Informational Bioelectromagnetics

Informational Bioelectromagnetics

David A. Copson, Ph.D., P.E.
Professor of Biophysics
University of Puerto Rico
Mayaguez, Puerto Rico

MATRIX PUBLISHERS, INC.
Beaverton, Oregon

To Val, Wendy, Laurie, David, Tiliń, Tilań, Ann, and Jan

© **Copyright, Matrix Publishers, Inc., 1982**
All rights reserved. No part of this book may be reproduced or utilized in any form or by any means, electronic or mechanical, including photocopying, recording or by any information storage and retrieval system, without written permission from the publisher.

10 9 8 7 6 5 4 3 2 1

Library of Congress Cataloging in Publication Data

Copson, David A.
   Informational bioelectromagnetics.

   Bibliography: p.
   Includes index.
   1. Information theory in biology. 2. Electromagnetism—Physiological aspects. 3. Electromagnetism—Physiological effect. I. Title.
QH507.C66   574.1'88   81-14253
ISBN 0-916460-09-6   AACR2

Illustrations: Scientific Illustrators
Printing: Pantagraph Printing
Editor: Merl K. Miller

Matrix Publishers, Inc.
11000 SW 11th St., Suite E
Beaverton, Oregon 97005

Contents

Preface xi

Introduction 1

1 **BIOLOGY CONCEPTUALIZED — THE INTERSYSTEMS APPROACH AND RADIATIONS** 17

   Information / Intersystems / Radiation

2 **THE NATURE OF SOME INTERSYSTEM COMMUNICATION** 61

3 **DNA INFORMATION, GENE MANAGEMENT AND RADIATION GENETICS FROM RECOMBINANT GENES TO SEQUENCING** 91

   Recombinant Genes / Protein Assembly or Synthesis / Radiation Input

4 **CONFORMATION, TRANSFORMATION AND DEFORMATION IN ATOMIC INFORMATIONAL BIOLOGY** 133

   Information-Rich Molecules / Informational Direction / Development and Other Evidence of Information Management in Biology / Feedback Influence of ATP, ADP, and AMP, the Adenylates / Direct Information Role Again for Amino Acids / The Immunoresponse, IFF, Identification, Friend or Foe? / Information Molecules in Evolution / Counterflowing Information and Biology's

Central Dogma / Information Theory, Tests and Bioinformation Management / In Aging the Information Management is Faulty / The Centenarians / Channel Death / Best Cells Survive / Error Catastrophe / Senescence and Aging / Effect of Free Radicals / Screening Out Noise / Code Redundancy and Degeneracy / The Riddle of Death / Ionizing Radiations / Psychological Factors / Nutrition / Fountain of Youth / Cognition Processes and Membranes / Information Features in Morphogenesis / Genome Expression and the Count / Concentration Gradient Information / Periodic Patterns and Wave Notation / Radial Growth and Polarization / Chemotactic Morphogenesis / The Factor of Mobility in Differentiation / Bioelectric Gradient Information Controlling Morphogenesis / Growth Determining Cellular Interactions / Chronon Theory / Development Information From Activity / Modeling and Modern Biology

5 WAVES AND ELECTROMAGNETIC ENVIRONMENT — POLARIZATION AND INFORMATION DENSITY REGIONS   205

The Electromagnetic Environment / High Information Density Regions, Antinoise, and Biological Squelch / Security of Information / Spectral Frequencies and Biomolecular Receptivity / Polarization / Time Constant and Half Life / Biomolecular Information from Spectroscopy / Energy Regions of the Spectrum / Quantum Fitting as an Informing Function / Waves and Particles

6 RADIATION BIOLOGY AND BIOSPECTROSCOPY   235

Polarization and the Dielectric Constant / Absorption and Emission Spectroscopy / Raman Information / Electron Spin Resonance / Spin Lattice Interaction / Nuclear Magnetic Resonance Spectroscopy / Probing Structures and Hazards Thereof / Special Interactions / Stark and Zeeman Effects / ESR, NMR, and Biohazards / Biohazards of Rotational Absorption / Biospectroscopic Studies

7 THE ELECTROMAGNETIC SPECTRUM   281

Miniformed Knowledge / Spectral Sensing / Spectral Reconnaisance / Spectral Units / Ionizing and Nonionizing or Polarizing Radiations / Landmarks in Electromagnetics / Dual Nature of Radiation / Commensurate Interaction / Wave Nature and Information / Radiation Polarization and Sensory Information Processing in Animals / Polarizers

Contents

8 THE RADIATION BIOLOGY OF IONIZING RADIATIONS 323

Agonizing Radiations / Types of Radiation / The Problem of Dose Integration from Various Types of Radioactivity, Units and Exposure / Environmental Ionizing Radiation / Radiation from Other Sources / The Entrance of Radioisotopes / The Radiation Biology of the Radiation Syndrome / The Course of Histological Changes in the Gastrointestinal Tract After Acute Exposure / The Nervous System / Impaired Embryonic Development / Blood System / Long Term Effects — Atomic Bomb Casualties / Surviving Exposure

9 UNRAVELING THE INFORMATION ROUTES OF THE IONIZING RADIATIONS 359

Activation / Collisions / Nuclear Reactions / The Chart of the Nuclides / The Decay Scheme / The Chart of the Nuclides as an Atomic Age Contribution / Atomic and Thermonuclear Explosions / Mass Defect / Availability of Isotopes / Dosimeters / Informational Links, Activation Analysis / Autoradiographs / Tracers / Isotope Dilution

10 INFORMING PROCESSES, ENERGY TRANSPORT AND REACTION DIRECTION WITH HYDRATED ELECTRONS AND FREE RADICALS AND SUPEROXIDE DISMUTASE 393

Operating Radicals / The Unpaired Electron and Reactivity, Photolysis and Radiolysis of Water / Electrons and Holes / Frequency Effects / Origin of the Free Radicals / Superoxide Dismutase / Linear Energy Dissipation / Direct and Indirect Action / Bound Water / Compartmentation

11 IMPEDANCE AND MEMBRANES, CORE CONCEPTUALIZATIONS IN BIOLOGY, DIELECTRIC ANALOGIES AND INFORMATION PROPAGATION IN TISSUES 415

Communicating and Compartmenting Membranes / Discovery of the Membrane Impedance Change and the Unit Membrane Concept in Biology / Role of Frequency and Other Electrical Parameters / How the Membrane, Molecule, Individual Cell or Tissue is Involved in Information and Energy Transmission / Relaxation Defined / How the Molecule, Single Cell or Tissue Becomes an Information and Energy Sink / The Frequency Response / Orientations — The Relaxation Response Extended to Include Systems / Phase

Relations, the Capacitor Model and the Dipole Ensemble / Types of Dipoles / Energy Conversion / The Impedance Conceptualization / Transfer Functions and Models in Biology / Measurements and the Temperature Effect with Microwave Parameters / Anomalous Dispersion / The Circular Frequency Graph / Behavior of Condensed Mixtures Compared with the NH_3 Resonance / Molecular Excitation and the Quantum Pump Principle of the Maser

12 DIELECTRIC LOSS IN MIXTURES, SOLIDS, AND COMPLEX MATERIALS   485

Absorption Mechanisms / Combining the Absorption Effects / Sensible Heat Production / Penetration / Microwave Heating / Semiconduction / Historical Landmarks / The Widening of the Scope for Microwaves / Sounding Bounds / Frank Injury

13 BIOLOGY AND ELECTROMAGNETIC RADIATIONS — NONIONIZING   503

Critical Mass / Terrorwatts / The Electromagnetic Environment / Electromagnetic Status, Characterization of the Exposure / Types of Exposure / Acceptable Units / Impinging EMR / Absorbed Energy / Instrumentation / Maximum Permissible Levels

14 DIRECT INFORMATION TRANSFER FROM THE ELECTROMAGNETIC ENVIRONMENT   523

Postulations and Evidence for Direct Information Transfer / The Microwave Cold War / Waves and Systems / Information and Noise in the Communication / Power Density and Exposure / Absorbed Energy Versus Exposure — Philosophy / Exposure and Absorption Models / Special Microwave Excitation and Bond Breaking / Continuum of Biological Effects / Low and High Power Densities / The Electric and Magnetic Components / Russian Work and Threshold Levels Give Interpretation Problems

15 INTERSYSTEM THEORY — DIRECT ELECTROMAGNETIC EFFECTS   545

Informing Forces for Coupling and Uncoupling / The Direct Mode / Direct Communication, a Function of Penetration / Radiant and DNA Information Interact / Time and Space Action Spectra / Conformation and Morphology / Neural Direct Communication / Exposure, Dimensions, Networks, and Thermia / Spatial Thermic Conditions and Dimensions / Shock Wave / Arranging Athermic Conditions / Modification of Growth and Development / Tumor Cells — Abnormal Development / Microwave Hyperthermia and

Tumor Thermotherapy / Stimulating Natural Immunoresponses / Lymphocytes and the Immunoresponse / Blood Pattern in Chronic and Acute Exposure

16 FIELDS, WAVES, AND PENETRATION INTO BIOLOGICAL PREPARATIONS — A BEGINNING BASIS FOR THE HAZARD ANALYSIS FOR VARIOUS FREQUENCIES 589

Model is Part of the Circuit / Composition and Units for the Energy Flow / Reflected and Plane Wave Experimental Preparations / Modulation Due to Electromagnetic Fan / Effect of the Magnetic Field / Capacitive and Inductive Coupling into an Animal / Inadvertent Radiators / Types of Fields Found at Various Distances From the Source / Polarization of Waves and Orientation of Biological Specimen / Absorption as a Function of Shape, Size and Layering / Absorption in Tissue Based on Equivalent Circuit Parameters / Reflection and Protection / Penetration Studies May Not Reveal the Internal Peaks of Energy / Heating Effect, Voltage Gradient, and Frequency / The Hazardous Conditions at Low and High Frequencies / Experimental Conditions in Microwave Devices

17 THE AUDIO EFFECT 611

Competitive Antagonism / The Hearing Mechanism / Structures Receptive to EMR / Rapid Rise in Temperature (RRT) and Shock Wave Mechanism / Polarization Force / Summary of Audio Effect

18 THE INFORMATION CONTAINED IN GEOMETRICAL RELATIONS, SPACE AND TIME CONSTANTS, AND SENSE MODES 625

Pulsations and Continuous Wave / Pulses — The Dimensional Response and Geometrical Fit / Time Constants and the Delayed or Flywheel Effect / Protective Functions of Skin / Selectivity Enhanced by Changed Dielectric Loss, Microwave Thermal Stress / Dimensional Responses / Brain Enzyme Inactivation / The Internal Peak and Thresholds for Biological Response / Biological Molecules Highly Susceptible to Absorption of Rotational Energy / Energy and Information Transport in Parallel / Specific Molecular and Mobility Effects / Resonant Effects / Power Density Levels for Direct Information Transfer / Pearl Chain Effect, Independence of Peak Power, and Basis for Transfer of Information / Relaxations in Molecular Biology / Natural Protection / Sense Modes for Information Input / The Bird as a Biopole in the Information Field / Animal Sense Modes / Perception and Behavior

19 CATARACT — NOISE AND ANTINOISE IN VISUAL INFORMATION FLOW   651

Multiradiation Bioeffect / Lens Records Physiological Insults / Comparative Radiation Cataractogenesis / Metabolic Lesion / Other Ocular Changes and Aging Cataract / The Exposure Model for Microwave Cataract / Flywheel Effect Postulated Based on Importance of Peak Power and Repair Time Constants / Juxtaposition of Internal Peak and Site of Primary Lesion / Migration of Internal Peak with Wavelength / Possible Coincidence of Peak Amplitudes / The Resonance Condition / Time Constants May be Unfavorable for Repair / Self-Generating Reaction / Biological Warnings / Microwave Oven / Diagnostic Indicators

20 QUANTITATIVE INFORMATION   673

EPILOGUE   683

Perspectives

APPENDICES

- A   Abbreviations   699
- B   Useful Factors   703
- C   Glossary   707

INDEX   737

Preface

The uncharted reefs in the course of biospecialization are dangerous, mainly because of lack of appreciation of the vision of the organism as a whole as coupled with its environment. Information and its inseparable action, communication, both provide easily followed direction markers through biology, reducing the probability of this perilous oversight. An interrelationship among the objectives of self-organization and cybernetic analysis, models, and automatic data processing is also established. Similarly, electromagnetic radiations must be recognized for what they really are, rich bearers of information for susceptible molecular systems.

The study of the nature of forcing and informing biofunctions is the next phase of the electromagnetic age in biology. It is based on the enormous effort of biophysicists in the decade beginning with Hugo Fricke in Cleveland in 1924 through the U.S. and U.S.S.R. involvement in the 60s and 70s with the biological effects of electromagnetic radiations. This book introduces the concept of these functions—an introduction which would not be possible without the dedicated efforts of distinguished scientists listed in the 900 references. During the gestation period of this book, its subject was brought to public attention by a clever exposé of an alleged cover-up by the United States of the exposure of our Embassy, Ambassador, and foreign service people to Soviet microwaves for 16 years in Moscow[1]. With 34 years of personal exposure to microwaves in mind, the author regarded this exposé as extremely interesting, coming as it did at the end of many years of writing and rewriting these words. Even more important however, was the growing realization that these years of study, teaching and writing on the subject of microwaves and EMR bioinformation transmissions, formed a propitious introduction to the world opened up by Dr. S. J.

---

[1] P. Brodeur, *The Zapping of America: Microwaves—Their Deadly Risk and the Cover-Up*, W. W. Norton, New York, 1977.

Webb in *his* " exposé " of exquisitely sensitive, bioelectromagnetic communication between precisely-controlled external millimeter waves and the very machinery of life in cells.

As a charter member of the Biophysical Society formed in 1956, the author saw how many, multidisciplinary natural scientists and engineers viewed that organization as being unfettered by rigid definitions as to its purpose, while, in fact, being oriented toward the study of radiation. This is because these scientists had matured and are still maturing through one of the branches of science into an integrated whole. This book mentions examples like Kenneth Cole, Alan Hodgkin, Norbert Wiener, Claude Shannon, W. Ross Adey, Maurice Wilkins, James Watson, Francis Crick, Sidney Webb, Andrew Huxley, Erwin Schrödinger, Georg von Békèsy, and other actual or potential Nobel Laureates.

Then, in 1956, in the long shadow of the atomic bomb, there was a great urgency about understanding the effects of radiations on cells and tissues. Then the singleness of destruction yielded to scientific inquiry. Radioisotope applications, radiotherapy, and diagnostics became indispensable. Next, other electromagnetic forces moved in to share the stage—some of them with new and strange ways of interacting with organisms. Currently, these are being seen as powerful sources for imaginative diversity in the sciences.

Appreciation of these developments has led to new agencies in the United States, including the Energy Research and Development Administration and the Bureau of Radiological Health. It is clear that the role of electromagnetic radiations is best appreciated through their usefulness as tools in the life sciences. Yet we must be fully alert to their hazards. These broad areas of awareness are admirably served by looking at electromagnetic energy as a flood of information around and about the living organism. Inundated in this sea of fields and waves, the organism establishes its rapport with its surroundings, deriving a constant input of essential data which enables it, in the final analysis, to cope with that environment.

To proceed logically, the book begins with a description of information and its management. This leads to viewing controlled and regulated systems at various levels as cybernetic models. The information theme is found to dominate the cybernetic one to such an extent that it offers a sort of plain language alternative in dealing with these systems. That is why it is the theme favored throughout the book. The second section deals with electromagnetics in detail, covering all the radiations—ionizing, nonionizing, polarizing, nonpolarizing, harmful, and beneficial. The third part is then able to deal with the interactions of radiations and biosystems, emphasizing microwaves in particular, since these are an unseen threat to the organism.

Information management furnishes sensitive guidelines for a biology text, since it can be used to emphasize analysis and the translation of experimental results into knowledge of the nature of biosystems. Energetics, mathematical analysis, axiomatic analysis, or some other approach could be used, but it is believed that these would find fewer receptive associations.

Preface xiii

On the other hand, the power of mathematics is fully brought out in connection with the analytical sequence of recognizing and coordinating the problem and applying the appropriate methods. When Louis de Broglie wrote his *Revolution in Physics*, he wrote out all his equations in words. Similarly, all of Banesh Hoffman's *Albert Einstein*, which details the work of the great mathematician contains but one equation—a record which we have not been quite able to equal although there has been an attempt to gain in communication what is lost in conciseness.

Given adequate briefing at the assumed level, it is felt that readers will seek out further appropriate references on subjects which seem to require more depth. The college-level physics and mathematics assumed is minimal and the book is self-contained for the theory necessary to understand the molecular biology portions. It is therefore written for a fairly wide range of readers having requirements in interdisciplinary biology. On the upper level side, the plain language exposition is often rigorous enough to allow a ready takeoff into appropriate computation and investigation, especially when supplemented with appropriate work in the laboratory.

Specific students who will be able to use the book and subject matter to advantage include biology majors with a need for amplification in the physical aspects or who wish to expand environmental sensitivities beyond chemical and traditional biology into electromagnetic compatibility. It answers the extradepartmental needs of physics majors, psychologists, engineers with research or secondary interests in biology who want to use their special training, and seniors and graduate students in medical technology, biochemistry, bioengineering, and biomaterials science. Taken with a "principles" text on biological processes, it is innovative for core curricula in liberal arts and, when combined with materials science and dielectromagnetics, it is a springboard for an interesting approach to engineering.

The author's current experience with the acceleration of medical and dental training and the premedical program shows that the selection of subject matter here is especially favorable for this project, which has high national priority. Interdisciplinary courses serve this objective well and these chapters in biophysics, covering the electromagnetic environment, bioinformation and intersystems that lead to man, bridge two basic sciences that must be considered essential.

The concepts related to the radiation transport of bioinformation and the even treatment given ionizing and nonionizing radiations lead to a comprehensive presentation of the electromagnetic spectrum. This permits an assessment of our incompatabilities with the little-known electromagnetic environment which is composed of these radiations. It is therefore directed to the concerned reader, who is interested in the biology of radiations, including nonionizing ones, with special emphasis on the emerging microwaves and their neighbors in the spectrum.

Whether or not America is being "zapped" (Brodeur 1977) by these radiations is a sobering question. Any threat with the potential for making the planet uninhabitable is worth a calm, thorough analysis. It isn't necessary to take the view of sinister manipulation; the radiations themselves become deadly due to what seems to be public acceptance of their proliferation. It is this march toward unseen "terrorwatts" and "agonizing" radiations that leads to the alarming message of Brodeur which suggests rather vividly that we have arrived at the electromagnetic frontier.

Microwaves lead on to dielectric processes and, in the sense of future possibilities, to biomolecular engineering so that a look at the periodic system of the elements is required. Similarly, when radionuclides are brought up, that other unique contribution of the atomic age, the Chart of the Nuclides, is explored. The spectrum, elements, and nuclides are examples of scientific conceptualizations that embrace a vast amount of electronic age knowledge "miniformed" in an easily-digested manner. It is on this note of conceptual flagstones, forming a path into biophysics, that the book begins.

In the author's experience, and within his capabilities, informational pathways have been used to lure biologists over the interdisciplinary bridge, whereas abstract rigor might have directed them elsewhere. Such a plan requires implementation with well-understood biophysical cases. However, the author cannot accept responsibility for the awesome contrast between the simplistic measurements and analyses possible now and the actual complexity of some of the systems being considered. This situation makes cautious unions among physics, physical chemistry, and strong suggestions from quantum mechanics absolutely essential.

Biological models must, of necessity, look fuzzy to physicists. Sometimes the probabilities give opportunities for statistical conclusions at an indicated probability and level of confidence, but care must be taken to assure that the effects follow the causes within the biolimits, and mathematical rigor itself is no guarantee of validity. The most plausible hedge against error here is to follow the informational pathways in biology, applying theory as appropriate, because this route affords ways of simplifying the impressive complexity of the biosystems. In writing on this subject, it has been essential to weigh the probabilities and even "guesstimates" have their place in careful hands, if only due to the imminent electromagnetic hazards to the biosphere.

The text owes much to its parents, the first and second editions of *Microwave Heating*, as it collects teaching experience and quiet analyses made among electromagnetic and other waves characteristic of this area.

General and cell physiology provide excellent avenues of introduction, and biological control theory and bioelectromagnetic specialization follow very well. Bioelectromagnetics enters the world of sensory information processing via these same informational pathways. The advantage is that an understanding of the fears and hazards connected with these radiations is at once correlated with their natural functions.

The virtue of the information theme becomes obvious when the study is complex, as in dealing with the nuances of immunorecognition.

To illustrate, A. Ebringer[1] recently described in the *New Scientist* the mechanism for cardiac valve damage in rheumatic fever. He showed that a *Streptococcus pyogenes* upper respiratory infection can lead, through a series of events, to antibodies which are unable to distinguish self, in this case, the damaged cardiac cells, from invading antigen, because of the practically identical atomic configurations at the end of one of the surface antigenic features on the streptococcus. The difference may be in one hydroxyl group. In addition it is a purely misinformational, or an identity-recognition problem, measurable in time, that makes one individual susceptible while another is not. Evidence of the recognition system involvement comes from the resistance shown by "O" blood types and the susceptibility of the "A" type. The "A" group surface antigen terminates in N-acetyl-D-galactosamine, while *S. pyogenes* has a surface antigen that ends with glucosamine. The difference in the message represented by a mere hydroxyl group is not enough submolecular difference to allow the challenged antibody to tell the difference. The "O" group, however, has a quite different ending, L-fucose, and the "B" group has galactose on the red blood cell surface antigen involved. These cross-reacting antigenic actions and the misinformation involve the associated glycoprotein. The result is that the body, via its immunosystem, challenges its own substance repeatedly, and cardiac damage occurs as a sequel to rheumatic fever.

The informational theme continually directs attention to communication from master controls in information-rich molecules by signals between systems. This leads as easily to hormonal regulation globally, as to conformation of macromolecules at the interaction site.

The romance of informational biology goes far into the past for its origins, and obviously well into the future for its hopes. The first crude attempts in ancient primitive molecular biology resulted in certain molecules with unusual capabilities for information management. Today, recombinant genetics thrusts into the hands of timid man nothing less than the management of life forms as his biological destiny.

It is the "science of science" and the mark of a good teacher to involve the reader or student in the operation of science. The first recorded case of electromagnetic radiations directed at humans is the Moscow Embassy case, and Pandora's Bizzare, which was, in part, the response of the United States, provide such an opportunity here. It is the effect of being on the scene, as humans, in the last two decades of the 20th Century, that must enable us to meet a formidable environmental challenge and to reap the rewards from understanding that lend relevance.

---

[1] A. Ebringer, "The Link Between Genes and Disease," *New Scientist*, 19, 1121, Sept 21, 1978.

Introduction

"As long as automata can be made, whether in the metal or merely in principle, the study of their making and their theory is a legitimate phase of human curiosity, and human intelligence is stultified when man sets fixed bounds to his curiosity."

Norbert Wiener

Informational biology forms a thread of life that winds through areas such as information-rich biomolecules and the transport of information in parallel with energy by electromagnetic radiations. It also traces these radiations through control systems and their coupling to organisms within specific electromagnetic environments. This may have positive or negative results. Electronic information processing has made a lasting impression on biology so that a text must now treat the reader as being concerned with the interface between man and machine. Informal computational language is therefore an advantage in smoothing the transition into today's biology. The simple models discussed here will soon demand nonlinear functions. To solve these functions, biologists may consider it crucial to have their friendly computer nearby. Thus, by beginning a biological study with information flow and implications for data processing, a special understanding of biological operations may be gained with an accent on modern methods.

The reader may acquire a researcher's *sensitivity* from the search for basic processes and models. One of the most productive results is the recognition of various *levels* of detail in explaining biophenomena. The relevant basics of information theory may be associated with important processes in molecular or *atomic* biology and on up to the electromagnetic environment in which biomolecules exist. Terms such as "information biomolecule" and "coding of bioinformation" are being widely adopted. Here, these concepts lead naturally into control and regulation, or cybernetic analysis, when we consider the equivalence between feedback and information.

Many of the elements of biological spectroscopy are being recognized for what they are, that is, aspects of molecular biology. This is bringing a rich diversity of tactics to bear on some of the current research. Emphasis on electromagnetic radiation tends to expand the reader's sensitivities toward the spectra in many biological operating systems, some of which exhibit resonance.

Many information processes in biology that are presented have open areas, but it is the modeling method that helps to expose the areas of incompleteness. The black box to white box transition, as knowledge is gained, is a method in itself. The fact that what happens in the black box is not yet understood does not mean that the existence of the particular control loop is challenged. The bypassing of detailed biological obstructions in this way is a strategy in the management of information. The black box is not seen as a termination but rather as a convenient means for proceeding to the output operations on the other side of the box function. What happens in the box begins as a functional concept, proceeds to a transfer function, and is ultimately expressed as a suitable equation. In the management strategy, the various means of investigation are associated with the gaps in data and with the methods of experimental biology to offer a sound rationale for studies. This produces the ultimate in research conditions where experiment assists theory and theory will, in turn, then assist experiment.

The foregoing leads naturally to the modeling process which is very much at home in biological research. It is often necessary to visualize the field of biological operations without the constraints imposed by insufficient information. Instead, the model maker must include all he has learned from experience in his efforts. Thus, a system model, on some chosen scale, of the biological subject is obtained. No harm can result from studying a model of a homeostatic loop, even if some of the elements remain to be uncovered. Simulation studies using the model can yield important results. They also have the decided advantage in that the model can absorb a great deal of punishment which the human race is spared.

The interfacing of humans with data processing machines has acquired new significance as the capabilities of the machines increase. The biological and physical disciplines merge successfully in studies of small systems at all levels of organization—molecular, cellular, or organismic. Now they are joining as well with other disciplines in the analysis of problems in large-scale systems such as populations or complex operations. The problems encompass environment, health and medicine, energy distribution, transport, communication, and humanities. Working models of these systems often include a *human reasoning* or *operating link* with the expertise necessary to answer machine queries when the computation reaches a crossroads and demands attention, or to take command of the system components. The human involved must exhibit two special abilities: he must be expert in the speciality in which the demand for reasoning and decision arises, and he must also be skilled at operations in the region of the interface between his own brain and the

Introduction

performance of the machine (Fig. I-1). However, like someone directly banking data or using feedback for soothing pain, he may be more like a *part* of the data handling circuit than is the biologist looking at the small system model. Neither analysis proceeds without the human operator, but the large scale model, and especially the interface operations, require that we pay close attention to the *coupling* of man and machine.

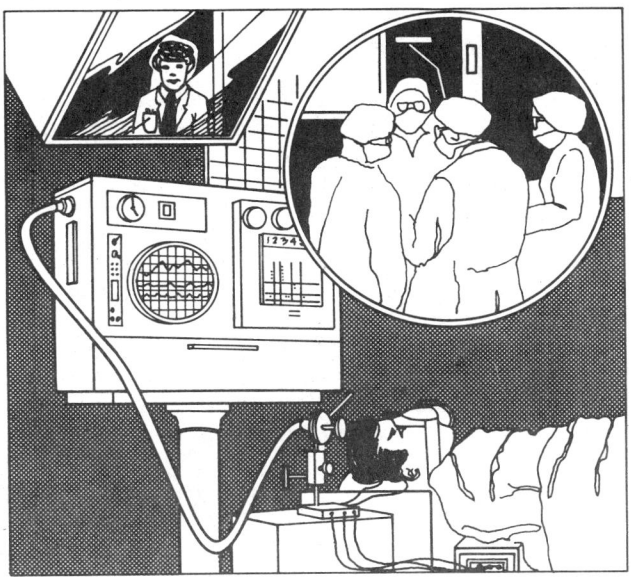

Fig. I-1. Arrangements for coupling of surgeon during complex operation into data processing links. Numerous man-machine interactions occur at this interface.

Theories such as those of John von Neumann in *The Computer and the Brain*, J. Singh in *Great Ideas in Information Theory, Language and Cybernetics*, and W. R. Ashby in *Design for a Brain* have fallen short in their attempts to explain man's interfacing function. They may have diverged on simplistic grounds while expecting neurophysiology to provide all the necessary explanations. More progress in this area may teach us what we need to know about how a creature is taught, decision making, intuition, and sensory experience. Even in the "Computer Age," the model for one tiny part of a nerve fiber is a very large computational program, but it would be dwarfed by the model for a complete neural cell with dendritic branches. Beyond this, the complexity is too great for accurate solutions to black box transfer functions.

Where there has been success, it was with limited models within the nervous system. Feedback has been recognized as being very important in regulatory electronics. It is also a reflex action in nervous systems. Cybernetics is a formal method that demonstrates the analogous nature of neural and electronic feedback. Now the feedback in neural systems is composed of incoming and outgoing information. This results in the response of the regulator to sensory messages which, in turn, brings the system into control. The study of the information contained in the messages, its role in reducing uncertainty, its precision, coding and related matters are all considered in information theory. Biological feedback and information are concerned with the accumulation or deficiency of nutrients, reaction products, or changes in physical conditions of interest to the organism. Regulation is then enzymatic, neural, and hormonal, the result being homeostatic control or response to system changes.

In order to access information and energy, chemical molecules use physical processes to combine according to the relationships established in the *Periodic System of the Elements*. Examples are seen in processes such as vision, photosynthesis and photoactivation, all of which involve radiant energy. The microform for study of all the radiations is the *Electromagnetic Spectrum*. It consists of all the informational pathways through which remote or local systems intercommunicate.

Even though the computer has a low intelligence quotient and depends on the human operator for real thinking power, there remains an inevitable, possibly embarrassing, comparison between ordinary individuals and smart machines. This shows up at the interface with any sizeable computer when its output of data is at a rate beyond the assimilation power of the operator. It is most obvious with operations which require on-line decisions such as in computer-assisted surgery or diagnosis. There is a burden on educators to redress the balance, but even with optimum memory utilization, and maximum reaction speed, the decision maker is likely to be the weakest link in a fast moving circuit, and the best of them are already in critical demand in heavily automated operations.

As Wiener expressed it, "The world of the future will be an ever more demanding struggle against the limitations of our intelligence, not a comfortable hammock in which we can lie down to be waited upon by our robot slaves." Thus one of the great future problems which we must face is that of the relation between man and the machine, of the functions that should be assigned to these two agencies." There is already a requirement for a therapeutic input for on-line computer language to assist in the training of the operator. It introduces relaxation and recovery into the highly efficient systems that require unusual on-line attention from the human link. Wiener's prediction is supported in those time-sharing systems which use a kind of man-computer commensalism. In this, the labor is divided according to the best performance areas of each partner at the interface. The intent is to gain, by flexibility and full utilization of both man's and machine's great resources, what could never be gained by the use of each partner alone.

The legitimate objective is to facilitate data banking in *learning* by favoring selected input, patterns for recognition, and efficient retrieval. This means striking known, universally recognized standard keys to set up the necessary couplings. The interface process in terms of decisions per unit of duty time is then directly dependent on the amount of data and knowledge. Speed in decision-making may result in faulty decisions, so in order to widen the pool of skilled human links in operating circuits, positive forcing functions are essential. This is a kind of channeling that is built on training, experience, and conditioning. Many intuitive reflex-type actions can be screened out and formed into a "sub-thought" level—the idea of *not* thinking of what you are doing, unless such high level conscious attention is necessary.

In biology, as in other disciplines, there are two approaches for optimizing data banks, conserving brain capacity and memory spaces while still providing the enormous base for associations demanded by recognition processes:

1. Specialize in areas with limited data requirements
2. In training, exploit certain bionic expanders or scientific conceptualizations
   a. *Regulation*, stability concepts, feedback, cybernetic analysis
   b. *Impedance* matching or balancing to assist coupling.
   c. *Self-organization*
   d. *Propagation* and waveforms
   e. *Amplification* and phase
   f. *Identification* techniques
   g. *Redundancy* reduction versus control of reliability
   h. *Multi-path*, "centrifuged" knowledge
   i. *Flag pathways* that indicate directions
   j. *Polarize*, nonpolarize, and relaxation responses
   k. *Resonance* for peak of action spectrum and "eased in" flow of information.
   l. *Transduction* and energy discrimination
   m. *System theory*
   n. *Closed and open loops*
   o. *Entropy*
   p. *Radiation* between systems

Impedance describes the generalized resistance to flow without dissipation of energy and its matching immediately suggests that the situation is transformed from one of resistance to one of transmission. However, impedance has a great deal of meaning in biology through a long series of great discoveries leading to the nerve axon mechanism and beyond. It has meaning wherever electric, dielectric, hydraulic, sound or other pressures influence the flow of the corresponding energy in biological systems. The essential part for information systems is that the impedance depends on the frequency of an applied field-forcing-function. The result is a capacitive or compliant part in the total energy admitted, which, in turn, explains phenomena that are difficult to explain in any other way. This concept, and the associated need for matched impedance in coupled systems, is taken up in Chapter 11. Amplification by masers and lasers as quantum pumps is also explained in that chapter. Dielectric pumps and amplification in the audio system are discussed in Chapter 17.

Propagation and amplification provide regulators for cybernetic systems in the sense that virus multiplication and cell division occur, in general, in response to appropriate conditions and feedback demand for *more of the same*. This eliminates the need for bulky stockpiling. Similarly, the DNA blueprint outlines growth and development and there is no need for maintaining models for them. In this, *self-organization* is also evident. Many biological systems undergo *polarization*: morphologically as in hydrophilic or hydrophobic regions of phospholipid membranes, operationally in nerve impulse transmission, and socially in the behavior of populations. In that status, they may be susceptible to forcing functions. Alternatively, the function that does the polarization may be an electromagnetic one.

The author remembers a signalman on board a ship in the navy. One time during close order maneuvers, held under the admiral's eye, the signalman was straining his every nerve to read the flag messages that were being hoisted at a frantic rate on the flagship. When the watch officer touched him to offer coffee, he nearly jumped overboard. There was tremendous action with imminent danger and his polarized system reacted in the manner in which it was poised.

Biosystems display recognition and identification of friend and foe as in antibody-antigen reactions in immunology, in the oxygenation process in hemoglobin, and in numerous retrieval procedures. Regulation, programming and memory are seen in regeneration schemes, respiration, volume and size control, and aging processes. Some twelve other *conceptualizations* form a basis particularly for Chapter 1 while those in this list are used throughout the book.

Some alert readers may see something metaphysical in the idea of information transmitted in radiation like a Supreme Being up above tapping out some Morse code. But, they must remember that microwaves, radio and other radiations can carry *all* the information in a telephone or other message by virtue of their natural or artificially modified characteristics. This capability is shown in the title of the book by Mischa Schwartz, *Information Transmission, Modulation, and Noise*. The radiation's wavelength or frequency and its energy are primary natural characteristics. Changes such as frequency modulation, amplitude modulation, pulsing or imposed changes in the radiation are artificial and, thus, they all constitute signals (Fig. I-2).

With visible light, the result is seen in visual phenomena such as the eye registering light interactions with the visible object, phototaxis,

Introduction 7

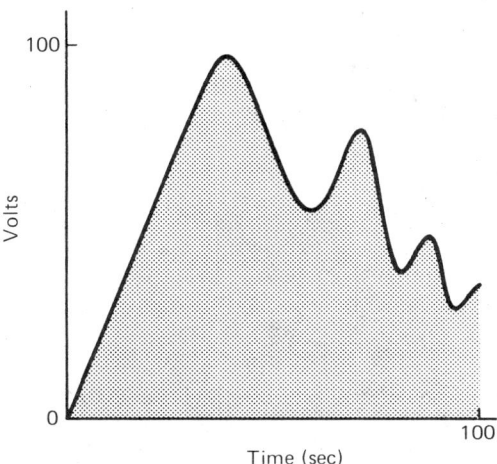

Fig. I-2. Time-voltage diagrams. Voltage amplitude can change between 0 and 100 in 100 seconds in the time/voltage diagram showing a system for carrying information.

phototropism, or photochemistry. The concepts in photochemistry further suggest how the features of all the other radiations capable of interacting or chemically reacting permit them to do so.

Man may *operate* and, at the same time, be the *subject* of a model of interest. It is from this latter point of view that the management of human bioinformation has developed. First, we must realize that we are dealing with specific information, and that measurement, instrumentation, and control or cybernetic analysis are means for filling the spaces or gaps in the knowledge available in the Life Sciences. As the physical analysis of life processes goes forward, we also notice that this analysis also clearly indicates the gaps in our knowledge. While many of the mathematical appraisals of biological areas such as ecology, morphometrics, sensory systems, regulation, experimental design, and genetics may have seemed to suffer from incompleteness, it is the very *definition of problem areas* which has been most helpful. As mentioned before, these deficiencies are most obvious when the model for the system is attempted. The indicated gaps then become logical subjects for investigation, with a consequent saving in scientific effort. It is the pressure of the application of the mathematical model or principle which forces the analysis to stop or circumnavigate at some step. Even if the value of the model in explaining a system were dubious on other grounds, it would be highly valued for merely delineating the problem. This is to say that the subloops required for cybernetic analysis are forced to surface. As seen, man may turn out to be both the subject and object of his "fate" to serve the computer or have it serve him. This result has broad significance in biological education and specialization. At the interface, the result can be to put man *in the circuit* as a decision-making step.

Readers' interest in electromagnetic radiation biology may range from lethal or deleterious effects and survival at one extreme, through therapy and diagnosis, to biomolecular analysis and intermolecular communication at the other. Then, convergence would mean that the latter should explain the former. It is hoped that such a completeness is evident in the treatment here of the questions of *ionizing versus nonionizing radiations* for it is known that many wish to explore this question in greater depth than most texts allow.

At the fundamental level, it is premature to come out "for" or "against" interacting radiation and biosystems as in the case of nuclear power, microwaves, and X-ray diagnosis. The search is certainly *for* significant biological mechanisms. First, however, is the invaluable observation of the bioeffects themselves. These are the positive or negative disturbances of the system state, or shifts in conditions, associated with the causing function or situation. Other significance may be involved, such as clinical, or hazard, in the case of biomedical uses of various radiations, but these should not prepolarize anyone one way or the other. Microwaves provide a case in point. These waves will apparently provide a leading method of food preparation in the future, not to mention the saturation of our airways with their silent and unseen presence. A powerful trend of this nature requires an appropriate moral concern on the part of knowledgeable scientists. A definite effort is made here to put the biological hazards in perspective. These hazards must be balanced with the notion of the electromagnetic radiations as tools in the Life Sciences. Hypotheses seldom yield to absolute laws, but at least, experience and judgment should be brought to bear on the matter. The acceptance of new technology turns out to be progressive. In the busy biological milieu, symptomology will probably remain the first observation and the leading stimulus to investigation. Care should be taken that this doesn't lead to narrow-mindedness in current research patterns. A central dogma in food science is the denial of athermic effects of microwaves while these same effects constitute a life's work and meaning to molecular biologists in microwave spectroscopy (Copson, 1975).

With microwave and ionizing radiations the *profile of the energy source* is examined without bias, just as if one were looking at sound, chemical, or mechanical energy sources. If, from these observations, some restriction on applications appears to be warranted, it should be a matter of putting the burden where it can be controlled. As with the creators of other environmental hazards in manufacturing, those (manufacturers) who cannot tolerate the heat should not be in the kitchen, to paraphrase the late President Harry S. Truman. To continue the analogy further, housekeeping is the name of the game—even when the disturbance is electromagnetic. The scientific world is also a political and an economic one, and the interdisciplinary biologist is called on to establish *maximum safe levels* of electromagnetic energy for the expo-

Introduction 9

sure of industrial and general populations when exposure is accidental, as opposed to intentional (in anger or visciousness). These levels should then *control* the *design* of electromagnetic generators and machinery or devices. They also act as a valve for progress in scientific, industrial, and military developments in which the utilization of the energy is involved. A declaration that some proposed device will operate under unsafe conditions will either counterindicate the device or stimulate the safe level solution. Thus, the influence of the scientist is a powerful one.

The strangest contrasts that emerge in these studies may be due to divisions and differences in the scientific world. For example, this book tends to bring out the fact that the U.S.S.R. has a maximum permissible microwave energy level applicable to the population but the U.S. does not. The teachings of Ivan Petrovich Pavlov, a Russian physiologist, have greatly influenced Soviet and other research. Those in the U.S.A. who stress the special significance of central nervous system functions and consequent behavior are certainly reacting to his work. The neural phenomena in the brain are closely involved with memory and learned behavior. Informational logic widens the applicability of the work and relates it to intersystem effects and biochemical changes (in the Pavlovian tradition). A special significance would be attached to athermic microwave effects as opposed to exclusively thermic ones. Yet East-West differences are no more evident than here. The U.S. scientists object to the sparseness that may make U.S.S.R. data difficult to *duplicate.* This basic law of scientific investigation being thus flaunted, their results are held in some disrepute. Yet the U.S.S.R. results must be taken seriously. The explanation, over and above communication problems with language and culture, evidently is that, on one side of the fence, in the East, long term effects were studied and the bioeffect searched sensitively. On the Western side, it has been the short term or acute hazard effect that has attracted the most attention until recently. It is likely only in the long term that behavioral changes from electromagnetic waves could demonstrate a consistent trend. Whether these behavioral effects might then be used in sinister manipulations is still another story.

In this book, an effort is made, perhaps uniquely, to bring both long and short term nonionizing radiation effects together in balance with many other aspects in addition to hazard. That any quantitative biological exposition of such effects can be systematically made today is certainly a tribute to those on both sides of the fence. Their polarized approach however, has had a negative influence. One has to wonder if the electromagnetic environment is equally hazardous in the two political regions. The attention paid to this matter from health and hygiene aspects in the U.S.S.R. has evidently been much greater than in the U.S. . If the appearance of anomalies in exposed persons stimulated wide scale studies in the sixties in the U.S.S.R., where were those symptoms in the U.S.A.? Even today as this is written, permissible standard levels of exposure adopted in the U.S. will, according to the

U.S.S.R. standards, impose a health hazard on exposed persons. The U.S. maximum of 10 mW/cm^2 is very comfortable for radar and experimenters to abide by, since detectors easily sense field only one thousandth as strong or approaching the U.S.S.R. standard of 10μW/cm^2, which is really an EMR pollution guide because it applies to the whole body and chronic long term exposure. Coincidentally, the U.S.S.R. has not, as yet, had the stimulus afforded by wide adoption of microwave ovens as is the case in the West.

From February 1976 on, the press carried news stories about the microwave irradiation of United States Embassy personnel and children in Moscow. The Russians might, in this manner, revitalize certain of their electronic eavesdropping devices in the Embassy and certainly could interfere with outgoing and incoming radio transmissions. In any case, there were biological results on the ambassador, his aides, and their families that are still being investigated, and it is clear that the Soviets were attempting to modify the electromagnetic environment for political reasons. Therefore, this environment has become one that can be accidentally *or* intentionally manipulated, and the electromagnetic age of biology has come with a flourish; its monument—the nine-story Moscow embassy ruined by fire in August 1977, a fire that seemed most severe in the secret areas in which Soviet "firemen" showed a keen interest.

The Moscow events had the effect of declaring open season on U.S. and other foreign service personnel. From 1975 to 1980, five ambassadors from the U.S. were killed and the rate of attacks climbed from two per year in the mid-1960's to 24 per year recently. The 50 hostages in Iran and the takeover of the U.S. Embassy occurred in a ten-year wave of 213 terrorist attacks on U.S. diplomatic missions in many countries from 1969 to 1979. It is obvious how electromagnetic radiations might be used to create additional instabilities. To the growing demand for military tactics like defensive perimeters, strong rooms, and high security must now be added protection against zapping so that the performance of these missions is in jeopardy. The pressure on them suggests that their work must be extremely unsettling to the opposition.

The Moscow microwave incident was hushed up to allow U.S. intelligence maximum flexibility in dealing with this menace. As a consequence, Ambassador Stoessel and the Embassy staff were apparently not alerted to the biological hazards. He was, in fact, reassigned to Bonn when he became ill, and measures were taken to counteract this additional stress of duty in Moscow. The principal fact seemed to be that only lymphocyte counts were cited in medical releases, which suggests a rather inadequate survey. Probably, one effect of classifying the incident was to invite a limited medical appraisal as all that could be done without compromising security.

East-West contrasts were visible in the management of the incident, public interest, and U.S.S.R. military policy. In the West, the large fraction of the population with microwave ovens, more than 10%, assured the existence of an influence group, probably knowledgeable and possibly concerned about EMR hazards. In the U.S.S.R., this would be a small, perhaps negligible, matter. The Russian approach was a "General Staff" one. Both nations understandably want their armed forces to retain the most options in the use of radiations for weapon guidance, target acquisition, and military/governmental communications. While the Russians could simply exempt the intelligence and military from the maximum permissible levels, in the U.S. where the "General Staff" concept is banned, the military-intelligence communities knew they had to come to grips with the problem and set maximum permissible levels that gave flexibility but also safety. This would then illustrate the disadvantage faced by a responsible nation when a public health factor is used as a military, political ploy.

Of course, to be on the safe side, a clause was inserted under Section 360 A (b) of the Radiation Health and Safety Act of 1968 which permits DOD to claim exemption when any standards made for consumer products threaten the free use of defensive and offensive weapon systems (in which case the U.S. Congress will not interfere).

This said, an open discussion of the menace can proceed so that microwaves and their neighbors can be revealed as tools as well as hazards in the life and behavioral sciences. It is a rare opportunity for a microwave specialist when international incidents forcefully introduce the subject. Then research accomplishments with these unique and useful forms of energy may be recounted against a background of the sudden need to know. Much will have been gained from the notoriety if the momentum of investigation is maintained toward the goal of attaining the *critical mass* for understanding of electromagnetic radiations. In this area today's ceiling become tomorrow's floor as these newer forces in nature become commonplace.

Lasers and masers add precise directivity and concentration to the electromagnetic capability. It suddenly becomes possible to visualize in true Flash Gordon, Buck Rogers, or Star Wars fashion, a weapon system able to lase or mase targets. At present this would require a large installation on earth, for instance, a generator facility in Havana to threaten Florida and the East Coast, or an offensive satellite constructed to destroy spaceships and other targets. These unseen EM weapons can be made very destructive. While not so dramatic, the EM environment in densely populated areas poses a threat of its own as civilization moves deeper and deeper into the EM age. A hospital in California, for example, found that, instead of vital signs of patients, it was measuring fields from a nearby transmitter site in its instrumentation. More optimism is expressed in the emerging concept of using EMR to stimulate or inhibit metabolic processes, control cells and perhaps help them reregulate their destinies when confronted with a viral, tumor, or parasitic challenge.

Bioinformation Systems and Management

Even the smallest biosystem is a large scale system in the sense of the flow of information required to sustain it. Stored information guides the development of the organism and an individual depends on processing the data he receives in order to determine his actions. While extending the study to information would seem an extra burden, it is actually a reducing factor when information flow and management are the subjects of interest. It is a category, useful for mental associations, one that has certainly withstood the test of time.

Organismically, bioinformation establishes and maintains the dynamic state of the organic constituents. It belongs with management and systems study. A constant biological status is achieved only by systematically changing or recycling the constituents. This is sometimes described as the changing renewal of the components. It is no longer sufficient to stand in awe of this complex process and it is now very profitable to have a look at the programming, in the cybernetic sense.

There are two distinct levels in the study of bioinformation, the forming level and the managing level (Fig. I-3). At either level, the

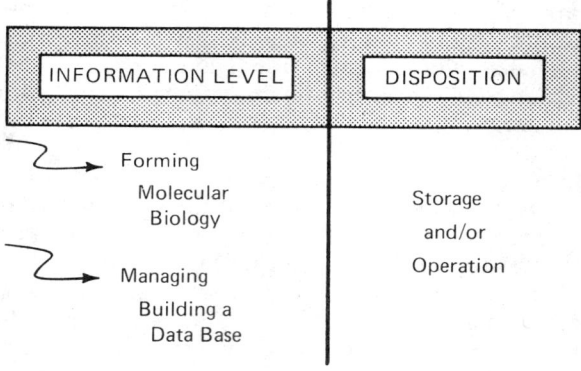

Fig. I-3. Levels and disposition for information in information management.

function of bioinformation is to reduce uncertainty. This reduction is in the sense of an inductive plan or a well-designed questionnaire. In the formative sense, information serves the needs of molecular biology. What happens at the organismic level, that is, how the disposition of information determines the operations of an individual, is clearly information management. The change from one to the other probably occurs at an analysis point. Thus, the building of a data base will, at some point, lead to its use in decisions and operations. The construction, selection, and scheduling of bits of molecular information and forming them into messages are forming features. The process, while highly effective, is not without its errors and omissions. Yockey cannot explain how DNA can blueprint an individual even when half of the "code" is erroneous. He is impressed enough by this unknown mechanism to suggest that it might be the process which, upon failure, produces aging and death. Later, knowledge flows into the sensory net to reinforce the individual's data base and prepare him for actions. It is optimistic to regard this flow as exclusively formative in effect, but if it is, it is also well-managed and hopefully is totally committed to the reduction in uncertainty. It is unfortunate that the information he receives may, at times, even increase this uncertainty. This is a good reason for emphasis on information management.

Teaching is concerned with the rate of acquisition of information and so, has always been a type of information management. Perhaps the rate of acquisition is a critical factor if, as concluded by specialists such as Streufert (1970), the decision maker should receive only one informational item, defined as a single dimensional fact (e.g., the location or amount of some influencing force, but not both), every three minutes, which is an eternity in some high-action situations. He was referring to the "push-button" naval commander receiving intelligence on the location or strength of an opponent, but his comments emphasize that education, information, and data processing are best designed to reduce uncertainty. Cybernetically, a functional organismic unit thus reinforced can reach a conclusion, make a decision, and prepare an action.

An effective rate of acquisition is formed by the individual's ability to receive many units or bits of information and to filter, accept, or reject them in the process of framing an action or a storage. It is easy to see how the reduction-of-uncertainty-function persists as a managing concept through these steps (Fig. I-3).

The educational burden in the developing biological sciences is maintained within reasonable limits with the help of an intrinsic management. As carcinogenesis is adequately blueprinted as to origin and development, the alternate theories become merely historical. Although there are serious scientists who have come out in favor of complexity because of the intellectual exercise and excitement, the student deserves the encouragement afforded by his finding simplicity in nature as observed by Szent-Gyorgyi (1964).

Cellular information includes the basic data in the macromolecular genetic code for scheduling development, describing unit organization, transport and transformation of products, and response to external stimuli. Basically the information in a single fertilized human cell specifies the final product—an individual with about a million billion cells. Operating data exist in every cellular, chemical, or energy gradient, and ionic disposition, so that mere displacements are meaningful. The cell *learns* to make its specific products, gains experience and retains impressions like traces of biomolecules, scars from viral attachment, and behavior patterns. The environment continuously contributes information as illustrated by the immunological response which produces antibodies, or the presence of ingested food which causes the appearance of appropriate enzymes.

The organism is seen as coupled, tightly or loosely, to his environment. This impression is one of an individual with an information net consisting of some ten thousand million neurons acquiring visual messages and interpreting them at a flow rate of about one hundred million messages per second. It is a very significant moment indeed, therefore, when an infant first opens his eyes and begins to bank his data. The audio system meanwhile will be coupled into the environment via more than 30,000 terminal receptors leading to the hearing center in the brain. Learning is a procedure in which the capacity to process the data is developed. The individual exercises some control over the closeness of his environmental *coupling* by his behavior. The data are screened and unwanted signals are squelched by learned behavior. Thus, a person on entering a noisy party room can approach and communicate with another person while unconsciously excluding the background noise.

The cell's program may be changed by mutation or transfer of genetic material, in which case, new information is stored. It also maintains the total information genetically in a redundant manner even if the cell is specialized for a limited operation which does not require *all* the possible information. This is perhaps a capability held over from the ability to regenerate anatomic parts. Retrieval of information is under genetic control which may be exerted by inhibitors, anti-inhibitors, or redirection devices (operons) according to external signals. There is information management in the life of the individual, ontogeny, and biological reproduction connects the process within the life of the race, phylogeny, so that information and learning are equally important to both the individual and the race. One result for the race would be natural selection.

We then have: 1) the body of biological information in circulation and, 2) the planning and operating data resident in the organism. Many biological questions concern the nature of this information itself. It would seem natural in this study to cross interdisciplinary bridges freely because information analysis is basic to all disciplines. There is particular interest in the theories developed in communications, information, cybernetic analysis, and biosystems.

Introduction

BIBLIOGRAPHY

Ashby, W. R., *Design for a Brain*, Chapman and Hall, London, 1952.
Blesser, W. B., *A Systems Approach to Biomedicine*, McGraw-Hill Book Co. New York, 1952.
Copson, D. A., *Microwave Heating*, Avi Publishing Co., Westport, Conn., 1975.
Hoffman, B., *Einstein, Creator and Rebel*, Viking Press, New York, 1972.
Lindsay, P. H., *Human Information Processing*, Academic Press, New York, 1972.
Lofgren, Lars, *Recognition of Order and Evolutionary Systems in Computer and Information Science*, 2nd Ed., (Editor: J. Tou), Academic Press, New York, 1967.
Miller, G. A. H., "A New Cardiac Catheterization Laboratory at the Brompton Hospital," *Bio-Medical Engineering*, Sept. 1967.
Schwartz, M., *Information, Transmission, Modulation, and Noise*, McGraw-Hill Book Co., New York, 1959.
Singh, J., *Great Ideas in Information Theory, Language and Cybernetics*, Dover Publications, New York, 1966.
Streufert, S., "Information and Decisions," *Naval Research Reviews*, Vol. 23, Sept. 9, 1964.
Szent-Gyorgyi, A, "Assumptions of Simplicity in Nature," *Science*, Dec. 4, 1964, p. 1278.
vonNeumann, J., *The Computer and the Brain*, Yale University Press, New Haven, Conn, 1958.
Wiener, N., *Cybernetics*, MIT Press, Cambridge, Mass. and John Wiley and Sons, New York, 1948.
Wiener, N., *God and Golem, Inc.*, MIT Press, Cambridge, Mass., 1964.
Yockey, H. P., "An Application of Information Theory to the Central Dogma and the Sequence Hypothesis," *Journal of Theoretical Biology*, Vol. 46, 369-406, 1974.

Films

Medical Electronics, McGraw-Hill Book Co., New York. 30 min, color. #689355
Miracle of the Mind, McGraw-Hill Book Co., New York. 28 min, color.

1

Biology Conceptualized — The Intersystems Approach and Radiations

"I sail on the wings of the wind but the hand on the helm is mine."

D.A.C.

War is not without its legacies. Ours, from World War II, Korea, and Vietnam, is a host of new observers of the living system. They are multidisciplinary scientists and engineers who form new work designs around conceptualizations that are fascinating extensions of the scientific method. One way to examine this powerful new force for creative synthesis and thinking is to make an "information series," selecting 12 informational concepts for consideration. Taking some liberties in going from general to specific, this would look like the following list:

1. Information
2. Bioinformation Management
3. Knowledge and Data Bases
4. General Systems Analysis
5. Modeling and Programs
6. Cybernetic Analysis
7. Operations Research
8. Feedback Information and Communication
9. Servomechanisms and Self-regulation
10. Automata
11. Closed and Open Loops
12. Organismic and Digital/Analog Computing

When the concept of *information* is carefully examined, we may find some unexpected developments. In the first place we are not so much concerned with information theory as we are with informational pathways, electromagnetic radiations, the management of information via specific meaning, and its recognition in biology. In our view, information, like most everything else in this world, varies in quality, with highly specific, noise free, strongly coupled information representing the high quality kind. This is a pragmatic view, stressing the logical intersystem connection provided by radiations. This concept leads easily to the electromagnetic spectrum and its wealth of information, as well as to local systems, dominated by vision and photosynthesis, but also exemplified by all the photobiological processes. Molecular addressees in the spectrum for the radiant information, such as chlorophyll and melanin and others, in fascinating periodic or rhythmic, phototactic, photomorphogenic, and phototrophic systems, are biochemicals, and often beautifully colored. However, the spectrum is broader and its biology much more diverse even than these.

The realization that *information* and its corunner, *energy*, with their carrier, *radiation* form intersystem channels or connections, permits a view of associated systems rather than of isolated ones. Then these associations form the wave structure for an electromagnetic, biophysical environment. Designing brains may or may not be too optimistic, but it is best to make one's own evaluation and not be denied any of the excitement that is properly a part of this age of automata, electromagnetics, and the enormous power in the atomic nucleus. Not all conceptualizations like *Origin of the Species* and *Natural Selection* can be immortal and most of those in the 12 listed will wax and wane, for they are "breakthrough-dependent" and sensitive to the amplification of new methods. When this happens, they seem to bloom. Then slightly different concept-theories emerge and there is sometimes imperceptible, but real, narrowing of the multidisciplinary gap.

INFORMATION

Information is so essential a part of data processing that many of its faces have been rather formally defined. For our conceptualization, the most useful is that in the *Oxford English Dictionary*. It tells us that to inform means to give form to, to give shape to, to impregnate with some specific form, and so information involves an energy interaction that reshapes and reforms. To inform means to give determinative character to, to stamp, impress with some specific quality, impart some active quality and make it what it is. It means to direct, guide and discipline in some particular course of action. Information is the action of informing matter or capacity to do so and the knowledge that is to be communicated. We can thus observe that radiant information connotes interactions with the material and chemical systems involved, and certainly deals with the form factors that are involved in radiation action. There is a connotation of involvement in which the impact is a changed system. The changes are in the form factors which we observe in these studies and they require energy. Strangely, mental processes, as opposed to photosynthesis and so forth, use only a tiny fragment of the universe energy, and here, energy and information are related to entropy. Thus a matter that concerns us the most, shaping our minds, turns out to involve only a small part of universe entropy and this happens in a small subsystem, the biosphere.

Information is basic enough to even involve motivation and drive when the action brings promise of a desired result. This is a self-organizing drive in which interaction produces feedback that motivates the biosystem. In the probabilistic view, information is measured as a matter of uncertainty. Information is such that a priori uncertainty yields to a posteriori certainty. Information is best considered in information theory as data that can be expressed continuously with time in a sequence of numbers to form a time series. There is a screening connotation in information in which only specific results selected by energy transducers are passed along. Still another feature in harmony with our concept is the causal classification which refers to information content received from sources external to a specific living system which can cause an event according to the laws of nature. This, combined with the systematic definition that information can change the state of systems, provides us with at least a literary framework.

A model for this concept is furnished by phototaxis. Shining a brilliant light spot on a culture of purple sulfer bacteria causes them to accumulate from the dimly lit portions to the spot within 10 to 30 minutes. Once trapped, the *change* in the light they receive when they touch the boundary of the illuminated area triggers a reverse motion. The specific information is contained in the exact wavelength of light that also activates their photosynthesis. Thus it is received by the chlorophyll and carotenoid light receptor pigments, each of which "knows" its own desired wavelength. The advantage of this response is that the bacteria are actively polarized or oriented, and able to localize, within a spectrum of energies, in the specific *energy* and *intensity* they need for most efficient photosynthesis. They do not actively seek the light, but their random swimming causes it to find them, and then they use their reverse response to *hold* in the pattern (Fig. 1-1).

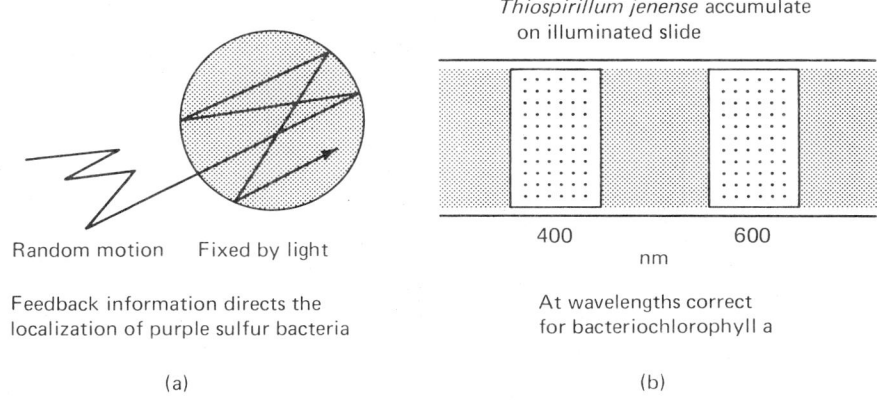

Fig. 1-1. Phototaxis in photosynthetic bacteria that guides the organisms into a "hold" pattern is an example of intersystem information.

Wave mechanisms are not limited to electrical systems, and information arriving in phototaxis continues to propagate within *Rhodospirillum rubrum* as a chemical wave of metabolite concentration. It communicates such information as switching on and off for response to stimuli, refractiveness to accommodation, periodicity, amplitude, duty cycle, and summation of subliminal stimuli. This kind of communication is even more natural in chemotaxis and illustrates the cooperativity of a biomembrane. Only 70 molecules of lethal Colicin E, on the *E. coli* membrane—that is, attachment of the ligand to only a few dozen receptor sites—means that a wave will spread out over the membrane in the form of a redistribution of membrane elements in the fluid lipid phase (Blumenthal 1975, Singer 1975).

The self-propagating waves were found in chicken muscle fibers in culture with a "wavelength" of 10 μm and velocity of 60 μm/sec. (Coleman et al, 1972). Cyclic AMP in cells responds to transient hormone messages and similar systems produce propagated oscillations. Redox systems are involved when, as in the cytochrome system electron transfer structure, the components of the multienzyme system oscillate between the reduced and oxidized states. The sarcoplasmic reticulum releases Ca^{++}, and many systems such as glycolysis, respond to pacemakers. Oscillations occur in membrane transport, and cyclic AMP is released by specialized amoeba in rhythmic messages to form the fruiting body in *Dictyostellum discoideum* when it encounters unfavorable nutrient conditions on the forest floor.

Radiation and information have the same model in that both have a source and a receiver, and a channel that connects the two. Similarly, both answer the definition of being something that may be generated, measured, stored and transmitted. Information theory dates back to Nyquist and Hartley in 1924 and 1928, the latter providing us with the logarithmic measurement for ordinary information content.

Transportation of information requires energy, and, between systems which are far apart, that would have to be radiant energy because information, as in vision, cannot be sensed without energy to move it to the sensor. But information can be generated and communicated by living cells and organisms. Alpha and delta waves in electroencephalogram patterns, the general emission of microwaves from active biomolecules, cardioelectric rhythms, and countless other examples of quantum generation exist. Energy cannot be created except by converting masses. Otherwise it can only be converted or transmitted. The brain has the capability of information storage and management, data banking and processing functions. The information it manages may have passed between systems as a result of radiation or other energy and material interactions.

Knowledge and Data

Knowledge and data are obviously closely related. Knowledge is a mental or otherwise recorded situation report (SITREP) on the world, formed from information acquired by sensory intelligence agencies as an output of their programs for learning. Certainty and uncertainty rest on knowledge, and thus are what is measured in information. Knowledge is of a more superior nature than data since it may be presumptive to say that one's data represents knowledge. On the other hand, data is more neutral, and it can be that information is data which is to be communicated, where the data might even be musical notes conveyed by a singer. A singer's vital signs and their transmission, with proper instrumentation, would be data of a different sort, but the song would then be incidental.

The foundations of knowledge are found in codes, symbols, signs and their information, all acquired by programs and experience. Philosophers may use knowledge to mean epistemology, which is the nature, grounds, validity, and limits of knowledge. Differences in meaning are acknowledged, but it would be difficult today to cover eons of civilization in histories of knowledge without the concepts of information. In the verb forms, *to know*, as in the result of being informed, the two concepts of knowledge and information converge compatibly as they do in modern, informational, computer, library and retrieval science.

A neuron receives information and stores it by molecular means, but it may be unable to associate it to form knowledge. Thus information sees the axon as a gate where, under the management of the nervous system, incoming information is self-organized by association with the necessary synapses, organized for a network searching function, and aimed at a consensus. In these transmissions, the quality of information may be designated by the success with which the desired signals can be separated from noise. Thus the signal to noise ratio suggests how much good data may be obtained from noisy transmission.

Noise is actually another concept that surfaces whenever there is a need for communicating signals in its presence, and sound or electrical systems are merely examples of this concept, which is common to all communication systems. Noise in electrical systems has a voltage, just as the signal does. If the noise voltage can be decreased to one-tenth of its original value, then the signal could be similarly reduced, all other things being equal. Quite appropriately, the noise is measurable in decibels and decreasing the noise in a typical microwave information system (radar) by only 3 db would mean we could cut the necessary power by one-half.

Insects, birds, bats, and frogs manifest signals in navigation behavior. Female moths attract males from incredible distances by pheromones, glandular communicating odors. In the silk moth, it is a 16-carbon alcohol, bombykol. The signal strength here is measured in molecules per second, the effective strength being about 0.5 mol/sec. Salmon also sense the natural odor of their family stream, always returning there to spawn. The music of crickets, coqui tree frogs and long-horned grasshoppers is essentially data transmission for attracting and screening suitable females, establishing territories, and other social operations. Publicized navigational accidents which can be literally lethal to whole populations of migrating species can arise when the signal to noise ratio does not permit the correct information to be received.

Information Management

The word "management" has two uses in *information management*. First, it suggests that information may be manipulated and handled in different ways. Evolutionary diversity for example, might be guided toward desirable results and information may be lost by unnecessary destruction which is avoided by good management. It also implies feedback and the ways that humans may differ from animals through the matter of feedback, by "knowing that one knows." This may be a human type as opposed to some simpler form of conceptualization and recognition for animals. Also, the hand on the helm in evolution would connote a feedback so that people might improve.

Management implies the executive and administrative functions also seen in operations research. Biologists are familiar with the management of infectious diseases, and of medical problems involved in bringing a disease under control. The theory of epidemics gives a popular mathematical model for distribution of an agent through a population where there are three states: susceptibles, infectives, and removals (by death or otherwise). Many time-dependent nonlinear processes follow, and are described by a system of differential equations. Goffman (1970) uses this model to analyze the growth of science.

Endemic diseases are always present in a population but do not cause high, abnormal rates of sickness. Then there is a periodic recurrence of childhood diseases, measles for example, increasing in incidence every year or two. With various models and conditions for new infections, deaths, type of contact, latency, probabilities, immunities, and velocities of propagation, the real populations may be simulated with the hope of evaluating the status and danger of various epidemics (Bailey 1967). There are also the matters of planned inheritance, population management with renewable resources in mind, and diagnostic cytogenetics which includes a sampling of the embryonic complement of chromosomes to detect aneuploid cases likely to be mongoloid or otherwise abnormal.

The busy output of brain waves visible in an EEG implies a communication and information center in action. That the action is orderly is shown by the precise frequencies, for example, about 10 cps for alpha waves, as messages are calmly routed to addresses and replies processed over networks for communications. The order is self-organized and results depend on the action itself. Questions are easily referred to managing centers which can refer, interrogate, correlate and operate on probabilities, calling on all the senses as need be, and establishing the necessary level of alertness—no more or less than necessary. What emerges is a sort of magic fabric of memory, communications, and retrieval woven into electrical and chemical patterns.

Biology Conceptualized

To accommodate this management of information, the brain commits large resources to obtaining sensory information. Vision requires a very large proportion of these resources. It must operate from imaged records, a temporary copy, to feedback signals to the intelligence agencies in visual control. One-third of all the nerve fibers entering or leaving the central nervous system are carried by the optic nerve. To keep the system smooth, the movement of eyes is blanked out as necessary to make one picture blend into another while, at the same time, the brain automatically classifies the information based on comparisons, newsworthiness, correlations, statistical probabilities, and the meaning of associated information bits. Normally, no two sequential results can be identical because there is a time factor with new combinations to be expected in the next moment.

Instead of standby generators and alternate power supplies, the brain matter is composed of blood vessels, collateral circulations, and provision for hooking up necessary nets for blood power on demand. Complexity brings peril, for with these energy and information requirements, disaster is always only one circulatory failure away.

Machines, *too*, grow in complexity and responsiveness. A good example of this is the Viking probe. It is capable of depositing itself on Mars only seconds from target time, with telemetered signals moving at light speed in light years of time. Its singular capability, like that of living systems, is the adroit management of considerable information.

> Brain structures, like such well-designed spacecraft, are important, but mainly for the information they contain and process and their continuity through off-on periods. In its magic fabric that is constantly being renewed, the brain images or reproduces the pattern of sensory reception in its own medium. It is certainly busier than a modern telecommunications center, more like all of them combined and operating on multichannels with over 10^{10} neurons in the cerebral cortex. The connections are personalized, connecting us with our individual and racial history, erasing here and repeating there, calling on our feelings, fears, and values. Then sleep, like a master switch, closes down a part of the operation, leaving only a skeleton activity of unconscious connections and dreaming activity in which motor centers stay asleep unless the sleeper takes a walk. There is never a need for a "hard copy facility," so coding and decoding *are* the operations that are determined. The volley of pulses of synaptic transmitters allows decisions and discriminations as the receptors integrate over the forces and pulses being transmitted. As in electronic devices, it is far easier to do all this with electrical *charges* than by moving *masses* of chemicals around at velocities that may approach the speed of light. The data passing in nerve impulses may have meaning only in their groupings according to frequency and modulations. The information is in the selected multineuron channels like red and blue giving violet, and memory enters via the channel selection. Neurotransmitter volleys and regenerating or other enzyme reactions are excellent for transmitting meanings via chemical gradients, grouping, and velocity modulations.

Systems

The "legacy concepts" are interwoven, especially with systems. This may be called systems science, systems analysis, systems theory, or general systems theory and involves the engineering of parts into systems where something special is contributed through cybernetics, computers, and automation. As a natural outgrowth of technological complexity and situations like the man-machine interfacing, systems study is of recent origin and, like the other conceptualizations, is antidotal for the dangers of overspecialization. At the same time, it has a certain philosophical rudeness that offers new dangers in the sense of human impersonalization if not interpreted correctly. It provides a powerful force toward organization in any size enterprise where a *team* effort is involved and the man-machine interface is apparent. As if the complexities of the living system were not sufficient, it was interfacing that signaled the requirement for systems organization and brought out support. Missile systems, weapon delivery systems, and air traffic control systems all presented new orders of complexity. The broad look of systems coupled with team expertise tended to make these more manageable, and their costs seem less formidable. The multidisciplinary approach that is implied could then be profitably extended toward the environment and this returns the concept to living systems. Experience with man-machine systems remains to prevent overindulgence in environmental protection and the concept gives us *overseers* such as cost effectiveness in which an optimized result is expected with a minimum cost.

Systems study means considering a problem, presenting various solutions from different points of view, and dealing with "entireties" especially. It is most likely seen where orders of magnitude in growth or complexity are proposed. In biology, the natural level of hierarchy affected is most likely in the higher strata of living organization, more of necessity than reason. One objective is economy of effort because generalizations learned for one system may be amplified to multisystems. Mass action is an example. If this can be recognized in a new study, the older rules surrounding it may be applied without having to rediscover the applicability. Similar systems subject to common rules may be lumped.

That the whole is greater than the sum of its parts is hardly a new concept, but that in itself does not express an analytical and organizational idea that is problem oriented. Sales of automobiles depend on the complete ride and not solely on the color, wheels, or motor, etc. This systematic thought ushers in cybernetics applied to car design, and it would be a misguided sales manager in Germany who tried to sell an American "soft" ride to Germans who prefer to feel the road more. Classically, scientific analysis requires categorization and then the parts of various classes may be viewed for their dependence. It may be simpler to deal, at least mathematically, with noninteracting parts. If the action in the parts can then be reconstructed into the system, little more than classical science is required. Systems come in along with the difficulties,

Biology Conceptualized

especially the appearance of nonlinear relations, where parts interact in complex ways. The utility of the scientific method is thus simplified and the computer is the tool required. The whole system equation in a simple system may relate a family (same form) of equations in the subsystems by a summation of their linear processes. These equations are likely to be simultaneous, nonlinear, differential or algebraic ones. As the analysis proceeds to something more complex than can be handled by either algebra with many equations or ordinary single differential equations, the computer becomes increasingly important in making the supporting math approachable. This trend develops into a surge as the power of math is succeeded by the much greater power of computers until it may *over*power the experimental information and conceptualizations available for the multidisciplinary effort. At this point, there is likely to be considerable noise from the opposition, and retrenching in the systems camp, while awaiting a breakthrough. The systems people and their equivalents in the case of the other concepts are dependent on mathematical advances in probability, statistics, and the correlation and smoothing of measurements. Some approaches which have provided assistance are:

1. Cybernetics, using feedback in loops.
2. Nets as applied to neural networks.
3. Sets as a group of axioms used to mathematically formalize the relations found in subsystems.
4. Theory of chain-distributed and centralized *compartments* which separates system elements, analyzes transport and bounds.
5. Graph theory which helps analyze problems with surface or topological features using matrix algebra.
6. Theory of games used to predict outcomes by making a game out of a conflict in which the players study *moves* that give them advantages.
7. Decision making and automata.

An example of systems approach is the division of systems into closed and open subgroups. Then the open system, of which the clearest example is an open flame oxidizing fuel in a rich oxygen atmosphere, is observed to differ significantly from closed ones. Most biological systems are open and most physical ones are closed. The concept of an open system leads to analyses such as that of dynamics and the recycling of renewable units which is characteristic of living systems.

Models

A model is an imaged theory, crystallized from the observer's analysis of a system's configuration. In a cybernetic model, the feedback is visualized in a set of blocks illustrating the flow of information. The form of a working model must be suited to the intended investigation. It may be a graphical representation of flowing information, a set of differential equations, or block diagrams with input and output. A forcing function F, for example, may be represented by Newton's second law in differential form

$$F = M \frac{dv}{dt}$$

in a mathematical model.

The common objective is clarity and that the model be true to conditions without oversimplification. This type of analysis provides a neat lesson in dividing the subject into manageable units which then receive closer study. Norbert Wiener viewed cybernetics in this manner, developing the analysis by logical divisions into black boxes. He adds broad emphasis on feedback control. Then, what is managed is the behavior of controls via detection of their efforts. When the black boxes are connected, information flows and feedback permits discrimination by the controls (Fig. 1-2). In this way a large information system is divided into small loops or a system may be synthesized from many loops.

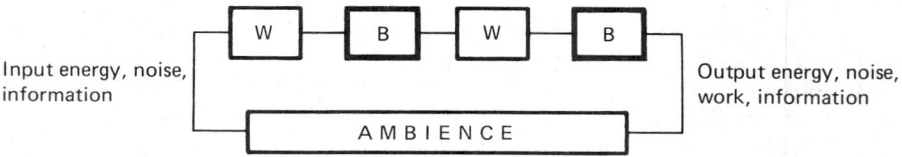

Fig. 1-2. Coupling of ambiental situation and elements of a system with white and black boxes. Organic units have their rules for action. The energy and information input would be useless without directions for their use. With these directions, the system can be self-organizing.

The informational pathway is the independent one. Therefore, it gives the meaning of feedback and is the actual input and output of black boxes. While energy and materials are used, they are usually in excess or a part either of the functioning of the power supply (dependent variables) of neural operations or of the structure. Thus, they are only independent variables when independent of other input variables, which might be the case if, for example, energy happened to be the critical or limiting factor and not in adequate supply. A problem comes when the white box ignominiously lumps the distinguished work of great scientists together, sort of taking their contributions for granted. A nerve impulse may only look like a line and arrow, but it represents the life work of many outstanding persons from Fricke to Young, Hodgkin, Huxley and on to Cole and others. A working concept leads to a model which is often built and tested by computer.

Programs, somewhat like those in computer software, have a broad significance in informational biology. J.Z.Young, one of this select group of neuroscientists has, in his current work, *Programs of the Brain* (Oxford University Press, 1978) carried the information theme into biopsychology in a manner that adds immeasurably to the bioelectromagnetic meanings developed here. A program refers to a coded sequence in time for goal-oriented, informing actions and brain operations where the goal is self-perpetuation of an individual's programs. It is like *mind*, used as a verb, as in minding your business or mind your ways, or less concretely in the usual sense of mind and brain. There is a programming in life as in social programs rather than "program" in the sense of agenda. Here the basis for informing is found in communicating *molecular codes, brains*, and their *manifestations in speech and record*.

Cybernetics

When Wiener felt the need of a new word to express the "essential unity of the group of problems centering about communications, control, and statistical mechanics, whether in the machine, or in living tissue," which he and Rosenbleuth had recognized, he gave it the name cybernetics for the Greek term which, with English letters, would be Kybernes. He then credited Maxwell with the field's first article on governors (1868), and noted that the word governor is a Latin corruption of the Greek word. Then, governors are associated with control devices well-known for assuring a steady performance in ships' steering engines.

Modern cybernetics resembles an analytical philosophy in the range of laws of thought and consciousness, and today, having joined forces with computers, it concerns the storage and processing of information in symbolic and logical form with binary notation. This statement has a limitation in that the binary logic may not readily encourage or recognize deviations due to inspiration or creativity like flashes of light in the creator's brain. These events are not black and white, and binary logic recognizes only 0 or 1, on or off, yes or no states. Inspiration occurs at a resonance point where circumstances provide an "eased in" flow of information, somewhat like a natural frequency for the inspired system. Cybernetics orders the system, minimizing deviations and so, seems in conflict with itself until it is seen that it is exactly the ordered, cybernetic system that creates with this order, opportunities for the significant deviations *to be significant*. They stand out boldly from the stabilized system, ready for selection processes to brand them as useful. There is also an element of purpose so that goal or mission orientation is an attribute of cybernetics.

The First International Congress of Cybernetics in London in 1969 formalized this multidisciplinary science. Perhaps Plato would have been surprised, for he regularly used the term in the literal sense of steersmanship as well as for the science of governing men in society, that is, government and politics. Addressing Socrates at one point, he said, "The cybernetics of men as you Socrates often call politics." It is also likely that Ampère would have perked up an ear since he used the term formally in his encyclopedic work on human knowledge in 1943. Thus, governing in the sense of regulating and guiding is a close relation of cybernetics.

Wiener tended to popularize his new field, hoping thus to assure its growth as a deserving area of study. The popularity has attracted many intellectuals in social science as well as in natural science so that to be a cybernetician is not necessarily to be a mathematician. The results are uneven, being proportional to the degree of naïveté or astuteness brought to bear on the model. Wiener extended World War II fire control and guidance systems analysis to peaceful situations in living systems. Here were analogous concepts of systems organizing themselves and controlling active loops using information, feedback, and activators. The anti-aircraft problem involved microwave ranging and detection (radar) with responses in the aiming of weapons. The living system receives sensory information which enables it to respond and steer activities of all kinds.

Cybernetic analysis in biology or the humanities may not be mathematical, but instead may operate using plain language. Cybernetic steering of evolutionary diversity by feedback (along worthy lines) seems sound. A religious viewpoint, which explains the loop nature of moral behavior as life in a circular system with feedback, is evidently close to the Golden Rule. With ouput being morally bad, the loop returns a hopefully corrective sampling of the bad result. This is cybernetic phrasing of "do unto others as you would they do unto you." Perhaps the appeal of cybernetics here is due to its intrinsic *mission*-orientation, and to the more easily understood logical loop with feedback as compared with the theological model. If so, it is probably due to the more reasonable and humble nature of this specific cybernetic task as compared with the enormous responsibility assumed by religion.

Cybernetics has a learning connotation relating cause and effect, with no feedback evident in those human laws stressing retribution, but operating clearly in those that suggest "turning the other cheek" or good for evil. Cybernetic analysis in the world of models and machines may also operate independently of abstract mathematics, being concerned with concrete devices. It demands equal rigor, for errors can be disastrous, as would be obvious in the case of the kidney machine that fails or the iron lung that is helpless even within its known operating specifications. A religious perspective is offered by Wiener in his little book *God and Golem*, which refers to the fabled prophet automaton, Golem, created by Rabbi Loew in Prague. Golem, who had superhuman answers, has a counterpart today in computerized mating and other decision assisting machinery. Wiener's message which probably evoked the title of his book was

> "Render unto God the things that are God's and to the computer the things that are the computer's."

Both mechanical and living cybernetic systems function on information which is also their feedback. This operates the loops and presents a requirement that the information be detected or sensed and communicated to give guidance. The dependence on information is illustrated in Ashby's law in which this cybernetician states that the amount of correct guidance that a control device can achieve is limited by the amount and quality of the information that it receives and digests. Detection and sensing are usually not enough; discrimination is required. Ordinarily, a sensor is some form of transducer or device which receives energy in one form and transmits it in another. Thus, visible light energy, by transduction in a photocell, may continue on as an electrical signal, or electricity, via piezoelectric crystals, may be transformed into sound. Another step is involved in the cybernetic system. There the transducer, by separating one form of energy from another, acts as an energy and information discriminator. A cybernetic device will then receive information only via compatible transducers, rejecting those forms not subject to the transduction. Noise, for example, might be so discriminated against. If cybernetics is to operate between distant systems, it must base its loop on radiation information or on action-at-a-distance. This is reminiscent of its original application in fire control against the bombers that threatened to erase London from the English countryside in World War II.

It is probably accurate to suggest that the Soviets have an abiding interest in control and regulation. So, the work of their mathematician A. N. Kolmogorov is often cited in the beginnings of cybernetics. Perhaps they should have encouraged him to publish as widely and boldly as Norbert Wiener did. Nevertheless, today, a mathematics section of the Academy of Sciences is known as the Institute of Cybernetics, Kiev, Ukr. U.S.S.R. and a comprehensive journal called *Cybernetics* is published there in Russian and translated into many languages. There are other such organizations in the Academy of Sciences including the Institute of Engineering Cybernetics as well as similar organizations in the satellite countries. The informational basis for a cybernetic system is very evident in the Soviet work, for example, in the algorithmic schemes of Kolmogorov. The algorithmic system recognizes commonality in the structure of problems, and devises an algorithm for their solution which uses a language such as MAD (Michigan algorithm decoder) that is understandable by both man and computing machine.

It may be that cybernetic ideas may be most fruitful in the studies on senescence or aging, especially if the life span of humans can be raised from today's levels (68.3 years for males and 75.0 years for females) to those "200 plus" lifetimes that are now predictable. The secret seems to lie in control over endocrine changes in a brain system which self-organizes according to the information from its own control operations. Other animals seem to be even more locked into a deadly control system. A death mechanism operating after procreation is seen in the death of a species of Australian mouse, when, after copulation, it receives a surge of pituitary ACTH. The same hormone kills the Pacific Northwest salmon who fight their way up their native rivers and then die within two weeks after the spawning effort. Perhaps here a positive feedback mechanism designed to support "all out" effort leads to system instability, rapid aging, and the final moment between life and death.

Since information is an intersystems connection carried by radiant energy, the eye dominates the banking of sensory data. Neuro-cybernetic operations occur as one enters a dark cinema and portions of the eye control system can set the scene for describing a feedback information pathway. The mission is to find seating. If the entrance is during a quiet scene in the film, many feedback loops may be operating. A delicate touch to the left may indicate that the seat is occupied, then to the right to find the next row and so on in the ensuing oscillation or search. The sensors are *vision*, which is accomodating itself by feedback and adjustment of the iris to open the pupil and admit what light there is available as the retina signals that it is in the dark, and *touch*, which may feed reversing information, that is, touch something to the left, move to the right. This is a kind of negative feedback where positive feedback would instead mean moving further to the left.

V. L. Parsegian who wrote *This Cybernetic World*, likes to use the example of holding a weight by a spring. The weight can easily oscillate up and down illustrating simple harmonic motion and periodicity, which is as fundamental as night and day. Motion of the hand *opposite* to that of the spring decreases the amplitude or dampens the action. Then, moving the hand *with* the direction of the weight can send the system into violent motion illustrating phase relations. These can be of special, even critical, importance when taken with some function that is forced into a system at its natural frequency of oscillation. Every wheel has such a frequency for vibration and it may be necessary to add weights to the wheel so that the amplitude and phase will be controlled and it will rotate smoothly at ordinary speeds. Rotating systems often have elements that may show excessive vibration at some frequency, necessitating operation on one side or the other. A dramatic example was provided by the Tacoma Narrows suspension bridge in the state of Washington. Vertical vibration was introduced by the force of the wind giving variations in simple harmonic motion. With the wind force applied in phase, and strong enough, sufficient energy passed into the oscillating structure to bring it down. Stiffeners in the new bridge prevent this destructive amplification. In phylogeny, or the history of races and populations, society may feel oscillations in numbers due to epidemics, famine, ice ages, wars, floods, and earthquakes, giving a rise and fall with time. To some sectors, these oscillations may mean opportunity—to others, disaster. Bubonic plague caused by the bacterium, *Pasteurella pestis*, which moves from rat to flea to man, decimated populations until the twentieth century arrived with its control procedures, especially tetracyclines, streptomycin, and chloramphenicol. The Middle Ages were marked by an epidemic of plague that killed one in every four persons in Europe. It is the virulence of the disease that varies with time so that even though sylvatic plague still makes dozens of persons ill in summer in the Rocky Mountain areas, it is not the

Black Death in which hemolysis causes a breaking of red blood cells and a dark skin.

Proportional control is applied to life situations so that reactions are commensurate with those required according to the information received. Normally, farmers base the size of their planting on weather and similar variables so as to obtain stability via normal crop size. Populations oscillate in response to economic conditions, immigrating and later emigrating. In history these changes were often the result of cannibalism, and periods of stability alternated with the man-eating ones. Modern methods bring advanced forms of food production into the system which dampen the oscillation toward cannibalism, but not all populations are allowed to migrate so there is another dampening. Proportional control is seen in the person-machine situation as in driving a vehicle. A broad system-state may be defined as that which fits over the case, such as calm and fear states which would be coupled to the estimate of danger.

At any time there is a mutation rate in genetic systems and some new departure from the phenotype (sum of traits due to all inherited genes) may be encouraged by the situation which led to its adoption or selection. In the stabilizing situation, one would expect the departure to be easier to recognize, because of the way it stands out.

Biocybernetics and biofeedback would appear to be subordinate concepts to two broad generalizations:

1. That conceived by Claude Bernard, a 19th century French physiologist, the *milieu interne*, and
2. That of Walter Cannon, in the 20th century who used *homeostasis* to describe the mission of regulation in the internal environment.

Many mechanical control systems have been derived from feedback principles, but living ones are much older, as well as orders of magnitude more diversified than those thus far managed. One must assume that it is often impossible to trace the biological loop. However, the possibilities using the feedback ideas are great. Proprioceptors or sense organelles of muscles and their central nervous connections respond to mechanical stimuli, to pressure, or to the action of stretching to give smooth, kinesthetic, feedback-controlled movement in walking, writing, reaching, or just maintaining posture (Fig. 1-3). The signals describe the muscular distension so that, for example, it is not necessary to watch the hand that is trying to brush a mosquito from the middle of the back. Arrows may be used very conveniently to indicate the direction of information flow or specific energy committed to control purposes in such a system (Fig. 1-4). A plus or minus arrow entering a comparison circle represents the algebraic operation intended.

Production in a biochemical system depends on the demand for its products. The difference between production and requirements can then be made the basis for a control action ϵ in the system. If there is a difference, demand may use feedback β to signal an increase or decrease

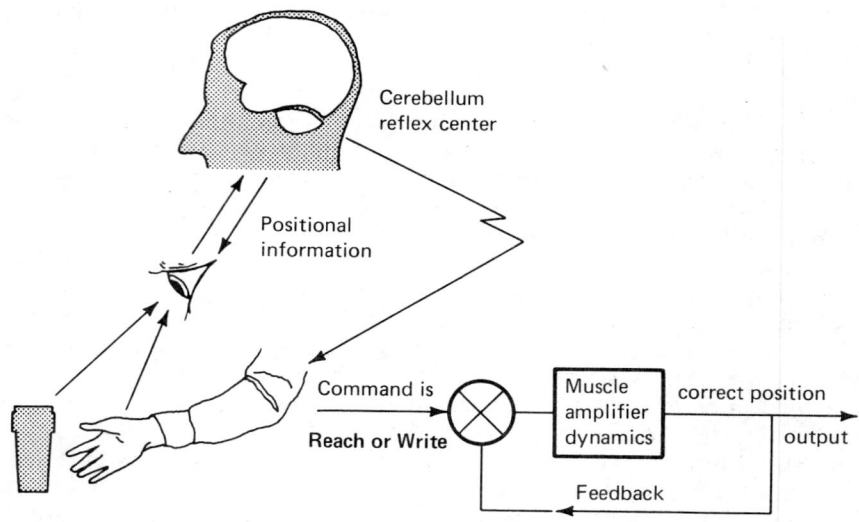

Fig. 1-3. The reaching reflex.

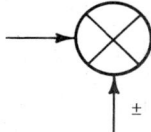

Fig. 1-4. Arrows entering comparison circuit have sign ± representing the algebraic operation introduced.

in production to stabilize the supply at a reference level ρ. In negative feedback, the *signal is arranged* so that the difference between it and the input gives an error signal of control action in negative form

$$\epsilon = \rho - \beta \qquad (1)$$

that commands the controller to bring the controlled quantity closer to the desired level. Comparison points are centers for distinguishing between negative, the common form, and positive, the *enhancing* form of feedback. In negative feedback, the comparison point subtracts, and in the positive form, the error signal is the sum of the input and feedback signals, as an increased output causes a larger error signal. In terms of information flow, positive feedback means that energy and signal, which reenter the forward flow add to the signal change. In negative feedback, the signal change producing the feedback is also reduced by it (Fig. 1-5).

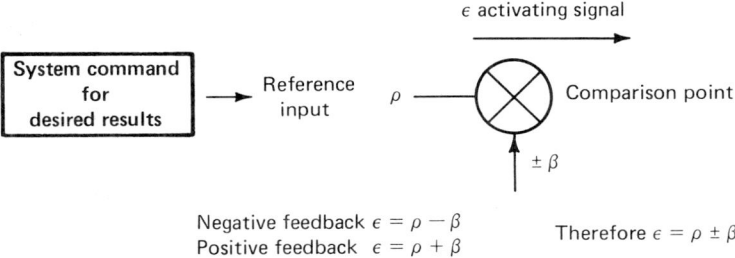

Fig. 1-5. Features of negative and positive feedback in the region of the comparison point.

The electrical circuit carries the feedback energy which usually need not be very large, often a voltage, voltage difference, or rate of change. It is, however, the *information* in the feedback loop that is compared, manipulated, and used to operate the controls. Breaking the information loop means opening it somewhere in the flow path, especially at the comparator, using electrical energy to do so; the result is an interruption in cycled information. Open loops have control over behavior that stems from closed loop characteristics. Thus, the two need not be greatly different. Individual loop gains are multiplied to give the total gain. Although remaining without dimension, this product combines the individual loop gains in the loop. Loop gain, if it is adequate, provides compensation for disturbances. J. Milsum (1966) saw the servomechanism as a control system which, slavelike, followed the reference input, while the regulator or biostat maintained *itself* at the reference level, even when disturbed. Then the former would be like neuromuscular systems and the latter like homeostasis. The gain is often described as the ratio of the response of the system without feedback to its response with feedback.

If, in *abnormal* thermoregulation, a fever made a person produce and keep even more body heat, there would be positive feedback. Normal thermoregulation with negative feedback would make a person produce more heat *in case of a chill* and cause him to conserve what heat is already present.

The actuating information proceeds to the amplifier controller or, more specifically, to the control block which connects to the controlled plant and modifies the controlled variable according to the information it receives via feedback. Consequently, some action is taken on the variable system and the result is an output from it. The need for feedback is the anticipation of some disturbance d which may bring a neutralizable influence to bear on the output forcing it away from control status (Fig. 1-6). Negative feedback "degenerates" d by producing a system response that opposes it.

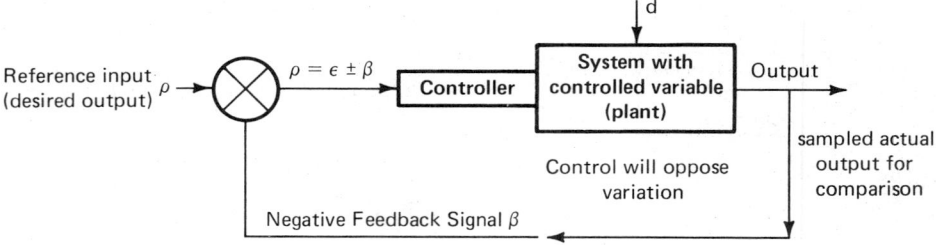

Fig. 1-6. Information flow diagram with blocks, showing disturbance.

Let us return now to the problem of getting seated in the theater. A primary requirement is adaptation to the darkness and in this, activity centers around control of the iris aperture or pupil (Fig. 1-7). The pupil sphincter (circular closing muscle) and dilator (radially contracting muscle) can cooperate in phase to achieve the desired result and a sympathetic binocular response can keep the second eye in harmony. The retina senses the darkness, and it must send a signal in the form of neural pulses to establish feedback. The process of transduction is needed in the flow line to put control information into a consistent form (electrochemical here) for the control system. Retinal nerve nets connect to the optic nerve and chiasm. Then there is synaptic action as the pulses pass from the optic tract in the pretectal nucleus. The information is relayed to complete the eye motor reflex. Both parasympathetic nerves (ciliary ganglion) and sympathetic nerves (dilators) are autonomic, giving reflex action in contrast to the voluntary control seen with the central nervous system.

The command given to this system is "Reduce the light error *variation* by closing or opening the iris," and the system closes except for the tiny aperture remaining after maximum contraction. The controlled output is the illumination of the retina. It is the result of various actions taken on the reference signal, summed at a comparison point, and effected at the controllers. The minimum information consisted of a function such as

$$F = a_1 \, [m(dv/dt)] \qquad (2)$$

deciding the contraction of dilator muscles plus another, such as cooperation in phase from the sphincter muscle (a_2) and constants such as one perhaps for dark eye pigmentation allowing slightly different dilation (−3). Then the pulse frequency information fed to the effectors is determined by the algebraic sum of

$$a_1 \, [m(dv/dt) + a_2 - 3] \qquad (3)$$

and any number of algebraic terms for the feedback could be introduced if the output of the comparator block should need additional factors. Factors are thus introduced meaningfully or for graphic purposes; the

Biology Conceptualized

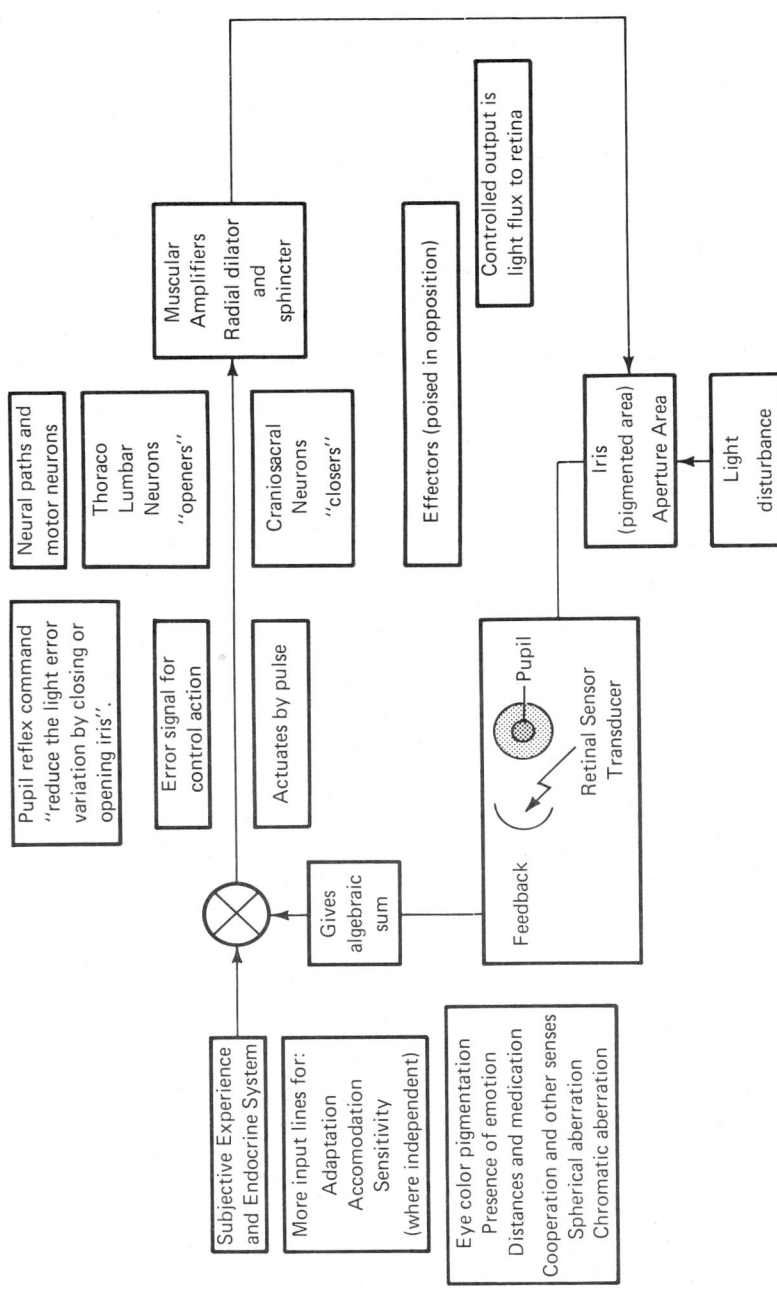

Fig. 1-7. Information flow and feedback control system for iris aperture or pupil operation, closed loop except for minimum opening. Information signals that go into the circle have a sign ± and the algebraic sum of these is the error signal. Establishing the pathway of information is the key in input-output and black box concepts because energy and materials are usually a) oversupplied or b) part of the structure, in power supply or neural operations.

subloops may be interrelated by their miniaturization (Fig. 1-8), a process which illustrates two features of the flow of information. First, the search for input and output in each loop has a way of promoting understanding of the system. Second, the dichotomy has a basic resemblance to many other systems such as electrical circuits governed by summations and other operations at each junction.

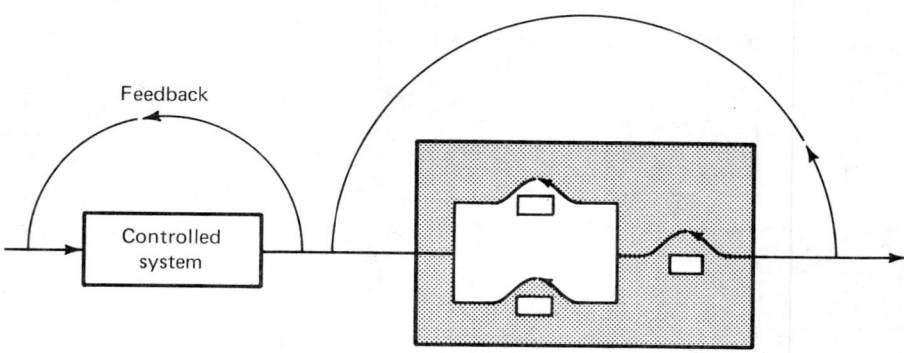

Fig. 1-8. Miniaturization via multiloops.

With the physiology and the feedback for eye motor reflex action in mind, the same system can demonstrate the failure of critical stabilizing loops and two widely used graphical presentations of this instability, the Bode and Nyquist plots. Experimental evidence, (Stark 1950) shows the model for instability. Then, loop gain and feedback equations may be used to establish the numerical basis. The center of the pupil is illuminated sinusoidally, that is, with a light that has a frequency of oscillation in intensity such that the stimulus for the reflex varies in a known manner. Then the response, the iris operation, will be seen to follow this varying intensity. The open loop gain G is

change in aperture area/change in light

This is open loop operation and lacks feedback. The response is expected to vary greatly if the input-output relation permits. The control action does not depend on the output. A home toaster is another example, with the control being a time of toasting that regulates the darkness of the toast. The signal traverses the pathway and is amplified in passing the loop by the value G, which represents the gain. A large gain, from many multiplications in the signal, would have the capacity for great compensation in the face of external disturbance. Thus a house amply supplied with furnaces could certainly gain on sudden drops in temperature outside. For the present case, the sinusoidal disturbance with the light will have its corresponding frequency. The responding system, the area made by the iris for light, will easily be able to follow the signals or will be *in phase* at low frequencies. At higher frequencies, the opening and the stimulus actions may get *out of phase*, thus setting the stage for instability as output lags behind input. For lag equal to a half cycle or 180°, and a gain closing toward a critical value of 1, there can be new, reinforcing feedback introduced and wild oscillation as a result (in terms of a normal stable system). Fig. 1-9 is a Bode plot in rectangular coordinates showing the gain amplitude

compensating light flux/disturbing light flux

as frequency-dependent and decreasing with higher frequency. A frequency of 0.5 *c/s* will move the iris less than a given light will at a lower frequency. In terms of phase, the Bode plot will appear as in Fig. 1-10. At about 1.3 *c/s* for the stimulus, the phase change is 180° or a

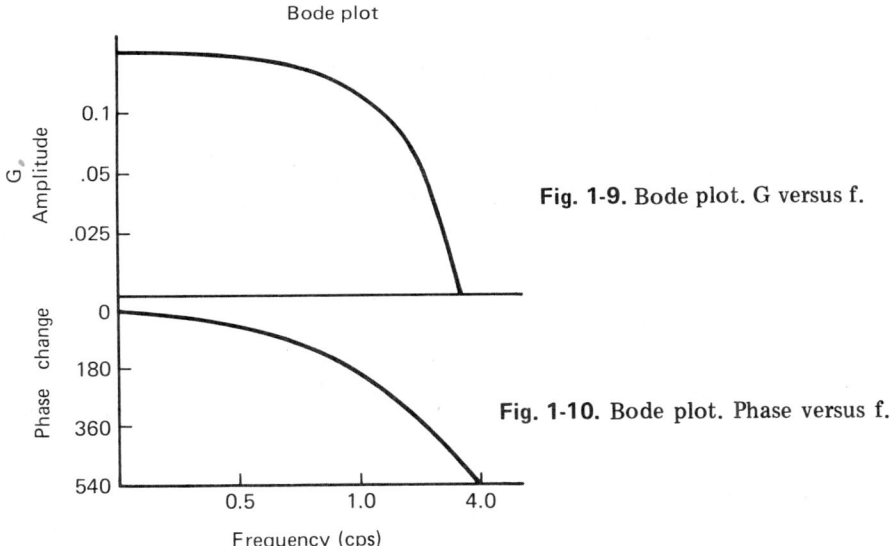

Fig. 1-9. Bode plot. G versus f.

Fig. 1-10. Bode plot. Phase versus f.

(From Stark, 1959 and Randall, 1962).

half cycle behind, where response lags stimulus. In terms of the direction of change, this means an increase in the light will increase the pupil area instead of shutting it down (positive feedback). Conditions for avoiding this crisis are graphically shown in the Nyquist plot of Fig. 1-11. The snail shaped curve is a light stimulus frequency versus response lag presentation in polar coordinates, so it needs an increasing gain value that radiates from the center and an angle measuring the phase between the light and pupil area. If the curve went to 180° and gain 1,

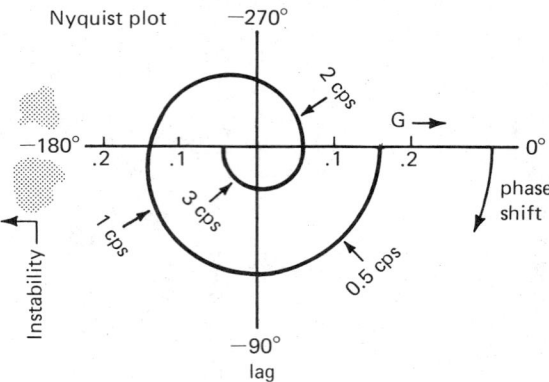

Fig. 1-11. Nyquist polar plot. Light stimulus frequency versus response lag and the instability region. (From Stark, 1959 and Randall, 1962).

there would be instability, but it reaches only slightly beyond 0.1 gain.

Instability can be produced intentionally by edge lighting the pupil at constant intensity. Then the stimulus will tend to be obstructed as the iris closes. This destimulates the system, which then oscillates by itself. Its oscillation will have the expected frequency that corresponds to 180° lag on the curves. The reason for oscillation versus stability is found in time factors in the information management. Sensory data will be out of phase with motor action timing and pulse communication.

Normally the visual system has feedback in a closed loop and the output determines the control action. Then closing one's eyes opens the visual loop and interrupts feedback, rather positively identifying the feedback as sensory information.

There is an example of the biochemical stage for information in the eye reflex since the synaptic transmission described must present signals due to neurotransmitter chemicals. Those like norepinephrine, dopamine, and serotonin control neuron to neuron communication, and illustrate both hormonal control, and, in general, electrochemistry in brain activity. In their absence, the motor reflexes can be inhibited. In serious cases, a disease such as Parkinson's can occur. Their availability is also a matter of youth and longevity and, along with their hormonal interaction, they may well determine how long a living human system can continue to function.

Biochemical information is clearly providing the feedback in the blood buffer system with carbonic acid and sodium bicarbonate, Fig. 1-12. This is an amphoteric system employing molecular feedback via a weak acid and its salt with a strong base in water. The command is "Reduce variation in pH of the blood away from pH 7.4 by ionization and dissociation." The controlled variable is dissociation because the pH depends on the hydrogen ion concentration. Carbonic acid ionizes only weakly. Sodium bicarbonate, the salt of a strong base and weak

Biology Conceptualized

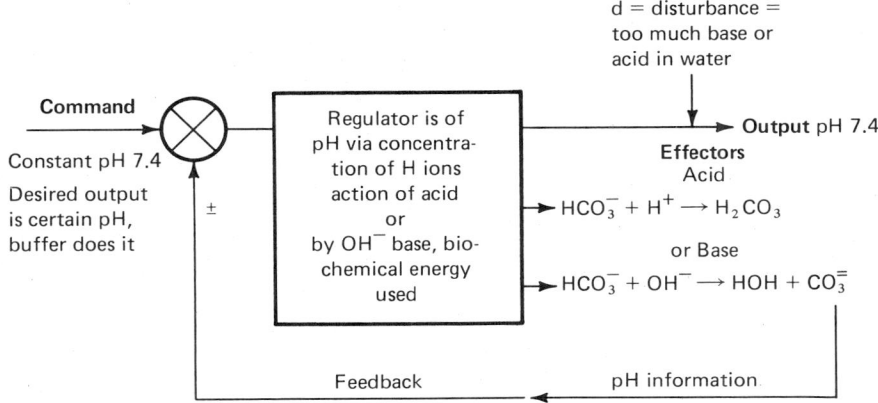

Fig. 1-12. Blood buffer system. Biochemical information controls reactions involving feedback based on molecular buffer action with weak acid and its salt with strong base. Feedback, output sampling, part of the output signal goes back to guide the regulation, actuating signal ϵ, $\Rightarrow$ error $\Rightarrow$ control action via algebraic sum. Amphoteric action with blood, sodium bicarbonate and carbonic acid.

acid readily ionizes. Adding acid produces more combined form of carbonic acid which removes added hydrogen ions. This keeps the pH constant when combined with the action of added base that tends to form water, which in turn captures the hydroxyl ion.

Operations Research

Almost everyone has faced the waiting-in-line problem, and some may take comfort from the work of operations research which deals, among others, with the queueing model. Experience has shown that the critical matters of choosing the size of a convoy steaming through submarine zones and waiting to unload in ports during World War II and the delivery of medical care to "patient" patients waiting in the doctor's office have useful common aspects. (Actually the success of the whole biological system rests on questions of supply and demand for the requirements of metabolic activity.) On an average, the waiting time for medical treatment may be related mathematically as a service time along with the arrival time of patients. Then these may be related in a ratio

service time/arrival time = traffic density

A ratio below one will frequently be better for the system. For the doctor's plans, one to one, or unity, is best, but the smallest traffic density is most convenient for patients. Any value below unity tends to reduce queueing. Looking more closely at the medical specialist in action, there are operations research effects changing the scene. Secondary administration in the form of an appointments desk or something more elaborate relieves specialists to concentrate on their specialty. The administrative person may have had training in health care as a delivery system, with input and output queue discipline such as priorities or "first come first served," and with the mechanics of service.

Systems that are operations researched are likely to have intrinsic regularities of pattern like those mentioned, sometimes with decision requirements in the sense of a game with conflicting forces. It is not all mathematical, especially in research where teamwork among rugged individualists (specialists) is desired. Without a smooth participation, the cooperative effort is unlikely to succeed and the kind of problem investigated by these methods requires a team effort plus diplomacy. Thus, while the research part of operations research is an embodiment of the machine-person relationship, it is now pluralized as machines-persons. To look at this as in this discussion is to be a generalist, but to participate is to be a specialist. The conclusion is that, without the biological, mathematical, and other subspecialists, there is no more chance for a real "game" than there is without ends, center and backs in a football game. At some point, human experience may require operations research for operating decisions and it is required in ethics too, in questions such as euthanasia, eugenics, country-wide vaccinations and drug testing. This will come when the need for special executive and administrative organization in the research involved is felt, to secure precise, specified thinking toward solutions.

Servomechanisms

The human body demands even finer control than that provided by ordinary feedback. Sensory organs dominated by eyesight collect much of the necessary information, aided by hypothalmic interest in thermoregulation (close adjustment in cerebellum) and the inner ear organs of balance. With all of the rigor in measurement, comparison, and correlation that these systems can display, they completely outshine the clumsy overall steering efforts of the masters they serve. The epitome of such fine control is the servomechanism which is the cybernetic model that self-actuates. This is a mission-directed system that can choose its own mission, evaluate its success, and respond in such a way that by itself it seeks a performance with zero defects.

The actuator in a servo is a servomotor which is fed by the output voltage of a servoamplifier. Servomechanism theory is illustrated in the eye-pupil reflex. Muscle contraction is the servomotor mechanism. In mechanical systems, potentiometers of a rotating type are used with electrical power in such a way that the movable or third arm can give angular information and control the amplifier which then turns the servomotor. The motor gives the torque for control action—moving a switch, closing a valve, etc. Another potentiometer "feels" how much the motor shaft turns and the potentiometer ratios can be made equal

Biology Conceptualized 41

at some system point, such as at the moment when the valve is closed. The feedback can then be mechanical and based on the radian measurement of angular position in the servomotor shaft. Such a system may be made to follow a program, actuating itself to add more action such as heat, light, or cooling according to feedback comparisons made between its actual status at a given time and the desired status at that moment.

There are also the saccadic eye movements which involve changes in the acceleration of the eye when it scans a field and acceleration which reverses, to correct sensed error quickly by reversing the forcing function. Servomechanisms seek a set point and dampen the variation.

Automata like Golem have robot qualities. They have a real manifestation in Turing machines which are controlled by messages on punched tape operating somewhat like a selection on a player piano. Their behavior and limits depend essentially on what the taped instructions can accomplish with computer control. The theory concerns problems of organization, structure, language, information and control. Natural automata are nervous, reproductive, and regenerative systems. Artificial automata include analog and digital computers, and telecommunication systems.

The status of a biomodel may be established or set up by a copying process, in the same way as RNA reads instructions on DNA resulting in appropriate protein syntheses, or by one like a Turing machine in which a tape of symbols can be read by a machine. Decisions may be included as a designed capability.

In terms of the number of functions, automata can be finite, just plain enormous, or infinite. Mathematical logic is used to pass to the "neural nets" of finite automata. These are cybernetic models which may lead to machines that simulate functions such as those of the kidney, lung, or heart, perhaps needing software programs plus hardware computers.

Digital Versus Analog Computers

Both digital and analog computing are seen in human systems and they shade continuously from one form into the other. Coming from counting digits, the earliest calculation a human learns, digital activity is evident in much of the nervous system action with the special exception of the end organs. Most human systems are analog as, for example, where nervous impulses control biochemical production. The hydrodynamics of biochemical distribution is also analog. To display digital output as a curve on an oscilloscope is an analog process. Thus, a digital operation may control an analog one as seen in genetic control of enzymatic activities. The genes behave like digital computers in many respects but the enzymes which they control function analogically.

Perhaps a more dramatic illustration is brought to mind from student days at M.I.T. Certain Ph.D. candidates devised a way to objectively measure the tenderness of a beefsteak. To do this they mounted a complete set of teeth and equipped it with force transducers that would respond to the biting and grinding force. Then, this digital variable could be displayed as a proportional, analog voltage on an oscilloscope thus recording the effort needed to manage the steak.

The management of information requires the biological system to constantly convert analog and digital signals. The analog signals are measured directly and form a continuous message. The continuous message is essential in communication of visual images with widely different tonal quality and a continuum of the colors in the spectrum. Analog signals take less bandwidth and are simpler, but digital ones are less noise-sensitive. Also, in mechanical systems, analog signals are easily encrypted, multiplexed, and compressed to conserve precious frequency space. Repeaters will amplify both the signal and the noise which accumulates when analog signals are being relayed.

Regeneration and antinoise measures are more successful in digital transmissions. These interface easily with computer systems, or with digital integrated circuits. They are, therefore, amenable to miniaturization of the associated equipment.

To go from one kind of signal to another, for example, to digitize an analog signal, it is possible to process it by pulse code modulation. This samples the analog transmission and represents the continuous voltage, for example, by a pattern of meaningful sample pulses. The same thing occurs if one converts an image of an animal into a pattern of dots for a child to connect (regenerate). The pulse samples have to be quantized, or given the nearest value in a preselected set of values which, in practice, may be a superposition of the analog signal samples over a grid of horizontal lines. The process then gives them their binary value so that binary pulses can be used to generate the quantized signal. Television pictures will normally need 128 quantized levels (for picture quality) or a transmission rate of 58 megabits per second. The bandwidth that must be dedicated to this transmission can be made smaller by transmitting fewer bits of information, which is often possible.

The receiver must then reverse the process, demodulating the signal, converting from digital to analog, reconstructing the analog message, and then correcting any errors.

INTERSYSTEMS

In intersystems, we look first at the physical aspects of the biosystem. Then it is examined as an open system with molecular radioreceptors, and finally as an intersystem whose parts are connected by radiation. Radiations are ideal for transporting information because they are many-state information media, providing a very large number of states in space and time. Each spectral band has some finite number of waveforms and each waveform is an information state of the radiation. Sometimes, the radiation is discrete, i.e., has a limited number of states, or it may be continuous, e.g., the field strength of a broadcast signal, but both cases are time dependent. Along with these time and space factors, there is also a meaningful variability in intensity and other characteristics when the nervous system processes information. These include:

1. The order in which the informing variations occur.
2. The limits, bounds, field, or peripheries.
3. Motion as in visual sensors operating in the scanning mode. Stationary objects may be invisible without conscious or unconscious scanning.
4. Wavelength discrimination. This is observed with colors and their wavelengths as well as with frequencies. A frog moves closer to blue objects than to green ones, and closer to lower sound frequencies than high ones.
5. Gradients in energy. Infrared sensation, for example, depends on a temperature gradient between the surface ambient and inward subcutaneous layers when this radiation is sensed as heat.
6. Recent experience of the receptors. Thus, a finger conditioned to *hot* will record a warm liquid as being cool.
7. Comparisons and integrations between multiple detectors as in the integration of color from the individual wavelengths.
8. Meaningful radiochemical changes. Thus light must obviously be absorbed before photochemical changes can occur. This concept applies to the other radiations which are capable of transporting energy to effect change in the status of radiochemicals. A common mechanism is electrochemical or charge transfer through chains of molecules. This often yields striking results in types of synthesis or in molecular radiation energy sinks.
9. Meaningful, direct information transfer via EMR (electromagnetic radiations) and receptive situations involving time and space constants, phase, and periodicity. EM fields then interact with living systems based on commensurate, dimensional, form or time factors.
10. Quantal fit, for example, for incoming EMR hearing information energy assures a receptive system via appropriate transitions.

Some biologists have emphasized the *self-perpetuating* features of open living systems, as if an isolated biosphere could continue indefinitely. Thermodynamics specifies that matter and energy are continuously imported and exported from an open system as well as altered in it. In the long run, after the intrasystem energy is exhausted, life is impossible. Admittedly, many lifetimes can be spent examining the energetics of the local living system alone, believing it has a "constant" or periodic energy supply readily available, and occupied, therefore, only with transductions, catalysis, and cycles.

Radiation and nuclear reactions can be the only signficant support for these open systems. Whether in solar thermonuclear chemistry which provides the spectrum as received by living systems on earth, or in terrestrial nuclear power, the supporting and informing energy source is radiant (Fig. 1-13).

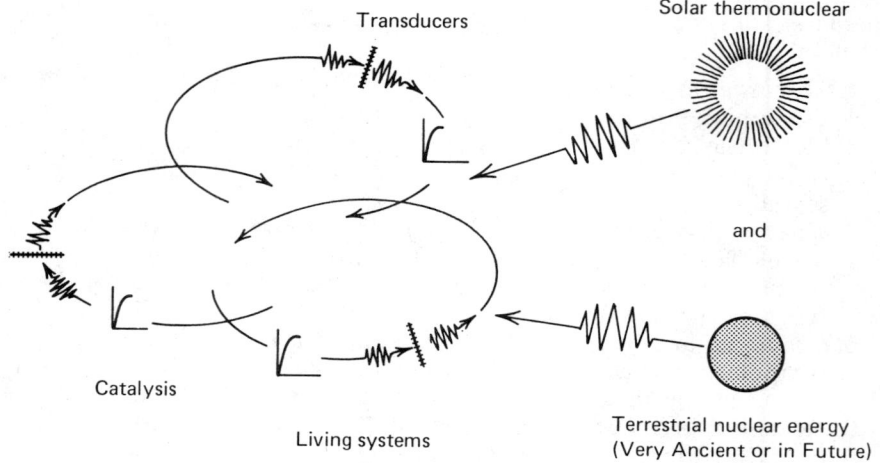

Fig. 1-13. Cycles of open living systems with initial and ultimate regeneration of consumed energy by nuclear power.

Molecular sensory reception of radiant information is illustrated by the rods and cones of the retina. The rods receive visible light, molecular changes occur, and a pulse is sent to the cerebrum. This is manifested by a reversible, fading or bleaching process in the rod-rich retinal boundary area especially seen in dim light. Rods receive information in an oblique direction as in scanning action. The photochemical importance of Vitamin A is indicated by the deficiency disease brought on by its lack, night blindness.

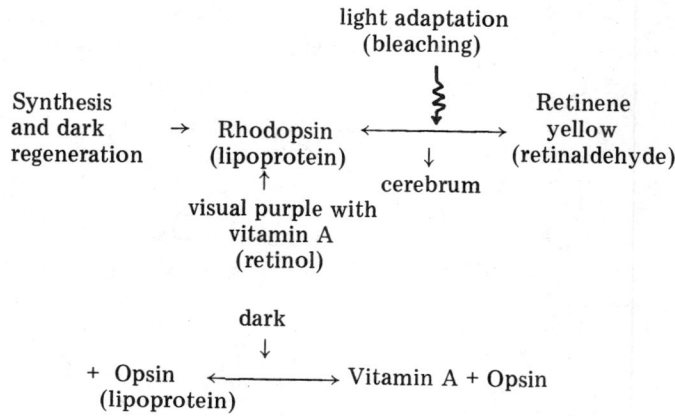

Synthesis and Bleaching of Rhodopsin

The chemical reaction by which cones reduce solar information is as follows:

$$\text{information in multi-band visible light} \xrightarrow{\text{absorption in}} \text{iodopsin, a protein-retinene that is photosensitive}$$

Thus vision is dependent on molecular receptors. The bull, having *none* of this molecular organization or, hence, color vision, cannot understand what all the "seeing red" business is about! The cones have a specific distribution on the retina. White is seen when all visible wavelengths are received, and black is seen when none are received (the light was absorbed *by* the object and thus it is seen as black).

In this optical system, rhodopsin is the molecular addressee. It is also responsible for energy reception in *Halobacterium halobium*, the halophilic bacterium which has an exceptional content of sodium in its wall. Bacteriorhodopsin uses radiant energy to transfer 250 protons per molecule per second across its membrane. This outbound conduction of positive charges leaves an electronegative cell, richer in energy, that can be used to make ATP and support cell synthesis. In this protonsynthesis, the membrane has a semiconduction structure that serves the light energy gathering process. When sunlight activates the pigment, it responds by transferring protons from the cell where they are readily found in the form of hydrogen nuclei.

The human body is only loosely coupled into the infrared spectrum with sensation coming from the end-bulb of Krause (cold) and Ruffini's end-organ (warmth) located in the dermis. Exactly what sensation is received depends on the previous immediate history, the species, the part of the body exposed, and the information in the radiant heat. Warmth receptors lie deeper than cold ones and are less numerous. The human has only 150,000 Krause terminations and, anteriorly, the arm has 15 times as many of these as cold receptors. There are about 16,000 Ruffini's terminations nonuniformly located. For example, there are none on the deep oral cavity mucosa. Some animals, such as owls and snakes, are particularly sensitive to infrared. Sensitivity to ultraviolet is well-known, the existence of erythema sunburn being a negative result and the conversion of 7-dehydrocholesterol to Vitamin A, a positive one.

$$\text{7-dehydrocholesterol} \longrightarrow \text{activated 7-dehydrocholesterol} \\ \text{Vitamin D}_3$$

Further exposure damage to skin results in solar carcinogenesis beginning in the ultraviolet.

Quantum fitting for self-luminescence in skin tissue means that certain ultraviolet energies produce special conformations in the chromophores. This is a darkened room phenomenon in which solid proteins, e.g., hair, teeth, nails, eyes, wool keratin, gelatin, and photosensitive chemicals *glow* after excitation at 325 nm, with some giving off a blue-white fluorescence at 403 nm. Some 94.3% of solar ultraviolet of this excitation (actually 250-265 nm) is blocked by the ozone layer so that bacteria are not naturally exposed to great concentrations of their absorption wavelength. For example, only minutes at 253.7 nm will destroy bacteria subject to those photoinactivation effects which, in one form or other, persist through the ultraviolet spectrum.

In these reactions both sensory and nonsensory information can exert a forcing function on receptive bodies via the radiant carrier. The resulting polarization is a transient stress (Law of LeChatelier) from which the receiver system can relieve itself by changing appropriately. It can then *relax* to its original values ready to receive more information. Meanwhile the "line is busy" or in harmony with other subsystems, the sense *adapts*, relinquishing sensory control to central-information-management.

Many sensory neurons with receptor terminations are radiant information processing units capable of interaction and often localized to indicate their function or positioned to *search* for signals. The superior colliculi, projection of the median brain section, connects to the retina, head, and spine and the inferior one connects to auditory and motor fibers.

Nonelectromagnetic, sensory, radiant information completes certain loops involving auditory data and gravity messages. Similarly, the sensory net receives input from the inner ear for balance detection. Additionally, nonsensory, radiant information has molecular addressees, often being received directly where the action is required. Photosynthesis extends into the far red or near infrared regions. When photosynthetic bacteria, algae, and green plants are included in the system, they have a direct intersystem connection with automatic insertion and data input.

Man, as well as all organisms then, couples both to his supporting medium, the boundaries of which give the definition for the living system, and to intersystem sources for radiant information. Radiant energy that physically moves toward life systems may not "see" the system boundaries in the way that other energies, e.g., chemical, see it. What it "sees" is a question of the mode of interaction. The human life span operates continuously as a system involving the behavior of all the tissue cells plus a communication and information network of about 10^{10} neurons when all of the system is operative.

A photosynthetic bacterium, being a complete, self-sufficient single-celled unit, is directly coupled to its radiant energy source. It is independent to a much greater extent than are parasites and saprophytes who derive energy either from hosts or from local sources which may suffer interruption with greater probability than the loss of the sun.

Entropy and Thermobiodynamics

With the foregoing as a background, it is convenient to look at one of the most fascinating controversies in biology – whether or not organisms feed on negative entropy. "Intersystems" tell how local systems may be connected to other parts of the universe. In this thermobiodynamic argument, the crux of the matter is precisely the relationship of the local system to its universe.

The content of information is measured by H, the message entropy in an ensemble of messages. This H has features of entropy in thermodynamics and statistical mechanics which are called S. Information is not solely a mathematical function so that it has to be described, not only in relation to uncertainty and order, but also in connection with its stockpile in DNA, in control, communication and elsewhere. Thermodynamic entropy with constraints appropriate in pure thermodynamics arises as a property from the probability distribution in electronic, atomic, and molecular energy. There are easily as many entropies as there are informations. When the latter is a mathematical function, it is rigidly prescribed for the kind of information (DNA to protein flow, sensory, feedback, and other types) being analyzed. The temptation to explain entropy by associating it with order in a system similar to that of ordered water molecules in ice (an organization that gives it less entropy) is justified, but relating that order to beauty is not nearly as helpful. Information is a matter of life and death when it is generated, stored, transcribed, and copied in living systems. Whenever an organism uses information in the sense of these pathways, it imposes a "biological tax" on such imports. Entropy is a thermodynamic concept and thermodynamic information is measured by the $\log_2$ of the *number of choices or alternatives involved*. Thus, with only two alternatives, the information is 1, and this gives the basic

$$\log_2 2 = 1 \qquad (4)$$

unit of information, the bit. Thus the on-off condition is equivalent to a bit in switching terms for an information, binary system. Then, through the use of symbolism in language and letters of an alphabet, information can be visualized as having some content. The nucleotide bases that are read by mRNA adenine, cytosine, thymine, uracil, and guanine will be seen as letters in the biological code where each letter will have a content of about one bit. Then the amino acid carried by tRNA will be the equivalent of a word and the protein molecule, a paragraph, with 100 to 200 bits or close to the maximized storage. From here it is a giant step to the "book of life" wherein a mammalian cell will use about 2×10^{10} bits comparable to the information content in some 100 sets of the Encyclopedia Britanica or the capacity of the human brain at about 10^{13} to 10^{15} bits or that in the DNA of the *E. coli bacterium*, about 10^7 bits. In vision, about 10^7 foveal units (rods and cones combining) process an average of 7 bits per unit for a total foveal burden of 10^{10} bits. This content is passed very rapidly and with few errors in coding to the brain.

Thermodynamic information is related to energy via entropy if operating temperatures are specified so that the operating energy requirement becomes known. Boltzman's constant, K, with units of ergs per degree (1.4×10^{-16}), can be helpful. Then if a "normal" temperature is taken ($T_{abs} = 298°$) the least energy for one binary act or choice between two alternatives ($N = 2$) can be calculated.

$$E = KT \log_e N = \underline{3 \times 10^{-14} \text{ ergs}} \qquad (5)$$

For sensory information obtained by the radiant mode, von Neumann estimated that the brain dissipates 25 watts, operating 10^{10} neurons in duty cycles of 10 times per second. This gives 10^{-3} ergs per binary decision. The difference between actual and expected energy here is 10^{+11} ergs. This estimated biological tax is the energy cost of information management in the human system.

The uniformity or information density is also associated with the radiation events and the information content. This is described when the probabilities for two binary (0,1) choices are equal (i.e., one bit per digit). But if 0 is 10 times more numerous, the information content of a 1 is

$$-\log_2 (1/11) = 3.46 \text{ bits} \qquad (6)$$

Neurons, for example, take more note of the *unusual*; the least probable happenings cause attention. For example, a bikini-clad girl at a formal party would excel in information content. For a 0, the information content is

$$-\log_2 (10/11) = 0.13 \text{ bits} \qquad (7)$$

and Shannon's *average* in this situation is 0.43 bit/digit. In addition, more information content comes with greater *meaning* in the radiant message (it may have a good signal to noise ratio). Meaning upgrades the information in terms of interneuronal information management.

If there is a question to be evaluated within a problem, the information enters into the judgment made, and validity in the judgment can then reduce the entropy of the problem.

Very large amounts of information are involved in coded messages received by the foveal rods and cones since they are arrayed in matrix form. A typical information burden is 2^{10} bits. This may be conceptualized by taking a checkerboard pattern with 10 x 10 squares that are either red or black, and imagining all the coded messages possible to obtain by varying the block colors between the two alternatives.

$$\text{Bits} = 2^{10 \times 10} = 2^{100} \qquad (8)$$

Introduction of the concept of uncertainty shows how entropy S and information can be related. For orientation we can say that when entropy decreases, the uncertainty decreases, since the system becomes more ordered in predictable states. Now uncertainty is subject to estimate by considering the statistical probabilities involved. This gives us a measure of information in terms of probabilities.

In biology, thermodynamics is introduced in the hope that its methods for managing energies and symbols may prove powerful tools to understanding.

1st Law: Mass and energy are transformable but cannot be created or destroyed. 100% interconversion is possible between energy forms.

2nd Law: Due to the bound energy in a product's molecules, devices can transform only a fraction of available energy. This law has consequences leading to a "direction" for time.

3rd Law: At $0°K$, molecules are in the most ordered, tranquil state.

We can show that entropy S connects the transforming energy and information in a most interesting manner. Of course, in the company of such venerable engineering concepts as entropy and thermodynamics, information theory appears as a comparatively new concept. Like biophysics, thermodynamics is best defined by what it does. However, in the former it is essential to assume a structural model for the biosystem. In the purely thermodynamic sense, no specific model (except for the randomness of distributed particles, such as gasses, crystals and isotopes) should apply constraints,

$$\Delta S = S_2 - S_1 \qquad (9)$$

and entropy involves a change of state. In physical thermodynamics and statistical mechanics, entropy is a measure of thermal energy capacity or, in an oversimplification, of the specific heat.

To be sure, entropy is a particular heat capacity involving the temperature as well as the storage and use of energy in reactions. It has units of calories per degree Celsius per mole. After a reaction, the heat capacity of the products is compared with that of the reactants and ΔS is determined to be either positive or negative. If the heat capacity of the products is greater, then ΔS is positive. The heat capacity of a system increases and so does its entropy in proportion to the activity in the modes of motion of molecules, that is, vibration, rotation, inversion, electronic, translation, etc. At any instant, the system with increasing, positive entropy will be found in more and more possible arrangements (complexity, disorder, multiorder, - chaos) as it exercises its degrees of freedom of internal motion. In addition to the arrangement of masses, there is a one way outflow of heat during some processes (exothermic) which contributes complexity in the arrangements as shown in Fig. 1-14.

The introduction to thermodynamics via the Carnot cycle and heat

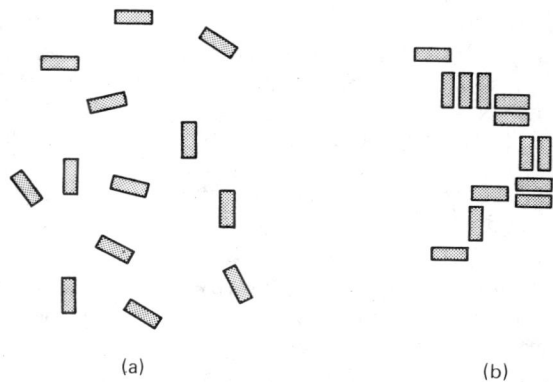

(a)   (b)

Fig. 1-14. Divergence. a) Universe "wants" to (must) expand or GAIN POSITIVE ENTROPY more probable arrangement for particles. b) Organisms "want" to organize, e.g., grow, PROTEIN (primary, secondary, tertiary, etc.—organized) structure, and live or make negative entropy or use POSITIVE ENTROPY UP as organism grows. This decreases the alternate ways for arranging the system. There is less inherent probability for the resultant state.

engines has seemed troublesome to some. Now information theory, with its certainties and probabilities, turns out to be surprisingly far-reaching, offering intriguing alternatives in the way of orientation to thermodynamics. Using such concepts as randomness, order versus disorder, states or forms a system can assume, coupling of sources, channel capacity, coded messages, and probabilities, it is found that information theory offers advantageous ways of expressing temperature, free energy, the functions of reactions and chemical potential. Biosystem states are selected from those worth describing and entropy is introduced to describe all the possible forms that the system can take.

$$S = K \ln \Omega$$

Here the constant K (Boltzman), or sometimes R in units of cal/degree mole, connects the thermoentropy or temperature associated with thermal energy with the various states. This logarithm can thus "reflect" system changes. If we think of defined systems with changes in status from A to B, and of small systems within large systems, while keeping homeostasis as a background idea, it is possible to see both many $\ln \Omega$ in little separate systems and their grand total, which merely counts them. However, an *increase* in $\ln \Omega$ describes changes going on in a system having constant energy and material. Next comes descriptions of entropy changes, heat changes, and reversible processes. In a suitably changing system, there can be more states of the kind suggested by energy modes in rotations, inversions, vibrations, and translations, all of which may be increased in products as compared with reactants. As opposed to this kind of increase it is seen that a need for greater *order* would put a constraint on the possible arrangements which would have the effect of decreasing entropy, that is, giving a negative entropy trend, negentropy, because *entropy increases (positive) posentropy, according to the alternate ways for arranging a system*. The increase in entropy may resemble disorder or randomness since it is associated with the more numerous arrangements, while an entropy decrease looks like a more ordered or organized system. Extraction of energy also mixes up the environment even as intrinsically the organism itself may become more ordered. Living processes nevertheless are using entropy up or adding to negative entropy. If all life stopped, and decay occurred, disorder would increase (positive entropy) and this would stop the negentropy trend. Therefore, this amounts to a thermobiodynamic description of death.

If a developing organism grows more ordered with *time*, there is a thermodynamic contradiction. As entropy increases, randomness moves in to replace order and probability exerts itself as improbable arrangements are replaced by probable ones. The order had existed in molecules, ions, crystals and their systems. Entropy always increases in actual cases as a sort of forward orientation for time (2nd Law). That is, an entropy decrease with time is *backward* according to the system. The arrow of time describes the irreversibility of physical events as entropy gives time a direction—toward more products and arrangements. Thus, without entropy, all processes would be *reversible* and the past would equal the future. Information has been introduced into thermodynamics, apparently with success and clear explanation. Time may well be introduced explicitly but, as of now, it is more like a thermoconceptualization based on the 2nd Law.

The answer to the apparent conflict of organisms living on negative entropy is not found completely in this time direction version of the 2nd Law, which would settle the matter arbitrarily. The thermodynamic truth is that the organism lives by negative entropy on an island in the universe *only*, while in a much larger unrestricted region, the whole universe, S remains positive. There is the possibility that this regional divergence represents continuous new valuable *production* within the living organism and is calculated to defy the heat death prognosis which specifies positive S and eventual progress to zero available energy. As the universe approaches a zero free energy status or maximum positive entropy (unavailable energy) the situation suggests catastrophe—like a battery running down to zero. The productivity counters this by continually introducing new synthesis which is tantamount to recharging the battery so that the process can continue as long as organisms do not "overdo a good thing."

In biological terms, radiant information changes the *Gibbs free energy* ΔG. In photosynthesis

$$\Delta G = \Delta H - T\Delta S \text{ (an endergonic process)} \tag{11}$$

and combustion heat is exchanged between the system and its surroundings. In the case of one mole of glucose, ΔH (its changed enthalpy) will be the well-known value 673 kcal/mole at absolute temperature T. Free energy changes positively here, concentrating energy in the plant product or "producing" glucose. Experimentally, an isothermal, equilibrium value of 686 kcal/mole is found for ΔG. Thus,

$$686 + 298°K \; \Delta S = 673 \tag{12}$$
$$\Delta S = -13/298 \tag{13}$$
$$\Delta S = -0.0436 \tag{14}$$

and the entropy change is calculated as negative and in kilocalories/mole. The negentropy is against the "arrow of time" but only on the "little island biosphere." while in the larger surroundings, entropy will always increase.

Free Energy and Entropy in Other Bioreactions. The second law may be stated as:

$$\Delta H = \Delta G + T\Delta S \tag{15}$$

where ΔG is the extractable or Gibbs free energy, sometimes ΔF, and again ΔH, the familiar change in enthalpy, is the kind of heat given out in the heats of combustion of compounds or the sublimation of ice. The role of T is that of a temperature or heat forcing function for the change in entropy ΔS, the unavailable energy according to the heat capacities *for products versus reactants*. Particular values may be introduced from handbooks, e.g., the behavior of ATP may be shown to produce energy up to $\Delta G = -7.7$ kcal/mole at pH 7 and 37°C by dephosphorylation hydrolysis.

$$\text{ATP} + \text{H} \xrightarrow[\text{pH7}]{37°C} \text{ADP} + \text{HPO}_4 + 7.7 \text{ kcal} \tag{16}$$

This available free energy is then used to drive a biochemical reaction. Glycine is bound to leucine in peptide units with a value of $\Delta G = 4.5$ kcal/mole, and to contract an actomysin unit in muscle may use up all of the 7.7 kcal/mole that is available from ATP. Then, to reform ATP, the 7.7 kcal must be reintroduced to eject a water molecule. This free energy is coupled out by electron transfer systems, such as the cytochrome system in which oxygen accepts electrons and $NADH_2$ is oxidized to NAD, or in intersystem photosynthesis where solar radiant information (molecularly) "orders" chlorophyll and intermediates to prepare energetic electrons with the accompanying energy and inject them into the system in the direction of ATP.

The Question of Isolating Systems. It is possible that those who work in pure thermodynamics or theoretical biology are justifiably upset at what seems to be an overworking of the idea of order. Perhaps they prefer to deal with fundamental units, mass, length, time, the fundamental properties of matter, temperature, and electromagnetics and would not see entropy as measuring disorder or telling about the universe. They may well object to the concept of "mixedupness" (McGlashan 1966) and an entropy increase in the universe, leading to the liquidation of all of us in ultimate disorder. They state that the universe is not a bounded, isolated body to which thermodynamics can be applied. One must admit that it is hard to say from what the universe might be isolated.

Nevertheless, biologists are committed to the indicated pathway because of the power of the thermodynamic method, their preoccupation with self-organizing systems, and above all, because of the growing realization of the fact that information moves in parallel with changes in entropy, informing on status changes in the organism. Other things being equal, it is readily understood that more information is required to specify a disordered system because of the random distribution of its elements. Thus, without reducing its precision, the concept of entropy is applicable as informational entropy, a more general expression than thermodynamic entropy. An alternate introduction is simultaneously provided into thermodynamics itself (Brostow 1972) as students of this subject will be gratified to learn.

Measuring Information. The measure of this information is in terms of the log of probabilities

$$S = -\sum_{i=1}^{n} P_i \log_2 P_i \qquad (17)$$

This shows that the information is a matter of probabilities in which the chance of a certain message getting through from among a larger number of messages is the sum of the products of the probability times the log of the same probability. To relate to the foregoing we note that since $\ln \Omega$ is a dimensionless number, the ways a shaken-up system can fall and be counted resemble a coin with two sides or modes of arrangement. Further explanation comes from the world of signals, symbols, and noise (interference). If the messages are those involved in radio, television, telecommunications or other media, then everyday words are involved. The procedures for communicating using people, machines, and electromagnetic wave behavior all introduce errors, and thus the probability involved computes information, because it says that many messages are beamed but the receiver selects only desired signals. Then, such selection involves probabilities or odds for success as in sequencing amino acids, card games, dice, elections, codons, and radiation decay. S has the valuable properties of conservation (as the sum of potential and kinetic energies is conserved), good definition, and it is also additive. In logs to the base 2, $\log 1 = 0$. If probabilities are expressed as

$$1/\text{number of messages} \qquad (18)$$

where all have equal chance, four messages give ¼ and log ¼ is

$$\log 1 - \log 4 = -\log 4 \tag{19}$$

Each probability is multiplied by the log of that probability and then all four are summed. Thus the result is the positive sum of all the denominators,

$$¼ \log 4 + ¼ \log 4 + ¼ \log 4 + ¼ \log 4 = \log 4 = 2 \tag{20}$$

or

$$= ¼(2) + ¼(2) + ¼(2) + ¼(2) \tag{21}$$

and the summation is

$$½ + ½ + ½ + ½ = 2 \text{ units of information,} \tag{23}$$
$$\text{the information entropy}$$

The fraction ¼ resembles a word ratio

$$\frac{\text{in terms of output, one component gains}}{\text{from an input of four contending components}} \tag{24}$$

This gives a probability of ¼ for any single component.

If two coins are flipped (states of a 2-coin system) at the same time, there are four possibilities HT, TH, HH and TT. Then the probability for each of these is ¼ and $S = 2$ bits. The same conclusion is reached by regarding information in output and input terms. First it is noted that description and specification can be designed to yield information progressively in small answers as in the manner of interrogation or in the rainy day game of animal, vegetable or mineral, AVM. Each question gives a yes, no; plus, minus; all, none type of answer. The player's mental image, which is some selected object in one of these three categories, progresses toward certainty or complete specification. His information content and noncertainty are changed by every query answered. Thus, noncertainty or doubt is a measure of the information needed to specify the identity of the object.

The answering unit (person or machine) will have an *input* of information—the question. It will have a probability that it will remove all uncertainty. A very obvious

$$P_I = 1 \tag{25}$$

would produce no new information. In contrast, a question like, "Is it nonliving?",—in starting the AVM series—produces a maximum amount of information because with an affirmative, two categories out of three are eliminated. Thus, the *output* probability is a maximum one. The last answer in the game fully specifies animal, vegetable, or mineral and removes all uncertainty. Information flows in during the series of questions and answers. These questions are of a binary type designed to give a one bit, one binary digit unit answer, yes or no. Logarithms to the base 2 are then convenient to use in keeping with the established custom for measuring the quantity of information.

A single question, for example, "Is it mineral?", has two possible answers just as flipping a coin may produce heads or tails with a statistical chance of 0.5. There is a difference if the instrumentation involved

Biology Conceptualized

is varied. For a real sample, a perfect mineral analyzer would give a maximum output of information with an output probability of 1. The bits of information added, or uncertainty eliminated by the answer, lie between these two limits in probability. The quantity of information that might be added is exactly one bit, the output probability divided by the *a priori* probability.

$$I = \log_2 (P_{out}/P_{in}) = \log_2 (1/0.5) = \log_2 2 = 1 \text{ bit} \tag{26}$$

Note that base 2 raised to the "1" power yields 2.

RADIATION

Radiation from radioactive atoms is a random process. We do not know which atoms will decay at any given time; we only know that 50% of them will decay within one half-life period. When we count the activity from a sample for a short time, there is a probability associated with that count therefore that it is a correct measurement of decay rate. Counts may be made repeatedly over a specific interval of time and the frequency with which each count or interval type occurs, is observed. The chance that some count is correct is then calculated according to the frequency or number of times this same count is obtained. In the resulting Gaussian distribution (see bell curve Fig. 1-15), the calculated P_I is 0.20 for a count of 3 to 4 per interval. Thus,

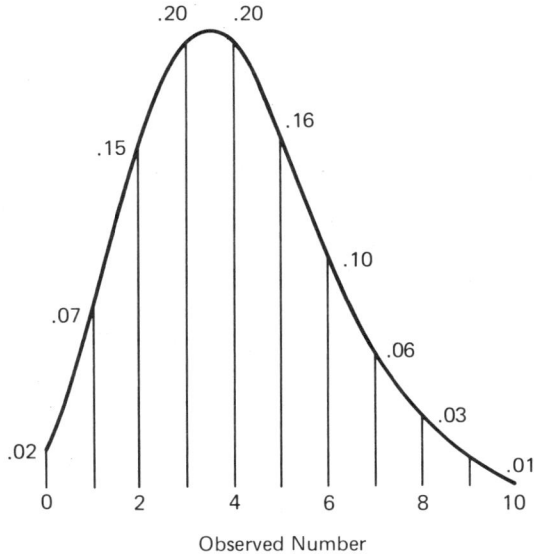

Fig. 1-15. Gaussian distribution of observations against the probabilities in the measurement of radioactivity.

in relation to a perfect instrument, the information added is

$$I = \log_2 (1.0/0.2) = \log_2 5 = 2.32 \text{ bits} \qquad (27)$$

which is a fairly small amount because the count of 3 to 4 is so probable. A much less probable count of 8 would carry more (almost 16 bits) information.

$$\log (1/.03) = \log 34 \cong 16 \qquad (28)$$

A very long counting analysis with a fine instrument (no noise) would produce the maximum output ($P_o = 1$) of information. In the limited analysis made however, there is a probability associated with each of the various counts that it is correct. The probabilities may be calculated using the formula,

$$P = e^{-N_o} [(N_o)^N / N!] \qquad (29)$$

Here P is the theoretical probability for a given interval, e is the base of natural logarithms, and N is the number of counts in each interval. The values obtained for P are similar to those shown in the graph (Fig. 1-15), when N_o, the average count per interval, equals four.

$$N_o = \frac{\text{total counts in all counting intervals tested}}{\text{all intervals counted}} = 4 \qquad (30)$$

The highest probability value, 0.20 in the curve, will be centered on the most frequently observed number of counts. If the number of intervals of each type is divided by the total number of intervals counted, an observed probability is obtained. This may then be compared with the theoretical.

In a communication like the AVM sequence, information flows with each message query beginning with query $i = 1$ among all of the queries n from the conclusions of Shannon (1948) and Wiener (1948). Again, the measurement of information shows that both information and entropy are related by similar functions. Both measure uncertainty in the system.

The entropy decreases as information reduces uncertainty. The greater the favorable possibilities from a question, the greater the removal or reversal of uncertainty and the greater the information communicated by the answer.

Messages make up communications which, in turn, organize information, speech, hearing, vision, sensory perception, and genetic control; the nervous system communicates information continually. Communication is the central feature of this system which has evolved to meet greater and greater information transfer requirements. The same features of organization are present. There is the black or white box with its input and output and rules for the behavior of the organic unit under input stimulation are included (Fig. 1-2). This organization may be either ad hoc or permanent. It would be ad hoc to fill a transient requirement. A very simple transfer function would be one which discriminates inputs of two classes, letting class one pass but excluding class two. This behavior would categorize the function in the manner of a gate or valve. The two possibilities give us a typical binary variable as opposed to the well-known continuous variable. If the message is in words consisting of coded triplets of amino acids, the information is that of DNA which holds the keys to information management in the living system.

PROBLEMS

1. Give the factors, input, output, and operation of writing as a sensory information management system.
2. Is a writing control system an open or a closed loop?
3. Diagram the biocontrol system for thermoregulation that involves muscle shivering, assuming that a control and comparison point is provided by skin and hypothalmic receptors.
4. Make a block "ideas" diagram for the flow of information in TV.
5. If 0.2 g of glucose are injected into an animal whose blood volume is 1 liter, and the blood sugar rises from 80 mg/100 ml of blood to a controlled level of 81 mg/100 ml, find the gain-feedback of the sugar control system assuming that 1 + gain-feedback =

    output without feedback/output with feedback
6. Name five biological concepts with an informational content in addition to the ten studied.
7. a. Define the probability of success which results when attempts are made to transmit messages to a receiver.

    b. If all have equal chance, four messages give ¼ and log ¼ is −log 4. What is the information content?

BIBLIOGRAPHY

Aleyer, S. L., *Simulation in Biology and Medicine*, U. S. Government translation of the Russian into English, CCM Information Corp., New York, 1965.

Ampere, A. M., *Essai sur la Philosophie des Sciences ou Esposition Analytique d'une Classification Naturelle de Toutes des Connaissences Humaines*, Bachelier, Paris, 1943.

Bailey, N. T. J., *The Mathematical Approach to Biology and Medicine*, John Wiley, New York, 1967.

Bellman, R. E., *Some Aspects of the Mathematical Theory of Control Processes*, Rand Corp., Santa Monica, California, 1958.

Bellman, R. E., *Adaptive Control Processes—A Guided Tour*, Princeton University Press, Princeton, New Jersey, 1961.

Browstow, W., "Between Laws of Thermodynamics and Coding of Information," *Science*, Vol. 178, October 13, 1972.

Chang, S. S., *Synthesis of Optimum Control Systems*, McGraw-Hill Book Co., New York, 1960.

Coeffegnel, E., *Le Concept d'Information dans la Science Contemporaine*, Cahiers de Royaumont, Goutier Villars, Paris, 1965.

D'Azzo, J. J., C. H. Houpins, *Feedback Control System Analysis and Synthesis*, McGraw-Hill Book Co., New York, 1960.

DelToro, V. and S. R. Parker, *Principles of Control Systems Engineering*, McGraw-Hill Book Co., New York, 1960.

DeRusso, P. M., R. J. Roy and C. M. Close, *State Variables for Engineers*, John Wiley, New York, 1965.

Distefano, J. J., A. R. Subberub and I. J. Williams, *Theory and Problems of Feedback and Control Systems*, McGraw-Hill Book Co., New York, 1967.

Doebelin, E. O., *Dynamic Analysis and Feedback Control*, McGraw-Hill Book Co., New York, 1962.

Dorf, R. C., *Modern Control Systems*, Addison Wesley, Reading, Mass., 1967.

Fjerdingstad, E. J., Ed., *Chemical Transfer of Learned Information*, North Holland Pub. Co., 1971.

George, F. H., *Cybernetics and Biology*, Oliver and Boyd, London, 1965.

German, J., "Studying Human Chromosomes Today," *American Scientist*, Vol. 58, 2, March-April, 1970.

Glushov, V. M., *Introduction to Cybernetics*, Trans. Scripta Technica, Ed. G. M. Kranc, Academic Press, New York, 1966.

Goffman, W., "An Application of Epidemic Theory to the Growth of Science (Symbolic Logic from Boole to Gidel)," *Progress in Cybernetics*, Vol. 3, Ed. J Rose, 1970.

Goldman, S., *Information Theory*, Prentice Hall, Englewood, New Jersey, 1953.

Graham, D and D. McRuer, *Analysis of Nonlinear Control Systems*, John Wiley, New York, 1961.

Gupta, S. C. and L. Hasdorff, *Fundamentals of Automatic Control*, John Wiley, New York, 1970.

Hartley, R. V. L. "Transmission of Information," *Bell System Technology Journal*, Vol. 7, p. 535, July 1928.

Helvey, T. C., *The Age of Information, An Interdisciplinary Survey of Cybernetics*, Educational Technology Publishers, Englewood, New Jersey, 1971.

Horowitz, I. M., *Synthesis of Feedback Systems*, Academic Press, New York, 1963.

Hubel, D. H. and T. N. Wiesel. "Receptive Fields, Binocular Interaction and Functional Architecture in the Cat's Visual Cortex, *Journal of Physiology*, Vol 160, pp. 106-154, 1962.

Hyvarinen, L. P., *Information Theory for Systems Engineers*, Springer-Verlag, Berlin, New York, 1968.

Jelinek, F., *Probabilistic Information Theory, Discrete and Memoryless Models*, McGraw-Hill Book Co., New York, 1968.

Jury, E. I., *Sampled Data Control Systems*, John Wiley, New York, 1958.

Karbowiak, A. E. and R. M. Huey, Eds., *Information, Computers, Machine, and Man*, John Wiley and Sons, Australasia and New York, 1971.

Kolmogorov, A. N. and V. A. Uspensky, "On the Definition of an Algorithm," *Uskekhi Matem. Nauk 13*, 4, (82), 1958.

Kuo, T. C., *Analysis and Synthesis of Sampled Data Control Systems*, Prentice-Hall, Englewood Cliffs, New Jersey, 1963.

Kuo, T. C., *Automatic Control Systems*, Prentice-Hall, Englewood Cliffs, N.J., 1962.

Legéndy, C. R., *Progress in Cybernetics*, Ed. J. Rose, 1970, 315.

Lettvin, J. Y. et al. "What the Frog's Eye Tells the Frog's Brain," *Proc. Institute of Radio Engineers*, Vol. 47, pp. 1940-1951, 1957.

Lindorff, D. P., *Theory of Sampled-Data Control Systems*, John Wiley, New York, 1965.

Marder, L. *Time and the Space Traveller*, U. of Pennsylvania Press, Phila., Pa., 1971.

Maxwell, J. C., *A Treatise on Electricity and Magnetism*, Dover Publications, New York, 1954, reprinted 1968.

McGlashan, M.L., "Thermodynamics," *J. Chemical Education*, Vol. 43, 226, 1966.

Merriam, C.W., *Optimization Theory and the Design of Feedback Control Systems*, McGraw-Hill Book Co., New York, 1964.

Milsum, J.H., *Biological Control Systems Analysis*, McGraw-Hill Book Co., New York, 1966.
Mishkin, and L. Braun, Jr., *Adaptive Control Systems*, McGraw-Hill Book Co., New York, 1961.
Newton, G.C. and T.R. James, *Analytical Design of Linear Feedback Controls*, Ed. A.G. Leonard, John Wiley, New York, 1957.
Norman, D.A., *Memory and Attention*, John Wiley, New York, 1969.
Nyquist, H., "Certain Factors Affecting Telegraph Speed," *Bell Systems Technical Journal*, Vol. 3, 324, April 1924.
Nyquist, H., "Certain Topics in Telegraph Transmission Theory," *AIEE Trans.*, Vol. 47, 617, April 1928.
Oestreicher, H.L. and D.R. Moore, Ed., *Cybernetics Problems in Bionics*, Gordon and Breach, New York, 1968.
Orians, G.H., *The Study of Life, An Introduction to Biology*, Allyn and Bacon, Inc., Boston, 1969.
Pagazzini, J.R. and G.F. Franklin, *Sampled Data Control Systems*, McGraw-Hill Book Co., New York, 1958.
Parsegian, V.L. et al, *Introduction to Natural Science Vol. I, The Physical Sciences*, Academic Press, New York, 1968.
Parsegian, V.L. et al, *Introduction to Natural Science Vol II, The Life Sciences*, Academic Press, New York, 1970.
Parsegian, V.L., *This Cybernetic World of Men, Machines and Earth Systems*, Doubleday, New York, 1972.
Pervozvansku, A.A., *Random Processes in the Nonlinear Control Systems*, Translation by Scripta, Technica Inc. by I. Herzer, Academic Press, New York, 1965.
Physical Science Study Committee, *Physics*, D.C. Heath and Co., Boston, 1960.
Pierce, J.R., *Symbols, Signals and Noise*, Harper and Bros., New York, 1961.
Prigogine, I., *Thermodynamics of Irreversible Processes*, Thomas Publishing Co., Springfield, Ill., 1955.
Randall, J.E., *Elements of Biophysics*, 2nd. Ed., Year Book Medical Publishers, Chicago, Ill., 1962.
Rose, J., Ed., *Progress of Cybernetics*, Vols. I, II, and III, Gordon and Breech, New York, 1970.
Savant, C., *Basic Feedback Control System Design*, McGraw-Hill Book Co., New York, 1958.
Schrodinger, E., *What is Life?*, Macmillan, New York, 1945.
Schwartz, M., *Information Transmission, Modulation, and Noise*, McGraw-Hill Book Co., New York, 1959.
Shannon, C.E., "A Mathematical Theory of Communication," *Bell System Technical Journal*, Vol. 27, 379-423, July 1948, 623-56, October 1948.
Shinners, S.M., *Control System Design*, John Wiley, New York, 1964.
Simon, H.A. and A. Newell, "Information Processing in Computer and Man," *American Scientist*, Vol. 52, 281-300, September 1964.
Singh, J., *Great Ideas in Information Theory, Language and Cybernetics*, Dover Publications, Great Barrington, Mass., 1966.
Smith, H.W., *Approximate Analysis of Randomly Excited Nonlinear Controls*, M.I.T. Press, Cambridge, Mass., 1964.
Smith, O.J.M., *Feedback Control System*, McGraw-Hill Book Co., New York, 1958.
Stark, L., "Stability, Oscillations and Noise in the Human Pupil Servomechanism," *Proc. IRE*, Vol. 47, 11, November 1959.
Stark, L, *Vision: Servomechanism of Pupil Reflex to Light*, O. Glasser Ed., Medical Physics, Vol. III, Yearbook Publishers, Chicago, Ill., 1960.
Thaler, G.J. and M.P. Paster, *Analysis and Design of Nonlinear Feedback Control Systems*, McGraw-Hill Book Co., New York, 1962.
Turner, R.P., *Impedance*, Tab Books, Blue Ridge Summit, Pa., 1976.
VanTrees, H.L., *Synthesis of Optimum Nonlinear Control Systems*, M.I.T. Press, Cambridge, Mass., 1962.

vonBertalanffy, L., *General System Theory, Foundations, Development, Applications*, G. Braziller, Inc., New York, 1968.

vonNeumann, J., *Theory of Self-Reproducing Automata*, University of Illinois Press, Urbana and London, A.W. Burks, Ed., 1966.

Watanabe, S., *Knowing and Guessing—A Quantitative Study of Inferences and Information*, John Wiley, New York, 1969.

Wiener, N., *Cybernetics*, M.I.T. Press, John Wiley, New York, 1st. Ed., p.194, 1948.

Wilts, C.H., *Principals of Feedback Control*, Addison-Wesley, Reading, Mass., 1960.

2

The Nature of Some Intersystem Communication

Why feel alone, marooned
As a featureless neuter?
With infinite companionship
In that national computer.
Our vitality and security,
Indiscretions and maturities,
In neat programmation
Without slur or approbation.
Our currencies and tendencies
In efficient binary notation.
Contented, assured,
Indexed and stored.
"Status ready"
For the terminal instruction
"Retrieve and accumulate
In heavenly registration"

D. A. C.

The outlook expressed in Chapter 1 correctly suggests a greater involvement of life scientists in the "substance" of information (and intelligence too). While brain designers may be pausing and regrouping, their searching questions have had their effects.

1. They accentuated the hunger for more hard experimental evidence to support further advances.
2. With automatic computation in mind they tended, perhaps, to be more conscious of the numbers involved in their physical studies; looked for a sharper scale of relative sizes and amounts because they were "recounting the neurons" and other cells, even atoms, and comparing living and mechanical responses for rates, time and space factors.

3. They may have influenced the trends toward conscious control over vital signs – meditation and positive thinking on blood pressure and other responses so that methods for dealing with tension and pain, for example, now involve biofeedback. Blood pressure, muscular tension and heart rhythm are monitored so as to make them behave under stress.

Behind it all is the realization that probing the informational areas inevitably carries researchers into the real operation of the nervous system and the brain. Designers had to quickly face up to philosophical questions of *mind and brain*. It may be argued forcefully that some of their conclusions formed after filtering out unacceptable terms and dogmatic statements give more benefits than would the elusive designs they seek. Rosenbleuth (1970) for example, rejected the idea of unconscious mental processes. Particularly, he did not believe in the stroke of genius process often reported in sleep. He felt that chances of finding a fruitful correlation among the infinite number of possible combinations by pure luck, when most of them would be trivial, is minimal, or nil. Fruitful correlation is most likely after much serious work, plus a great deal of reflection. Some suppose that there is a scientific sensitivity to be developed. After this *training*, unconscious mentality can flag the consciousness with a kind of attention and interest alerting process that otherwise stays dormant in the insensitive, a concept that has its merits. Arturo Rosenbleuth sees the degrees of consciousness as all belonging to one kind of mind which can vary only in awareness, from sharp to diffuse. To be unconscious, thus, is to be absolutely without awareness, so there seems to be *no need to have an independent unconsciousness or even a mental entity that is not a matter of switching, associating neurons, and making correlations.* Anything helpful that goes on during such lapses in consciousness as sleep is due to the "*memory* bridges" that span these gaps. It is certain that unconscious *nervous action* occurs and even recall of this action with all the necessary causes and results may take place, but this is not equal to unconscious mental events or processes with awareness. The latter are not compatible with unconsciousness. Real events yes, mind entities, no.

The background given may suggest here that the beautifully regulated and self-activated designs that maintain homeostasis while not denying change, are often possessed by masters unwilling to take advantage of their own systems. A better way perhaps is to suggest that there may be flaws in the training which society offers. With certain notable exceptions, mostly muscular and related to sports, there is insufficient drive toward taking advantage of the enormous powers that be.

Telepathy and spiritualism must be denied and the communication described here is strictly behavioral, verbal or by similar means. A person's own mental states and events provide exclusively that of which he can be directly aware. This may be necessary to indicate here, if information is taken to suggest metaphysics with someone "up there" sending out a Morse code or if it is too difficult to dissociate information from the separate concept of the energy that carries it. However, microwave carried telecommunications and the descriptions in this chapter will help to explain this factor.

Within the organism, the messages carrying sensory or other information undergo many transformations including self-organization. Yet these messages have features in common with the outside world of messages and communications. Thus a prerequisite for an adequate view of DNA information is to see how communication theory has developed formalisms for the *internal structure* of messages and language which are not random. Codes and intersymbol influences are reasonably well understood as is the medium or channel. There is sufficient evidence also for intersystem communication by electromagnetic fields between the systems in organisms.

Possibly the statement of Marshall McLuhan that "The medium is the message" is a terse statement of one mechanism in the communication of information. The first requirement for transporting information or storing it is some physical medium such as a sound wave, tape, cable, microwave link, or radio connection with an organization similar to that shown in Fig. 2-1. Selection of a transmission *symbolism* such as

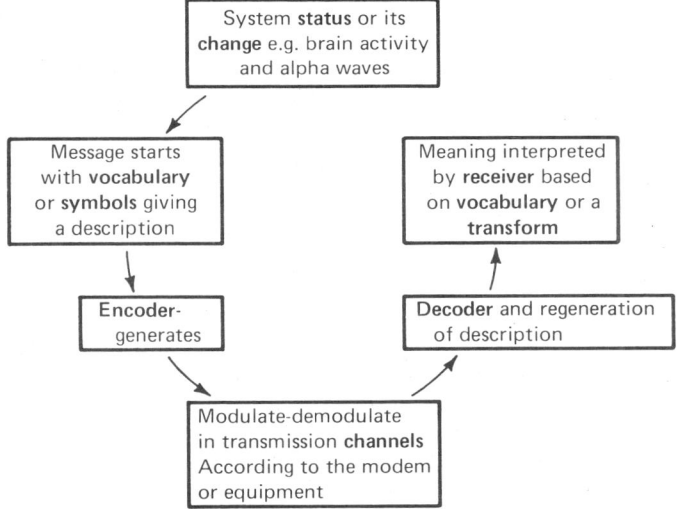

Fig. 2-1. Organization for communication analysis.

a 26-letter language coded in dots, dashes and spaces for letters and words, completes the idea of an appropriate *channel*. A dash has a special duration, longer than a dot, so the channel is made up of symbols passing with some duration of time for each. The ratio, information/T is the rate of transmission divided by the total duration and, as this value increases, it brings up the question of the capacity of the channel, which is limited when T is long and it is, of course, measured in bits per second. The sender can also be a person speaking, with the receiver, the ear of the listener.

A vocabulary is involved and it is made up of the alphabetical set of symbols. The AVM vocabulary may be binary encoded to give three categories, animal 10, vegetable 01, and mineral 11. It will be noted that a one in the first column gives a choice of vegetable or mineral and a one in the second could make it animal or mineral. Complete specification of mineral is only made by encoding a one in both columns. With this three word vocabulary, a very limited repertory of messages may be transmitted. Actually, devices are available that operate on variously sized vocabularies but they require standard speech and limited expressions. The IBM 7772 audio response unit, ARU, can have a vocabulary of 1,000 words or an unlimited vocabulary if there is enough storage space in the form of drums, discs, or photographs. On receipt of a coded inquiry, the audio response unit will search for a reply and deliver it in a female voice which uses the 1,000 American English word vocabulary. "Words" can also mean the code words that stand for the word as we know it, e.g., 11 stands for mineral. The symbol set in molecular biology can be nucleotides in nucleic acids or amino acids and proteins, progressing from letters, through words, to messages.

Elaborate codes such as that in the ARU obviously create requirements for economy, that is, for taking word and code lengths into account. Coding should be such as to make best use of channel capacity in transmitting information. The code symbols are selected carefully giving short codes to the more frequently used expressions. The quality of redundancy, while essential in communication, may act counter to economy, but redundancy protects against errors even if it overloads the system. A message is redundant if it is in other than its briefest form. Most cable or telegram (telecommunications) users will have faced the necessity of expressing their thoughts in the fewest words to reduce cost. Just as it is often necessary to repeat a key word (arriving tomorrow, repeat tomorrow) to guard against error in an important cable, an organism duplicates glandular, structural, chromosomal and informational functions. This is especially true for development, cardiac, pulmonary, cerebral, and reproductive system information where the redundancy is a form of insurance for vital functions. Similarly, all of the cells in an adult individual contain the entire genetic code for the individual, as if any *one* cell might be called upon to regenerate the whole organism.

A more general way of looking at redundancy is as a strong intersymbol influence. For example, n after *tio* is an obvious next letter in English as is h after t, p, and c, u after q, and t after gh. Noise is the villain in communications. It must be outwitted in channels and this is done by using redundant transmission to produce reliability. Following this theme along its many ramifications and possibilities gave Shannon his theories for information in telecommunications. Suppose that the telegram had been garbled to read "crriving tomorrow." Redundancy would allow ready recognition of the deviation in the language pattern and permit accurate interpretation of the message. Shannon can now enjoy the prospects as molecular biology and other biologies use the potential of his theory in the quantitative testing of their fundamental theorems.

Biological Software

It is convenient that, in this broad sense, the language of communication is in numbers, such as the binary and decimal sets. The economy offered by numbers in genetic communication is striking. In the nuclear repertory of DNA, some few thousand numbers are available for the communication of development information in lower organisms; complex organisms will require an order of magnitude more data. The qualities of flexibility and economy are shown in the way the developing organism expands by means of his *self-organized* and *self-generated* data. Only the expansion schedule or adult plan appears to be required, saving millions of numbers in the gamete. Since a *minimodel* on a cellular scale is not available, the developmental plan must come from an informational *outline* with self-generated detail. It must be admired how communication of this organismic detail occurs with all its *form factors, redundancy versus error, provision for modification on receipt of environmental information and other data transactions*. This self-scheduled expansion must proceed according to a computational program with its *biological software* which is the program of instructions for producing an adult. Great stability is demanded, even in intergenerational characteristics, so that much of the process must be repetitive and subject to strict product control.

Boolean Algebra, Numeric and Alphanumeric Logic

In order to penetrate more deeply into these studies in communication and the realm of black boxes as well as into the automata they represent, we need some sort of numeric or alphanumeric logic because there are so many numbers from so many boxes to fit together (Fig. 2-2). George Boole (1847) offered a suitable *special algebra* which Shannon, Wiener, and others have adapted to telecommunications—the latter being nonexistent in Boole's day. This logic should now be related to *binary arithmetic* for an operating basis, and finally, with the help of a few *electromechanical devices*, a rather formal *algebraic symbolic logic* will be obtained. Although forgotten for many years, the constructions and applications of Boole have emerged in the last two decades into a fashionable and useful study under the stimulation of automatic data processing.

Boole consciously associated his work with the mechanics of brain operation for he called his book *An Investigation of the Laws of Thought*, and he presented a mathematical construction for logic.

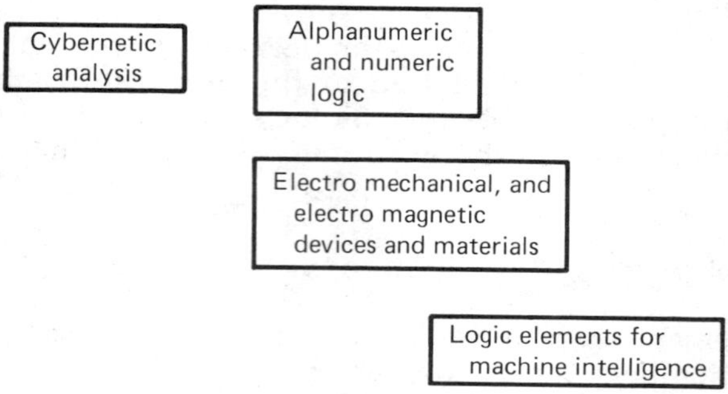

```
┌──────────────┐   ┌──────────────┐
│  Cybernetic  │   │ Alphanumeric │
│   analysis   │   │  and numeric │
│              │   │     logic    │
└──────────────┘   └──────────────┘

              ┌──────────────────────┐
              │ Electro mechanical, and│
              │    electro magnetic    │
              │  devices and materials │
              └──────────────────────┘

                         ┌──────────────────┐
                         │ Logic elements for│
                         │ machine intelligence│
                         └──────────────────┘
```

Fig. 2-2. Realm of black boxes and automata.

Shannon's analysis in 1938 came from his graduate work at the Massachusett's Institute of Technology where he has now returned as Donner Professor of Science (Fig. 2-3). His analysis provided a needed relation-

Fig. 2-3. Dr. Claude E. Shannon whose landmark work in communication theory continues its chain of development from Boole, Kolmogoroff, Wiener and others into the present expansive period of applications. He has now joined the M.I.T. faculty.

ship between electromechanical devices and Boolean algebra. He called it "A Symbolic Analysis of Relay and Switching Circuits." Thus, in information measurement

$$H = -K \sum_{i}^{n} P_i \log P_i \qquad (32)$$

This is Shannon's first theorem from which he went on to a theory of communcation, which is truly an impressive case of mathematical modeling, stating as it does remarkable theorems of information with persistently apt logic. K is usually one and log is to the base 2.

Norbert Wiener, also at M. I. T., and the Russian A. N. Kolmogoroff were contemporaneously advancing the broad aspects of communication theory in cybernetics and mathematical prediction analysis. Wiener received his Doctor of Science Degree from Harvard at the age of 18 and, among other singular accomplishments, is closely identified with the expansion of ideas about communication. To him, cybernetics meant the science of communication and control whether in machines or in living organisms. His book *Cybernetics - Control and Communication in the Animal and the Machine* written in 1948 was followed by *I Am a Mathematician* in 1950. As the title suggests, the latter was biographical but it consolidated the earlier cybernetic excursions into other disciplines. On his death in 1967, Wiener, an "Institute" Professor at M. I. T., was an acknowledged genius whose contributions were much broader than any of the Institute's departmental boundaries. Many publications since then have attempted to clarify the more difficult or obscure portions of his writings but some, such as *God and Golem*, written in 1964, are quite readily followed. The latter is an essay which anticipates many of the philosophical arguments which have been upcoming on the religious significance of automata and learning machines. It also shows a whimsical side of Wiener that was evidently not obvious in his mathematics.

It is quite clear that these scientists recognized the biological content of their work. Wiener, for example, felt that cybernetic analysis should be applied to biology before social science because of the promising status of experimental biology. Shannon, however, had some misgivings and appears to have been content with restrictions that limited the implications of his communication theory developments. He broadened out only slightly in his 1948 paper, "A Mathematical Theory of Communication." He has been an interdisciplinary scientist without admitting it and has attracted others to cross the bridge to deal with him where he knows the footing is soundest. On the other hand, his former association with the Bell Telephone Laboratories provided a ready use for his contributions without the necessity for crossing his primary boundaries of interest.

Two-State or Bistable Devices

An essential feature of the development of the living system is the kind of automatic control that has been discussed. Control and regulation keep the organism on an acceptable pathway as it proceeds toward goals such as reproduction. Regulation is often a matter of choosing between "on and off" states or of ordering "go or no go" for enzymatic processes. This system regulation may take the form of imposing and withdrawing inhibition of some sort that resembles switching or activation and deactivation. These actions often resemble the designs made with two-state or bistable mechanisms with symbolic electromechanical devices, and the logic which connects them. Symbols of special interest in logic are

$$A'$$
$$A \cdot B$$
$$A + B \tag{33}$$
$$\overline{A \cdot B}$$
$$\overline{A + B}$$

In Boolean algebra, for example, one can examine sets and classes of events, objects and the logic of propositions, and chains of propositions or statements concerning them in the sense of formal deduction. The algebraic logic involves pairs, C is a matter of how A and B, a pair, behave; or $C = f(A,B)$. The propositions describe logical inclusions and exclusions among the groups. A part of a disposition of points on a paper may be circumscribed and the circle will include some and exclude others. For example, $C = A^1$ is a subset of events or things which are not in A. More formally, from a Venn diagram (Fig. 2-4), A is

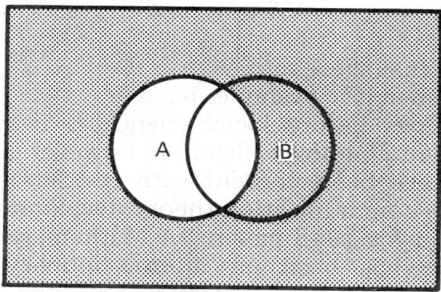

Fig. 2-4. Venn diagram. A excluded, non A or A'.

excluded from the shaded class. These useful illustrations are due to John Venn, a 19th century logician. Thus a logical statement A has its negation A^1 which, when taken together, give us an exclusive *pair* of statements (one or the other occurs). Negation means that A fails to happen or fails to be true so that we have non A or negative A. If A is a set of persons who wear rings, A^1 is the set of "non class ring" people. This gives the proper impression that a logic built around two-state or bistable devices might appear, where the circles are used to symbolize various concepts in logic and to represent statements.

The OR Operation, Class Sum

If another event C is considered and C equals A or B, it means that C occurs if both A and B do or when either A or B does. Thus, C is true when either A or B or both (logical sum, $A+B$) are true. The technical term for the *or* operation is disfunction. This produces a kind of logic with electrical meaning or a particular electrical situation may be taken to illustrate the numerical logic in the *or* disfunction, denial or contradiction. The simplest example is a parallel circuit which furnishes an alternative pathway for joining two points (Fig. 2-5), which may be circuit junctions. Current will flow through either branch A or B. A specific electronic device is the OR gate or buffer (Fig. 2-6) in which

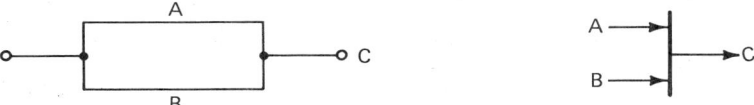

Fig. 2-5. C occurs when either A or B does.   **Fig. 2-6.** The OR gate.

either A or B reaches C. The gate has the effect of disjoining A from B. The Venn diagram for the set logic will look like Fig. 2-8 described as $C = A$ or B, also $C = A+B$. Figure 2-7 relates the logical sum to an actual

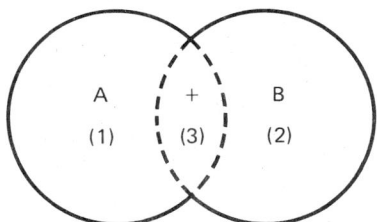

Fig. 2-7. a) Class of intelligent people; b) class of rich people.
Members may fall into 3 classes: (1), (2), or (3). The sum set is $C = A + B$ *either* intelligent or rich (heavy outlines). The product set (members in common) is shown only as the overlap in dashed outline both intelligent *and* rich.

class sum. The logical sum is so called to distinguish it from the arithmetic sum. If A and B had no members in common to both, however, the terms of the logical sum would be the same as the arithmetic sum.

The Conjunction or AND Operation

Conjunction is another operation called the *and* operation. If still another event C is considered such that $C = (A$ and $B)$, C is the logical

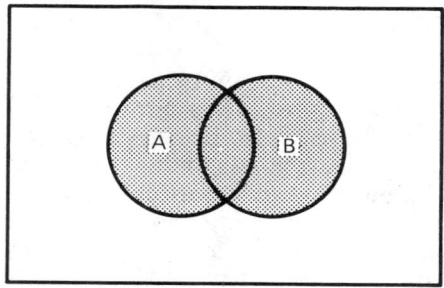

Fig. 2-8. C is A *or* B.

product of *A* and *B*. We obtain *C* if both *A and B* occur. The series electrical circuit illustrates this if two switches *A* and *B* are included (Fig. 2-9), and when the switches close there is a series current in the

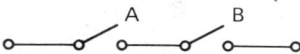

Fig. 2-9. Series electrical circuit illustrates the *AND* operation.

circuit. This is written $C = A \cdot B$ or the product AB as in the multiplication that the operation represents through the logical *and*. The set of encapsulated bacteria is the *logical product* of the two sets, bacteria and encapsulated organisms. To visualize the subset, the Venn diagram (Fig. 2-10) is for *C = (A and B)*. An electronic manifestation would be

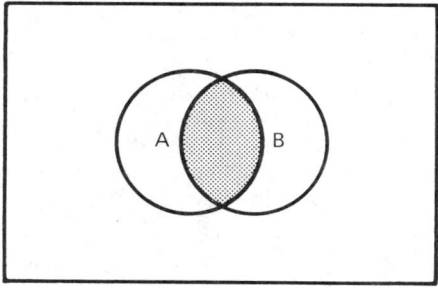

Fig. 2-10. C (A *AND* B).

the *AND* gate having two inputs to *C* which form a union or join *A* and *B* (Fig. 2-11). The *and* operation is a logical rather than arithmetic product. If two sets have three and two members respectively and have only one of these in common, the logical product has one term, while the arithmetic one has six.

Intersystem Communication 71

Fig. 2-11. Two inputs to C.

The NAND and NOR Operations

The *NAND* operation means not "and." Here, $C = \overline{A \cdot B}$ and the logic is illustrated in Fig. 2-12(1). There is also a *NOR* operation meaning "not or." This is written $C = \overline{A+B}$ and symbolized as in Fig. 2-12(2).

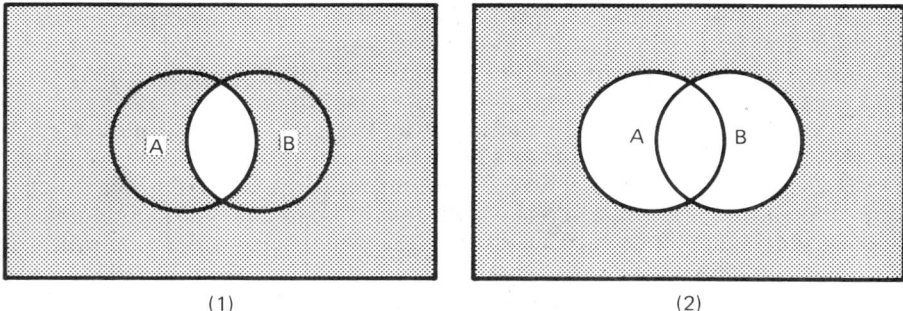

Fig. 2-12. (1) and (2). The *NAND* operation (1). The *NOR* operation (2).

Compound Devices

When these operations are used, they are usually combined in a numerical logic to describe set relations and to obtain algebraic polynomials. Two *and* gates may be combined, for example $C = A$ and $B + A$ and B for a compound device. Usually the laws for manipulation of these symbols are well-known. The cumulative law $A+B = B+A$ is applicable so that sums come out identically even if the order is inverted. The same holds in multiplication A and $B = B$ and A. The distributive law combines the addition and multiplication operations, $B(A+C) = B$ and $A + B$ and C. The associative law covers the relations $A+B+C = A+(B+C) = (A+B)+C$ and $C \cdot B \cdot A = C(A \cdot B)$. However, in contrast, the operations are idempotent, giving $A \cdot A = A$ and $A+A = A$ for a proposition A. The three operations are involute and dual. They are involute because $(A^1)^1 = A$ and a double negation returns us to the original proposition. They are dual because $C = A+B$ with the equivalent dual $C^1 = A^1 + B^1$. This is to say that non C happens if neither A nor B does, the same logic as C occurring if either A or B does.

These algebraic set relations illustrate the several types required for manipulations. We can conclude that they bring numerical analysis to bear on logical propositions. As for what one might expect to gain, at least a few pertinent results may be selected. Taking one "bionic" view, an understanding may be gained of how organismic communication functions can be so *compressed* that we have such dramatic results as the super "microminiaturization" of the adult elephant concept in its gamete. The other side of the coin is the minimization problem so visible in electronics. Algebraic manipulation of the logic behind electronic circuit elements is a means for studying the reduction in the number of elements by revealing equivalences in them. While not always predictable, the results have often been almost as dramatic as the elephant/gamete strategy. Perhaps the "bug" so important in electronic espionage is aptly named as it suggests miniaturization, biology, and information. In any case we have tiny, pill-sized radiotransmitters, light emitting diode (LED) or liquid crystal (LC) electronic wrist watches, tiny hand calculators, and desk computers which have replaced the bulky devices seen in earlier generations of the devices. Thus, the algebra of combinational relay circuits has taken on considerable standing in the sciences and practically all important electronic devices and materials like semiconductors have been carefully dissected so as to reveal their logic elements. The reconstruction of logic elements into an operating device is essentially the creation of machine intelligence and packing more functions into an integrated circuit (IC) results in miniaturization.

The brain designers selected simple, mathematical, logical models for purposes of discussing the human nervous system. The surprising result in that case, and in the electronics as well, is something that is actually an equivalent of formal logic.

Many variations are possible in Venn diagrams to represent the logic of biological situations. Other diagrams such as Euler's circles are also useful. Figure 2-13 represents an illness based on a chronic airflow

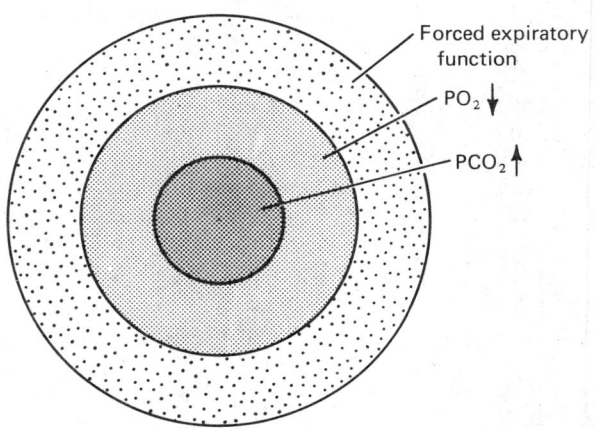

Fig. 2-13. Euler's circles for logic of respiratory illness.

obstruction (Lewis 1973). It would have been possible to state a simplistic high partial pressure of CO_2 accompanied by a low pressure for O_2. The ratio of ventilation to blood flow points up certain inequalities. Members of the class in the outermost circle represent those patients who were examined and found to not vary enough in the ventilation/blood flow ratio to renew blood gasses. The middle circle shows a class of patients with decreased oxygen tension $\downarrow PO_2$ and normal carbon dioxide tension PCO_2 related to a fairly uneven ratio for ventilation/blood flow. Then the inner circle shows patients as a class, having large inequalities in the ratio and both low $\downarrow PO_2$ and high $\uparrow PCO_2$. Thus all patients with $\uparrow PCO_2$ have a forced expiratory function and those with $\downarrow PO_2$ are also inside the function circle. Then premises may be formed and the possibilities examined. For example the "function" patients have two classes which are not identical but one is included in the other. As long as the conclusion follows of necessity from the premise, various speculative diagrams become possible in the same field of measurements (Fig. 2-14). Thus the function group could enclose the $\downarrow PO_2$ in the small inner circle instead of having it encircle the $\uparrow PCO_2$ class. The $\downarrow PO_2$ class could overlap the $\uparrow PCO_2$ one, or its

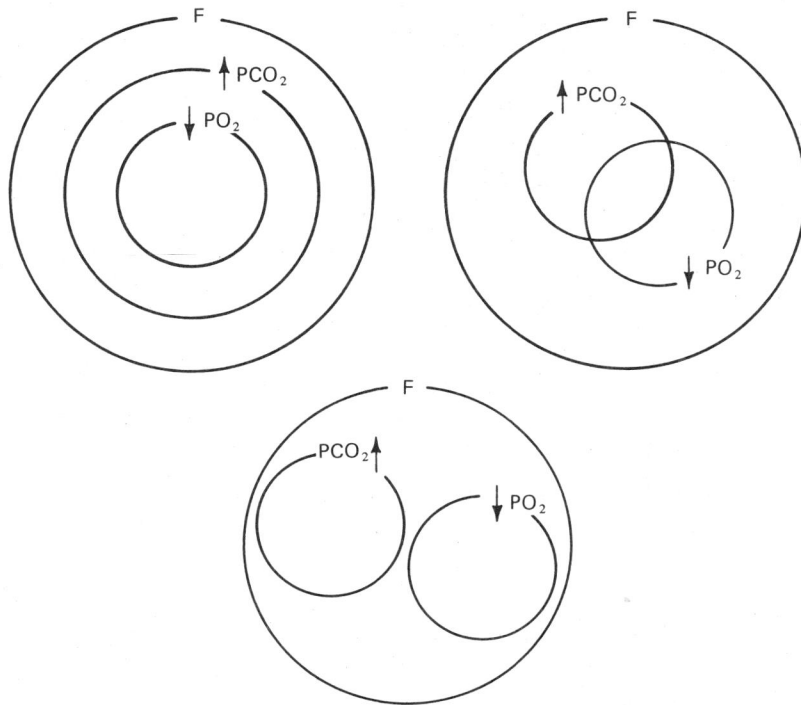

Fig. 2-14. Completely different premises and conclusions diagrammed.

members might, under other circumstances, be completely excluded from the ↑PCO$_2$ class.

Binary Operations

Symbolic logic thus deals with formal concepts and rules for deductive science and uses symbolism for deductive inferences. It may be assumed that with all the deductive inferences which a living system must make as it deals with its information and self-regulates, symbols would be essential. There is a quality about symbolism that must also be important in the living design. Roman numerals and Arabic numbers represent two notations to use in problems. Try to carry out an elaborate algebraic series with Roman numerals as compared with our usual Arabic number notation. One is extremely difficult, the other is far more comfortable. Another virtue is that symbols make visual operations feasible with almost mechanical actions. Thus it is *not* necessary to think of what we are doing and the more acts we can safely delegate to this kind of mechanical action, the more higher faculties are reserved for decision and judgment.

The pair which is recognized formally in Boolean algebra in $C=f(A,B)$ may be regarded as a "pulse" or "absence of pulse" result from a logical decision in electromechanical circuits. This is a dominant form of communication, just as it is in neural communication in organisms.

Such a general factor in systems as this "go or no go" process requires an appropriate counting method to go with it. There is one which was contributed by very ancient cultures. It is the binary or base 2 number system. For ordinary purposes we usually use the decimal system to base 10, but base 5 (quinary), base 8 (octal), and other systems exist and find appropriate application. As one result, already noted with 2 as the base of logarithms, a unit of information may be called a binary digit or bit. In the decimal system, such a unit could be called a decit. The term bit was adopted by Shannon over equally possible alternatives such as binit and nat, for natural logarithms.

In measuring the information content, the probability times $\log_2$ of the probability gives the number of bits. The logarithm of the number is the power to which 2 (the base) must be raised to equal the number. $2^1 = 2$ (2 must be raised to the first power to equal itself).

In the binary system the idea is to grow accustomed to operating with only two symbols, 1 and 0, which are the two binary "words." Adding would be direct,

```
    0           1           1
   +0          +0          +1
   ──          ──          ──
    0           1          10
```

The last may look unusual. It is explained by the economy of symbols (that is, only 1 and 0 are available) and some reexamination of decimals. It is just as reasonable as passing to the second place when you pass 9

Intersystem Communication   75

and go to 10. A point to mark place is used or understood for numbers in the decimal system. Places are counted off from it. The decimal 5315 can be separated as shown,

or
$$5315 = (5 \times 1000) + (3 \times 100) + (1 \times 10) + (5 \times 1)$$
$$5315 = 5 \times 10^3 + 3 \times 10^2 + 1 \times 10^1 + 5 \times 10^0$$

Place determines which exponent of 10 is chosen and the point position is understood to the right of the last 5. Negative exponents would come to the right of the point. An algebraic type of representation produced this group of polynomials. A series in exponents of 10 begins at the right.

As if counting had to be done with only two fingers, the available symbols in binary numbers (0 and 1) end at 1 so that repetition of symbols must occur every time the count reaches 1. Thus, 2 becomes 1+1=10, 3 is 10+1=11 etc. as shown in Table 2-1. Powers of base 2 are used to locate places in the binary number. Following the method used in the decimal example a binary number, 001110, would be

$$1 \times 2^3 + 1 \times 2^2 + 1 \times 2^1 + 0 \times 2^0$$

with the zeros to the left, as usual, not changing the value. The change

TABLE 2-1. Representation of Numbers

Decimal Base 10	*Binary Equivalent*	*Logarithm Base 2*
0	0	$-\infty$
0.5	0.1	-1.0
1	1	0.0
2	10	1.0
3	11	1.59
4	100	2.0
5	101	2.32
6	110	2.59
7	111	2.81
8	1000	3.00
9	1001	3.17
10	1010	3.32
16	10000	4.00
17	10001	4.09
32	100000	5.00
100	1100100	6.64
128	10000000	7.00

to decimals is 8 + 4 + 2 + 0 = 14. The last digit to the right fixes the point as in decimals. The places in binary form are:

$$2^5 \qquad 2^4 \qquad 2^3 \qquad 2^2 \qquad 2^1 \qquad 2^0$$

The power-of-2 solved gives the value and in the *place* indicated by the *exponent*; that value either exists (1) or does not exist (0). It would also be seen that a decimal 8 gives the binary number 1000 and that this binary 1000 is changed via 2^3 back into 8. Decimal 16 will give binary number 10000 and via 2^4 one returns to 16. Decimal 17 is 10001 in binary from the sum of 2^4 and decimal one which is 2^0 or binary one. The binary number is easily placed over the corresponding power of 2. For example, for binary 101010,

$$\begin{array}{cccccc} 1 & 0 & 1 & 0 & 1 & 0 \\ 2^5 & 2^4 & 2^3 & 2^2 & 2^1 & 2^0 \end{array}$$

The power-of-2 is solved,

$$\begin{array}{cccccc} 1 & 0 & 1 & 0 & 1 & 0 \\ 32 & 16 & 8 & 4 & 2 & 1 \end{array}$$

Then these two, the binary number and the solved power are related

$$32 \quad 0 \quad 8 \quad 0 \quad 2 \quad 0 = 42$$

to give the decimal equivalent of binary 101010. Only a 1 digit in the binary number gives the use of the solved-value of the power-of-2 in a column. By inspection several decimals can be made into binaries.

$$\begin{array}{llll} 1 = 2^0 & 8 = 2^3 & 32 = 2^5 & 128 = 2^7 \\ 2 = 2^1 & 16 = 2^4 & 64 = 2^6 & \end{array}$$

Any other number up to 128 will be within 2^7. The decimal 29, for example, is between $2^4 = 16$ and $2^5 = 32$. It can be located by a simple process

$$\begin{array}{llll} 2^4 = 16 & 2^3 = 8 & 2^2 = 4 & 2^0 = 1 \\ 29 - 16 = 13 & 13 - 8 = 5 & 5 - 4 = 1 & \end{array}$$

Thus

$$29 = 16 + 8 + 4 + 1$$
$$29 = 2^4 + 2^3 + 2^2 + 2^0$$

All the powers should be included.

$$29 = 1 \times 2^4 + 1 \times 2^3 + 1 \times 2^2 + 0 \times 2^1 + 1 \times 2^0$$

Then the value of 29 in binary form is a copy of the coefficients 11101.

Using the rules for addition 0+1 = 1+0 = 1, 0+0 = 0, and 1+1 = 10, the two binary numbers already examined can be added. The rules were applicable to forming binary numbers above one. Because there are no other numerical digits, the same rules provide a strikingly brief means of addition. Two of the binary numbers already studied are added,

$$\begin{array}{r} 11101 \\ 1110 \\ \hline 101011 \end{array}$$

There is a carry in the fourth digit not obvious in the rules. The three 1 digits in that column give a 1 as the sum with a carry going over to the fifth column. The sum is equal to

$$\begin{aligned} &1 \times 2^5 + 0 \times 2^4 + 1 \times 2^3 + 0 \times 2^2 + 1 \times 2^1 + 1 \times 2^0 \\ =\ & 32\ +\ 0\ +\ 8\ +\ 0\ +\ 2\ +\ 1 \\ =\ & 43 \end{aligned}$$

Computers

Internally a computer uses binary but programming uses the usual systems (decimal). Translation to binary is automatic within the machine from tape or card data. Internal instructions are stored in binary form. To address the binary instructions directly, it is convenient to use a suitable code. This machine code informs the machine of the expected arithmetic. This means for example, that a stored number can be ordered into the *accumulator*, where a second number may be combined with it arithmetically and the result stored at a known address.

The idea is to write a program which instructs the machine to go to an address, take out the instruction or number stored there, carry out some needed operation such as addition, store the result and put the original numbers back where they came from. The instructions are represented by code numbers which the machine "understands" and it carries out the instructions, always referring both to operation codes and to addresses where the numbers have been stored. These are written down on a program flow sheet by someone skilled enough in the work to have learned the code and who understands the peculiarities of a particular machine. Alternatively, the programming may be automatic from simpler instructions, for example, verbal, as an option in the sense of greater computer automation or by special programming language, for example FORTRAN (Formula translator) or ALGOL (Algorithmic language) which is actually an automatic person to machine translator. Another is COBOL (Common business oriented language). The task of learning to use the special codes and arithmetic and adapting them to a certain computer is programming and the product of this specialty is a designed computational procedure.

Information in Microwave Radiation

The machine transmission of information must operate economically and the equipment must be both reasonable in scope and efficient. If the living system has the same constraints, then it should show many commonalities, which it does. There is an element of surprise in information because anything that causes a "So what!" or "Old stuff!" response is not meaningful in terms of information. The sequence of signals involved is predictable and known so there cannot be transmission of knowledge. There may, however, be emotional pleasure as in hearing a much-loved song of a bird. A close study of meanings, unknown harmonies or unpredictabilities in the song would convey information.

A quantity of bits of information can be transmitted by telephone microwave link in a sending time t via the changing amplitude of the signal during that time, see Fig. I-2 in the Introduction. It is limited by practical factors in the design, especially by the energetics of the system. The time constant TC, the response time or the minimum time it takes for the system to change its voltage, has some value based on the energetics, or what it takes electromagnetically to put through a signal change. Voltage change with time, the reciprocal of TC, would be bandwidth and thus brings in the microwave frequency.

Suppose that the variable amplitude can show only eight steps or different levels that are meaningful for signals. With a fixed sending time t sec., and a limited set of eight steps, there is a definable capacity or a constraint on information transmission for any time t. This is a large number 8^{30} if there can be eight different amplitudes in each one of, for example, 30 sec. intervals or

$$\text{combinations in } t \text{ sec} = v^{t/TC} \tag{34}$$

where TC is the time constant. Of course, the sending time can be increased and a logarithmic change in information is then expected. Or, with logs to base 2, to be consistent with bits,

$$\text{Information in time } t = (t/TC) \log_2 v \tag{35}$$

Thus, in 2 seconds of sending, and thus, two intervals,

$$2 \log_2 8 = 6 \text{ bits} \tag{36}$$

Then the information capacity of the microwave line is

$$\text{Information}/t = (1/TC) \log_2 v \tag{37}$$

in bits/sec.

The transmission of information is then accomplished by sending changes in voltage amplitude of the microwave (amplitude modulation, AM). So each changed period, e.g., second number one, second number two, etc., is like one question in the AVM game with the voltage received as the answer. There is a difference; the AVM answer can only be one of two, yes or no. Thus, certainty as to the answer in one interval would have to come from a small "questionnaire." Is the signal voltage in the *upper* half of the 8 steps (Is it living in the AVM game?) or the lower half (nonliving)? Then among the lower 4, for example, one would have to work toward specification of the exact level of the signal by the same process. It is seen how the binary yes or no reduces the steps in finding the exact level from eight to three, and yes can be 0, no can be 1, and the testing of the signal can be completed in terms of only 2 digits and by only 3 queries specifying the answer among 8 steps.

Thus, if it is desired to *code* the level of amplitude in each sending interval, a good way to do it would be to describe it with these two binary digits (like identifying the answer in the AVM game in advance) and then use only 3 possibilities instead of 8. It would then take only 3 answers to specify which one among 8 levels is intended

	Answer	is
Upper half; 0, 1, 2, 3	can be 0 or 1	0
Upper half of rest; 4, 5	can be 0 or 1	0
Upper half of rest; 6, 7	can be 0 or 1	1

Thus 3 bits disclose the necessary information that the voltage is at level 6, and 0; 0; 1, three numbers sent, specify the whole voltage pattern and, thus, the information in the interval. The intervals are combined for a longer message which will need $t/TC \log_2 v$ bits. The larger the number of steps v in the signal pattern, and the larger the number of t/TC intervals, the greater the information content in bits that must be transmitted.

When the communication specialists began to unravel the intricacies of this kind of information flow through cables, relays, switches, and other circuit elements, they were not too far from the telegraphy of Morse. They were soon deep in the world of automated transmission, computer storage and control, speech compression, time and signal readout, and the related semantic and linguistic studies as well as automatic translation and security measures to protect the integrity of private communications. When any hope of rewarding progress existed, organismic communication was examined. Bionics is a quite deliberate pathway or science of systems, charted by Major Steel in the United States Air Force at the 12th Annual Aeronautical Electronics Conference at Wright Patterson Air Force Base on 2, 3, and 4 May, 1960 to describe the study of biological systems in order to improve thereby mundane physical ones (Geradin, 1968). A reverse process seems to be in motion in communication theory in some areas of study such as prosthesis, proving that it is only fair to regard bionics as reversible and to look at many biophysical phenomena as reversible bionics.

The expected benefit from the work of these communications specialists is an increased understanding of the communication networks which we know permeate all organisms. The relevancy of communication theory is that it was laboriously developed to deal with situations that are often analogous to biological ones. What follows then is a sampling of the vocabulary used in this field. It has far wider significance for the inevitable expansion of computers into all of modern biology which in turn depends greatly on the use of symbolic notation.

Symbolic Information

Everyone is involved with symbols from the moment they begin to bank data as a child, but in the present context the big moment is when we first learn to let x equal something.

Symbols carrying information between entities are subject to the rule of economy or a minimization of the number of them which must be communicated. An illustration is the selection of the Morse code single dot for the letter E, the most common letter in the English language. Information flows, however, via groups of letters, words and expressed thoughts, so that economy, and its close relative redundancy, would be served by analyzing expressions more broadly than by single symbols. While nonredundancy is the briefest expression of a message in symbols, it is gained by making full use of the meaning conveyable, via intersymbol influences which are vital in language but apparently missing in the protein "idiom." Perhaps the most striking of such influences would be the time during which some conceived influence acts. In the present transmission, for example, it is intended to be clear that the subject is communication and the application of its concepts for the flow of information between systems in biology. As long as this understanding, redundance, or influence persists there is no need to restate it. In fact any new word that appears is immediately suspected of being a part of this subject. In this manner symbols also extend the influence due to some understanding or prearrangement. Constraints are being introduced here, one being the necessary rule of clearly changing topics when one is finished. All organisms must learn the many chemical, physical, biological, and social constraints of the system in which they operate. For us personally, these are rules and laws and those observations which experience has uprated to full guidance status. Just about everything and everybody come into this category when their actions affect us or the sytem under study. In adjusting, we consciously filter out the significant constraints and so maintain a high adaptive efficiency.

Many of these influences which reach us with information are in categories like states of systems which go together and a collection or ensemble of such like things can be visualized. A source of messages can produce an ensemble of all its possible messages. In measurements one could measure a large number of identical systems for their output at one time in an *ensemble* method. No variation in the density function would make the process stationary. Measuring one system over a long time could give a time average. If the result is independent of the time interval chosen, then the process is stationary with time.

Stationary statistical processes then are those in which the probability distributions are the same independent of the time at which they are measured. Thus we have constant, "dependable" experimental conditions like voltages that do not vary no matter on which day they are measured.

Ergodic processes give the same statistical results whether ensemble or time averaged and can always be stationary. An ensemble or bank of constant voltage batteries of different output would provide an ergodic but nonstationary process. One source measured against time would be stationary with time.

A grouping of symbols such as those of Fig. 2-15 is a simple ensemble in which a row of playing cards represents a sequence. Many operations

Fig. 2-15. Sequences in an ensemble.

are done on ensembles both in biology and in communications. The states in which a system can exist may be finite, as in a step-change regulator with steps 1, 2, 3, etc., or continuous and infinite as in the time states on the face of a clock (Fig. 2-16). "Stationary" prescribes a fixed order for a sequence in an ensemble that does not change as symbols become more distant from the origin. Thus, in a temporal sense, it is homogeneous and the statistical properties are not changed when the starting point is advanced to any point of the original process. This book is a certain sequence of words in the ensemble of all sequences of equal length. Lullabies, soul, rock, and symphonies share a common element, the ensemble of musical notes contains them.

A constant voltage source such as a battery could be tested and the resulting probability distribution obtained at three different times. If the measurement begun at T_1 produces the same results as that at T_2

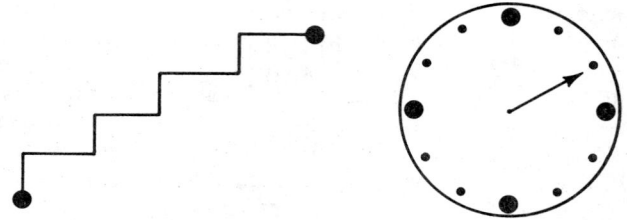

Fig. 2-16. Finite states and infinite states.

and T_3, the process is stationary in time. If this applies to many such equal voltage sources, it would be ensemble statistics. There, identical density functions of the ensemble at three different measurement times would make the process stationary. Many astronaut candidates, as an illustration, might be prepared for EEG recordings and the traces related to tasks they were required to perform. Statistics would be applied to the recordings in this ensemble of astronaut signals.

In linguistic and chemical examples, the structure of a language with letters, symbols, and words or that of a protein with amino acids is studied to learn if it is stationary. For language it is concluded that this is not strictly true because symbol probabilities as to what letter follows another are not rigid throughout an ensemble of messages. Certain statistical math is appropriate if the process is stationary. Many characteristics of language will be obvious. As shown above, *the* is the most common word in English, and the letter *e* appears most often. Common words such as "a, the, or, and" tend to be short; otherwise everything we read would be much longer.

An ensemble average may be taken for example, of all the opening symbols or letters in words, or the second symbols, etc. Common combinations can be described in this way, for example, "in," and "for." The card sequence (Fig. 2-15) can be used to show what is meant by a time average, in this case the lumping of suits in a sequence and averaging. Thus, in the first two sequences or rows, the probability for clubs or diamonds is ½. Taken from the first column, hearts have the probability of ⅓. In an ensemble, this time average may equal the ensemble average, in which case the originating source of information is called ergodic. This one is nonergodic, ½ ≠ ⅓. It is convenient to model for communication study in terms of ergodic sources. It will be consistent statistically—symbols will recur unvaryingly, and they will be in a fixed order in respect to digrams, trigrams, or the equivalent. When the probability of a specific symbol is constant through a message, the numerical analysis result for the source and the output are specified.

For a coin, heads and tails are statistical states for that system. These symbols, H and T, occur with a frequency that approaches a limiting value and this process is *stochastic*. Tails will come up randomly, but ultimately its frequency will be close to 0.5. Thus a stochastic process, phenomenon, or variable is a random or chance one and is described by states at various times. These states are determined by both chance and by the history up to then. A scene in a film will depend on chance selection by the observer and on the scenes that went before. In a Markoff process, the future depends upon its history only. The result of flipping a coin depends only on that toss and not on the last one and is not a Markoff process. Word formation on the other hand is Markoff because of the correlations and intersymbol influences.

We often measure one system and yet know that the statistics will hold for an ensemble of such systems at any one time. The process measured is then ergodic and it will always be stationary. Measuring the constant voltage sources and setting up the density probability functions would give a stationary ergodic process. With *different* voltage batteries, however, the output may be constant and the process stationary, but the ensemble statistics would not be that of an ergodic process.

Energy flows with information to construct the order in a message and energy is required to extract information. Therefore, it is not surprising that the word ergodic comes from the Greek expression for energy pathway. It is compounded from the word that was accepted as a unit of energy in the C. G. S. system, the erg, and the word for way or road. It is a term in statistics or in general mathematics. A stochastic ergodic process has the property that time averages taken from an observed record may be used as an approximation to the corresponding ensemble or population averages. In the long run, an ergodic system will return to states which are similar to previous ones. An ergodic "path" has the most direct bearing on the origin of the word, for it is like an ideal billiard ball in its rectangular "box" going on forever and touching near every location. This implies a recurrence, an energetic trajectory, and reaching all points in a confined space with equal frequency.

A language is less ergodic after several letters in a long word have used up the intersymbol influence which is strong initially. As an easily analyzed stochastic source (code-like), language is assumed ergodic. While intersymbol influence is exerted on letters, this is only over a limited (8 letters in English) number and, after that, the probability of a certain letter's occurring is random. This is then an ergodic ensemble of symbols whose occurrence is probabilistic but whose influence on each other tends to disappear exponentially.

The influence of emotions is to emphasize the nonergodic. Thus there could be distinct pleasure in the blaze of neon artistry of Main Street, U.S.A. to someone unable to read. When systems communicate by radiation as a bat does with sonar, there is an infinite possible ensemble of sequences of signals like echoes. It is obvious that the message processes are quite the opposite of random ones, but are guided instead by these internal features of communication.

Novelty increases the density of information. To the brain decoder, a millisecond of lower voltage in a neural signal or a discontinuity, can alert the cell population to an unusual event. The brain is also sensitive to the ratio of signal to noise, so that much more information flows when the ratio is small.

Physical Features of Information

For the physical manifestation of the systems discussed, communication and control systems are described and identified with information theory. Organismic information transmission and processing is seen in relation to capacity, input and output as a form of direction and energy, generators as sources, users as addressees, transducers as discriminators between electrical, magnetic, mechanical, and chemical forms, thresholds and junctions as selectors or switches, and the qualities of information that make its processing so fundamental. Nervous systems illustrate these points rather well, for example, in axonal and dendritic transmissions. Then it is seen how the information capacity is related to the diameter of the fibers for accepting a charge and to the antinoise measures available in the biosystem. Molecular information is seen as being modulated and demodulated, especially by key enzyme systems which provide connections, switches, and branching functions. In sensory information processing, selective narrow-band-pass filters in the central nervous system allow memory and attention connections to be conserved by blanking out whole volumes of the attention sphere that are at one moment not relevant.

Dendritic impulses live only for their mission distance, otherwise their "all or none" character would trigger a somatic impulse. Instead they act as synaptic computer input from the dichotomy of branches out to their distal filaments. The excitation *threshold* at a junction is a logical cancellation point and mechanism for route selection if the input to the parent root is not satisfied. The dichotomy may give "AND" or "OR" switching action, and summations from numerous entering roots may accumulate the threshold charge to the parent. For two inputs, pulse A "OR" pulse B may excite pulse C,

$$C = A \text{ "OR" } B; C = A + B$$

Alternatively, the necessary input for C must include pulses from both A "AND" B,

$$C = A \text{ "AND" } B; C = A \times B$$

Then this decision may be related as in the study of Scott (1973) to the impedances of the two possibilities and to the diameters of the root fibers. For root A to excite parent C which has a threshold, the "OR" switch depends on the inequality,

$$\frac{\text{diameter of A}}{\text{diameter of B}} > (\alpha/1-\alpha)$$

where α is a factor less than one and about ½. Otherwise, the switch is an "AND" gate and it is observable that the diameters also control excitation for root B. Similarly, a threshold charge may be accumulated from n roots for any number in a junction,

$$\text{Threshold} = (\alpha/1+\alpha)[n + (\text{diameter B}/\text{diameter A})^{3/2}]$$

and again provision is seen for connecting into any daughter root, as well as into the parent. The logic may be strengthened by form factors in the dendrite and the possibilities of a kind of "responsive growth" where the morphology of a dendritic root shades into molecular aggregations and selections.

The information theory of hearing is physically manifested in the communication via transducers and transfer functions when the transducers are cochlear or other membranes, hair cells, Organ of Corti, and ossicular chains. The physiology of hearing is incompletely understood as will be obvious when an attempt is made to explain the hearing of microwaves (Chapter 17). The information capacity of about 5000 to 8000 bits/sec. represents a flood of auditory information for a listening capability of only about 25 bits/sec. The value of informational biology is seen here. Whether the investigation is psychological as that of Jacobson (1951), or biophysical as in the case of Rink (1973), the conclusion can be framed in the same units (and approximately the same numbers) as just given for information capacity for auditory transmission. In one case, the difference limens for simple discrimination of frequency and intensity of pure tones with masking distance effects and complex stimuli, distinguished in a second, are measured; in the other, physiological phenomena of neural noise lead to computation of cochlear patterns, spike counts in the space-time domain of auditory events, randomness of interspike intervals in the primary auditory neurons and consequences in hearing acuity.

The latter analysis of the channel message set and its entropy, error probabilities of auditory output and input, and the effect of different step-size in spatial and temporal processes and their directions, allows neural noise in the system distal to the brain to account for the information bearing quantity, identified by Rink in the primary auditory-neuron discharge pattern—the spike count obtained in a certain time interval. They are assumed to occur randomly by a Poisson process when forcing functions are applied. Then two results are evident:

1. Biophysical results support the psychoacoustic data in that neural noise constraints in hearing desensitivity are peripheral rather than cerebral.
2. Masking and saturation are related to neural noise such that Rink's fractional differences, a parameter involving spikes and intervals, are maximized and the information rate is minimized except for on/off, one bit, information in a pulse-interval; a possibility too when microwave input overrides the sound system.

For these studies, the *morphology* of the auditory nerve system in fiber numbers (approximately equal to 30,000) *lengths*, important in the sensory envelope, *process time constants*, important in the temporal activity of the envelope, the *transfer functions* for middle ear transmissions, *basilar membrane* and *stapes displacement* (space) and time constants, and *synaptic integration* for mechanical-neural encoding, form a basis for the information capacity in the channel, which is the biological result as the acoustic input sets a meaningful distribution of space-time surface displacements on the cochlear membrane. Finally, the output information I in a signal for time t is,

$$I = (Lt/\Delta x \Delta t) I(\Delta) \text{ bits}$$

where L is the length of the cochlea, Δx and Δt are increments on the space-time surface where spike density may be considered, and Δ is the fractional difference between spike counts from neighboring intervals which relate to channel error and neural noise and should be maximized.

Physical Basis for Information Management

The CNS is responsive to long-term environmental deviations by developing an information management, by what Conrad (1974) calls the selection, by probing, of circuits composed of effector molecules. The information-processing capacity of molecular systems can be related to the CNS by using an array of discriminating enzymes on the neuronal membrane. The response can include retrieval of information from memory, DNA or RNA sequences which connect the genome with its internal information with information from the ambient, through life spans, into ontogeny and evolutionary changes. The phenotype depends on these segments which determine the specific conformation of molecules. This morphology or architecture has, thus, specificity, provision for amplifications, and a means for gradual change. It is important in the genome that access or inhibition manages the information flow and that the *transforming principle* in DNA be expressed. Physically, the unit structure of the CNS must provide both the net, and local responsive membranes with selectivity, and the molecular information in enzymes, as well as their catalysis and activation. Activation is followed, in turn, by the energy expression in a fired neuron, but feedback stimulates the synthesis of nucleic acids with codes for the enzymes.

Information management requires the selection of nets, evaluation of the output and direction of DNA transforming principles. Thus, to the word information and its variations in Chapter 4, transformation must be added as a "management technique." The net connects interested regions on a basis of their need to know or the demand for their products. DNA can operate the system by making transportable enzymes available on a self-organized basis or in response to external signals.

Segment information is far more meaningful as leading to conformation than codon or nucleotide base information. In the net, the excitable enzymes express the transforming principle by establishing a *responsive space, time surface* on membranes. These surfaces permit selection through electrochemical sensitivity (EM band combined with enzymatic sensitivity), filtration of data, recognition, and via nets, a comparison, connection, or recourse to memory. Internal EM signals are ubiquitous from relaxing molecules, asymmetries in dendritic roots, cellular morphology (local fields), and conductive processes. The DNA segment can then be appreciated as an effector group of "letters" or an informant unit that makes *reality* out of a *possible* change in status.

In identification the net in action responds to an EM field, with alerted, activating enzymes in the space, time, surface-field assuming the burden for the neuronal spikes that follow. An integration of the position and temporal information would be required for transmission efficiency.

The information is manageable by molecular processes that vary the field by continuous amplifications or cancellation in answer to evolutionary or contemporary pressures. Recombinant genetics in a dynamic sense, transformed genes, and DNA molecules in glial cells are essential. The neuron is seen in this manner as a transformer with decision power from its membrane enzymes manageable as to population in ontogeny or phylogeny.

PROBLEMS

1. Show electrical circuits illustrating the AND operation with two inputs to the outlet.
2. Add the following binary numbers:

   0+0, 1+0, and 1+1
3. With respect to poliomyelitis at least four classes of persons may be recognized: those who have been immunized, those not immunized, those who have been challenged by the disease, and those who remain unchallenged by the disease. Show how the four classes may combine to produce eight different classes.
4. Give the binary equivalent and logarithm base 2 for decimal, base 10

   3, 7, 10, and 128
5. What is the sum of binary numbers 11101 and 1110?
6. Define: stochastic, ergodic, sequence, ensemble, stationary and noise.
7. What three-question sequence can be used to locate the third step in a set of eight possible voltage amplitudes counting 0 as one step? What three binary numbers would be sent to give the information if the upper is 0, and lower is 1?

BIBLIOGRAPHY

Beckenbach, E. and C.B. Tompkins, *Concepts of Communication—Interpersonal, Intrapersonal and Mathematical*, John Wiley, New York, 197-. 14 authors. Treats information for the brain and nervous system including trauma in three chapters and communications of a socio-bio-mathematics nature in areas of psychology, mathematics, psychiatry, neurology, engineering, anatomy and physiology.

Bolschelet, E., *Introduction to Mathematics for Life Scientists*, Springer-Verlag, New York, 1971.

Boole, G., *An Investigation of the Laws of Thought*, London, 1954, (1847) original.

Broadbent, D.E., *Decision and Stress*, Academic Press, New York, 1971.

Conrad, M., "Evolutionary Learning Circuits," *Journal Theoretical Biology*, 46, 167, 1974.

Daniels, C. and F.S. Wood, *Fitting Equations to Data*, Wiley Interscience, New York, 1971.

Day, R.H., *Human Perception*, John Wiley, New York, 1969.

Eugene, W., *Stochastic Processes in Information and Dynamical Systems*, McGraw-Hill Book Co., New York, 1971.

George, F.H., *Cybernetics and Biology*, Oliver and Boyd, London, 1965.

Gerardin, L., *Bionics*, McGraw-Hill Book Co., New York, 1968.

Hohn, F.E., *Applied Boolean Algebra*, 2nd Ed., Macmillan, New York, 1966.

Jacobson, H., "Information and the Human Ear," *Journal Acoustical Society of America*, 23, 463-471, 1951.

Kolmogorov, A,N., *Grundbegriffe der Wahrscheinlichkeitsrichnug*, J. Springer, Berlin 1933. Tranlated Edition by N. Morrison, *Foundations of the Theory of Probability*, Chelsea Pub. Co., New York, 1956.

Kolmogorov, A.N., "Logical Basis for Information Theory and Probability Theory," *IEEE Trans. on Information Theory*, IT-14, 1968.

Lewis, B.M., "Laboratory Medicine," *Postgraduate Medicine*, Vol. 53, April 4, 1973.

Norman, M.F., *Markov Processes and Learning Models*, Academic Press, New York, 1972.

Oestreicher, H.L. and D.R. Moore, Eds., *Cybernetic Problems in Bionics*, Gordon and Breach Science Pub., New York, 1966. Symposium of USAF at Dayton Ohio. Gives electric signs of expectancy and decision in the human brain. Strong cybernetic presentation.

Rose, A., *Computer Logic*, Wiley Interscience, New York, 1971.

Rosenblueth, A., *Mind and Brain*, M.I.T. Press, Cambridge, Mass. 1971.

Schrodinger, E., *What is Life?*, Macmillan, New York, 1945.

Schwartz, M., *Information Transmission, Modulation, and Noise*, 2nd Ed., McGraw-Hill Book Co., New York, 1977.

Scott, A. C., "Information Processing in Dendritic Trees," *Mathematical Biosciences*, 18, 153, 1973.

Shannon, C.E., *A Symbolic Analysis of Relay and Switching Circuits*, M.I.T. Thesis, 1938.

Shannon, C.E., "A Mathematical Theory of Communication," *Bell System Technical Journal*, Vol. 27, 1948.

Shannon, C.E. and W. Weaver, *The Mathematical Theory of Communication*, University of Illinois Press, Urbana, Ill., 1949.

Shannon, C.E. and J. Macarthy, Eds., *Automata Studies*, Annals of Mathematics no. 42, Princeton University Press, Princeton, New Jersey.

Simon, W., *Mathematical Techniques for Physiology and Medicine*, Academic Press, New York, 1972.

Venn, J., *Symbolic Logic*, 2nd Ed., Ben Franklin, New York, Reprinted 1972.

Venn, J., *The Logic of Chance*, Chelsea Press, London, 1962.

Whitehead, A.N., *Introduction to Mathematics*, Oxford University Press., New York, 1911.

Wiener, N., *Cybernetics or Control and Communication in the Animal and the Machine*, John Wiley, New York, 1948.

Wiener, N., *I Am a Mathematician—The Later Life of a Prodigy*, Doubleday, Garden City, New York, 1950.
Wiener, N., *God and Golem, Inc.*, M.I.T. Press, Cambridge, Mass. 1964.
Young, J.F., *Information Theory*, Wiley Interscience, New York, 1971.

3

DNA Information, Gene Management and Radiation Genetics from Recombinant Genes to Sequencing

> Not all parts like, but all alike
> informed With radiant light, as
> glowing Iron with fire.
>
> Milton in *Paradise Lost*

RECOMBINANT GENES

The biological management of information is in the hands of genetic molecules whether in ontology, which is responsible for the individual, or phylogeny which is likewise for the race. At least this was true until recently. On 20 August 1976 in San Francisco at the Americal Chemical Society meeting, the work of Dr. Khorana at Massachusetts Institute of Technology was announced in which he and his colleagues had synthesized a gene and allowed it to recombine in *Escherichia coli* cells where it carried out its function. It is now clear that synthetic genetic instructions can be inserted into living sytems and the DNA information management has, by the power of biochemical synthesis, been *shared* for the first time in a direct way with humans. The shift is potentially for *any* genetic information management from the individual organization to humans. This is what worries those who regard work on recombinant genes as dangerous and something that may have to be forbidden.

There are now three ways for this process to occur:

1. By recombinant genes—natural transfers of operational genes carry stored data from one organism to another giving the donor's features to the receiver at the discretion of the operator.
2. By chemically synthesized transfer genes in limited cases where the gene can be synthesized and will function when transferred.
3. By enzymatically ordering the reversing process in which the product makes the gene it came from. As in 1. (above) the gene composition remains unknown so it cannot then be synthesized.

These recombinant genes can instruct the recipient cells to make biochemicals needed but not now being made—in the sense of biomolecular engineering. They can also be used to trace the pathways of information management in health and normal genetics in biosystems, or of mismanagement, e.g. in cancer, as a rate function under genetic control. This is done by deliberately mutating a gene and then using it to learn the effects. Now that manipulation is feasible, it is necessary to know what should be deleted or added. Done carefully, the method of chemical substitutions can alter message segments in the cipher according to a controlled plan. This may be safer than recombining natural genes into a new nucleic acid environment where the result may be unpredictable with monstrous possibilities.

Today then, the question, "What is a gene?" can be answered with precision with an example, the tRNA (tyrosine) one, while, at the same time, noting that others may be larger with perhaps up to 1,000 or more base pairs in human cells instead of the 199 in the synthesized gene.

The recombinant gene technique consists of splicing the new gene into the plasmid, circular bacterial nucleic acid (see, for example, the section in Figure 3-3), or combining it in phage which is then permitted to infect the organism. Passing new genic information management to man puts him in a "forced evolution" role, especially due to his new capability for changing species which have evolved since deep in history. He now sees no species barriers. This situation strains his understanding of genetics and, by accident, scientists may mingle genes into a "witches brew" of new organisms. This doesn't seem likely to be a problem if proper laboratory confinement is practiced. Degrees of simple care are those of aseptic techniques which are learned in general microbiology. The student comes to respect the organisms he uses and to treat them as potential pathogens in the sense of *excessive prudence*. Beyond this are the barbed wire and the security of biological warfare projects. Extraordinary care is a matter of which organism to use: 1) One that is supersensitive and cannot survive outside the experimental area. *E. coli* is the best understood organism for these recombinant purposes and many would prefer not to substitute; 2) Change to an organism not so intimately associated with every human being, and already known to vary in virulence. The security is highly sophisticated involving cytogenetics, virology, species similarities and ecological interactions and many places will not have the expertise demanded.

To understand these questions fully, one must begin with the biochemical background for these informational molecules. Radiation is an important element. It was used to probe the DNA structure and reveal the double helix with confirmation by other tests. DNA is also a molecule which is particularly sensitive to radiation. This appears to account for early evolutionary changes and continues to provide a means for altering the information in the molecule. Induced mutations, followed by selection on a tremendously accelerated time basis are common when adequate radiation is available. The situation is delicate however, for overmutation is more probable followed by inactivation or inability to divide. Numerous radiation inactivations may be followed by reactivations, as almost magical repairs are *induced* in radiation altered molecules as a characteristic which is tightly coupled to the applied radiation wavelength.

The Biological Management of DNA Information

Genes in chromosomes in the nucleus are on DNA molecules which carry the operating information. They communicate individual and species characteristics in a dependable and orderly manner. DNA behaves as the source of *information* in an organism and provides for its genic storage, coding, communication and utilization in enzyme and other protein synthesis. According to the classical description of Beedle, molecular control is expressed by having one gene associated with one enzyme, or more accurately, it is associated with one polypeptide chain. With this Central Dogma and Sequencing Theory, the intellectual foundation is formed for molecular biology. Information management is the basis for both, dealing as it does with the registration, transmission, and reception of information in DNA, RNA, and protein.

To recapitulate, nucleotide bases (ATCG) are like letters with a value of one bit each; the triplets will represent words having an appropriate number of bits; the protein molecule will be a paragraph with some 100 to 200 bits; the cell will involve some 2×10^{10} bits; and the brain will have a capacity upward of 10^{15} bits. The DNA of the much-studied bacillus *E. coli* is estimated to have information equivalent to 10^7 bits while the eye depends on 7 bits per foveal unit of the rods and cones for a total of about 10^{10} bits for the foveal burden for these units.

Immobilized Enzymes. A shift in information management occurs when a biological component is removed from its natural system. Man is the new management and the component is isolated because it gains him some advantage. Thus, information in linear amino acid sequences is all that is necessary for the structural complexity and much information for the operating level is deposited in molecular conformation and functional linkage. The catalyzed synthesis may be duplicated and reaction products obtained in a semi-artificial manner.

Specificity Information: Retention of specificity is the test for these conclusions and it is most evident in the behavior of immobilized enzymes which can operate independently and quite specifically when separated from the natural system. The burden for control, regulation, inhibition and activation now changes to the new management and the success of the process depends upon how fully this responsibility is understood.

In its best manifestation as in the case of immobilized enzymes, it is possible to utilize the enormous speed (hundreds of thousands of molecules per second) of bioenzymatic catalysis in isolated, productive systems. The rapid inactivation of enzymes is something different and does not have the burdens from control which remain in these operating systems. Taking a lesson from the natural system, the immobilized enzymes are captured in structures such as:

1. Absorbing columns
2. Covalent binding and supporting matrix
3. Cross-linked on their substrates
4. Held in suitable gel
5. Entrapped in capsules
6. Held on walls of counter-flow devices
7. Exchanged and bound to membranes, for example in constructing ion selective electrodes for urea using urease
8. Used in combinations of these

The result is that the immobilization permits the substrate to flow by for continuous catalysis and, most significantly, the enzymes are retained for reuse. Since they are often expensive, this reduces the cost and takes advantage of their natural tendency to survive the enzymatic process intact. In their natural system, the enzymes are usually positioned on membranes or membrane subunits, cristae in mitochondria, where the significance of the position is seen in the order of the catalyzed process within a larger scheme of multienzymes and reactions. Immobilization is then a form of coupling and a part of macromolecular binding and affinity systems.

Then isolated, synthetic reactions may be used for some step in a process such as the preparation of penicillin by enzymatic engineering. The enzyme isolates may be used in insoluble form or compartmented for the soluble form but, in any case, advantages may sometimes be gained from versatility in the enzyme reaction conditions and from their resistance to decomposition. In such processes the structures may be reversed, for example by inverting a membrane to introduce directivity and control in an experiment.

Restriction and modification enzymes employ the specificity in dense enzymatic information to manage natural or man-made changes in DNA. In one type of operation, the restriction, host, DNases cleave viral DNA after an infection. Appropriate nucleic acid modification, in turn, by the virus can defeat this destruction of its program. The specificity in these DNases is for special nucleotide code *sequences*, that is, for places where the enzyme finds identical sequences in the double stranded DNA. The scission it then causes may be permanent. In the laboratory the specificity is transmitted to probes to identify the fragments and *sequencing* of the nucleotides is made possible as will be discussed later in this chapter. The DNA structure puzzle can then be pieced together from this sequence information as many scission enzymes are now being made available. Conversely, the assembly of new sequences by DNA ligases makes management of DNA information a reality as the cohesive fragments can be shuffled. Restriction enzymes act specifically due to subunits in their recognition region which recognize a certain repeating sequence in DNA. The danger of self-destruction is normally inhibited by having only "friendly" or modified sequences in a host DNA. A typical modification could be addition of a methyl group to a base in the endangered sequence. This host-protective ploy is also available to the attacking virus which can thus "learn" to refine its attack. The restriction enzyme's substrate target is a point of symmetrical sequences such as TC and CT in the parallel strands of DNA.

An application for restriction enzymes in eucaryotic cells is found in the diagnosis of the "informational disease" sickle cell anemia by restriction enzymes in the embryo. The catalog name for the endolase or DNase is HpA 1 and it is used to cleave the fetal DNA. Its informational specificity causes it to single out its target nucleotide sequence from the parent beta globin gene and separate it. Radioactive copies are used as probes for interrogating the excised fragment. A positive diagnosis permits the anticipation of this congenital anomaly.

DNA emerges as the *information management macromolecule* which not only provides for cellular structure but also for function, as well as for information storage in gene units. The control is exercised through the information encrypted in its segmented nucleotide sequences and ultimately in its base pairs. The information storage and its retrieval on demand by the organism are followed by certain crucial operations, for example, protein synthesis and self-replication, all of which are essentially the bases for biological science. Input information from the environment and stored instructions (to which the organism must respond within the constraints of its system) set off the chain of events. Molecular information biology begins in this manner as DNA provides a molecular flow of information into all the levels of bioactivity.

There are four cryptographic DNA units which are *nucleotides* or compounds of a nitrogen base, a five carbon sugar and phosphate (Fig. 3-1) and one type with uracil is only in RNA. This nitrogenous base gives a letter and triplets form the words of messages read out by messenger RNA.

The hyodrogen bond is a conveniently weak atomic interaction which can easily break or open for replication purposes. A strong covalent bond would not serve the purpose since such a bond usually requires enzymatic assistance in its forming and unforming. The hydrogen bonding is a common way for setting the shape of macromolecules and for positioning at the molecular biology level. Weak opposing or attracting forces among molecules can determine the ordering or form factors in cellular components (Fig. 3-2).

The position of cellular components may be established by weak interactions such as hydrogen bonds, ionic bonds and van der Waals forces. They can insure proper contact between interacting molecules which also use distinctive surface features to assist in the acceptance or rejection of interacting units or substrates. By bonding or responding to forces, the cellular components assume a comfortable "status". This generates a release of free energy which is then available to break the bonds. The hydrogen bonds, however, are broken by energy only slightly more than that available at the thermal level. Polarity in molecules separates those capable of hydrogen bonding (for example, polar water to polar water which can form four H bonds) from those likely to form van der Waals bonds instead (such as many water *insoluble* molecules).

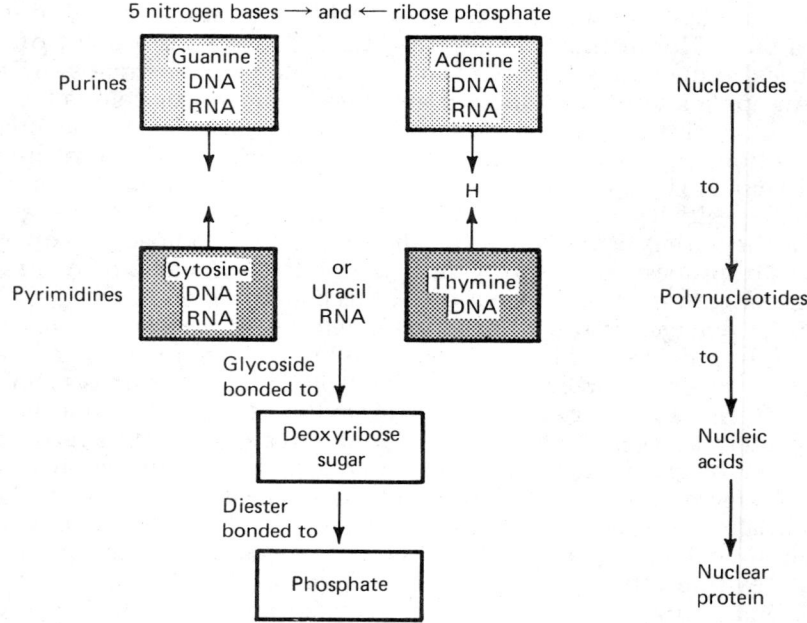

Fig. 3-1. Nucleotides in DNA and RNA. At the next level to that shown, nucleotides associated in polynucleotides form nucleic acids which are then conjugated with protein loosely so as to release easily on hydrolysis.

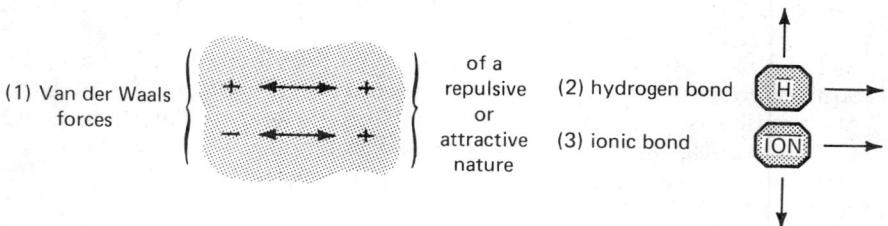

Fig. 3-2. Ordering forces.

The many H bonds hold the DNA molecule together securely. Form factors and bonding factors split biomolecules into those which can and those which cannot interact (hand and glove fit).

The base included in the nucleotide identifies it as A, G, C, or T. These are the cryptographic *symbols* in the genes and the *sequence* in which they occur spells out the desired *message*. Thus, the DNA helix is a script which uses the language of base pairs (mated ATs and CGs). Three base nucleotides, taken together, form a *codon* "code-on," where they are neighbors in the DNA "tape." Each such triplet is sufficient to command the ultimate appearance of a certain amino acid in its appropriate site on a protein chain where it can undergo demand synthesis into an enzyme or other molecule. The DNA nucleotides are joined by hydrogen bonds between base pairs to form a polymer of nucleotides in a chain. The DNA coil is double membered and helical in form so that it has the essential capability for transformations such as uncoiling or opening to permit the attachment of corresponding bases. Each member or side can then behave as a template pattern for new DNA in replication. Just as the symbols in any message form may be analyzed, one can study the unusual constancy of the symbolic codon appearances in a species. This leads to the use of the guanine-cytosine percentage in DNA (GC%) to separate genera. It can also be shown that the purines equal the pyrimidines or A+G equals T+C in DNA and A equals T and G equals C. Further, the genic code applies to all organisms, that is, it is universal. The DNA vertical or stacked members are laterally fastened into a helix by hydrogen cross bonds as if one began with a ladder and twisted it at each rung. Thus, the rungs are joined together along their length by the hydrogen bonds (Fig. 3-8). A gene such as the one for tRNA (tyrosine) has 199 base pairs. The human fertilized egg will have about 5×10^9 nucleotides to establish.

The information needed for development varies in different species. The DNA will differ in its total mass accordingly. Genic information thus has an equivalent DNA mass for a corresponding amount of code which is constant for a given type of cell. In the haploid and diploid conditions this will vary by a constant factor of two and the relation continues through the other ploidies. These equivalent DNA masses are measured to be on the order of 10^{-9} mg of DNA. The specific mass of DNA is quite a fundamental characteristic of a type of cell since the number of bits of information required to form the individual is expected to be specific for that individual. Another way of looking at the information is to measure it as a length on a genic chart—or map of the genes which shows the characteristics for which they are responsible. This is a graphic ordering of information in DNA-specific distances (Fig. 3-3)

The genic maps come from studies of effective areas on the chromosome as these areas become involved in typical genetic behavior. Some genes have about 1,000 base pairs and a complex organism like a mammal will have thousands of genes per chromosome in comparison with the few needed in a simple virus. Whereas *Drosophila* need a few thousand kinds of protein, man needs a few tens of thousands.

An important result of DNA's information management is genetic control of cell processes and some manipulation of organisms is feasible if the genic material is susceptible. It can be manipulated by chemical,

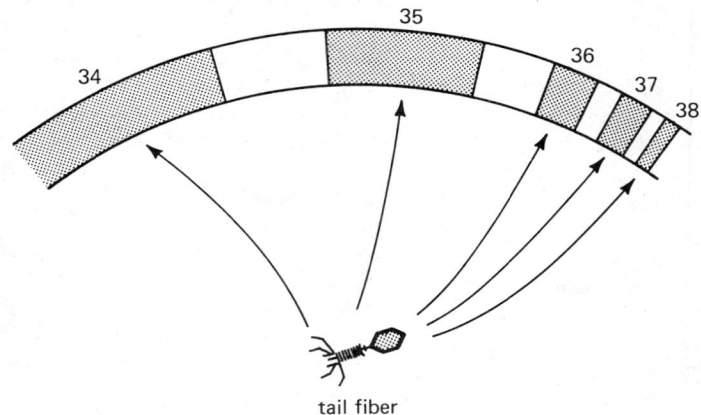

Fig. 3-3. Portion of genetic map of T_4 bacteriophage, phenotype shown in Fig. 3-26 micrograph. Numbers refer to DNA cistrons that code for amino acid sequences in peptides. (From Edgar and Wood, 1966).

physical, or natural mutation experiments or direct genic interference, and by DNA transfer in recombination, transduction, transformation or other invasion. A well-known example is a biochemical mutant which has a metabolic block. This may be, for example, a strain of microorganism, unable to survive without the addition of some ordinarily unnecessary nutrient (an amino acid, for instance). This is the "one gene, one enzyme" theory where changes at one locus produce results that depend on the types of alleles there. Further manipulation of such a strain by culturing with others not so constrained may produce a restored capability in some organisms. This management of DNA information which is under steady refinement, inevitably summons up interest in organism behavior genetics. The basis for these opportunities lies in molecular bioengineering in which DNA information is managed for desired results, an example being recombinant genes. Microorganisms naturally possess a vast potential for good or for harm to man as suggested by comparing *Saccaromyces cervisiae*, the bread yeast, with *Mycobacterium tuberculosis*. The control of DNA information in this area implies that the results might be weighted to man's favor. Viruses pass in cycles between host cells. In the evolutionary time frame, cytoplasmic inheritance, as shown by chloroplasts and mitochondria can carry to host cells, DNA or RNA, contributing new information to the recipient.

The crossover of chromosomes is a principle means for incorporating nuclear data into the life stream of new organisms. It resembles editing of a tape and splicing in a new section and is a common behavior in chromosomes. Physical agents like radiation can create chromosome fragments that, in turn, also cross over. Crossing over results are, in turn, a means for studying the chromosome and the genetic map (Fig. 3-4).

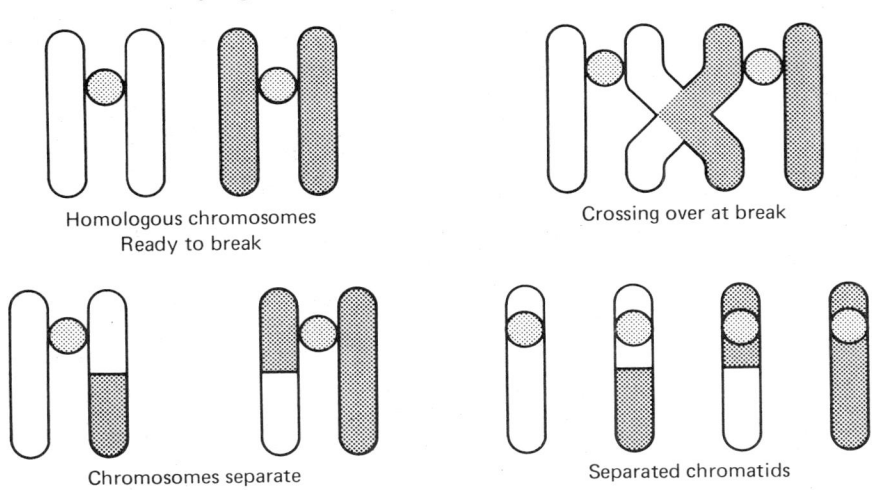

Fig. 3-4. Crossing over in meiosis.

Shuffling of entire segments along the chromosome causes a recombination of alleles. By denying absolute linkage, it assures that information will not be inherited in a fixed sequence forever. Exchange occurs in the 4-strand stage between 2 non-sister chromatids.

Messages Communicated by DNA. The uniqueness of the individual begins at the molecular level with the code on DNA which resides in the sequence of codons and makes up the exclusive genic information which is to be operating in a given individual. Messenger RNA is like a strip of pressure-sensitive plastic substance or modeling clay which, when pressed over the stereomorphic indentations on the DNA will retain the impressions. Actually, the mRNA nucleotides are complementary to those on the DNA, fitting together by virtue of their identity and the influence of the cipher system. The RNA can then form a printout of the codon line of nucleotides containing the complementary version of the DNA genic code. This message is addressed to a ribosome center in the cytoplasm or endoplasmic reticulum. This center will have a few ribosomes dedicated to the synthesis of some required protein. The information carried in the message is required so that the center will know how to proceed. The protein requires amino acids which will come to the ribosome assembly point via transfer RNA. Since the energy source for the process is the energy-rich phosphate bond in ATP molecules, phosphorylation is a first step in peptide synthesis. The tRNA has a hand-in-glove capability for fitting on the mRNA, that is, it is "opo" coded on, or has the necessary opocodons to fit on the mRNA and to use its sequence, start, and stop information as an assembly line pattern while discharging its cargo—a specific, required, amino acid. Once the function of an opocodon is appreciated it may be more desirable to call it anticodon—a term in wide use today. With this step, the original DNA message has been read and acted upon with the demanded amino acids now being properly located in a growing protein or subprotein unit in the ribosome assembly region. DNA control is seen to have informational features such as molecular recognition and a means for the transfer and release of subunits. The amino acid is carried on specific terminations at the other face of tRNA which is thus doubly specific and "sequences" the peptide.

DNA genic information is self-sufficient to begin this process of transformations and if, by some chance, some of it finds its way into the correct environment by accidental virus infection or intentional transfer, it will have the capability for getting its message through (Fig. 3-5). The information on the other hand, is sensitive to loss or deletion

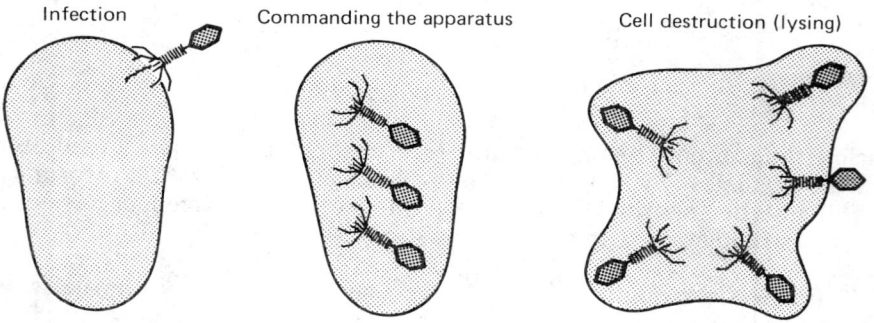

Fig. 3-5. Virus infecting, as in the case of T_4 bacteriophage in its host *E. coli*.

by unfavorable physical influences such as ultraviolet, ionizing radiations, unfavorable pH or similar chemical changes.

In the process of photoinactivation after ultraviolet exposure, for example, deletion of information occurs in many organisms if they are not exposed to photoreactivating light shortly thereafter. This is a complex example of information and energy transported by radiation between systems. Without the second message, the organism is lost. The whole DNA molecule can also disintegrate in the presence of its specific enzyme deoxyribonuclease in order to recycle its chemical components. This occurs under appropriate conditions which resemble a self-destruct situation.

Molecular Informational Access: The two functions of nucleoprotein; 1) design and, 2) synthesis control, may be separated further with respect to the fibrous, single-stranded, or straight DNA sections as opposed to globular control DNA which is in bundles according to the banded Crick model. A chromosome will have the latter in alternate banded or striped regions, that is, its bundled DNA. The two different regions are designed for molecular informational access—that is, the bundle is a dense concentration of information for more active requirements. The interband portions are coded for protein synthesis, with control functions for this process coming from the much more detailed DNA information which is in the bundled band. Chromosome histone protein is structurally responsible for maintaining this alternating pattern. The development of advanced species may reveal differences in the bundle characteristics for the more complex needs of advanced organisms require a more active access.

Determining Structure

In the hands of skilled persons, such as M. H. F. Wilkins, a New Zealander by birth who worked in England at Kings College, X-ray crystallography can be used to reveal the design of molecules such as DNA. During the period of discovery of the DNA structure in the late fifties, J. D. Watson and F. H. C. Crick were building molecular models while Wilkins was actually probing with X-rays. The result was confirmation of the suspected helical structure, a result which earned this threesome the 1962 Nobel Prize in physiology or medicine. When DNA is in the gel form under the microscope, fibers can be readily drawn out in a way that suggests a springy molecular design of great regularity, rendering it susceptible to X-ray study. The molecules were obviously arranged in a manner resembling a flexible crystalline lattice. The interaction of X-rays with such a configuration is a scatter, diffraction, or reflection pattern from the atoms, molecules and planes on which the rays are incident. The rays behave as coherent, reinforceable or cancelable waveforms scattering from the atoms in their pathway. Information in the outcoming waveforms may be resident in the wavelength and wave phase, either in phase with the incident radiation or out of phase with it. This may be referred to as interference information or patterns and it exploits the wave nature of X-rays with their wavelength and frequency features.

In this type of study, one objective is to photograph the information to show, as in Fig. 3-6, the vertical stacking outline of the macromolecule. A helical form is deduced from the X pattern of X-ray reflections. In these dark places, the photographic film has been darkened by the reflected X-rays. Thus, the main information is of atomic concentrations as revealed by the intensity of the spots. If some magic lens were available, the scattered radiation could be gathered into a form which would be an actual image of the structure. A picture would require collection and refraction of the scattered beams—the function of a microscope, which is not as convenient for X-rays. Lacking this, the method utilized radiation features—monochromaticity, incident angles, and interacting distances as well as simultaneous chemical and computer

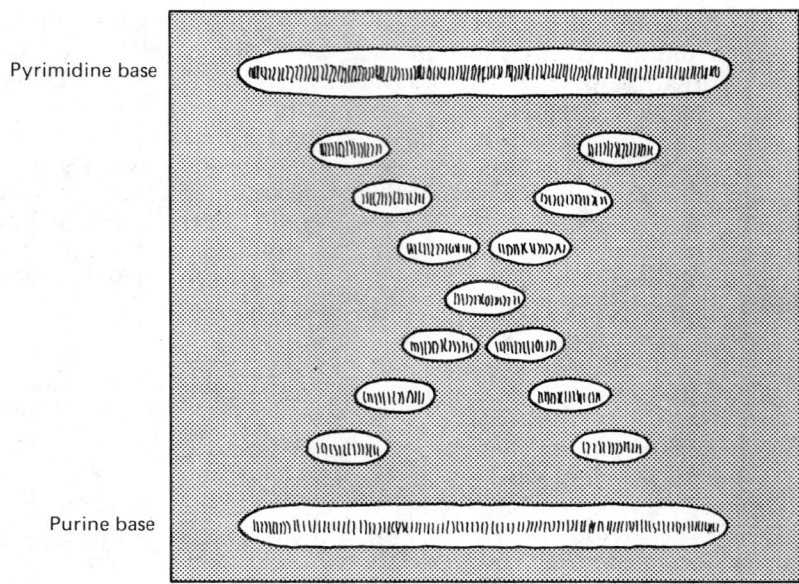

Fig. 3-6. Diagrammatic version of X-ray photograph of DNA proves the stacked helix configuration and shows the distances between bases and other units and the planes in which they are located. Measurement of the amplitude and phase of the scattered waves (variation in intensity as a function of direction) gives the structure in X-ray diffraction.

analysis. If, from the other evidence, a helical molecular configuration were already suspected, the calculated model could be and was compared with that observed by X-ray scatter from DNA preparations, giving absolute conformation, especially when DNA in a more natural state gave essentially the same helical pattern.

Figure 3-7 shows an arrangement in which the sine of θ, the wavelength, and experimentally interacting distances will yield information about the intermolecular spacing S. The nature of the important interference effect is illustrated in the lower portion of the diagram which shows wave reinforcement and cancellation. These relations are combined with the planar ones in Fig. 3-7 (a). Here, the incident X-rays are initially parallel and scatter at the same angle θ with the planes. Electrons associated with molecular structures at the circled points of incidence will interact with the oscillating field energy of the X-rays, taking up vibration energy and reradiating it. These emerging secondary rays in the direction determined by θ will constitute a scattered or diffracted beam with the intensity determined by the phase relations. *In* phase waves would produce the maximum intensity in the diffracted radiation. If the angular and space relations are correct, parallel incident rays will reinforce and produce this maximum.

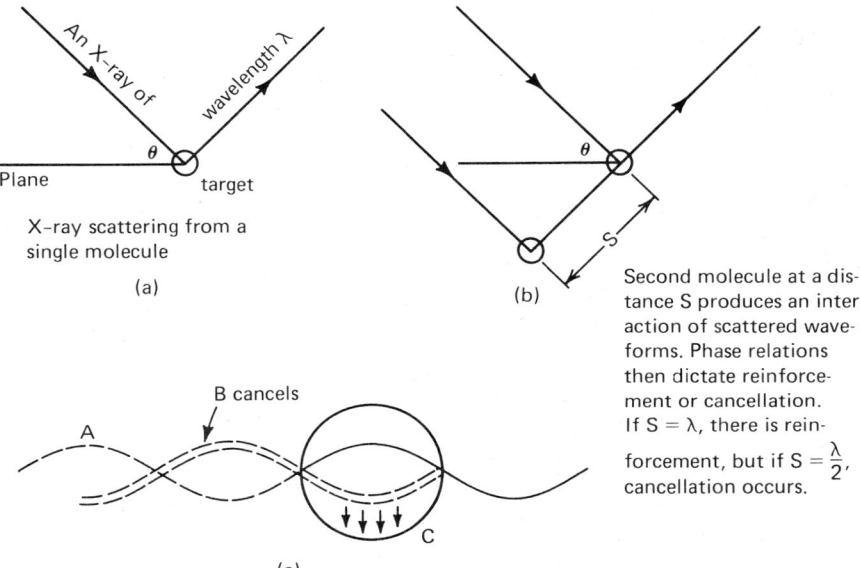

Fig. 3-7. X-ray scatter crystallography. The measurement of wavelength and angle θ are related to the spacing S between molecular units in a biomacromolecule. When two waveforms interact as at arrow, there is equal and opposite effect or cancellation. However, two in-phase waves result in reinforcement as at C. Thus, known wavelengths of X-rays used to irradiate two target bases at the angle θ will diffract and interfere by reinforcement in the resulting wave. The wavelength of the resultant wave and its phase information (in or out of phase) allow the distance S to be measured.

In other words, the directions in which the diffraction maxima occur are associated with the spacing of units in the crystal. These intensities must, in turn, be related to the phase information by indirect methods. The objective of the mathematician is to locate the coordinates x, y, and z of the atoms that scattered the X-rays. There are so many possibilities that must be ruled out, that the educated guestimate is essential. This arises from the previous evidence and judgment of the diffraction specialist. The heavy, electron-rich atoms "dominate" the diffraction effect so that information and clues are available here as well. They make sharper photographic spots than do lighter elements in the image captured in the photographic film. Also, some element, a heavy one for example, may be deliberately replaced and the resulting differences, in an otherwise identical sample, are then searched for information about the other molecular locations. The technique of crystallography here yields an electron density map with photographic darkening associated with drawn contours all within atomic dimensions. In addition to the structure of nucleic acids, vitamin B_{12}, penicillin, myoglobin and others have been mapped in detail and virus structures are effectively examined.

Mention of interference patterns would not be complete without a brief account of holography. Here the scattering of light from objects informs the eye about those objects via the intensity and phase relations when a reference beam such as a laser is caused to interfere with the light. The photographed pattern can be made into a 2- or 3-dimensional hologram which can be illuminated to produce an excellent image.

The radiation features and interactions which make X-ray crystallography operate will be far more obvious after the discussions of the electromagnetic spectrum.

The important intramolecular distances determined for DNA were:

1. The interrung distance or the distance between the nucleotides, or between stacking levels of purine and pyrimidine bases which is 3.4 Å (see ladder-like aspect in Fig. 3-8(c)).
2. One ten-nucleotide turn or helical period or major groove which is therefore 34 Å.
3. The width of the ladder or of the molecular helix which is constant unless spiraling occurs and is 20 Å.

When the X-ray diffraction information is combined with other experimental observations, e.g., ratios of bases, chemical compositions, nearest neighbor analyses, and bond angles, the DNA helix can be accurately modeled as shown in Fig. 3-8 a, b, c, and d. If placed in the

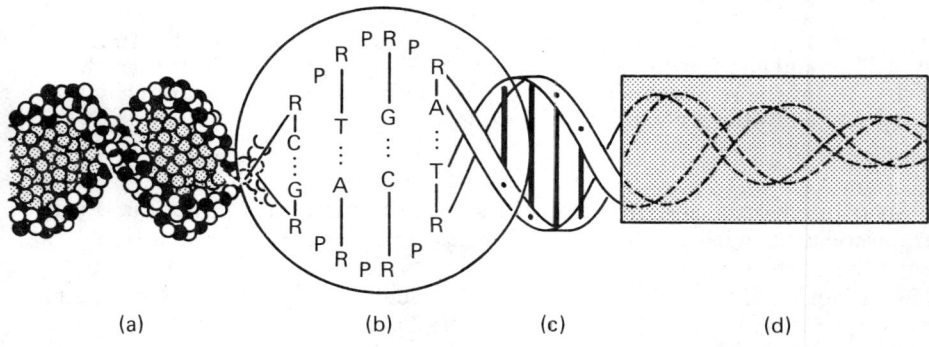

(a)   (b)   (c)   (d)

Fig. 3-8. 4-view of DNA helix. The DNA helix appears like two waveforms 90° out of phase but the chains are actually interwound. a) fleshed-in version of the macromolecule showing the bases hydrogen bound, inside the two chains of sugar phosphate groups; b) molecular outline shows the structure but does not reveal the stereochemical features. CTAG are bases. R is the deoxyribose sugar and P is the phosphate group; c) "twisted ladder" view of DNA spiral. Four doubles shown, one hidden with 10 rungs per turn; d) diagrammatic view of an electron micrograph.

horizontal plane, the helix resembles a pair of waveforms as they are usually pictured. These hypothetical waveforms would be about 90° out of phase. However, the DNA chains are actually interwound, and joined by the "twisting" bases and their hydrogen bonds.

The forward and reverse directions of a single DNA chain may be defined as leading from the 3^1 position of a phosphate forward and from the 5^1 position in the opposite direction (Fig. 3-9). This direction of chain advance is then reversed for the other DNA chain in the helix (Fig. 3-10).

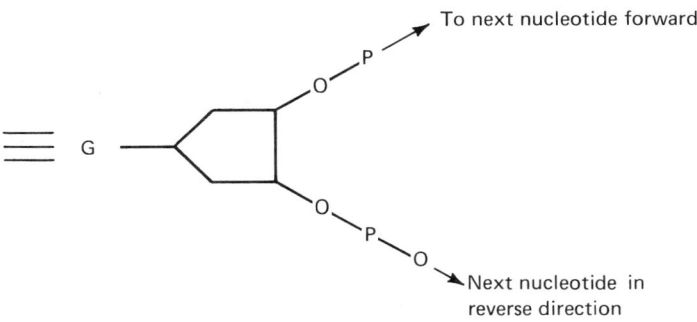

Fig. 3-9. The 3' (upper) and 5' deoxyribose positions with attached phosphate groups establish the direction in the DNA chain.

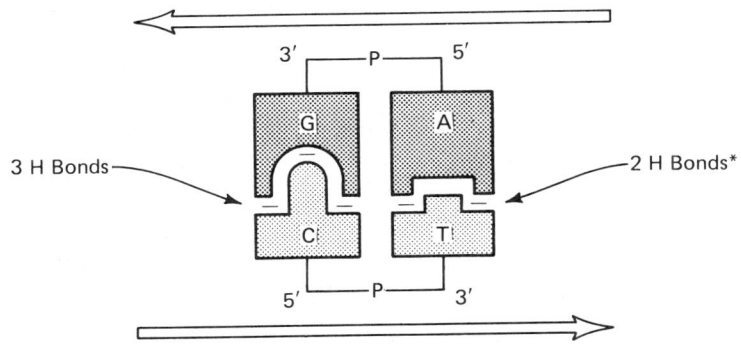

Fig. 3-10. Oppositely polarized DNA strands can be demonstrated by analyzing their nearest neighbors with radioactive tracers. Two levels of bases AT and GC in a helix section with hydrogen bonding and opposite polarity of phosphate groups (end for end). Diagram suggests why A = T and G = C or the proportion of adenine always is the same as that of thymine and guanine equals that for cytosine.
[+]Due to relative structures at GC and AT base pairs. A and G are larger molecules than C and T. By specifying base pairs, chains become complementary.

Two or three hydrogen bonds join nitrogen or oxygen as the case may be (Fig. 3-11 a, b, c). T and A are H-bonded at about 2.9 Å between thymine oxygen and hydrogen and adenine nitrogen and hydrogen (Fig. 3-11b). The C and G are H-bonded at about 2.8 Å distance between cytosine H, N, and O with guanine O, H, and H (Fig. 3-11c).

The "string" of hydrogen bonds is located in the partitioning line position. This is where the helix separates to replicate and the opened-up pattern of AGTC template elements provides an excellent assembly arrangement for making a copy of the separated DNA chain using new nucleotide components from cellular supplies (Fig. 3-12 and 3-13).

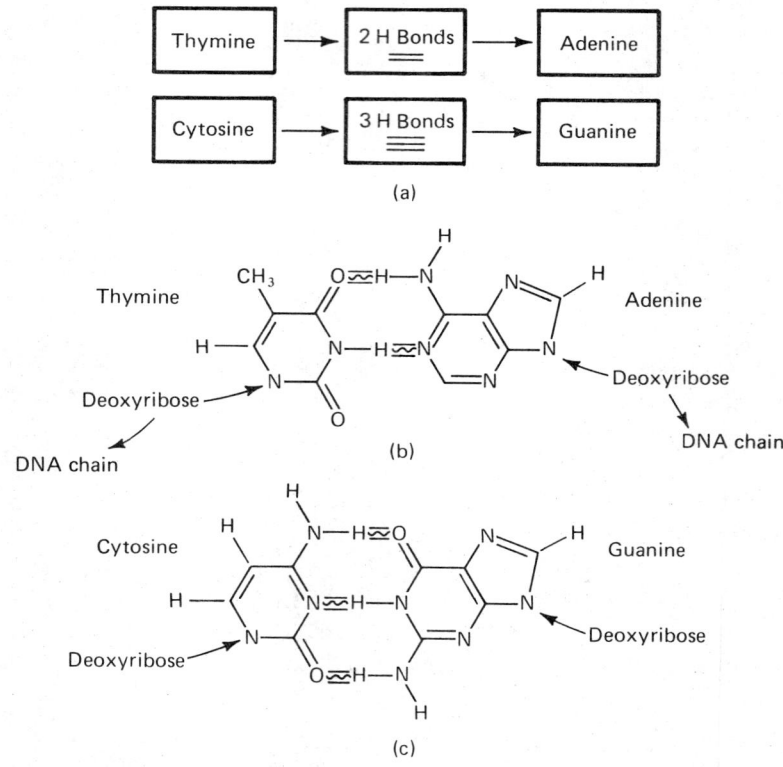

Fig. 3-11. Weak hydrogen bonding of (a) two types allows the identification of the mate in each base pair (b) and (c). The atom of hydrogen links two electronegative atoms.

Self-organization is a valuable characteristic gained by these unique structural details. Order is managed via the information carried in these information-rich molecules in sequences like CGATCTGC and orders are effected by the DNA capacity to replicate on demand. Information

DNA Information, Gene Management and Radiation Genetics

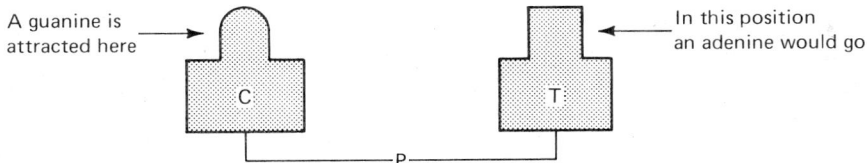

Fig. 3-12. Replication refills the opened H bonds. The steps in replication are:
  a. Base pairs separate.
  b. The helix uncoils.
  c. New nucleotides flow onto the old strand according to the H bonding and complementary to the existing AGTC bases (Fig. 3-13).
  d. DNA polymerases arrange the phosphate ester bonds that link the nucleotide chain.
  e. Two chains emerge and each coils up to form a helix.

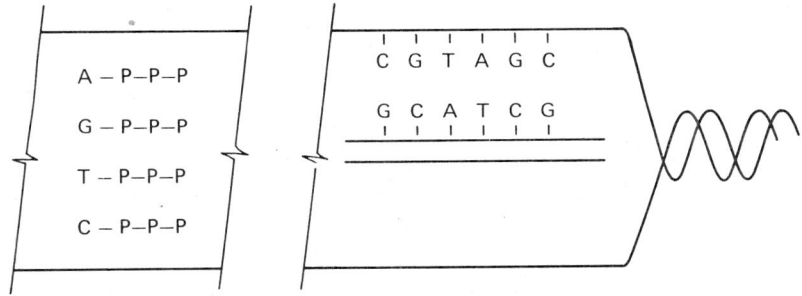

Fig. 3-13. Flow of new information units, nucleotides, encrypted in complementary base pairs, accomplishes the replication process. Kornberg and his group synthesized DNA in vitro using primer DNA template and DNA polymerase.

printout is provided by RNA which can utilize the information readout to form enzymes and other protein material. These in turn establish the growth, operation and replication which will constitute the life of the unit with self-organized metabolism and energy routes.

Cipher Characteristics

In breaking the DNA cipher, an early step was to establish that the codon is a triplet, i.e., it has three nucleotide symbols instead of two or some larger number. Next, given a message of, for instance, three codons, the cipher can exist in 3^3 or 27 permutations of the three codons. With a longer message, e.g., five codons, the variations could have grown rapidly, that is, $5^3 = 125$, a capability not necessary here.

Another step in understanding the nature of the cipher is to observe that it is nonoverlapping and that each codon triplet stands on its own and does not restrict what might be the nucleotides in the next codon in the message. This is critical if unnecessary constraints on protein synthesis are to be avoided. A hypothetical overlapping codon (not used) might be CUU - UUG - UUU, which would be translated via the UCAG code into leucyleucylphenylalanine. A study of this overlapping codon sequence shows that the presence of symbols in one triplet of the sequence (that is, UU), predicts the next triplet to be UUG. The CUU triplet influences the next two triplets. Since a codon results in an amino acid later in the peptide, it can be seen that overlapping acts as a constraint on the peptide. The nonoverlapping code section might be CUU - GGU - CUU and the triplets in this case would produce leucyl-glyclleucine. Without intersymbol influence, the arrangement may in a sense be more flexible while less redundant (Fig. 3-14).

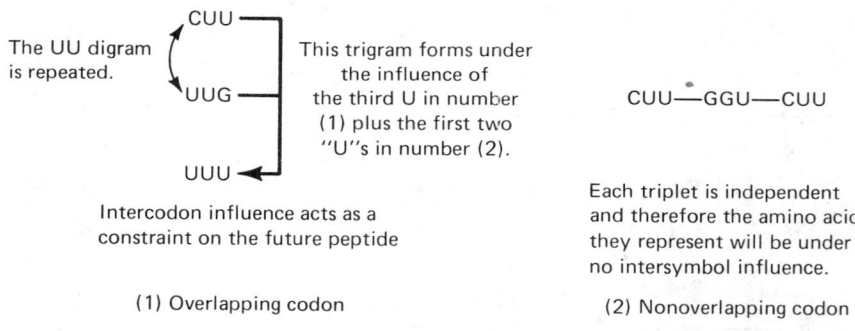

Fig. 3-14. Intercodon influence in Overlapping Cipher.

One to One Relation for Codon to Amino Acid

The code feature wherein one codon results in a single amino acid required a unique kind of chemical detective work to support it. Suppose that a special mRNA contained only one nucleotide, that is, pure U, or C, or A, or G. From the code, it would be found that only the corresponding monoamino acid peptide is possible in each case. This would be polyphenylalanine for UUU, polyproline for CCC, polylysine for AAA, and polyglycine for GGG. The triple U, C, A, or G codon carries the same message to the protein each time and only one kind of amino acid is laid down in each case.

Degeneracy

Another feature deals with the information value in each nucleotide of the triplet. This leads to an evaluation of the degeneration in the code. Reference to the code reveals that both UUU and UUC result in the laying down of phenylalanine in the peptide; in other words, to the operation of the same opocodon in tRNA. Therefore these are degenerate triplets. Since the codons are varied for one amino acid, it is called a logical degeneracy because synonym codons are related in a logical manner. In this case only base three varies and the first and second bases are the same as expected. The key information, therefore, is in bases one and two.

Experimental exceptions observed in the cryptographic readout system show further that the middle symbol is the single key informational one, since both nucleotides one and three may occasionally have substitutions for a given amino acid. The codons for methionine and tryptophan are exceptions. These have single codons and, thus, are nondegenerate or consistent portions of the cipher. Otherwise degeneracy in the coding is the rule and a given AA may be solicited by up to six codons. Mutations involving a less informational base in the codon, such as the third base, are not as effective and would not be noticed due to the degeneracy in the code.

Redundancy and Recognition Schemes in the Code

The theory of information assigns a specific set of characteristics to degenerate codes. While these are not directly related to this biological system, it is apparent that the ability of several codons to direct an amino acid dampens the oscillations which are due to mutations. The latter look essentially like noise in the information flow and redundancy appears as a special feature for reducing the effect of noise or for protecting the communications against it. Thus a damaged or mutant gene need not necessarily lead to a changed amino acid or error in the protein, and a somewhat higher level of mutating radiation would be needed to produce a new characteristic.

Coding is a communication device which, to produce a new characteristic in the present system, causes a flow of information via physical and chemical reactions. The operation of synthesis on the ribosome brings the message information into contact with the units upon which it is to operate, including two enzyme systems, the tRNA synthetase and the one that completes the results of mRNA meeting tRNA at the assembly site in order to produce the insertion of amino acids into the protein. Identification or recognition schemes, for example the one that permits selection of the correct amino acid at the assembly site, are incorporated in the information supplied and play a large part in the effective flow of information.

The base sequences and stereochemistry of the nucleic acid units are related to the functional structures of proteins of the many types needed, from enzymes to membranes. Thus DNA helices are interrelated with single stranded base paired loops in mRNA and bobbypin loops in tRNA (Fig. 3-15), as well as tertiary protein configurations in the final product. Nonerror codes are still possible through these different communication vehicles thanks to the way base sequences can be varied

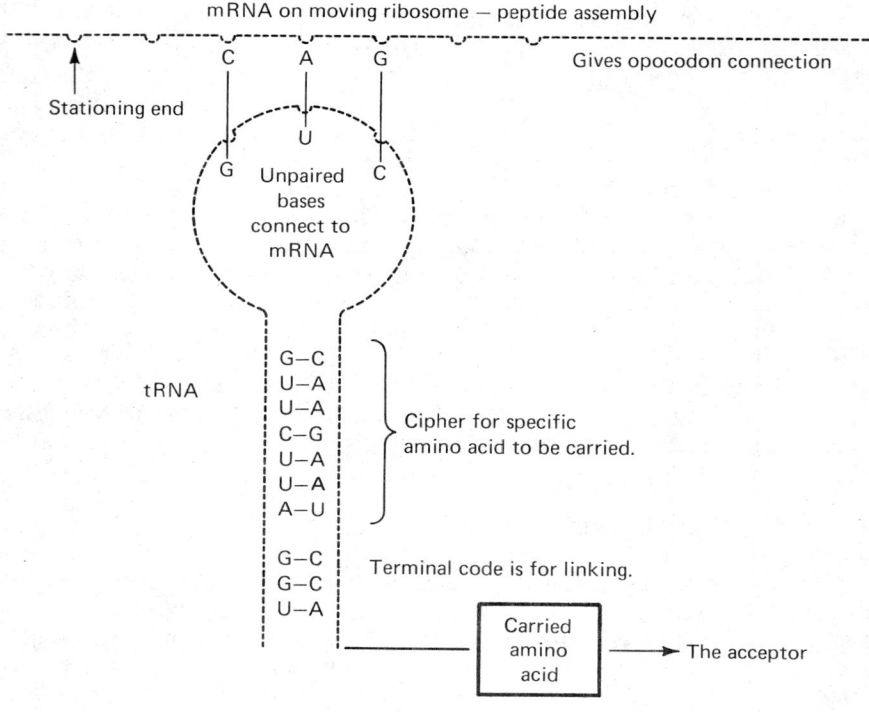

Fig. 3-15. Bobbypin loop secondary structure in RNA. A section of base pairs and triples with the amino acid coded equivalent is shown. mRNA, ribosomal RNA and smaller tRNA arise from coded instructions in DNA. To make a double strand out of a linear structure, it can be doubled back which gives base complementarity even with single stranded RNA. The loop may appear in various configurations on the tRNA molecule.

with the help of the redundancy in the genic code. In other words, base pairs can be made from varied base sequences and amino acids will still be laid down correctly.

The tRNA leads or carries specific amino acids to the mRNA on the ribosome. Thus there is glycine tRNA, leucine tRNA, etc. There are about 40 tRNAs for the twenty amino acid components so some have multicodon representation as a redundancy measure. Gene sequences need not be repeated because this multicodon measure, multiploidy neutral mutations in homologous protein, and repair provide redundancy, $(1-R)$ where $R = H/H_{max}$, relative entropy; R also is $H/\log m$, the equal alternative m).

Colinearity

A model protein synthesis system can be set up with known components to show that the ATGC code for amino acids has colinearity with

DNA Information, Gene Management and Radiation Genetics

the twenty amino acid code in polypeptide sequences. The gene-controlled synthesis by protein-*constructing* genes reveals this colinear feature via sequences of DNA and RNA bases just as a simple code made up by dividing the English alphabet will illustrate the same system. The letters A through M and N through Z give thirteen equivalents for columnar code if they are so listed. They are colinear, that is, A is read into the code as N, B equals O, C equals P, etc. In DNA, the nucleotide code is colinear with the polypeptide that results from the synthesis. Reading from one end of the constructing gene to the other corresponds to the amino acids laid down from one end of the peptide chain to the other. An appropriate genic change or mutation gives a corresponding modification in the gene-controlled protein synthesis in the way of a substituted amino acid which may well be undesired.

Nearest Neighbor Analysis

Tracer techniques using radioactive phosphorus-32 or other isotopes have been very useful when combined with other methods for showing the linear arrangement of DNA units. The radioactive ^{32}P can easily be incorporated into the triphosphate nucleotide of thymine, for example (Fig. 3-16). When this is linked to deoxyribose to form the nucleotide

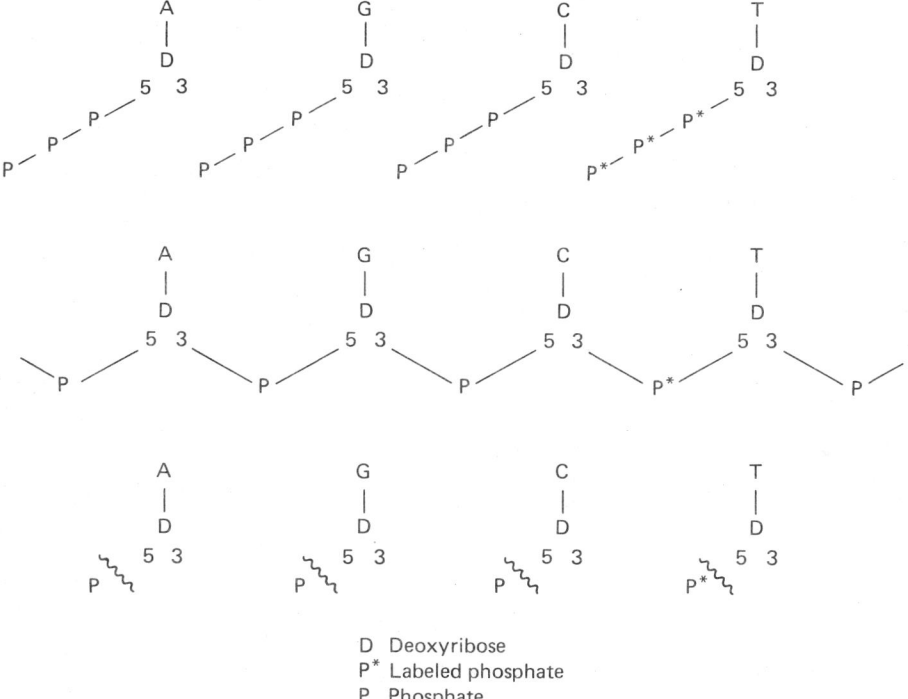

Fig. 3-16. Nearest neighbor analysis with radioactive phosphorus.

T, and it, in turn, is added to a building DNA chain, the tracer will be seen to pass to the cytidine phosphate. A tracer on the phosphate at the 3^1 position of the deoxyribose in the cytosine nucleotide after these steps means that cytosine nucleotide was the neighbor of thymine nucleotide in the DNA as it was laid down. In this way, by many "nearest neighbor analyses," the linear arrangement of the code can be worked out. Sequencing, in turn, facilitates the determination of the residues in a sequence.

Operating requirements for RNA in its various forms include a special diversity in secondary structure. One form, shown in Figs. 3-15 and 3-17a, is called the bobbypin loop. The spatial and stabilization requirements in certain forms of organisms such as viruses, as well as in higher organisms at certain appropriate times in their vital activity, lead to the formation of these base-paired loops in RNA. The redundancy in the code system may be one reason for this structure since it allows the logical synonym triplets to be exploited to full advantage. A two-nucleotide code would lack this redundancy and could not easily use base-paired loops, as for example in mRNA. The triplet code can make this possible and still pass the information via base sequences for protein assembly. The loop structure is a convenient configuration for this operation. The RNAs can follow base pairing procedures but since they contain unusual bases such as 5-ribosyluracil, loops can be formed at such places. In the helix parts, the usual base pairs form.

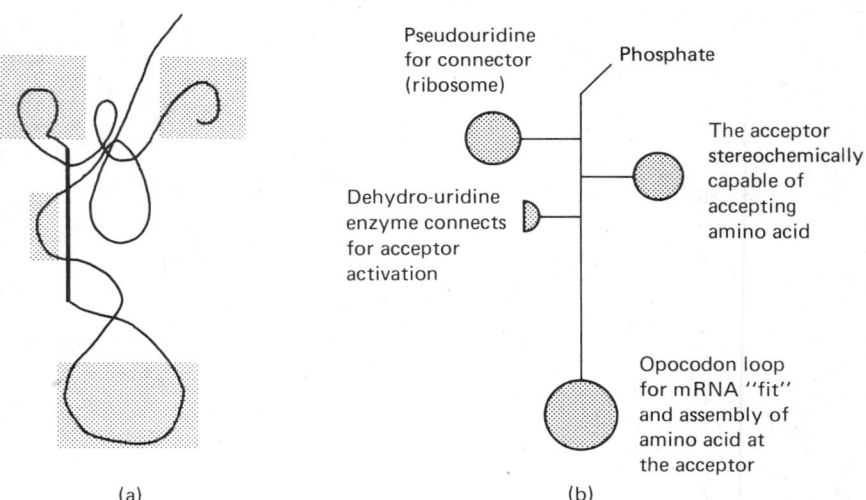

Fig. 3-17. tRNA has specific places where stereochemical action centers. a) diagrammatic representation of tRNA molecule; b) loop action centers.

A loop forms in mRNA opposite a point mutation site which leads to a deletion in DNA (Fig. 3-18). Depending upon degeneracy, the base deleted may or may not seriously affect the information transcribed. A third base change is not likely to be harmful. However, the looping constitutes a reregistration of the *reading* and *read* nucleic acids, in the sense of sliding two scales out of registration. This may shift the readout from thousands of bases and could be catastrophic at the protein level. The amount of disturbance depends on the extent of deletion. Repair of simple, single-base deletions by inserting a correcting base nearby, which would normalize the situation, is a natural possibility.

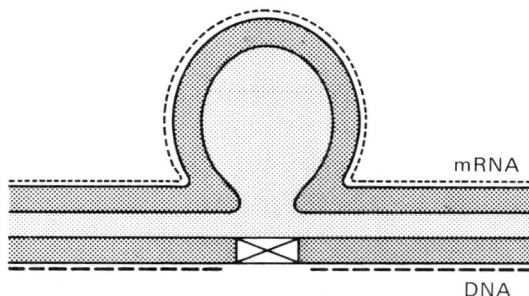

Fig. 3-18. Loop in mRNA opposite a deletion in DNA results in a reregistration of the reading frame.

Sequence Analysis of Isolated Peptides

There is a more direct technique than that of nearest neighbor analysis. After suitable analytical characterization and preparation, polypeptides may be cleaved, the peptides fractionated, and the isolated peptides sequenced in such a way as to reconstruct the primary structure, assign disulfide bridges and the like. These procedures may be automated using suitable strategies, as in the Edman degradator and the Sequenator, solid phase sequence analysis, and mass spectrometry. Another device, the Mini-15, with proper programming, gives a solid phase sequence analysis of protein and peptides with single or dual columns.

In order to cleave a certain bond such as Try-Met selectively, the polypeptides or proteins are likely to be reacted with, for example, 50% acetic acid and NBS, N-Bromosuccinimide. The result is then determined with the ninhydrin reaction. Degradation by enzymes gives the sequence in chains as well. This sequencing is then applied to an enzyme system to probe the significance of a functional group and to identify such a group in an enzyme. Then the mechanics of recognition and specificity for enzymes of high information density such as DNA-specific enzymes may be examined.

The linear sequence of amino acids in a protein accounts for the higher levels of organization. This sequence is coded in nucleotide sequences via the universal code, and thus, is derived from the information in DNA. The amino acid sequence gives the operating specificity insofar as this depends on the protein conformation. Therefore, this sequence is sufficient as far as information is concerned. Sequencing has this information as its objective, and the critical step is obtaining a workable protein preparation for studying the sequences. The selection of the native conformation among the 10^{100} possible ones in a protein with 100 amino acids is also a formidable task.

The conforming protein uses many aids in folding to a thermodynamically comfortable position within local constraints. Some structures aid in thermal stability. Others furnish periodicity, cooperativity, amplification, and motion. Others are forbidden on energy grounds or compatibility of neighboring bases such as cytosine-guanosine in eukaryotic mRNA. Still others will relate to host enzymes, metabolic pathways, alternative substrates, recognition sites, and ligating enzymes.

PROTEIN ASSEMBLY OR SYNTHESIS

The protein synthesis operation tends to resemble an automatic computer-controlled assembly system. The mRNA coding and tRNA decoding result in a supply of amino acids at the ribosomal or rRNA assembly site. The ribosome functions as a holding device in motion for the assembled peptide (Fig. 3-19). While in production, a specific ribosome is "adopted" by the peptide (or vice versa) for the duration

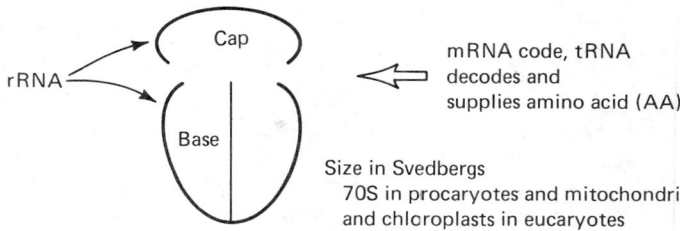

Fig. 3-19. The ribosome as a positioning unit or "jig" for the peptide. It is of large molecular weight (in millions) and very numerous in an active cell. A specific ribosome is "adopted" by the peptide while it is being produced, and they are fixed together for this period. The mRNA acts as a conveyor with coded instructions for a dynamic assembly in the moving jig holding a growing peptide chain. Thus ribosomes are relatively small stations in motion along larger mRNA lines. Ribosomes are nonspecific in the sense that a virus commanding a cell can adopt the ribosomes available without difficulty. Thus viral takeover occurs at this level of synthesis. The *recognition* seen at the conjunction of tRNA, rRNA and mRNA is weighted toward the information rich, communicating mRNA. The tRNA specificity is toward its amino acids, their transport, and deposition.

DNA Information, Gene Management and Radiation Genetics   **115**

of assembly. Ribosomes (Fig. 3-20) may be located either on the endoplasmic reticular membrane or free in the cytoplasm. The mRNA acts as an addressable information source spaced on a conveyor with coded instructions for the dynamic assembly in the moving peptide jig on the conveyor. mRNA tape conveyors (Fig. 3-21) are large when compared with the relatively small ribosomal stations in relative motion along

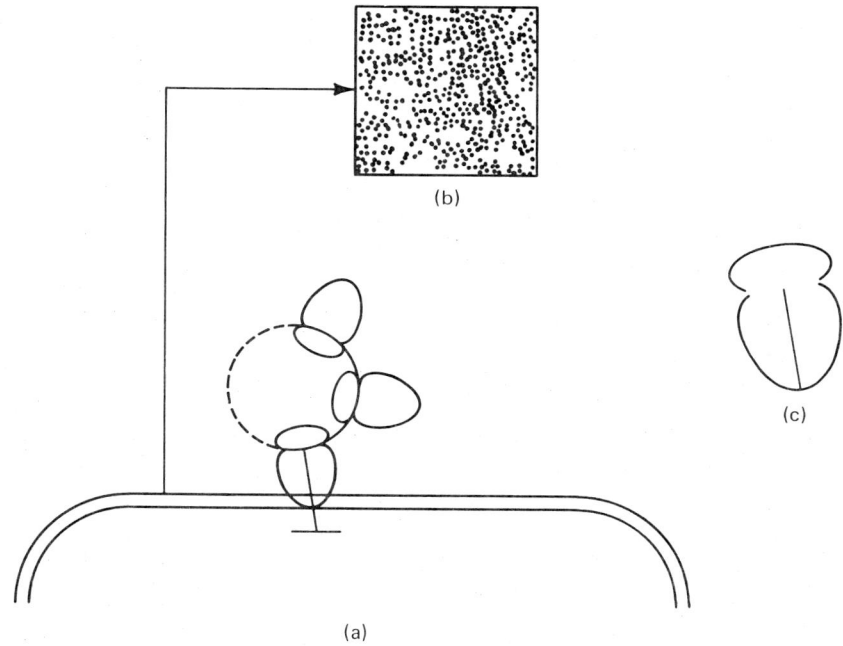

Fig. 3-20. Ribosomes occur in membrane of endoplasmic reticulum (a) and (b) or free in cytoplasm (c).

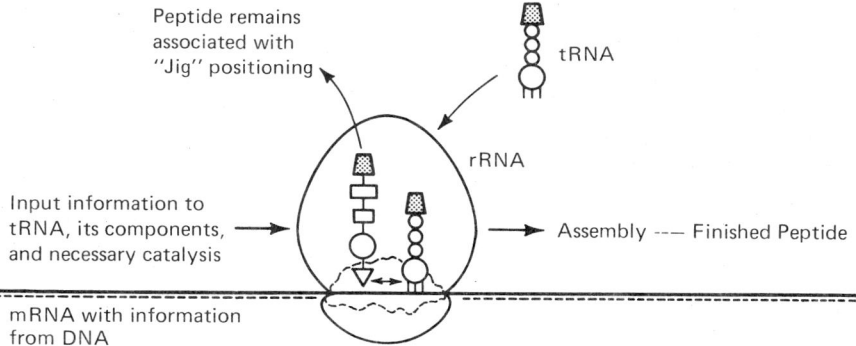

Fig. 3-21. Single ribosome shown as part of whole peptide assembly system.

their length. The whole peptide-forming system requires input information, tRNA, components and enzyme assistance. Operations at the load position are shown in Fig. 3-22. Here, mRNA, tRNA, and rRNA assume positions for decoding and unloading amino acid components relative to the ribosome holding device for the protein in process. An essential supply of amino acids is necessary for successful completion of the product along with assistance from bonding and synthesizing enzymes, guanine triphosphate, glutathione at the rRNA site, and energy-rich phosphate bonds from ATP to bind amino acids.

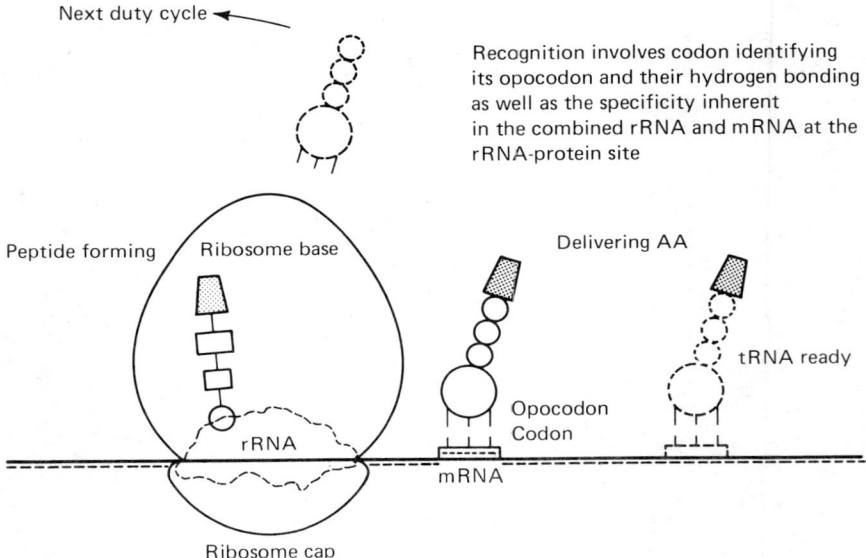

Fig. 3-22. The load position for ribosome, computational code control and peptide product handling in presence of complete component supply, enzymes, guanine triphosphate, and glutathione at rRNA site. tRNA is identified, and discharged of its amino acid to the forming peptide in the ribosome base jig.

When the amino acid approaches the tRNA-mRNA binding site, the mRNA code is read and tRNA is able to recognize the presence of the proper building block and conduct itself to the assembly site. Work may be done in multiples of this unit in the form of a polyribosome (Fig. 3-23). The essential feature of this loop is the ribosomal access time to the mRNA "tape" for reading instructions. The product flow involves entering and leaving tRNA as it deposits both its charge and recycled rRNA as the ribosome goes from the initiator side of the protein-forming loop to the terminator side where its product is discharged. Ribosomes number in the tens of thousands, thus constituting a dominant proportion of the cell mass. About 120 amino acids per minute per assembly site are formed into protein.

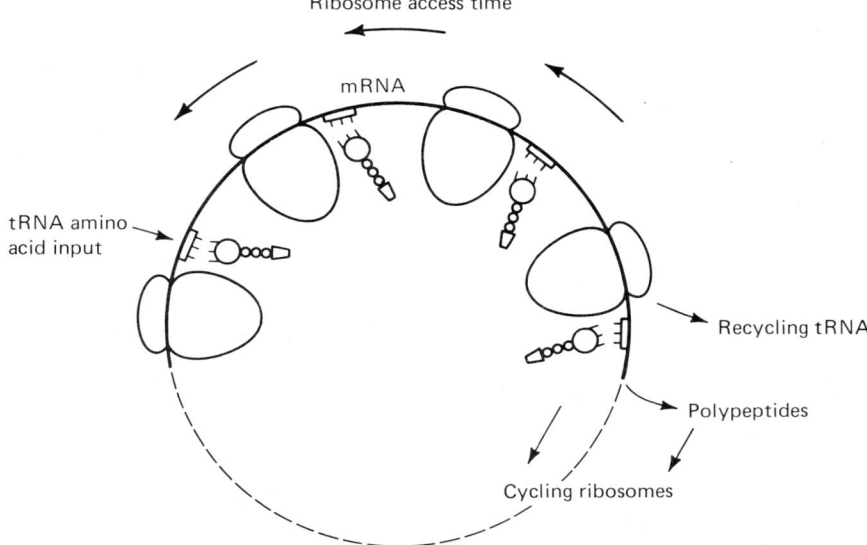

Fig. 3-23. Loop representing ribosome access time to code on mRNA. Multiribosomes arranged for integrated protein synthesis and combined into a loop by mRNA.

The central role of the ribosome begins with the switching action in fertilization. *Informosomes,* or mRNA protein complexes, are formed in oogenesis and remain coupled to polyribosomes within a constraining envelope of protein. Fertilization then releases the units and couples them into the protein synthesis which begins embryogenesis (Fig. 3-24).

In the genome or total chromosomal code, the logic resides in some 3 million nucleotide bases which is a sufficient number for about one million triplet code groups in lower forms. The protein may have about 300 amino acids. The one million triplets should produce an equal number of amino acids less a factor for inert or repeated units. Thus the one million triplets, divided by an average amino acid number of 300 in

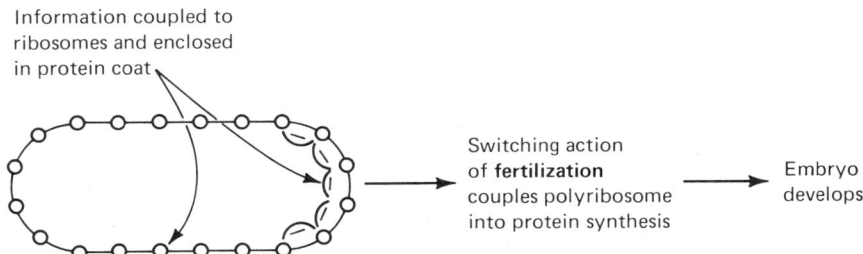

Fig. 3-24. Informosomes are mRNA protein complexes that await fertilization for switching on. Thus they link oogenesis with embryogenesis.

protein, produce a figure of about 3000 types of protein as an upper estimate in simple cells. The protein synthesis is seen as an assembly system with parallel lines where several thousand assemblies occur simultaneously (Figs. 3-25, 3-26, 3-27). The overall rate would be on the order of 300 serial amino acid input operations per assembly per protein. A line of symbols in DNA or RNA contains the triplet bases with its instructions. In DNA the four nucleotide bases are guanine (purine), adenine (purine), thymine (pyrimidine), and cytosine (pyrimidine). RNA has uracil (a pyrimidine) in place of thymine so that adenine pairs with uracil in RNA base pairs. These carry the alphanumeric logic in the Nirenberg code which thus encodes four base letters into 20 amino acid letters. The universal UCAG genetic code is shown in Table 3-1 and Table 3-2 lists the 20 amino acids with their structural formulas.

The human mitochondrion DNA is an exception to the universality because *its* code list with 64 word-codon-nucleotides is opposite 20 amino acids as shown by Frederick Sanger sequence studies. When electrophoretic separation is combined with autoradiography and tracers by his group, B. G. Barrell, A. T. Bankier, and J. Drouin, the triplet UGA is found to be the codon for tryptophan, not for termination. The codon AUA is for methionine and not for isoleucine. Thus universality is dated beginning on the date of adoption of mitochondria into the host cell in evolution, at least according to the cellular inclusion theory.

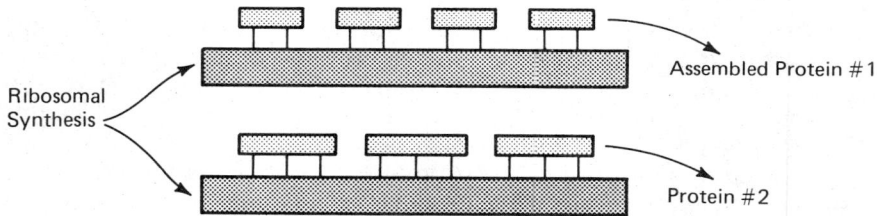

Fig. 3-25. Protein linear assembly is repeated in parallel with many lines producing many proteins. The features are phased production with provision for great amplification.

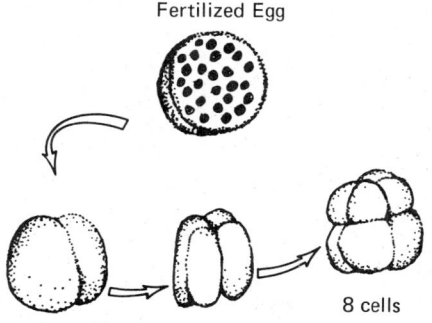

Fig. 3-26. Cellular geometrical growth.

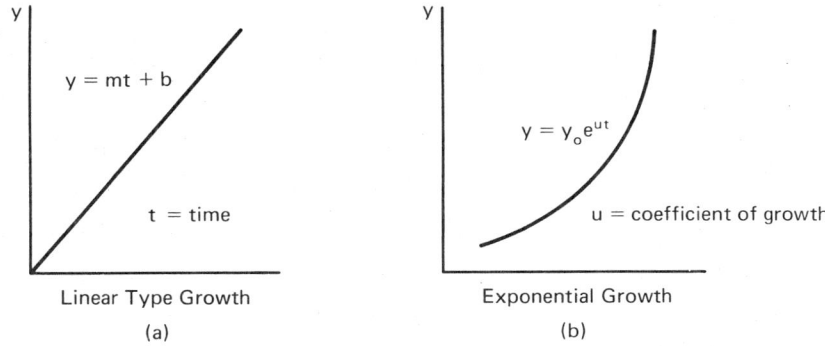

Fig. 3-27. Linear and exponential growth. Proteins are assembled on a linear rate basis as opposed to the geometrical progression seen in the multiplication of organisms. However, many assembly lines in phase operations can support the rate demands during geometrical growth.

The *beginning* of a synthesis is established by the Met-RNA initiator and the *end* is set by the terminator. In between, the combination of the previous step and coded instructions establish subsequent steps. The initiator codon establishes the phase of translation and without it there would be uninhibited phases. Initiator codons instruct "read this word first," thus establishing the phase for later reading. Three kinds of chemically different RNA and DNA codons occur—5′ terminal, 3′ terminal, and internal.

The evidence points to a remarkable universality in the UCAG genic code. One codon specifies the same amino acid in all species and in all forms of protein. However, several kinds of tRNA exist for the same amino acid, in some cases, and they respond to different sets of codons. Species may have evolved changes in codon recognition in order to acquire more storage capacity for information. Such changes can exert secondary influences or pressure on development by controlling the *rate* of protein synthesis. The translation may vary for some environmental reason. Gene mutations can cause errors or noise in the code so that the operations involving mRNA, tRNA, AA_tRNA, synthetases, and ribosomes may be modified. The code has an error reduction feature. An incorrect base limits amino acid substitutions chemically. The program probably varies in a sliding way to reflect growth needs, that is, the codon translation operation is adjustable. Genic disease is a manifestation of the presence of noise in the signal or error in the code. For example, in sickle cell anemia, hemoglobin has one error affecting two out of its 574 amino acids and, therefore, does not carry oxygen well. Specifically, DNA gets translated to valine instead of to glutamic acid in position 6 in the two beta hemoglobin chains. In evolution, those offspring having one parent clear of the error (heterozygous) would be selected, other things being equal.

In some *viruses* the mechanics of nucleic acid injection show that the infective transfer is very fast. In a small volume such as the head of a

TABLE 3-1. The UCAG Genetic Code

		Second Letter			↑One base changed = changed AA	
	U	C	A	G		
U	UUU Phenylalanine UUC Phenylalanine UUA Leucine UUG Leucine	UCU Serine UCC Serine UCA Serine UCG Serine	UAU Tyrosine UAC Tyrosine UAA Term UAG Term	UGU Cysteine UGC Cysteine UGA Term UGG Tryptophan	U C A G	
C	CUU Leucine CUC Leucine CUA Leucine CUG Leucine	CCU Proline CCC Proline CCA Proline CCG Proline	CAU Histidine CAC Histidine CAA Glutamine CAG Glutamine	CGU Arginine CGC Arginine CGA Arginine CGG Arginine	U C A G	*Third Letter*
A	AUU Isoleucine AUC Isoleucine AUA Isoleucine + AUG Methionine	ACU Threonine ACC Threonine ACA Threonine ACG Threonine	AAU Asparagine AAC Asparagine AAA Lysine AAG Lysine	AGU Serine AGC Serine AGA Arginine ACG Arginine	U C A G	
G	GUU Valine GUC Valine GUA Valine ++ GUG Valine	GCU Alanine GCC Alanine GCA Alanine GCG Alanine	GAU Aspartic GAC Aspartic GAA Glutamic GAG Glutamic	GGU Glycine GGC Glycine GGA Glycine GGG Glycine	U C A G	

First Letter (left side)

← Start code reading with symbol at 5' position

→ Finish reading codon at the 3' position

+ N-formyl methionine tRNA initiator, initiator in *E. coli*.
++ GUG codes for N-formyl methionine at beginning of mRNA.

Note:
mRNA carries these triplet codons read from the DNA script and the sequence of nucleotides present is directly translated into the indicated amino acid. One codon means one amino acid. N-formyl methionine is a tRNA initiator. The idea is that certain codons such as AUG result in the beginning of a readout and others indicate the end. Thus UAG and UAA signal operation complete, terminate the assembly and remove the finished product from the ribosome. The alpha amino end of methionine has the substituted formyl group added by an enzyme after methionine has joined its tRNA. Ordinary methionine can enter at other places in the peptide. (Modified from a table originally developed by F.H.C. Crick.)

bacterial virus (Figs. 3-28 and 3-29), the DNA may be coiled up neatly on a spool or alternate structure. It then squirts nucleic acid out of the tail into the host cell. A fraction of a second for 40,000 nucleotides in a

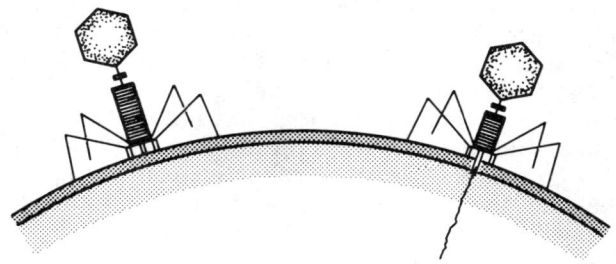

Fig. 3-28. Viral nucleic acid injecting host cell.

TABLE 3-2. Amino Acids for Synthesis of Proteins

A rare apparent universality in biology, the 20 amino acids in protein and their codification by four bases in nucleic acid. Both are found essentially in an identical way all through nature. All the amino acids have a common alpha carbon with attached carboxyl and amino or imino (Proline) group. The rest of the molecule or the "R" portion varies with each acid to provide specificity useful in information and recognition processes. Proline is seen as an exception which does not have this common feature. The neutral form is shown for the amino acids. Thus the carboxyl and amino groups appear as CO and NH rather than as COOH and NH_2.

Amino Acid	Abbreviation	Formula
Glycine	Gly	$-CO-CH_2$ $\quad\quad\;\;\mid$ $\quad\quad\;\;NH$ $\quad\quad\;\;\mid$
Alanine	Ala	$-CO-CH-CH_3$ $\quad\quad\;\;\mid$ $\quad\quad\;\;NH$ $\quad\quad\;\;\mid$
Valine	Val	$-CO-CH-CH \langle \begin{array}{l} CH_3 \\ CH_3 \end{array}$ $\quad\quad\;\;\mid$ $\quad\quad\;\;NH$ $\quad\quad\;\;\mid$
Leucine	Leu	$-CO-CH-CH_2-CH \langle \begin{array}{l} CH_3 \\ CH_3 \end{array}$ $\quad\quad\;\;\mid$ $\quad\quad\;\;NH$ $\quad\quad\;\;\mid$
Isoleucine	Ile	$-CO-CH-CH \langle \begin{array}{l} CH_2-CH_3 \\ CH_3 \end{array}$ $\quad\quad\;\;\mid$ $\quad\quad\;\;NH$ $\quad\quad\;\;\mid$
Phenylalanine	Phe	$-CO-CH-CH_2-C \underset{\underset{H\;\;\;H}{C-C}}{\overset{\overset{H\;\;\;H}{C=C}}{}} CH$ $\quad\quad\;\;\mid$ $\quad\quad\;\;NH$ $\quad\quad\;\;\mid$
Proline	Pro	$\quad\quad\;\;CH_2-CH_2$ $\quad\quad\;\;\mid\quad\quad\;\;\mid$ $-CO-CH\quad CH_2$ $\quad\quad\;\;\backslash\quad\;\;/$ $\quad\quad\;\;\;\;N$

Amino Acid	Abbreviation	Formula
Tryptophan	Trp	—CO—CH—CH$_2$—C—C(H)=CH—...—NH—...HC=C, N-H, C-H ring
Serine	Ser	—CO—CH—CH$_2$—OH | NH |
Threonine	Thr	—CO—CH—CH⟨OH, CH$_3$⟩ | NH |
Methionine	Met	—CO—CH—CH$_2$—CH$_2$—S—CH$_3$ | NH |
Asparagine	Asp-NH$_2$	—CO—CH—CH$_2$—CO—NH$_2$ | NH |
Glutamine	Glu-NH$_2$	—CO—CH—CH$_2$—CH$_2$—CO—NH$_2$ | NH |

BASIC

Amino Acid	Abbreviation	Formula
Histidine	His	—CO—CH—CH$_2$—C=CH | ... H$_2$N N NH \\ // | CH
Lysine	Lys	—CO—CH—CH$_2$—CH$_2$—CH$_2$—NH$_3$ | NH |

Amino Acid	Abbreviation	Formula
Arginine	Arg	$-CO-CH(NH-)-CH_2-CH_2-NH-C(=NH_2)NH_2$ ACID
Tyrosine	Tyr	$-CO-CH(NH-)-CH_2-C_6H_4-OH$
Aspartic Acid	Asp	$-CO-CH(NH-)-CH_2-COO$
Glutamic Acid	Glu	$-CO-CH(NH-)-CH_2-CH_2-COO$
Cysteine	Cys	$-CO-CH(NH-)-CH_2-S$

string 42 micrometers long is all that is required. In one *E. coli* virus or phage, the DNA is 50 micrometers long and quite an ingenious compression is necessary to produce a viral head only 0.095 micrometers by 0.065 micrometers.

In *evolutionary* processes, the determination of how amino acid sequences in protein are associated with DNA and heredity is a main, but not the sole, key to information pathways. There has always been great interest in the changes that have occurred in these sequences due to the alternate pathways of evolution as shown by the work in the field of micropaleontology. Variation in biochemical structure is a clear way for populations to present new phenotypes as an experimental trial *historically* in the game of natural selection. Information has certainly flowed via these sequences to give a diversity in organisms which is subject to analysis. The cell itself is an ensemble of automated molecular systems which use the logic described above.

A study of code variations reveals possible causes and the information management is in response to the laws of natural selection. How

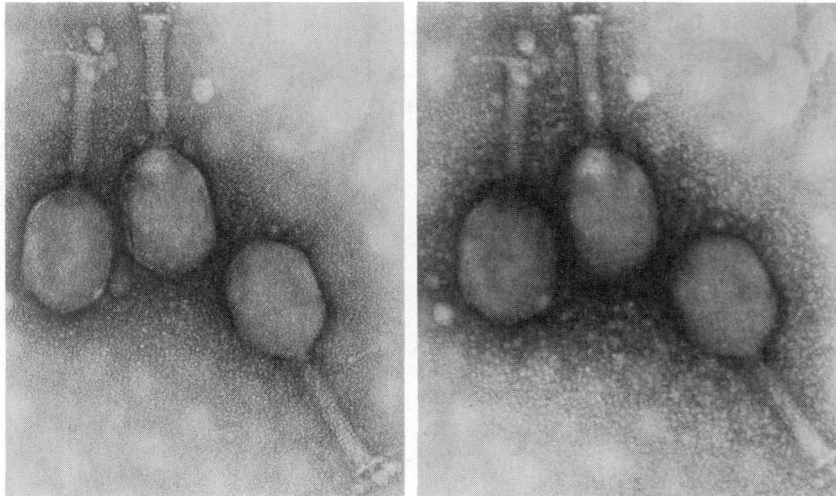

Fig. 3-29. Electron micrograph. *E. coli* T_4 phage penetrates host cell injecting about 40,000 nucleotides in a fraction of a second. Specimen: $T_4 D$ phase stained with 0.5% uranyl acetate; Magnification: 80,000X direct, 480,000X total; taken at 100 kV (left), 200 kV (right). Courtesy Perkin-Elmer Corp.

the code can vary, around what mean can the program deviate purposefully, and how do the code, concentration gradients and other patterns steer synthesis and respond in feedback loops are existing questions. As of now, the emergence of phenotypes via molecular codes can only be described in the classical physical manner. Further study may give a quantum mechanical description.

DNA Molecular Information in Taxonomy

The subscience in biology of molecular taxonomy has developed out of the information resident in DNA base compositions. The base composition of a sample of DNA may be given as the mole percent guanine and cytosine GC. If GC equals 50, then adenine and thymine AT would also be 50. Just as the total amount of DNA reflects the complexity in organisms, variations in GC reflect taxonomic differences. The existence of equal GC values does not mean identical peptides because the code is degenerate and GC can vary without having the peptide vary, but differences of more than 10% in GC are quite meaningful in tracing the evolutionary and metabolic closeness of organisms. For example, the DNA percent GC now takes its place alongside phage typing, serotyping and antigenicity, fermentation tests, pigmentation, and motility in such procedures as the separation of major subgroups within a group of related microorganisms. Very often, as in the case of the genus *Bacillus*, DNA base compositions reveal differences never supposed to exist, so that reclassifications can be logically made.

RADIATION INPUT

Chapter 1 showed how information concepts provide an alternative introduction into thermodynamics. Then, in Chapter 2, a groundwork was prepared for communication concepts within living systems by means of the alphanumeric symbolism and background of deductive inference and logic. Here, in Chapter 3, the conclusion may be inescapable that the information approach is an alternative entry into the study of genetics with the pathways being those outlined in Chapter 2. Usually this subject is approached from the molecular or the Mendelian side. The role of DNA is that of information manager, and the study of genetics from a combined, molecular/information aspect allows a concentration on the steadily unfolding *modern* picture in this field. It is especially useful to follow the features of information theory that can be identified in genetics and to use the molecular structures to understand the very informative radiation interaction.

If DNA is the master molecule in biological control as it certainly is, then it meets *its* master in radiations. In the case of the much-studied microorganism *Escherichia coli*, the main cause of inactivation of cells by radiation is genetic damage caused either directly or by some route that leads to such injury. Ultraviolet radiation is an efficient way to stop DNA synthesis in bacteria. This energy forms dimers that prevent normal operation in the master molecule. Switching to a second radiation of longer wavelength results in normalization with recovery. Thus radiation frequency is a matter of life and death. Deep in history there is the primordial scene where the first living system quite probably existed in an environment heavily contaminated with radiations and atmospheric storms. Mutations by radiations are highly efficient in accelerating genetic changes which are the essential basis for evolution.

Radiations are electromagnetic communications that can also play the role of *noise* in the channels. Then genetic information must be communicated through noisy pathways without suffering too much signal loss. Lightning is an atmosphere instantaneously saturated with disposable charges. As far as living systems are concerned, one would expect a surge of noise in the channels of communication with such electromagnetic disturbances making redundancy essential if the organism is to carry on. Information-molecular genetics reveals how this is accomplished. Noise exists in many forms depending on what type of communication is occurring. The organism may reject the noise or communicate internally through it. Conflict, overloading, and saturation sensory effects are possible results from various degrees of noise introduction. The unexplained effects of ball lightning may fall in this category, giving bizarre results that lead to some of the best ghost stories available today.

Ball lightning is the probable cause of atmospheric luminous objects which are often sighted and reported with the utmost seriousness but that defy positive explanation. It varies somewhat in that although usually associated with storms, it may be found inside structures. It may be a moving luminous sphere inside a flying aircraft or, as in a documented story from England, an energetic, glowing object that struck a lady in her kitchen and burned her dress and pants. In a report of this incident by a physicist, M. Steinhoff, in Nature, April 15, 1976, the lady related how a blue sphere of light appeared above her stove during an exceptional thunderstorm and came after her, striking below the belt and disintegrating her clothes at the point of contact. Instinctively, brushing the little ball about the size of a softball away, she suffered redness and swelling on that hand and a burning inside her gold wedding ring as the ball evaporated with a bang. The cloth was shrivelled, but not burned, around the hole and the polkadot pattern was bleached. It is photographed in the account of the incident which was the woman's first experience with the phenomenon. It is certain that with houses being struck by the electrical charges in the neighborhood, exposed living systems would have had to be affected.

There are many manifestations such as the ball lightning in the surface-charged atmospheres of electrical storms. On a ship, the outlines of antennas will be traced with a glowing discharge which is sometimes called "St. Elmo's fire." Ball lightning may be like the Kapissa (Russian physicist) model with ultrahigh frequency electromagnetic waves generated by the storm—then high energy would be concentrated in the maxima of standing waves. This energy may well be sufficient to ionize the air and make it glow. Sustaining energy would flow in from the super-charged atmosphere. To us, the fact that ball lightning is an inside phenomenon provides the distances and reflective surfaces that would add to the credibility of this model.

Lightning represents a sudden massive noise input into internal communication. Charges and ionized particles are released and circuits experience surges and momentary saturation. These are large instantaneous variations in local forces, pressures, and in the motion of free, conducting electrons and ions. This is the picture of noise in channels and is not limited, therefore, to sound effects. The noise is a form of electrical "rattling" of vibrating charge carriers superimposed on normal motions. After the lightning dies away, average conditions return, broken continually by the lower levels of "domestic" thermal noise. Presumably, certain conditions may arise in which electromagnetic fields are greatly amplified in storms so that the ordinary micro or milliamperes and microvolts become amperes and volts of sufficient strength to be disruptive until finally, at the critical levels of contact, tissues suffer paralysis and individuals are killed.

Normally the internal communicating, sensory, and metabolizing systems will be involved with phosphorylations and energy transfers through valence bond variations, charge transfers, excitations, and redox reactions. At ordinary levels of activity in electrical circuits, random or uniform electron emissions and electron flow produce internal circuit noise or *shot* noise. This adds to the other type, *thermal noise*, to give the total internal noise. Since the "rattling effect" is temperature dependent, it is thermal noise. *White* noise is a term taken to mean that all frequencies are included with no cutoffs just as white light means all of the visible spectrum is included. Then, any frequency peaking would reinforce certain spectra and be significant in noise analysis.

Sudden surges in the sensory communication system could saturate the signal lines, spilling over into alternate pathways so that sensation might occur in the hearing mode through stimulation in another mode. "Microwave hearing" suggests itself as a case in point. Here, one postulation is that these electromagnetic waves spill into the hearing channels from other receptors, so that the microwave is sensed without an apparatus designed for such sensing. A better explanation is a direct sensing by a distortion of the normal audio receptor system.

The nature of the noise is a function of the type of living system involved. A microorganism may need communication to serve its needs for data on light intensity, pH, oxygen, carbon dioxide, concentration of energy sources, nearness to a neighbor, and temperature. Animals depend on the sureness of their sound, sight, touch and smell signals. Organization man, however, is more diverse and his language is advanced. In English, he has a communicating capability in bits corrected for redundancy as he struggles with the noise that engulfs him. Successful communication internally in living circuits depends on getting over the noise barrier to reception of needed signals so that numerous methods must be used to outwit noise. All the sensory systems are straining when on full alert to filter out the incoming noise. Some of the incoming signals will be in the form of radiant energy and this radiation information will see barriers based on the living system's response time constants. If these are too greatly out of balance, coupling is defeated. The bandwidth of the incoming radiation is inversely related to the time constants of the responders. Incoming radiation may be a signal via DNA telling the organism not to reproduce and only a few hundred *rads* (radiation absorbed dose) carry the message. Otherwise a gene or a point in the DNA can be reformed. Alternatively, it may be like noise in the organism. In this case, rejection or filtering mechanisms are expected. In organisms, this would amount to protective or repair processes leading to recovery from the disturbing noise.

Radiations on the planet Earth did not begin with the atomic age. In prehistoric times, radiations must have been a barrier to life on the planet but as the radiation cooled with time, life became possible. At the same time, residual radiation would have accelerated the evolutionary changes so that adaptation could occur. This development of life had to involve some protected storage of information against radiation damage so that stored information could be used in molecular reproduction. Some of the known protective mechanisms such as those of aminothiols with their sulfhydryl and amino radicals, cysteine, cysteamine (beta, mercaptoethylamine MEA), and cyanides may have been important during this period. If life is taken to have existed for about 3×10^9 years, there had to be precursors for nucleic acids in a pre-DNA era as well as a "motivating" influence toward these reproducing molecules. Purines, pyrimidines, amino acids (and sugars) can be assumed to have been formed out of simpler substances such as methane, ammonia, water and hydrocyanic acid in a radiation environment especially at moderate levels, possibly initiated by lightning stimulation. Electrical discharges cause amino acids such as alanine and glycine to form in vitro. Something is present as information or electromagnetic energy that directs even as it energizes.

Bases continue to be ready receivers of radiation. X-rays or gamma rays destroy the DNA bases in the presence of oxygen but the early earth had a reducing, protective, atmosphere. Pyrimidines are more sensitive than purines, and thymine (which is in DNA but not in RNA) is the most sensitive of the bases. DNA is certainly the main contributing factor leading to radiation injury and cytolysis. Conversely, irradiation alters the population of mammalian cells. Moderate doses of X-rays delay DNA synthesis, higher levels inhibit it, and small doses may increase synthesis, so that, obviously, there is information for the organism in the *level* of radiation received. If the double helix is irradiated, it loosens its "spring," allowing tighter folding and the breaking of the strands by alpha particles. In the hydrogen bonds there is a resistance against breaking in ionizing radiations but these are sensitive to ultraviolet so that hydrogen bond deformation, strand breakage, and folding are "frequency dependent" responses. As the DNA folds up in solution (smaller axial ratio) its viscosity decreases and a crosslinked structure would be preferred in the absence of oxygen. Alternatively, oxygen would form peroxides on the ends of any breaks, effectively preventing repair so that with oxygen, DNA tends to break. Again, these results differ according to the kind of radiation received.

To be mutagenic, the energy of the radiation must be correct for particular bonds in the structure. Whereas ultraviolet easily affects the hydrogen or normal scissioning bonds in the DNA helix, ionizing radiations affect the hydrogen bonds of globular protein. Scission of peptide bonds in polypeptide chains is a case of highly accurate modification. This happens without alteration of the two adjacent amino acids forming the bond. The time constants in the receiver system depend on the stage of activity, and DNA occurs in a highly organized system with mitotic and intermitotic phases. The chromosome-splitting time is its most vulnerable one and breaks in one strand are always more hazardous than double strand ones. The nucleus is most sensitive at the end of the resting stage and the synthesis of DNA is also most vulnerable during this stage. The mutagenic action may be deletion of bases on DNA or base substitution which then leads to the wrong registration of opocodons when the message is read. This may be followed by reregistration, looping in the mRNA or deletion of a whole gene, giving a result like albinism. Thus, this is usually irreversible and equivalent to the insertion of new information in the race since the changed instructions are faithfully reproduced and executed in the successive generations.

Inactivation is a drastic case of radiation absorption rather similar to "overmutation." The same molecule that can mutate can be inactivated under more radiation. The result may not be obvious until division, since the time when genetic information is being manipulated in mitosis with spindles is a sensitive time, (a fact used to advantage in radiotherapy). Thus, the errors could have been caused prior to mitosis, a time which shows the effect of accumulated error. If the exposure to X-rays is doubled with *Drosophila*, then so are the mutations. The absence of a threshold would make any quantity of radiation mutagenic and the effects would be accumulated. Their manifestations are seen in point deletions in chromatids and chromosomes and in exchanges and rearrangements of genetic material. Photoinactivation at 2,200 — 3,000 Å has a special effect at 2,537 Å. Between 3,130 and 5,490 Å, photoreactivation occurs with a peak at 3,800 Å and requires some absorbing body. In *Streptomyces griseus*, the photoreactivation is at 4,300 Å. A suitable absorbing body or molecule would be DNA or a nucleoprotein which is inactivable and reactivable. The best mutation wavelength coincides with the inactivation one, at an absorption peak for DNA. In addition, there is an effect in the medium that is passed on to the organisms. Mutation is greater with a medium that has been irradiated before inocculation.

One can study the problems of aging, radiation effects and degenerative disease in terms of loss of information. Radiation causes *errors* as well as favorable mutations and marginal systems cannot tolerate errors! They are not able to operate when the system is too lossy or noisy. In development, differences of an order of *magnitude* in the information content occur when passing from one level of organization to another. Redundancy enables the information to get through a noisy channel.

1. Alternatively, the messages carry large amounts of internal instructions,
2. Or the information is in great excess for molecular convenience (permutation).

In primitive times, living systems needed replenishment of free energy from intersystem radiation on a continuous or cyclic basis. After radiation cooled on earth, this replenishment was due to the sustaining force of solar thermonuclear reactions sending the spectrum constantly to earth. Any structural change may have been significant. Lack of oxygen in deoxyribose may have given DNA a packing density evolutionary advantage over RNA which is now the only genetic material in some viruses.

Algae remaining green near intense radiation suggests that the chloroplasts could have survived and "flowered" on the primitive earth, yet they can concentrate radioisotopes up to 10,000 times their concentration in the water in which the algae grow. The primitive radiation environment would have favored certain cell aggregates over single cells. Similarly, cell aggregates—multicellular packets and clusters of bacteria— would have been favored over thinner constructions such as pairs and chains.

1. Forms with protectors like —SH would also have been selected, or those with enzyme systems capable of destroying radiation-induced toxins, for example, the system H_2O_2 and the enzyme peroxidase.
2. Multinuclear forms and higher ploidy ones would have DNA protection.
3. Oxygen sensitivity could have come about in this manner: The primitive cells in a reducing atmosphere were resistant to radiation damage and they could prosper and develop. Then, with oxygen increasing in the atmosphere and radiation dying down, aerobic cells could emerge.

PROBLEMS

1. How is information management taken over by the human operator in recombinant genetics?
2. Why is *E. coli* considered a hazardous mode for recombinant gene research?
3. How are enzymes immobilized for use in catalysis in separate systems?
4. To how many cistrons does the T_4 phage refer for details of its tail fiber morphology?
5. Is cross-over in chromatids a normal operation?
6. At what positions in the deoxyribose sugar molecule of DNA bases are the neighboring bases bound?
7. What kinds of nucleic acid are involved in protein synthesis?

BIBLIOGRAPHY

Anfinsen, C.B., Jr. *et al*, Eds., *Advances in Protein Chemistry*, Academic Press, New York, 1970-1978, Vols. 24-32.

Edgar, R.A. and W.B. Wood, "Morphogenesis of Bacteriophage T_4 in Extracts of Mutant-Infected Cells," *Proc. National Academy of Science, U.S.*, Vol 55, 498-505, 1966.

Franklin, R.E. and R. Gosling, "Elucidation of the DNA Structure by X-ray Crystallography," *Nature*, Vol. 171, 740, 1973.

Hamburgh, M., *Theories of Differentiation*, American Elsevier Pub. Co., New York, 1971.

Jovin, T.M., "Recognition Mechanics of DNA-Specific Enzymes," *Annual Reviews of Biochemistry*, Vol. 46, E. Snell, Ed., Annual Reviews Corp., Palo Alto, Calif., 1976.

Kendrew, J.C., "Information and Conformation in Biology," *Structural Chemistry and Molecular Biology*, A. Rich and N. Davidson, Eds., W.H. Freeman, San Francisco, Calif., 1968.

Kornberg, A., "Biological Synthesis of Deoxyribonucleic Acid," *Science*, Vol. 13, 1503-1508, 1960.

Neuroth, H. and R.L. Hill, Eds., *The Proteins*, 3rd. Ed., Vol. 3, Academic Press, New York, 1977.

Nirenberg, M.W., "The Genetic Code," *Ill. Sci. Amer.*, March (offprint, 153) 1966.
Paterson, D., *Applied Genetics*, Doubleday and Co., New York, 1969.
Spirin, A.S., "On "Masked" Focus of Messenger RNA in Early Embryogenesis and Other Differentiation Systems," *Current Topics in Developmental Biology*, Vol. 1, A.A. Moscona and A. Monroy, Eds., Academic Press, New York.
Watson, J.D., *Molecular Biology of the Gene*, W.A. Benjamin, Inc., New York, 1965.
White, H.B., B.E. Baux, and D. Dennis, "mRNA Structure Hairpin Loops with Protein Structure," *Science*, Vol. 175, 1264, March 17, 1972.
Woese, C.R., *The Genetic Code: The Molecular Basis for Genetic Expression*, Harper & Row, New York, 1967.
Yanofshy, C. and P. St. Lawrence, *Ann. Rev. Micro*, Vol 14, 311, 1960.

4

Conformation, Transformation and Deformation in Atomic Informational Biology

> Given the code,
> The tools,
> And a list of rules,
> The formless emerges,
> ... a shapeless block
> ... or stairway to stars.
>   D. A. C.

INFORMATION-RICH MOLECULES

It is one of the virtues of English and of several other related languages that they provide for the information aspect of the stereochemical behavior of molecules as suggested by the title of this section. Conformation and deformation are features in, and operations performed by, information molecules, particularly macromolecules functioning in operations such as catalysis, regulation, immunization, and transport. Transport refers to those biological operations involving movement of energy and materials so, obviously, transport phenomena are ubiquitous in biology. Macromolecules have high organization levels which can make them information rich. In immune reactions for example, the lower limit for antigen recognition is a molecular weight of about 5000. In this they behave like automated, portable, intelligent, machinery with access to meaningful ambiental messages.

The word "information" will be seen to connote a flow of internal management directions as opposed to a random meeting of reacting elements. Outside information management includes the introduction of inhibitors, competitors, blocking agents, activators, and so on as in the case of drugs and chemotherapy. Intersystem management then involves the transport of information by radiant energy to interfaces as in vision, photosynthesis, phototaxis and photobiology in general.

In this area of atomic biology, the features of individual atoms, or

TABLE 4-1. Information in Atomic Structures. Location of Some Information-Dense Regions

Spatial Conformation	In higher levels of informational molecule (e.g., quaternary) configuration, especially macroproteins and protein complexes with lipid and polysaccharide, subgroup behavior, and membranes
Amino Acid Sequences	Messages in communication, substitution cipher, digram, trigram, and other segments, side groups, angularity, shape, helix, etc.
Carboxyl Group Amidation	Reaction coupling
Presence of Conjugated Small Atoms and Molecules	Transitions, unpaired electrons, coordination
Polymerization	Repetition, redundancy translation of information into form factors
Alternate Folding Patterns	Space Charge exposures, exposure of prosthetic groups, *spatial* changes in covalently linked groups
Isozyme Configurations	Alternate form of isoenzyme having closely related function, affecting development, regulation, redundancy changes in direction of synthesis, switching, discrimination of substrates, selective repression in synthesis of subtle protein differences, flexibility in meeting changing requirements
Reconfiguration	Enhancement, repression, regeneration of state
Renewal	Accumulation of favored characteristic and loss of recessive factors, cumulated errors
pH-Influencing Atoms and Molecules	Reaction and allosteric effectors

their chemical characteristics in the solvent or surroundings, control the atmosphere in the interatomic spaces where the action takes place (Table 4-1). Some "influential" atoms such as transition metal elements (e.g., the iron in hemoglobin) have an extremely commanding authority over their atomic regions.

Dielectric Constant of Water as Moderator

Certain solvent characteristics are also revealed as determining features, especially the dielectric constant ϵ'. This constant for water is

Atomic Informational Biology

about 80 as compared with values of about 4 for a lipid. Such a large value insures that the charges on solutes will be forced to keep their distance. However, in the interatomic matrices in macromolecules and depressions, the solvent influence and ϵ' are subject to transformation. This reflects the quite different hydrocarbon surroundings as compared with the intermolecular conditions. Reactions become feasible in these special locations then, due to the closeness of the reactants. Water does not separate them as it does outside. One of the features affecting reactions therefore is the degree of regulation exercised by the solvent as a *dielectric* mediating the charges. Another is the proposition that these special locations constitute *active* or operating *sites* for reactions (Fig. 4-1).

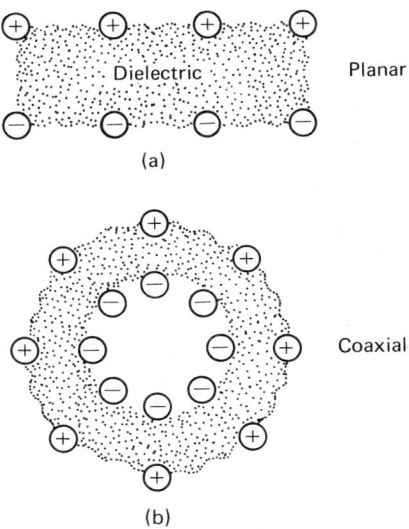

Fig. 4-1. Dielectric functions as a planar or coaxial charge spacer in atomic biology. It may have water-solvent features between molecules or alternatively hydrocarbon-type characteristics in the interatomic spaces of macromolecules.

The information density involved in such regulation has been considered to be such that 125 atoms can store about one bit. In the case of DNA, 50 atoms store this amount. Following through with the calculations on the number of atoms needed to store man's whole treasurehouse of information, normally the writings in about *24 million* volumes of an encyclopedia, the conclusion is that physical laws allow equivalent information constructions or densities all within a cube $1/200$ inch on a side, which is essentially a mere trace of substance. The estimates that have been made for DNA are taken as proof of the incredible information density in that molecule. On these dimensions, a macromolecule which has perhaps a molecular weight of 10,000 to 100,000 and which is both a protein and an enzyme will have enormous, highly organized information and communication capabilities.

Significance of Lysozyme

Lysozyme is a case very much in point. Among the mechanisms of resistance to disease are those within certain cells, e.g., phagocytes. One type of phagocyte is the granulocyte found in the blood and bone marrow. It is a motile cell which has lysosomes on its membrane. Lysozyme, which hydrolyzes a mucopolysaccharide in the cell wall, is one of the enzymes in this organelle. Along with other enzymes, it gives the granulocyte the power to digest and destroy invading pathogens. Granulocyte activity is greatest when the infection is most acute. It illustrates the phenomenon of *Cooperative Effort* in that recent phagocytosis so modifies a phagocyte that it increases its capability tenfold.

Lysozyme is distinguished by being first among the few enzymes which have been successfully analyzed in terms of its informational behavior. Lysozyme from egg white, which is a rich source, has been analyzed. Perhaps an even more interesting source would be tears which, thus, protect eyes. There are several thousand enzymes in cells, and so far, some dozens have had their catalytic action explained. The studies reveal, above all, that the speed of enzyme-catalyzed reactions, which is some 10^8 to 10^{11} times as fast as that of nonenzymatic ones, is associated with conformational and deformational changes in the molecules. Since these are special capabilities in protein with their several levels of structural organization, it is anticipated that the general mechanisms have been at least partly discovered. Another concept is that most proteins use their special *form factors* to participate in one way or another as informational and/or enzymatic structures. In evolution, there has been a reluctance to discard a highly ordered structure once it has evolved. This is apparent in the identities found between proteins in differently evolved radiations (Fig. 4-2). Widely different organisms thus possess identical amino acid sequences and form factors in their proteins. These identical substances are used in such organisms for different purposes, showing that a "make do" policy was in effect when the biochemical evolutionary change or mutation occurred. No doubt, highly coordinated molecules with recognition features were too valuable to be replaced, considering the burden of evolutionary pressures.

The persistence of this informing organization is related to the invariance between generations—the true copies that continue to be generated. It will be observed that what is in the process of formation here is a background of cybernetic ideas or, as stated by Lars Löfgren, "The discussion will point on a natural extension of cybernetics from communication and control in animal and machine, towards a study of invariances of evolutionary characteristics over animal and machine (or over animal formation and theory formation). The control concept requires externally defined norms to control after, whereas such norms are automatically generated in a natural selection process."

Atomic Informational Biology

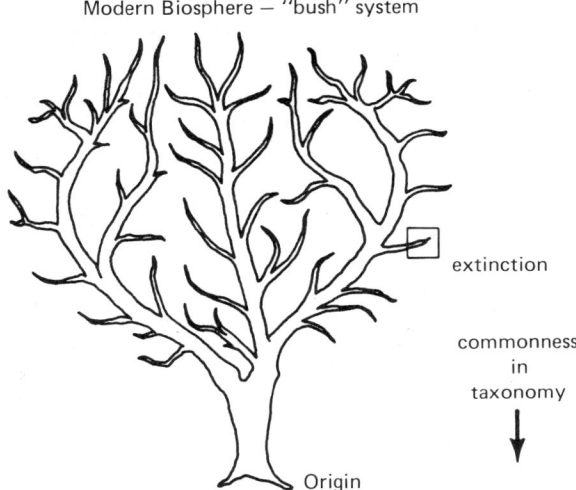

Fig. 4-2. Evolutionary radiation. Valuable information-rich molecules are conserved during evolutionary change because they furnish powerful assistance in meeting evolutionary pressures. Identities are thus found between proteins in different evolved radiations.

The information in lysozyme resides in the chain of amino acids, its meaningful distribution of these links back and forth, bending at each atom, its disulfide linkages across the space between chains, the surface and internal electric charge distribution, polarity of groups, alignment with respect to solvent, and especially the location of its active sites; the information exists in the presence of a depression whose dimensions permit only the substrate to conform and be admitted for catalysis (Fig. 4-3).

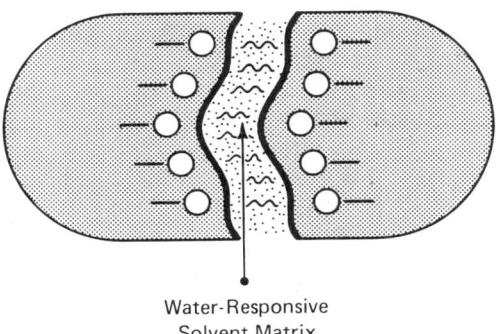

Fig. 4-3. Lysozyme Information. The location of active sites and dimensions of available depressions allow conformation of substrate to the region of catalysis, as seen in enzymes such as lysozyme.

Portions of lysozyme tracing only the chain and linkages (Fig. 4-4) form merely an outline. The real fleshed-in molecule is very different,

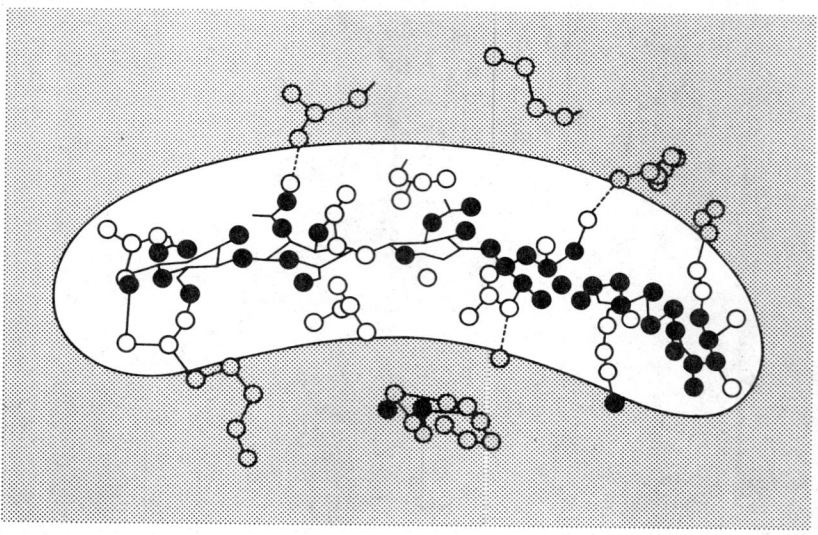

Fig. 4-4. Molecular outline shows atomic configuration of the chain and linkages in one orientation only. The "fleshed in" module would appear differently to the connecting substrate. See Fig. 4-5 in this connection.

being fully packed in the atomic sense. The drawing, however, shows the bends and twists and internal groups, while the model demonstrates what the protein would look like to an approaching substrate molecule.

The substrate hexasaccharide recognizes an opening with which it can conform and which extends across the lysozyme molecule with appropriate amino acids on the receptive surfaces. This is the basis for the specificity and is the active catalytic site. There is also a control point for regulation of the enzyme according to the internal and external requirements for its suppression and activation. N-acetyl-glucosamine and its dimer chitobiose are molecules capable of conforming to the crevice across the molecule at the other active site. They inhibit the enzyme action since they carry information from the ambient to the effect that action is not desired.

The suppression due to the presence of a regulator molecule is obviously the reverse of the *cooperative effect* mentioned in connection with the mechanism for phagocytosis. As a result of this effect, a recent meal makes the phagocyte much more eager to devour more bacteria.

Another example is that which makes it easier for hemoglobin to attach to a second oxygen than it was for it to attach to the first. In

Atomic Informational Biology

this case, the cooperation can be explained on an atomic basis within the hemoglobin itself. By virtue of the electron resonance reactions based on the presence of the unpaired electron in the transition metal (i.e., iron) of hemoglobin, and the x-ray crystal analysis of the molecule, hemoglobin has become quite well understood (Fig. 4-4a). If one of the four heme groups binds an oxygen in the lungs enroute to the myoglobin which is waiting for the oxygen and anxious to get rid of its CO_2, then the second oxygen finds its way to another heme group with remarkable ease. It is so easy in fact that the action may be looked at as enzymatic. The heme groups are too far apart at 25 to 37 Å to explain this cooperation in terms of electromagnetic forces. Thus, the impetus has to be stereochemical and, specifically, related to a deformation of the heme group in which the iron moves relative to the porphyrin ring. It appears as a deformation in which the earlier oxygen makes the displacement by the force of its own attachment. In the sense of an allosteric enzyme (where allosteric refers to binding control-functions), hemoglobin has its substrate, oxygen, its coenzyme, the heme group, and its allosteric effectors. These are hydrogen ions and 2,3-diphosphoglycerate. In the nonoxygenated form, H^+ ions are released as hemoglobin loses its salt bridges when heme-heme interaction energy is supplied to loosen them. The effector (DPG) is stereochemically complementary to an active site between the beta chains which is closed off by oxygenation. Thus the allosteric effectors are regulators tuned to the nonoxygenated status of hemoglobin, that is, the *ready* status for its function.

Subunits, Forcing and Informing Structures

INFORMATION RESIDES IN CONFORMATION OR DEFORMATION PRESENT OR OPERATING IN TERTIARY, QUATERNARY, QUINARY AND HIGHER LEVELS OF PROTEIN ORGANIZATION.

The deformational change which occurs when an oxygen is attached is a distributed or communicated disturbance. Deformational changes follow in the other atomic positions due to variations in their bond angles, and their rotation about certain bonds. This amounts to a form shift in the subordinate group of the protein in response to internally transmitted information or form-factor signals. The signal either tells the affected atoms to assume a position favorable to the attachment of oxygen or it deforms the molecule, forcing it to assume a new favorable conformation for the same. Since the hemoglobin structure is compact, the signal distribution to the attachment site is assured, where any constraints against the oxystate are cancelled. The multilevel organizational structure (Fig. 4-4a) which is general for enzymes, favors this kind of reactive deformation on receipt of appropriate molecular information. Thus, it is a way of responding to the demands made among substrates, enzymes, and coenzymes. On the other hand, the first signals "open the door," making subsequent oxygenation obviously easier, up to saturation of the molecule, because the deformation for oxygen bonding is already begun. As Max Perutz has said, paraphrasing the Bible, "To him who hath shall be given."

Studies of the subunit behavior in the pyruvate dehydrogenase complex and other dehydrogenases show the coupling occurring between metabolic pathways. The super enzyme system for complement in the immune response is a fine example (Fig. 4-16). In metabolism the multi-enzyme coupled systems, with common intermediaries, show cooperative effort based on their energy requirements, being complementary (exergonic coupled to endergonic). In these reactions the time and distance relationships are important so that the action may be tightly coupled as in a close-order football play. The displaceable subordinate groups of the enzyme are arranged in such a way that they are self-organizing. There are simple physical systems in which a set of existing influences causes structures to assume an orientation reminiscent of steel balls falling into their holes in a tray when it is shaken, or of cubes having one *magnetic* face which assume a preferred growth type of orientation when shaken up together. In the latter case, a magnetic face on each cube makes it self-organizing and the agitation is a low order information input.

At least another level past that shown in Fig. 4-5 is visible in the degrees of freedom for the protein. The globular structures can cluster and, in virus protein, the subunits associate in a meaningful array. The geometry and symmetry are meaningful and sensitive to operations that can take advantage of such form and shape. One may find from forcing functions that these operations offer even higher levels of organization. Linearly coded DNA instructions set up conforming structures that are

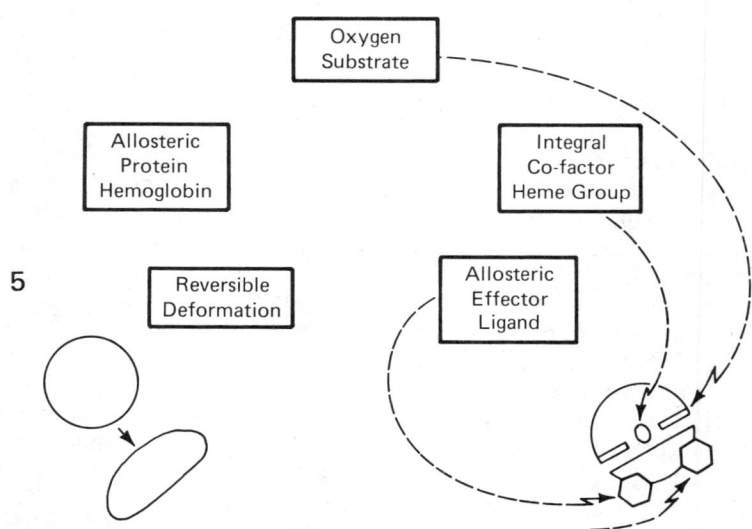

Atomic Informational Biology **141**

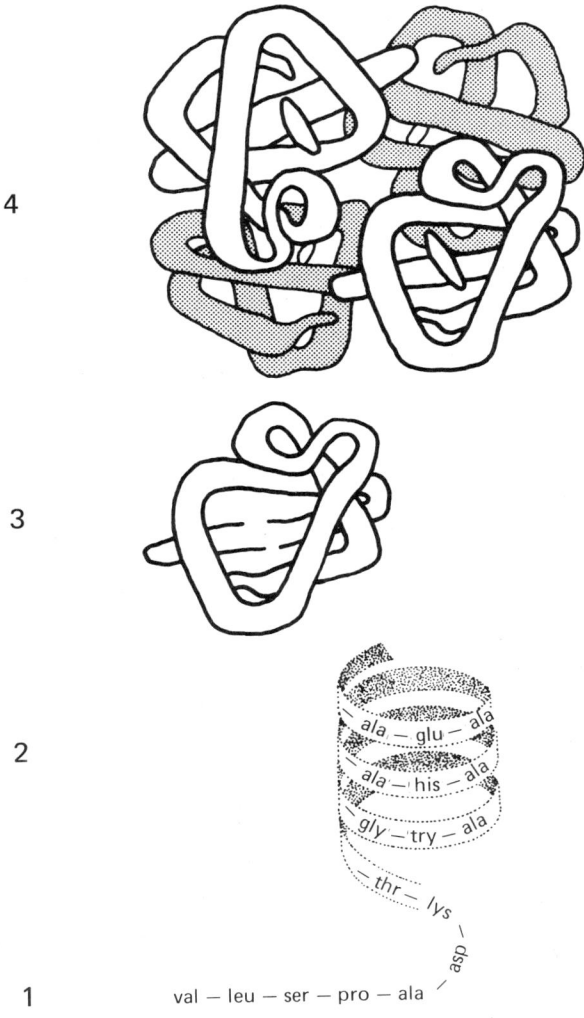

Fig. 4-5. Organizational Level 1—Portion of polypeptide. An alpha chain in hemoglobin.
Organizational Level 2—The alpha helix.
Organizational Level 3—Tertiary structure. Folded helix.
Organizational Level 4—Quaternary structure. Hemoglobin subgroups combined.
Organizational Level 5—Quinary-protein organizes into a system and responds with subgroup deformations as required.
Hemoglobin molecule. Heme-heme interaction in Hb uses the porphyrin rings and deformation of the ions and subgroups to regulate activity. Conformation is shown when oxygen binds, *displacing* the region of the combining site and signaling the cooperative binding site. The reaction is regulated by allosteric effectors H^+ and 2,3-diphosphoglycerate.

not linear but three dimensional. Unless there is some management implied in the helical shape of DNA, linear encoding produces conformational encoding. It is probably adequate to explain 3D forms from 2D codes by the bonding cycles of various amino acids relative to each other, wherein many changes in direction are easily made within an amino acid sequence. Therefore, information in the genotype is able to make conformation in the phenotype.

Conforming Bounds

The basis for conformation is response to communicated information, but the responses often show distributed effects such as cooperativity and amplification. Except for procaryotes, i.e., bacteria and blue-green algae (cyanobacteria), cells appear more and more as highly compartmented, fibrous, tubulated structures with compartments designed for specialized activities without interference and tubules or fibers for transport and reformation of the cell substance. In neuroplasm, these changes probably utilize axon and dendrite microtubules with appropriate conformational shift and ionic messages. The overall status presented contains the bounds and semipermeabilities as operating constants, on a dynamic status. Membrane dynamics suggests this picture of a conforming bound in lipid bilayers. Small ionic shifts, for example, give conformational change. Ca^{2+} on pH-sensitive, surface sites varies the whole organelle structure when its concentration varies by only a small fraction of one pH unit. The plasma membrane can also carry out pinocytosis or a reverse process in immunoglobulin production in which the membrane detaches a section of itself and launches the Ig.

To achieve the full flexibility of the bound, the structure can call upon the degree of order in the phospholipids, their packing density, the pH, the mono and divalent cations, the hydrophobic and hydrophilic phases or sides, the size of openings, surface charges as attractive or repulsive polarizations, and active, transporting, protein complexes within the ordered bound. The main shift is from fixed to fluid lipid or from an ordered mosaic to a fluid phase with varying degrees of association (Fig. 4-6).

Calcium Ion Signals. The purpose of the calcium ion is to communicate information via pulse, charge, or reactivity when the addressee is a Ca-dependent function. A vast amplification may be forthcoming. In the brain, for example, there is a tendency to see the electroencephalographic signal as a clinical convenience provided by the body, when actually it is the wave that results from the integrated brain activity. This activity is like a magic fabric of great significance to the individual and only incidentally of interest to other individuals as far as is known.

Atomic Informational Biology 143

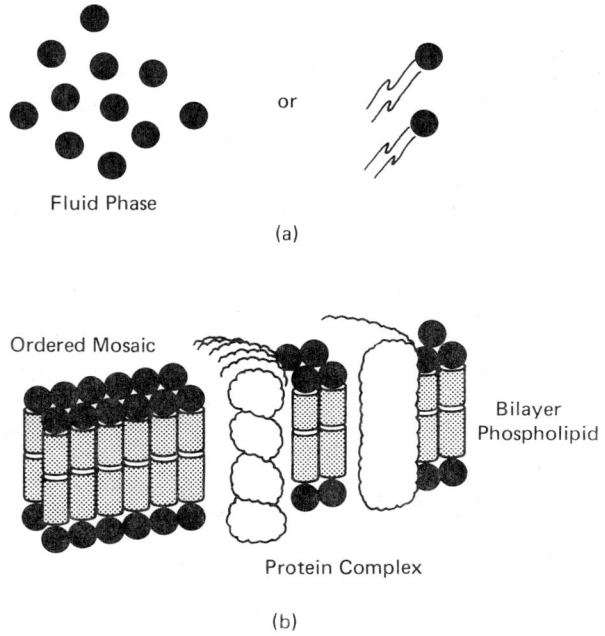

Fig. 4-6. Fixed and fluid conformation of bound.

The fabric is woven of electric gradients, nerve impulses and spikes, transient cation pulses, their migrations, their reversals, reinforcement and cancellation, neurotransmitter substances in electrochemical designs operating as excitors and inhibitors, specialized intramembranous substances and conforming membranes. The brain units link together, plug in, or form a network for the neurochemical actions and transmit information which communicates the substance of the fabric's network from which the magic of memory, meaning and information retrieval emerges. To move from one stable biophysical status to another with low E fields requires clever cooperation and ready amplification. These assure that a small input can manage a larger output, a conformational change and feedback, either positive or negative. All of these functions link back to DNA information by genetic control over mRNA messages relating to synthesis of enzymes for the neurotransmitters. Information management is by inhibition—activation which can control access to instructions as when DNA polymerases open the double helix for bacterial replication (polymerase III), repair (I and II), or for mRNA to position and read a protein synthesis code. When the DNA forms, the first turn is a conformation set up by the directional bonds and first turns on secondary structures may also take more energy than subsequent coils.

Evoked Biopotentials and Structure

Membrane surface macromolecules reversibly bind the calcium ions forming a glycocalyx which extends the membrane influence as much as 2000 Å. The calcium could switch on prostaglandins as an amplification of energy when the calcium causes conformational changes in its binding and effects across the membrane.

Electric and electromagnetic fields definitely interact with brain tissue establishing electric gradients and a flow of ions across membranes quite different and weaker than that due to common excitation. The electric gradients seem to be overshadowed by the surface effects and calcium has a key part. The electrical gradients may integrate into an evoked potential and have consequences in the behavior and neurochemistry observed. Calcium may signal the membrane to shift its axial conformation in order to prepare for changes across the membrane. Polyanionic glycoprotein is a surface counter-ion base or receiver for cations. Calcium is more useful than other divalent ions or monovalent ones because of the power in its bond, being more like hydrogen ions. These are delicate activities involving 0.1 $\mu v/\mu m$ and need amplification and cooperation in the structures to extend their influence along and across the membrane. The cationic flux can be associated with dielectric constants four orders of magnetic greater than those seen ordinarily, and this is at frequencies less than 1.0 kHz. EMR would see a high cerebral impedance and attenuate rapidly in the material. One would expect that electric gradients and neuronal surface states would be affected and would establish excitability levels and transductive coupling.

Whether or not helix or ribbon structures are formed or whether the structure is more irregular in the carbohydrates of cell membranes rests on whether or not they have a regular sequence of sugars. Sialic acid terminates glycolipids and glycoproteins in the membrane bilayer. Then, carboxylate and oxygen functions in this sugar-carbolic acid mean a cooperative binding of calcium in the bilayer as well as it does on some cell surfaces. The ionic radius of calcium gives an intercavity dimension which may hold several oxygen atoms for coordinating. In molecular models there can be C-H bonds of sugar rings to form hydrophobic basins. Calcium may attach and detach from apolar species in such molecular models.

Conforming to Communicate

In a system, state B can be reached from state A by several routes. The information and energy may be different in the forward ascending and back descending routes. This is called hysteresis (Fig. 4-7). It implies the presence of energy barriers barring one state from the other by ordinary direct line routes. The intensity of an emitted radiation may vary with the temperature of a molecular emitter. The differences may be due to changes in phase or order on the two routes as well as to conditions such as cations, pH, temperature, and the disordering process.

Amplification is a magnifying response that builds up a feedback energy signal, e.g., as Ca^{2+} informs muscle troponin to initiate contraction and one troponin passes the message to 7 actin conforming subunits and dense meromyosin acts with triphosphate action and the tropomyosin-troponin complex. The functional linkage is information communicated to signal the conformational shift. Brain communication is like a "now hear this" announcement in which a local signal is quickly passed to all surrounding units having a need to know, even at a distance. Sensory receiver molecules need to amplify a weak stimulus to polarize membranes and dispatch the sensory pulse.

It is obvious that information rich molecules, both small and large, communicate by moving so that they can bind at effector sites. This leads to their being regarded as ligands—often being linked to protein in allostery. Allosteric proteins have information roles through their control and recognition functions. This begins when they single out their own monomers for polymerization which is seen here as a molecular amplification. Ligands have their conformational roles and they respond to other ligands when the stereochemistry ordains it. The response is almost independent of background noise, even that due to the concentration of similar proteins.

INFORMATIONAL DIRECTION

Information management operates in an isodirectional or linear manner and in many cases it cannot double back on itself. Thus it is an informational requirement which sets up metabolic pathways in an amphibolic manner but keeps the processes orderly. Although the same reactants are involved, catabolic enzymes differ from anabolic ones in the important cases. Pyruvate may be the catabolic product of sugar or amino acids which then becomes the precursor of the same sugar or amino acid in anabolic reactions. Information is then needed in the form of different specific steps and key intermediates to *sort out* the routes and bridges—to mark the way. Then, essentially, there is a route for building and "debuilding" the metabolite. That most important of all processes is another example. DNA makes protein but protein, in a sense, can never make DNA, and the reason for this is found in the partner that travels with information, the energy.

These reactions are so firmly established that the energy relation is wrong if you go in the opposite direction. All enzymes and intermediates have to follow this rule if they "wish" to activate the reactants. Catabolic reactions yield energy by degradation of source material and anabolic ones receive and then store energy in bonds. By separating the two routes, the two enzyme systems can run at proper rates to maintain control. The processes may even be physically separated, one in mitochondria (catabolism and microbodies) while the cytoplasm is left for anabolic sequences. Then they can proceed independently with their own information. With these compartments, control is different than in bacteria where it is completely enzymatic and informational (Fig. 4-15).

DEVELOPMENT AND OTHER EVIDENCE OF INFORMATION MANAGEMENT IN BIOLOGY

In energy-metabolism reactions, organisms demonstrate the role of information management. There are definite constraints on the simultaneous movement of informational molecules in reverse directions on the same route. To avoid this problem, organisms use alternate pathways for energy in metabolism. Thus, for synthesis of substances, pathways are chosen which do not insert noise into the degradative routes. Then organisms are able to move metabolites along one pathway to derive energy or needed intermediates while, at the same time, moving precursors along the alternate route that synthesizes this same metabolite. The correct enzymes of each pathway are informational agents as are the key intermediates (such as acetyl coenzyme-A). The latter are bridging compounds that may exist in different cycles, e.g., citric acid and glyoxylate cycles, so as to provide a metabolic crossing point between pathways and cycles in metabolism. Common intermediates between cycles and routes, and energy-matched redox compounds in electron transfer chains, are examples of coupled units. The enzymatic agents manage the division of energy pathways under genic control which is very approximately described as the one enzyme (or one polypeptide)-one gene system, so the informational path is well established. Local control also includes small molecules such as AMP and ATP, and feedback based on mass action. ATP is a coenzyme, or coupler, and with its high energy phosphate, is a temporary energy shunt which engages the enzyme only during the reaction, or in phosphorylation as preparation for it. Coenzymes are involved in electron or energy transport, or in the movement of hydrogen atoms and chemical groups. Prosthetic groups, on the other hand, remain on the enzyme.

Coenzyme NAD nicotinamide adenine dinucleotide (also coenzyme I or DPN), the derivative from the vitamin niacin, has a central degradation management while coenzyme nicotinamide adenine dinucleotide phosphate NADP (also coenzyme II or TPN) has a synthetic one in redox reactions of metabolism.

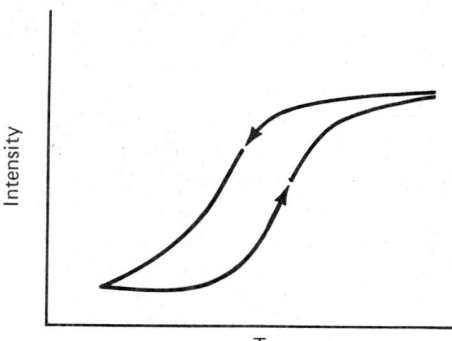

Fig. 4-7. Hysteresis curve.

Atomic Informational Biology

One of the most important mechanisms, which is also attended by feedback, is mass action, where the accumulated mass of a product of a reaction inhibits its own further production. An advanced refinement of this is suppression—control of a reaction sequence, or a cycle by the number one enzyme of the sequences, a final-product-sensitive-type enzyme acting as a "go, no-go" switch. This enzyme is inhibited by the final product of the sequence until the latter falls below the threshold concentration and deinhibition of the sequence occurs (Figs. 4-8, 4-9, 4-10, 4-11, 4-12).

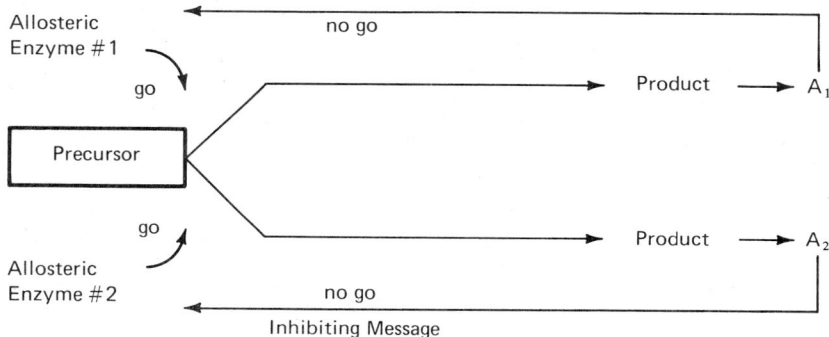

Fig. 4-8. Selective suppression in the case of a precursor common to 2 pathways which lead to 2 different amino acids. This acts as feedback regulation on top of mass action. The feedback acts in the same direction as mass action passing any enzyme steps in between to modulate the key enzyme step.

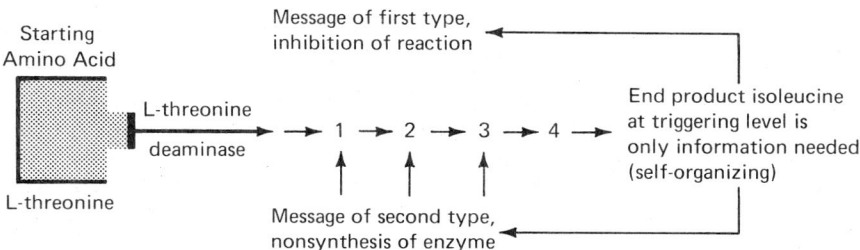

Fig. 4-9. L-isoleucine synthesis and feedback. End product inhibition of first enzyme in *E. coli* synthesis. This situation is sometimes known to create an *essential* amino acid situation. The organism such as *E. coli*, for example, finds itself short of methionine because the amino acid products inhibit threonine and a master inhibition occurs affecting methionine also. Thus one enzyme can be inhibited by several end products (8 in the case of *E. coli* glutamine synthetase). Product inhibits first enzyme by binding. The intermediates 1, 2, 3, and 4 could also be inhibited by the product.

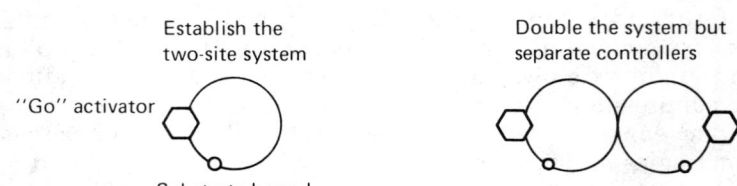

Fig. 4-10. The two-site system in allostery keeps the controllers out of each other's way, leading to the successive possibilities of binding, oscillation, and shape modulation.

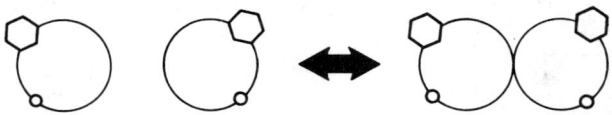

Fig. 4-11. Let the combined systems oscillate between the bound and unbound states.

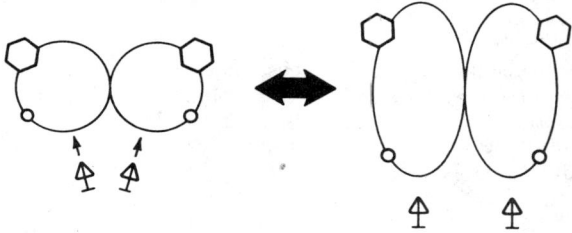

Fig. 4-12. Provide for conformational changes or deformation displacement as a modification with a new signal ↑.

The last figure includes deformation in controlling operations. Deformation may occur by breaking covalent intramolecular ester links and other protein connections. Since these give the molecule its conformation, the result is a relaxation or deforming which may "let the molecule down." Other deformations may involve:

1. Intermolecular covalent bonds in a polymer.
2. Peptide bonds.
3. Photolysis of disulfide bonds of cystine.
4. Chain breakage following destruction of amino acids.
5. Intermolecular crosslinks.
6. Optical rotation of space conformation of helix.
7. Smaller axial ratios and reduced viscosity as proteins or chains undergo scission as in irradiation.

Gamma and even ultraviolet can specifically affect the helix structure or the rate and extent of its formation. Excitations may follow radiation absorption with emissions from protein, as in skin, serving to relieve the situation and so, to protect.

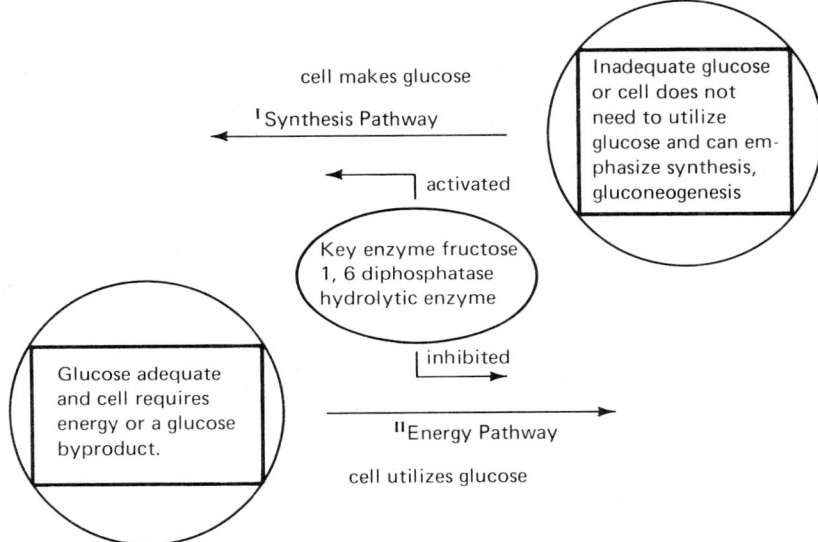

Fig. 4-13. Metabolic crisis management as shown by separated synthesis and degradation directions for glucose.
I Fructose-6-phosphate + ATP → fructose-1,6-diphosphate + ADP
II Fructose-1,6-diphosphate + water → fructose-6-phosphate + H_3PO_4
This is the sequence in the pathway that illustrates the two directions. Enzymes in the reactions are energized by ATP and AMP to make reactions go in right directions. Separate pathways permit their simultaneous operation.

In the all-important synthesis of glucose (Fig. 4-13), the dual flow of information is via inhibit-deinhibit action of the key enzyme, fructose 1,6 diphosphatase. The existing choice is with which pathway glucose is going to be associated. The switching depends on the requirement. The hydrolytic enzyme, fructose 1,6 diphosphatase is the switch and mass action controls it. If there is a "threshold or better" supply of glucose sugar already, then the enzyme is depressed and the switch toward synthesis is in the no-go position; the cell chooses pathway two. With less than the necessary glucose concentration present, the enzyme is deinhibited and pathway 1 is placed in operation. Thus there is a feedback from the glucose concentration. Thermodynamically, the two pathways are favored in their respective directions—the equilibrium constant in one direction is so large that it prohibits the reverse one.

The enzymes for each side favor opposed directions. The consequences include a *crisis management* capability. The metabolism can marshall cell reserves in a fast response to need, rather than having to wait for slow equilibrium to occur before setting up the reverse pathway for degrading fat or polysaccharides. Demands cannot occur, therefore, at an embarrassing moment when reserves are depleted as long as the feedback control is in effect.

In fatty acid and lipid metabolism, the pathways are between the triglycerides (Fig. 4-14) and their fatty acid side chains of even numbered carbon atoms. This system shows the use of two kinds of enzymes,

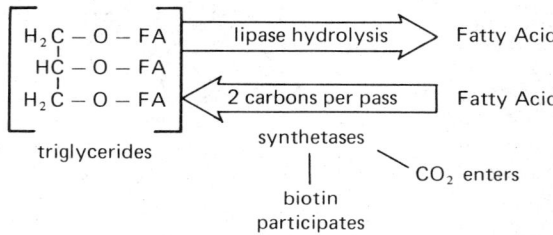

Fig. 4-14. Fatty acids and their dual pathways.

lipases and synthetases, to separate the pathways rather than using the cyclic activation of a single one via inhibiting factors. Many of these concepts may be seen in the Fig. 4-15 master plan for information and energy utilization in metabolism as applied to organisms in general.

When a kind of sugar appears at a membrane, a message is sent to DNA. The operations that follow result in effector enzymes but the response is based on information acquired in about 3×10^9 years of heredity. Membranes provide this external to internal transmission and it becomes more highly ordered with compartmentation in eucaryotes.

In evolutionary development it turns out that it is essential to increase informational macromolecules. This is an exception to isodirectional information and means that a loop is necessary so that requirements can change DNA when necessary in response to the needs of higher organisms, until in man, a central nervous system (CNS) involvement is made. The control of cell activity is referred to conclusions reached in the nervous system and exerted by communicated reflex action and hormones. The allostery shows a behavior in which enzyme inhibition

Fig. 4-15. Information flow with energy pathways in metabolism. Under the corresponding control are metabolic cycles that synthesize or degrade, cooperative interactions between coupled cycles which provide push plus pull effects with the aid of multienzyme systems and key metabolic intermediaries.

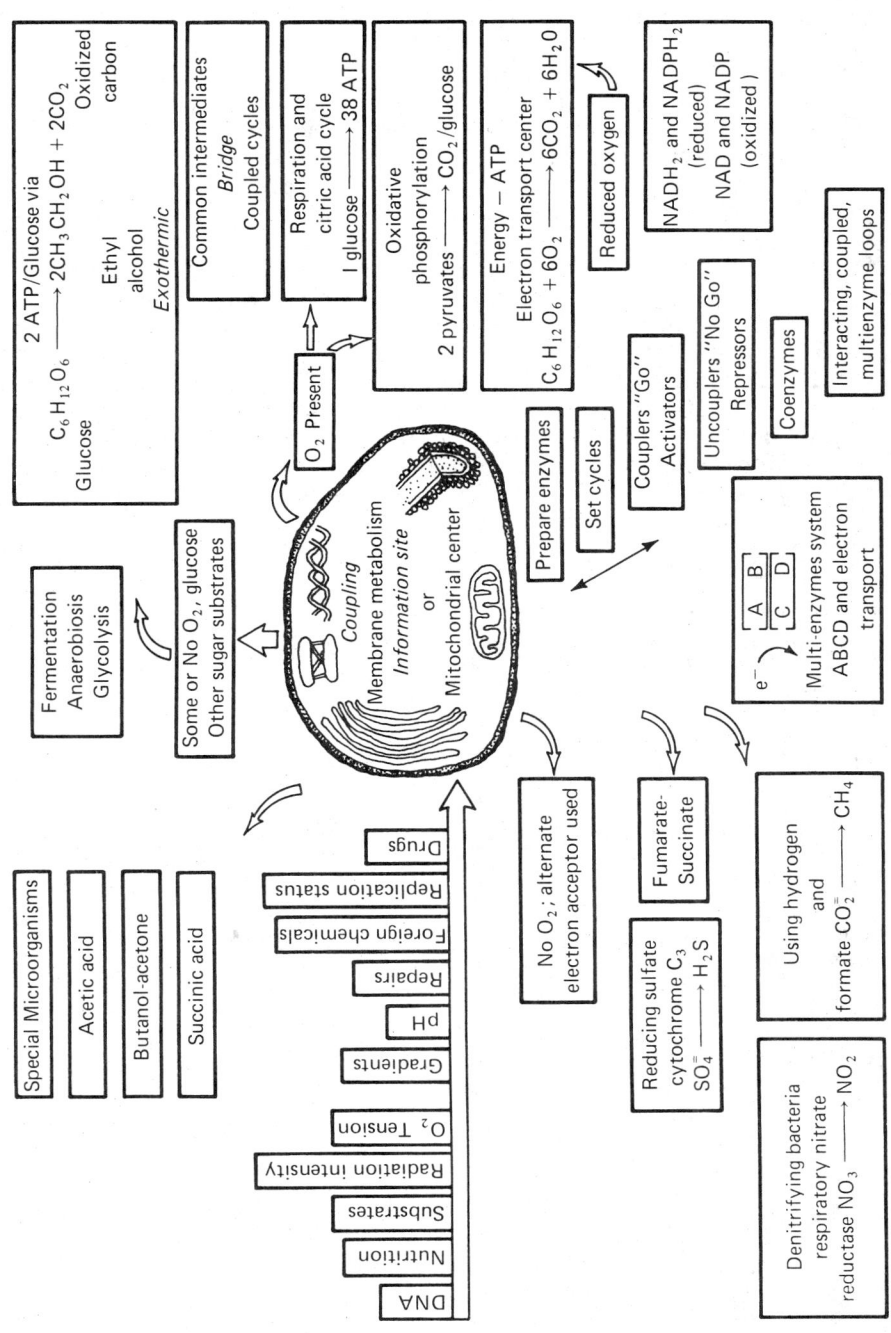

occurs by small molecules arranged to counter mutual interference (Fig. 4-7). Conformationally, *catalysis* is seen as a union of enzyme specificity information with form factor or shape signals. These ordering procedures combine with feedback, in the cases where the first enzyme is sensitive to the end product of the reaction, to dampen the production of the unwanted product that is giving a negative signal at an allosteric site. This is seen in the sequence in which threonine goes to isoleucine in bacterial cells and accumulated isoleucine can then exert a pressure on the enzyme which carries threonine to the first product, α-ketobutyrate. The plan is also observed in glycolysis in erythrocytes, where an excess of the product phosphoglycerate stops the phosphorylation of glucose. Similarly, aspartate transcarbamylase production is dampened by cytidine triphosphate in pyrimidine base synthesis. The enzyme needs two conformational sites to accommodate both its substrate, aspartate and its inhibitor cytidine triphosphate (CTP). These sites must communicate to achieve the regulation. It turns out that both of these requirements (2 sites and communication) are heat labile features (in biochemical analysis) of the enzyme's higher organization but the substrate aspartate can, by force, go into position when the enzyme is in the "go" status. The "go" status can be dampened by reinforcing the "no go" features of conformation and this is what CTP does (Fig. 4-16).

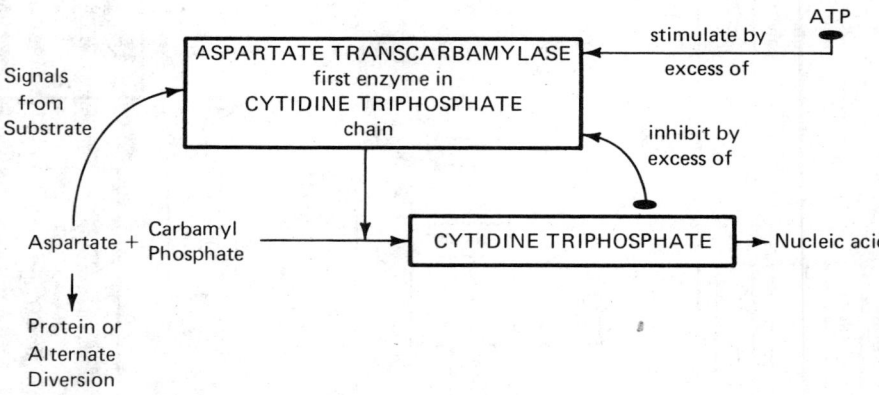

Fig. 4-16. Pyrimidine base synthesis and its stimulation by ATP, and inhibition by cytidine triphosphate in the regulation of aspartate transcarbamylase.

"Searching for the switch" is a valuable research concept in metabolic studies, and enzyme repression or induction is an outstanding means for control.

An isosteric control is seen when the inhibitor is sterically similar so that it can competitively antagonize the enzyme. Both allostery and isostery involve enzymatic control by the conformation of an inhibitor

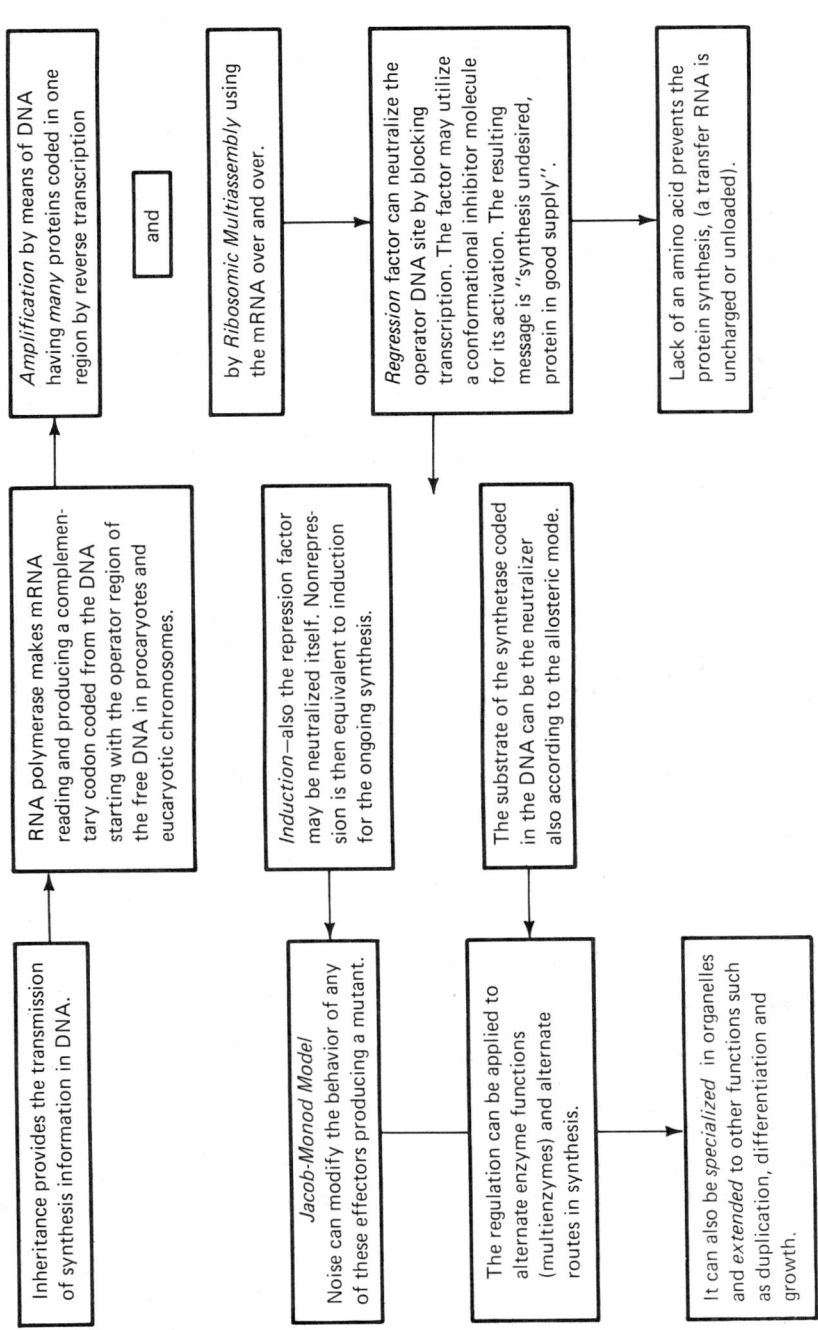

Fig. 4-17. Informational regulation in protein synthesis.

to an active site which is distinct from that reserved for the substrate, while simultaneously there is an associated displacement in the informational or ordered structure of the protein of the enzyme.

Induction or noninduction of enzymes would control products by the presence or lack of the enzyme itself and this control comes in response to the presence or absence of the products and substrates. If the enzyme is not used due to the good supply of its products, its synthesis is repressed. The appearance of its substrate or its energy source can cause the synthesis to occur (Figs. 4-17 and 4-18).

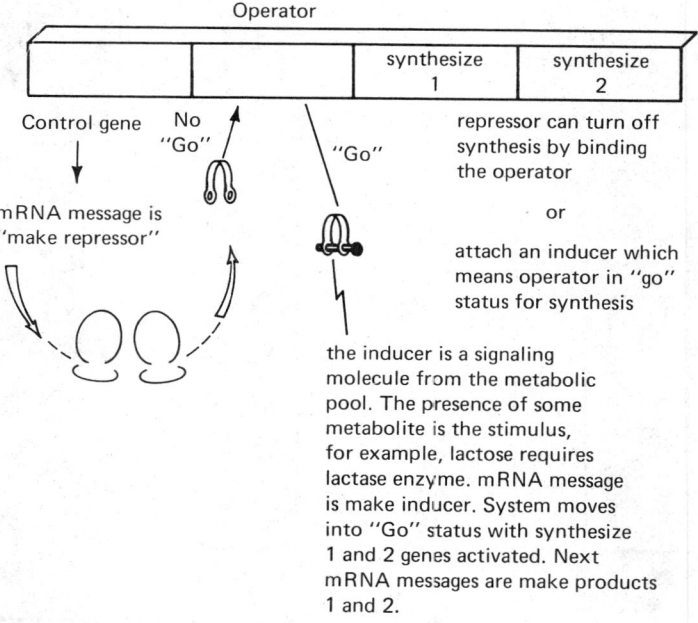

Fig. 4-18. Operator under control of repressor and inducer in synthesis balances waste activity in producing unnecessary enzymes as long as there is zero error. Competitive analogs can introduce noise— the system may then be immobilized.

An operon may have several genes which are expressed in the synthesis of a molecule whose intracellular concentration controls induction or repression. A regulatory gene may produce a substance that can combine with the synthesized molecule. The feedback is from that molecule with the cooperation of the regulatory gene substance back to the operon genes.

FEEDBACK INFLUENCE OF ATP, ADP, AND AMP, THE ADENYLATES

The role of ATP as an energy-rich molecule includes a subrole in the information flow, in the feedback plan, in repression and stimulation and control in general. The participation is extremely wide in all cases where energy is converted or coupled into metabolic cycles. Cycles behave in such a way as to support the oscillation of the adenylates between three forms. Both the presence alone as well as the concentrations of these compounds influence the enzyme catalysis in metabolism.

Energy management requires a response to the presence or absence of alternate substrates, carbohydrates, nitrogen, the need for synthesis of operating molecules, and the disposition of temporarily-in-excess material into "stores" status. Diversion of the intermediate, acetyl coenzyme A, away from the citric acid cycle occurs if there is enough energy at hand and little nitrogen for synthetic purposes and it is done by the repression of citrate synthase by ATP. Otherwise the synthase gates the intermediate into the cycle. When the adenylates are predominantly in one form, the others are scarce. Thus, if citrate accumulates, it can direct the key intermediate, acetyl-CoA, to the fat cycle. It does so by stimulating the joining of acetyl-CoA carboxylase and its substrate under conditions where AMP is in short supply, that is, where there is much ATP. Plentiful ATP gates acetyl-CoA away from the citrate cycle and the AMP shortage pushes it toward the fat cycle.

The withdrawal of the rich ATP energy deposits occur as this adenylate oscillates toward its lower energy forms to support synthetic, electromechanical, transport and (photo) chemical activities since the cycles which regenerate ATP support these activities using carbohydrate, fat, and protein fuels to do so. The intermediates, standing at the cycle interacting points, manage the intermovement of material and the adenylates by their oscillating intracellular concentrations, repress or encourage these energy dispositions. They act as enhancers and/or repressors on the key enzymes such as phosphofructokinase, isocitrate dehydrogenase and fructose diphosphate phosphatase. The high concentration of the triphosphate may stimulate one reaction direction at the same time as the lack of monophosphate is inhibiting the opposite reaction, thus achieving a push and a pull cooperation. Conversely, the monophosphate, by its presence, means that a low energy level for cell work exists (little ATP). Thus it activates energy sources (glycogen) to rebuild ATP from the acquired energy, whereas accumulated glucose-6-phosphate would mean *make* glycogen and store energy. Activation means to exert the effect via the preparation of the necessary enzyme.

DIRECT INFORMATION ROLE AGAIN FOR AMINO ACIDS

Information theory is useful in predicting the conformation of molecules. Molecules such as lysozyme and myoglobin are dissected by conformation analysis right down to the atom, its bonds, distances, and angles. The search is for conformational cues or clues regarding how a macromolecule slips into its functional conformation. The alpha helix and certain other protein structures apparently are self-folding rather than having design provided by the ribosome, since they fold separately from it. Nucleation would be the use of assisting structures such as the helical one and stable structures. For example, myoglobin would be built on the law of minimum conformational free energy with a given set of bond angles providing this desired structure. *In a direct utilization of the theory in which the bits present in specific amino acid words are programmed by computer to yield most probable conformation, those amino acids, especially near the helical part are found to carry the alpha helix message.* This is the immediate result of their evaluation in terms of probability via the source of information for the design. It can be shown in the same manner that the backbone of an amino acid interacts with its side chain in such a way as to *self-inform*, as the amino acid forms or deforms the helix. Side chain to side chain information gives the fine directional adjustment, and side chain to neighboring amino acid interactions establish the alpha helix firmly. These predictions are conveniently made by computer, trying various conformations out to see whether the helix arises as it is expected to, from various constituents, sequences, and angles for products such as lysozyme, ribonuclease, myoglobin, subtilisin, and papain.

In these studies, alternate arrangements of molecules contributing to stability in a macromolecule may be referred to as *conformers* and the potential energies associated with *syn* and *anti* conformers (referring to 2 possible orientations of a group about, for example, a glycosidic bond) are interpreted in conformational maps. Here equal energies for the molecules give "contour" lines, and illustrate regions of abrupt change in value or deepest minima for the energy leading to conformation and stability.

The rigidity in polynucleotides is due to the *anti* orientation of the *nucleotides* rather than to the *nucleosides*. The flexibility comes from P—O bonds and puckerings in ring structures are called the C_n, -endo or exo conformations referring to the carbon number in, for example, a C_3, -endo or C_2, -endo describing the conformation at an important carbon.

Chemoreceptors involve conformation in important information pathways. Conformation is seen in their shape and classification as specific for certain sensory messages, olfactory, taste, etc. The conformation determines what receptor the molecule will join. Stimulation then initiates a coded message through sensory nerves to the reception center.

THE IMMUNORESPONSE, IFF, IDENTIFICATION, FRIEND OR FOE?

Immunology probably began with the early smallpox epidemics in Europe which used to kill 20% of the population in their twenties and disfigure 10% more. Jenner, perhaps the first immunologist, observed how milkmaids, while never contracting the disease, nevertheless bore scars on their hands. Cowpox virus infection gave the milkmaids immunity to the immunologically similar smallpox virus. The pathogen in infection has a probability of success which may be expressed as follows:

$$P = \frac{\text{viral pathogen virulence} \times \text{size of invasion} \times \text{parasite and host nutrition}}{\text{genetic and exposure immunizing factors} \times \text{host resistance and defenses}}$$

Good nutrition gives the host protection while a good supply of nutrients for the pathogen gives it virulence.

Gamma globulins clearly show a protein synthesis that is based on information input from the ambient as these proteins constitute antibodies in the immunoresponse. Immunoglobulin M, IgM, comes first in both ontology and phylogeny so there is recapitulation, and after a challenge, IgM is synthesized first but is apparently preparatory for the IgG synthesis which comes next. The immunoresponse must encompass all possible challengers to the organism's defensive identification and security system. For economy, the responsibility for matching antibodies with *all* of these antigens is shared between somatic mutation and germline genetics. Thus the challenge can input information into all the reservoir of data represented by the higher levels of protein organization and the organism can therefore limit the sets of necessary protein constructions that have to be blueprinted in advance. Speed and flexibility of response would be represented by somatic mutation while allowing germline genetics to be far more parsimonious of genic information. This essential economy would come about if only a *limited number of responding cells* needs to mutate, since not *all* the cells would need or have the antibody capability.

The features of the immunoresponse include an interrogation, on challenge, which results in an antibody recognition of the challenging antigen, self-discrimination as friendly substance, and organization for defense. There is cooperative effort or amplification in which many antibody units are deployed at the right time. Another cooperative effort relates to the final disposition of the foreign body. The antibody action, precipitation and agglutination may be a prelude to ingestion of the antigen by phagocytosis. Antibodies serve as a surface engagement while the latter assures destruction. There is also a phase consciousness, in that antibodies not employed in the time-space situation of the challenge, will turn off automatically. Once again the information or defensive intelligence resides in molecular levels of organization, in the function of subordinate units in macromolecules, and in enzymes and their control systems.

For the antibody and disease resistance mechanisms it is a survival situation as far as the challenged organism is concerned. The invader can destroy the life of the host. Thus macrophages and phagocytes are equipped to engulf and digest intruders, with the assistance of antibodies. Complement reinforces this antibody assistance and adds its own lysis to the general defense. The challenge is the basic step that stimulates the synthesis of antibodies and it may be made by foreign protein and cells. It is the input from the ambient, which, during a long evolutionary existence, has come to involve an identification process to separate friend from foe. Interrogation essentially resides in a recognition capability at the active sites on antibodies. These bodies and sites have an incredible identification success in resolving incoming target information, being highly specific in response. *Yet as a system, they can mount an opposition to a very wide diversity of invading substances by cooperating among themselves to dispose of them.* The antibody may be specific for a particular target cell or substance but the antibody reaction, being cooperative, is much more general with defensive capability against thousands of targets. In carrying out its task, the antibody has a highly useful ability to conform around the stereochemistry it encounters. Thus recognition depends on structural uniqueness and exposure of reacting antigen sites. This exposure is a matter of atomic arrangement, conformation and reciprocal configuration in the reacting antibody. The "tuning fork" configuration of immunoglobulin G, IgG, has two active sites which can operate in such a way that the angle between tines can change, possibly giving a sort of mechanical arm effect. The significance of recognition is great because the organism must not destroy its own substance (tolerance or no self-rejection). Thus the process is arranged so that immunogenicity is proportional to the degree of identifiable strangeness. To distinguish friendly material from invaders, the antibody uses a process similar to that of protein translation from the DNA code. The antigen, or challenging substance which can elicit the antibody, is "read" by the defensive mechanism which then constructs the antibody according to this information. Typical antigens are macromolecules with all their special features for identification, and as invaders are likely to be very different, that is, a toxin, or a foreign body "not fitting in" or possibly threatening in nature. An organism's own macromolecules cannot call up this response unless the immunosystem has been damaged by agents such as radiation or radiomimetic chemicals or has not been fully developed or maintained nutritionally. Similar, or friendly-like molecules pose a greater problem and no doubt a serious hazard. Thus viruses can usurp the host's machinery and need not even carry their own, only their nucleic acid plans. However, friendly-like molecules, in general, may not be dangerous due to their similarity to the host's own molecules. The identification is the initial action and it depends on cooperative behavior between haptens, adjuvents, and determinants which are intermediates within the recognition and antibody elaboration system. Collectively these antigenic determinants modify the recognition conditions. Haptens, for example, modify the minimum weight limit for recognition of an antigen, being themselves a small molecule on a protein. Determinants and adjuvents are aids to antigenicity although not immunogenic themselves. Reaction also depends on the *intact* atomic features of the reactants. Intracellular antigenic features, external appendages, and walls make antigenic determinants as seen in the human blood groups, flagella, etc. If the determinants are damaged, there is error, or non-recognition. For example, the system may not recognize its own damaged macromolecules and therefore may take them for a "foe."

There are many determinants in an individual's red blood cells that produce a pattern in the sense of a fingerprint. When the host produces an antibody, it is equipped against a determinant regardless of the circumstances of the next meeting. Each bacterial cell carries a unique classifying group of determinants. However, since there is a limit on the total determinants, immunity can approach the full challenging capacity. Transplant rejection is a large scale operation of the immunoresponse which must be dampened by anti-recognition measures if the transplant is to succeed.

Radiation is used for immunosuppression in human organ transplants. In a kidney transplant, for example, about 600 rads are delivered in 4 doses on the first, third, fifth, and seventh days after transplant as a preferred way to avoid rejection. The prognosis is good, especially if the organ has come from a near relative. Even transplants from cadaver sources are usually successfully made. The recognition system can sometimes distinguish between the sexes. For example, a male with his distinctive Y chromosome accepts skin grafts from the female who has only the "his and hers" X chromosome. However, the female will reject a male skin transplantation.

Human blood groups are based on IFF (Identification, Friend or Foe) reactions. There are four types of blood, each of which will agglutinate (give antigens against) the other types (A, AB, B, and O). Type O produces anti A and anti B bodies and so transfusions must be from O type. Type AB produces neither, so transfusions can be from the blood of any of the four types. The other two accept their own or type O. That is, all types can take O but O-type takes *only* O. Rh positive blood produces a certain antigen factor Rh, Rh negative does not and Rh positive is a dominant germline character. It may be dangerous for the fetus if the child is positive and the mother negative. She may reject the baby when her antibodies enter the fetus through the placenta (erythroblastosis). The mother herself becomes immunized to the Rh factor in her fetus. Her antibodies (anti Rh agglutinins) cross the placenta and usually damage the fetus. Thus the mother, father and fetus should all be Rh compatible.

The immunoresponse begins early in fetal development as a preparation for entering a world full of hazards. As the process is learned, it replaces the no-response condition found in early embryos. The memory or recognition features are apparent in this development in that the process actually improves with development. The significant step is learning to identify "self" and to turn "on" only when the challenge and interrogation so indicate. Antibody recognition power is proportional to differentiation. More differentiation in development means that more foreign bodies can be identified as foes. Thus, there is a threshold in development *time* beyond which induced acceptance of grafts is impossible. Prior to that time, however, acceptance can be conditioned by injecting the foreign strain material. This is a step change from low to high discrimination. Much remains to be learned about the overall process and its development but the effort that has been focused on it is a triumph in itself and many outstanding scientists utilize methods such as those shown in Table 4-2.

татSLE 4-2. Methods in Immunology and Their Uses

Fluorescent Dye Labels
  DIRECT
  INDIRECT
	Explain immunoresponse
	Detect antibodies and identify
	cells under UV microscope

Radioactive Tracer
  Radioimmuno-
  electrophoresis
	Follow mechanisms and separate fractions

Serological Reactions
  Utilize the immuno-
  reactions themselves
  and their specificity
	Study recognition, identification
	Fixation of complement
	Sensitization
	Anaphylaxis
	Cross Reactions

Amino Acid Sequence Analysis	Structure, coding, active siting, and evolution
Antagonists, Analogous Compounds	Enzyme behavior of antibody and antigen, allergies in which the immunoresponse over-responds
Sedimentation Rates	Separation of parts of the responding system by weight differences
Enzyme Digestion	Identify active immuno fractions and locate active sites on the bodies
Transplants	Study limits of recognition, homo and hetero grafts
Population Surveys	Show presence or absence of antibodies, for example, against polio; racial differences, e.g., Semites with genetically incorporated resistance and Negroes with susceptibility to tuberculosis
Electron Microscope	Visualize structures, phagocytosis, observe lysis through loss of bacterial wall integrity
Electrophoresis	Order of mobility of immunoglobulins, seperate IgA, M, G, E, D
Heat Stability and Lability	Study functions separated out by heat treatment

Studies have shown that the system for antibodies is localized in the blood plasma cells, lymph nodes, spleen, thymus, Peyer's patches and bone marrow cells where antibodies or precursors are active. Within these, primary immunoresponsibility is cooperatively shared by:

1. **Macrophage.** These can ease lymphocyte responses by secreting activators and nonspecifically scavenging foreign cells which they identify by adsorbing B and T cell immunoglobulin.
2. **B system lymphocytes from bone marrow.** The B cell response is an immunoglobulin with information for identifying certain antigens—thus forming antibodies. Examples would include responses to somatic and flagella antigens in the humoral (lymphoid) system.
3. **T system lymphocytes from thymus.** Their response is to reject foreign tissue and cancer cells as a cellularly managed activity. For many antigens, these cooperate with B cells in antibody formation. They transport immuno-materials to the challenge site where they amplify in numbers and refine the identification and receptor process already underway with B cells.

Conformation is seen as the antigen couples to any matching surface. The lymphocyte expresses recognition by amplifying that feature which attracted the antigen. This means that it can rapidly transform itself into a clone of antibodies, remembering which immunoglobulin they are to secrete. IgM is an immunoglobulin or specific antibody class made by B cells which identifies and receives certain antigens.

The antibody reaction with its cooperative features can lead to neutralization, agglutination, precipitation of a foreign body or, finally, phagocytosis or liquidation of the cell. The complement fixation process is an immunoresponse from serum which leads to lysing's being superimposed on the antibody reaction, which itself can lead to the neutralization and clumping of foreign bodies. Finally there is phagocytosis or liquidation of invading cells. Complement does its work using informational protein, enzymes, and interacting sequences in a manner which is now familiar. The first reaction aids the second and so on until there is a sequential lysing of the foreign cells. The antibody-antigen engagement comes first and these units then complement the result (Fig. 4-19). In a sort of team action, the first complement prepares the scene for the second and so on through nine, these complements functioning as proenzymes, enzymes, binders, and time and place-sensitive organizing agents. The latter sensitivity functions in support of recognition, producing a switching on and off action for proliferation of the mechanism according to need. Once the system is fully tuned, the numbers amplification or cooperative effort provides for an escalation. The final coup d'etat is the lysing of invading cells via an effective attack of complement nine on the membrane or cell wall integrity.

The information burden is on the antibody and only the challenge is required by the antigen. The immunosystem presents an ensemble of defensive lymphocytes and the engagement depends on conforming surface features between antibody and antigen. The amplification

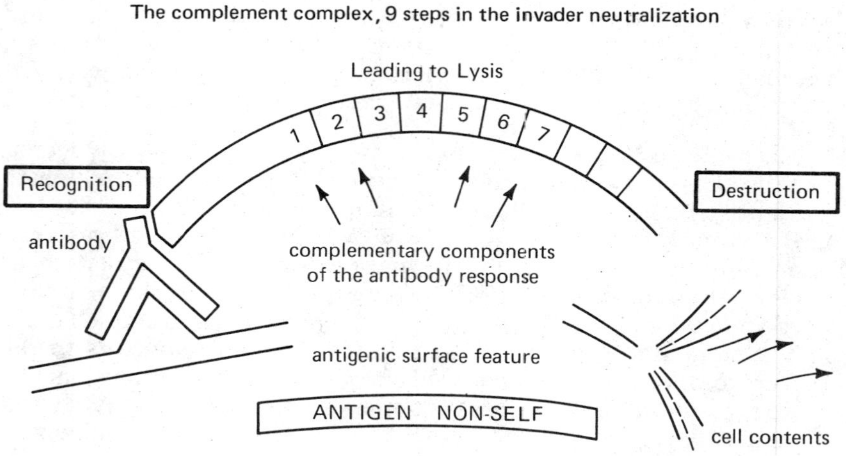

Fig. 4-19. The super enzyme system of complement—a system of coupled enzyme sequences.

assures adequate numbers for defense and is a mitogenesis initiated by antigen binding. Lymphocyte responses to EM radiations are striking and one of the first clinical observations to be made. Thus, microwaves are mitogenic for lymphocytes as discussed in Chapter 15.

INFORMATION MOLECULES IN EVOLUTION

Primordial cultures living in their ambient broth were characterized by a fortuitous grouping of amino acids and life components which probably still exist today but are now overseen by more evolved information structures. These are seen in unambiguous, well-informed, ordered, genetically controlled biochemistry, built around informational DNA. The primitive culture depends on inefficient excesses in both the quality and quantity of nutrients and precursors. As evolution moves up and pressures develop, the excesses cannot be afforded and the deletion of some of the precursors *imposes a greater demand in the way of information molecules to direct a more orderly utilization*. To reach the triplet codon-opocodon system with DNA and RNA involved with synthesis of proteins and, thus, the derivation of enzymes, it was likely that there was considerable searching, redirection, terminations of unproductive systems and endless modification in the way of biochemical evolution. An early defect, in the sense of efficiency, must have been that the coding was ambiguous. One codon probably designated several amino acids; so the primitive cultures no doubt suffered from error in protein synthesis. Once in a while, then, the culture must have hit upon a lucky new protein structure with useful features for operations in the primitive broth. The enzymes duplicating DNA are a case in point because any changes in the action of these enzymes would have mutation consequences as we know them. The existence of variations of this kind is the essence of mutation out of which forms better able to meet the ever increasing demands created in evolutionary processes could arise.

PRIMITIVE CODE VERSUS MODERN — EARLY CODE REQUIREMENTS DID NOT SPECIFY WHICH AMINO ACIDS ENTERED THE PROTEIN POLYMER — PRODUCTION OF THE CORRECT PROTEIN PRODUCT THUS REQUIRED SOME FORTUITOUS ELEMENTS AS COMPARED WITH HIGHLY REFINED CODES SUPPORTED BY ADEQUATE INFORMATIONAL MOLECULES

This degeneracy in coding is redundant but its information never gets through from RNA to protein. Today, nevertheless, only methionine and tryptophane lack degeneracy by having a single codon. The primitive code had redundancy and a high survival value which have persisted. The triplet code had many intrinsic advantages that supplanted doublet and other coding systems and became universal. The third letter involved in purine and pyrimidine pairs affects the two or three hydrogen bonds in the codon-anticodon complementarity and makes an advantage possible for one number of bonds over the other in reducing readout error. It was always a case of pitting one life system against another and the primitive system worked—in the primitive environment.

At some early moment, the primitive broth thus lacked the highly ordered, information-rich nucleic structures with which the current concepts of mutation are related. Information had to flow from protein toward DNA or toward the evolutionary precursors of DNA. This is the reverse of the Crick hypothesis, but is the only way more ordered structures could develop in succession.

$$\begin{array}{c}\text{Information Flow Forbidden} \\ \text{by the Crick Hypothesis}\end{array} = \begin{array}{c}\text{Central Dogma of} \\ \text{Molecular Biology}\end{array}$$

$$\text{protein} \not\to \text{protein}$$
$$\text{protein} \not\to \text{RNA}$$
$$\text{protein} \not\to \text{DNA}$$

Thus it would not be surprising to find extranuclear genetic information or cytoplasmic genes in full operation today since their advantages would have favored their retention. This cytoplasmic inheritance occurs in many ways such as in viral invasion, bacterial infection of the host cell, sperm action, translocations, in transforming DNA, in chloroplasts, and mitochondria. Thus, the offspring is a product of multigenetic sources.

DNA can be isolated from mitochondria, and lambda particles are symbiotic DNA in *Paramecium aurelia*. This symbiosis causes an independence from the main duplication to evolve so that the DNA can continue to be part of the symbiont operations while only needing to be sure that some of its units are included whenever the parent cell makes a cytoplasmic division. Alternatively, the organelles themselves can utilize their own cytoplasmic genetic apparatus to insure continuation. This may apply to mitochondria, chloroplasts and kinetosomes and may even occur in isolated organelles. The exact nature of the DNA involved is uncertain but mitochondria divide transversely by a splitting which is similar to bacterial fission and the DNA probably needs its own protein assembly system.

Lambda is a cytoplasmic informational substance that contains DNA, RNA, protein, polysaccharide and phospholipid. In *Paramecium*, its presence probably provides the folic acid essential in growth stages for the parent.

Enzymes certainly have informational features including recognition even though they are highly specific but those DNA polymerases which can cause mutations have high information content. It can be shown that if a polymerase synthesis gene suffers mutation, then a virus with which it may be associated will suffer further mutations (mutagenic mutations) or specific mutations in additional genes. The ultimate argument in this direction is that information can flow from protein to DNA, the reverse of the Crick hypothesis, or more accurately from *enzymes* to DNA. The evidence begins with results such as those with DNA polymerase. The question is whether the somatic changes follow Mendelian rules only or follow a more *plastic* course and are able to incorporate information during development. *Somatic mutation* adds flexibility and appears essential in instances such as the immunoresponse which is so dependent upon environmental information input.

The extragenetic influence is seen in exogenous growth-*rate control* information that is changed when organ cells are transplanted into tissue culture. There they divide about every day, while in the intact animal, they divide under their normal information input only every month or more.

COUNTERFLOWING INFORMATION AND BIOLOGY'S CENTRAL DOGMA

Crick (1970) warned against the counterflow of information and Yockey (1974) used Shannon's seventh theorem to support this hypothesis, but he calculated the information entropies without giving the degeneracy in the code its due (Ebringer, 1975). This is curious since he commented on the significance of degeneracy and on the real problem which is the channel capacity of the code in informational entropy units. *Information theory cannot be used to evaluate the central dogma* but such matters as channel capacity, matching of polypeptide sequences, calculating the bits for forming sequences, and the information content of genetic mutations can be examined in a most fascinating manner with the aid of the theory.

INFORMATION THEORY, TESTS AND BIOINFORMATION MANAGEMENT

Immunology with its recognition and operating features based on immunoglobulin phenotypes and questions of antibody diversity can be assessed in terms of the theory. Here the different theories for explaining the wide defensive scope of antibodies and transmissions of information about antigens can be examined in terms of bits per response. Probably an information entropy of 220 bits per immune response is obtained if antibodies come from a selected population according to an Instructive Theory, 46 by Germ Line Theory, and 23 by Random Somatic Mutation Theory. The signal source is an antigen which probably uses 30 bits per response. Then, according to Ashby's Law of Requisite Variety (from a Shannon theory), the random somatic mutation theory of antibody diversity cannot explain the phenomenon's being contrary to information theory as explained by Shannon and Ashby, (23 < 30) (Ebringer, 1975).

Very precise biochemical data are obtainable from information theory. Many proteins whose sequence and conformation are known such as papain, subtilisin, myoglobin, chymotrypsin, lysozyme, and ribonuclease can be tested for helix forming power and helix forming information. That is, the teachings of information theory are useful in evaluating helix-forming information in sequences of amino acid in protein chains when the information datum is available for each unit. These studies prove the critical importance of information in single amino acids for establishing helical regions. Then the studies show how conformation with neighbors constrains the operation of this information in cooperative action. Otherwise single residue information may predict a helical region by error. These predictions made by probability are straightforward and admirably helped by the computer. A first structure will elicit the information for the second (Robson and Pain 1971). Thus, less information is required since only denial information is needed following residues where the helix is not intended. This conserves capacity for other levels of structural complexity. Similarly, tests by the theory show that side chain-backbone interactions and interside chain ones, contribute the structural information rather than other stabilizations. Expression of the information is found in the energy of the bonds forming the structure (side chains to backbone) with predictions tending to be borne out by energy differences between helical and nonhelical structures. The information flows from first steps toward second ones, seemingly (to us) in contradiction with the denial of intersymbol influences in these same messages by Yockey (1974). Apparently the amino acids set up a loose helix which establishes a local, rich, informational milieu excluding some of the influence of the watery environment and utilizing hydrophobic-philic polarization. This, in turn, builds the information management system by exposing operating groups externally, at frequencies set by the turns. The structure is then steerable into subsequent conformations which can put these groups into action status elsewhere and can also create an interacting unit for other dynamic configurations including energy waves of known and complementary modulation.

IN AGING THE INFORMATION MANAGEMENT IS FAULTY

> "If however, there exists the possibility of indefinite prolongation, the termination of a life or even the refusal or neglect to prolong it involves a moral decision of the doctors. What will then become of the traditional prestige of the medical profession as priests of the battle against death and as ministers of mercy? . . . What if every patient comes to regard every doctor, not only as his savior but his ultimate executioner?"
>
> Norbert Wiener

From the standpoint of carrying out a *program* of maturation and decline, the limited life span of the organism is simply a result of the program. However, the evidence is that the usual limits are not universal and so there is no single program for aging. Instead there is a whole set of circumstances involving biological information management, socio- and psychobioeffects, regeneration, repair, synthesis, and information errors. In tissue cultures of cell strains, it can be shown that age is a matter of how many unused divisions a cell retains at a given stage of time. Thus fetal lung cells in culture will live during a lifetime of fifty divisions and sometimes more, but usually the decline is within more or less ten doublings of the cell around this "age" of fifty. Old age in such tissue culture is a matter of excessive generation time, interruption of mitotic activity, and accumulation of cellular debris which all lead to failure of a "passage" or subculture.

Most information estimates of the value of a developed individual are ballpark ones and are enormous. The tentative limits are between 10^5 and 10^{12} bits for the descriptive essentials in a cell or individual. This information is derived from essential atomic orientations, molecular structure, chromosome volume and the genotype catalog by Quastler (1953).

In practice the adult can decrease these requirements which strain credibility by absorbing information from his nurturing process. Twinning is a monozygotic process for identical twins so morphogenetic information has to include redundance in the germ cell and the real differences among adults at levels below the whole organismic one are proof of the extrazygotic contributions. At Princeton, Murray Eden used the teachings of information theory to study probabilities in a morphogenetic model. Surprisingly little information in the germ cell is needed to project the overall form of the organism after the passage of a certain number of divisions.

In terms of cell renewal, about 10^{11} red blood cells are made every day for a total lifetime generation of 10^{16} cells. An even more active germinative tissue is that of the epidermis, which is replaced every four to eight days or about 5,000 times in ten years. The burden on information management is to allow enough divisions in a lifetime to supply all the cells needed. The best way would be for the body to have an infinite capability until death stops the action due to causes other than a lack of replacement cells. In the absence of such a capability, some limit such as fifty divisions must be recognized. It would be possible to operate within these constraints of life by keeping some of the progenitor stem cells in reserve when the daughter cells divide and differentiate to replace the cells that are lost. The management then involves decisions as to which cells to specialize (with the prospect of extinction) and which to hold in reserve, as well as how to distribute the stem cells where needed, ready to be activated in division or differentiation. The reserve concept, even in the scope of fifty divisions, could then serve replacement needs.

If the culture is taken from an older source, the number of divisions seen will relate to the age of the source material and will steadily decrease with the increased age of the source but *all* cells need not die according to a program. Immortality is a possibility, at least in tissue culture of cell lines. Erlich ascites tumor cells are maintained and grown in incubators all over the world today as a standard laboratory strain and the HeLa cells originated from Helen Lane's cervical cancer in 1951.

Cell strains which resemble normal somatic tissue could become immortal by changing to cell lines as in tumors or by virus transformation (growing a sister cell line from a cell strain). A correlation between "longevity" in cultured cells and longevity in the source animal suggests that given the metabolism provided in cultures, humans could attain a life-span of 110 years. More importantly, spanning as opposed to immortality is due to factors carried by transplanted cells and tissues from the host, something like a going away "gift" of information on previous history, metabolic requisites, and environmental relations. Tumor cells and tissues presumably do not get such a gift and continue their wayward ways independent of the habits of the host source.

In budding organisms such as yeast, the parental cell can continue a program of new daughter cell generation only as long as there is a virgin region on its surface which has not already been used for budding and does not show a scar. This program lasts for some 20 to 25 generations in new cells. This limitation is avoided in bacteria by compartmentation of the whole parent cell into daughters by binary fission. Fungi that grow by hyphal extension or only from the tips of the filaments show aging in that the oldest cytoplasm is found farthest back from the growing tip. All these microbes however, can thus count their divisions because the total fits into some area or volume. Other cells such as dividing spermatozoa produce a certain size or bundle which limits the capacity. Management then involves a mechanical "belongingness" which would self-limit cells in the wrong place or, like a benign tumor, would direct itself to control abnormal growth.

THE CENTENARIANS

Some people live to an age that is well on the way to the *second* hundred mark. It is only recently that the annual convention for U.S. Civil War Veterans or Veterans of the War Between the States finally lacked survivors to attend and, in the U. S. today, there are three centenarians per 100,000 people. Reports of centenarians come from Abkhazia in the Caucasus in Eastern Europe and they are apparently surviving due to dietary habits and to the chemical makeup of the soil in the region. In Vilcabamba, Ecuador some nine citizens over one hundred years old (120 to 150) live usefully in a beautiful tropical valley on modest diets and notable intake of unrefined rum. Their baptismal records are quite acceptable and the records for others show that, in the past, other life spans have been as great. An exception to this good documentation is the Hunza in Pakistani Kashmir, where the scene is one of very remote, high valley and cliffside life with vigorous oldsters whose ages are far less certain (due to language and record problems) than is their good condition.

The people involved in these places are alert until they die whereas this isn't always true elsewhere. They die from diseases introduced from outside and from accidents. They seem to prefer to drink from their river than take water from other sources and eat standard country meals of grain, bean and potato soup, about one ounce of meat per week and plenty of vegetables. Most of the long span types seem to have come from country places rather than from cities. Smoking is quite a common practice. In fact in Ecuador, they carefully produce their own tobacco. Many oldsters attribute their longevity to a devoted spouse, but studies tend to implicate good lifetime habits and to suggest the possiblity of an oldster personality rather than an inherited tendency. Elsewhere, this personality may involve enjoyment of life but with accents on the subtleties ordinarily contributed by higher education and tranquility. Moderation is a key, as is a curiosity and persistence about what things lie ahead. In dietary habits, regularity seems to be a factor with high protein, low calorie selections in early years contributing favorably. Eggs do not seem to be shunned and neither do milk or alcoholic beverages in the later years. In the U. S. life expectancy is now 69.3 years for males and 77.1 for females even as experts predict spans of from 200 to 400 years. In the whole world one might estimate some 5000 or so centenarians, mainly in the Caucasus.

CHANNEL DEATH

It is tempting to conclude an informational reason for long life spans or at least one involving management of biological information. In particular, the hypothesis would be that these persons enjoy a net *zero error synthesis*. If one looks at the continual renewal process that is a way of life for so much of the tissue structure, it is apparent what effect error can have. With constant turnover, errors can cause the accumulation of error products and so accelerate decline. Behind this result is an accumulation of errors in protein enzyme synthesis. Thus aging would appear as a steadily worsening defect level in this synthesis. As the informational content which provides for specificity, communication, recognition, and catalytic behavior is reduced by damage, errors accumulate and aging results.

Thus the really senior citizens would seem to have a low cumulative error, either from lack of such damage or from a high *redundancy* in the information source. This would be specifically a tendency toward repetition in the DNA genetic structure to offset losses and errors. The decline in information management can then be stated in fairly specific terms. There is a known loss of DNA in brain cells, which are nondividing, which makes aging appear as information attrition.

Attrition in protein forming information means that errors tend to increase as noise from sources like radiations, thermal input, chemicals and analogs enters the system. Below some minimum of protein information, the consequences to the living system are lethal. On the way to this crisis, the slope of decline would pass mathematically from a simple linear effect to an exponential decay. The change may be related to channel capacity which, it will be recalled, is the information valve in the connections that communicate information. In some region this will become less than that necessary for adequate protein synthesis. Either too much noise or too much competition for the channel could produce the fatal crisis or "channel" death. Certain noise input has been studied carefully and is often related to radiations which are mutagenic or to radiomimetic chemicals but this noise input may also be due to biochemical "mischief" such as that caused by analogs, competitive inhibition and antagonism, particularly among the amino acid analogs. Exactly what would be involved in conserving channel capacity is unclear but redundance is protective and *the channel must include membrane integrity* of genetic information. Not all information for protein synthesis is genetic; therefore the nongenetic or extranuclear sources could be enhanced. One of the most understandable of these that will be developed here is the biopsychological concept of extending the morphogenetic and developing period.

According to the attrition view, the whole structure of renewable systems as well as nonrenewable or nonregenerating ones, is gradually weakened as DNA information is lost, and the cumulative error increases. Good management would mean that the systems that hardly divide, such as muscle, would have a high information security, guaranteeing correct metabolic synthesis and repair while those with dividing cells would have a high redundancy in the processing of genetic information, protecting DNA from the accumulation of renewal errors by repetitions and other measures. Repair is counteracting but is under management control itself. A high repair capability would mean a high threshold for the damage to be permanent. In other words the threshold for permanence of the injury or loss of DNA information is at the maximum of the repair capacity. Thus for longevity, a durable repair process is desired. The eye lens crystalline fibers store defects as a function of age and these, plus some increasing inflexibility, working against accommodation, produce senile vision problems. Mitochondrial changes appear late in the aging process. The membrane integrity is critical if inadvertent mixtures are not to be permitted (enzyme release). It is known that other membranes enclose enzymes capable of catalyzing cell self-destruction. Recognition failure may also play a role by favoring body rejection of its own substance.

The ability to process or metabolize sugar is known to decline with age along with other hormonally controlled processes. Thus diabetes may occur as an old age disease. Similarly, in the older person, additional vitamin C is needed due to failures in its utilization and subsequent problems in fatty acid metabolism.

BEST CELLS SURVIVE

On the opposite tack from cumulative error via constant renewal, there is the favorable outcome of the survival of the best cells. That is, replaceable cells will tend to be made up of the best selected survivors of mutation and genetic damage. Thus, the weakest tissues would be those having the nonregenerating cells, the nerves and muscles, and protective measures for them must exist in individuals enjoying long lifespans. Typical damage to macromolecules due to physical agents, and possibly to aging, are crosslinking or gel formation and scission of chain type molecules. If the chain degrades by scission, it may rejoin or repair the broken ends in an improper manner. From an information aspect, the repair may produce a hair pin loop (clover leaf structure) or a sliding fault on one of the chains as in the case of DNA and RNA. Where previously the bases were properly paired on each side of the double strand, it is easily seen how one side could get out of register, or out of phase in a gene region, if slippage, bunching or looping up happened. This would be like a loop of film meeting an obstruction in a projector and bunching up before the point of blockage. These loops have been observed to be stable reregistrations of the code frame but in the loop itself, bases lack opposites or are unpaired. The result is haphazard repair of DNA in such regions after damage from any cause and eventual deletions of the faulty genetic substance. This causes a loss of DNA in the cells of the tissue which may be attributed to aging and other changes. The loss of DNA may be specific so that some essential function such as mRNA synthesis or transcription is effectively prevented.

ERROR CATASTROPHE

The cumulative error in enzyme production and functional form may have some frequency-of-expression rate. If this becomes truly excessive it may produce an *error catastrophic situation* in which most enzymes are defective and synthesis is chaotic. The mutations involved would certainly be lethal and the causes would include synthesis errors in altered enzymes and faulty proteins in the nucleus and cytoplasm, both of which would again seem to be involved in a genetic process. In aging via these events, DNA polymerases may be the central misfunctioning entities.

The error catastrophe hypothesis leading to eventual loss of critical information is more specific and less respectful of DNA repair capability than an "aging hypothesis" based on information mismanagement. Management is quite naturally connected with the endocrine system's exerting control over immune reactions and neurotransmitters, and information from the environment in which the organism finds itself should logically influence the hormonal secretions. Figure 4-20 shows a group of endocrine and interrelated systems which are directly related to aging in that if they remain healthy, apparently so does the individual. The thymus gland is involved in resistance to disease as well as in growth. It is largest when growth is most active and then atrophies with age. It receives material from the bone marrow system in connection with the recognition or rejection of foreign bodies and it delivers cellular substance to the active areas of the body for such operations. As the thymus organ shrinks in aging, fewer of these substances are produced and the individual has accordingly less resistance. The thyroid functions in active resistance and cardiovascular health, and the struggle against aging appears to settle on how to maintain this health and resistance as well as on mastery of diseases by learning all about them. A leading matter that is involved seems to be brain electrochemistry and in this, the neurotransmitters have a central role. These connect neurons chemically and include norepinephrine, dopamine, and serotonin in particular as well as acetylcholine, glycine, aspartic acid, and glutamic acid.

With respect to resistance, thymus, and bone marrow cells the key seems to be effectors versus the simple action of an antibody. T cells can be accumulated so that production is less necessary with age. Therefore, less efficient thymus glands need not be as serious. T and B cells are primed memory cells and cooperate to make effector in response to antigenic challenge. B cells in turn produce antibody generating plasma cells. T cells make effector cells for cellular reactions and helper cells for blood responses. Memory means that cells have learned to respond and therefore are more efficient on the next occasion due to the immune information system. Lymphocytes form the large antibody

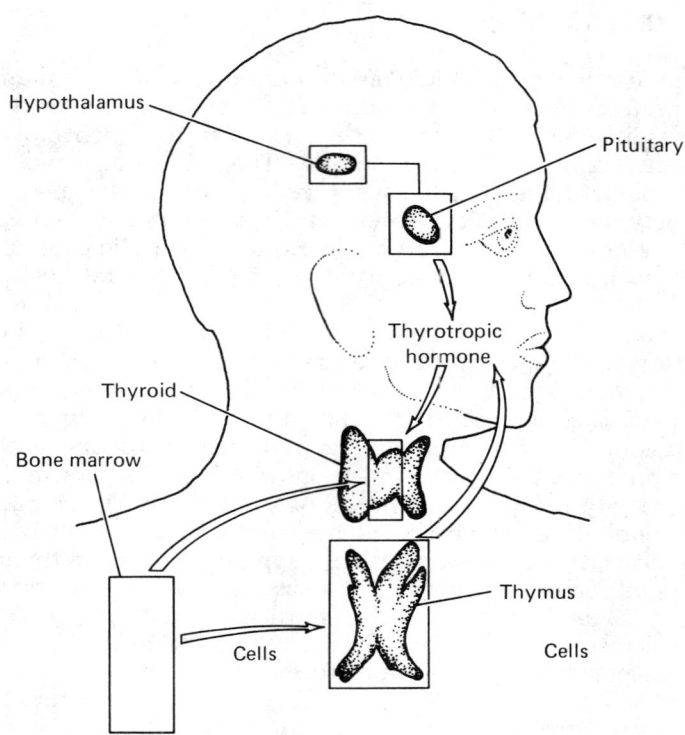

Fig. 4-20. Glands and systems interrelated in maintaining health and resistance versus aging.

group which functions in hypersensitivity and rejection. Autorejection of self is a real danger because the recognition information seems to decrease with age, leading to the possibility of a gradually amplified rejection mechanism, incompatability, or intolerance which could produce all sorts of information-related reactions like malignancy, age diseases, rheumatoids, amyloidosis or just plain chaos. Hashimoto's thyroidism in which antibodies work against thyroglobulin and hemolytic anemia in which serum antibodies attack RBC are other possibilities.

In this event, the older organism would be more sensitive to:

1. Changes, especially injury. This is exemplified by an accidental meeting of enzyme substrates which would include immunoresponses of normally compartmented elements;
2. Disease, in which infection challenges immunosystems with transformed or transduced host cells that now look like foreign cells;
3. Mutation, and

Atomic Informational Biology

4. If B or T cells transform into foreign cells themselves, they would not be guided by learned information that originally taught them which cells were "own."

A progenitor spleen donor cell that is antibody forming can be propagated in tissue culture. The antibody forming ability is immunization to an antigen. The culture can then be examined during its lifetime which depends on its personal history and is finite in span. The clone that develops from this antibody secreting cell line may, after about seven transfers, lose the antibody producing ability apparently as a matter of aging. After about 300 hours and seven transfers, the antibodies decline to practically zero probably as a result of having fewer effective antibody cells due to age (Fig. 4-21).

From studies in Parkinson's disease, in which a deficiency of L-dopamine precipitates aging, it has been shown how this substance stimulates the thalamus and affects motor responses. It also decreases in the hypothalamus and depletion affects pituitary function, so that all told, some million neurons come under the influence of dopamine, and

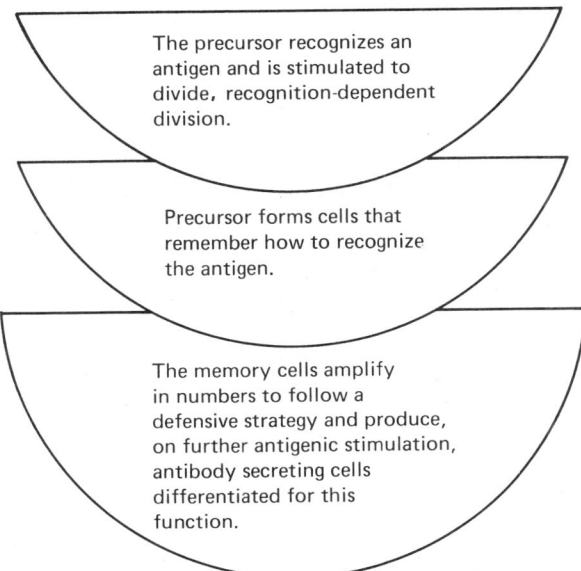

Fig. 4-21. The process "proliferate then differentiate" is recognition and memory dependent. If the cells differentiate enough, they cannot proliferate as a matter of age.

its role in precipitation of aging suggests it is some sort of pacemaker. In this sense, aging begins to look like a brain-controlled pattern of endocrine changes. This pattern also responds to further control by feedback with the associated possibilities of instabilities caused by imbalances in the essential substances.

There are some dramatic examples of acute imbalance. Pituitary ACTH kills a species of Australian mouse after copulation in a sudden death mechanism. Pacific Northwest salmon drive up their native streams, then succumb in two weeks after spawning by the same overdose mechanism.

As man peaks in development at about age 19, it is thyroid hormonal control that accounts for his psychomotor reactions and maximum strength. An "aging pituitary" hormone at the other end, changes the permeability of the thyroid membrane creating a problem for the thyroid hormone supply. This can be demonstrated by a pituitary extract (bovine) which decreases oxygen consumption and blocks the thyroid, thus inducing age. Yet in some way, the complex information system involved in the immunoresponse is sensitive to time as is the cardiovascular system. Associated phenomena under endocrine control also provide visible effects as in the transparent look of aging skin, due to the more fragile collagen fibers, with subsequent loss of elasticity due to crosslinking of these fibers (somewhat like the tanning process). There is a distinct sex difference in this biophysical effect. In terms of skin condition, a woman of thirty is said to be equal to a man of eighty.

SENESCENCE AND AGING

At the menopause, a woman's reproductive system shows senescence as does the thymus gland in young adults, so biologically, *senescence* and *aging* are different matters. The longevity of an *individual* and the average life span of *populations* are determined by the rate of aging and certainly by environmental stresses. *Molecular* systems such as enzymes undergo a form of aging ending with death but this is not seen as a decline. The life of a *single cell* may end with its division so that its life span is sometimes a matter of minutes. However, the cytoplasm transferred to the daughter cell may establish a kind of immortality through the generations. Aging is also *clonal* where the whole line of cells undergoes the process rather than an individual cell.

More defective enzymes may turn up in the aging organism so that if these could be identified, the diet could be especially adapted to correct for the deficiency. This would take the form of enrichment so that the products of the enzyme's action would appear in the diet just as they would have to be in a nutritional mutant. One mutant of *Neurospora crassa* involves abnormal leucyl tRNA synthetase. In this case it mismanages an amino acid insertion in the peptide chain so that many proteins are defective. Tests show that the error spreads as the information machinery is damaged. DNA is inaccurately replicated, leading to mutation and genetic senescence, or senescence due to the action of a gene. This sort of catastrophe or point mutation is clearly visible in sickle cell anemia where a congenital error is caused in the insertion of a single amino acid during the transcription for hemoglobin. It is obvious that this disease could be called an informational one. It results in sickled red blood cells which can become lodged in capillary junctions. This leads to deprivation of oxygen in multiple locations or massive tissue strokes. It is the snowball effect of the faulty management that can be serious. When older cells contain defective protein, rapid degradation of the protein can be expected, but if the errors build up within the protein synthesis system, the consequences can be widespread. Thus the snowball effect causes proliferation of defects, with errors in the master molecule DNA being most serious. It is not the error in protein synthesis that causes senescence, but information management in general, and the analysis has two main thrusts: a) control of the damage; and b) programmed senescence involving all the elements of (a) but with a direction.

EFFECT OF FREE RADICALS

The damage may appear in nuclear mutations and, therefore, in DNA or in mitochondrial DNA. Free radicals which are always present with ionizing radiations, but which also occur as metabolic products, can menace membrane integrity. When this damage results in forbidden unions between enzymes and substrates, the damage could be suicidal. The free radicals can cause cross-linking in nucleoproteins and macromolecules. Cross-linked DNA and protein become comformational variants with their normal functions inhibited. If helices turn and flex to "steer," any stereochemical inhibition would be misdirecting. A dramatic form of error expression would be in nonrecognition of the organism's own tissue, which results in the rejection of cells. Coding operation errors in protein synthesis may produce errors in metabolic system enzymes which, in turn, lead to faulty transportation, communication, secretion, and energy transduction, making the organism death-prone.

SCREENING OUT NOISE

Exactly how cells that seem free of error can purge themselves of defects in the cell strain is unknown. Thus germ cells must have some way of selecting *one* out of four sister cells; the others die out, so only one meiotic product actually becomes an egg. This selection can then be refined by asymmetrical division that tends to purify the stem cell strain while tangentially spinning off the faults in the differentiated cells. Meiosis, with its crossing over stage, is a suitable process for shuffling genes in such a way as to develop weaknesses which then destroy the cells with defective information. Certainly, the germ line is immortal and thus there is ample time for purging the strain even if each organism has a finite lifetime. The process is aided by any means which penalizes those cells having errors. This requires a monitoring information molecule that can be activated when something goes wrong so that further DNA readout can be inhibited. The sooner the defective cells self-destruct, the better for the culture, since their influence would then extend for only a few operations. Rapid screening thus protects the survivors and can even insure their immortality. The expression of the genetic information necessary for debugging the cell's program can be an analysis and filtration of error at the level of molecules. This process would work against damage to DNA from free radicals, cross links and chain scission. Defective protein is discriminated against in *E. coli* which can recognize missing or erroneous peptide chain links as well as ribosomal or tRNA products with mutant defects. DNA damaged by UV undergoes enzymatic or photo repair which corrects errors such as the formation of dimers between thymine bases and the mechanics of radiation repairs are wavelength specific.

In this sense older tissues have a greater antinoise burden than do young cells. In the genetic disease *Xeroderma pigmentosum*, lack of a squelch mechanism against the expression of faulty DNA information is lethal, since the skin becomes hypersensitive to solar radiation. However, in separate culture, the cells do not die earlier. The biosquelch process appears very durable and it is obviously more necessary in older persons and in those activities in which errors tend to accumulate.

CODE REDUNDANCE AND DEGENERACY

There is considerable evidence in favor of programmed aging. Redundance suggests it, as does the difference in longevity in cells which may vary from minutes to years. Degeneracy in the genetic code is evident since several codons usually code for one amino acid. The instructions a cell may receive for the synthesis of its proteins may have gap problems or information missing. For example, the cell may lack some instrument for effecting the operation, such as tRNA with a specific codon. This degeneracy would function as a biological squelch forcing some other codon to deliver meaningful information to bridge the gap. What is in operation here is a control loop for information itself, and the loop feeds on degeneracy. Then, as the options are used up, senescence occurs due to the fault in the decoder. The more demands made on a system, as in active growth, the more attrition there can be in protein adequacy. This situation suggests that a slow rate of growth affords better chances for deprogramming.

Redundancy is seen in the fact that genes exist in multiple copies. The genes that code for rRNA are repeated something like five to ten times and a similar code is found for the genes that code for tRNA in bacteria. In *Drosophila* it is 100 to 130 times while in vertebrates the redundance may be from 250 to 600 times. In humans the redundancy is greater than in animals like cows and mice. It could be concluded that redundant DNA information is most probably the problem kind in aging. The loss of information via damage to DNA is more serious in renewing tissue while in nonrenewing tissue the limiting factor is the accumulation of membrane damage. Another conclusion that may be made is that deprogramming the aging process is similar to promoting the sustaining program and, in this, there is a parting of the ways between cells and organisms. This is because the sustaining program must be the emphasis for the cellular level while for organisms, the purging of defective, but renewable strains, can mean individual survival.

THE RIDDLE OF DEATH

Evolution should favor longevity after all the effort expended in ontogeny. If individuals reproduced at age 60 instead of at 20, longevity genes would be selected, but this is not the case, so that genes that harm in senescence can accumulate. The result is that defective genes which contribute their effect at a late age are favored. Selection favors strength at the age of reproduction even if this results in a reduction in longevity. The mutant may have both of these double effects, which are then amplified. Thus, the greater the fertility, the greater the senility. In this case one would expect a generalized break-up rather than a progressive decline in tissues and organs. In terms of life support and limited resources on the evolutionary scale, short life spans would possibly spread the resources better through a population giving it an edge, while short generation times favor the fast expression of evolutionary pressure. Again the advantage is against populations with longevity. It would be tempting to conclude as far as longevity is concerned, that what is good for the individual is bad for the population except that, in the long run, the population is composed of individuals and what is good for them is good for the population.

While these comments on senescence are oriented toward the noise information aspect, there are many other information factors involved. Particularly important are the effects of the individual's interaction with the social ambient in which he is aging. Probably the old Eskimo practice of putting their aged "out on the ice" to conserve their short supplies of food and materials is an interaction likely to accelerate all the other decline factors for biopsychological reasons. There is also undoubtedly some contribution toward longevity from inheritance and certainly from nutrition. However, the information factors and the delayed action physiological insults point out the importance of reducing the accumulation of defects if the full life span is to be experienced. The centenarians probably have a set of inherited and acquired factors, some no doubt psychological, which have the overall effect of deprogramming the aging process. Exceptions are those like the Vilcabamban woman who would not care to repeat the distinctly difficult 95 years behind her. Thus it is inevitable that it is an individual matter. It is the quality of life that counts in the overall motivation.

In general it seems to be true that the smaller the mammal, the smaller its lifespan and the greater its metabolic rate. Man seems wasteful, using three times more lifetime energy than other animals. Based on body weight his span should be about thirty years. As far as cephalization is concerned, the larger the brain size, the greater the extension of the life span so that this factor seems to compensate for body weight via superior metabolic control. Bats (*Chiroptera*) live twenty years or three times the expected life span based on brain size, probably because of their habits of hibernation and diurnal torpor. Evolution, which inevitably favors rich information centers, should favor the information processing afforded by the large brain, but it takes longer to give birth to large-brained animals and therefore fewer numbers result.

Marion's tortoise *Testudo sumeiri* can live to 152 years and the Carolina box tortoise *Terrapene carolina* lives to 123 years or more. Fish may be considered immortal, at least in the sense that many can outlive man, e.g. the sturgeon. This is to say that they do not die naturally and, during their lives, they may maintain their growth mechanisms. If these functions persist, then any problems from leftover growth mechanisms that might encourage senility are avoided. Nevertheless, even in fish, the forcing function of mortality increases with age so that many simply die while still growing up or suffer too many accidents. Some coelenterates have spans of 35 years and bivalves may live for 70 to 80 years. In protozoa the existence of clonal age is a distinct possibility. Amoeba in continuous growth seem immortal although on restricted media their life span becomes finite.

Apparently the best advice for a child is to select long-lived parents, especially the mother. Mothers who have children at older ages produce offspring with shorter lives. This tendency seems to be reversible across the generations in some species, causing a shift toward younger mothers. Inbreeding has a depressing effect in that it may reduce longevity since the inbred organism has almost the same genes. In this case, the hybrid tends to represent vigor. Unmarried persons live less time than married ones and the average life of widowed and divorced persons is less than for unmarrieds, so that widowing and divorcing introduce social effects.

Man has fewer X chromosomes since he has the X-Y sex determining pair, versus double X for the female, but it seems unlikely that longevity could be logically connected with the X chromosome.

IONIZING RADIATIONS

Low doses of ionizing radiations provide no clear evidence of damage until suddenly the span of life is found to be shorter. *Drosophila melanogaster*, after being exposed to 80 kilorads at an age of ten days, has 50% of the normal longevity and 40 kilorads shorten the span to 50 as opposed to 70 days for the controls, in terms of percent survivors

(Lamb 1977). However, radiation exposure does not necessarily shorten life, although there may be more malignant disease. Even chromosome aberrations and mutations will not always reduce longevity. Polyploidy as in diploids, tetraploids etc., or the double X karyotype plus polyploidy will not insure a normal span. The early conclusions about radiologists having a shorter span were troubled by the problems of very large doses of radiation when the dangers were uncertain, inadequate data on the actual doses received, and the lack of control groups. In the Marshall Islands incident, when 175 rads of gamma radiation were received by the islanders who had failed to leave, the exposure has not given aging evidence, e.g., accelerated greying of hair, more tightening of the skin or differences in blood pressure. The sample is rather small however, and the tests insensitive. The massive studies of Hiroshima and Nagasaki are similarly hampered by biased populations, since most young people were away from these cities in the armed services and the men who stayed were often unfit for duty. While there have been more cases of leukemia and other malignant disease, there is no clear evidence of an aging effect. It may be because of our extensive studies of ionizing radiations that they are considered one of the most positive life shortening agents known. There are many resemblances to the natural process, but the mechanisms are far less certain.

Vitamin E

Free radical damage to unsaturated fatty acids, as in membranes, is linked to aging through the accumulation of membrane faults which may arise from free radicals of external (radiation) sources or internal sources such as metabolism. It has been suggested that vitamin E may aid in preventing this type of damage. In any case, membrane damage can be invoked to account for accumulation of lipofuscin pigment. This color in cells precisely chronicles age in animals. The enzyme release hypothesis, where damaged membranes permit mingling of normally separated catalysts and substrates is another membrane effect. A vitamin E deficiency and hyperoxia both increase lipofuscin and accentuate free radical damage. Vitamin E may be indicated in later life where degeneration is occurring.

PSYCHOLOGICAL FACTORS

There is no doubt that psychological and sociological factors play a strong role in determining the life span. There is, for example, the scene where Miguel Carpio, a Vilcabamban who is 123 years young, can no longer see women well enough to identify the sex, but who now uses the much more certain "Braille" method. There is a strong motivational effect, but its importance varies among individuals. The biophysical and informational aspects are extremely important. Death may strike if senility means uselessness, but this may be simply a transient mental attitude. There is almost the suggestion that such thoughts are an invitation to self-destruction which, if misunderstood, could lead to consequences in immune system recognition errors and rejection of self, cross links in collagen, blood vessel stiffness, etc. There may be an aging center acting as a programmed pacemaker like the ovaries which age to menopause with a built-in counting system, or there may be many such centers. Longevity often looks like a game in which it is essential to conserve the ability to relax any homeostatic mechanism after a stress, in order to guard the precious recovery systems. Alternatively, it would be necessary to avoid the challenge wherever possible. Game strategy would suggest that systems may be preserved and protected by adapting, or taking advantage of, any physiological change that is on your side, such as the wisdom and understanding that comes with maturity and better immunorecognition. Similarly, concurrent stress situations might be anticipated so that they are never allowed to couple and produce physiological instabilities. One example might be in thermal regulation. If a person reacts more quickly or changes faster when exposed to a temperature stress, and recovers more slowly than the norm, it can mean double trouble.

Sustained activity, often expressed in useful work, improves the cardio-pulmonary oxygen supply in the heart muscle so that myocardial infarction is possibly less a disaster to the cardiac muscle. There seems to be a definite place for stimulation, especially in relation to sex and alcohol, taken regularly and in moderation, respectively. Marriage is very important if there is a good mate. Diet is a factor that is usually possible to control.

NUTRITION

The U.S. is probably in transition in terms of dietary habits, stature, and activities, but compared with the little Shangri-Las of longevity, the caloric intake in the U.S. is consistently higher, with an actual average of 3300 for all ages, and a recommended level of 2400 calories for persons above 55 against 1800 calories per day typically in the Shangri-Las. In those places the diet composition varies greatly although usually consisting of about 1700 to 1900 calories, coming from milk, vegetables, meat and fruits. The carbohydrates are usually from bread while low-fat cheeses are also important in the diet. Coffee seems to be well tolerated.

The restriction of food intake means a smaller stature and a minimum food intake is a stress which is also felt at the cellular level. Chalones are enhanced by hormones leading to mitotic delays. Concentration of these substances sets the mitotic program or, at least, its time scale, therefore determining cellular longevity according to one theory. Nevertheless, a restriction of food to about one-half the usual amount suggests an avenue for combining cellular and organismic level control of the aging process.

FOUNTAIN OF YOUTH

Deprogramming should probably exert some influence on pacemakers or zeitgebers which probably include the nonrenewable tissue, including nervous tissues. There is a loss of 25% of the brain cells from 10 to 100 years of age. If the biology of fish is considered, then it is again attractive to try and deprogram by extending the growth period. This is an effort to convince the organism that it is still developing and to inform the pacemaker that the time scale is not approaching that "finished growth" setting. This procedure combines logically with the restricted food intake designed to maintain the weight so that the growth program is extended in time. Maximum diet would be a reserve process for getting out of trouble, establishing a good operating status, or carrying out regeneration or repair. Research on the biology of aging has involved some good work with animals but the psychological factor appears as one that is both critical and impossible to manage with animals. Thus, it is not merely an extended growth program but the understanding that there *is* a growth and improvement process underway that is too much to expect of animals. In deprogramming, it would be logical to expect to live to one hundred rather than to program death at sixty or eighty. The Shangri-Las define youth as extending to eighty or ninety and connect it with sex activity, so that most have large families of four to six children. It seems to be important to include work in the pattern of life, to be independent and free, and to keep things enjoyable, light, and not overly serious. Mental tranquility without tension and life in the right location would appear to be significant.

The Shangri-La types who live to 140 or 150 could be explained if city life pushes children toward maturity too rapidly. Exposure to the media, especially TV and attitudes in school tend to force development like plants in a greenhouse. If the children could stay young in mind and the environment selected accordingly, then development would be extended. To deprogram in this manner requires an effort to set a slow development in the children by letting them rest and recuperate with a great deal of school-free time. Thus if they are taught to be young it may set a standard that will help against aging also. For an adult, deprogramming would involve doing what children do, which is play.

It is obvious that aging is not the simple summation of the effects of disease, injury and the residue of wear and tear, but that longevity is a matter of attaining and maintaining functional maturity which means to stay as a bud, because to become a flower means the end is in sight.

COGNITION PROCESSES AND MEMBRANES

The recognition-dependent division seen in Fig. 4-19 is only one instance of identification followed by a communicated command. The transferring biomolecules such as hemoglobin recognize their addressees and deposit cargo such as oxygen. The transferin molecule delivers its charge of iron in the correct heme manufacturing location. The active and passive transport of solutes via membranes requires that the cell molecular locus or action center "read" the concentration and act on the measurement information. The active transport system responds to mass action on the receiving side of the membrane. When the concentration level reaches the threshold, the "no pass" signal closes off the active mechanism. The passive transport system depends only on the difference, or gradient, in concentration however, and will continue to pass solutes through the membrane as long as a gradient exists. In the active system, information management requires that an entering substance be identified via membrane recognition. Other solutes are shunted aside for slower or non-transport. Thus the membrane barrier behavior of closely related molecules should reveal those chemical features which constitute molecular information. This is seen to be the case with molecules that are related in that they are in competition for "pass." There is a resulting nutritional significance in this behavior at the biochemical level. Too much L-alanine on one side of a cell membrane prevents the acquisition of the closely related L-valine and vice versa. Thus, the presence of one group may inhibit the passage of another related group even though both groups are essential compounds for the well-being of the cell.

The cell membrane is also functional in presenting its active sites to antibodies, viruses, as well as to the determinants of cell division. *Membrane variables include permeability, pore size, hydrophobic and hydrophilic behavior, chemical constitution, shape or configuration, appendages, packing sensitivity, cohesiveness, electrical charge, capacitance, dielectric "constant," and pressure.* The regulation of cell division is conveniently localized on the membrane. Binding sites for serum factors exist although these active sites may remain repressed during the resting stage. Unblocked active sites, reconfiguration of membrane protein plus proteolytic enzyme action signal resumption of division which leads to the complex sequences of cellular reaction involved.

Viral identification of its host depends on the mutual recognition between the viral connector mechanism and the host cell attachment site. In bacteria this site may be membrane protein or lipopolysaccharide which couples to the viral tail fiber connectors. In order to infect the host and adopt its machinery, the virus must first go through this recognition and coupling process. If this process is lacking, the virus must mutate to acquire identification. Polio virus recognizes a lipoprotein site on the cell membrane for attachment while the flu virus recognizes a red blood and mucosal cell membrane glycoprotein. The specificity of the virus for its host results from these recognition sequences. Uncoupling mechanisms, if they were known, could play a role in the development of badly needed antibiotics against viral disease.

The molecular forces involved in recognition will vary with the units' conforming and coupling. Hydrogen bonds controlling the complementary bases function in the codon-anticodon recognition. A repressor recognizes either its starting point for transcription or the operator as a sequence of nucleotides. Highly developed organisms demonstrate various ranks of recognition. Differentiation and developmental information is in the nuclear DNA, especially after gastrulation, and when this is passed on to differentiated cells, the effect is to elaborate a specific array of required enzymes. Probably about 250 or more are needed. About 1,000 different kinds of cells with these arrays will constitute the higher organism. Differentiation will require that only those genes contributing to the specialization will go unrepressed and be allowed to produce the enzymes. Thus the inducers must, by molecular recognition, select the genes or cistron units of genes that they are going to turn on. To do so, they must have a polynucleotide prosthetic group capable of the essential recognition of the gene molecules.

The operon has two zones of about the same size which are the acceptor and structural information regions. The synthesis of one or more proteins is planned out in the cistrons of the structural zone. The acceptor region is really an identification site, recognizing protein enzymes and repressors, and accepting blockage as a "no go" status. The promoter region is close by the acceptor zone in the operon. One operon has an array of acceptors so that it is able to receive information from among factors such as hormones, suppressors, inductors, and other signals. A single regulator macromolecule can simultaneously signal several operons if it wants to couple them and they may have varied actions producing a complex response, such as a complex synthesis in differentiation.

INFORMATION FEATURES IN MORPHOGENESIS

The large scale system of the fully differentiated and developed organism is a complex of smaller systems. These in turn involve processes of synthesis, degradation, energy utilization, and response. The individual systems are described in terms of the states in which they exist and the timing and phasing of their processes. Systems can change their status under time rules which take the status of other contributing systems into account, i.e. they are synchronized in development. The regulation of these processes, their initiation, termination, and synchronization all require communication between cells and tissues and a number of determining actions which are described in this section.

At an early embryonic stage, the supply of nutrients, amino acids, fats, and carbohydrates, will dictate the rate of growth and the rate itself is manipulated to differentiate the organism. This is in the sense of differential growth which is fundamental in the forming of parts and structures. After the birth transition, the composition of the external nutrient supply becomes critical, since it is not independent of the maternal supply. Maternal milk is designed to satisfy the development demands and maintain the optimum rate of growth. The typical commands to cells would be to attach or connect, as embryonic implantation, induction of protein synthesis (supplies permitting), growth to some state, and division. A coupling of time and other signal loops is essential, as evident in commands such as "grow until a certain size," or "divide to a certain quantity," because when the objective is reached, the organ will have been finished. The time for completion must be phased into the transitions planned into the organism so that only necessary constraints are imposed on metabolism. The degree of coupling will always depend upon the urgencies involved and the disturbance signals that are processed. The coupled systems that support these operations will react to input signals and search about the set point with an oscillating frequency related to the degree of coupling. The coupling of cycles produces a nonspecific regulation by itself, because the united loops gain interdependence. The couples function with key intermediaries such as aspartic acid, ATP and the nitrogen supply in metabolism. This interlocking control, as compared with more passive behavior such as mass action, contributes response speed and selectivity. These are based on enzyme control; more, less, suppress, go, no go, commands and switching nets all happening within process time constants of a few minutes.

In developing ribosomes in bacteria, these assembly sites fasten to the developing mRNA, matching their growth to the RNA polymerase action. Here the transcription and translation are coupled. The mRNA has not finished its work until the last translation in a ribosome. Thus a signal to degrade it would involve both its presence in the ribosome and the completion of its translation or a blocking action on these steps. It may be commanded to degrade or to stabilize according to its situation. In the event that it is to degrade, this process is coupled to nuclease action along the chain and also to separation from the protective effect of ribosome association. There seems to be no conflict between *simultaneous cell growth and function*, and the more general conclusion is in favor of coupled growth (for example, more DNA, functioning mRNA and ribosomes) and function. The very growth of many cell lines with their retained functions belies this conflict. The reason for the conflict being said to exist is probably related to the different nutrient requirements for growth and for function. This is especially true of vitamins and hormones. The existence of switching nets and a series of commands is suggested by the illustration of cell division in bacteria (Fig. 4-22).

GENOME EXPRESSION AND THE COUNT

The teaching of the genome in organism development is expressed in the accumulation of structural units up to a certain specified number. The genome is the group of all genes in a haploid set of chromosomes.

Atomic Informational Biology

Cell Division in Bacteria

1. Cell membrane message is "initiate division".
2. Form duplicate chromosome thread.
3. Attach chromosomal material.

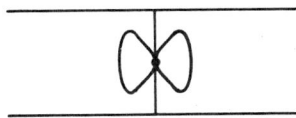

4. Prepare to execute division on signal *from* division regulator gene *to* duplicator gene.
5. Weaken, then thicken wall at point of separation.
6. Grow centrally if a coccus or first distally if a rod.

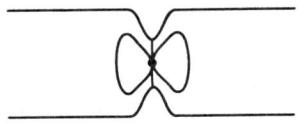

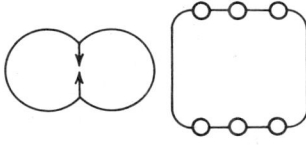

Elongation

7. Divide wall and separate.

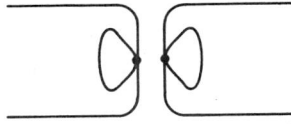

Fig. 4-22. Cell division in bacteria.

Thus all the genes in the cell give the genome. How does the organization measure progress and recognize completed products, terminations, and changes in direction? This is one set of questions. About 300 amino acids finish a protein, for example, and its end is signaled with a terminator codon. The tally along the way is based upon a one-to-one program from the DNA. The result can be consistent only with the rates of catalyzed and highly coordinated processes. For instance, in the 50 minute process time for one bacterial division in minimal media, some 2,000 different genes may be read. It takes about one half this time for many microorganisms to divide during their logarithmic growth phase. A molecule of mRNA or protein is synthesized in one to sixty seconds.

Then the count must be maintained on the whole set of simultaneous reading actions which go along with the linear correspondence observed for the DNA codons with the translated protein product. The resulting picture is of groups of genes being read and transcribed at the same time. The synthetases are revealed as high-speed matching or coupling agents playing a *key information role* along with their well-known specificity and permanence through the catalyzed reactions.

The expanded dimensions for the role of the enzyme come from a realization that it has recognition and verification functions which are not relinquished to be a function of chemical bonds of the template. Then, with its own information, the enzyme both *orders* and *executes* its special operation. The reactions catalyzed by synthetases, with translations by tRNA, demonstrate this determining role with high precision.

Fairly tight coupling is evident for bacterial cell development and the timing of organelles and other parts. For example, the cell will have one fiber of DNA or its chromosome, possibly 1,000 kinds of protein in various concentrations, a related number of mRNA types, about 60 kinds of tRNA, plus ribosomal RNA, mucopeptides, and some types of polysaccharides. The coordination of their production is designed to meet the demands on schedule without shortage or excess. Penicillin exploits this situation because it will antibiotically inhibit both the mucopeptide and cell wall construction. Other polymers will then be in excess and the cell will lyse. When a waste product accumulates due to the failure of its degradation process, the whole operation goes into biostasis or the result may even be bacteriocidal. The main difference in timing for multicellular forms is the demand for even more coordination.

Process rates may be repressor-controlled, both for on and off status but questions arise as to how this influence is exerted and how a characteristic rate can be assigned to some organismic reaction. The success with protein synthesis studies would seem to warrant additional *conceptualization* along these lines. The idea of a program with binary control can be presented. It would permit either a bistable or multistable synthesis, by assigning go, i.e. *with* enzymes, or no go information in the genome for any number of desired rate levels demanded by the organism. Then derepression can be stepwise, within a range of rates, and can be responsive to the concentrations or to other influences from the repressor.

CONCENTRATION GRADIENT INFORMATION

Concentration gradients, by their very existence, can exert directive influence over compartmental differentiation or over certain molecular behavior by assuming an understanding of the gradient by the units. Compartmentation is used in a broad sense to categorize groupings within meaningful boundaries. In the region of compartments being differentiated, or of an organ in development, molecular or higher units may concentrate along a single or multiple gradient toward or away

Atomic Informational Biology

from the organ. A vectoral distribution is contemplated with length, mass, and possibly time, dimensions.

Special meaning may also be assigned to solutions in the vicinity of their *critical* concentrations. Beyond these meaningful levels of concentration, a specific instability will occur. In the polymerization of macromolecules, for example, a rate process or monomeric reaction may ensue which will be time dependent. Then, in the dynamic field of differentiating structures, with boundaries established, the sudden polymerization may be regarded as a *disturbance* with spatial features. Remembering the interest in accounting techniques, and that units must be distributed according to the specified pattern, it is attractive to evaluate the utility of the concentration information. Without implying a quantum mechanical explanation, it may be shown that quantizing is an obvious possibility. This is because the development proceeds in a series of discrete steps, usually with an observable symmetry. Thus, no third arms or half toes are acceptable but exactly how the count is made is obscure.

It was shown how incident light waves become chemical and membrane waves in *Rhodospirillum rubrum* and that muscles show self-oscillating waves and hormone to cyclic AMP signals are propagated pulses.

PERIODIC PATTERNS AND WAVE NOTATION

Turing, of automata fame, taught the feasibility of a process associated with periodic patterns and wave notation. Possibilities then exist for the application of well-known wave characteristics. Taking Turing's plan somewhat pragmatically, a count for repeated periodic structures such as segments, vertebrae, ridges, and layers could be visualized. The system accumulates and distributes components according to specified rates, and at a discrete value. An instability is created such that a "last straw" reaction produces a disturbance that passes through the medium in the manner of a wave pattern, as in radially spreading ripples on water. The model may lend itself to wavelength processes, for example, concentric maxima related to the concentration of a substance may be formed. Each maximum then may generate a focal point for a local spurt of synthetic activity to initiate the part involved. There is also a wave number, or waves per unit of distance. If these are possible, a unit count as well as the quantity and distribution may be suggested for periodic structures like symmetrical features. An associated wave behavior is valuable also for other informational contributions including frequency, amplitude, and phase relations.

RADIAL GROWTH AND POLARIZATION

The radial basis gives a set of growth directions, some of which may be growth suppressed while others are preferred. In the limit of this type of selection, one preferred directional or axial growth may arise. The result is a polarized growth pattern (heads and tails) which is clearly in evidence as a natural guide to positioning in many forms. Other directions may not have the necessary *receptivity* to the wave disturbance in the media. They may be suppressed by the emerging patterns, interregional influences, susceptability to preferred directions or the presence of repressors. In any event there is a binary choice between repression or nonrepression superimposed on the radial pattern to provide preferred radii of development in morphogenesis.

In order to return to another state (the most important one being the nongrowth state), it may be theorized that this state is weighted, or dominant. This would be a binary no-growth state, functioning to limit the change as necessary, and it would prevail whenever an activation of the growth state is lacking. At times, the reverse type of control would be desired in which the action *goes*, unless suppressed.

Amplitude-dependent reactions, when they exist, would add significance, proceeding only when and where (e.g. at maxima) the oscillation or other process establishes the necessary concentration. The wave frequency can also be related to enzyme bursts of activity to account for amplitude maxima for a process which is spatially distributed.

A molecular expression of amplification is found in the action of the enzyme reverse transcriptase which is an RNA-dependent DNA polymerase. This amplification is essential if the synthetic machinery of the cell wants to prepare for an imminent demand such as that arising in rapid metamorphosis, differential growth, a probable fertilization, or any step change. One procedure in preparation would be to multiply ribosomal RNA genes for greatly increased production of protein. The message must come from DNA so that it must amplify its transcription capacity many times. It does this via the RNA intermediate (Fig. 4-23) with the help of the reverse transcriptase enzyme. With this mode, transcriptase polymerizes RNA at a rate of about thirty-three bases per second.

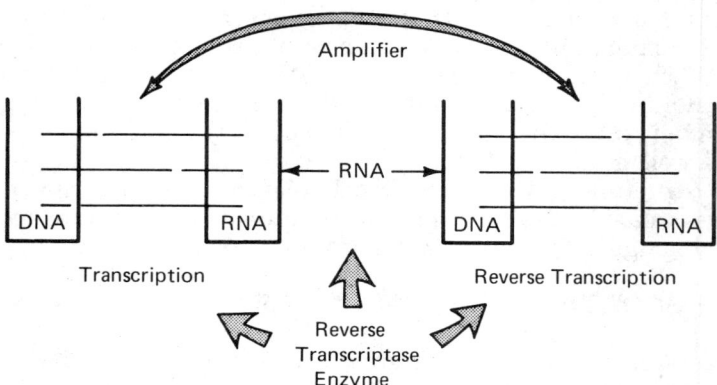

Fig. 4-23. DNA to RNA to DNA, an amplification at the informational molecular level.

Phase relations for multiple waveforms can carry interwave information. Phase, being the difference in time when the maximum amplitude passes for one wave with reference to another, establishes timing. These are mechanisms by which set controls and biological clocks can express themselves through concentration gradients, catalysis, and propagated disturbances. Their organization can be ordered by communication of information from its storers and carriers.

Biology provides structures, notably the regenerating ones, in which these conceptualizations may be subjected to experiment. The periodic growth patterns seen in circular, spherical, helical, and segmental forms allow development to be examined. These patterns often reestablish their numerical integrity and order after removal of a unit, and in so doing, can demonstrate how repair is controlled and the information in memory utilized (prediction effects). Memory is related to electrical (possibly oscillating) activity which affects coding proteins, neuron membrane states, and functional amines which control transformations in the nervous tissues of the brain and elsewhere. The propagation of the beating of cardiac cells over to new cells which take up the rhythm has been demonstrated in vitro, clearly showing the wave mechanism for information transfer. As John von Neumann pointed out in his last work before bone cancer took his life in 1957, nerve impulses are well known electrical disturbances which occur in pulses of about 50 millivolts with a duration of about a millisecond before dying out. However, this is an apparent propagation which actually involves ionic shifts in position on the membrane. It is a propagation of, and by, an ionic disturbance, so it is electrochemical.

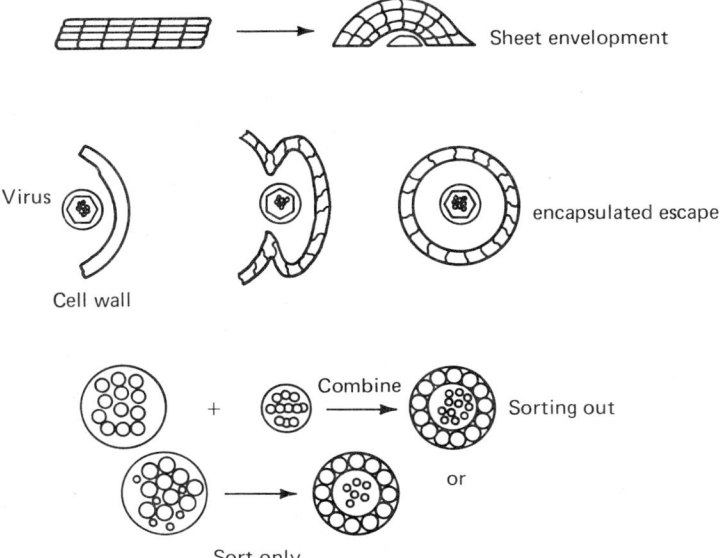

Fig. 4-24. Some migratory operations in cellular differentiation.

CHEMOTACTIC MORPHOGENESIS

The study of *Dictyostelium discoideum*, the slime mold amoeba shows how this organism can react chemotactically to the ambiental information. Should the supply of nutrients become low enough, an intercellular signal causes the organism to clump and start differentiation. Occasionally they enter a cycle of growth and multiplication with the formation of a plasmodium, and a mature upright structure. They can move a little and the stalk is differentiated with the elaboration of cellulose fibers (Fig. 4-25). The change is associated with the production of acrasin or c-AMP by an occasional secreting cell which initiates the chemotactic response. Signals from informing cells are amplified as more and more secreting cells join the swarm. Togetherness brings a unity of information and an organization. Then cellulose is elaborated to give rigidity to the stalk and spore form for reproduction. A simultaneous phototaxis leads the slug toward the open light where chance winds and animals can distribute the spores.

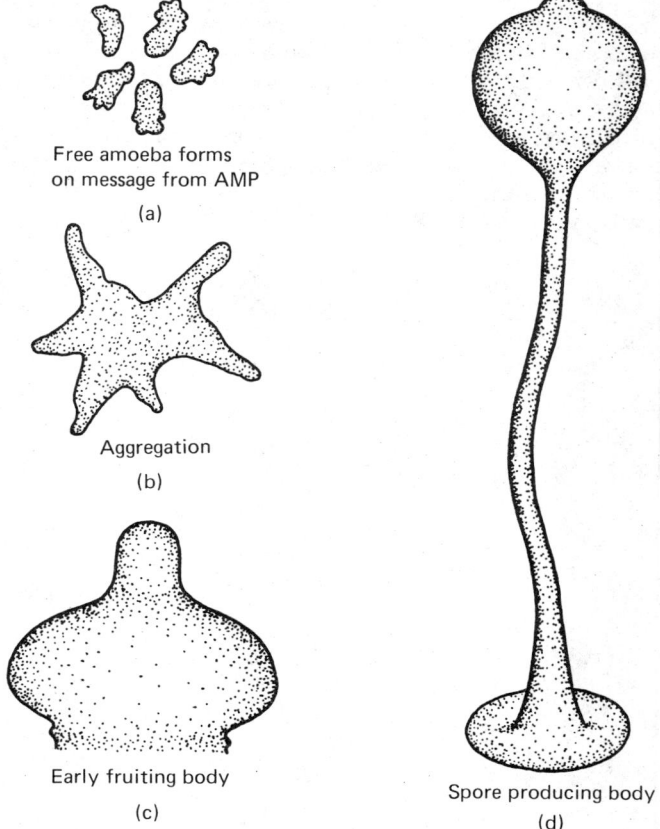

Free amoeba forms on message from AMP
(a)

Aggregation
(b)

Early fruiting body
(c)

Spore producing body
(d)

Fig. 4-25. Differentiation of amoeba.

THE FACTOR OF MOBILITY IN DIFFERENTIATION

The separation of cells and organisms by mobility is probably the most fundamental division into types that can be made. The importance of mobility is apparent in disease situations such as metastatic cancer or invasive tumors. Given cellular mobility, the developing organism can select a "move" or "stay" status, a partial motion under constraints to make special organizations, or a directed motion under intersystem influence as in phototaxis. Similarly, partial motion can be coordinated in planes and geometrical configurations to produce migrations of groups of cells, as in sheets, or a mode of exit for viruses from cells (Fig. 4-24). One of the available constraints is cohesion, which can be applied to one end and removed from another to create relative movement.

When two kinds of embryonic tissue in culture, such as liver and limb bud cells, adhere together, the liver cells will surround the bud cells or the tissues will sort out assuming the spherical, most comfortable, (with maximum adhesiveness) equilibrium status possible. The sphere will have the least surface area exposed and the strongest internal adhesion, thus providing at least these two construction features. To cause the limb bud cells to stay central, they should have the greater mutual attraction, which they do. Putting the less adhesive ones outward thus makes for a greater total strength than if they were centrally disposed.

After fertilization in sea urchin eggs, cleavage units, or blastomeres, continue with protein synthesis, apparently due to a store of mRNA in the developing unit. The supply is adequate to carry the development until gastrulation. In a bird, the information that sends a developing *limb* toward the wing development goal or, alternatively, in the direction of a leg, comes both from the limb layer and the under layer, that is, the wing bud mesoderm is affected by information from the next embryonic layer.

The nucleus in specialized cells contains all the genetic information that other cells have in an organism, so they could initiate embryogenesis. However, this does not happen. They maintain their specialization as intestine, liver or the like. Some of the repression is of a "Do Not Go Unless Deinhibited" type and semipermanent, using histone, most likely in DNA linkage. Others may be "Go Unless Inhibited," or transiently uninhibited by hormonal messages like that of estrogen to acid chromosomal protein informing it to initiate mRNA synthesis for specialized cytoplasmic components. These components then develop exclusively, even though the DNA information for all of the cytoplasm is available.

Fertilization is a primary switch which triggers an amplification via release of mRNA that had been masked or inhibited. Then a steering directivity becomes evident in gastrulation when the destiny of a cell is set with respect to its specialization, unless countermanded at a future time by a tumor type of deviation.

BIOELECTRIC GRADIENT INFORMATION CONTROLLING MORPHOGENESIS

It can be shown that a small electrical field, polarized with respect to a healing bone fracture, will accelerate the regeneration of tissue and healing. Many influences that may deposit or remove charges such as flowing water or elevated temperatures will favor plant regeneration by causing local electronegativity with respect to the neighborhood. Thus, repression can be exerted by making the local region electropositive. In a plant with a cut stalk section, both ends may be made electronegative with respect to the middle. Regeneration begins when tentacle ridges appear with their distal ends directed toward the most electronegative locus. Thus a directivity is available in which distal and proximal growth is governed by an electropotential only slightly above that of the natural level of the plant. Similarly, the position of regeneration is varied by fixing the polarity of the bioelectric field applied. This effect is one of command by a superimposed, polarized (i.e. oriented "heads and tails"), slightly stronger electric field. To reverse the direction of regeneration in the plant, the orientation is made *opposite* to that of the bioelectricity present. To counteract lagging or stopped regeneration, the biofield is reinforced by a *stronger* applied potential. The regenerates are oriented toward the electronegative side.

The development of polarity in ordinary regeneration is spontaneous. Cut ends become electronegative with respect to the middle. The natural gradient is toward electronegativity at the distal end. Nonregeneration will be associated with less electronegativity, for example, at the proximal end, or it may even go to electropositive, meaning that the repair has failed. Multiple negative centers will introduce corresponding special directional effects in regeneration. The associated oxidative metabolism is mobilized at the negative ends of electrical surface gradients. These gradients thus have both informative and communicative functions.

GROWTH DETERMINING CELLULAR INTERACTIONS

The influence of one cell on another is great in tissue culture as well as in the whole animal. If a single cell is dependent for survival on a limiting nutrient, a cooperative effort is exerted in a whole population so that the dependence is eliminated by having enough cells. The informative molecule from the dense population is the compound affecting density itself. Serine, for example, is a limiting nutrient, required for good cloning of cultured cells when there are 10 to 100 cells per milliliter. This is the density at which the population turns toward internal support of this nutrient requirement to allow for added cells. The switch is population density related, and the action is the contribution at higher densities of serine or the derived glycine from the population. Cellular metabolism intermediates, such as cystine (and cofactors such as folinic acid) change their limit status in the same way, but the population differences required to show the switching effect (10 cells/ml limit, 200,000/ml nonlimit for cystine) are wider apart. Population density dependence is connected with the vigor with which the cell type approaches the synthesis requirements. This vigor, which connotes a sort of conditioning, is quite variable, so the dependence applies to many systems such as differentiation, expression of cell function, and viral synthesis.

There is also a contact pressure effect in the contacting membranes as the growing culture of cells bunches together; the intercellular influence depresses the synthesis of DNA, RNA, and protein and growth. Polyribosomes in the cytoplasm disappear and these effects are reversible on transfer of the culture. To avoid this inhibition, other cells of different ploidies (heteroploid from normal liver, intestine, conjunctiva and HeLa) are used and they readily pile up in multilayers. Thus, variable contact inhibition and cell density or population maxima are cell-type related factors (Fig. 4-26).

(a)   (b)

Fig. 4-26. Some cells demonstrate contact inhibition forming perhaps only a monolayer in culture (a) while other types are released from this inhibition (b).

In a tissue culture it is difficult to recreate all the factors of the organ environment as well as the normal communication of stimuli. It is something like asking the toe to mimic the man. Table 4-3 shows the type of function that seems to need organ-type population pressures for expression, and thus fail unless the population in culture is dense enough.

TABLE 4-3. Population Dependent Cell Functions as Seen in Cultures

Culture	Function Needing Dense Population
Iris and Retinal Epithelium	melanin
Adrenal and Pituitary Tumor	hormone production
Fibroblasts	collagen production
Chondrocytes	cartilage
Myoblasts	myofibrils

Contact Pressures

Contact pressures, inhibitions, and interactions, interpassage of nutrients, and cooperation all have significance in cancer cells which dedifferentiate. The situation in a malignant cell is in conflict with that of a regulated cell, but the basic controls are still studied. Some questions are whether a cancer cell commits malignancy by circumventing control via special synthesis of control neutralizing agents, or does it *omit* control to proceed on its proliferation, or is malignancy a failure to communicate in the sense of identification of unwanted variations in components?

Contact inhibition refers to the termination of the growth of cells which are crowded and it is observed in tissue or cell line cultures at saturation densities (cells are touching each other). Uninfected animal cells in a medium with serum, salts, amino acids, and vitamins will often persist through an infinite number of passages to form a distinct cell line. Contrary cultures may transform into tumor producing ones, and it is seen that the rule for contact inhibition is then being violated. A normal cell line like 3T3 from the mouse embryo is inhibited by contact and will stop dividing, even in a monolayer, and is stable against transformation to the tumor producing state, unless infected by a tumor virus. When infected with SV 40 DNA tumor virus, it relinquishes information management to the virus and is no longer contact inhibited.

The removal of some normal 3T3 cells from a culture would be equivalent to injuring the culture. There is a repair response in which cells are reformed around the damage but information instructs the cells to reform only at the damage, not at a distance. The message to grow comes via the saturated medium of growth and control substances whose concentration gradients are recognized by the cells within it. The medium provides both this gradient and a separating force that prevents contact inhibition. In vigorous growth, the momentum pushes cytoplasmic processes, edges, and membranes around neighboring cells intimately, until the lack of growth and control substances brings this engulfment to a halt. An injury is like scraping away the inhibition at edges and

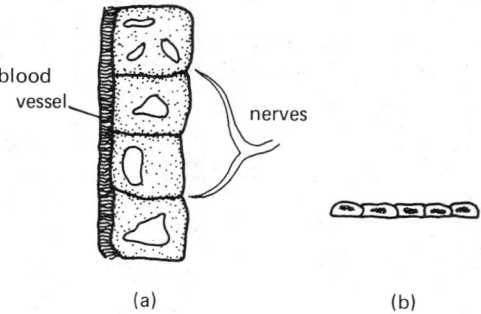

Fig. 4-27. a) The cell in the organism is coupled into its environment; b) the cell is excommunicated in culture.

flooding with these substances so that repair is stimulated. The repair is amplified by causing the nutrient layer flowing over the culture to be almost motionless right on the cell surfaces, so that it is exhausted more quickly there. This is exactly what is predicted by boundary layer theory with the thickness of this layer depending on the velocity of the stream of the medium. A discontinuity at an injury gap thins out the boundary layer, adding nutrition by working against depletion or stagnation. This encourages specific growth at that location.

The reason that tumor cells lack contact inhibition or that tumor-transformed cells lose it, is that they grow without the benefit of the normalizing substances in the medium, such as hormones, and thus their growth is abnormal.

Many cellular processes are implicated in circadian rhythms in plants and other organisms. Some of these are shown in Table 4-4. Not all plants show these periodic production or behavior characteristics but various plants have indeed shown these types. The rhythmic changes may be temperature, time, or radiation sensitive, that is, they have components that depend on temperature or radiation under system stabilization by means of clock action. In these rhythms, communication signals and energy availability are coupled to the intrinsic energy levels in pigment systems and photoreceptors. Resonance wave phenomena, natural frequencies, and the optimized capture of radiant energy play a significant role.

CHRONON THEORY

Circadian rhythms are named from the Latin "circa" and "dies" giving a meaning of "about a day" but the rhythms in natural cycles

TABLE 4-4. Circadian Rhythmic Processes in Plants and Photoperiodism

Photosynthesis
Luminescence
Division, growth
Chloroplast deformation (*Acetabularia*)
Luciferase, amylase, and catalase
Flowering
Zonation
Uptake
Translocation
Phototaxis
Migration
Odor Release (e.g., nocturnal)
Germination
Spore Release
Movement of cotyledon, leaf, petal, and rhizome

usually range from 19 to 28 hours. Electromagnetic radiation signals associated with daylight and darkness inform organisms of the 24 hour period, allowing more standardized rhythm, which is then expressed in synchrony of activity cycles or common metabolic phases. In terms of time and space, circadian rhythm means spacial registration transformed into temporal registration, with the scale provided by sequential RNA transcription, a time process coupled to DNA spacing of codons.

Codons combine to make a gene or cistron in a length of DNA molecule (a chromosome) adequate for describing a protein unit. Then a replicon may be defined as a poly-cistronic length of code, the chromosome having about 1000 of these. The time for a complete transcription and recycling is gated *to the longest replicon* which therefore has a "ticking off" function and could be called a chronon since it measures approximately one day. A circadian chronotype is an organism's time profile for completing chronon transcription. Also important is the fact that RNA made at different phases in the rhythm is itself different, being a copy of the unique DNA that happens to be exposed by decoiling at that time. Animals use the resting time during the comparative safety of darkness for regenerative or supportive phases. Leaves of plants droop to conserve moisture but light signals them to attention. Other zeitgebers, or time givers, interact with radiant information in establishing the rhythms. *Basic metabolism responds to the consequent clockwork.* Single cells and microorganisms take up the rhythms when conditions, including communication, favor synchrony. Zeitgebers like stimulants and sedatives are substances that chemically reregister the metabolism phase. This is illustrated by caffeine or theophylline used as a wake up signal in coffee and tea but light or electromagnetic signals may dominate over these same zeitgebers later in the day. Paramecia reset their biological clocks if they sense the radiation during what should be the darkness time and metabolic events may be moved up as the clock is set ahead. How much advancing is done is dependent on how much radiation is received (the intensity) in the signal and whether the metabolism is advanced or retarded follows from the phase of darkness affected, for example

> too early — retard
> first part — advance
> middle   — may not be informing at all or give varying effects
> in light phase — no effect

Since the presence of a definite nucleus seems essential for the rhythms, more is involved than the DNA control itself. The metabolism or DNA transcription behind glycogen synthesis, mitosis, neutrophils, lymphocytes, eosinophiles, cholesterol, serotonin, and corticosterone, is time-phased throughout the light period in the rat while adrenal corticosterone, urine histamine, serum urea, urine urea, norepinephrine, and alkaline phosphatase in the serum are measureable functions that dominate in that order during the dark phase. The action behind the clockwork is the changing character of RNA as DNA uncoils, as a function of the length of the replicon which gives a time scale, and the "interpretation" by the chronotypic enzymes that result.

Noise from competing zeitgebers enters the information scheme to complicate the phase relations. Both microbes and higher organisms are sensitive to such input. In the sense of electromagnetic wave reinforcement, in-phase or synchronous zeitgebers are probably harmless or else serve an unknown purpose. On the other hand, cancellation by equal and opposite signals may lethally confuse the organism. Insects and plants may show tumors when signals are 180° out of phase. The tomato wants it to be warm when it is light and cool when it is dark. Crossed signals always result in the development of tumors.

Chapter 15 examines in more detail the effect of electromagnetic signals on the functions and phases of organisms.

DEVELOPMENT INFORMATION FROM ACTIVITY

Signals from the ambient can include information about the relation between the organism and its environment. Such information is probably crucial in determining any development that is related to the behavior of sensory-motor systems. There are also the elements of psychological biology in which motivation, determination, persistence, drive, and such vital factors influence the entire individual via signaling systems, secretions, and nervous system responses. At a very primary, uncomplicated level (which is difficult enough to obtain) it may be shown that an animal's own motion changes his information input or feedback. The input from the surroundings in relation to the animal's simple motion in those surroundings keys development of coordination in muscles and nerves. This is called *plasticity*, as distinguished from a rigidity in which all such information would be provided by genetic coding. In careful experiments in which developing animal pairs are environment-stimulated in one case and left passive in the other, such dependence is demonstrated. For example, one animal is in harness rotating a beam pivoted at the center; the other end of the beam carries a passive animal who doesn't move but only "goes along for the ride." Normal development is subsequently seen only in the active animal, while the other is retarded (Fig. 4-28).

There is a self-generation factor in such experiments wherein the motion or response offers special reafferent stimulation or feedback-accompanying-movement in the sensory loop, special information processing, and error recognition. The passive animal is deprived of this experience and can neither develop at the full rate nor adapt. Related effects accompany signals for growth and maintenance, continuation of normal coordination, as well as adapting to modified visual and audio information input.

If plasticity is taken in the larger sense (opposite of germline genetics or hereditary rigidity) then it would include cellular dependence on

Fig. 4-28. Two animals on a rotating beam—one active, one passive.

accidental DNA input. These events are ambient related, as are the plasticity experiments with motion and animals, with their external input of information. There is evidence that infective agents such as bacteria, may begin as harmful invaders, and then, after a long period, become symbiotic to an extent that the host nucleus becomes dependent on the presence of the bacteria for carrying on the normal functions. There is evidence that both mitochondria and chloroplasts demonstrate cytoplasmic inheritance. They have their DNA and self-duplicate. Somewhat in the same manner, cytoplasmic inheritance, transformation, and transduction operate with agents of inheritance (Table 4-5).

TABLE 4-5. Genetic Recombinations Via Agents of Heredity

Transformation	*Streptococcus pneumoniae*—Its capsule holds the key to pathogenicity. The noncapsulated strains of the diplococcus can be transformed into pathogens by giving them DNA from the capsulated strains.
Transduction	Temperate viruses make lysogenic bacteria out of nonlysogenic ones by transferring DNA into the host's genetic material.

Light deprivation in cats during development significantly inhibits their normal progress as measured by the amount of pacing they will do in a maze to reach a reward. Pacing is an estimate of the searching needed to find the correct route and cats raised in normal light easily outperform their deprived mates.

MODELING AND MODERN BIOLOGY

Cell dynamics is a term for the growth, division, processing, and development at the hierarchy of a cellular organism. The basic model sequence of input-black box-output guides the attention to the significance of a process. Other terms would do as well such as stimulus-cell-response if they manage to convey the substance of the system that is being projected into the abstract. These terms should also suggest immediately that a search is being undertaken for a forcing function, transfer function, and output effect in the system being explained. The fractionation of the problem into what are estimated to be answerable units is also conveyed and a special logical formalism is anticipated to accompany the terminology.

It must be remembered that a model by itself has no lasting importance and the intent is to have it serve accurately as a reflection of the real organization. Specialists in various biological fields perform this conceptualization best but contributions from theorists are essential too. Problems of large scope are likely to be approached as a *team* effort in which the individual scientist brings his own specialty to bear in what is essentially a staff study. Mathematics by itself is not expected to provide a complete plan any more than is the model itself. Yet, rigorous formal thinking is required to explain the more complex problems and for the designing and correlation and optimization of experiments.

The value of a model lies in its comparisons with the real system. An understanding of a good model usually leads to an understanding of the real system. The teaching value of the model rests on its careful, analytical preparation. With it, cause and effect experiments are possible which define essential relationships. Then the model is subject to refinement until it meets the established criteria. This might mean a proper, quantitative response to the input stimulus. With it, one may be able to express transfer functions as well as forcing functions which are usually functions of time. A natural outcome of having a model is the possibility of *computer simulation* which may be far less hazardous and less costly than work with real living systems. Even models that only approximate the real system have great value. Naturally the model is always at a disadvantage when the real system has been under intensive Darwinian product development for over 2×10^9 years.

There are both natural and artificial models. A classical biologist can point out features in certain species that make these species unusually appropriate as natural models. Observation of squid propulsion leads to questions concerning its nervous system and the squid axon is found to measure one millimeter in diameter versus 0.01 millimeter in the average human single nerve. The depolarization and impulse propagation, and ion displacement work using this model shed light on nerve transmission in general. The octopus has a memory that is of great interest to psychobiologists and the shark liver concentrates Vitamin A to an interesting extent. If one is studying eggs, the ostrich egg may be more convenient that the gamete of a gnat. Bioelectric voltage may be studied in the electric eel which has a very high potential.

Tissue culture or propagation of cell clones can model organ development and function because many cell lines retain this function through several generations. These probably meet the miniconcept of the model process, at least in the sense of a small scale model.

It is essential to know when a model has served its purpose if this does, in fact, happen. The persistence of the Carnot cycle as the introduction model in thermodynamics with its questionable value, unreal aspects, and unfathomable concepts is a warning that models may have a limited useful life unless they apply with unusual force in some limited field.

A convenient strategy is to model one organ by another. In the case of prolactin hormone studies, for example, an organ such as the fish urinary bladder is known to be influenced by this hormone and perhaps by a cortical hormone in connection with its fluid control, osmoregulation, and sodium retention. The real objective may be the mammary gland function but the bladder is observed to provide a much less complex organ structure in which the same hormone functions.

Work with a differentiating model may lead to conclusions regarding the program in this development. The result is a switching network exercise in which the series of organismic commands is expressed as a program under master control. The eastern salamander, for example, turns toward the water after a terrestrial existence to remetamorphosize, the first change having been that from the tadpole stage. In the water existence, it will reproduce. The series of changes may be expressed as,

1. Make skin slime secretions.
2. Change pigmentation.
3. Make water breathing structure.
4. Grow larger.

This is, admittedly, an inadequate attempt to program the salamander's water drive, but it will demonstrate how the series might develop. Missing are all the necessary subroutines and an ability to sequence or place the commands in their proper order.

In the case of control systems there is evidence that regulation must, in fact, model the system it regulates. If it does not, the regulator may come out to be unnecessarily complex. This cannot happen with a satisfactory model or optimal regulator which models its regulated system, because both follow the same mapped version of events. In a large scale neural system, the brain must model its environment because the brain is a regulator which regulates its environment. Hopefully, with further information-flow-study this may become feasible and the brain may be evaluated rather than treated as an unchallengeable standard in its field. The brain should be found to regulate the body's ambiental interaction via an understanding process built upon a series of interaction models.

In circadian rhythms it has been observed how direction is exerted by radiation, as in photosynthesis. A direct example of this at the macromolecular level would be helpful because organisms are immersed in radiation with many different electromagnetic characteristics other than those of photosynthetic light.

A great many biological experiments have been designed to reveal radiation effects on organisms where the radiation may carry information and energy influences on the course of events. At strong levels, these tests often result in *inactivation* which may be selective, or *saturation*, which is not. Activation is an alternative but the energy "message" must arrive in the midst of noise and be interpreted—a difficult proposition especially when high levels of the incoming radiation will certainly mask the activation. It can be shown that radiation at 546 nanometers wavelength activates the lactate in the lactate dehydrogenase system LDH, and increases the yields in the lactate-to-pyruvate direction while depressing the reverse process in the manner of a switch. The exposure point in time represents the signal pulse or position while the amplitude is the length of the exposure as modulated by any intensity changes due to polaroid filters and distance from the light source. There is a maximum action at a particular wavelength which transports the information signals as well as carrying selected quanta for the action.

The programming of biological change, its control, the forcing and transfer functions will be exercises in information management so that information is a line of life through biology. This continuous line may be followed gainfully by many biologists if only because educationally, multidirectional studies provide a proven route to true mastery. It is an unmistakable conclusion that the essence of control is in the behavior of signals at discriminating gate-type points and at switching points where alternative directions may be selected by an organism.

Information flows in parallel with energy. Physics stresses the energy thread while biology follows the information one primarily, but they are but two sides of a coin. Biology emphasizes control as well as states in a system, and physics is wedded to the description of states and the procedures involved in passing from one state to another.

PROBLEMS

1. How are identities between proteins in differently evolved radiations explained?

2. How does the hemoglobin molecule reveal functional linkage?
3. What are the advantages of a two-site system in allostery?
4. What are the features of coupled cycles?
5. Can the sex chromosome be responsible for long life span?
6. What are zeitgebers?
7. What two metabolic pathways are separated for crisis management?

BIBLIOGRAPHY

Abercrombie, M. and J. King, *Advances in Morphogenesis*, Vol. 9, Academic Press, New York, 1971.
Adey, W.R., *Functional Linkage in Biomolecular Systems*, F.O. Schmitt, D.M. Schneider, D.M. Crothers, Eds. Evidence for Cooperative Mechanisms in the Susceptibility of Cerebral Tissue to Environmental and Intrinsic Electric Fields. Raven Press, New York, 1975.
Belamy, D., "Thymus in Aging," *Nature*, Vol. 237, May 5, 1972.
Birshtein, T.M. and O.B. Ptitsyn, *Conformation of Macromolecules*, Wiley Interscience, New York, 1966.
Burnet, Macfarlane Sir, *Auto-Immune Disease*, F.A. Davis Co., Phila. 1972.
Busch, H. and D. Smetana, *The Nucleolus*, Academic Press, New York, 1970.
Christensen, C. and A. Palmer, *Enzyme Kinetics*, Saunders, 1967.
Comorosan, S., M. Cru, and S. Vieru, "The Interaction Between Enzymic Systems and Irradiated Substrates," *Enzymologia*, Sept. 1971.
Conant, R.C. and W.R. Ashby, "Every Good Regulator of a System Must be a Model of that System," *Int. J. Systems Science*, Vol.1, no.2.
Cumming, B.G. and E. Wagner, "Rhythmic Processes in Plants," *Annual Review of Plant Physiology*, Vol. 19, 1968.
Crick, F.H.C., *Nature*, London, 227, 561, 1970.
Davies, D., "A Shangri-la in Ecuador," *New Scientist*, Vol. 57, 831, Feb. 1, 1973.
Durapinski, J.B.G., Ed., *Research in Immunochemistry and Immunobiology*, University Park Press, Baltimore 1972.
Eagle, E., "Metabolic Controls in Cultured Mammalian Cells," *Science*, Vol. 148, April 2, 1965.
Ebringer, A.J., "Information Theory and Limitations in Antibody Diversity," *Journal of Theoretical Biology*, Vol. 51, 293, 1975.
Ebringer, A., "A Comment on Code Degeneracy—Information Theory and the Central Dogma," *Journal of Theoretical Biology*, Vol. 53, 243, 1975.
Ehrat, C, "Circadian Rhythms," *Spectrum*, Summer Issue, Argonne National Laboratory, University of Chicago, 1977.
Elden, H., Ed., *Biophysical Properties of the Skin*, John Wiley, New York, 1971.
Garod, D., "Cells Need to Stick to Move," *New Scientist*, Vol. 57, Jan. 11, 1973.
Goodwin, B.C., *Temporal Organization in Cells*, Academic Press, New York, 1963.
Goraig, E., "Synthesis of Specific Stabilized mRNA when Translocation is Blocked," *Genetics*, 331-336, Feb. 1972.
Gordon, B.L. et al, *Essentials of Immunology*, F.A. Davis, Phila. 1971.
Held, J. "Plasticity in Sensory Motor Systems," *Scientific American*, Vol. 213, Nov. 5, 1965.
Hiragg and Yanofsky, "Regulation of the Tryptophan Operon," *Nature*, Vol. 237, May 5, 1972.
Holliday, R. and G.M. Tarrant, "Altered Enzymes in Aging Human Fibroblasts," *Nature*, Vol. 238, July 7, 1972.
Inman, F.P., Ed., *Contemporary Topics in Immunochemistry*, Plenum Press, New York, 1972.
Jacob, F., S. Bruner, and F. Cuzen, "Regulation of DNA in Bacteria," *Cold Spring Harbor Symposia on Quantitative Biology*, Vol. 28, 1963.

Jacob, F. and J. Monod, "Genetic Regulation in Protein," *Journal of Molecular Biology*, Vol. 3, 3, 318, June 1961.
Jeon, K.W., "Development of Cellular Dependence on Infective Organisms, Microsurgery in Amoebas," *Science*, Vol. 176, no. 4039, June 9, 1972.
Johnson, R. and B.L. Strehler, "Loss of Genes Coding for Ribosomal RNA in Aging Brain Cells," *Nature*, Vol. 240, 412, Dec. 15, 1972.
Kendrew, J.C., "Information and Conformation in Biology," *Structural Chemistry and Molecular Biology*," A. Rich and N. Davidson, Eds., W.H. Freeman, San Francisco, 1968.
Korn, R.W. and E.J. Korn, *Contemporary Perspectives in Biology*, John Wiley, New York, 1971.
Lamb, M.J., *Biology of Aging*, John Wiley, New York, 1977.
Leaf, A. and J. Launois, "Search for the Oldest People, " *National Geographic*, 143 n. 1, Jan. 1973.
Lee, D.H.K. and D. Menard, *Physiology, Environment and Man*, Academic Press, New York, 1971.
Lofgren, Lars, *Recognition of Order and Evolutionary Systems in Computer and Information Sciences II*, J. Tou, Ed., Academic Press, New York, 1967.
Mandelstam, J. and K. McQuillen, *Biochemistry of Bacterial Growth*, John Wiley, New York, 1968.
Monod, J., Pierre Changeux and F. Jacob, "Allosteric Proteins and Cellular Control Systems," *Journal of Molecular Biology*, Vol. 6, 4, 306-329, April 1963.
Nazlin, R.A., *Biochemistry of Antibodies*, Plenum Press, New York, 1970.
Rees, D.A., "Fundamental Principles of Cooperative and Transductive Coupling," *Functional Linkage in Biomolecular Systems*, F.O. Schmitt, D.M. Schneider, D.M. Crothers, Eds., Raven Press, New York, 1975.
Reichert, T.A., D.N. Cohen, and A.K.C. Wong, "An Application of Information Theory to Genetic Mutations, and the Matching of Polypeptide Sequences," *Journal of Theoretical Biology*, Vol. 42, 245-261, 1973.
Reiner, J.M., *The Organism as an Adaptive Control System*, Prentice-Hall, Englewood Cliffs, New Jersey, 1968.
Robson, B. and Pain, R.H., "Analysis of the Code Relating Sequences to Conformation in Proteins: Possible Implications for the Mechanism of Formation of Helical Regions," *Journal of Molecular Biology*, Vol. 58, 237-259, 1971.
Robson, B. and Pain, R.H., Ed., "Studies on the Role of the α-helix in Nucleation," *Conformation of Biological Molecules and Polymers*, E.B. Bergman and D. Pullman, Academic Press, New York, 1973.
Rose, S.M., *Regeneration: Key to Understanding Normal and Abnormal Growth and Development*, Appleton-Century Crofts, Ed. Div., Merridith Corp., New York, 1970.
Rose, S.M., "Bioelectric Control of Regeneration," *Developmental Biology*, Vol. 28, 274-279, 1972.
Shannon, C.F. and W. Weaver, *The Mathematical Theory of Communication*, U. of Illinois Press, Urbana, Ill., 1979.
Shields, R., "The Fallacy of Cells in Contact," *New Scientist*, June 24, 1976.
Sivak, A., "Induction of Cell Division: Rose of Cell Membrane Sites," *Journal of Cell Physiology*, Vol. 80, 167, October 1972.
Sager, R., *Cytoplasmic Genes and Organelles*, Academic Press, New York, 1972.
Sundaralingam, M. "Concept of Conformationally Rigid Nucleotide," *Conformation of Biological Molecules and Polymers*, B. Robson and R.H. Pain, Eds., Academic Press, New York, 1973.
Szent-Gyorgyi, A., *Bioenergetics*, Academic Press, New York, 1957.
Szent-Gyorgyi, A. *Introduction to Submolecular Biology*, Academic Press, New York, 1960.
Temers, H.M., "RNA Directed Synthesis," *Scientific American*, p. 25, Jan. 1972.
Turing, A.M., *Phil. Trans. Royal Society B*, 237, 37, 1952.
vonNeumann, J., *Cerebral Mechanisms in Behavior*, L.E. Jeffress, Ed., (Physical Laws), John Wiley, New York, 1951.

vonNeumann, J., *The Computer and the Brain*, Yale University Press, (completed posthumously) New Haven, Conn., 1958.
Watson, J.D, and F.H.C. Crick, *Nature*, London 171, 964, 1953.
Widdowson, E., *Harmony of Growth*, Lancet, p.901, May 1970.
Wiseman, L.L., M.S. Steinberg, and H.M. Phillips, "Experimental Modulation of Intercellular Adhesiveness Reversal of the Tissue Assembly Patterns," *Developmental Biology*, Vol. 28, 498, 1972.
Yockey, H.P., *Journal of Theoretical Biology*, Vol. 46, 369, 1974.

5

Waves and Electromagnetic Environment — Polarization and Information Density Regions

"Her constant beauty doth inform
stillness with life and day with light."
Tennyson

THE ELECTROMAGNETIC ENVIRONMENT

The information aspects of radiation biology and an impressive list of biological actions, some vital, some beneficial, and some harmful, and all connected with the spectrum (electromagnetic action spectra) must be included in the study of biology. The radiations are physical agents which interact physically and chemically with biomaterial. Electromagnetic theory may be introduced in this study as necessary to explain these interactions. Thus, orientation is provided on wave behavior and mechanisms so that one becomes aware of an electromagnetic environment with which organisms may or may not be compatible. Building upon such a foundation, the study can progress to vital relationships between wave energy and the change in status of the systems with which it interacts (Fig. 5-1). From the behavior of the electromagnetic waves, the study may be profitably extended to analysis of creative uses of the electromagnetic spectrum, information flow from biospectroscopy, ultrasonic waves, to magnetic influences, gravity, and direct electrical power (dc or ac) dissipation in tissues. When all the interactions are considered, along with present day emphasis on radiation, radiobiology emerges clearly as a major subarea of biology.

Ionizing and nonionizing radiations should be carefully distinguished. Ionizing radiations are energetic waves of very short length—angstrom waves—beginning with the ultraviolet spectrum. This is the usual meaning for radiations which characteristically produce ions in tissue by the interaction of wave energy such as X-rays, gamma rays, and radiation particles such as alpha and beta rays with the nucleus or electronic clouds. Nonionizing radiations are longer waves (such as

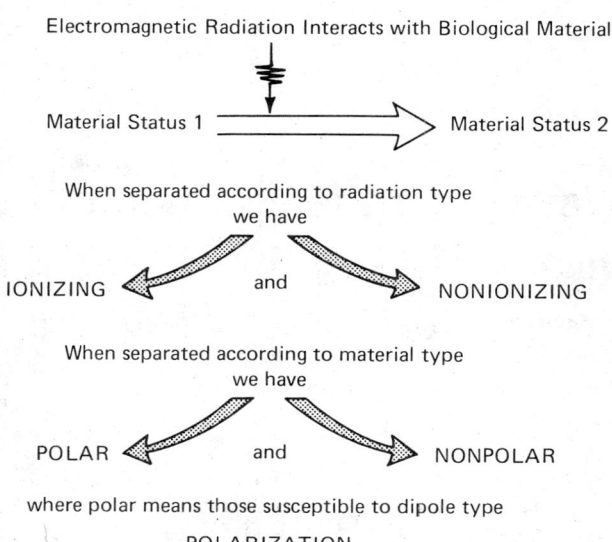

Fig. 5-1. The electromagnetic radiation action is a radiation-material interaction.

microwaves—centimeter and millimeter waves) and are radiations of equal biophysical importance. Many volumes of radiobiology have been written which give the impression that all radiations are ionizing, or that only ionizing radiations are biologically significant. For example, microwaves are now coming into much wider use and, probably due to the lack of attention in the past, their biological interactions have suddenly become a matter of deep concern to public health officials. Since nonionizing radiations will be treated in some depth, the reader will be able to understand microwaves and their spectral neighbors, infrared, ultraviolet, and visible light, all of which are of critical importance for the survival of all species of organisms through such special interactions as photosynthesis, vision, black body radiations, and thermal transfer (Table 5-1).

TABLE 5-1. The Main Radiation Divisions

Nonionizing	*Ionizing*
Radiowaves	Far Ultraviolet
Microwaves	X-rays
Infrared	Gamma Rays
Visible Light	Alpha, Beta Rays
Near Ultraviolet	Neutrons and Other Particles
Sound	

HIGH INFORMATION DENSITY REGIONS, ANTINOISE, AND BIOLOGICAL SQUELCH

It is always useful to identify sensitive sites or, conversely, resistant structures, for purposes of control and understanding of radiation interaction with tissue. The information content will be concentrated in high information density regions such as the prosthetic groups of highly specific enzymes, coenzymes, allosteric enzymes, reduction-oxidation centers in key intermediaries, electron transfer system couplers and uncouplers, and of course, in a less specific way, in the whole indispensable genetic apparatus. How these parts look as targets and how ploidy (the haploid, diploid, tetraploid conditions) and other dispositions of information affect the redundancy are, therefore, important questions.

Atomic information is stored in and transferred in and out of (1) the stereochemical configuration, local solvent factors, charges, and bonding states or the possible bonds the atoms can form as well as (2) the nature of the element itself, such as hydrogen, phosphorus, sulfur, nitrogen, etc. Then, these sources and their numerous possibilities provide the total information as in the animal, vegetable, and mineral game (see Chapter 2).

In terms of the information density, it is clear that, among all the possible states in which the system can deploy itself, some will be much more meaningful biologically. For example, the fraction of absorbing, reacting, sensing, functional or active states among all possible states is most significant.

When radiations interact in biological systems as they do in causing mutations, the geneticist requires analysis of the effect. The genetic example is apt, because the incoming information carries some kind of message to stored genetic information. This message is frequency dependent and specific for an addressee target, or else is nonspecific in the sense of a bullet in battle that has "someone's name" on it. Stored data will be modified by deletions or additions. The radiation can hit a gene in the sense of a target or there may be a sensitive volume in a nucleic acid molecule.

A chromosome would be a much larger target. The chromosome would then respond by rearrangements and redistribution of information and probable repair. This results in chromosomal mutation. The message received may communicate conflicts or improvements, i.e., deleterious or favorable mutation. Transmission may be chemical, by ionization of a base in the time frame of replication, or physical, by rearrangement of genic fragments. In any case, these reregistrations in the copying, and possible errors or changes include the introduction of meaningless noise, inversions, insertions, and substitutions of gene characters. This resembles a typographical error in copying the genic message.

Suppose the message has w words with ℓ letters so that the arrangement carries ℓw symbols. The information in a letter is H_ℓ and letters influence other letters so that I_ϱ is the information measurable from intersymbol influences,

$$H_w = \ell(H_\varrho + I_\varrho) \tag{38}$$

The message information is given by

$$H_m = w(H_w + I_w) \tag{39}$$

where I_w is the information found in interword influences. Then,

$$H_m/\ell w = D \tag{40}$$

where D is the information density in bits per letter. Informational molecules in biosystems depend a great deal on intersymbol influences, digrams, trigrams, sequences and nearest neighbors, so that l_ϱ and l_w are significant. As a noise, an appropriate unit of radiation can delete one letter. The information density is $H_m/\ell w$ if the target biofunction is determined by *all* the letters. In the usual case, there is an expendable fraction due to :

1. *Redundance*, where equivalent data reside in some parallel structure.
2. *Antinoise*, where there are biological squelching effects, or
3. *Repair* possibilities where an adequate time between exposures is provided.

The essential letters are $\ell - \ell_x$ and the bioinformation density can be corrected to indicate real functional density.

$$(H_m/\ell w) \times [(\ell - \ell_x)/\ell] \tag{41}$$

Here, bits per letter, multiplied by a fraction of essential letters, gives bits in essential letters or essential information. Expendable letters are represented by ℓ_x, and if the appropriate unit of radiation action is incoming, such as an ionization in the target, it can trigger a dysfunction with a probability of

$$P = \ell - (\ell_x/\ell) \tag{42}$$

where

$$\ell_x/\ell < \ell w \tag{43}$$

This reasoning evaluates the expendability of the fractions and of the receptive, sensitive, or target area, which is smaller than that of the whole structure (Fig. 5-2).

SECURITY OF INFORMATION

The antinoise elements may represent protection as does hypoxia in radiotherapy of tumors. The core of the tumor is often found to be regionally hypoxic and the low oxygen tension provides good protection so that this part is radioresistant. Conversely, normal tissue may be temporarily ischemic in hypothermia or circulatory arrest. It would then be protected by the antinoise effect during therapy. Redundance will also protect the data. Suppose that ℓ_2 letters are resistant or supported by equal data present for correcting all errors, artificial or

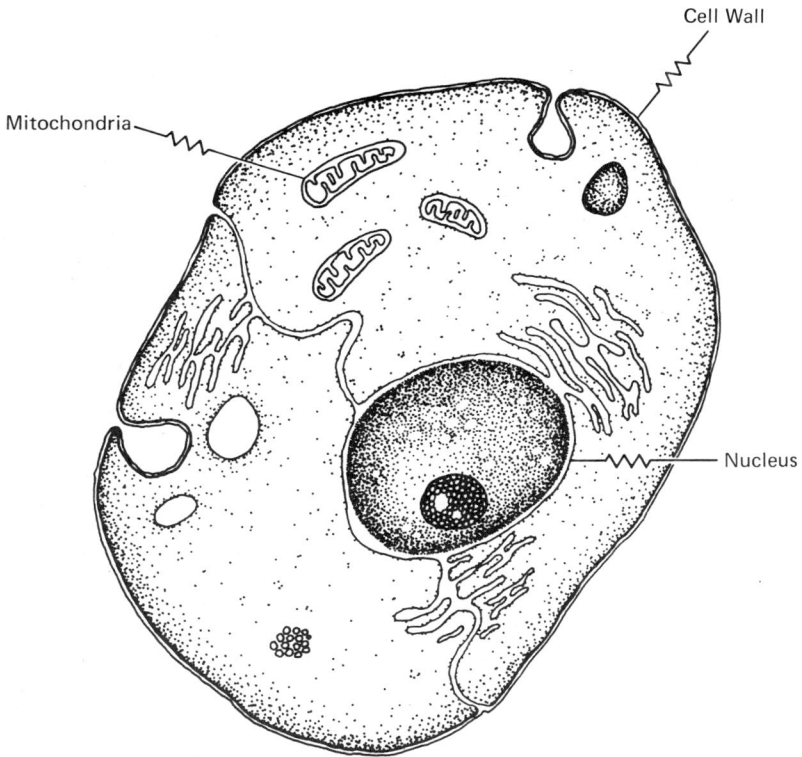

Fig. 5-2. Information density. In terms of sensitivity to the input of noise, a cell represents a target whose size is smaller than the whole structure.

natural. Now ℓ_2 letters are error correctable and the bioactivity is secure. Then,

$$D_{sens} = (H_m/\ell w) \times [(\ell - \ell_2)/\ell] \tag{44}$$

represents the information density in bits per letter for the radiation-responsive or sensitive portion. Normally, organisms gain information security with their ploidy, or the number of sets of chromosomes, N, 2N, or 4N. The exposure required to produce the same percentage inactivation decreases inversely with the ploidy above haploid. The organisms or organelles in their medium or structure are an information ensemble, receiving their communicated information within the system. Coding accidents, disturbances, and genetic transformations all entail some internal noise and error. The external error due to deleterious radiations and other foreign input may be added to this. The effect of this kind of error is loss of functional information which leads to conflict. A threshold exists beyond which information is no longer expendable. Losses beyond this threshold mean lethality for the organisms in the ensemble. To support these conclusions for genetic functions, many experiments show that diploids (2N) survive much better than haploids (N) among yeasts and other organisms exposed to X-rays.

The analysis for the process of decline or aging and for other deleterious changes and their agents is similar to the above. Other radiations need not necessarily be described in terms of sensitivity, damage and protection. Alternatively, the analysis may be in terms of receptivity.

In vision, photosynthesis, and the activation of photochemicals such as 7-dehydrocholesterol to vitamin D in visible light, the process of receptivity is involved. Here, the result is so important that skin color and ultraviolet absorption for this favorable activation have actually determined the distribution of races on the continents of the earth. Blondism, for example, invites skin cancer in the tropics but is an asset in cool, foggy regions.

SPECTRAL FREQUENCIES AND BIOMOLECULAR RECEPTIVITY

Biomolecules are often peculiarly receptive to specific frequencies. This leads to fundamentally useful processes such as photosynthesis, photoinactivation, photoreactivation, vision, and unusual transmissions or reflections. In the ionizing radiations, the first requirement is for wave energy sufficient to cause ionization, and only secondly for a specific frequency that may be involved in some sort of discriminatory action among biomolecular species or tissues.

Frequency, on the other hand, is of such great importance in nonionizing bioevents that it becomes of real value to locate specific interacting frequencies. If one is found, great selectivity and discrimination are gained in the application of wave energy to biomolecules at this frequency. The other side of the coin is the receptivity of the atoms, molecules, and biological systems to the radiations.

These interactions can be divided into two broad groups on the basis of polarity (Fig. 5-1). Nonpolarity is equivalent to nonreceptivity in the case of microwaves and also, to some degree, for other radiations, both ionizing and nonionizing. With ionizing radiations, nonpolarity may protect a molecule through a partial absence of the capability to respond to the sudden appearance of radiation-carried charges in its structure. It also may mean neutralization of ions formed by immediate recombination.

Cells which divide frequently are radiosensitive. The mitosis operation is a distinct polarization, and is accompanied by the production of electrical impulses. Into this scheme, ionizing radiations and charged particles of high energy can and do introduce drastic changes. The radiation may "know" which molecules it can affect, the frequency and quantal wave energy being the means by which this information is communicated (Fig. 5-3). If the ionizing wave information is communicated to the genetic substance itself, it finds a very ready receptivity. The interaction can result in deletion of information (a noise effect) or sometimes in a substitution of new characteristics (positive mutations).

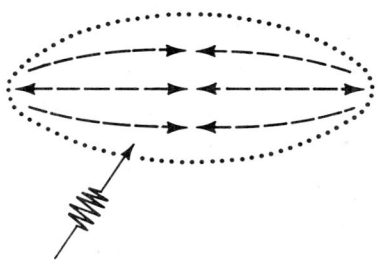

Fig. 5-3. Mitosis, with its complex active structures, polarity, and charges, is a time of radiosensitivity.

Biomacromolecules offer an easy target to ionizing radiations. A relatively bulky alpha particle (helium nucleus) will cause a macromolecular chain to separate and, with hits on both sides of the DNA helix fairly near each other, a break will occur (chromosome breaks). The alpha particles with their high density of radiation can easily saturate a large molecule with ionization events, causing profound changes in the genetic system. Protein synthesis will be disordered when the information stored in and transferred from the nucleoprotein is obliterated by the noise or radiation input. Sometimes these events can be characterized as forming a bipole which is the polarization of a biological system in terms of alternate states exhibited in response to an influence or disturbance.

Vision is an example of vital information transfer by light radiation. Microscopic observation, for example, requires a preparation and a source of light radiation. The incoming light is nonspecific until it interacts with the preparation. On doing so it becomes cognitive energy which is meaningful to the retina, where it forms a retinal image. Subsequently this is translated by cerebral data associations into a virtual image, but the primary interface is seen to be in the objects observed.

In photosynthesis, corresponding beneficial events occur in chromatic interfaces in the bands, cups, chloroplasts, and networks of chlorophyll, leading to energy transduction and the production of cellular carbohydrates. In this case, the life system is receiving stabilizing energy through a recognition process in the energy sink. The recognition process is resonance with molecular absorption. That is, the photon magnitude is commensurate, with each type of chlorophyll having its correct wavelength for excitation energy by resonant transfer.

POLARIZATION

Polarization phenomena constitute a fundamental response of molecules to radiation. Therefore, an understanding of them is of great value since they provide an area of common response whereas commonality is otherwise difficult to establish. All that is needed to use this model is a displacement of the molecule or its subunits in response to the spectral energy influence. There are recurrent electrical charges on nucleoprotein in its electron transfer regions so that the instant that radiation-induced charges appear in the molecule, or within reacting distance from it, a polarization response resembling a parallel wave response can occur.

The sensitivity of proteins to ionizing radiations is based on the fact that only a few ionization events are needed to inactivate a molecule with a molecular weight in the millions. In proteins, an important polarization at secondary bonds such as hydrogen bonds and salt linkages may occur when they are exposed to wavelengths between the infrared and radio wavelengths (the nonionizing microwave region).

For *strong* bonds and *small* masses a resonance or excitation occurs in the infrared from 2 to 25 microns, and with *weaker* bonds and *large* masses, the resonance is expected at longer wavelengths.

The principal ionization event is the formation of an ion pair, with an average use of 34eV of ionizing radiation energy. This is a polarization in which inner electrons are ejected, leaving a more positive atom. In ordinary ionizations as in electrolytes, outer valence electrons are removed by a few electron volts. Neutral atoms are unpolarized in the sense that the nuclear charge, centralized in a sphere of radius about 10^{-10} m, is equal to that diffused in the electron cloud with its sphere of radius about 10^{-6} m, so that a dipole structure is indeed difficult to define.

If the water molecule is viewed as a polarized body, with charges localized at the corners of a tetrahedron, then its polarity assists in the forming of four H bonds with other water molecules.

Multiradiation Effects

Spectroscopy provides excellent examples of radiation interaction in which molecular changes occur due to the absorption of energy. However, the results may not be fully utilized to explain *macro*biological changes since the chief interest is to analyze and describe molecular properties. As a result there is a rather interesting source of avoidable error. This is illustrated when protein is studied to see how ionizing radiations alter molecular weight. Ultraviolet absorption may be used to show changes in the spectrum of the protein. The ultraviolet itself is a radiation with its own effects, one of which is a scattering absorption by large aggregates as in the present example. While the experimenter would like to neglect these changes as nonionizing or unimportant, he cannot do so. Sometimes the whole change in a protein, e.g. bovine serum albumin, is attributable to the ultraviolet analysis rather than to the original or test irradiation (Alexander, et. al., 1956).

Polarization and Displacement

The requirements for identification of displacements in molecules as a unifying spectral response may be examined in three main categories;

rotations, vibrations, and translations. Translation, as a motion, involves displacement of the entire structure with each point moving in the same general direction as in the attraction or repulsion of a whole body, and is equally distributed. Rotation and vibration will be made clear by examples. Displacement is a useful general concept of broad application and most fundamental nature as shown in Table 5-2. The motion may also be capable of causing a structural change involving shifts in the internal order of atoms and nuclei. Translational motion may mean kinetic energy shifts and may result in temperature change. Similarly, vibrational and rotational kinetic changes may be seen as absorption of infrared and microwave radiation.

TABLE 5-2. Selecting Forcing Functions which Demonstrate Polarizations, Displacements, Reversing Polarizations and Resonances in which Transduction of Spectral Energy Occurs

1. Molecular, Atomic Dipolar Rotations, Vibrations and Translations	Bond twisting, torsions, stretching, bending and reorientation of molecular subunits.
2. Conduction, Semiconduction, Dielectric Polarizations and Orientations at Interfaces	Charge migration or mobility, dipole motion, and local realignment of charges.
3. Piezoelectricity	Crystal or biological transducer such as pressure sensor responds to intersystem information by striction or the production of current. A forcing function generates a voltage on opposite faces of the piezoelectric or if a voltage is applied, then the opposite response—a force or vibration is produced. See heart pacemaker application in text. Sensitive response makes piezoelectric quartz crystals useful for sensing pressure effects with the needle in the record groove of a disc player—an example of direct information transfer.
4. Electrostriction	An applied, electric field, forcing function deforms a dielectric acting as certain elements of strain as opposed to the piezoelectric vibrator. The latter is a bar or other shape with attached electrodes which excite a resonant frequency.
5. Magnetostriction	Ferromagnetics experience a dimensional strain or distortion according to the field direction and magnitude in magnetization.

6. Ferroelectric Hysteresis	Polarization of ferroelectrics in which the response to the forcing function depends not only on that applied field but on the recent "experience" of the material as well. Magnetic hysteresis would be similar but in the magnetic sense in ferromagnetics. Can be compared with dielectric and electric hysteresis. A material showing electric hysteresis can be represented by a hysteresis loop which shows the polarization P, or electric displacement as a function of the applied electric field E. The effect is to vary the magnitude of a polarization during increases in the magnitude of the forcing function as compared with decreases in it. Internal friction is responsible and the word is from the Greek word for lag. Some solids show elastic hysteresis under stress.

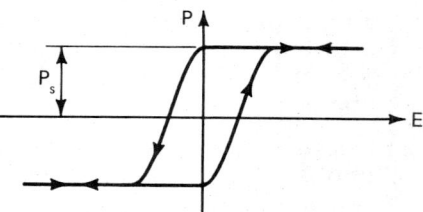

Hysteresis loop showing the relation between the resulting polarization P of the ferroelectric crystal and the externally applied electric field E; P_s is permanent or spontaneous magnetization.

7. Magnetic Domain Wall Resonance	There is a volume, probably microscopic within a ferromagnetic or antiferromagnetic which has aligned magnetic moments in parallel or antiparallel to a given direction. Without the magnetic force, this alignment is parallel to some major axis. The crystal symmetry establishes many domains and their preferred directions.
8. Ferromagnetic and Magnetic Resonance	Resonant absorption of applied EM radiation in paramagnetic materials to indicate the internal magnetic fields of ferromagnetic substances which are important in UHF applications. Ferromagnetic atoms have magnetic spin systems which absorb at resonant frequencies if they are subjected to alternating fields at their natural frequencies. The ferromagnetic or magnetic domain is a region in the material that shows parallel alignment of atomic or molecular moments. A uniaxial crystal has oriented magnetic fields in its various domains.

9. Spin Wave Resonance	The crystal lattice experiences a sinusoidal wave whose variation is seen in the angular momentum due to magnetism, mainly the electron spin angular momentum.
10. Ferrimagnetism	Atomic magnetic moments of neighboring ions in solids are forced into magnetic order or antiparallel alignment. Ferrimagnetic and ferromagnetic materials spontaneously magnetize but there are important differences in saturation magnetization of magnetic moments.
11. Antiferrimagnetism	Neighboring electron spins align antiparallel in metals, alloys, and transition elements. The magnetic moments of atoms align and realign or spiral so as to cancel the net total moment.
12. Nuclear Magnetic Resonance	Many atomic nuclei with unpaired spin or orbital magnetic moment separate into two or more levels when oriented in strong magnetic fields. They absorb energy at certain radio frequencies that satisfy quantum conditions for the energy separations. Absorption is recorded as a pattern of resonance lines, for example, for paramagnetic salts.
13. Cyclotron Resonance	Charged particles orbiting in a uniform, static, magnetic field can gain kinetic energy from resonant coupling of electromagnetic fields. The resonant frequency of the applied EM field is related to the orbital frequency of the charged particle in the magnetic field.
14. Gas Collision Frequency Resonance	A resonance based on the number of collisions per unit time for a colliding gas particle in the kinetics of gases.
15. Ion Pairs and Polarons	Emissions of charged particles can result for example, in a polarization of the residual ion with one charge and the emitted particle with an opposite charge ($A \rightarrow A^+ + e^-$). Polaron is the polarized orientation of water molecules that produces the aqueous electron (Chapter 10.)

By applying the characteristic motions to the many species of substances suggested by the list shown in Table 5-2, the listed displacements can be unified to a great extent by association. Many motions in this list form a basis for discussion elsewhere in the text. It is also useful to describe the motions as modes in which an interaction occurs. Polarization is an even more fundamental motion if we choose to make it so (Tables 5-3, 5-4, and 5-5). Then resonance would be a reversing polarization which is characterized by the "easing in" of a particular form of

TABLE 5-3. Resonance Situations Involving "Eased-In" Energy

	Resonance	Comment
Mechanical	Observed in wheels, swings, bridges	Small applied force produces large oscillations
Chemical	Benzene ring	Molecular structure may be stabilized in resonant forms
Radiation	NH_3 resonance in microwave region	Resonant molecular system enjoys an eased-in energy
Electrical	I is maximum	Surprisingly high V in circuit
Acoustic	Resonant column of air, energized by a musical note	Resonant sound effect
Biological	Chlorophyll	Molecular and other photochemical processes
Dielectric	Water molecule, frequency with optimum forcing function	Absorptive polarization

TABLE 5-4. Radiation Spectra and Polarization Involved

Frequency	Polarization
Radio Frequencies	Nucleus, Magnetic Dipole, NMR from H component of electromagnetic field
Radio and High Frequency 10-100 MHz	Nuclear Spin System Polarizations, Dipolarizations by RF Field
Microwave	Molecule rotates about a dipole axis
	Electron Spin
	Transitions of Magnetic Dipoles
	Also Electric Dipole Transitions
Microwave	Nuclear Quadrupole Resonance; Ellipsoid-like nucleus orients in electric field
	Debye Effects
	Electron Transition
	Dipole Orientations
	Crystal Defect Polarization
1-3 cm (microwave)	NH_3 type rotation
	Inversion Polarization
Infrared, Visible	Vibration, Electric Dipole, Atomic or Molecular (from E component).
	Resonant Polarization
	(Reversing polarization)

energy, Table 5-3. By studying polarization, it is possible to relate the energy causing it to the structure, which thus absorbs radiant energy. Starting with the simplest structure, a dipole, this has been extended to many other structures such as quadrupoles, octopoles, and their *implied sets* of structures.

Piezoelectric behavior is of particular interest in the intersystem passage of information. If skin or pressure sensors are subjected to a forcing pressure, they may act as electrical generators similar to substances such as quartz, zinc salts, sodium chlorate, boracite, tourmaline, calamine, topaz, tartaric acid, and sucrose. In skin transducers, the electricity stimulates nerve impulses to continue the communication. Cardiac pulses could activate a piezoelectric generator in such a way as to give a capacitor-stored charge of electricity, sufficient to generate a pulse, if needed to power the heart in case of cardiac failure. This device is a particularly interesting form of *pacemaker* since it uses the heart's own energy to power itself, and can be "massaged" externally if such emergency treatment is necessary.

TABLE 5-5. Resonance Polarizations as a Fundamental Behavior in Biophysics

Resonance Polarizations	*Selected Biological Research Involved*

Nuclear Magnetic Relaxation in which τ, the relaxation time is measured as a function of the applied field and *nuclear magnetic resonance.*	Information studies
	Recognition
	Transferrin function
	Heme-Heme Interactions
	   Oxygenation and Configuration
	Biological function of protein
	   bound water
	Other analysis of reaction products
	   in molecular biology

Electron Spin Resonance
ESR
   also called
   Electron Magnetic Resonance
   EMR
   and
   Electron Paramagnetic
     Resonance
   EPR or E(P)SR
Many systems have paramagnetism
   and Free Radicals, reactive
   species with the unpaired
   electron
	Photosynthesis
	Free Radical Analysis
	In vivo analysis, for example
	   Chloroplasts in algae
	Enzyme Mechanisms
	   Free Radical Behavior
	
	Biological Effects of Ionizing
	   Radiations
	
	Protection from Ionizing Radiations

The squeezing of a piezoelectric body to produce a current was defined in 1880 by the Curies, Pierre and Jacques. The word piezoelectricity is taken from the Greek verb "to press." Quartz illustrates this phenomenon very well because the information interaction is localized in the conformation of the lattice or in the molecular asymmetry. In Fig. 5-4, molecular charges are shown as two equilateral triangles with sides connecting like charges and having a common center. Although the coincident centers produce electrical neutrality, displacement as in Fig. 5-4 (2) forces the faces together, resulting in anticoincidence at the center which, in turn, leads to charge separation and electric current. When the reverse action takes place, an electrical voltage impressed on the faces causes a deformation which can alternate repeatedly to produce a vibration at some particular frequency. Sonic vibration, for example, can then be coupled into fluids for sonic measurements in blood or in the sea for sounding (sonar).

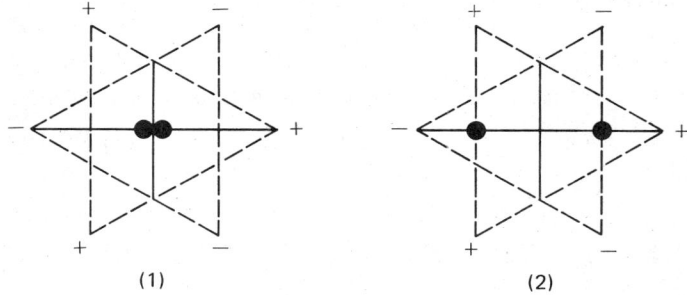

Fig. 5-4. Piezoelectricity as a charge separation when a force distorts the symmetry of charges.

Secondary and tertiary information levels are inherent in the piezoelectrics. For example, the frequency limits for the vibrating crystal are a function of its size, with large sizes exhibiting slow vibrations and a lower frequency. The displacement of the crystal can be managed in a characteristic mode;

1. Flexing of a plane crystal controls the frequency with the range of 0.4 to 100 kHz.
2. An extensional or longitudinal mode of displacement, outward from the center, gives 50 to 1000 kHz.
3. A shearing displacement with the upper face to one side and the lower face toward the opposite side, gives a change in height (like slanting a rectangle to one side) in a quartz crystal. The vibrations resulting are in opposite directions around a nodal plane at the center of the crystal, for frequencies from 1 to 125 MHz.

Polarization due to electrical fields is used to study molecules as systems of electrical charges. The fields disturb these charges, causing an electric moment per unit volume, or polarization. Debye explained this model and identified controllable external fields and/or neighboring molecular systems as causing the disturbance. He was interested in a special model for formal treatment. If there is simple *displacement D* of a unit charge in a cylindrical model containing the lines of force from the disturbance, then D has components from E and P where E is the intensity of the disturbance and P the polarization.

When E is the *electrical* intensity, a dielectric material (as opposed to conductors and semiconductors) has a property ϵ called its dielectric constant and $D = \epsilon E$. Here the dielectric constant is clearly seen to be a constant of proportionality between the displacement or bioaction and the field or forcing function. As a material descriptive constant, it governs those parts of biochemical reactions that depend upon the interaction of local charges in reaction sites. The ϵ value for water is often about 80 but, in the restricted region in which action occurs, the dielectric constant is not 80 but will depend on the influences of local charges. Its variation can then influence the possibility of reactions via the role of water as a solvent. Charges indicate the existence of local fields, and ϵ depends upon the forces exerted by such fields according to

$$D = \epsilon E \tag{45}$$

Thus, this constant contributes to reaction control and regulation.

Polarization can occur by orientation, the latter being a measurable response most obvious in a rigid system of charges (Fig. 5-5). As the electrical disturbance loses energy, the orienting system *gains potential energy*. This is characteristic of rotating structures and the disturbance is characteristic of radio and microwave frequencies. Moving higher to infrared, the displacement is a vibration and, similarly, it may involve whole molecular structures or groups free to move about some more rigidly held point.

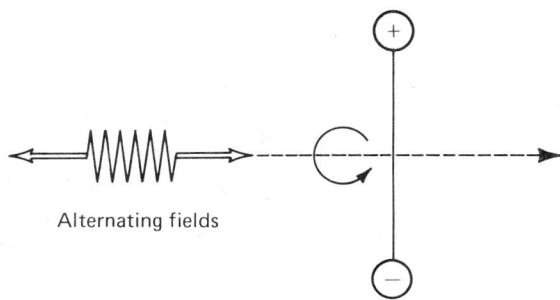

Fig. 5-5. Dipole orientation at radio and microwave frequencies.

Exponential Absorption and Polarization

In the interest of information management, it would be convenient to describe the interaction of electromagnetic energy with matter in a manner characteristic of the whole spectrum and modifiable for specific parts. Perhaps the nearest one can come to this is to observe that the disappearance of radiation energy in material is always exponential. The amount absorbed in a small distance x depends on the amount available at the start of this pathway.

$$dE/dx = \alpha E \qquad E_o \Rightarrow |\Delta x| \to E \qquad (46)$$

The constant α is the absorption coefficient which lumps all of the mechanisms by which radiation can be absorbed. The larger α is, the more these mechanisms are operating to account for the absorption. If any single general mechanism exists, it is likely to be in a suitable interpretation of displacement or, even better, of polarization. Dielectric polarization relaxation is a process common from microwave down to low radio frequencies. Whatever is polarized is displaced, but it is a little more difficult to see all displacements as polarizations. For one thing, displacement requires only one part of an object to be present, while a dipole must have two, separated by a distance (or at least two polarized parts are required for polarization). Yet even in the passage of an alternating wave train of energy, the oscillating, repetitive effect, and the spreading, elastic disturbance it causes, *do* resemble polarization in the sense that the surface of still water is polarized if an object drops on it (Fig. 5-6). The term polarization may be defined broadly enough to include what happens when a single dipole orients its positive and negative poles in an electromagnetic field as well as what occurs in an ordered medium being traversed by an alpha, beta, gamma, delta or other ionizing ray. The ionizing radiation disturbs the nucleic and/or electron clouds in its pathway, forming an ion track of disturbed and probably polarized matter, one feature of which is its ability to undergo recombinations, which is analogous to depolarization.

Absorption Coefficients and Characteristics

Without a proper generalization, the situation demands a more diverse appreciation of individual mechanisms, that is, an analysis of the lumped coefficient α. Many of these mechanisms are readily described in terms of available conventions and notations but an overall mastery is seldom achieved by one scientist.

$$E = E_o e^{\pm \alpha M} \qquad (47)$$

is the process of exponential change in a fundamental unit M by the various mechanisms α. Absorption means decay in the magnitude of E so α is negative. Growth, on the other hand, would give a positive α. M may be spatial (x) or temporal (t). The coefficient α may be determined by such processes as dipole polarization as in microwave rotations, inversions, vibrations in infrared, vibrations in ultraviolet and ionization as in the X-ray spectrum. In practice, the total effect is usually composite, being made up of subprocesses characteristic of primary, secondary and other degrees of interaction.

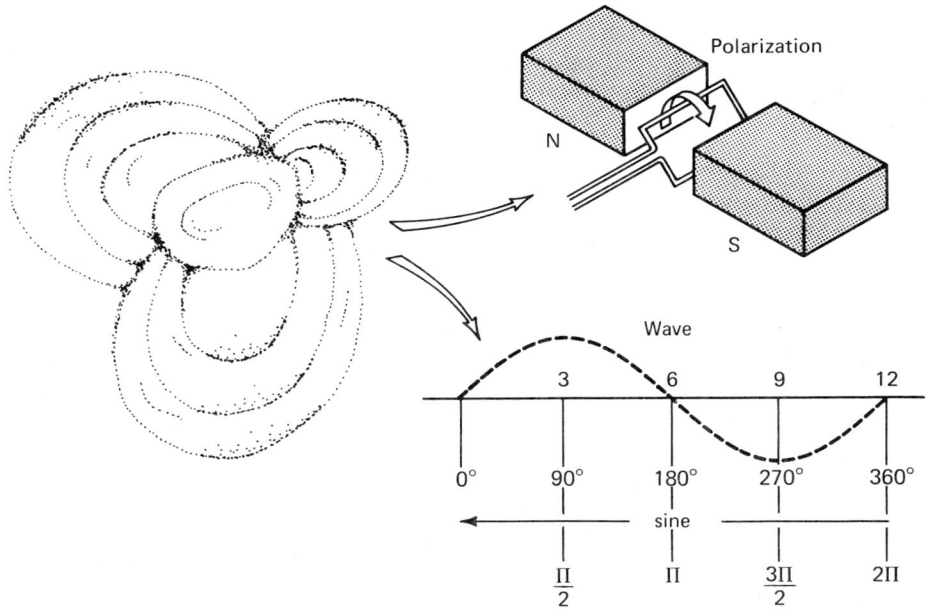

Fig. 5-6. Polarization pattern in water. The spreading elastic disturbance in the surface of a liquid is a polarization phenomenon. Such basic disturbance patterns may characterize the interaction between radiation energy and the matter with which it interacts. Insert shows the usual sine wave notation where the voltage at any time is $e = E \sin \theta$, θ being the angle shown with its equivalent values in radians and time at which the loop cuts lines of force.

With microwaves there may be components from dielectric absorption, electron spin resonance (ESR), nuclear magnetic resonance (NMR) and I^2R loss (conduction, resistance). In the X and gamma ray absorption process, α is determined by the components from the photoelectron effect, the Compton effect, and the production of electron-positron pairs in the medium. Within its limitations, the exponential relation is extremely useful as illustrated by the manipulations that follow. The natural logarithm factor (base) specifies from

$$E = E_o e^{-\alpha x} \tag{48}$$

that

$$\ln (E/E_o) = -\alpha x \tag{49}$$

Conversion to log, base 10, is obtained by dividing by 2.3 and

$$\log(E/E_o) = kx \text{ where } k = -2.3 \, \alpha \tag{50}$$

k, or sometimes ϵ, is the extinction coefficient in optics which is proportional to the optical density

$$k = OD/x \tag{51}$$

with units in cm^{-1} if x is in cm. In a spectrometer, k is measurable so that the material may be identified by reference to tabulated values for extinction coefficients.

Mass Absorption Coefficient. An important property of radiation can be shown by separating linear and mass effects. The mass absorption coefficient α_m equals the linear absorption coefficient α_ϱ divided by the density ρ. That is, when

$$E = E_o \exp(-\alpha_\varrho x) \tag{52}$$

$$\alpha_\varrho/\rho = \alpha_m \tag{53}$$

$$\alpha_\varrho = \rho \alpha_m \tag{54}$$

The decrease in radiation energy per unit length of travel is proportional to the mass of the absorber, that is

$$(\text{energy/cm})/(\text{g/cm}^3) \tag{55}$$

equals, in terms of mass and linear units involved

$$\text{cm}^{-1}/(\text{g/cm}^3) = \text{g/cm}^2 \tag{56}$$

and also, the number of centimeters of absorber traversed times its density equals g/cm^2.

This surface density, density thickness, or filter factor as it is variously called, relates different materials according to their absorbing ability. It is important in shielding against ionizing radiations with *equal* g/cm^2 values (i.e. equivalent thicknesses) of concrete, lead, steel, water, etc. It also permits calculations to compensate for radiation-counter windows, air, and source covers in handling radioactive materials. As the units indicate, it is measurable by weighing a sample of known area.

Therefore the absorption in a material depends upon the number of molecules in the path of the radiation. When we determine the concentration of a substance spectrometrically, this is really the matter of interest. The material may absorb 10 per cent of the radiation entering for each centimeter traversed. Then α is 0.1 and, after one centimeter, 90 per cent of the radiation remains to continue along its path of absorption. Each centimeter will take its toll, giving the exponential decrease in E (Fig. 5-7).

Optical Density

If k is the extinction coefficient in cm^{-1} then the concentration for a certain x (such as the width of a sample holder in a spectrometer) comes from the change in the energy

$$E = E_o\, 10^{-kcx} \tag{57}$$

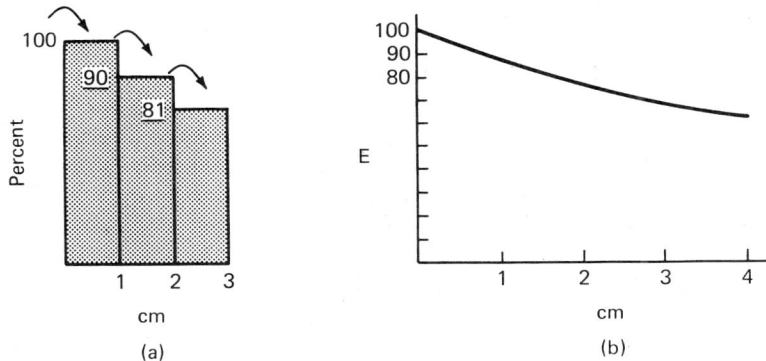

Fig. 5-7. Change in a distance x depends on amount present at the start of x. Assume a 10% change.

with the concentration in the sample in moles per liter

$$kcx = (cm^{-1})(moles\ cm^{-3})(cm) \tag{58}$$

and the exponent simplifies to moles per cm^3. Similarly, the value of k in cm^{-1} may be found for a certain concentration C and certain distance $x = b$. Also

$$kCb = -\log\ \text{transmission} = OD = 2 - \log\%\ \text{transmission} \tag{59}$$

where $-\log T$ gives the decimal value of transmission. If all the radiation is transmitted, there is no solute present. In this case there is 100 per cent transmission, or a decimal (coefficient) value of unity. Then the optical density OD (sometimes D) = $2 - \log 100 = 0$, because the log of a number is the power to which you must raise 10 to equal the number, and $\log 100 = 2$. Thus, there is no solute ($2 - 2 = 0$). Let $OD = 2$; then

$$OD = -\log 0.01 \tag{60}$$

$$OD = -(-2) = 2 \tag{61}$$

Here there is a 1 percent transmission at high concentration

$$OD = 2 - \log 1 \tag{62}$$

because

$$0.01\ N = \tfrac{1}{100}\ N = 1\%\ N \tag{63}$$

$$OD = 2 - 0 \tag{64}$$

$$OD = 2 \tag{65}$$

or almost all the energy is absorbed. These conventions are, unfortunately, hardly universal and must be validated for the range of concentration and wavelength of interest, especially at the concentration limits.

At times, the expression used is absorbance, A

$$A = \log_{10}(1/\text{transmittance}) \tag{66}$$

which, in many cases, is proportional to the concentration. The absorptivity a is the absorbance divided by the product of the concentration c of the test substance in grams per liter and the length b of the sample path in centimeters

$$a = A/bc \tag{67}$$

Transmittance is the ratio of transmitted to incident power and, multiplied by 100, would represent the percent transmittance

$$T = P/P_o = 10^{-kbc} \tag{68}$$

$$\% \text{ transmittance} = (P/P_o) \times 100 \tag{69}$$

Then, in a cuvette containing a sample, these values may be taken for the solvent and the solute as well as for unit thicknesses. The inverse relation for per cent transmittance and absorbance is shown in the scales of Fig. 5-8.

If the absorption coefficient is vanishingly small as with the usual transmission distances for telecommunications, then variations in the atmosphere, bounds, and medium interact with the frequency to produce such effects as radar echoes and other EM applications. The exponential change will be an attenuation that is described in dB/km or the equivalent. Rain, fog, snow, hail, radio horizons, temperature, tropical, temperate, polar regions, and the troposphere will be some of the variables. The bounds will contribute reflections, reinforcement, and cancellation.

Of great importance is the possibility of distributing the disturbance over various distances which has the effect of diluting a local dissipation. Sometimes shock waves are produced by a sudden step change in wave propagation. Then the shock wave carries the energy from a small interaction region to relatively vast areas for less violent interactions. These are seen in high pressure test equipment, nuclear explosions, and sonic booms but there are many suggestions of shock waves in biological systems especially when attempting explanations for one wave mode preempting the usual one in sensory information processing systems. Signals in a microwave mode received in the auditory system may thus be a matter of distribution of energy away from the sensor rather than an alternate mode of reception.

TIME CONSTANT AND HALF LIFE

A time constant τ can be defined for a decay process. It is equal to

$$\tau = 1/\alpha \tag{70}$$

where α is the decay constant. The latter can then have time dimensions, for example $\sec^{-1}$. The constant τ is used to describe first order pro-

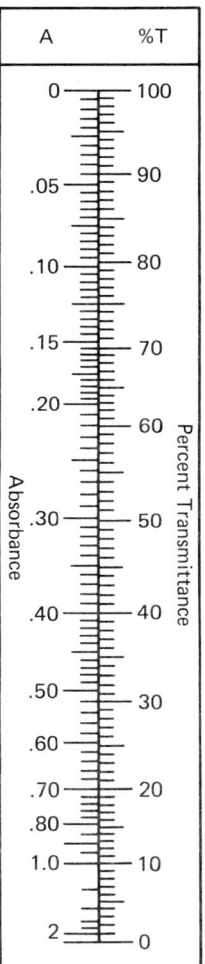

Fig. 5-8. Conversion scale for absorbance to percent transmittance.

cesses in radiation such as dipole polarization, instrument response, and radioactive decay (Fig. 5-9). The basic form of the relationship is

$$y = y_o e^{-\alpha t} \tag{71}$$

Here y_o is the initial quantity and y is the quantity remaining at time t. If *one time constant is taken for t where, by definition, τ is equal to the reciprocal of α*, and α is in reciprocal time units such as one per second or sec^{-1}, then

$$y = y_o e^{-(1/\tau)\tau} \tag{72}$$

$$y = y_o e^{-1} \tag{73}$$

$$y = y_o \frac{1}{e} = y_o/2.718 = 36.8\% \, y_o \tag{74}$$

Again, e^{-1} gives a "remaining fraction" (i.e. y/y_o) of 0.368. There are several objectives here. The first, in the general time constant case, is to point out the special "36.8% left and 63.2% lost" point in the exponential curve for a systematic absorption, orientation, decay or instrumental process. Another is to express a negative exponential change in terms of time and time constants. Thus five time constants will allow 99 per cent of the possible change to occur. The value of α for microwave absorptive polarization, for example, may be determined via certain approximations from Maxwell's equations modified for the material of interest. Table 5-6 gives the exponential functions for x and e^{-x}.

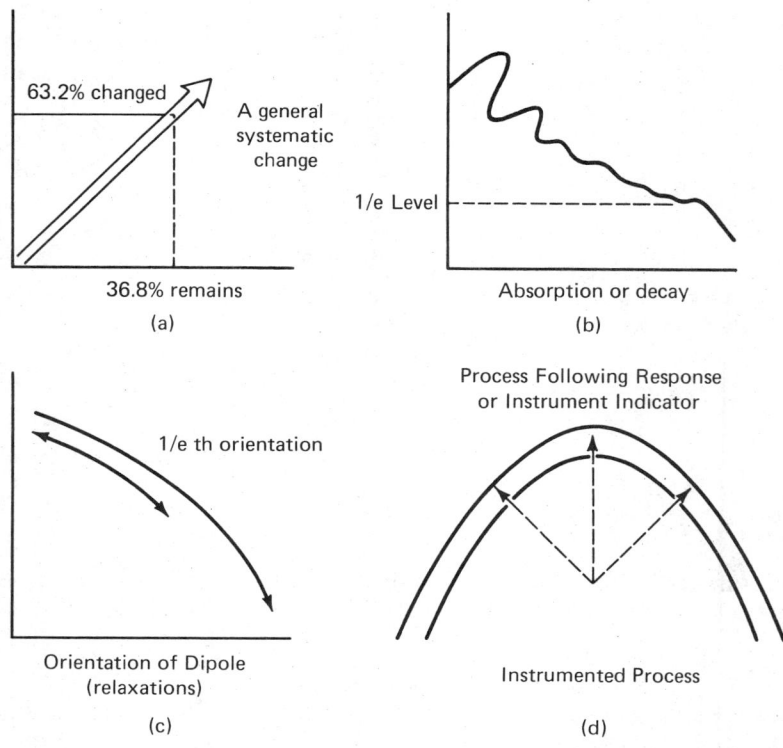

Fig. 5-9. First order processes in which the time constant or relaxation time describes a stage of particular interest and permits a ready comparison between different systems.

TABLE 5-6. The Exponential Functions for x and e^{-x}

x	e^{-x}	x	e^{-x}	x	e^{-x}
0.00	1.000	1.30	0.2725	2.60	0.0743
0.05	0.9512	1.35	0.2592	2.65	0.0707
0.10	0.9048	1.40	0.2466	2.70	0.0672
0.15	0.8607	1.45	0.2346	2.75	0.0639
0.20	0.8187	1.50	0.2231	2.80	0.0608
0.25	0.7788	1.55	0.2122	2.85	0.0578
0.30	0.7408	1.60	0.2019	2.90	0.0550
0.35	0.7047	1.65	0.1921	2.95	0.0523
0.40	0.6703	1.70	0.1827	3.00	0.0498
0.45	0.6376	1.75	0.1738	3.05	0.0474
0.50	0.6065	1.80	0.1653	3.10	0.0450
0.55	0.5770	1.85	0.1572	3.15	0.0429
0.60	0.5488	1.90	0.1496	3.20	0.0408
0.65	0.5220	1.95	0.1423	3.25	0.0388
0.70	0.4966	2.00	0.1353	3.30	0.0369
0.75	0.4724	2.05	0.1288	3.55	0.0351
0.80	0.4493	2.10	0.1225	3.40	0.0334
0.85	0.4274	2.15	0.1165	3.45	0.0317
0.90	0.4066	2.20	0.1108	3.50	0.0302
0.95	0.3867	2.25	0.1054	3.55	0.0287
1.00	0.3679	2.30	0.1003	3.60	0.0273
1.05	0.3499	2.35	0.0954	3.65	0.0260
1.10	0.3329	2.40	0.0907	3.70	0.0247
1.15	0.3166	2.45	0.0863	3.75	0.0235
1.20	0.3012	2.50	0.0821	3.80	0.0224
1.25	0.2865	2.55	0.0781	3.85	0.0213

The half life $T_{1/2}$ of a process is another special change where

$$y = 0.5 \, y_o \tag{75}$$

Then

$$y = y_o e^{-\alpha T_{1/2}} \tag{76}$$

$$y/y_o = 0.5 \tag{77}$$

$$y/y_o = e^{-0.693} \tag{78}$$

$$T_{1/2} = 0.693/\alpha \tag{79}$$

$$\alpha = 0.693/T_{1/2} \tag{80}$$

Also

$$\tfrac{1}{2} = \exp(-\alpha T_{1/2}) \tag{81}$$

$$2 = \exp \alpha \, T_{1/2} \tag{82}$$

$$\ln 2 = \alpha T_{1/2} \tag{83}$$

$$0.693 = \alpha T_{1/2} \tag{84}$$

Then if either the decay constant or half life is known, the other may be found. A well known example is the decay of a radioactive sample.

Nuclear reactions depend on the opportunity afforded a bombarding particle to interact with a nucleus in its pathway. This situation is governed by an interception factor σ and the cross section, measured in barns, a unit which equals 10^{-24} cm. The product of σ and N, the number of nuclei per cm^3, is the absorption coefficient for the radiation. Thus for a thickness of absorber x, the fraction of particles changed as a result of the interaction would be y/y_o where y_o is the incident number of them. Then

$$y = y_o e^{(-\sigma N x)} \tag{85}$$

BIOMOLECULAR INFORMATION FROM SPECTROSCOPY

Information about the molecules in biophysical systems is accessible through their interactions, which is the mode of access used in spectroscopy. At this level of analysis, a great deal of information on the nature and function of biomolecules is available. Recent bioresearch makes it increasingly clear that this is the logical place to begin a serious study of radiation interactions. Spectroscopy begins with radiofrequencies and continues through all the spectra to the highest frequencies in which devices are available. Sometimes a multifrequency situation will forcefully indicate that more needs to be known than can be obtained by the operation of one analyzer alone. If there is a radiation experiment calculated to show biofunctional changes at a certain wavelength, observed as actual structural changes in the molecule at a second wavelength, then the spectroscopy at both wavelengths needs to be understood. In addition to the protein absorption in the ultraviolet already mentioned, this situation frequently arises with photosensitive materials analyzed by visible light after ionizing irradiation. In a sense, spectral analysis occurs at both wavelengths, and both the irradiator and the analyzer are instruments for spectroscopy.

Organizational Hierarchy. Thus we have a field of study (bioaction spectra) which rather conveniently presents the theory of this instrumentation and the molecular biophysics susceptible to analysis by it. From this level, one can build through the organizational hierarchy by adding system, cellular, organismic, and finally bulk isotropic and anisotropic effects. Considerable confusion may occur when the fundamental molecular level is not included. A well known scientific hassle over whether microwave effects should be called athermic or thermic is an example.

ENERGY REGIONS OF THE SPECTRUM

The resonant frequency of a biomolecule with respect to its forcing function is perhaps more its true "fingerprint" than any other identification. At this frequency the molecule is known by its characteristic absorption of the electromagnetic energy. It is quite like an abnormality in the curve for absorption versus frequency—a point of abrupt change —which is usually called anomalous absorption, dispersion of the dielectric, or some other descriptive constant.

The visible, infrared, and ultraviolet radiations all cause many bioenergetic changes and an extensive formalism has developed for their description. The energy E as proposed by Max Planck in 1901 in these as well as in all electromagnetic radiations, is equal to the product of his constant h and the frequency. E quanta = hf where h has units of erg·sec. That is, $h = 6.6 \times 10^{-27}$ erg·sec/quantum and 1.6×10^{-12} ergs = 1 electron volt. Quanta are wave units which fit events and structures in the atomic and molecular world and, thus, occur in discrete packets. *The match between these energy packets and their molecular sinks is the logic in the quantal connection.* In Table 5-7, five sample wavelengths are shown with their equivalent quanta. The spectra (infrared, ultraviolet, ionizing radiation, and visible) will have quanta (absorbed energy) ranging from 0.1 to 12 or more. The lower frequencies or the radio and microwave bands have smaller associated energy, ranging from 10^{-2} to 10^{-6} eV. The energy level of the hydrogen atom in the lowest orbital state is 13.5 eV. This is the ordinary ground state. Absorption of this much energy would correspond to the work needed to pull the electron away from the proton nucleus. More than the separation energy would give the separated electron added kinetic energy. When appropriate wavelengths irradiate this atom, they can be absorbed to produce an excitation transition to a different energy state (above the ground state). If the two states are E_1 to E_2, then E_3 must be the one satisfied by the equation

$$E_3 = E_2 - E_1 = hf = h(c/\lambda) \tag{86}$$

TABLE 5-7. Wavelengths and Their Equivalent Quanta (eV)*

Å	eV	Region
10,000	1.24	IR
7,000	1.77	VIS
4,000	3.1	VIS
2,000	6.2	UV
1,000	12.4	UV

*The logic of the quantal connection is that each of these energy packets matches some specific molecular sink.

The excited atom can then lose its excess energy in the form of an emission of radiation. This second transition can occur after a microsecond or less. Usually, the absorbed energy will match the difference E_3. If not, as in a case like Compton scatter with X-ray absorption, the surplus portion of the initial radiation can continue on at a reduced energy. This radiation may easily be powerful enough to ionize the molecules it meets, continuing the ionization radiation process until the energy is dissipated either thermally or by the ensuing reactions. In the hydrogen example it is seen that there is, in addition to those already given, an *electronic mode* for the interconversion of energy.

The fast electron from gamma interactions with matter results in a local injection of 55 eV on the average which is capable of disrupting any molecular bond if it correctly interacts.

QUANTUM FITTING AS AN INFORMING FUNCTION

The radiation, according to its discrete quanta, is then selective for actions or events with which it can logically relate, that is, fit and inform. This is true for polarizations or displacements involving quantized motions in an atom, such as angular momentum of the nucleus or electrons in orbit. It is true for photoabsorption processes in the visible spectrum where quanta may be called photons. Photons of the logical size will be above the threshold level for tautomeric changes in rhodopsin and retinal flashes so that information may be passed to the optic nerve. In photosynthesis, only the energy associated with specific wavelengths such as 880 nm in bacteria, or 700 nm and 400 to 500 nm in plants are correct for exciting chlorophyll to release energy-burdened electrons and set off the chain of events leading to ATP and reduced coenzymes.

Forces such as those in momentum are familiar from Newtonian physics, but atomic behavior involves somewhat different, or quantum, phenomena. The time periods of interest are less than femto seconds (fs = 10^{-15} sec) and reality in this domain is vested in the energy levels of the atoms rather than in their precise locations which are uncertain according to Heisenberg's principle. The masses are orbiting electrons, atomic nuclei, small molecules, and the related particles. The distances are atomic, that is, about 10^{-8} cm or 10 nanometers. Thus the space occupied by these particles is enormously disparate, the nucleus appearing like a golf ball centered in an electron planetary system revolving at a distance of two football fields away. Thus, density is essentially nuclear and *there* it is about 100 megatons per cubic centimeter or 1.6 gigatons per cubic inch. At this packing density, the substance or weight in a human body could be expanded from material less than the size of a dot. The larger world is an assembly of such tiny dimensions, enormous densities, fleeting times, occurring in a space desert and following a correspondence principle. Measurements in this world deal with the average ensemble characteristics of the small particles. Radiation biology deals with the conditions for interaction between electromagnetic fields and these entities. With interactions quantized, the energies are prescribed at specified levels, such as E_3 which is the difference between E_2 and E_1 in the example above. The magnitudes encountered are not continuous in the scale from low to high but, instead, there is a discrete set of quantal steps.

WAVES AND PARTICLES

Again, interest may center on wave behavior when electromagnetic radiation is discussed. The wave is seen as a sinuous track, an elastic disturbance in the media, or the periodic rise and fall in amplitude of the wave quantities such as electric and magnetic field strengths (see Chapter 7). Sometimes it is convenient to emphasize this wave character of radiations giving at once the benefits of both known wave behavior and mechanics. The latter suggests phenomena such as reflection, transmission, absorption, refraction (and its indices), diffraction, reinforcement, cancellation, and interference, the phenomena of interacting waves.

When interest is turned to the particle nature of waves, it is clear that entities like high velocity electrons in beta rays, alpha particles or helium nuclei which can be accelerated and take on kinetic energy are involved. The radiation pressure of light from a laboratory-type arc lamp on a blackened disc suspended from a quartz fiber (radiometer) is about 7×10^{-5} dynes/cm^2, a number to be expected from particle theory and calculations. Again, in the electron microscope it is a stream of particles, electrons emitted from their source, which *scatter* from portions of the observed preparation. At this primary interface, nonspecific electron energy is converted to a "related," almost cognitive, energy capable of forming its own image on a sensitive plate. This favors particle form but relates to both, that is, a specific wavelength and scatter or collision reactions. In light microscopes the result is similar, but the interaction is customarily in *wave* terms, refraction, transmission, etc., giving a distinction that is comfortable to microscopists.

Current research in biology relies in a growing way on quantum mechanics as the theory to predict the status of a system and the behavior of its components in the atomic and molecular domain. The status is described in quantum numbers which are lumped in the energy states E_1, E_2 or E_3 etc.

If an energy state meets incoming radiation of the correct frequency, according to $E_1 = hf_1$, then the energy in f_1 can be absorbed if the atom is to increase its energy, or emitted if it is to be the other way around ($hf_3 = E_2 - E_1 = E_3$). The behavior of components of the system—the quantization or specification of spaces, orientations, momenta, response to magnetic and electric fields, excitations, and transitions—all are determined by the special rules of quantum mechanics. They permit a characteristic absorption pattern, emission pattern, or typical transition so that the status of the biological system is revealed through its radiation interaction. In all cases the supporting concept is information flow to or from molecules as shown by molecular spectra. A well known emission of quanta is the yellow-orange light given off by sodium in the flame test. This light has a clearly defined wavelength of 590 millimicrons. The radiation energy emitted informs the observer that the emission represents a discrete difference between the excited and ground states of the sodium atom. Activation energies for many bioreactions range from 10 to 100 kcal/mole so that many photoreactions are possible at corresponding wavelengths from 3,000 to 30,000 angstroms. In photosynthesis, absorption of 3 quanta at 6800Å or 41 kcal/mole each, is theoretically capable of fixing one molecule of CO_2. (One kcal = 4.186×10^{10} ergs.)

Another level of inquiry is opened up when it is considered that energy may be interconverted between the given modes, that is, from one to another of the rotational, vibrational, translational, and electronic modes. These particular interactions suggest the many "black boxes" that still exist in the understanding of the transfer and storage of energy by molecules. Not much is known about gas phase systems in biological cells and systems, for example, or about electronic and vibrational transfer of energy in them. Yet gas phase reactions or systems in which the radiation interactions might occur are a feature of mammalian lungs and insect spiracles, specifically the alveoli and tracheoles, in which the exchange of oxygen and carbon dioxide can occur. Also, cells have gas vacuoles and the important bladders and intestinal systems have variable volume gas spaces. There is a recurring and often lethal involvement of the latter in radiation biology. A great deal is known, however, about some modes of interaction such as dielectric absorption, the relaxation process, and resonances. These conceptualizations are developed in the next chapter. Polarization, as a concept developed in this chapter, remains one of the most important unifying characteristics of the radiation response as progress continues to be made toward casting the needed theoretical framework.

PROBLEMS

1. How does ploidy affect the ionizing radiation resistance of yeast?
2. What is the information density in a cell?
3. List seven types of resonance.
4. What is the logic of the quantal connection?

BIBLIOGRAPHY

Alexander, P., M. Fox, K.A. Stacey, and D. Rosen, *Nature*, Vol. 178, 846, 1956.
Benumof, R., *Concepts in Electricity and Magnetism*, Holt, Reinhart and Winston, New York, 1961.
Clark, G.L., *The Encyclopedia of Spectroscopy*, Reinhold Pub. Co., New York, 628-651, 1960.
Clayton, R.K., *Molecular Physics in Photosynthesis*, Blaisdell, New York, 1965.
Davies, D.W., *The Theory of the Electric and Magnetic Properties of Molecules*, John Wiley, New York, 1967.
Day, R.H., *Human Perception*, John Wiley, New York, 1969.
Debye, P., *Polar Molecules*, Dover Pub., Great Barrington, Mass. 1930.
Goodrich, F.C., *A Primer of Quantum Chemistry*, John Wiley, New York, 1972.

Hedvig, P. and G. Zentai, *Microwave Study of Chemical Structure and Reactions*, CRC Press, Cleveland, Ohio, 1969.
Ingram, D.J.E., *Spectroscopy at Radio and Microwave Frequencies*, Plenum Press, New York, 1967.
Kamen, M.D., *Primary Processes in Photosynthesis*, Academic Press, New York, 1963.
Pollard, E., *Physics, An Introduction, 1969*, Oxford U. Press, New York, 1969.
Pollard, E.C. and Kraus, K., "All-Or-Nothing Character of DNA Degradation in Bacteria After Ionizing Radiation," *Biophysical Journal*, 1973.
Poole, C., *The Theory of Magnetic Resonance*, Wiley Interscience, New York, 1972.
Rojansky, V., *Electromagnetic Fields and Waves*, Prentice-Hall, Englewood Cliffs, New Jersey, 1971.
Squire, C.F., *Waves in Physical Systems*, Prentice-Hall, Englewood Cliffs, New Jersey, 1971.
Uberall, H., *Electron Scattering from Complex Nuclei*, Academic Press, New York, 1971.
Vonsovskii, S.V., *Ferromagnetic Resonance*, Pergamon Press, Oxford, 1966.

Films

Crapuchettes, P., *Wave Action and Microwave Heating*, Litton Co., Electron Tube Div., San Carlos, California, 1970.

6

Radiation Biology and Biospectroscopy

"Since nature designs everything from atoms, we should be able to create with foresight any feasible kind of material and device if we understood the Periodic System in all its implications."

Arthur von Hippel in *The Molecular Designing of Materials and Devices*. The M.I.T. Press, Cambridge, Mass. 1965

The subject of radiation biology involves radiation interactions, the effects of radiation on biochemicals, water, macromolecules, cells, enzymes, tissues, radiosensitivity and repair, acute, chronic or delayed effects, low and high level exposures, radiation sicknesses, protection, treatment, radiotherapy, and genetic changes. Other branches, such as radiology, health physics, and work with isotopes involve specialization in a related area. As a second consideration, biospectroscopy is the use of radiations, mainly of the electromagnetic spectrum, in order to extract information on the interactions of these radiations and on the structure and function of biological substances from their spectra. It is apparent that, due to the tremendous number of kinds of action spectra, this second kind of study is more fundamental and unifying than the first and may, if one can generalize on the associated instrumentation, include many of the principles belonging to radiation biology. Thus the radiation used may come from any subspectrum—from radio to gamma rays—and carry the corresponding name, for example, microwave, infrared, ultraviolet, and others (Table 6-1). Spectrometers utilize a source of radiation energy in the spectrum of interest, controllers, a sample or preparation (which may well become a secondary radiation source itself), sensing, or detecting and readout devices which are characteristic for the spectrum selected. Tracer studies are, basically, modifications of the specimen with isotopes or other indicators which help to identify the parts of the biosystem and the changes occurring. The

TABLE 6-1. List Showing Diversity In Selected Spectroscopies

Nuclear Magnetic Resonance or Nuclear Magnetic Relaxation (Overhauser Effect) form of NMR with other superimposed saturated NMR	Fast Neutron
	Raman
	Mössbauer
	Kerr
	Zeeman
Microwave	Proton Magnetic Resonance
Radio Frequency	Atomic Fluorescence or Atomic Fluorescence Flame
Electron Spin Resonance	
Infrared, Near, Far	Internal Reflection Spectroscopy
Visible	Phosphorescence
Ultraviolet	Low Temperature Spectroscopy
Vacuum Ultraviolet	Astronomical
X-ray	Attenuated Total Reflectance
X-ray Diffraction	Absorptiometry
Soft X-ray Spectroscopy	Fluorimetry and Low Temperature Fluorimetry
Gamma	
Meson	Quenchofluorimetry
Thermal Neutron Scattering	Phosphorimetry
	Quenchophosphorimetry

similarities between radiobiology and spectroscopic studies are in the interactions studied and, often, in the objectives.

Too often we examine mechanisms and effects in this field of radiobiology without the benefit of knowing intermediates and reaction products. This comes about because so much information remains locked into living systems. Spectroscopy now provides abundant clues as to this kind of information, so that the two fields are interdependent. Now the problem is to interpret all the information that is forthcoming against incomplete understanding of some bioprocesses, for example, the semiconduction or biosolid state in enzymology and, in general, the bioelectricity of tissues and membranes, and "manipulation" (for either good or evil) of DNA information—for example in good radiotherapy or bad carcinogenesis.

All cells, from small to large, have about 1 $\mu f/cm^2$ capacity for membranes which may have dielectric loss mechanisms and components with the characteristics often found in solid dielectrics. The cell interiors, on the other hand, have conductances in the range of those of biological electrolytes, of the order of 100 ohm cm. Therefore, *explanations* may be sought in semi-conduction-solid state theory.

In this chapter, biospectroscopy is examined to provide an understanding of its importance in the biological sciences, as a fundamental way of looking at the biology of radiations, and finally, to support the concept that radiation information avenues may be followed into biomolecules and their operations.

Radiation Biology and Biospectroscopy

We also refer once again to polarizations, now in relation to dielectrics and the dielectric constant both for their connection with microwave spectroscopy and with the subjects explored in the next two chapters. First, however, the intriguing concepts of biomolecular engineering and dielectromagnetics will be mentioned briefly. The former recognizes the possibility that biomolecules may be combined into materials in which the characteristics and performance are predictable from a knowledge of the molecular properties, particularly of their constants such as the dielectric constant, the semiconductivity, and related electrophysical properties. The term dielectro-magnetic engineering actually arises from the two areas under discussion, dielectrics and electromagnetics. The dielectrics part of this is explained by the quotation that opens this chapter and suggests the kind of information in the Periodic System that Dr. von Hippel had in mind—a kind of electrical or dielectrical reaction to applied forces and processes, varying with the element. The author once observed that with a fully classified and utilized electromagnetic spectrum, one might select frequencies to satisfy defined requirements in a process. Then, by resorting to the device which delivers this specific energy in the form required, one could obtain some truly sophisticated solutions to process problems. The term dielectromagnetic engineering combines aspects of von Hippel's enthusiasm for the Periodic System of the Elements with the author's respect for the electromagnetic spectrum. It is then obvious that neither sentiment is justified fully without the other (Fig. 6-1). The conclusion for educa-

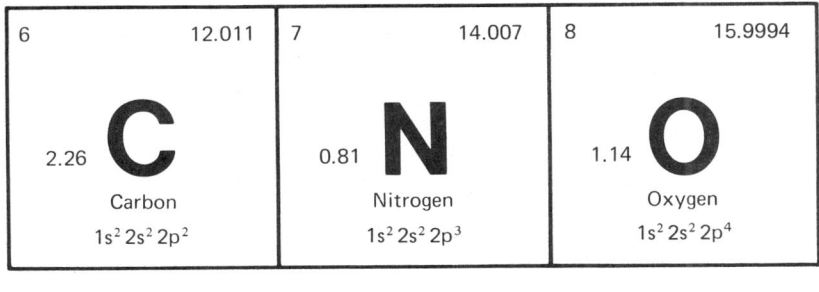

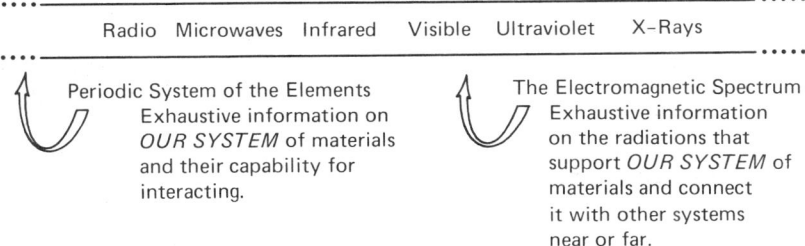

Fig. 6-1. Function of Periodic System information.

tion in the future is that the study of materials and electromagnetic energy interactions and especially of dielectric processes will need much more emphasis. As if to connect these lines of thought we have the fact that it was the early work in spectroscopy following Nils Bohr's dynamic model of the atom that established the basic physical terms of the Periodic System of the Elements.

A good beginning may be made by viewing polarization as a displacement, and the dielectric constant as an ensemble molecular property in the polarization model. Special entities that emerge from the associated "electronics" include the free radicals and, of course, their biological functions. The resonances (electron spin and nuclear magnetic) appear then as special situations in which radiation energy is eased into the interaction. Time-dependent processes bring in the nature of relaxation and its effects. The possibility then arises that access to information in these radiation methods may lead to hazards as well as to biological tools.

It is no accident that polarization is a recurring theme in these sections dealing with radiations, for this behavior is as fundamental as are positive and negative charges, on-off switching in electronic circuits, repressor-depressor control of reactions, and the unit of information, the bit, which derives from the on-off choice between system conditions.

$$\log_2 2 = 1 \text{ bit} \tag{87}$$

What may not be obvious as we look at spectroscopic avenues into biological molecules and their functions, and involve ourselves with polarization notation, is that these ideas have broad implications in the evolution of society. Similarly, we are ever more interested in electric fields and waves among organisms, and the radiation energy and information complex serves to unite these concepts quite effectively.

It is the genetic deviation from some constant pattern in the shape and form of biological units such as molecules and cells that permits evolutionary experiments to begin and natural selection to operate. If 100 standard blocks are contemplated, they reveal little of this adventurous possibility, but if ten blocks among them have pegs to which another block can attach itself, then the assembly is richer in information and offers more possibilities as these units are made to interact in various environments. The deviant state with pegs may be said to complete a form of polarization of pegs and unpegs and, alternatively, the constant form would represent depolarization. Similarly, molecules such as the sugars may boast of little differences in structure—perhaps only a hydroxyl group is displaced—but, at once, an optical activity may be added. To the trained observer this represents information on the form-related functions of this molecule. There are many other examples involving information in or from molecules. The orbits of electrons, excited states, energy emissions and absorptions, distances between nuclei, rotations, vibrations, inversions, angular momenta, quantum size for radiation interaction, and mode of interaction, inter-

```
        H                         H
        |                         |
        C—O                       C—O
        |                         |
   OH—C—H                     H—C—OH
        |                         |
    H—C—OH                    H—C—OH
        |                         |
        H                         H
```

    levrorotary glycerose   *dextrorotary glycerose*

bond distance, bond nature, and interatomic forces are all examples of this. The form factor deviations in or among any units in an organism or population may be adventitious and followed up. They are avenues along which change can occur as natural pressures and selections develop in the history of the species.

    There is an uneasy element to this in that we often recognize constancy or perfection of form as beauty, and yet, by this standard, it is the imperfect unit that offers the greater possibilities. However, this may be no more than we have come to expect as we recognize stability as easy and change as difficult. Nevertheless, it is the deviation in the coded information in molecules that makes innovations possible and this leads to the progress represented, hopefully, by human beings. While stability is essential, and guaranteed as we have seen by cybernetic-like control and regulation, adventure is also required and it is inevitable as DNA master control biomolecules are exposed to mutations and alternations in genetic information. Every cybernetic control uses a searching and probing in which the system seldom returns to the exact same status. Thus these changes have the blessing of nature. The electromagnetic radiations interact quite specifically in mutations, sometimes by intention and sometimes by accident, thus communicating information. Normally, the visual subspectrum would be credited with the lion's share of this activity as vision brings us beauty which leads to ideas and photochemistry brings energy for all the necessary processes. Yet as we see in this section, when we study some of the great ideas in information spectroscopy, all the spectra have their contributions. In extrasensory perception, many persons feel that communications occur in some physical or metaphysical manner, even at great distances. While we have such communication as that represented by electroencephalography within the environment and the intraorganismic waves that support it, it is intimate. The possibility of certain interactions between organisms cannot be dismissed. An alpha wave reader used as a means for feeding back brain operating data for use in personal meditation and psychic training is now an accepted device. The tracing is char-

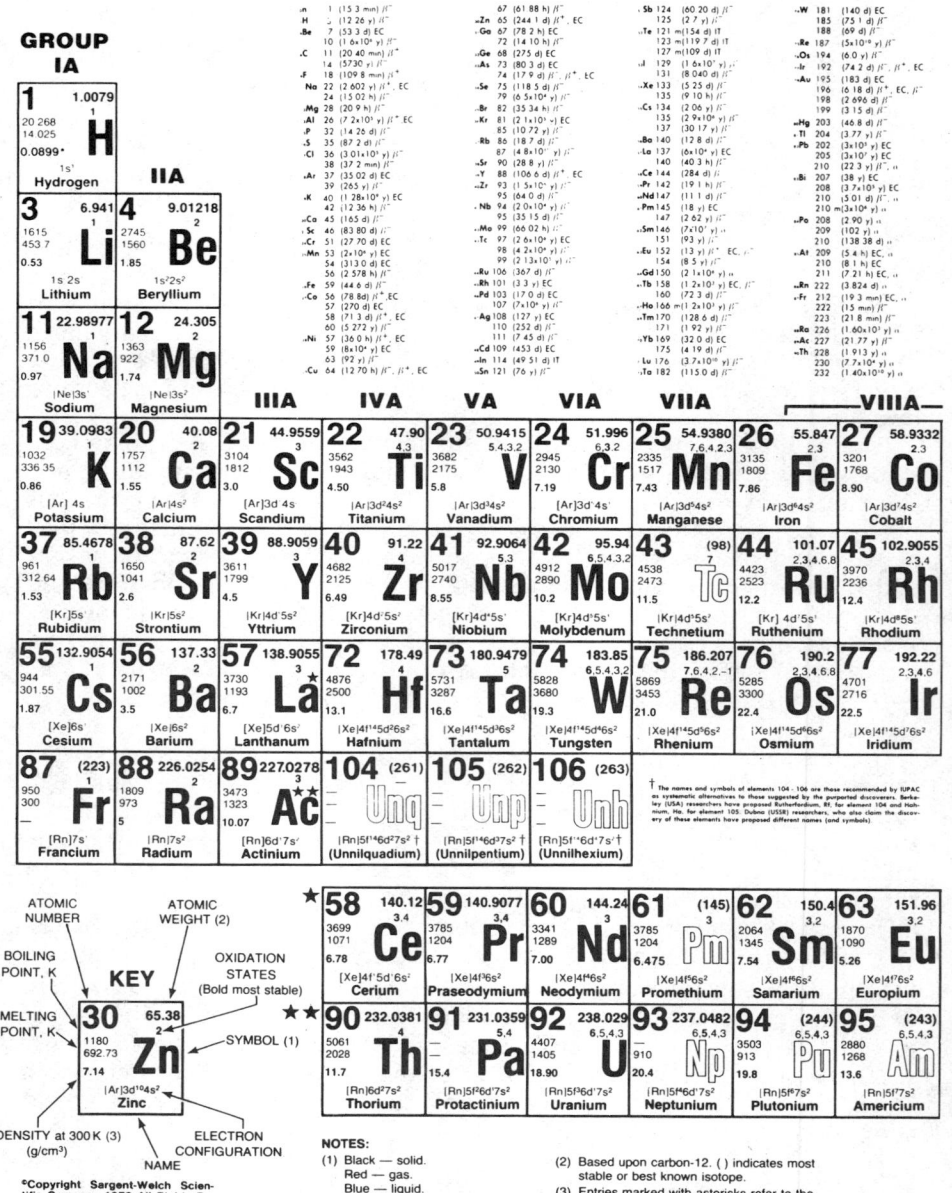

OF THE ELEMENTS

Isotopes

Pa	231	(3.28×10^4 y) α
U	233	(1.59×10^5 y) α
	234	(2.44×10^5 y) α
	235	(7.04×10^8 y) α
	236	(2.34×10^7 y) α
	238	(4.47×10^9 y) α
Np	236	(1.1×10^5 y) EC, β^-
	237	(2.14×10^6 y) α
	239	(2.346 d) β^-
Pu	238	(87.75 y) α
	239	(2.41×10^4 y) α
	240	(6.54×10^3 y) α
	242	(3.8×10^5 y) α
	244	(8.3×10^7 y) α
Am	241	(432 y) α
	243	(7.37×10^3 y) α
Cm	242	(163.2 d) α
	244	(18.12 y) α
	247	(1.55×10^7 y) α
	248	(3.5×10^5 y) α, SF
Bk	247	(1.4×10^3 y) α
Cf	249	(351 y) α
	251	(900 y) α
Es	252	(472 d) α
	253	(20.47 d) α
	254	(276 d) α
Fm	255	(20.1 h) α
	257	(100.5 d) α
Md	258	(55 d) α
No	259	(58 min) α
Lr	260	(3.0 min) α
104	261	(65 s) α
105	262	(40 s) α
106	263	(0.9 s) α

Selected Radioactive Isotopes

Naturally occurring radioactive isotopes are designated by a mass number in blue (although some are also manufactured). Letter m indicates an isomer of another isotope of the same mass number. Half-lives follow in parentheses, where s, min, h, d, and y stand respectively for seconds, minutes, hours, days, and years. The table includes mainly the longer-lived radioactive isotopes; many others have been prepared. Isotopes known to be radioactive but with half-lives exceeding 10^{12} y have not been included. Symbols describing the principal mode (or modes) of decay are as follows (these processes are generally accompanied by gamma radiation):

- α alpha particle emission
- β^- beta particle (electron) emission
- β^+ position emission
- EC orbital electron capture
- IT isomeric transition from upper to lower isomeric state
- SF spontaneous fission

	IB	IIB	IIIB	IVB	VB	VIB	VIIB	VIII
								2 He 4.00260 4.215 0.95 (at 26 atm) 0.1787* $1s^2$ Helium
			5 B 10.81 3 4275 2300 2.34 $1s^22s^2p$ Boron	6 C 12.011 ±4,2 4470* 4100* 2.62 $1s^22s^2p^2$ Carbon	7 N 14.0067 ±3,5,4,2 77.35 63.14 1.251* $1s^22s^2p^3$ Nitrogen	8 O 15.9994 −2 90.18 50.35 1.429* $1s^22s^2p^4$ Oxygen	9 F 18.998403 −1 84.95 53.48 1.696* $1s^22s^2p^5$ Fluorine	10 Ne 20.179 27.096 24.553 0.901* $1s^22s^2p^6$ Neon
			13 Al 26.98154 3 2793 933.25 2.70 $[Ne]3s^2p$ Aluminum	14 Si 28.0855 4 3540 1685 2.33 $[Ne]3s^2p^2$ Silicon	15 P 30.97376 ±3,5,4 550 317.30 1.82 $[Ne]3s^2p^3$ Phosphorus	16 S 32.06 ±2,4,6 717.75 388.36 2.07 $[Ne]3s^2p^4$ Sulfur	17 Cl 35.453 ±1,3,5,7 239.1 172.16 3.17* $[Ne]3s^2p^5$ Chlorine	18 Ar 39.948 87.30 83.81 1.784* $[Ne]3s^2p^6$ Argon
28 Ni 58.70 2,3 3187 1726 8.90 $[Ar]3d^84s^2$ Nickel	29 Cu 63.546 2,1 2836 1357.6 8.96 $[Ar]3d^{10}4s^1$ Copper	30 Zn 65.38 2 1180 692.73 7.14 $[Ar]3d^{10}4s^2$ Zinc	31 Ga 69.72 3 2478 302.90 5.91 $[Ar]3d^{10}4s^2p^1$ Gallium	32 Ge 72.59 4 3107 1210.4 5.32 $[Ar]3d^{10}4s^2p^2$ Germanium	33 As 74.9216 ±3,5 876 (subl.) 1081 (28 atm) 5.72 $[Ar]3d^{10}4s^2p^3$ Arsenic	34 Se 78.96 −2,4,6 958 494 4.80 $[Ar]3d^{10}4s^2p^4$ Selenium	35 Br 79.904 ±1,5 332.25 265.90 3.12 $[Ar]3d^{10}4s^2p^5$ Bromine	36 Kr 83.80 119.80 115.78 3.74* $[Ar]3d^{10}4s^2p^6$ Krypton
46 Pd 106.4 2,4 3237 1825 12.0 $[Kr]4d^{10}$ Palladium	47 Ag 107.868 1 2436 1234 10.5 $[Kr]4d^{10}5s^1$ Silver	48 Cd 112.41 2 1040 594.18 8.65 $[Kr]4d^{10}5s^2$ Cadmium	49 In 114.82 3 2346 429.76 7.31 $[Kr]4d^{10}5s^2p^1$ Indium	50 Sn 118.69 4,2 2876 505.06 7.30 $[Kr]4d^{10}5s^2p^2$ Tin	51 Sb 121.75 ±3,5 1860 904 6.68 $[Kr]4d^{10}5s^2p^3$ Antimony	52 Te 127.60 −2,4,6 1261 722.65 6.24 $[Kr]4d^{10}5s^2p^4$ Tellurium	53 I 126.9045 ±1,5,7 458.4 386.7 4.92 $[Kr]4d^{10}5s^2p^5$ Iodine	54 Xe 131.30 165.03 161.36 5.89* $[Kr]4d^{10}5s^2p^6$ Xenon
78 Pt 195.09 2,4 4100 2045 21.4 $[Xe]4f^{14}5d^96s^1$ Platinum	79 Au 196.9665 3,1 3130 1337.58 19.3 $[Xe]4f^{14}5d^{10}6s^1$ Gold	80 Hg 200.59 2,1 630 234.28 13.53 $[Xe]4f^{14}5d^{10}6s^2$ Mercury	81 Tl 204.37 3,1 1746 577 11.85 $[Xe]4f^{14}5d^{10}6s^2p^1$ Thallium	82 Pb 207.2 4,2 2023 600.6 11.4 $[Xe]4f^{14}5d^{10}6s^2p^2$ Lead	83 Bi 208.9804 3,5 1837 544.52 9.8 $[Xe]4f^{14}5d^{10}6s^2p^3$ Bismuth	84 Po (209) 4,2 1235 527 9.4 $[Xe]4f^{14}5d^{10}6s^2p^4$ Polonium	85 At (210) ±1,3,5,7 610 575 — $[Xe]4f^{14}5d^{10}6s^2p^5$ Astatine	86 Rn (222) 211 202 9.91* $[Xe]4f^{14}5d^{10}6s^2p^6$ Radon

The A & B subgroup designations, applicable to elements in rows 4, 5, 6, and 7, are those recommended by the International Union of Pure and Applied Chemistry. It should be noted that some authors and organizations use the opposite convention in distinguishing these subgroups.

* Estimated Values

64 Gd 157.25 3 3539 1585 7.89 $[Xe]4f^75d^16s^2$ Gadolinium	65 Tb 158.9254 3,4 3496 1630 8.27 $[Xe]4f^96s^2$ Terbium	66 Dy 162.50 3 2835 1682 8.54 $[Xe]4f^{10}6s^2$ Dysprosium	67 Ho 164.9304 3 2968 1743 8.80 $[Xe]4f^{11}6s^2$ Holmium	68 Er 167.26 3 3136 1795 9.05 $[Xe]4f^{12}6s^2$ Erbium	69 Tm 168.9342 3,2 2220 1818 9.33 $[Xe]4f^{13}6s^2$ Thulium	70 Yb 173.04 3,2 1467 1097 6.98 $[Xe]4f^{14}6s^2$ Ytterbium	71 Lu 174.967 3 3668 1936 9.84 $[Xe]4f^{14}5d^16s^2$ Lutetium
96 Cm (247) 3 — 1340 13.511 $[Rn]5f^76d^17s^2$ Curium	97 Bk (247) 4,3 — — — $[Rn]5f^97s^2$ Berkelium	98 Cf (251) 3 — 900 — $[Rn]5f^{10}7s^2$ Californium	99 Es (252) — — — — $[Rn]5f^{11}7s^2$ Einsteinium	100 Fm (257) — — — — $[Rn]5f^{12}7s^2$ Fermium	101 Md (258) — — — — $[Rn]5f^{13}7s^2$ Mendelevium	102 No (259) — — — — $[Rn]5f^{14}7s^2$ Nobelium	103 Lr (260) — — — — $[Rn]5f^{14}6d^17s^2$ Lawrencium

NSRDS

SARGENT-WELCH SCIENTIFIC COMPANY
7300 NORTH LINDER AVENUE, SKOKIE, ILLINOIS 60077

Side 1

Catalog Number S-18806

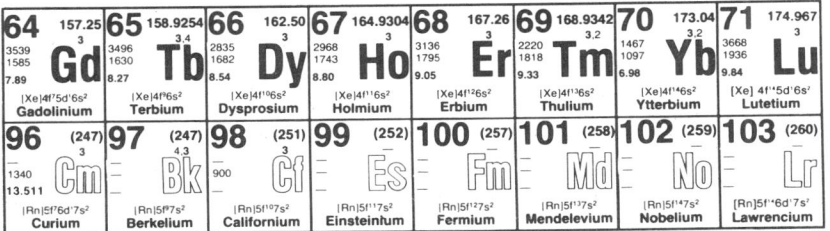

Fig. 6-2. Periodic Table of the Elements (by permission of Sargent-Welch Scientific Company).

Informational Bioelectromagnetics

TABLE OF PERIODIC PROP-

Percent Ionic Character of a Single Chemical Bond

Difference in electronegativity	0.1	0.2	0.3	0.4	0.5	0.6	0.7	0.8	0.9	1.0	1.1	1.2	1.3	1.4	1.5	1.6	1.7	1.8
Percent ionic character %	0.5	1	2	4	6	9	12	15	19	22	26	30	34	39	43	47	51	55

DATA CONCERNING THE MORE STABLE ELEMENTARY (SUBATOMIC) PARTICLES

	Neutron	Proton	Electron*	Neutrino*	Photon
Symbol	n	p	e (e^-)	ν	γ
Rest mass (kg)	1.67495×10^{-27}	1.67265×10^{-27}	9.1095×10^{-31}	~0	0
Relative atomic mass ($^{12}C = 12$)	1.008665	1.007276	5.48580×10^{-4}	~0	0
Charge (C)	0	1.60219×10^{-19}	-1.60219×10^{-19}	0	0
Radius (m)	8×10^{-16}	8×10^{-16}	$<1 \times 10^{-16}$	~0	0
Spin quantum number	1/2	1/2	1/2	1/2	1
Magnetic Moment†	$-1.913\ \mu_N$	$2.793\ \mu_N$	$1.001\ \mu_B$	0	0

KEY

- SYMBOL: Zn
- COVALENT RADIUS, Å: 1.25
- ATOMIC RADIUS, Å (7): 1.53
- ATOMIC VOLUME, cm³/mol (8): 9.2
- FIRST IONIZATION POTENTIAL, V: 9.394
- SPECIFIC HEAT CAPACITY, J g⁻¹ K⁻¹ (3): 0.39
- CRYSTAL STRUCTURE (2)
- ACID-BASE PROPERTIES (1)
- ELECTRONEGATIVITY (Pauling's): 1.65
- HEAT OF VAPORIZATION, kJ/mol (4): 115.30
- HEAT OF FUSION, kJ/mol (5): 7.322
- ELECTRICAL CONDUCTIVITY, $10^5\ \Omega^{-1}\ cm^{-1}$ (6): 0.166
- THERMAL CONDUCTIVITY, W cm⁻¹ K⁻¹ (3): 1.16

© Copyright 1962
© Copyright 1964
© Copyright 1965
© Copyright 1966
© Copyright 1968
© Copyright 1979
© Copyright 1980

SARGENT-WELCH SCIENTIFIC COMPANY
7300 LINDER AVENUE, SKOKIE, ILLINOIS 60077

The names and symbols of elements 104–106 are those recommended by IUPAC as systematic alternatives to those suggested by the purported discoverers. Berkeley (USA) researchers have proposed Rutherfordium, Rf, for element 104 and Hahnium, Ha, for element 105. Dubna (USSR) researchers, who also claim the discovery of these elements have proposed different names (and symbols).

Catalog Number S-18806   SIDE 2

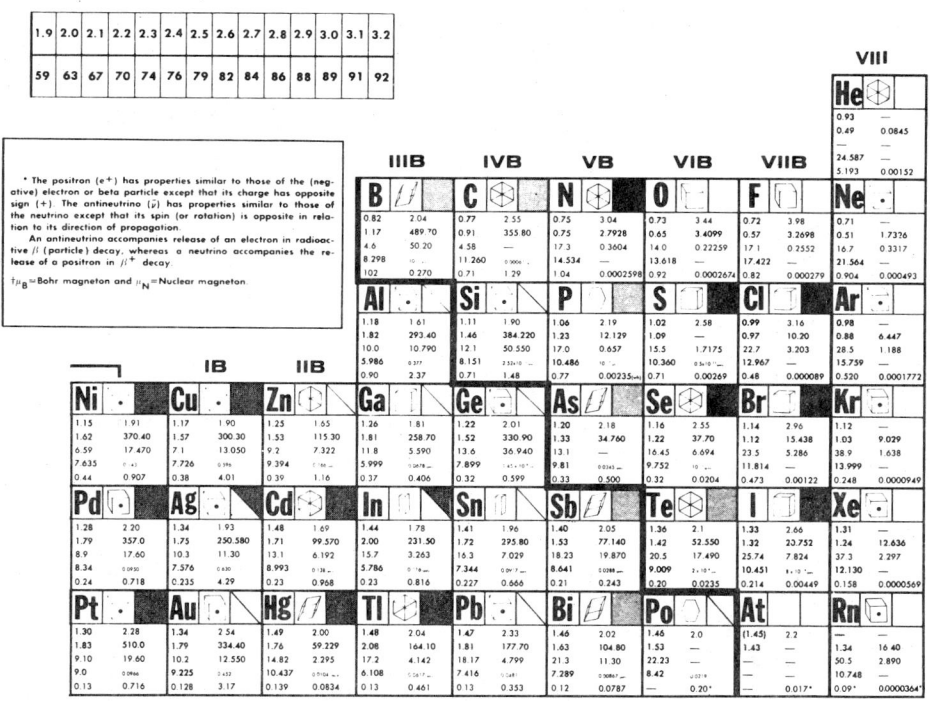

Fig. 6-3. Table of Periodic Properties of the Elements (by permission of Sargent-Welch Scientific Company).

acteristic in frequency for animals, children, or adults. The visual receptivity afforded by opened eyes alters the tracing as if to orient the adult to alertness and attention (Fig. 6-4). However, it is not necessary

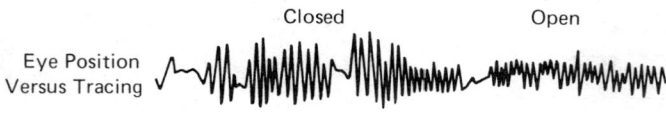

Fig. 6-4. Electroencephalographic tracing and eye position.

to use the feedback machines as the other methods of studying meditation such as Yoga do quite well, but the clinical feedback training can be effective against chronic pain as a kind of conditioning. Later, there will be instances where organisms may be considered as immersed in electromagnetic fields, and their compatibility in such environments is considered. It would be foolhardy to argue against interaction even if it is difficult to demonstrate.

The interacting systems—hormonal, neural, and muscular—behind the EEG and MEG records of brain electromagnetic activity are broad in both scope and significance as shown in Fig. 6-5. Newer detectors, especially the Squid field monitor, and superconductivity permit a separate assessment of the magnetic field as distinct from the evoked or stimulated *potential* which *gives* the EEG. Squid detectors are used with primary flux transporters, notably the second derivative gradiometer, to detect magnetic fields not previously mapped. The noise level of the detector at 10^{-14} tesla is less than magnetic fields of brain, eye, heart, skeletal muscle, and lungs which range from 10^{-8} to 10^{-13} tesla and far less than the 50 microtesla steady field of the earth. If there is any message in the weak electromagnetic fields, it might come more logically from the MEG type, which represents more localized, more specific neural activity.

While the significance of such communication is uncertain, the fields are capable of informing on magnetic contaminants in the lungs of asbestos miners and arc welders. Strong fields, MCG are generated by polarization currents of cardiac muscle during its rhythm. There is also a magnetoculogram which records a steady field of about 10^{-11} tesla around the current due to the EMF of about 100 mV from cornea toward retina.

One of polarization's simplest and most responsive molecules is water. A group of these molecules can orient around a drop of impurity in a classical exhibition of polarization based on some feature such as hydrophobic and hydrophilic forces. Ordinarily, the water molecules will be randomly distributed dipoles with the necessary elements—the positive and negative centers of charge—spaced at a water bond distance. The resulting orientation around the particle is the beginning idea for a formation which can indeed become an information channel in the form of a *membrane*, in turn distinguishing between permeable and impermeable molecules based, for example, on their solubility in lipid or water, molecular size, shape, charge and other form factors. An informative structure has been added via polarization and input of

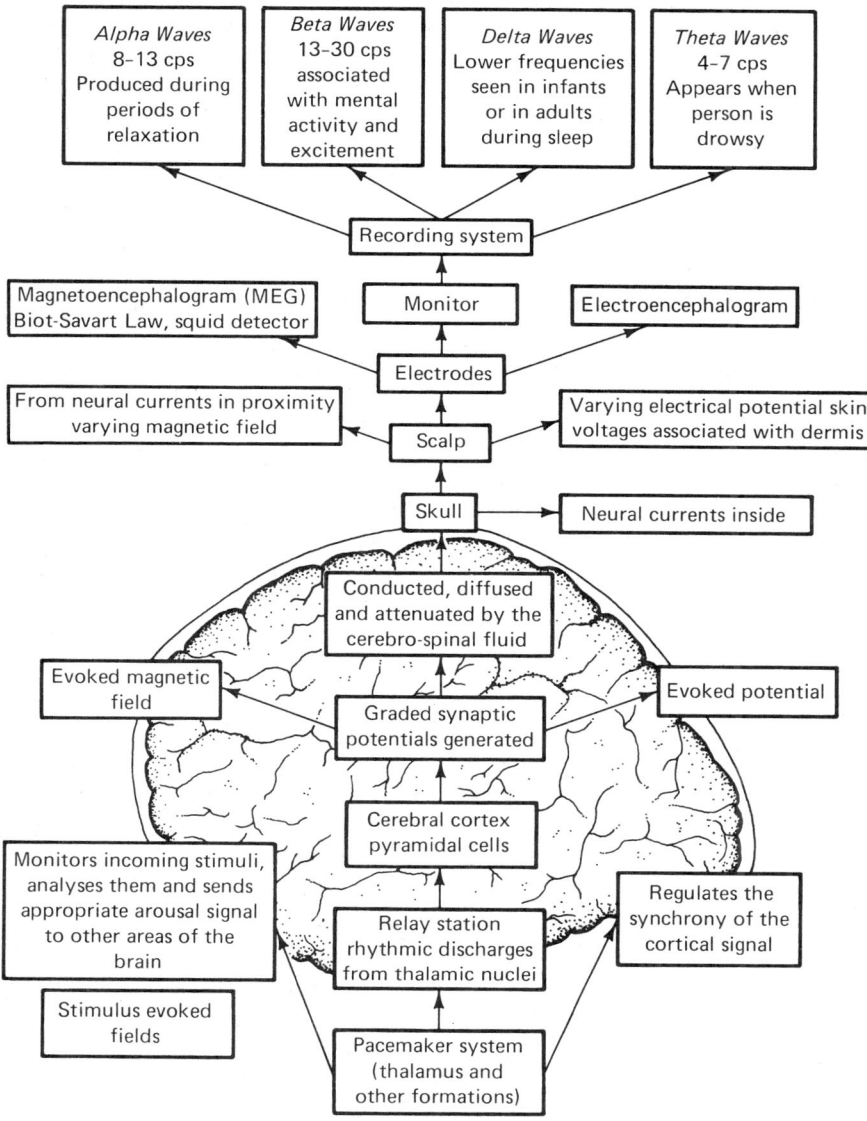

Fig. 6-5. Brain waves.

energy, and part of a network for biocommunication has been anticipated.

In a much broader sense, the oriented state characteristic of polarization has points in common with biomolecular systems and social ones. It may be orientation of a paramagnetic transition metal in hemoglobin in a magnetic field so that precessional energy at the natural frequency can excite easily. It may be a politically, religiously oriented population which will, due to the susceptibility brought on by polarization, much more easily receive some message or instruction relating to the group behavior. Thus social movements may be directed toward the polarization or depolarization of populations. Education should teach how to resist dangerous forms of polarization such as extremely dogmatic communism, or fascism, electing instead the more random structure in a democracy for its overall virtues in the long run.

POLARIZATION AND THE DIELECTRIC CONSTANT

Polarization also provides a line of development toward that elusive relationship describing a commonality among all radiation interactions. When an electromagnetic force acts upon a susceptible ensemble of nuclei and electrons, corresponding dipoles are induced by the electric and magnetic components of the radiation. The dipole may be formed by any two susceptible entities separated by a distance. The more general term, *displacement*, refers only to a single susceptible unit and its discernible movement in response to the electromagnetic force (Fig. 6-6). If an electric field E influences the electrons and nuclei of biologi-

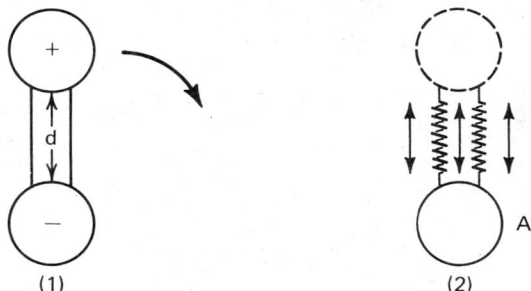

Fig. 6-6. Displacement is a more general response than polarization. (1) Example of dipole made up of charges separated by a distance. Typical movement indicated by arrow is polarization rotation; (2) displacement of a single body A in response to an alternately repelling and attracting force function.

cal atoms and molecules, then these entities may be taken as unit charges between electrode plates perpendicular to a field. Then the displacement D is larger than P due to the presence of induced charges on the plates. P is the polarization, or the electric moment per unit volume, induced by the field force E, the electrical intensity, acting on the unit of charge in the volume. All biological atoms and molecules will have inherent polarizability. Even nonpolar substances can have an induced polarity under correct circumstances.

Radiation polarization $P = P_e + P_m$ combines the effects of both components of the electromagnetic radiation field. $D_e = \epsilon E$ where ϵ is the dielectric constant. This constant introduces molecular or material properties and thus implies all the significance that can be attached to the structure and composition of the material. It is also an ensemble property and the number of molecules present will be reflected in the moment per unit volume, or *polarization*. This results in a dielectric susceptibility because the material behaves as though permeated by charges in a certain concentration that can be displaced by electric forces. The polarization is the product of the number of molecules acting per unit volume and the electric field. Substances of a polar nature may be tested in a gas or in a nonpolar solvent to find the dielectric susceptibility. The dielectric constant is measured and the susceptibility appears as a proportionality constant between the dipole moment per unit volume and the electric field.

Another essential definition is that of the free radical, either in atomic, ionic, or molecular form. Much of the ionization that gives ionizing radiation its name involves the highly reactive species called free radicals. These in turn are most conveniently identified and followed in their behavior by means of the spectroscopies, especially electron spin resonance (ESR). Later it will be shown in detail how water, ever present in biology, is subject to radiolysis or photolysis in ultraviolet, and that radiolysis yields three special radicals

$$HOH \rightsquigarrow e^-_{aq} + \overset{\otimes}{H} + \overset{\otimes}{OH} \quad (88)$$

the hydrated electron   the hydrogen atom   the hydroxyl radical

and the G yields, or the radicals formed per 100 eV of ionizing energy absorbed, are about 2.8, 0.5, and 2.5 respectively to give an idea of relative abundance. Then, molecules with an unpaired electron are known as free radicals as shown in Fig. 6.7. This odd electron gives the structure a magnetic moment and a paramagnetic susceptibility. The result in terms of electrons is an unpaired spin. Usually this is the only situation producing a molecular magnetic moment. There may be two unpaired spins yielding a species called a biradical. Orbital magnetic moments are said to be damped when magnetic susceptibility arises from the spin. Then the interactions affect orbiting electrons.

Thus, under the correct instrumental conditions, paramagnetic susceptibility gives electron spin or, equivalently, electron paramagnetic resonance (EPR). Paramagnetic susceptibility or "willingness" of the particle to be influenced is thus related to the molecular magnetic

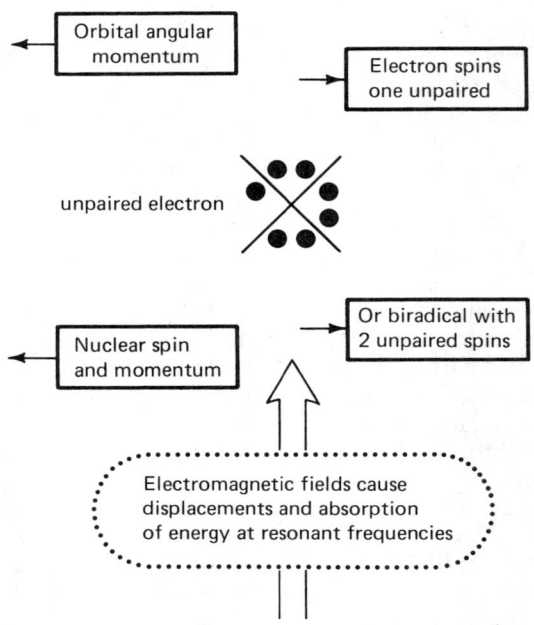

Fig. 6-7. Electron spin resonance. The odd electron gives a magnetic moment and paramagnetic susceptibility. The resulting free radical has an unpaired spin.

moment or to the orienting force for a magnetic dipole. This is one of two magnetic susceptibilities for molecules, the other is diamagnetic. This latter kind is defined as a tendency for molecules to orient themselves at right angles to, rather than along the lines of force. As a general molecular property, it is measured by carefully weighing a powdered sample between magnetic poles with and without the field. Some molecules, for example, the aromatic hydrocarbons, benzene and napthalene, have a large diamagnetic susceptibility.

Ionizing radiations not only produce the three common free radicals but these, in turn, continue the process by forming organic free radicals in a chain of events. The persistence of these often dangerous, highly reactive, oxidizing and reducing species has serious consequences for exposed organisms as explained later. Now we may further note that a very broad implication for free radicals in biology is being recognized over and above their production by radiation. Paramagnetic atoms with odd electrons are almost everywhere in the biosystem and are even more common than enzyme systems. Enzyme mechanisms in redox systems must allow for electron transfer and if two electrons transfer, one at a time, a free radical will exist momentarily after the first electron passes. Thus flavoproteins and cytochromes are portions of an electron transport system. Cytochrome has the heme, or iron containing porphyrin ring linked to protein. An electron transport system arranges for the acceptance and the donation of electrons. In the cytochrome system for example, the acceptor is oxygen and the donor is $NADH_2$, nicotinamide adenine dinucleotide, providing a route to the temporary energy shunt ATP. Redox occurs by the gain or loss of an electron in the coordinating metal atom; in cytochrome, this is iron which is known as a transition metal. Then, redox systems are dynamically coupled to form enzyme catalysis chains in which odd electrons "transfer" charges.

All the enzymes with functional metallic centers have paramagnetic properties because of the transition metal. Thus copper (Cu^{2+}) is found in ceruloplasmin enzymes which come from various organisms and react with substrates such as hydroquinone and catechol.

Anhydrase has zinc, iron is in cytochrome oxidase, and xanthine oxidase has molybdenum plus iron. Then, even multiple metallic paramagnetic centers occur and, by ESR, their components may be studied in the act of catalysis. The unpaired electrons are in orbits around the metal in these transition metal molecules. Thus ESR shows how they behave fundamentally in catalysis and electron transfer, as if the metal had a tracer attached. The free radical concentration itself becomes a variable which is dependent on the type and history of tissues. None are found in certain neoplasms while many occur in the obstructive type of jaundiced liver. Thus, operable or nonoperable jaundice can be distinguished by the push of a button on suitable instruments using ESR. Very low electromagnetic energy is sufficient in ESR studies so that the cells or biosystems need not evidently be disturbed unless they react to the strong magnetic fields imposed.

If, in a certain disease, an implicated (causative) chemical has a conveniently low electronic ionization potential to form free radicals in the body, then use may be made of this fact to make early disease conditions recognizable via ESR. Some carcinogens do exactly this, for example, paradimethylamino azobenzine and thioacetamide. Under ionizing irradiation, an enzyme may produce a free radical. This in turn may lead to other free radicals in a free radical chain reaction. Those such as the perhydroxyl radical, $\overset{\otimes}{HO_2}$, can be lethal for the enzyme. Then on the protection side, ESR can show how certain compounds, such as sulfhydryl (—SH) containing substances, protect against damage by breaking the chain of events after the first free radical is produced by irradiation. Thus, among the substances studied by means of their paramagnetic systems using ESR, we find organic radicals, either in solution or as free radicals from ion pairing and solvation, enzyme systems, electron transport in metabolism, drugs and their action, and carcinogens.

Melanin, a normal pigment that gives the suntan color, shows ESR signals due to its unusual chemical properties. It is a stable free radical and it, or its variations, are implicated in skin and other cancers and in abnormal pigmentation such as amelanotic melanoma. There, while not even detectable as a pigment, it does give the ESR signal. The demonstration is unambiguous and illustrates a biomedical, spectroscopic, identification.

The concentration of free radical intermediates is determined by first calibrating the ESR signal intensity against standards. In a given bioenergy transforming system, ESR is used to find out how much of the enzyme-substrate complex is in free radical form and how this situation leads, in part, to the enzymatic changes observed in the substrate molecule. Application to bacterial photosynthesis gives the time constants (usually in milliseconds) for the appearance, relaxation, or decay of the ESR signal after illumination or darkening. When the magnitudes are known for these relaxation processes, reaction steps may be compared. The free radical production and concentration can then be associated with the appropriate steps—those which are equally fast, constant, or slow in time. The moments of the nucleus in the free radical may be artificially varied by substituting a heavy isotope, typically nonradioactive deuterium, for hydrogen. *Chlorella* algae, for example, may be cultured in heavy water. The ESR absorption band in this case gives a narrower band when photosynthesis is studied. The form of the ESR signal or hyperfine structure has, in such a case, reacted to the changed magnetic moment of the nucleus in the "heavy" free radical, presenting new information in the process.

ABSORPTION AND EMISSION SPECTROSCOPY

In spectroscopy, elements signal their presence by either the absorption or emission of radiations. The spectrum produced specifically identifies the element while the strength of the radiation energy gives the concentration within a certain concentration *quenching* limit which is characteristic of the compound. The rainbow is an apparently continuous spectrum. On the other hand, the yellow sodium flame test illustrates a discontinuous spectrum lacking the other colors. Mercury gives blue-green, potassium red, strontium cerise, lithium red, and barium green.

Spectroscopy involves prisms, apertures, slits, and ruled gratings. Thus, it involves constructive and destructive interference, diffraction, and resolution as does optics. Diffraction is the bending of some radiation from its normal path when passing an edge, aperture, or diaphragm (Figs. 6-8, 6-9) and it leads to the formation of smaller waves. When radiation impinges on the surface of a reflection grating it forms numerous scattered waves that can constructively reinforce, or interfere and thus will either intensify or diminish. Diffraction gratings or apertures are rulings or grooves, often on glass. The aperture and the glass are then coated with sputtered aluminum and chromium, and

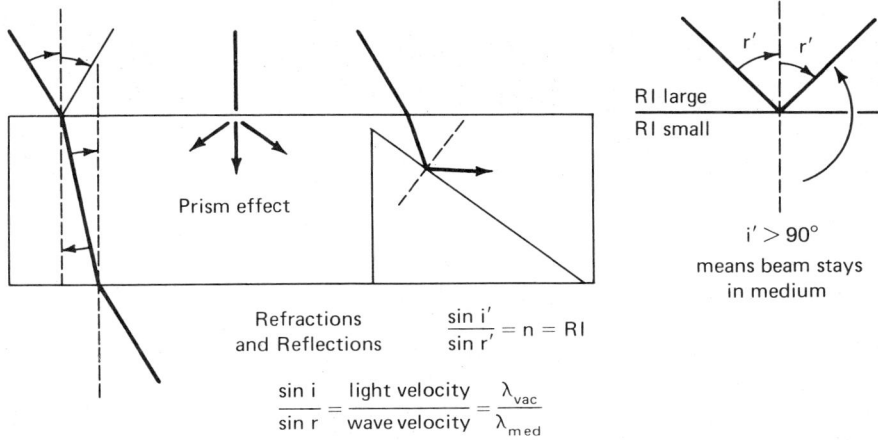

Fig. 6-8. Reflection and refraction.

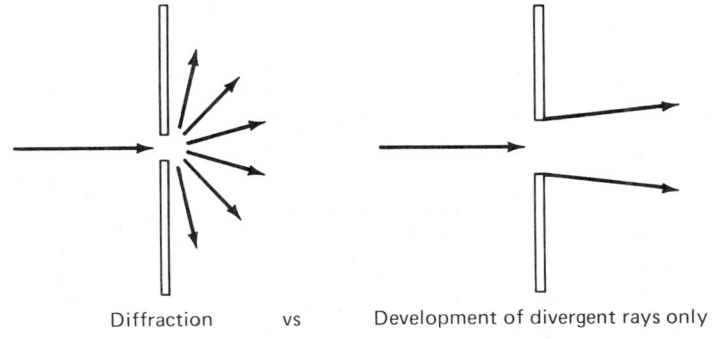

Fig. 6-9. Diffraction and divergence.

duplicated in plastic. Resolution depends on the total number of rulings in the width of a grating rather than on the wavelength. In the prism, resolution depends on the refractive index and, thus, on the wavelength. The monochromator in a spectrophotometer has slits for admitted and emitted light, mirrors, and the diffracting prism or grating.

Bands are defined in terms of the wavelength that produces them, for example, 3883Å to 4606Å for the violet band system of CN (Fig. 6-10). The lines may be resolved into other closely-spaced lines. At 5821Å, the long wave, (red side), there is the cesium line which is, therefore, near infrared. On the short wave (ultraviolet) side, 2400 is the photographic limit because in film, the gelatin absorbs the radiation, thus losing its sensitivity for lines. Recourse may then be made to photoelectric cell detectors. If the atmosphere absorbs the radiation as it

Fig. 6-10. Violet band system of CN in angstrom units.

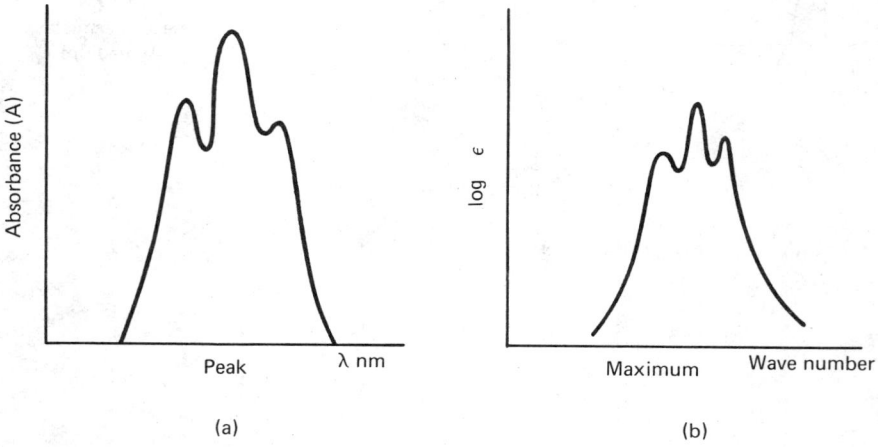

Fig. 6-11. Presentation of hypothetically equivalent absorption spectra as wavelength (a) and wavenumber (b). Absorbance $A = -\log T$ transmission. The extinction coefficient epsilon has concentration units, that is, $m^2\ mol^{-1}$ or liter $mol^{-1}\ cm^{-1}$ which is liters per mole per cm, depending on the system of units. Wave number or waves/cm or K (Kaysers), e.g. $cm^{-1} \times 10^{-3}$. This number varies linearly with energy.

does below 1800Å (that is, with decreasing wavelengths), the instrumentation may be evacuated to give vacuum ultraviolet. Spectral *lines* are derived from *free*, vaporized atoms while *bands*, or a continuous spectrum, are derived from *bound* atoms as in polyatomic molecules.

The spectral readout may be presented in a number of ways (Fig. 6-11). In a medium of refractive index n, where $n > 1$,

$$\lambda = c/nf = \lambda_{vac}/n \tag{89}$$

the difference in λ,

$$\lambda_{vac} - \lambda_{air} = (n-1)\lambda \tag{90}$$

which ranges from 1 to 2 Å in the visible spectrum.

If specks of sample are added to a flame, spark, or to an arc, they are vaporized, and the excitation of liberated atoms produces the emissions as electrons change orbit. A prism or grating separates the light and disperses it into subspectra of various colors on a dark field. The grating of fine rulings or slits produces diffraction and interference and the lines of bright color are formed in the optical slit (image maker). Against a scale of wavelengths, the element is identified by its spectrum of lines. The exciting radiation may vary into the X-ray region as opposed to the other side (toward light). Then the slits, made by ruling suitable material, are not commensurate with the wavelength. Crystal lattices are substituted, acting as three dimensional slits or gratings for the fluorescent emissions. X-ray readout is by a geiger counter or the equivalent.

In atomic absorption, the light from a hollow cathode ray lamp, which is the illuminating radiation and varies according to the cathode material, lights up the grating. The width of the grooves in the grating is selected for the wavelength involved. The cathode beam traverses a flame into which a fuel and oxidant are introduced. The absorbance is proportional to the concentration and samples are compared to standard curves. Only tiny amounts of solution are needed and these are introduced through the burner nozzle. Along with precision, this is the great virtue in spectroscopy; micrograms or picograms are often sufficient. Spectroscopists also delight in pointing out the difference between their chemical analysis and the customary "wet" ones done in chemistry.

In mass spectroscopy there is an interesting departure from electromagnetic energy excitation and the use of the word spectrum (Latin: *spectrum*—appearance; Greek: *skottel*—to see; French: *spectre*—ghost) is not actually correct. High energy electrons impinge on the sample and ionize it. The ions move through a potential and are magnetically deflected through arcs which separate them by mass and charge on the way to the cathode. Readout is by photograph or ion detector which provides for interpretation of the ions and fragments from minute amounts of sample. By interpreting the fast-moving electrons in a wave mode, this can be called a spectroscopy. In general, they would be able to excite molecules, produce radiative emission or nonradiative neutralization to ground, break molecular bonds in susceptible molecular ions, or ionize the positive remnant part of the molecule. Alternatively, a negative ion may form if the electron is held in the molecule.

Figures 6-12 and 6-13 show a prism spectrometer as it might be used in a biospectroscopic study. The English scientist, David Keilin, made such a study on yeast cytochromes a, b, and c, and the absorption bands at three wavelengths were found for the reduced states of the compounds. Then, when he aerated the sample, the absorption bands disappeared showing that the cytochromes had changed to the oxidized state. To Keilin, the spectral changes meant that he had demonstrated the presence of cytochromes in yeast cellular respiration, a significant determination, since the presence or absence of the cytochrome system is an indication of which substrates are compatible with yeast.

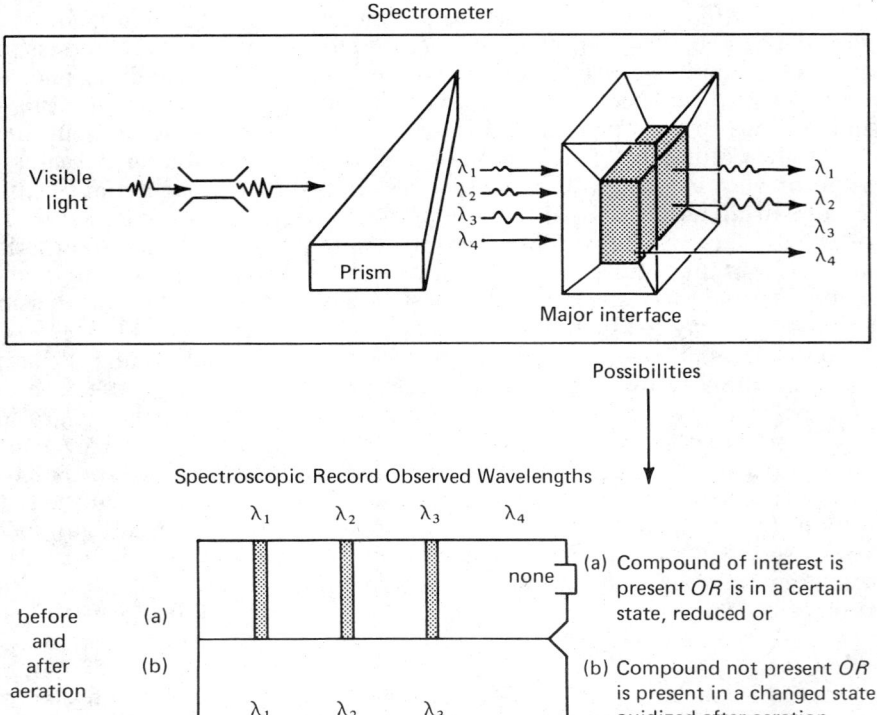

Fig. 6-12. Study of a system by its absorption of visible radiation. The absorption band means that the absorbing substance is present. If it is not there, neither is the substance, or the substance has been changed, possibly in some very significant way, such that it can now transmit rather than absorb the light at the three wavelengths. In the latter case (change), the two different states for the substance have been demonstrated as they have in the case of yeast cytochromes a, b, and c in reduced and oxidized states in yeast. $\lambda \approx 550$, $\lambda_2 \approx 562$, and $\lambda_3 \approx 600$ nm.

Molecules at ordinary temperature usually exist in their singlet ground state. Absorption of the necessary photons at wavelengths selected for the spectroscopy (visible, ultraviolet, infrared, microwave) raises the molecule to an excited singlet or vibrational energy state (Figs. 6-10, 6-13). The phosphorescent emission bands are longer than the fluorescent wavelengths (toward the red side) because the first excited triplet state is of lower energy than the first excited singlet state. Going over the alternate route and the more selective phosphorescent spectroscopy is called *system crossing* and it is encouraged by

Radiation Biology and Biospectroscopy

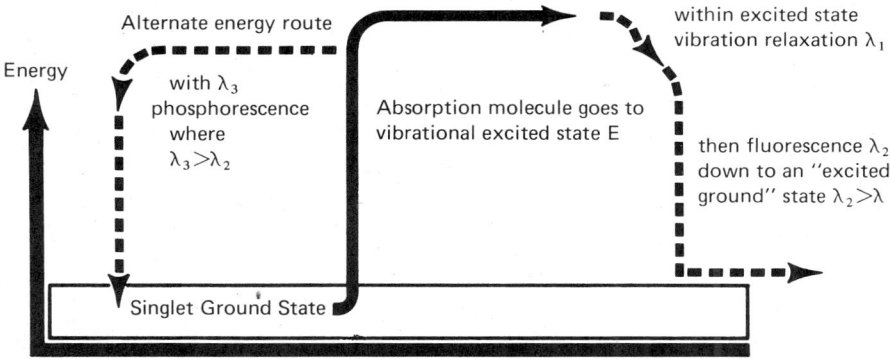

Fig. 6-13. Spectrophotometry or photospectroscopy with ultraviolet fluorescence is dependent on solvent dielectric constant and the structure of the solute, concentration, temperature, and liquid versus solid state. There is also an important radiationless transfer possible for both routes in which the excitation energy transfers from lowest excited to ground state, a nonfluorescent change depending on quenching factors. The results may be a visible fluorescence from an invisible excitation or such radiationless transfer. From singlet ground to an excited singlet and the reverse transitions appear as "mirror image" processes in many cases. Phosphorescence is a delayed emission in terms of microsecond relaxation times of such excitations when compared with the faster fluorescent change.

making the sample react with certain organic preparations or by suitable synthetic steps. When solvents are varied so as to quench or cancel the phosphorescence, it is called *quenchophosphospectroscopy*, which is suitable for the analysis of functional groups. Thin layer chromatograms, pherograms, and cellular, living tissue are appropriate sample presentations.

In absorption spectroscopy, the radiation is subject to five or more interactions, some before the sample and some after it (Fig. 6-14). First, the radiation enters a slit which functions like the iris or pupil of the eye, controlling the intensity; then it enters the section that produces specific wavelengths or spectra. This may be a prism, grating, crystal or color filter. The monochromatic radiation is then passed through the sample, which may be held in a cuvette. After the radiation reacts with the structures being studied, the results must be registered in a manner typical of the particular radiation. Visible light can sensitize the film or photocells can give visual signals. Thermosensitive detectors such as the bolometer may be used. Registration may also be on a chart record or spectrogram, a galvanometer, potentiometer, or cathode ray oscilloscope.

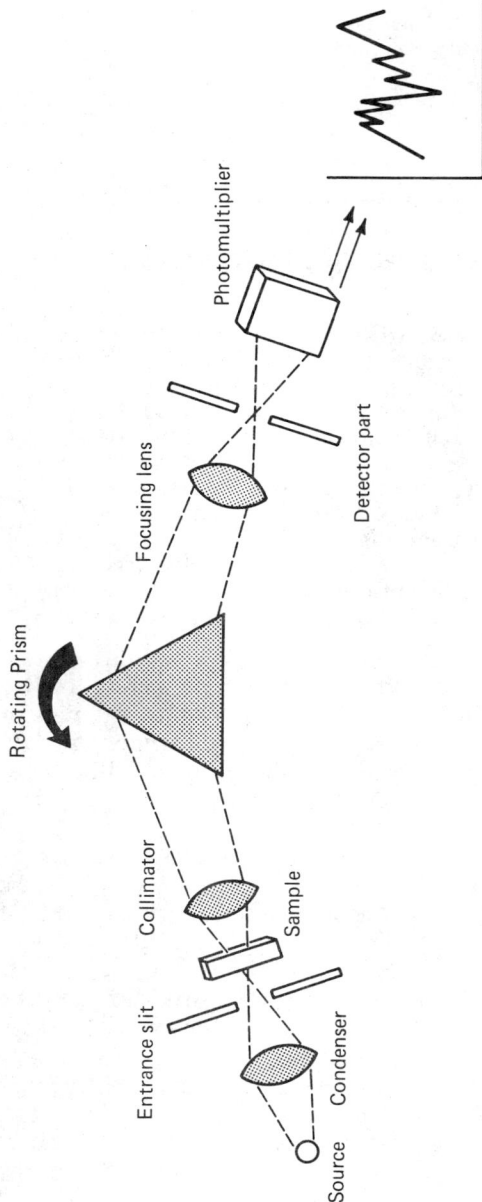

Fig. 6-14. Absorption spectrometer. A general arrangement of the various parts. This instrument has a light source, a sample cell, a wavelength scanning device (in this case, a prism), a detector-recording system, and some optical components like lenses and slits. The light source in an absorption spectrometer should ideally be a "white source," that is, a source emitting the same intensity for all wavelengths of the light. As much light as possible is gathered by the condenser lens and focused at the entrance slit. After passing through the sample cell, the light is rendered parallel by the collimating lens, and then the parallel beam is allowed to go through the prism. The prism disperses the light according to wavelength onto the slit which serves as the entrance to the detector. The detector itself is generally a photomultiplier tube which has a photocathode that is sensitive to the desired wavelength range.

RAMAN INFORMATION

"The universality of the phenomenon, the convenience of the experimental technique, and the simplicity of the spectra obtained enable the effect to be used as an experimental aid to the solution of a wide range of problems in physics and chemistry. Indeed, it may be said that it is this fact which constitutes the principal significance of the effect. The frequency-differences determined from the spectra, the width and character of the lines appearing in them, and the intensity and state of polarization of the scattered radiations enable us to obtain an insight into the ultimate structure of the scattering substance."

Sir C. V. Raman, Nobel Lecture
Stockholm, 1930

There are two ways of analyzing the theory behind the spectroscopies. One is that due to Newton's classical understanding and the other is the quantum explanation. Both ways are useful and each forms part of the total picture. It must be remembered that the development of the modern spectroscopies parallels and possibly joins with, the modern concept of the atom, which dates from the Nils Bohr model, so that here one deals with no less than the greatest basic ideas in science.

In terms of the eigenstates, the equations in quantum mechanics for absorption and emission are the same.

$$E = E_1 - E_2 \tag{91}$$

where E is the energy of the emitted radiation.
The emission of energy is by eigenfrequency

$$E = hf \text{ (Planck's quantum equation)} \tag{92}$$

and the difference between two energy states or eigenstates E_1 and E_2 describes the emitted energy

$$\Delta E = E_1 - E_2 = hf \tag{93}$$

which has the wavelength

$$\lambda = hc/E \tag{94}$$

On the other hand, absorption of energy E will raise the energy of a molecule from the lower eigenstate E_2 to the higher state E_1. The wavelength required is λ.

Planck's quantum equation led to the birth of the quantum theory. Then Albert Einstein, while studying the photoelectric effect, determined that the energy of the photoelectrons depends only on the frequency and not on the intensity of the radiation as most people thought. He saw how a photon or quantum was an energy hf packet for a given frequency and that this packet had enough energy to eject outermost electrons from their nucleus in the surface of the photoelectric metal. Shown as a packet of energy, the quantum would look like that shown in Fig. 6-15.

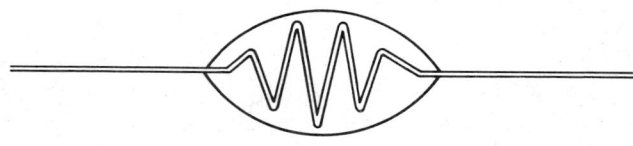

Fig. 6-15. The quantum shown as a packet of energy.

The vibrational, rotational, and electronic modes are types of excitation or motion in quantized levels, experienced by molecules when radiation of the correct eigenvalue is absorbed, with more quanta needed for higher levels. The absorption peaks for these modes in spectroscopy convey molecular information on the dissociation energy, types of bonds (covalent, ionic, etc.) in the structure, symmetry, homopolarity (e.g. O_2) and heteropolarity (e.g. HCl) of dipoles, and probably much more if only all the wealth of detail could be interpreted.

The Raman effect is used in a particular kind of vibrational or rotational spectroscopy. Raman excitation produces spectroscopic signals or lines, and the frequency of the signal gives the rotational or vibrational energy of the excited molecule, specifically the frequencies of oscillation of the chemically bonded atom of a transparent gas, liquid, or solid molecule. With interpretation, these are the blueprints of the geometry of the molecule and of the molecular forces binding its parts. When monochromatic ultraviolet or visible radiation collides with sample molecules, it may be scattered as well as transmitted and absorbed. Normally, scattered radiation is seen with small particles and this is known as the Tyndall scatter. Scatter from molecules is called Rayleigh scatter and, in both instances, the scatter frequency equals the incident frequency. What Professor C. V. Raman in India showed in 1921 was that there were also scatters in which the frequency shifts and the changed frequencies then correspond to frequencies of rotation and vibration transitions in the scattering molecules.

To obtain the Raman shift, the substance is excited by a monochromatic light whose wavelength does not coincide with the resonant frequencies of the substance under study. The frequency of the radiation will then be between two frequencies at which the molecules resonate. Other scatter will be identical with the incident frequency but it is the fraction containing lines of higher and lower frequency than the original that is of special interest.

Scatter is associated with photon-molecule collisions. Most of these are elastic and the scatter has energy and frequency equal to the original radiation. That is, the total kinetic energy stays constant as explained in Chapter 10. Inelastic collisions occur if the internal energy of the molecule is changed during the collision process. Collisions may also be mixed (show features of both), in which case they are said to be imperfectly one or the other. If the molecule is raised to an excited state by the collision, discrete vibrational or kinetic energy changes occur and the colliding photon loses an amount of energy equal to the energy level difference of the molecule. Then the scatter has a lower frequency. On the other hand, a number of the molecules may be in excited vibrational or rotational states before the collision and can increase the energy of the photon by a discrete amount. This gives rise to the scatter photons of higher frequency. The original methods are very straightforward and produce much new information on molecular structures.

With Raman spectra dependent on frequency shifts, it is easy to see how polarization of the stimulating radiation and its resultant in the Raman lines would be significant. While polarized light has not been mentioned very much, polarization certainly has, and polarization of radiation is already familiar in polarized cameras, filters, insect behavior, and eye lenses. Here it is noted that a molecule may be presumed to have certain preferred orientations or internal directions where it and its electrons, in a state of sort of induced dipole moment, might suffer displacements rather more easily in an applied electric field. These preferred directions are not necessarily the same as the field directions or orientation as in vector directions. Thus, with Raman lines, the polarizability of the lines is considered in connection with the polarization of the radiation. That is, the radiation may be depolarized, polarized perpendicularly to the axis of the sample, or parallel to it in the experimental arrangement. Thus the Raman spectral lines have built-in horizontal and vertical components, or an innate reaction to the state of polarization which affects the intensity of these two components and, thus, the character of the lines. As a ratio, the horizontal to vertical contribution may be between 0.1 and 0.5. Below 0.1, the line has a polarization, above 0.5, the ratio indicates that the line is depolarized. As we try to fathom the highly-coded data compressed into the composition, distribution, spins, and bonds of the molecule, these tricks of the trade are invaluable, but it is readily understandable that they are most meaningful to those skilled in their interpretation.

This is the problem, because the spectra have much more information than anyone can fully decode, and this problem must be attacked on a wide scale and by looking at the process carefully.

In laser Raman methods, the exciting radiation becomes a laser source which immediately gives flexibility due to the laser's properties, especially controlled radiation. It improves the resolution in a synergistic effect with the photographic emulsion, allowing the observation of spectra formerly hidden due to overexposures of plates and related resolution problems. Overexposure of exciting lines had tended to make ghost effects, indistinct, and overlapping spectra. Some of this is pressure broadening of the Raman lines (excess sample vapor) and some is increased line background from overexposure, but the result in any case is poor resolution. The laser method uses a narrower excitation line and thin, fine grain emulsions to reduce this interference, so that the spectrogram has considerably more clear detail. Laser microprobes are sampling procedures used in emission spectroscopy in which elements are detected in trace amounts in tissue (absorption emission spectroscopy and Raman) using the fine focus of the beam for intracellular and *in situ* analysis. To prevent destruction of exposed tissue, the beam may be pulsed instead of continuous. The coherence of the laser radiation is its prime feature but the emissions cover the photo spectrum, for example, for gas lasers, from 2358 to 120.08 μm. Thus, lasers enhance the Raman effects, "pinpointing" the Rayleigh peak absorption so that spectra at both sides can demonstrate the Raman effect by avoiding that absorption band.

An infrared Raman laser can be tuned to any desired wavelength so that it becomes a carrier of information such as the presence or absence of some particular pollutant molecule in the air. Where the frequency shift of incident radiation is interpreted in terms of molecular structure and behavior in Raman spectroscopy, in communication or detection, the frequency shift, for example in a carbon monoxide gas laser, is caused by or informed by a source or target material, depending on whether it is a communication device or a detector.

ELECTRON SPIN RESONANCE

The free or organo-radical with the odd electron acts as a special indicator whose localization and energy are revealed by ESR. Odd electrons are the particles out of which bioenergy transport is made and they function as well in chemical bonding through their valence status in and between atoms. Thus, unpaired electrons are associated with extra energy and activity in a molecular system and their positions are revealing for chemical structure. The presence of an odd electron means that an electron shell in the Bohr atom is incomplete and this status is present in paramagnetic centers like those in the metal transition elements and organic radicals. Thus ESR can aid in studying any biosystem such as electron transport schemes, metal coordination compounds, enzymes and organic radicals and ion-pairing radicals whose behavior is a result of the paramagnetic center. These results are based on the fact that electrons possess charge, mass, angular momentum, and magnetic moments. They are also constantly interacting with their neighboring atoms, molecules and lattices. Here then is a fundamental picture of structure at the level of finest detail which could be examined with the help of an intelligible probe of sufficient sensitivity.

Polarization in Electron Spin Resonance

The electron spins and magnetic moments are normally depolarized, that is, they are not oriented in a particular direction, but a strong magnetic field polarizes them. This produces a first splitting in which groups are formed both parallel to the applied field and opposite or antiparallel.

The polarization, or alignment, in ESR under the influence of an alternating microwave field is thus one of spins and moments. In samples placed in applied radiation fields, electromagnetic energy can be absorbed by such spinning and orbiting systems. The odd electron in molecular or atomic orbit generates the magnetic moment as a result of its spin. There is an ensemble of such electrons in tissues which can respond to externally applied magnetic fields. In equilibrium without external field influences, the spins and moments are unaligned. When the magnetic field is applied, they form a parallel or antiparallel orientation. (The lower energy subgroup orients parallel and the other antiparallel.) These two orientations represent a separation of the electron population into two subgroups, splitting, by quantum description, into the two possible configurations $+\frac{1}{2}$ and $-\frac{1}{2}$ (Fig. 6-16).

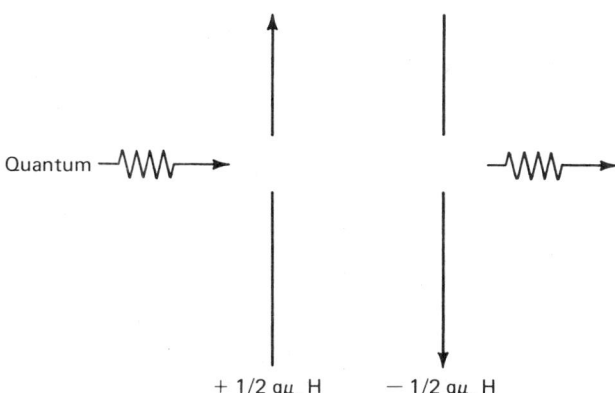

Fig. 6-16. Splitting into subgroups. g is the gyromagnetic ratio or splitting factor. Energy of the two quantum mechanical levels $+\frac{1}{2}$, $-\frac{1}{2}$; applied magnetic field H. Resonant frequency quantum perturbation of electron.

The two energy groups are separated by a *splitting* difference which arises from the magnetic moment due to spin momentum, applied magnetic field intensity, and gyromagnetism coming out of spin and orbit angular moments. Instrumentation for ESR supplies this magnetic field (Fig. 6-17) to separate the groups and it also applies the preces-

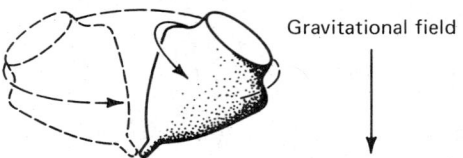

Fig. 6-17. One may rigorously describe an electron as a tiny gyroscope having a magnetic moment along the axis of angular momentum. In toy gyroscopes, the precession is caused by the force exerted by the gravitational field of the earth.

sional electromagnetic energy equal to the difference between the energies of the groups. This electromagnetic energy is likely to be in the microwave region and capable of exciting unpaired electrons to the upper energy state in the electron resonance and orientation of magnetic moment that occur. (Unpaired electrons mean that there is a triplet state existing.) The presence of paired electrons instead of the odd electron would cancel out the spin effect as well as absorption. With the magnetic field and electromagnetic field both set into the instrument, the gyromagnetism can be observed and related to electron spin, angular momentum, concentration and orbital state of odd electrons, and bonding in the structure (Fig. 6-18).

When the particular states of the unpaired electron are being considered, they are therefore characterized by the magnitude of the spin momentum resolved along this axis of reference and hence, for the case of the single unpaired electron, would be characterized by

$$M_s = +\tfrac{1}{2} \quad \text{or} \quad M_s = -\tfrac{1}{2} \tag{95}$$

according to the orientation of the total angular momentum. If an external magnetic field is applied *across* the specimen, however, the electrons will now have a common axis of reference and will all precess so that they have a resolved component of either

$$M_s = +\tfrac{1}{2} \quad \text{or} \quad M_s = -\tfrac{1}{2} \tag{96}$$

in the direction of this applied field. General quantum conditions allow only quantum states in which the quantum numbers differ by unity; hence the two cases of

$$M_s = +\tfrac{1}{2} \quad \text{and} \quad M_s = -\tfrac{1}{2} \tag{97}$$

are the only ones allowed for these single unpaired electrons.

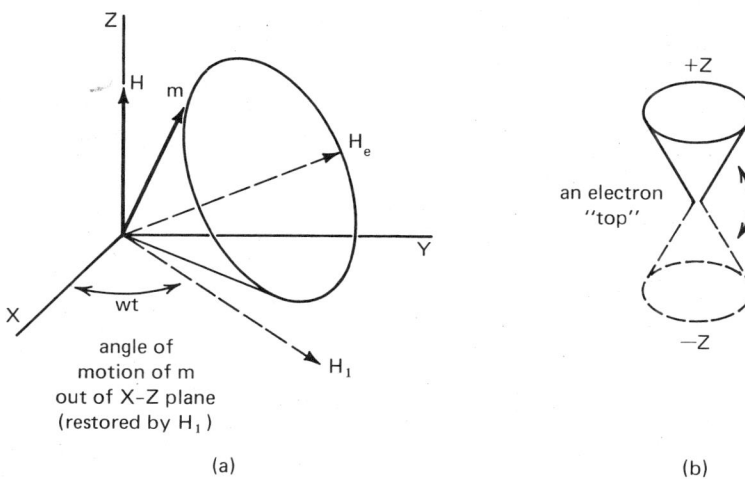

Fig. 6-18. Electron paramagnetic spin resonance. Paramagnetic means that the magnetic field in a substance is more than in a vacuum whereas, in diamagnetism it is less. Not all spins are paired, therefore the molecule *has* moment. H_1 (from oscillating component) precessional field at right angle to H pushes m back. Therefore motion = ±z. (resonance) m precesses around H_e (effective magnetic field). The electron then absorbs or emits ΔE between states plus or minus one half. We may effect the required torque by electron intrinsic moment tied rigidly to its axis or angular momentum such that when a strong magnetic field is applied to the electron, there is a torque to make the electron precess with a frequency which is both proportional to the applied magnetic field, and to the electronic magnetic moment.

The application of an external magnetic field divides the unpaired electrons into two groups, those with their spins aligned parallel and those with their spins antiparallel to the direction of the field itself. It does, moreover, also give different energies to the electrons in these two groups. Those which have their magnetic moments lined up parallel to the magnetic field will have their energies reduced, whereas those lined up antiparallel to the field will have their energies increased. The reason for this behavior can be seen in the analogy of a simple bar magnet and its behavior in the magnetic field produced between the pole faces of an electromagnet. Normally, the bar magnet will be placed in this magnetic field so that its north pole faces the south pole of the electromagnet, and vice versa. This is the position of stable equilibrium and a motion of the bar magnet from this position will be rapidly overcome, with the magnet relaxing to its original polarization.

When the two energy groups form, their energy is $g\beta H$, where g is a splitting factor, the gyromagnetic ratio from the spin and angular momentum, H is the magnetic field and β is a factor for considering spin and magnetic moment together. Called the Bohr magneton, it converts units of angular momentum into magnetic moment units. Sometimes we use μ_o, which is called the orbital magnetic moment.

The resonance is provided by the magnetic effect which gives the "first" splitting effect while simultaneously injecting microwaves at energy $g\beta H$ across the sample

$$f = g\beta H/h \qquad (98)$$

from which the necessary microwave frequency (the Larmor precession frequency) may be found. What this energy does is to put hot or resonant electrons with excitation $g\beta H$ into the higher energy group, so that this is a "second" splitting or reinforcement of the separation. The quantum value connected with the microwave frequency is equal to the energy difference between the two groups of electrons (Figs. 6-19, 6-20).

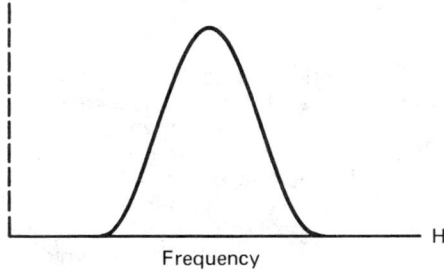

Fig. 6-19. Resonant frequency result with increase in H.

$$E = hf = g\beta H \qquad (99)$$

The injected energy goes into the odd electron at the level of the lower energy group boosting the electrons in it to the upper level and reorienting their spins. The absorption is revealed in microwave power detectors downstream from the sample and read out on a cathode ray oscilloscope or screen (Fig. 6-21). Substitution of superconducting magnets with higher fields permits the use of higher frequencies. With electromagnets, the field might be 3200 gauss and the frequency 9GHz and the wavelength 3 cm. Under those conditions, the g value may be found and, for organic free radicals, it may be two, compared with the free electron value of 2.0023. With a localized, odd electron, the g value difference depends on the interaction of spin magnetic moment with the orbital angular momentum. At 38 GHz, or 8 mm wavelength, the magnetic field would be 12,000 gauss. At 10,000 gauss, it is 43 MHz. Resonance absorption intensity is generally optimized in these combi-

Radiation Biology and Biospectroscopy

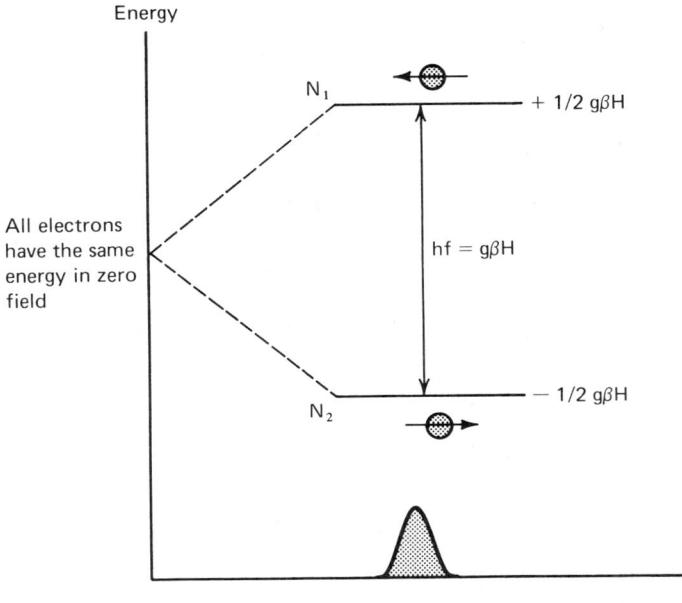

Resonance Condition hf = gβH
Resonance for free electron f = 2.8 × 10⁶ H

Fig. 6-20. Splitting effect into two groups N_1 and N_2 at precession frequency.

nations in the microwave region. The g value is 2.0023 for a free electron and this differs from an expected 2.0 due to the interaction of the electron with its own radiation field. The energy in the relation $E = hf$ for radio frequency—microwave ranges from 10^{-2} to 10^{-6} eV. This low energy is correct for extremely low energy separations or splittings which, accordingly, can be measured at these frequencies. The large magnetic field assures a good splitting of the electron population between a less numerous upper energy level and the lower one and, thus, a good absorption of microwaves. Other practical factors in the frequency selection would be sensitivity, sample size, and power. The electrons form a density gradient in a sense, being more numerous in the lower energy state. The difference is then optimized by the applied field.

SPIN LATTICE INTERACTION

No resonant electron spin absorption could occur if there were only paired electrons. Electron net spins balance. The instrumental values of H and E for resonance allow g to be calculated. The g is interpreted in terms of orbital states of the odd electron. This in turn tells about the

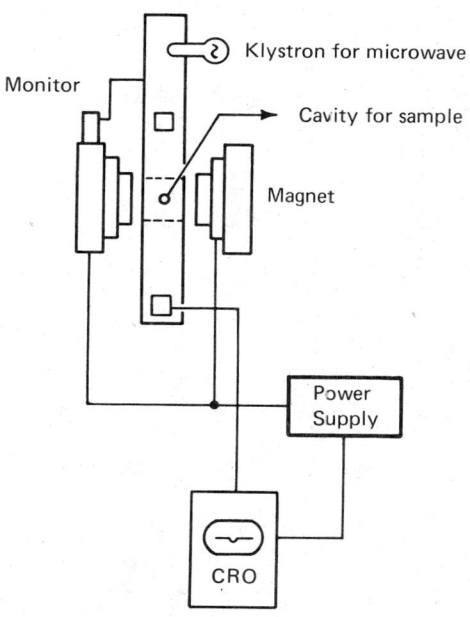

Fig. 6-21. Electromagnetic large microwave and magnetic fields are simultaneously applied to the sample in the transmission line and the resonant absorption read out on suitable oscilloscope or screen. The microwave radiation is fed down a waveguide to the cavity resonator, in which it is effectively concentrated, and, in this way, a high level of the oscillating microwave magnetic field can be applied to the specimen. In the simple transmission system, a certain amount of power from the cavity resonator is coupled out and fed to the detecting crystal at the far end of the microwave run. The value of the detected current flowing through the crystal is then a measure of the power existing within the cavity, a reduction will also be produced in the current detected through the crystal. The external magnetic field is applied across the cavity as shown. This field is normally produced by an electromagnet, since it is the magnetic field and not the frequency that is usually varied to obtain the resonance condition.

kind of bonds present and the structure of a biomolecule like myoglobin from g values at various crystal orientations (Fig. 6-22). The concentration of odd electrons comes from the area under a single resonance absorption line. Its width depends on the lattice structure around the electron. For example, spin-lattice interactions are energy exchanges that cool, dampen, or quench the "hot" electrons, somewhat shortening their excitation.

Typically, a single spectral line splits because the odd electron interacts with the magnetic moments of nearby nuclei and other electrons

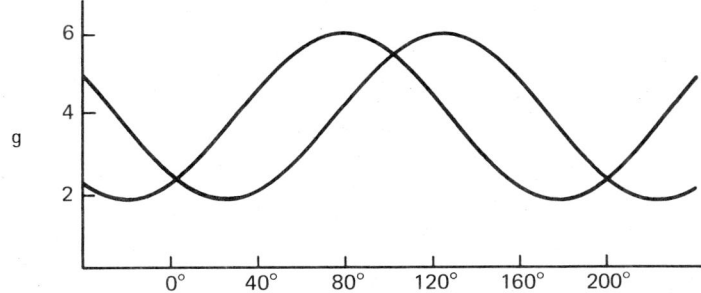

Fig. 6-22. g-value variation in myoglobin. Crossover points of g locate crystal axes in the 2 molecule/unit cell crystal. Myoglobin structure is analyzed by ESR thanks to variations in the g spectral line splitting factor with the angle of incidence of the external magnetic field. This field can be moved about the plane of the crystal and g plotted versus the degrees of angle in the shifted field. Crystal axes will be in directions determined by the ESR spectra from different ions and the resulting spectral coincidence points. The frequencies being observed, g factors are calculated as they vary for molecules in a unit crystal cell around the plane.

producing spin-spin splitting or hyperfine structure. These are interpreted as information about the immediate atomic surroundings since orbits indicate molecular energy structure and orientation. Spectral or absorption changes (e.g., line or band widths at resonance), gyromagnetic effects and other results are probed to reveal the nature of the atomic environment and thus give information on the behavior of matter, reaction schemes, and physical constants such as the heats of hydration. Other spin interactions are spin-orbit coupling, electron spin-spin coupling, and electron-nucleus coupling. Electron nucleus interaction is evident in paramagnetic systems and analyzed as hyperfine structure measurable in the ultraviolet, microwave and electron spin resonance spectra. Coupling such as this occurs in diamagnetic systems between the nuclear magnetic moment and electronic magnetic moment induced by rotation and forms the basis for rotational level measurements in microwave spectra. Coupling constants are associated with these interactions which also include nuclear-nuclear coupling with spin, orbital, and diamagnetic features. Nuclear spin-electron spin and nuclear spin-electron orbit coupling can be involved as indirect interactions. The spins may or may not be opposed. As we have seen, the inequality between particles such as electrons, is exposed and, even with a stable magnetic field applied, responses of these particles indicate that their magnetic susceptibility varies with their environment. Thus in NMR when protons resonate, even the electrons change motion, thus diamagnetically making a new magnetic field to oppose the main field, all in a meaningful way to those skilled in extracting the information.

Materials susceptible to electron spin resonance may be chelating compounds where the metal (e.g. iron in myoglobin), is in valence rings. They are also likely to be metal-protein complexes with energy manipulating functions or compound transfer functions based on chelation and properties of the special metal ions. Transferrin is such a complex. It feeds two ferric ions per trip in erythrocytes in the hematopoietic center of bone marrow. Its binding, structure, release, and especially its identification of the recipient cells to which its cargo is assigned are informational questions subject to electron spin resonance study. It is a sugar protein of molecular weight about 80,000, and is salmon-pink colored. Electron spin resonance is revealing because it is very sensitive to transferrin structural changes showing also that HCO_3^-, the bicarbonate ion, is needed to keep the ferric ion tightly bound in transit. Otherwise, it would form an iron hydroxide precipitate with the readily available OH^- ions in the water. Without the HCO_3^- ion, the color is clear (colorless) whether transferrin is with iron or apotransferrin, without iron. Electron spin resonance shows that the iron atoms are specifically oriented at transitional metal binding sites in the molecule and not loosely attached around the molecule. The preparations, when left to react in the ESR instrument and then tested again, show the changes or reactions that occur with changes in the preparation. These changes may be, for example, with and without HCO_3^-, with excess Mn, Cr, or Co ions etc.

NUCLEAR MAGNETIC RESONANCE SPECTROSCOPY

Where electron spin resonance utilizes the inbalance created by the odd electron and a net spin not balanced to zero, nuclear magnetic resonance is based on the magnetic moments of magnetic nuclei that come from an odd number of nuclear particles—protons and neutrons. In the same way as before, magnetic moments occur due to the spin (Fig. 6-23). Again, an intense magnetic field is needed to polarize the nuclei in the permitted quantum energy states that belong with $(2I+1)$ polarizations where I is the nuclear spin quantum number. Once again, an electromagnetic field is applied over the magnetic field but this time a radio rather than a microwave frequency is needed because the proton has a smaller magnetic moment. Thus, the nuclei absorb and can resonate with the emission of detectable radio signals. These are analyzed for their intensity and distribution in the spectrum.

Hemoglobin transport properties are investigated by nuclear magnetic resonance. In its methemoglobin form, in which the Fe^{2+} ferrous is oxidized to Fe^{3+} ferric form, hemoglobin can shift its reaction to alkaline from acid and its nuclear magnetic resonance signal shifts accordingly. NMR relaxation times reveal how hemoglobin, in its dissociation in buffers, is sensitive to the size of the buffer ions and, by this difference, it is revealed to be quite different in its heme group from myoglobin which is independent of the buffer used. The heme-heme interaction is a unique instance of cooperation. The oxygen load is bound at the

Radiation Biology and Biospectroscopy

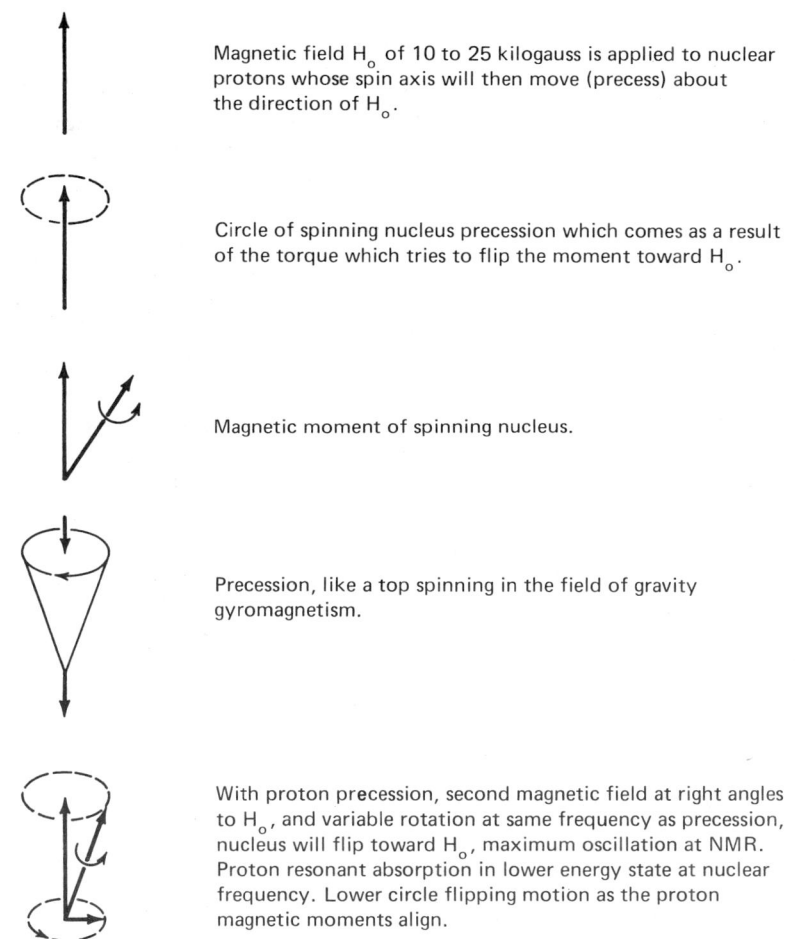

Fig. 6-23. Exploded diagram of nuclear magnetic resonance.

heme groups which are quite distant in the atomic scale for direct interaction. Yet when one oxygen is liberated to the recipient structure in tissue, this causes the remaining groups to want to leave also. The hemoglobin molecule has two similar but distinct chains—two alpha and two beta configurations stacked and interwoven together with four heme iron groups coming out to the surface of the large molecular bundle.

Arginine, at the peptide junction of alpha and beta chains, changes its charge from positive as the pH changes but its charge and location are essential for the heme-heme interaction. Probably, movement of arginine and its changes in form in hemoglobin occur just at the time oxygenation does.

Relaxation Effects (Nuclear Magnetic) and Biomolecular Time Processes

The constraints on protons of solvent water in a magnetic field allow only special polarizations. These H atoms relax in an equilibrium relaxation time after application of the external field. Thus nuclear magnetic relaxation can be *timed* so that this relaxation is added to the information gained as large metal proteins are added to the solvent. Then, the relaxation is itself a mechanism which yields information on the molecules and their behavior. With water molecules alone, the relaxation effect is due to the interaction of the two hydrogen protons and the time is about three seconds. In water, this time is dependent on Brownian movement and the magnetic field applied by one proton on the other one as it rotates. The Brownian movement or the frequency of molecular rotation due to this Brownian motion can be decreased. If it is, the correlated time during which proton-proton interaction occurs is increased. The relaxation time is then shortened. Proteins like transferrin with transition metal, paramagnetic ions, have larger magnetic moments than the protons, so the water protons come under a larger interaction as they feel the protein's presence. The relaxation time is still shorter because there are now two effects on it. Relaxation times thus allow logical connections to be made based on the function of the biological compound and its relation to the time constants for the specific bioprocess.

Water Binding

The protons in the water live a dual life as free solvent water hydrogens or as bound water hydrogens on the protein attached to ionizing groups. In essence then, the behavior of the relaxation in protons mirrors (reflects) this changing role of water. The nuclear magnetic relaxation or proton relaxation in proteins can be studied in connection with their response as the magnetic field is varied and bound water is visualized as it rotates on the larger and slower protein. The protein rotates more slowly with a lower frequency for the externally applied field. This provides the longer time for attached water molecules to interact and equilibrate after the disturbance of the applied field (Fig. 6-24).

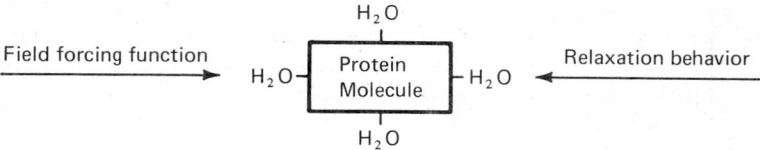

Fig. 6-24. Bound water. Free versus bound water on proteins can be studied by nuclear magnetic relaxation. Binding contributes functional effects as well as relaxation ones. Proton relaxation is a special type observable by nuclear magnetic relaxation.

A number of bound molecules of water of this interacting type can be found on any protein and there may be a dozen or more per molecule. On the other hand, loosely bound water moving around the metal binding site, is *not greatly relaxed* by the paramagnetic bound-metal iron in transferrin. These studies then move us toward an understanding of the water binding and relaxation in proteins. Steps in this direction have shown paramagnetic contributions to relaxation behavior due to one that depends on the frequency and associated with a water molecule bound at a distance of five angstroms from each Fe^{3+} ion, and, one that is heat activated (above +25°C.) in its relaxation behavior and due to a water directly coordinated to the metal ion.

Thus, using an instrument, the NM Relaxation spectrometer, the physical effects of radiations on biological molecules and systems, the structure, function and local relationships may be effectively investigated.

The rate acceleration for release of Fe^{3+} to erythrocytes from a passive delivery is in the ratio of 1 or 2 seconds to 10,000 years. Bionically, considerable interest exists in such enormous acceleration as well as in biological identification by transferrin of the reticulocyte recipient of the ferric ion for this procedure, since a similar one may also occur in body invader recognition, grafted tissue response, or autoimmunology. It may also find application in artificial devices where computer or other associations dictate the need for high speed and/or recognition features.

PROBING STRUCTURES AND HAZARDS THEREOF

The environmental hazard question that arises in connection with the use of radiations includes polarizing, nonionizing radiations. Basically this question concerns the compatibility of organisms with the electromagnetic fields required with the instruments and devices using these radiations. It develops that extraordinary skill has been used to design the devices, particularly where they are intended to probe into and ask questions of molecular structures. Such is the nature of the information in informational biomolecules. Only the proper query, made in these instances by probing electromagnetic radiations, can give access to the structural and functional design. Some important radiation interactions, such as those in Table 6-1, then become informational routes into molecular biology.

Some of the interactions have been discussed in connection with the specific analysis, for example, in the Raman effect or in the background discussion. The impression should have been gained that new discoveries, effects, or approaches involving the manipulation of the radiation or the samples are eagerly absorbed into this science with the beneficial result being the development of modified spectroscopies (Table 6-2). The reader will benefit greatly from following the historical development.

TABLE 6-2. Selected Radiation Interactions that Provide Avenues for Investigations in Molecular Biology

Bunsen Effects
Fraunhofer Lines
Kerr Effect
Cotton-Mouton Effect
Faraday Effect
Rayleigh Scatter
Stark Effect
Zeeman Effect
Overhauser Effect
Raman Effect
Mössbauer Effect

While these effects actually produce absorption in the sense of orientation in gas or liquid, the absorption is *between* resonant frequencies and related to scatter. In scanning electron microscopy, the preparation scatters the electron beam, bringing information that can be registered on a suitable screen.

Thus scatter usually involves a dispersion or a change of direction with some interesting side effects that may involve those secondary structures that may happen to be in the altered pathway of the scattered radiation. In a magnetic field parallel to the direction of propagation of a radiation such as light, a moment is induced in all molecules. This moment gives a rotation to the plane of polarization of the light passing through (Faraday effect). Diamagnetic and paramagnetic molecules have different mechanisms but involve induced molecular magnetic and electromagnetic field moments. Light photons are scattered by these processes at interband frequencies. The wavelength may or may not change. In *Rayleigh* scatter it does not, and the scatter is simple refraction. If the molecular state changes, the wavelength will be changed giving the *Raman* effect. The change of state is a definite absorption. the model for analysis of the scatter is a plane polarized wave of a given frequency moving on an axis as a vector potential. The scatter involves electric dipoles, angles of incidence, simple refraction, magnetic dipoles, and rotation of the plane of polarization.

SPECIAL INTERACTIONS

The Kerr constant K is due to the difference in refractive index $n_\parallel$ and $n_\perp$, parallel and perpendicular to the field, respectively, when a beam of polarized light is transmitted through a group of molecules.

$$K = (n_\parallel - n_\perp)/\lambda E \tag{100}$$

where E is the electric field strength and K comes from molecular polarization.

This effect was observed by J. Kerr in 1875 and is due to the fact that molecules can polarize in an electric field (Fig. 6-25). The optical properties of materials will thus depend on this polarization or alignment as opposed to the random situation existing in materials without

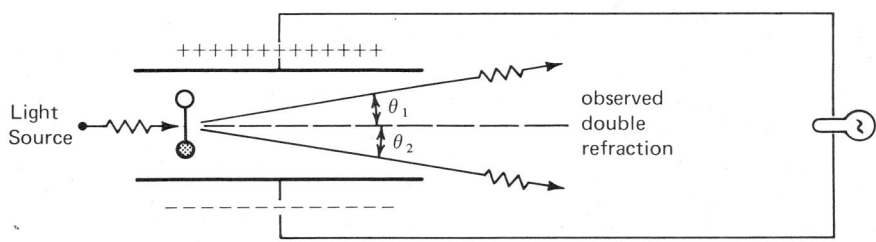

Fig. 6-25. Kerr effect. Molecular dipole in electric field between condenser plates is oriented. Light radiation is doubly refracted or spread by angles θ_1 and θ_2. In many respects this spreading resembles a polarization of the light.

an applied field. The Kerr effect is observed by transmitting light through a liquid held between condenser plates. The light "sees" a uniaxial crystal with its optical axis paralleling the electric field direction or lines of force. The electric lines of force connect the plates of the condenser. When light passes normal to this field through the liquid, it divides into two linearly polarized waves.

The Cotton-Mouton constant C is obtained from the magnetic analog of the Kerr effect. A magnetic field perpendicular to the beam of plane polarized light produces two refractive indices, $n_\perp$ and $n_\parallel$, perpendicular and parallel to the field where H is the magnetic field

$$C = (n_\parallel - n_\perp)/\lambda H^2 \qquad (101)$$

If Kerr and *Cotton-Mouton* effects occur on the refractive indices perpendicular and parallel to the field, there will be abnormal effects near to absorption bands and transitions. Birefringence is an optical result in an assembly of anisotropic molecules. The Kerr effect depends on this anisotropy, the magnitudes of the dipole moment and electric field, and on the temperature. Magnetic polarization by a field of the optically anisotropic molecules produces the Cotton-Mouton effect which then depends on anisotropy and molecular magnetic susceptibility.

Absorption of radiation can cause a transition from one electronic or rotational state to another. As in the Stark effect, atoms and molecules have a different response. For the smaller units, electronic energy (that is, of electrons *only*) is involved. In molecules, the electronic energy effect is well-separated from *rotational* energy. Stark effect studies are thus appropriate for each type as well as for the two together. Here, external fields are interacting with the electric dipole moments.

STARK AND ZEEMAN EFFECTS

There is a small shift in the frequency of the light emitted by an atom when it is placed in an electric field. An externally applied electric field causes the electron cloud surrounding the atomic nucleus to be moved "off center" with respect to the nucleus, because the two are oppositely charged (Fig. 6-26). This results in distortions of the elec-

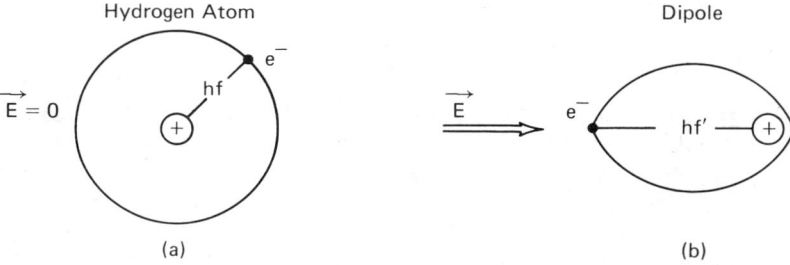

Fig. 6-26. Induced polarization. Stark effect.

tron orbits, and changes in the energies associated with them. Since the frequency of the light emitted by an atom is directly proportional to the energy lost by an electron in falling from one orbit into another, it follows that there will be a slight shift in the frequency of the light emitted in transitions between the altered energy levels.

In electronic interaction as opposed to rotational interaction, the Stark effect is small. In a field strength of 30 kV/cm or 100 e.s.u., the split is 10^{-2} cm^{-1}. With microwave rotational absorption of radiation, a strong Stark effect is observed. This is the effect that the alternating field in the microwave region has on the rotational energy of a molecule. Instruments used to study this produce a splitting of the population of molecules into subgroups with the corresponding different energy levels. Thus it is the energy state which is split in magnitude by a certain number of frequency or wave number units. It is convenient, as far as instrumentation is concerned, to observe the results spectroscopically.

It has been shown for molecules that there are electronic orbital and spin angular momenta plus rotation. A magnetic field interacts with the rotational angular momentum. It also couples with the nuclear angular momentum. If the molecule has the odd electron, the magnetic field also interacts with the electron spin. In the Stark effect, the electric field interacts with the molecular dipole moment and

$$Energy = \mu_e E \tag{102}$$

where E is the electric field and μ_e is the electric dipole moment vector. In the Zeeman effect, a magnetic field changes the molecular energy level where

$$Energy_{molecular} = \mu_m H \tag{103}$$

The applied magnetic field is H and μ_m is the magnetic dipole moment. The H field is seen to split the energy levels into two sublevels. The effect on orbital magnetic moments is the normal Zeeman effect; on spin, it is an anomalous effect. Rotational energy levels experience a change in spacing as a Zeeman effect.

In these interactions, the microwave absorption begins at the level of orbits and spins. Atomic orbital and spin angular momenta contribute to the Zeeman effect, being influenced by an applied magnetic field. Other contributions come from the spin-orbit interaction and the hyperfine interaction with nuclear spin and spin angular momenta. The spin Zeeman effects occur in free radicals and are observed by NMR and ESR. The Stark and Zeeman splittings or rotational energy level spacings result in subgrouping of the molecular population, and this is seen as line splitting or the formation of spectral lines. Lines are much more narrow in liquids and in gases than in solids (Fig. 6-27). This is motional narrowing as seen in nuclear magnetic resonance. In solids, one nucleus would affect another as in a rigid lattice.

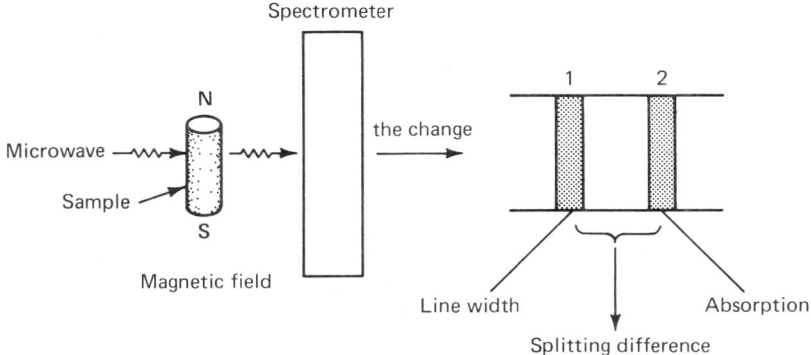

Fig. 6-27. Line width and splitting difference in microwave spectroscopy.

For example, the NH_3 inversion occurs in a magnetic field of 50,000 gauss, producing splitting, due to rotational magnetic moments equal to 45 Mc/sec. The inversion occurs as the N atom overcomes potential barriers between the two states $(E_2 - E_1)$ or the barrier formed by three H atoms which $\approx 1.24 \times 10^{-4}$ eV, placing the absorption line in the region of 1 to 2 cm waves. This is an instance of molecular polarization (Fig. 6-28). Slight changes are seen due to rotational states of the NH_3 molecule because these affect the hydrogen potential barrier. Thus it is also called a rotation-inversion line spectrum.

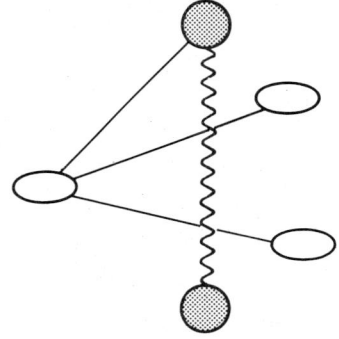

Fig. 6-28. Molecular polarization of NH_3. Resonance as reversing polarization.

The molecular transitions to new rotational energy states in the millimeter-centimeter spectrum for a diatomic molecule are described by the energy

$$E = N^2/2\theta \qquad (104)$$

where N is the angular momentum and θ is the moment of inertia of the molecule. Selection rules from quantum mechanics give the vector N only those discrete values allowed by the relation

$$N = (h/2\pi)\sqrt{J(J+1)} \qquad (105)$$

where J is the rotational quantum number from the scale of rotational states (Fig. 6-29). The moment of inertia is governing and it comes from the structure, i.e., the mass and distribution of atoms. If small atoms are drawn close, the rotational energy levels of separation will be greater than in large atoms at greater distances. Characteristically, microwaves cause transitions between rotational levels. Thus massive molecules with large θ have low frequency absorption peaks. Ice is heavier than HCl and ice absorbs with its rotational peak in the microwaves while HCl is in the infrared. This situation dramatizes how electromagnetic spectrum absorption behavior overlaps in the various regions. Other axes of symmetry can be added to this concept to allow

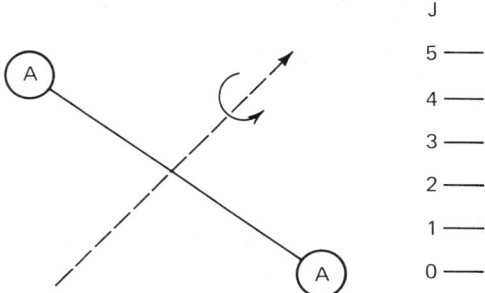

Fig. 6-29. Rotational energy transition for diatomic molecule (A,A).

for more rotational quantum numbers, vibration, distortion caused by rotating force, and band widening due to molecular interaction. The interactions provide components for absorptions and for serious consideration in molecular analysis. They may be decreased by studies in the gas phase. Bandwidth widening then would be associated with condensed phases and higher pressures. At stake is *resolution* in precise absorption studies. Interactions which produce bandwidth broadening may look like noise to the spectroscopist. Similarly, the communications engineer sees absorption of his signal en route as a loss. In molecular biology the microwave absorption is the fact of interest because of its possible effect on function.

ESR, NMR, AND BIOHAZARDS

ESR and NMR represent excited states in which unusual reactions might ensue in mixed systems at the time when reactions are studied by these resonances. Biofunctions could conceivably be placed in jeopardy by these super reactions. Further, the resonances occur in liquid and solid phases rather than in the gaseous phase alone. What then is the biological hazard from these basic absorption mechanisms in the microwave energy spectrum? ESR and NMR are used to study function, for example, transferrin, with normal products assumed to result. This, then, is like other radiation analyses in which it is assumed that the effect of the analyzing radiation on the preparation is zero.

The principle safeguard against danger is the need for the strong static magnetic field which must exist along with the microwave field. Similarly, in heterogeneous biological media, many interactions are possible which would dampen the effect beyond recognition as ESR or NMR. Nevertheless, under experimental conditions in large magnetic fields under crossed influence from microwave, organisms could conceivably present a proper sample considering all the possible organisms and all their possible molecular presentations. This situation would be difficult to arrange intentionally, let alone by accident.

BIOHAZARDS OF ROTATIONAL ABSORPTION

Turning to microwave rotational absorption, the situation is a little different because no separate magnetic field is required and the microwave energy is sufficient to cause the absorption. A number of restrictions lie in the way, however. The pure effects can only be observed in the gaseous phase. Within the concept of motional narrowing of absorption bands, possibly no lines will occur in liquids and solids, but the effect may persist even if it is difficult to measure it with present equipment. The millimeter and decimeter waves that are involved are common enough in microwave devices which therefore can expose organisms including humans. If the effect can occur, even though much dampened due to less than ideal conditions, what can be expected? The absorbing system will surely be excited and special reactions, excited products and drastic changes, possibly highly selective ones, can then occur. Exactly what results are possible remain unknown. Given a local gaseous space such as in the lungs, urinary bladder, gall bladder, the gastrointestinal tract, or in cell vacuoles or network structures, open sinuses, etc., and a suitable absorbing substance, this absorption could occur. The possibility of reversibility occurs as well as neutralization of excited products. Yet it is unlikely that the biosystem could return to complete normalcy. As to what products can be expected, this is an unknown area. Hundreds of compounds of C, H, N, S, O, and other bioelements have known rotational absorption in microwave fields under highly specific instrument conditions. These include radiation energy, frequency, orientation of fields, size and phase of sample preparation, etc. Nevertheless, there is a likelihood that some effect similar to the highly controlled rotational absorption will be observed in natural organismic states.

It may not be entirely a coincidence that the above mentioned organs which have gaseous spaces are often mentioned as sensitive to microwave absorption.

Deliberate creation of ideal conditions might well be attempted for any of these microwave absorptions in order to produce some, as yet unknown, therapeutic result based on excitation, resonance, or highly selective absorption at low power levels. The growing use of these techniques in analysis of bioreactions may well spin off some useful medical procedures but with highly specific instrumentation unlikely to be developed by accident or by haphazard exposure to microwaves.

BIOSPECTROSCOPIC STUDIES

The list of compounds associated with carcinogenesis in one way or another is very long and always growing and includes many substances uncomfortably familiar in everyday life. Many of these are easily measured spectroscopically. There are a number of old, established toxic substances ranging from rhubarb leaves with stored oxalic acid to pneumoconocoses, farmer's lung, bagassiosis, and byssinosis (moldy cotton) that are conveniently studied. Substances like nitrosamines are formed in the intestines upon ingestion of nitrate-nitrite cured meats (which are thus made bright in color). The nitrosamines may be carcinogenic and spectroscopy can give some evaluations although, of course, final action would depend on animal feeding tests. Suspected carcinogens include polymers, adhesives, pesticides, peroxy compounds, petroleum products, and food orange and red dyes. The latter are banned or are about to be but will persist for a few years until supplies are cleared.

A promising type of spectroscopic measurement has to do with rapid clinical diagnosis and detection on small samples and even on multiple readouts in a single sample. Urinary lactic dehydrogenase is implicated in bladder cancer, alkaline phosphatase activity in genital disease as well as in renal cell and prostate carcinoma. Catechol-amine-secreting tumors show spectroscopically detectable amounts of 4-hydroxy-3-methoxymandelic acid. For dopamine-secreting tumors, homovanillic acid may be measured and in childhood leukemia, leukemogens. Pure DNA is measurable in tumorigenesis and RNA in those viruses implicated in tumors. In cardiovascular disease, plasma beta lipoproteins may be related. The biochemistry of congenital disorders with inborn errors of metabolism such as phenylketonuria are subject to study. Other subjects include atmospheric pollutants (where rapid test results may be used for alerts), allergenicity, and penicillin sensitivity.

The information pathways in biology may be followed via biochemical tests in problems with emotions and behavior, memory, creativity, sexual activity, psychosis and schizophrenia with hallucinogens. The carcinogens, according to one point of view, activate an antimetabolite which has the effect of noise in the nucleic acid information system. This leads to faulty communication with the DNA and RNA which means loss of biological control and cancer.

PROBLEMS

1. What would dielectromagnetics mean as applied to biology?
2. What do the Periodic System and Electromagnetic Spectrum taken together illustrate?
3. In the broadest sense, what disciplines are included in the meanings of polarization?
4. How can biosystems be stabilized and yet evolve new systems?
5. How does opening the eyes affect the EEG tracing?
6. What is a free radical?
7. Which process is faster in terms of microsecond relaxation times, phosphorescence or fluorescence?
8. Would you expect the ESR or NMR to be especially biohazardous in the general environment?
9. What are the advantages of the MEG over the EEG?

BIBLIOGRAPHY

Assa, R., *IBM Res. Reports*, Vol. 5, no. 4, 1969.
Brittain, E.F.H. et al, *Introduction to Molecular Spectroscopy, Theory and Equipment*, Academic Press, London 1970.
Casy, A.F., *Spectroscopy in Medicinal and Biological Chemistry*, Academic Press, London, 1971.
Christophorou, L.G., *Atomic and Molecular Radiation Physics*, John Wiley, New York, 1971.
Birks, J.B., *Photophysics of Aromatic Molecules*, John Wiley, New York, 1970.
Clark, G.L., Ed., *The Encyclopedia of Spectroscopy*, Reinhold Pub. Co., New York, 1960.
Dewitt, C. and J. Matricon, *Physical Problems in Les Systemes Biologiques*, Gordon and Breach Science Pubs., New York, 1970.
Dole, M., *The Radiation Chemistry of Macromolecules*, Academic Press, New York, 1973.
Francon, M. *Holography*, Academic Press, New York, 1974.
Greenland, K.M., *Spectroscopy and its Instrumentation Translation*, Hilger, London, 1971.
Harrick, N.J., *Internal Reflection Spectroscopy*, Interscience, New York, 1967.
Hayflich, L. *Current Theories of Biological Aging*, Fed. Proc. Vol. 34, 1975, 51-5.
Hochstrasser, R.M., *Behavior of Electrons in Atoms*, W.A. Benjamin, New York, 1964.
Hucper, W.C. and W.D. Conway, *Chemical Carcinogenesis and Cancers*.
Holliday, R. and G.M. Tarrant, "Altered Enzymes in Aging Human Fibroblasts," *Nature*, Vol. 238, 26-30, 1972.
Ingram, D.J., *Spectroscopy at Radio and Microwave Frequencies*, Plenum Press, New York, 1967.
Lamb, M.F., *Biology of Aging*, John Wiley, New York 1977.
Meyer, B., *Low Temperature Spectroscopy*, American Elsevier, New York, 1971.
Mohyi, E. Abu-Zeid, et al, "Emission Studies of Pyrene Solutions," *Chemical Physics Letters*.
Myers, L.S., "Macromolecules of Biological Interest, Nucleic Acids, Protein, and Polysaccharides, Free Radical Biochemistry," *The Radiation Chemistry of Macromolecules*, E.M. Dole, Ed., Academic Press, New York, Vols. I and II, 1973.
Narahari, Rao, K. and C. Weldon Mathews, Eds., *Molecular Spectroscopy*, Academic Press, New York, 1972.
Orgel, "Aging of Clones of Mammalian Cells," *Nature*, Vol. 243, 441-5.
Sawicki, E., *Chemistry Analyst*, Vol. 53, 24, 28, 56, 88, 1964.
Strekler, B.L.,"Implications of Aging Res. for Soc.," *Fed. Proc.*, Vol. 34, 5-8.
Schlicke, H.M., *Essentials of Dielectromagnetic Engineering*, John Wiley, New York, 1961.
Ternberg, J.L., and B. Commoner, *Journal of the American Medical Assoc.*, Vol. 183, 339-42, 1962.
Ternberg, J.L. and B. Commoner, *Photometric Organic Analysis*, Wiley Interscience, New York, 1970.
Udenfriend, S., *Fluorescence Assay in Biology and Medicine*, Academic Press, New York, 1962, Vol. II, 1969.
Walker, S. and Shaw, H.S., *Spectroscopy*, Vol. 2., pp.12-13. Chapman and Hall, New York, 1962.

7

The Electromagnetic Spectrum

"But some officials expressed concern for the health of embassy residents and workers. High-intensity microwaves, like those used in electronic kitchen ovens can "cook" human cells. They can cause cataracts and raise levels of serum triglycerides, or blood fats in humans, predisposing them to heart attacks. The waves can also interfere with the operation of heart pacers."
  from: "American Embassy Personnel in Moscow Receive Microwave Radiation from Russian Transmitters" Science News in *Time Magazine*, 23 February, 1976

MINIFORMED KNOWLEDGE

Science and technology have brought us to a point where electromagnetic radiations influence us constantly and more than ever before. As seen in Chapter 5, it is only fair to raise the issue of our compatibility in this electromagnetic environment. Here the electromagnetic spectrum logically comes after that concept and others such as the waves, forcing function, information density, polarization, and quanta as taken up in that chapter. It is also positioned after the dielectromagnetics and informational-spectral probes for molecular biology discussed in the last chapter because the spectrum is a major objective in itself. It might have gone first without introductory sections. This would have allowed different aspects of it to be followed subsequently but that would not seem to be fair for one of the great central ideas in science. It is clear that an understanding of these radiations and their interactions has taken on the highest priority.

If it could be presented, the spectrum would be an encyclopedic classification of action spectra, frequencies, wavelengths, wave numbers, associated energies of the quanta, equivalent masses of photons, sources, applications, interactions, informing processes, administrative constraints on users, and biological hazards of all the radiations. At one time or another, biologists will have an interest in each of these characteristics and more, all of them constituting the completely classified spectrum. It is convenient to view the spectrum as a continuum of frequencies extending from low frequency, electric, alternator oscillations as used in power, light, and telephone application, to the highest frequencies produced by powerful particle accelerators and super energy cosmic rays from outer space. Between these limits which could theoretically go from zero to infinity, are the radio, microwave, infrared, visible, ultraviolet, X-ray, gammas and more energetic radiations according to increasing frequency.

The spectrum may be divided into many subspectra depending on the depth of treatment but some minimum such as the 5 shown in Fig. 7-1 (see back pocket) is appropriate. Then further definition and resolution is provided on the spectrum itself and the linear array of frequencies with a vertical development of the characteristics should convey the idea of a steadily expanding collection of knowledge in that direction. *Here then, are all the radiations that have the burden of carrying, containing, or retrieving specific information and the energies that are associated as a function of frequency. Both may be destined to reach transducer molecules or organs playing the role of selective receptors, to function in a lesser capacity as noise, with the energy to be finally degraded to thermal levels.*

Applications or manifestations are essentially intersystem interactions of this information-energy flow with particular systems located in the transport pathway. This action-at-a-distance capability is characteristic in that the systems need not be physically connected. Only a sampling can be given, since whole handbooks of chemistry and physics, encyclopedias, and international data books are needed to provide all the details.

SPECTRAL SENSING

One informational probing science that summarizes this quite well is spectral sensing or remote, environmental surveillance with analysis, interpretation, synthesis, and diagnosis, combined under the broad term *spectral information analysis*. Here the sensing of information may utilize one or all of the available frequencies across the spectrum. Multiple records by photograph or other graphs are used for analysis but the intersystem feature is vividly illustrated when this sensing is done between planets.

This is a special kind of spectroscopy at a distance, with the information quite clearly presented as a sensing record of the interaction between energy at some wavelength and the material of interest. Instead of molecular information, this sensing is often after *spectral signatures* of gross material features, just as an X-ray produces a gross comparison between tissues which can be interpreted by a specialist. The spectral signature is the wavelength band (see the Electromagnetic Spectrum) and from the air, photograph records will show the imprint of an oil slick best in near ultraviolet and infrared, while municipal sewage will be shown in yellow-green in the visible spectrum. Varying the water with rhodamine dye tracer will show currents or sources of contamination in the visible spectrum at 0.56 to 0.62 micrometers. Figures 7-2 and 7-3 show spectral sensing applied to thermal pollution of a river and harbor discharges. Both are single band information returns which would often be adequate for aerial surveillance for ecological control without additional ground data, and the band may be any wavelength for which sensors are available. This could be almost any portion of the spectrum, certainly microwave radar pictures, infrared, visible, ultraviolet, X-rays, or gammas, the only limit being an engineering one. Multirecord sensing, on the other hand, can produce more information by a convergence of evidence. This means that new data is synthesized from the sample that would be unavailable to single sensors.

> This is somewhat like a *superacute form of sensory perception*, as if combining the ultraviolet "vision" of certain insects with the infrared sensors of snakes and the visual acuity of humans.

The photographic film may be sensitive to the wavelength from beyond X-rays through infrared and, beyond these limits, photographs are taken of the instantaneous information return as with a radar screen. On the sensitive film, the radiation behavior is recorded in tone differences based on reflection, transmission, and absorption of the energy, any of which may be the signficant mechanism. Based on their coefficients, these would equal unity

$$r + t + a = 1 \tag{106}$$

With daylight, many photographers would depend only on r, (r=1), there being no transmission or absorption, but this records the required information on surface details and subtle differences in texture among all those represented in the object. Emissivity is important too, forming the basis for night infrared photographs.

Since *one* sensor wavelength will not usually give the total record needed, but only a portion of it, careful selection of a complementary probe wavelength can make a unique result. To obtain such complementarity may require analysis of the signal information as affected by the transmission path, scatter, attenuation, sensitivity of the sensor, and the use of discriminators and filters, possibly with computer aided analysis. The multiwavelength signals produce a series of records with the interpretation then coming from a comparison. The interpretation uses converging evidence to form a conclusion not based on any single record but arising out of the comparison.

Scatter and attenuation influence the signal strength, for example, short ultraviolet will be excessively scattered in accordance with Rayleigh's law while infrared will penetrate smog and haze to return infor-

Fig. 7-2. Spectral sensing. Nuclear plant cooling water raises the surface temperature of Connecticut River at East Haddam Neck. Plant at first arrow produces the warm stream (second arrow) which blends into the river. The warm stream is much lighter by infrared sensing and the river is lighter too, after the coolant enters. Courtesy, Bendix Co., South Field, Michigan.

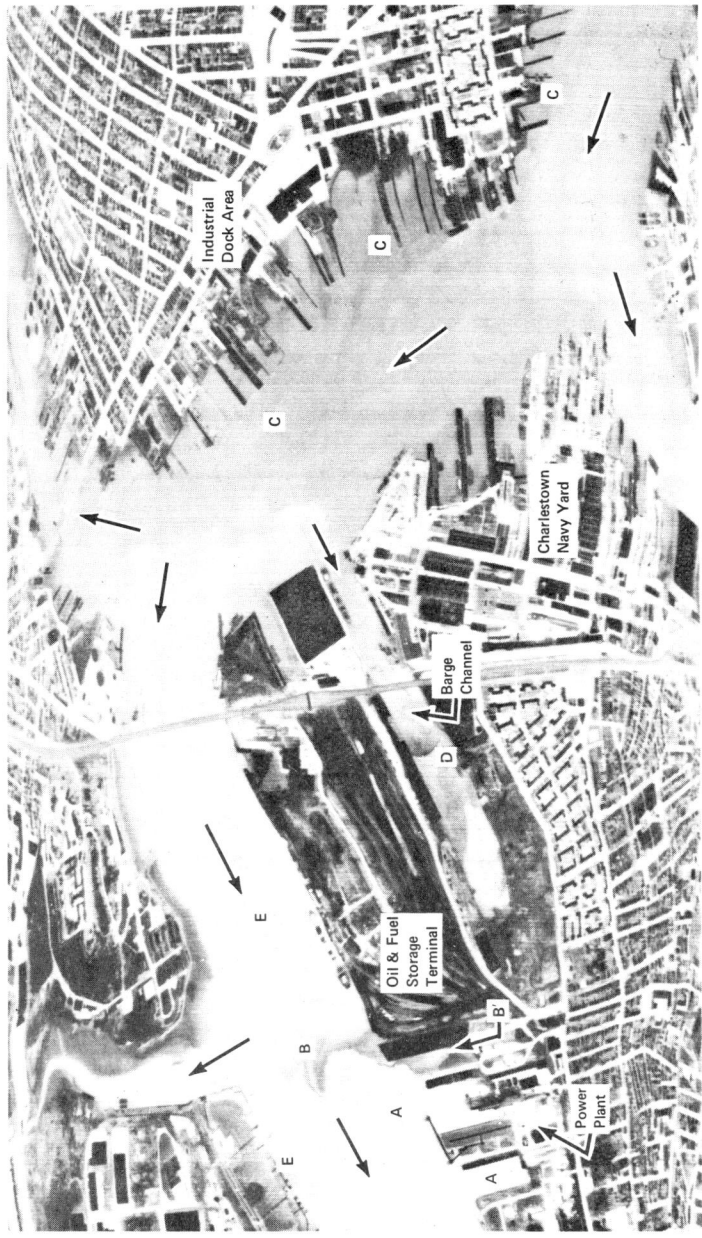

Fig. 7-3. Boston Harbor. Infrared sensing at 3 A.M. evaluates water quality in the inner harbor as being determined by a power plant, fuel terminal, large channel, Navy Yard and industrial dock area. A-thermal discharges, B-oil slick from B' fall storage area, C-diesel, paint waste, acid pickling baths, D-shallow, inadequately flushed channel, E-tanker heat sources (heat oil for pumping). Courtesy, Bendix Co., South Field, Michigan.

mation. Infrared "windows" in the atmosphere occur between 3 and 6 micrometers or between 3.7 and 5.5 and a wide window exists from 8 to 12 or 14 micrometers. Water photos must take into account that infrared is strongly absorbed during the day and emitted at night. Thus the reflectivity coefficient is important in daylight hours and the emissivity is important at night. Naturally, the light must be sharply focused on the film due to chromatic aberration. (Fig. 7-4). This occurs

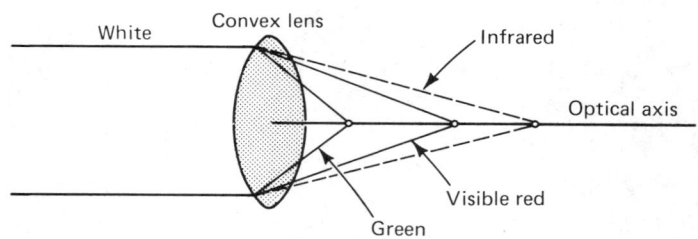

Fig. 7-4. Dispersion of focus with wavelengths. The wavelength changes and is subject to correction in optics.

due to dispersion of the focus as the wavelength changes and is subject to correction in the optics.

Convergence of evidence in spectral sensing is illustrated in aerial diagnosis of plant diseases (Fig. 7-5). The diseases due to *Puccinnia* in wheat and oats result in a texture change in the mesophyll (turgid to spongy) which changes the reflectivity of the plant to infrared. Diagnosis can be earlier by this means than by panchromatic photos because the texture change is not evident in the unchanged green color seen in that film. Plants reflect green from their chloroplasts, absorb blue and green in photosynthesis, and reflect infrared from the lower lying mesophyll. These methods have been used for diagnosing disease in artichokes, wheat, oats, cotton, peaches, pears, grasses and naval orange trees. The multistructural sensing gives the diagnosis.

Phytophthora is another fungal disease of trees that may be diagnosed early by panchromatic-25 A (film filter) plus infrared 89 A as the hyphae penetrate and clog the mesophyll pores, changing the texture. Infected trees in an orange grove may be identified by this method.

Phase variations of a wave may be recorded if the wave is made to interfere with an incoming, additional, coherent laser wave in holography. Coherent waves produce the clearest interference whereas incoherent waves would mix in a more complicated manner. An intense laser illumination that is temporally and spatially coherent is the hologram background. The quality is explained in terms of sharpness of interference patterns or fringes. The hologram is a record of amplitude

The Electromagnetic Spectrum

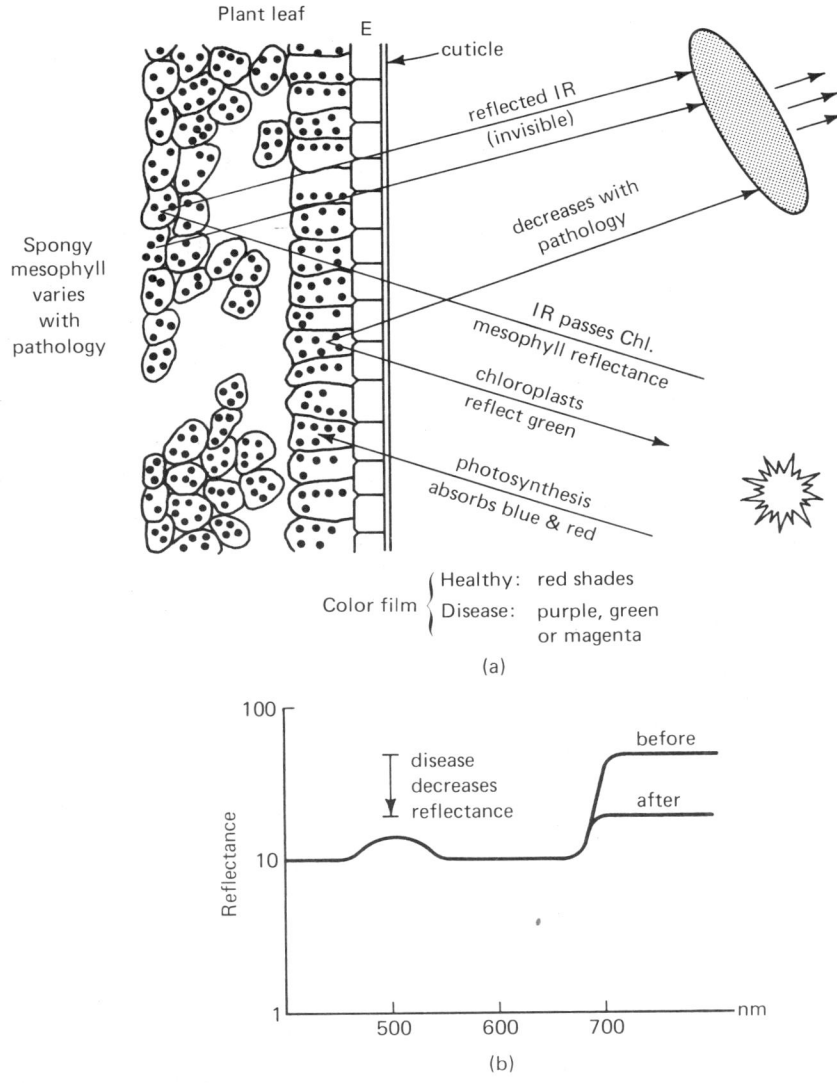

Fig. 7-5. EMR energy for information transfer from tissue infected with *Psorosis* or *Tristeza* virus, *Xyloporosis*, *Exocortis* or *Nematosis* (in older trees)

and phase from the exposed subject which is reconstructed into a view or image using the available diffraction pattern.

Unusual effects such as shock waves can be recorded. Potentially, a holographic microscope could gain resolution equal to the ratio of forming and observing wavelengths so that a hologram made with X-rays and observed with visible light would have a result like an electron microscope. A 3D motion picture is possible which would not require special glasses.

Other converging evidence is obtainable from ground facts when necessary. For example, water suspected of contamination may be sampled for *Escherichia coli* and suspended solids.

Since infrared imaging is by thermal sensing, and the subject can emit its own radiation, the technique has great analytical potential. Blood circulation "photographs itself" and shows abnormalities such as improper umbilical cord positioning which could add to birth hazards, as well as the position of breast cancers. Lasers provide light wavelengths, but with coherent radiation which then increases the utility of the wavelength. Motion, for example, can be followed in a warm body as in chest motion during respiration. Then irregularities are revealed to one trained in interpretation. The laser capability is extended to spectral repair as in the case of retinal detachment when laser surgery helps to repair the disorder. The process is wavelength-dependent with neodymium lasers damaging structures leading to the optic nerve as compared with ruby lasers. In addition to the superior properties at a wavelength plus the coherence in the wavelength due to lasing, we must also consider the significance of a microbeam. This beam can be made small enough to manipulate inclusions within cells, to modify or observe chromosomes as well as to change mitochondria and other components. If the action spectrum for a biomolecular absorption is the same as the laser wavelength, selective molecular interaction is added to the microsurgery potential. Holography is a means for producing realistic three-dimensional images and lasers have made this procedure practical. With it, ocular distortion in depth is diagnosed during the progress of glaucoma which changes the ocular pressure.

A multisensor diagnosis is obtained by combining lasers and acoustics which produce an interacting sound pattern in tissue. Then the reflections become visual patterns, returning internal information in the manner of X-rays but without the hazardous exposure of various structures inside a body.

SPECTRAL RECONNAISANCE

Microwave thermography is a clinical trial method for detecting cancers, especially of breasts. Converging evidence may be accumulated by infrared, centimeter, and millimeter wave imaging of the cancer which may be as small as a centimeter. Then the methods may be used in a scanning mode for mapping tissues in depth, locating areas of greater or lesser vascularization and inflammation throughout the body and brain. In microwave hyperthermia, thermography used in an active versus passive mode can be used to monitor the therapeutic temperature.

The Electromagnetic Spectrum

Microwaves or ionizing radiations may be used to destroy tumors leaving the residue to be absorbed. Some wavelengths may be far more selective than others for the tumor tissue and the metabolism of a region may be stimulated or inhibited from outside by suitable wavelengths that absorb selectively, without affecting normal tissue. Masers relate to microwaves as lasers do to the various light frequencies. There are instruments using the laser to fuse tooth-colored material directly to the enamel to reinforce it against decay, and UV is used (3200-4000Å) to polymerize composite resin to chalk or seal teeth for a similar result. Recent versions incorporate safety measures for the dentist and patient.

In time sequence sensing—like time lapse photography—the change in a scene becomes apparent. For example, infrared photographs show a bright image for the warm automobile hoods in a parking space showing which car or cars may have just been parked. Aerial photographs in sequence show warm areas associated with trucks and construction activity which is important around new missile sites in military intelligence. Certain wavelengths are ideal as well for "selecting" camouflaged objects and showing what they actually represent.

In the X-ray region, 12.3keV rays with a wavelength of one angstrom are suitable for probing crystal structures by X-ray diffraction as has been shown for the DNA helix. The crystal is made to describe itself along the lines of methods used in information theory and coding. Also in the X-ray spectrum, soft X-rays (of lower frequency than hard rays) permit the visualization of internal structure because they are differentially transmitted by denser and softer tissues. This selective transparency pattern is revealed in the sensitized grains of the radiograph after development. In biospectroscopy, radiation characteristics are matched with biological test requirements. The absorption in tissue for all radiations is inversely proportional to penetration. This relation leads to the use of harder X-rays in radiotherapy of deep tumors. Not only is the penetration adequate but it is possible to concentrate the absorption toward the end of the penetration. This fact tends to spare the intervening normal tissue, as well as the tissue beyond the tumor, from damage. Good geometric positioning and correct radiation energy are used to insure these results.

The ultraviolet portion also furnishes examples of discrimination or selectivity. An equal mixture of nucleic acid and serum protein will absorb over 90% of ultraviolet at 2600Å in the nucleic acid and only about 10% in the protein. At 2537Å, *E.coli* bacteria irradiated by ultraviolet are inactivated permanently unless they happen to be rejuvenated by exposure to visible light (photoreactivation) within a reasonable time. The information-energy complex, carried by the radiation via its frequency characteristics, is literally a matter of life and death to these microorganisms. A direct communication occurs between a specific wavelength and the organism's energy system. Thus, *E. coli* is photopreprotected by near ultraviolet at 3340Å against subsequent exposure to far ultraviolet at 2540Å which is approximately the same lethal wavelength as previously discussed. *The longer wavelength* (between 3000 and 3800Å) *retards oxidative respiration and phosphorylation*. This welcome pause in its energy operations allows the cell to regenerate missing components of the electron-transport chain in a sort of repair-in-the-dark mechanism based on selective metabolic inhibition.

In a manner somewhat like the self-photographed infrared images, radioactive isotopes like phosphorous-32 and tritium (hydrogen-3) can make autoradiographs of their distribution in tissues when they are placed on a film and allowed to expose it. The interaction is with the silver bromide (AgBr) crystals suspended in a film of gelatin. Affected grains are then reduced to silver in the developer and unaffected silver bromide dissolved by fixation. In other cases there is a primary or an intermediate interaction between the radiation and the structured matter in its pathway, such as the tumor, in contrast to differentiated tissue with X-rays; and again with radar targets in front of the microwave beam. Then the final image or secondary interaction is in a photograph, fluorescent screen (fluoroscopy), or phosphorescent display (radar scope) showing the results of the primary interaction. The autoradiographic readout of information is somewhat more direct from the radiation interaction standpoint because it does not have both primary and secondary interactions. The end result in these instances is information retrieval.

SPECTRAL UNITS

Probably the best way to think of the spectrum is as a collection of electromagnetic frequencies rather than of wavelengths, but remembering that these are related by the relation

$$\lambda = c/f \qquad (107)$$

Thus radiations can pass through a particular media without changing frequency but the wavelength λ and the velocity vary according to the nature of the medium. The velocity of the electromagnetic radiation wave in space is always c (3×10^8 m/s), that of light. In another medium, this velocity is obtained from the frequency properties of the material, that is, the dielectric constant ϵ' and the magnetic permeability by the relation

$$velocity = c/\sqrt{\epsilon'\mu} \text{, and } \lambda_m = velocity/f \qquad (108)$$

This is true for all waves and, while the discussion is about electromagnetic waves, it must be remembered that there are also sound and water-like waves. Electromagnetic waves are distinctive in that their "medium" is the electric and magnetic fields which form them, while both sound and water waves depend on a separate medium of propagation. Water-like waves are often chosen to demonstrate electromagnetic waves using a rope, hose, Slinky, or any elastic substance held at one

The Electromagnetic Spectrum

end and moved to generate a wave. In a medium with refractive index n the velocity $v = c/n$ and, in air, it is about equal to c where n is about one.

One can always calculate the frequency by measuring the wavelength with wavemeters. Portions of the spectrum are sometimes named by the characteristic length of the wave, for example, meter wave band radio, centimeter in microwave, micrometers (μm) in infrared, nanometers in the visible (sometimes millimicrons), angstroms Å for X-rays, and milliangstroms for gamma rays (see Fig. 7-1).

The various spectra may also be clearly indicated in frequency units such as Hertz, after Heinrich Hertz, or cycles per second, so that there are gigahertz GHz (10^9 c/s) and terahertz THz (10^{12} c/s) frequencies and others as suggested by the conventional prefixes included in Table 7-1. There is still another aspect to the terminology. Frequencies often

TABLE 7-1. Decimal Prefixes for Forming Acceptable Numbers

10^{12}	tera	T	Teracycle	10^{-15}	femto	f	femtoampere
10^9	giga	G	Gigacycle	10^{-12}	pico	p	picofarad
10^6	mega	M	Megatons	10^{-9}	nano	n	nanometer
10^3	kilo	k	kiloton	10^{-6}	micro	μ	microfarad
10^2	hecto	h	hectoliter	10^{-3}	milli	m	millimeter
10	deca	dk	decaliter	10^{-2}	centi	c	centimeter
				10^{-1}	deci	d	decigram

used to be referred to by the energy characteristics of their particular part of the spectrum. Some examples appear in Table 7-2. This policy of associating the energy is especially true beginning with the microwave and infrared spectra where the frequency is often given in wave numbers (cm^{-1}) for the resonant frequencies of compounds. Then visible light and ultraviolet are usually described in eV or photon energy. In any case, the relation for energy and frequency is again E (for example, in eV units) equals hf, where h equals Planck's constant. Conveniently

$$\text{eV} = 1.24/\lambda\mu m \quad \text{and} \quad J = (1.987 \times 10^{-19})/\lambda\mu m \quad (109)$$

From low frequency broadcast radio to cosmic rays, the energy in eV ranges from 10^{-9} to millions of electron volts. When the radiation energy increases through the thousands, millions and billions of electron volts (keV, MeV, and BeV), the frequencies pass from those of X and gamma rays to cosmic rays. Spectra are more and more commonly being described in "frequency" units (cm^{-1}) or wave numbers and this convention will tend to predominate for spectroscopy because the wavelengths of sharp spectral lines have been very accurate, even more so than c, the velocity of light, and relative photon energies are conveniently given by λ^{-1} rather than by frequency. This is the space frequency, sometimes $\bar{\nu}$ or σ, and where f or ν is the frequency in cycles/unit of time,

$$\bar{\nu} = \sigma = f/c = \nu/c = \text{cycles/cm or waves/cm} \qquad (110)$$

With 10,000 waves/cm,

$$\lambda^{-1} = 10^4/\text{cm}, J = 2.1 \times 10^{-20} \qquad (111)$$

and this equals 0.124 eV. Another notation is the angular space frequency in radians/cm, sometimes k, or phase constant, where

$$k = \omega/c = 2\pi f/c = 2\pi/\lambda = \text{radians/cm} \qquad (112)$$

TABLE 7-2. Energy Levels, Wavelengths and Responses

Energy Level and Quanta eV	Exciting Source Wavelength	Spectral Region	Selected Response
4.13×10^{-6}	30 cm	microwave	molecular rotations
1.24×10^{-3}	1 mm	microwave	molecular rotations
0.01	124 micrometers	infrared	molecular rotations
0.1	12.4 micrometers	infrared	molecular vibrations
1 (quanta individually detectable)	1.24 micrometers	infrared	electronic-molecular disturbance
1.24	1 micrometer	near infrared	bond flexing and displacement of electrons relative to nucleus in optical region.
1.8	680 millimicrometers	visible	photosynthesis
3.1	4000 Angstroms	visible	ionization of molecules (gas) in glass
13.5	900 Angstroms	ultraviolet	*$IP_{hydrogen}$

*Typical dissociation energies are 2 to 5 eV. Ionization potentials are on the order of 10 eV, for example, for hydrogen (H), it is 13.5 eV. This is the work needed to pull the electron away from the hydrogen nucleus.

The Electromagnetic Spectrum

If

$$\lambda = c/f, \text{ then } \lambda^{-1} = E/hc = 10^4/\lambda\mu m \text{ cm}^{-1} \tag{113}$$

and the wave number is linked to the energy of the waves. If, in the interaction under study, there is a natural tendency toward one or the other descriptive units (for example, photons, eV, Hz, or Gc/s), then best understanding dictates that it should be used. The last two are understandable in terms of the associated energy. The length of a standard wavelength such as the red cadmium line in air at 760 mm pressure and 15°C, is quite naturally given in linear units (6438.4696 angstrom units). Hertzian waves are often understood to be those having wavelengths *longer* than 2.20×10^6 angstroms (millimeter waves). The log of frequency or log of wavelength (same except for right-left inversion) may appear, at times, because here there are advantages realized in setting up numerical instrument ranges in different spectra or in the convenient plotting of absorption or transmission on log graph paper. The wave number λ^{-1} for an energy of one electron volt is equivalent to about 8,066 cycles per cm. This relation reveals the wave number as a frequency unit with an associated radiation energy. Useful conversions would be

$$1 \text{ Å} = 10^{-10} \text{ m} = 0.1 \text{ nm} \tag{114}$$

$$1000 \text{ Å} = 10^{-5} \text{ cm} \tag{115}$$

$$\text{Å} = 10^{-5}/10^3 \text{ cm} \tag{116}$$

$$2.2 \times 10^6 \text{ Å} = (10^{-5}/10^3) \times 2.2 \times 10^6 = 2.2 \times 10^{-2} \text{ cm}$$
$$= 0.022 \text{ cm} = 0.22 \text{ mm} \tag{117}$$

Radiation Sources

The radiations of the whole spectrum may be designated by these various names related to their dimensions, or by the generators or sources in which they originate. The spectrum is simply described as extending from a few c/s with rotating electrical machinery, to more than 10^{24} c/s associated with cosmic rays. Up to a frequency of about 3×10^{12} (which is about where microwaves end and far infrared begins), electronic oscillators, or tubes, semiconductor devices and resonant L and C (tuned) circuits are used to generate the frequencies. A fraction of a millimeter wave (e.g. 0.1 mm) is the smallest that can be so produced electronically. Some of these frequencies are involved in absorptions in matter by polarization, rotation, vibration, and translational modes.

Table 7-3 shows the spectral designations and units with their associated forcing functions. What is absorbed may, at times, also be emitted, according to the frequency and energy rule

$$frequency = (E_2 - E)/h \tag{118}$$

TABLE 7-3. Forcing Functions and Associated Frequencies

Frequency Name	Frequency c/s	Electron Volts	Forcing Function or Region on which it is Exerted
Radiowaves	3×10^5	1.2×10^{-9}	Free Ion and Electron Translation
Short Waves	3×10^7	1.2×10^{-7}	Nuclear Resonance
Microwaves	3×10^9	1.2×10^{-5}	Nuclear Resonance Free Electron Translation Electron Resonance
Far Infrared	3×10^{13}	1.2×10^{-1}	Rotational Excitation
Near Infrared	3×10^{14}	1.2×10^0	Vibrational and Electronic Excitation
Visible Light	0.6×10^{15}	2.4×10^0	"Bond" Breaking
Near Ultraviolet	1×10^{15}	4.1×10^0	Outer Electron Liberation
Far Ultraviolet	2×10^{15}	8.3×10^0	Middle Electron Liberation
Long X-ray	1×10^{16}	4.1×10^1	Inner Electron Liberation
Short X-ray	3×10^{16}	1.2×10^2	Ionization
Gamma Ray	3×10^{20}	1.2×10^6	Nuclear Reactions

where E_2 is the higher energy level and E is the lower one involved in the change. Thus these frequencies may come from energy changes in the structure of appropriate atoms and molecules, that is, in the electronic arrangements or nuclear resonances so that *tissue has an ensemble of EM generators*. If $\lambda = 1$ Å,

$$E = (6.625 \times 10^{-34}) \times [(2.998 \times 10^8)/10^{-10} \text{ m}] = 1.986 \times 10^{-15} \text{ J} \quad (119)$$

$$= 12.4 \text{ eV}$$

A mole of hydrogen in the ground state dissociates, for example, in an excited plasma

$$H_2 \sum{}_g \rightarrow 2 \text{ H atoms} \quad (120)$$
$$104 \text{ k cal/mole}$$

This energy of 104 k cal/mole comes via the energy of electrons. The electrons have this amount of energy which is not heat (athermic) since the reaction can occur at cryogenic temperatures as in the collection of free radicals.

Maser and laser sources play a significant role in the overlapping region of microwaves and far infrared on into the visible region. These acronyms describe the generation of microwaves and light respectively by the s̲timulated e̲mission of r̲adiation. A much less sophisticated example is the universal production of infrared rays by every warm body over 20°K and especially by the sun and glowing ceramic, carbon arcs, flames, filaments, and the like. The heat radiated from a black body

(5) $\quad S = \sigma(T_1^4 - T_2^4)$ (121)

(6) $\quad T_1$ = absolute temperature of source (122)

(7) $\quad T_2$ = absolute temperature of environmental medium (123)

(8) $\quad \sigma = 0.567 \times 10^{-4}$ ergs/cm^2/sec^2/deg^4 or (124)

$\quad\quad 0.567 \times 10^{11}$ watts/cm^2/deg^4 (125)

The human skin at 308 degrees absolute (Kelvin) radiates mainly in infrared at 10 micrometers. The frequencies from infrared through visible are found to be generated whenever atomic and molecular energy shifts occur, and these frequencies also come from the sun and from thermally excited atoms by molecular vibration. Ultraviolet comes from the sun, atomic collisions with electrons, and by gas discharge tubes such as mercury lamps. X-rays are typically produced by machines which accelerate electrons to great velocities and then suddenly stop them in a target such as tungsten. Gamma rays, of still higher frequencies, are considered as coming from radioactive disintegrating atoms such as cesium-137. In terms of radiations emitted by such materials, gamma rays are *of* the nucleus and come from nuclear transformations while X-rays are extranuclear in origin. The cosmic rays are produced by nuclear and atomic reactions not usually seen on earth but common in space and on bodies such as the sun which generates most of the EM spectrum. When the great nuclear reactions, fission, fusion, or thermonuclear occur, they produce almost an entire spectrum along with neutrons, protons and nuclear fragments.

It would appear possible to generalize about these many sources by observing that they tend to function through the acceleration of electric charges. In this sense there are indeed many generators. The hydrogen atom with its proton surrounded by six possible electron orbiting levels, or shells (Fig. 7-6), can generate energy equal to the difference between any two states, for example from 3 to 2 or 3 to 1 in the relation

$$E_3 = E_2 - E = hf \quad (126)$$

and the corresponding, generated frequency is

$$f = E_3/h \quad (127)$$

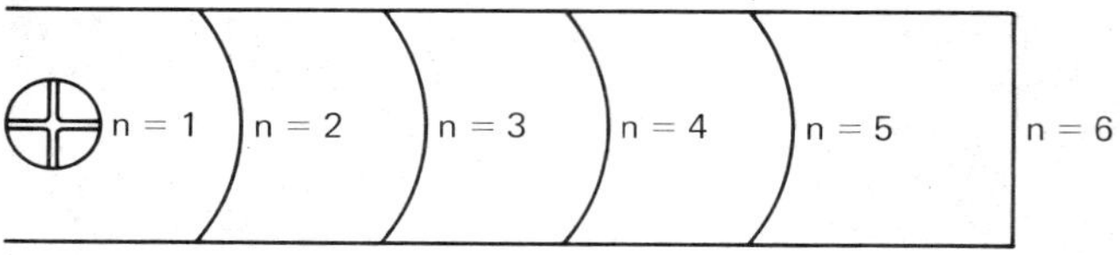

Fig. 7-6. Hydrogen atom with its proton surrounded by six possible electron orbiting levels or shells.

Now if all the atoms are included and it is considered that each one is unique with its set of possible configurations and energy changes, then the number of frequencies it is possible to generate is very great. The radiation is a field propagated away from the generating excitation. Therefore, all substance, biological material in particular, is composed of generators that can be sources for most EMR.

IONIZING AND NONIONIZING OR POLARIZING RADIATIONS

The behavior of radiation can be divided into two large areas: the ionizing and nonionizing types. Spectrally, the first ionization capability is usually assigned to ultraviolet but this is subject to some interpretation, certainly, for example, in photochemical reactions and ionizing-like changes that occur due to excitations below the visible, near infrared photosynthesis, for example.

These sources produce wavelengths which vary from an enormous 18,600 miles at the low frequency end at about 10 Hz to 0.03 Å at 10^{20} Hz (X-rays). This is the distance traveled from the source during one cycle or vibration. In each section of the spectrum, it takes several bands and corresponding generators to occupy the entire range of frequencies. Besides natural emissions, very common from the sun during solar storms for example, from solid state oscillators, and from masers, there are electronic generators such as magnetrons, klystrons, and traveling wave tubes. These are the workhorses of microwave spectral applications, for example for the many types of radar, communication beam relay links, cooking, diathermy, spectroscopy, dehydration, plasma production, industrial processing, and radio astronomy to list only a few of the myriad of microwave uses.

Overlap Regions

The sections of the spectrum are classified and subclassified for convenience in managing this vast presentation of information. In the borderline regions between bands, the generation, transmission, absorption, detection, and behavior of the radiations assume characteristics of both sides as may be noted for microwave and infrared. A millimeter wave band may be distinguished which goes from 0.1 mm to a few mm. From 1 mm to 0.1 mm this is an overlap region where microwaves merge with optical frequencies. The tiniest microwave at about 0.1 mm is, therefore, in the optical range which spectrum experts begin at 3 mm or 100 GHz. What the operation in these overlapping bands will actually look like is a matter of convenience in the technology, or depends on the side from which the technologist approaches the application. Millimeter wave operations in microwave are semi-optical and lasers, which involve light, compete successfully here with microwave generators or masers in many cases.

Similarly, in the microwave-infrared overlap region in radiation interactions, there are rotations of molecules by microwaves, then rotations and vibrations mixed, until finally interactions tend to be vibrational and to have thermal characteristics in the infrared region (Fig. 7-7).

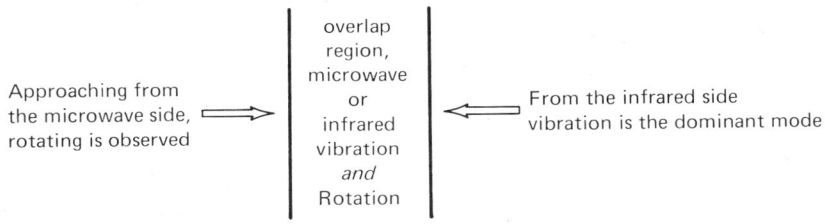

Fig. 7-7. A spectral overlap region.

These vibrations are intense and superficial, leading to great local absorption and little penetration so that heating of *surfaces* occurs.

In terms of incandescent sources, a wide range of wavelengths is possible according to the absolute temperature of the thermal source. If the peak emission is in the visible, then it will shift to infrared when the temperature is lowered. The thermal source may be an organism such as the human body, a well-known infrared generator but also a *good microwave source at least in the microwatt range. Thus when emitted wavelengths pass from micrometer to millimeter sizes, the emission is microwave with many infrared or optical similarities.*

Photovoltaic power absorbs solar energy by forming pairs of negative and positive charges via ionization. Gallium-arsenide cells collect an electron and a hole or positive charge for each photon absorbed. The charges move to the p-n junction, positives to positive side and negatives to negative. Solar energy may then be drawn off as electricity. The cell may have seven layers proceeding from the receptor side:

1. The p contact.
2. An antiscatter coating to enhance absorption.
3. The aluminum-gallium-arsenide almost transparent trap.
4. Gallium arsenide p crystal where the ionization occurs.
5. Gallium arsenide n crystal opposite the junction.
6. The n + gallium arsenide.
7. The n contact.

LANDMARKS IN ELECTROMAGNETICS

The primary behavior of radiation energy is expressed in its interaction with matter. This behavior is inseparably related to the mass equivalent into which it can, under proper conditions, actually transform according to the famous equation

$$E = mc^2$$

The great discoveries about radiation shown in Table 7-4 are actually steps in the understanding of this interaction.

TABLE 7-4. Landmarks in Electromagnetism

1800	Sir William Herschel	Infrared spectrum
1873	J. C. Maxwell	Electromagnetics
1887	H. Hertz	Basic work in electromagnetic waves
1895	W. C. Roentgen	Discovery of X-rays
1896	A. G. Becquerel	Discovered natural radioactivity
1897	J. J. Thomson	First observed the electron
1898	Pierre and Marie Curie	Isolated and identified polonium and radium
1901	Max Planck	Planck's constant introduced. Electromagnetic radiation emitted in discrete quanta.
1903	William W. Coblentz	Infrared Spectroscope
1911-1919	Lord E. Rutherford	Atomic and nuclear basic concepts and transmutation
1913	Nils Bohr	Planetary picture of atomic structure
1916-1956	Albert Einstein	Relativity theories of physics, electromagnetic radiation and gravitational radiation. The relation $E = mc^2$ which led to nuclear fusion.
1919	Louis deBroglie	Wave and particle duality of electromagnetic waves; $\lambda = h/\rho$ where $\rho = mv$ is the momentum of particles.
1925	Wolfgang Pauli, P.A.M. Dirac and W. Heisenberg	Quantum electrodynamics
1926	Erwin Schrödinger	Wave equations
1934	Irene and Joliot Curie	Artificial radioactivity and spontaneous disintegration of radioactive nuclei
1940	W. F. Libby	Radiocarbon dating
1940	Randall and Boot	Magnetron inventors
1941	D. W. Kerst	Betatron for acceleration of electrons

1939-42	Enrico Fermi	Atomic cell or uranium-graphite nuclear reactor, fission
1943	W. M. Dale	Substrate protection for enzymes when irradiated
1943	G. H. Tenny	Isotopic radiography
1943	G. Hevesy	Isotopic analysis
1947	W. E. Lamb	The Lamb shift, energy for electron shift in hydrogen by molecular beam action
1948	D. Gaber	Holography and microscopy by reconstructed wavefronts
1949	M. Calvin	Path of $^{14}CO_2$ in photosynthesis. Radioprotective chemicals
1957	C. N. Yang	Effect of weak interactions such as beta decay or parity laws. Mössbauer effect in the absorption and emission of gammas.
1963	R. Tousey	Airborne solar ultraviolet spectroscopy
1964	C. H. Townes	Maser action with microwaves

A spectrum may be obtained with the use of appropriate radiation and a spectroscope. For example, in 1800, Herschel passed sunlight through a prism spectroscope which spreads the colors according to their unequal diffraction due to longer and shorter wavelengths. He discovered that thermometers in one location, the infrared, adjacent to the red end of the visible spectrum, registered the highest temperatures (Fig. 7-8). This portion, the heat rays from 0.76 to 350 microns, is

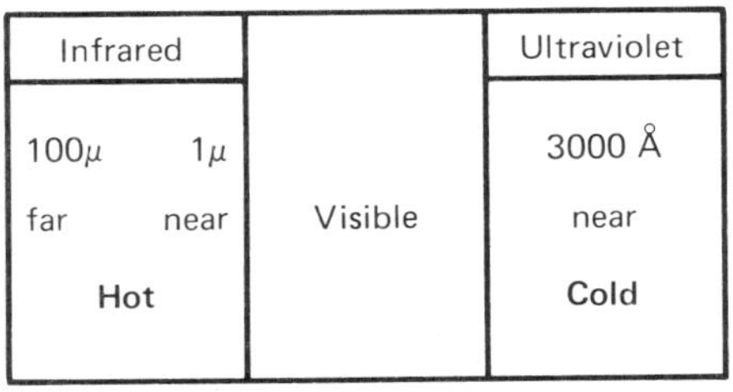

Infrared		Ultraviolet
100μ 1μ		3000 Å
far near	Visible	near
Hot		Cold

Fig. 7-8. Hot and cold regions of the middle spectrum with respect to thermal energy absorption.

always characterized by an interaction with matter that results in the dramatic absorption of thermal energy due to the great motion (vibration) imparted to the molecules within a very limited absorber range. A record of such an experiment might be called a spectrograph and studies of interactions between the energy and matter are, as we have seen, often in the realm of spectroscopy.

DUAL NATURE OF RADIATION

The work of pioneers in this field has shown that radiations have a dual nature (Table 7-5), that of waves and of particles. Sometimes it is

TABLE 7-5. Dual Nature of Radiation

EM Wave Nature	Particle Wave Nature
Diffraction, refraction interference effects	Single particles and beams
Energy associated with frequency $E = hf$	Acceleration of particles Energy associated with kinetics of particles $E = \tfrac{1}{2}mv^2$
$E = E_1 - E_2 = hf = hc/\lambda$	$E = eV = 0.5\ mv^2$ $E = h^2/2m\lambda^2$
Velocity = c	Velocity varies with particle up to c Velocity, $v = \sqrt{2eV/m}$
λ is associated with frequency, $\lambda = c/f$	λ is related to momentum, and mass m; $\lambda mv = h$
Power, $P = hf \times$ flux	Power = eV $\times$ flux = current $\times$ accel. potential
Maximum energy is fixed by the source	V is variable for further energy after source
Microbeams possible and limited by diffraction effects	Microbeams possible but beam diverges from electrostatic repulsion
Skin depth is less, e.g. red light penetrates 0.05 μm in aluminum	Skin depth is greater, e.g. e^- of 100 kV penetrate 50 μm
Incandescent luminescent or laser light sources	Hot cathode, cold emission electron guns, plasma electron beam, ion beam sources
Beams drill very hard metals	Pulses give large exposures

convenient to consider one aspect, for example, the particle nature, as when radiation exerts a pressure on a radiometer (e.g. sunlight = 4.46 $\times$ 10^{-5} dyne/cm^2, arc lamp = 7 $\times$ 10^{-5} dyne/cm^2), obviously in the manner of particles of photons or quanta producing a force (Fig. 7-9). These values are also obtained from electromagnetic theory.

Electrons are displaced from a photocell by impinging light and their motion constitutes a measureable current. When ultraviolet strikes a fluorescent surface, the result is visible light. Thus all waves carry energy and the forces in water waves during hurricanes are enormous. Sound is felt near loud drums and excessive sound energy can be lethal.

The Electromagnetic Spectrum 301

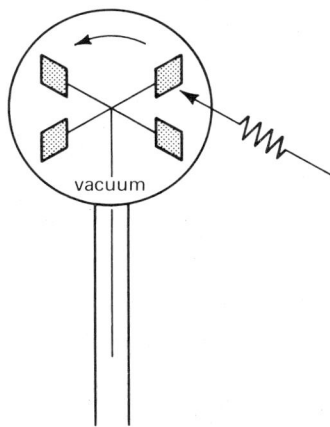

Fig. 7-9. Crookes radiometer.

Thus radiations like alpha rays (He nuclei) are called particles. Electrons are particles with a mass of 9×10^{-28} gram or 0.000549 amu (atomic mass units) (based on carbon = 12) and can be accelerated and take on kinetic energy. Again, the primary interest may be in behavior such as diffraction, which is wavelike.

All the radiations are altered in their courses by materials. Just as the Herschel spectrum was diffracted by a glass prism, spectroscopists use, for example, ruled gratings and crystalline prisms made of potassium salts with halides. Most bodies or arrangements of ionized material in strata like the atmosphere, as well as magnetic fields, will diffract or bend appropriate spectra. Atoms, after all, *are electromagnets with their electric current provided by the moving electrons. External magnetic fields can therefore influence them.*

Here the two faces of radiation are separately described but they exist simultaneously as a duality. The waves with their length, velocity, phase, amplitude, frequency, and energy characteristics can be *shifted* in frequency by a *gain or loss in energy*. Amplitude is the extent to which the wave form alternately swells and contracts according to the mutually perpendicular electric and magnetic vectors. The waves are transverse because the electric and magnetic fields move out transversely about the direction of propagation and the fields are in phase. The leading historical figure in defining theoretical electromagnetics and the equations which describe wave configurations is Maxwell (1873). The smallest wave energy would be like an e^- carrying current and the energy cannot be decreased beyond a quantum. However, the quanta are their own transmission lines. If the velocity of e^- is 5.9×10^8 cm/s, then KE for the particle is $\frac{1}{2}mv^2 = 100$ eV. The wavelength is 1.22Å.

Energy Changes

The interactions involve both the energy in the wave, in eV, and the number of molecules in the reacting region or eV/mole or per molecule or, alternatively, in kcal/mole (1 eV = 23 kcal/mole) or other energy/concentration ratio. However, the electromagnetic radiation has its exact equivalents in other energy units. Thus an eV equals 1.6×10^{-12} ergs or 1.6×10^{-19} joules and appropriate descriptions may be watts of power flow with time (1 joule/sec = 1 watt) and any other energy unit. Thus photons of light at 1.66 eV would have a wave number of 14,000 cm^{-1}, equal 40 kcal/mole, and be in the infrared. The interaction depends on this energy. Thus, a *colored* light from the flaming of *a metal* at a wavelength between 3800Å (3.8×10^{-5} cm—violet) and 7800Å (7.8×10^{-5} cm—red) passes rod and cone transducers on the eye retina to become optic nerve information for brain recording. This metallic light originates when an electron in the atom's electron cloud is able to accept thermal energy provided by the flame and move to a higher position energywise. On returning quickly to the lower lying state, it emits characteristic visible energy recognized by the eye's receptors. Stronger emitters of radiation energy than are seen in the visible range allow deeper excursions into an interacting atom's territory, even into the nucleus. These penetrations may change the construction of the atomic material by changing the number and arrangement of atomic and nuclear particles. The energy of the quanta is always determining. Thus if X-rays or gamma rays interact in material, the result may be mainly ejected photoelectrons from atoms hit, if the X-ray has energy up to 100 keV, or Compton scattering, characteristically like a projectile ricocheting, for radiation up to 1 MeV. Still higher radiation energy causes predominantly electron-positron pair production. Each kind is important in creating or studying a particular kind of interaction such as the X-ray picture as compared with the radiotherapy objective noted before. Photoelectrons receive both their energy of liberation (ω = work function) plus their kinetic energy from the light photons.

$$KE = \tfrac{1}{2}mv^2 = hf - \omega$$

This equation shows that photoelectrons have a frequency and, thus, a wavelength. If the action could be run backwards, electrons would return, strike the target and generate X-rays which is somewhat the way it is done in an X-ray machine using very energetic electrons and a tungsten target.

COMMENSURATE INTERACTION

Microwaves Commensurate with Biological Entities

There is an advantage in relating wavelengths to the interacting object's size in anticipating organismic or, as shown throughout the electromagnetic spectrum, gross interactions with matter. Small λs can probe into larger objects, but if larger λs are used on small objects, the objects are not "noticed." The waves simply pass the small object as if it did not exist in many cases. This certainly holds true for microwaves since it is usually considered that an object must be at least one tenth of the wavelength before interaction will occur. Then microwaves would be very important simply because their wavelengths happen to be commensurate with the size of important objects like macroorganisms. Perhaps, just as important, they are *unique* in the spectrum for having wavelengths commensurate with sizes of ordinary experimental laboratory equipment. The wave nature of radiation then provides this added dimensional relationship which also finds numerous examples in the transmission, detection, and reception of radio and microwaves (Fig. 7-10). In these operations the wave dimensions determine the geometry of guiding and detecting structures such as quarter wave antennas and waveguides. The wave passes according to its propagating direction and the magnitude of its wave number vector (number of waves per unit of distance). Its phase is described in terms of time, the character of the wave, and its frequency. The direction of propagation is perpendicular to both electric and magnetic vectors so that the vibrations are perpendicular to this direction (Fig. 7-11).

This is the concept of the principal wave, naturally called then the TEM or transverse electromagnetic one. Depending on the transmission, the waves can vary or *mode* infinitely, and each higher mode conforms to a distribution of electromagnetic fields determined by the material boundaries and Maxwell's equations. One set of waves is TE, transverse electric, the other TM, meaning that they have E or M field components in the direction of propagated energy. A TE wave has a transverse electric field.

The energy and wavelength may be related using statvolt and gauss units for E and H respectively. From their fundamental relation in terms of Planck's radiation constant h, and the frequency

$$Energy = hf = hc/\lambda \tag{128}$$

$$\lambda = hc/Energy \tag{129}$$

or in terms of wave number $\bar{\nu}$

$$\bar{\nu} = 1/\lambda; f = c/\lambda \tag{130}$$

$$\lambda = c/f; 1/\bar{\nu} = c/f \tag{131}$$

$$f = \bar{\nu}c \tag{132}$$

Therefore

$$E = h\bar{\nu}c \tag{133}$$

This electromagnetic energy in terms of electronic charge and the statvolt (electrostatic cgs unit) is

(a) Transmission lines

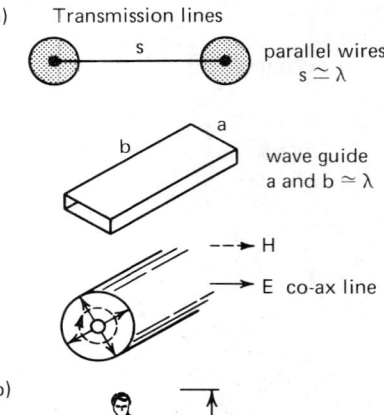

parallel wires
s ≃ λ

wave guide
a and b ≃ λ

co-ax line

(b)

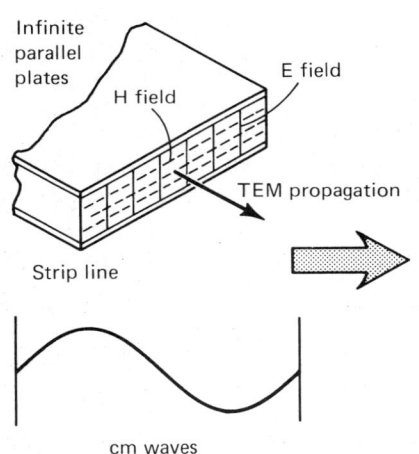

man or any animal

for interaction h ≧ 0.1λ. Quantum mechanics relations at atomic level.

(c)

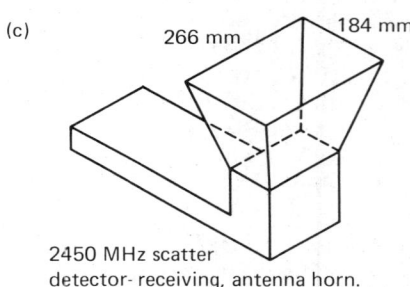

2450 MHz scatter detector- receiving, antenna horn.

(d)

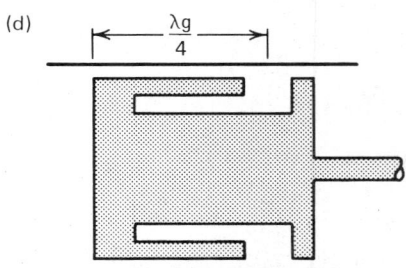

Plunger-tuner with slots and separated from guide walls. I = 0 in opening between $2\frac{\lambda g}{4}$ slots.

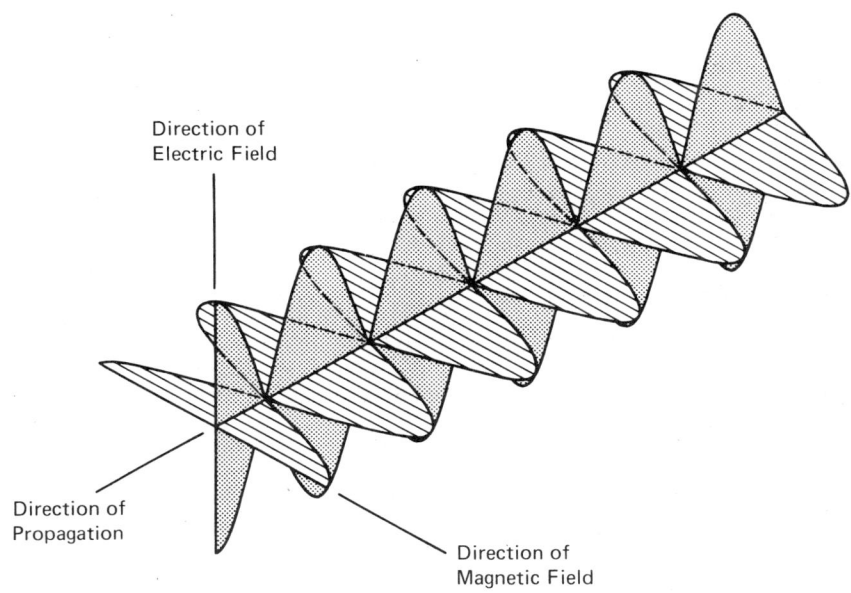

Fig. 7-11. Electromagnetic wave with mutually perpendicular, transverse, electric and magnetic fields, both perpendicular to the direction of propagation. This is the principal mode, TEM, or transverse electromagnetic wave which will transmit energy at all frequencies down to d-c passing along separated conductors like a coaxial line. Higher modes, TE or TM, will pass along lines with one conducting boundary such as a wave guide.

Electromagnetics explains basic features that naturally or artificially carry the information content between interacting systems. Features like quanta, polarity, range or penetration, amplitude, wavelength, and pulsation then assume the mission of serving a communication channel. Thus goal oriented, EMR widen in meaning as they link sources and receptors to make information couples. Units are essentially bits from the theory and, subordinately, units appropriate for pulse coding, amplitude, geometrical and quantum fit and others in consistent terms. An informational probe is a feedback of interaction results.

Fig. 7-10. Geometrical relationships for microwaves. a-transmission line where s, the distance between two wires, must be a suitable size, and a and $b \approx \lambda$ waveguide in which dimensions a and b are specified for guiding a certain size λ and coaxline; b-absorbing objects where h or some other interacting dimension must be about one-tenth of the wavelength or more for certain types of absorption; c-a short flared horn to act as a radiating antenna, receiver from a transmission line where opening has given dimensions when wavelength is 12.5 cm.; d-noncontacting tunable plunger for matching transmission line has $\lambda/4$ slots.

$$\text{Energy} = e^-, \text{ electron charge, times } V/300 \text{ statvolts} \qquad (134)$$

Thus an energy of

$$1 \text{ eV} = (4.8 \times 10^{-10})(1/300) \qquad (135)$$
$$= 1.6 \times 10^{-12} \text{ ergs}$$

If the voltage per unit length or statvolts/cm for the E-field is measurable, or known at some position, then

$$\text{E-field}/300 \qquad (136)$$

gives statvolts/cm or ESU. Then, the magnetic field H is also known because the magnitude of these E and H vectors must be equal in consistent units. Thus with an electric field maximum of 10^{-3} volts/cm

$$\text{E-field} = (10^{-3}/300) \text{ statvolt/cm} = 3.33 \times 10^{-6} \qquad (137)$$

and the E-field would be 3.33×10^{-6} ESU and the H-field would be 3.33×10^{-6} gauss. The wavelength is therefore

$$\lambda = hc/eV = \frac{6.62 \times 10^{-27} \text{ erg sec} \times 3 \times 10^{10} \text{ cm/sec}}{4.8 \times 10^{-10} \text{ ESU} \times (\text{volts}/300)} = \lambda \text{ cm} \qquad (138)$$

The radiation energy passes a unit area perpendicular to the direction of the wave number vector at a rate

$$S = \frac{cE \times H}{4\pi} \text{ ergs/cm}^2/\text{sec} \qquad (139)$$

As the wavelength decreases in size, the radiation energy eV is seen to increase. Ionization can occur when the latter becomes equivalent to the ionization potential for forming a certain ion. Thus ionization of different substances will require different types of radiations, i.e., ultraviolet, X-rays, gammas, etc.

This energy, in packets of photons and quanta, is necessary for interactions and the reaction requirement is matched by the energy in the packet, i.e., the quantum fit. Small photons at 60 Hz would equal, with h in joules now as in the mks system,

$$E = hf = 6.62 \times 10^{-34} \text{ joule sec} \times 60 \text{ Hz} = 4 \times 10^{-32} \text{ joules} \qquad (140)$$

which is a very tiny amount when it is considered that a joule is one watt sec.

On the other hand, a tiny wavelength would yield a much greater energy. With $\lambda = 10^{-18}$ meter

$$E = hc/\lambda = 6.62 \times 10^{-34} \text{ joules} \times 3 \times 10^8/10^{-18} \qquad (141)$$

$$E \approx 20 \times 10^{-7} \text{ joule} \qquad (142)$$

Guiding the energy, or radiating it in wave structures, is carried out in such a manner as to minimize losses so that the full energy may be received by detecting or end-use devices. Sometimes the energy in radiations is deliberately "lost" by absorbing it in tissue material as in microwave diathermy, freeze-drying, or radiotherapy. At other times it is accidentally absorbed with varying degrees of danger or frank injury. As transmissions in the whole spectrum increase in number with everyday advances in technology, and their concentration increases in density in populated areas, these accidental absorptions become a matter of official concern. This in turn tends to generate standards for safe human exposure levels and studies aimed at fitting the spectral radiations into a compatible relationship with man in his electromagnetic environment.

Microwave Ultrasound

The microwave spectrum is related to ultrasonics (as opposed to audible sound) in both generation and in the general characteristics being discussed. These coherent, elastic, sound waves propagate from an electromechanical transducer, for example a piezoelectric crystal (like quartz). Application of microwave energy to the crystal causes it to vibrate at the corresponding frequency, but in place of electromagnetic radiation, the output is microwave ultrasound. For this reason, ultrasound has been excluded from the electromagnetic spectra. The alternate rarefactions and compression in the conducting media, which are characteristic of this energy tend to produce results similar to the microwave ones in cases like molecular resonance absorption. The main difference is that the foregoing considerations are applied to elastic wave quanta and their material interactions instead of to electromagnetic quanta. Unlike microwaves then, sound can penetrate into conductors. In its vibration modes of interaction it will be able to reveal molecular structures and transitions as well if not more flexibly than microwaves, which are currently more easily managed from an engineering standpoint. The molecular sound experiments however, can conveniently be conducted on the condensed phase or liquid materials. The energy in sound is capable of disintegrating biological units as well as of analyzing them. On the microwave side, such destruction at similar saturation energy levels would be characteristically thermal phenomena. At such saturation energy levels, both microwave ultrasound and microwave electromagnetic radiations mask all of the usual molecular and microscopic interactions which are of great interest (athermic effects). Observable instead would be the disintegrated fragments in one case and a general temperature rise in the other. At sub-masking levels, more radiation energy management is in evidence. Consideration of the molecular and structural interactions at radiation interfaces may be feasible along with the consequent extraction of the cognitive energy and its interpretation.

Intersystem Coupling with Chlorophyll

The commensurate nature of wavelengths and potentially interacting systems is not quite as obvious in optics where wavelength is very tiny as compared to the lenses and systems that are involved. Other phenomena such as interference, diffraction, and polarization may be more meaningful. *The best radiation for intersystem transfer is the frequency showing the least intervening attenuation, which is to say that the displacement of molecules in the medium through which the wave train passes ought to be minimal until it meets an information couple.* This minimal loss in transit is always true for a propagating medium or transmission line and for waves in general with respect to good elastic media.

In plants and algae, the information couple may be highly specific. Thus a difference in two elements separates chlorophyll (a) from chlorophyll (b).

Blue-Green algae and plants with (a) have the CH_3 group
Chlorophyll $C_{55}H_{72}O_5$ with the porphyrin ring and
Mg-phytol side chain $C_{20}H_{39}OH$
Yellow-Green Chlorophyll or type (b) has the CHO group
Chlorophyll $C_{55}H_{70}O_6$ also with the porphyrin ring and
Mg-phytol side chain $C_{20}H_{39}OH$

CH_3 type (a) is recognized by blue 430 nm and red 660 nm, and CHO type (b) is recognized by blue 460 nm and red 645 nm. Figure 7-12 summarizes these relationships for various chlorophylls and their informing wavelengths.

WAVE NATURE AND INFORMATION

The "motion" in a propagated wave may be considered as that part of the energy which is actually transported through the waveform or "train" from origin to destination. A manifestation of this is the sinusoidal curve which thus gains the appearance of a wave in motion. Basic to the motion is the concept of vibration or pulsing of the fields in and out about the direction of propagation. If a special tag could be put on the end of a vector of the voltage amplitude and photographed, it would look like a vibration as in a vibrated rope tied at one end or a *musical* string, and the nature of the motion is *simple harmonic* where the displacement in the y direction according to a rotor diagram is

$$y = A \sin(\omega t + \gamma) \qquad (143)$$

If the special tag N were on the edge of a phonograph record and its spinning were viewed in the plane of rotation at a distance, it would describe this kind of motion (Fig. 7-13).

Here, y is the y coordinate of the projection of point N at P. A wave train is the whole "channel" which a wave uses in its medium where the

The Electromagnetic Spectrum

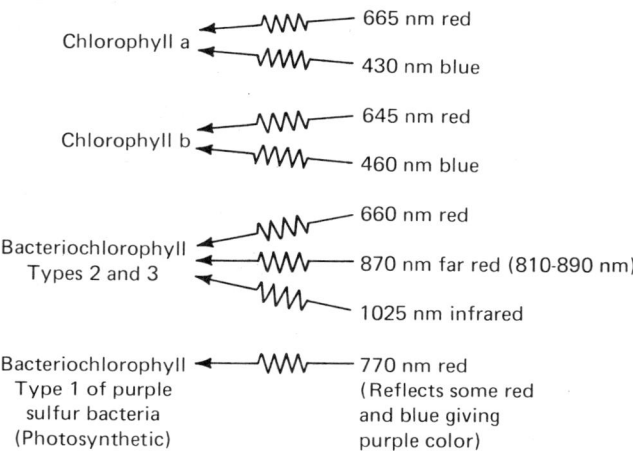

Fig. 7-12. Relation of various chlorophylls to their specific informing wavelengths for ether extracts of the pigments. λ differs if measurements are not made on ether extracts. The resonant difference between chlorophyll a and b rests on a molecular configuration that differs in only one group. Chlorophyll a is green because it absorbs red and blue.
With their accessory pigments, "antenna" and photoreaction center chlorophylls, the system is resonant, using conjugated double and single bonds which have pi e^- easily excited to move through the reception pathway.

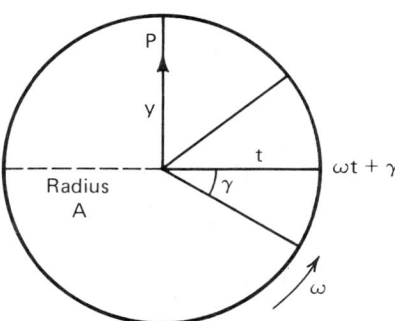

Fig. 7-13. Simple harmonic motion.

ends (the distance from generator to "addressee") may be separated by interatomic or interplanetary distances.

Phases of waves are best illustrated by pendulums set in motion, each one a little delayed or ahead of the other in their oscillations. A pulse is like the flipped loop sent along a rope when it is given a quick motion by the wrist. The number of waves depends on its duration.

Identical waveforms reinforce each other if they arrive at a bound in phase or are mutually extinguished or cancel if opposite. Reinforcement here would give four times the intensity because the latter varies as the square of the amplitude. Such intensification can easily be meaningful in the information sense when, for example, light rays arrive in phase and combine four times as intensely. Light intensity is a key to phototaxis, chlorophyll induction in etiolated (dark raised) plants, and adaptations.

As opposed to traveling or propagated waves, standing waves indicate *quantization* because the *nodes* give discrete wavelengths only (Fig. 7-14). Thus

$$\lambda = 2L, L, 2L/3 \tag{144}$$

and so on, where L is a length along the x direction of a stretched string and it gives the number of wavelengths that can be accommodated. The light photoelectric effect where impinging light displaces electrons can be explained by the waves containing quantized packets with one particle thus having a given effect. Taken together, all the packets will produce the gross effect of a wave.

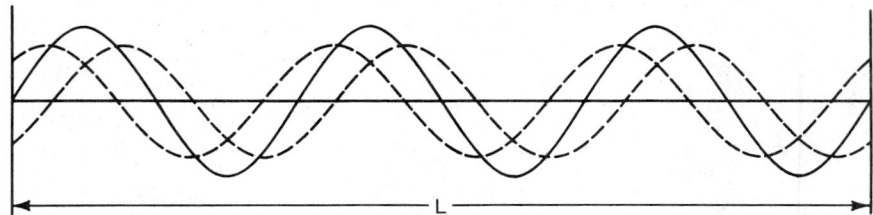

Fig. 7-14. Standing waves. Two waves, equal in length and amplitude but opposite in direction. They are separated from each other by a quarter wave and a quarter period and form a standing wave which is the sum of the two waves, in this case reinforcing each other. The amplitude is minimum at a node and maximum at an antinode. The sequences of nodes and antinodes is accommodated in the length L.

A meaningful pattern made by alternate reinforcement and cancellation would then be an attractive design as well as a probing method for interacting materials. It is then referred to as an *interference pattern* or *fringing* (Fig. 7-15).

The Electromagnetic Spectrum

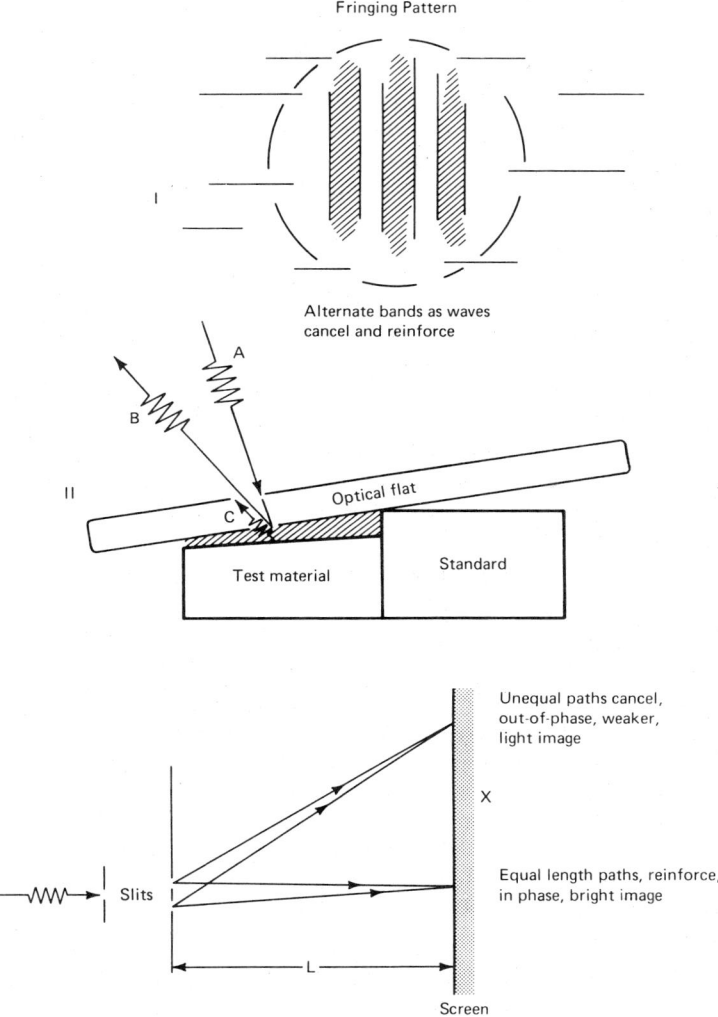

Fig. 7-15. (a) Interrogation of a surface by light. Interference or fringing pattern I as in light measurement of surface defects II. Crosshatched wedge above test material is exaggerated deviation from the standard surface. Monochromatic light λ beam A reflects in part B from lower optical flat surface and the rest continues on to reflect from test material, C. Looking down on the interference pattern formed (I), you might see, for example, three bands made when C cancels B and this would mean that the wedge height is equal to 3λ. Phase Change. At optical flat, lower face, wave B reflects with its troughs corresponding to wave A's crests.
(b) Interference.

Ubiquitous microwaves give interference information on physiological events such as muscular movements. There are two main kinds of detectors used to see such movement:

a. The 3-port circulator where the three parts are the generator-waveguide, the detector part, and the receiver. The detector reads out a waveform showing amplitude of movement, and the receiver takes the microwaves generated from the muscle. This version is 6dB more sensitive;

b. The magic T hybrid with an input from the *movement* directed to its T-shaped microwave hybrid circulator that divides test energy equivalently for reference signal and sample. Half of the microwaves reflected from the muscle, etc., strikes a detector.

The generated microwave, the reference signal, and the detector, circulate and sense an interfering signal that describes the movement and this type is simpler to operate. A variable attenuator and short circuit in the reference branch determine the amplitude and phase of the signal that is received for reference in the detector.

If spectral measurements are regarded as absorption, reflection, or scatter, then, alternatively, the interference devices must be seen as depending on the generation of microwaves by tissue. The more penetrating nature of microwaves then permits their origin to be much more distal than infrared, thus covering a much larger diagnostic volume than in thermography and on down to a significant wavelength fraction. Backscatter is the fraction of scatter known as reflection and is fairly easy to monitor with simple equipment (generator and receiver) down to microwatt levels of signal strength. In practice, the reflectometer can be an interferometer because the movement excursion is limited and amplitude varies insignificantly with almost constant surfaces. The information derived is supplemental to EEG, ECG and EMG or other biopotentials for correlation diagnosis in a no-contact mode. Phase differences specify the position of the test subject and following them gives the motion to be observed. A klystron is probably the simplest kind of signal generator to use for the source.

In order to understand coherence, we may return to the atomic generation of wavelengths which is actually incoherent since it occurs at various times from excited atoms. Waves do not generate *in phase* except from lasers or specially designed interference sources such as one which uses one light source illuminating two slits. Where interference proves that radiation is in waves that have phases, *polarization* is a phenomenon that shows the transverse field, a vibration characteristic of waves. Thus waves are transverse circularly *all* around the direction of propagation and plane polarizers are needed for polarization in only one plane. There are polar lenses which separate the wave into two polarized beams. The lens acts as a filter passing light in the two preferred planes. Thus polarization may give double refraction or birefringence. If one of the planes or beams is absorbed as it is in crystals such as tourmaline, the single polarized ray is dichroic and if all the crystals are properly stacked in a filter, the result is a Polaroid lens.

Proof of the angles that are set up in polarization is seen in two superimposed sheets of such filter material. There are two directions, one that is up and down (vertical) which lets the polarized light through. The other is a left and right (horizontal) orientation which absorbs the transmission from the first sheet through its own complete absorption. The information is seen in that the polarization either tells how the crystals are arranged or it furnishes clues to the identification and composition of materials. This is what happens with optically active substances. They rotate the plane of polarization in a left or right direction as a function of the light path or of the concentration of the solution.

An optically active kind of molecule between Polaroid filters would "depolarize," and correcting this change by rotation of the Polaroids is highly informative. The rotation needed to correct for the depolarization can be calibrated and used to find the concentration in a standard cell.

A transverse wave can be polarized because there are *many possible* directions that are still perpendicular to the direction of propagation. Thus substances of interest to an observer will scatter light en route to his eyes, and scatter is plane polarized if it is scattered or reflected at 90° (perpendicularly). If we could detect the atmospheric scatter of sunlight in terms of plane polarization, we could sense the *sun's direction* because the angle would always be 90° from the source or direction and partial plane polarization would also be information. This is exactly what the visual equipment of bees and similarly-endowed insects can do. Humans require instrumentation to do this. A refractometer for example, finds the refractive index of a material by reflective and polarizing optics.

Diffraction takes place at a slit or hole when radiation passes. An unimpeded source can give a sphere of outgoing radiation but obstacles in the path create interference patterns, secondary sources, and diffraction when the light, for example, is later observed on a presented surface. There are variations of dark and light areas due to restrictions in the wave front, and once again interference produces maximum and minimum impinging radiation on the surface. After radiation passes the slit, the waves can be in or out of phase and thus may interfere. The smaller the slit, the more the diffraction, until it is much less than a wavelength, in which case it tends toward zero. Alternate bands of intense and weaker radiation are received on the surface. For example, a hole makes a bright spot surrounded by dark and light circles of light. A slot makes a bright line with less intense bands falling off to each side. If the surface is the retina, diffraction makes an indistinct image with such features. Resolution of the diffraction pattern depends in part on the distance between slots. If they are too close, it makes the patterns merge. They should be spread a distance equal to or greater than that to the first minimum after the main pattern. In a grating, the slots may be simulated by microgrooves to give perhaps 20,000 lines per inch and then illuminated. The process can be repeated in the cross direction.

RADIATION POLARIZATION AND SENSORY INFORMATION PROCESSING IN ANIMALS

Thus, not only are the polarity of molecules, their components, and groups important in the radiation interactions, but the polarity of the radiation is equally informative. *It is the interaction of polarities that produces the total information.* A pigeon orients by sensory input and "homes-in" on its target even at long distances. Birds migrate by choosing a course based on learned or inherited guidelines and sensory information received en route over thousands of miles, the course being an orientation with respect to the radiant information input. They are not immune to "nonsense" input or disorientation and sometimes make disastrous migratory errors at obstructions like towering buildings. Ants orient by visual estimation of the angle which the sun makes with the course from their current location to their nest, always following a true course to their objective. They can be disoriented by misinforming them on the time (keeping them in the dark) in which case they cannot correct for the elapsed time and start off on the course that would have been correct earlier. The orientation of bees is a case of using their own polarization with respect to that of polarized light from the sun and sky. The "dart" angle in their communicating dance informs other workers in which direction to fly (the beeline) to locate flowers, and the number of darts per minute reveals the distance away (Fig. 7-16).

The blue sky is our visual interpretation of light waves that have been Rayleigh scattered (principally) once or more out of the direct solar beam. Blue is scattered more than red because of its smaller wavelength. The scatter is by molecules that make up the atmosphere and the sky is bluer near the zenith and whiter at the horizon due to light scattered from larger aerosol particles. In the ocean, hydrosols scatter in a similar manner and polarization occurs as well due to downwelling light that undergoes total internal reflection with elliptical polarization. Adding surface waves makes glitter in the light and its pattern is strongly polarized. Marine animals often have sensitivity to polarized light so they can find their direction under water. The polarization is usually at a maximum near the horizon and at right angles to the principal plane so it orients a sensitive animal. At sea, without waves, the horizon would be invisible. The blue sea color is due to a preferential scatter of blue and absorption of red, and blue also penetrates best to great depths.

The informing action is not unlike a biopole that is "magnetized" or oriented between poles of a magnet. Then after sensing its orientation, it could move in the direction of polarization or some overriding influence could lead away the polarized units. Honeybees (*Apis mellifers*) are sensitive to the state of light polarization and use light polarized by solar scatter from air particles which informs them of the sun's direction even when looking in a direction other than that of the sun. What they sense is the direction of the plane of vibration of polarized light. Unpolarized light is vibrating in all transverse directions around the direction of propagation as opposed to one directional plane of polarization.

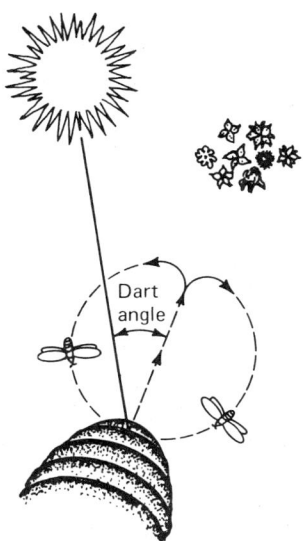

Fig. 7-16. Information in bee dance is a code for direction; arrow toward flowers indicated by a dart in the bee dance accompanied by abdominal motions; the bee dances around the half circle, makes the dart, then completes the other half-circle. The darts per minute give the distance to the target (slow means longer distance). The amplitude of the dart, or diameter estimates the nectar deposits. The bees detect the polarization of the light entering their eyes and thus the message brings an innate response mechanism into action.

What is polarized with polarized light is the transverse field component that vibrates perpendicularly to the direction of propagation. These components in depolarized light vibrate in all such planes while polarization selects only certain permitted directions for them. Thus it is a vibrating direction that is polarized. This then constitutes information within the radiation for any capable detecting or sensing system.

Figure 7-17 shows various forms of polarized and unpolarized radiation. Light from the blue sky is partially polarized and its greatest polarization is at right angles to the line connecting the observer and the sun. This makes the polarization vary with the apparent position of the sun so that it does not require a visible sun, only a patch of blue sky. The polarization is then a direction dependent variable (Fig. 7-18).

Honeybees are able to navigate directly to the hive. There they communicate their message to the other bees. To do this, the scout bee alights on a convenient vertical surface so that the observers can then receive polarization information while getting the message. At this point the uncertainty is complete for the flowers, or the target could be anywhere. The message is sent by a ritual dance in which the pattern of movement is the code which gives distances, directions and quantities (Fig. 7-18). Experimentally, mirrors and screens will force the insects to adjust their orientation (Fig. 7-19).

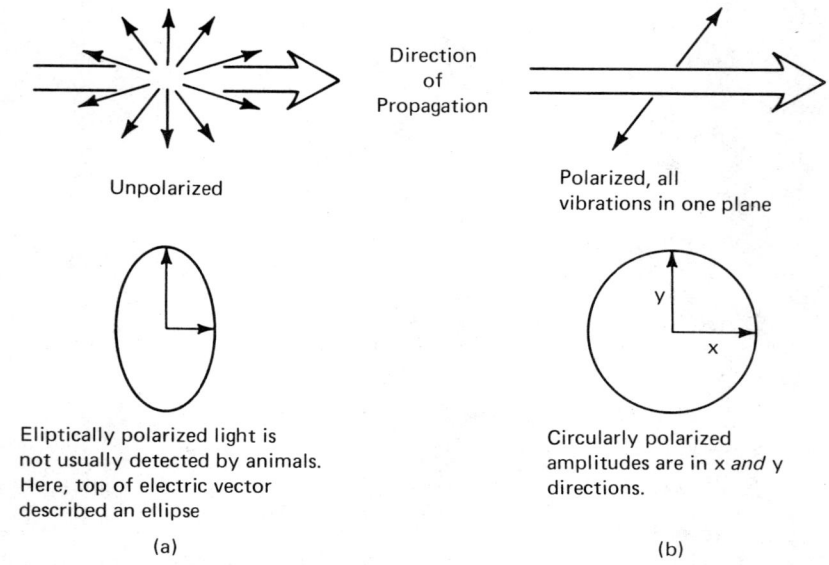

Fig. 7-17. Forms of radiation polarization.

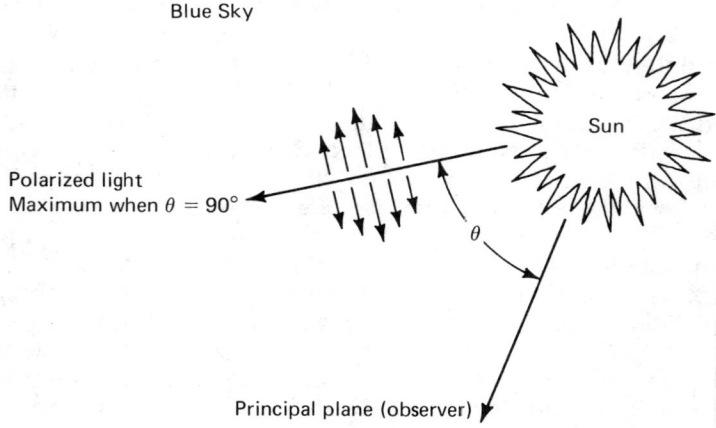

Fig. 7-18. Polarized light gives a pattern and an angle established at any time by the sun's position. It is sensed by honeybees and used for position finding because the polarization marks a particular direction which allows them to orient themselves.

The Electromagnetic Spectrum

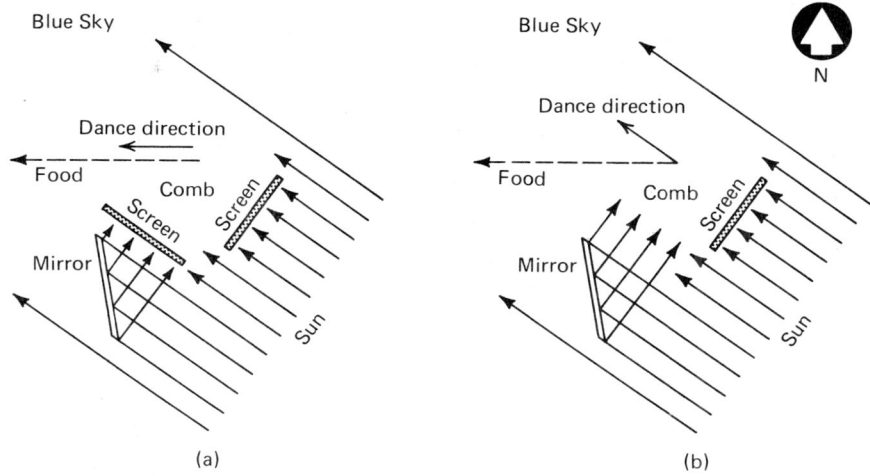

Fig. 7-19. Mirrors and screen force bees to adjust their orientation.

POLARIZERS

A polarizing crystal makes two plane-polarized rays. The radiation vibrates in planes which form the direction of both the propagation and vibration, and these are mutually perpendicular. Thus the polarizer is a discriminator for information, selecting those two directions for transmission. No matter what planes the incoming vibration may be in, only two component ones are passed on. Taking an incoming vibration with amplitude A, and angle θ to one of the allowed "transmit" directions, gives the two components shown in Fig. 7-20 of amplitudes A cos θ and A sin θ, mutually perpendicular. If θ is 45°, then the components will be equal.

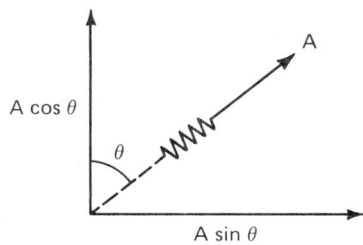

Fig. 7-20. Crystal calcite polarizes with two permitted directions (perpendicular). Vibration of amplitude A separates into components A cos θ and A sin θ giving double refraction or birefringence. Either component is a wave of polarized radiation for other uses.

Light will pass through a Polaroid filter sheet and be plane-polarized. At a second Polaroid filter, it may be reduced a little. If the second one is rotated enough, the light will be cut off. These crossed polarizers then permit an analysis of component molecules as in the analyzer instrument. If optically active molecules are positioned as an inserted sample in which the amount of rotation is proportional to the path length and concentration (for example, a sugar), then the second polarizer must be rotated to depolarize by an angle which is proportional to the concentration of the sugar. Then reference to a calibration curve will give the concentration of the unknown (Fig. 7-21).

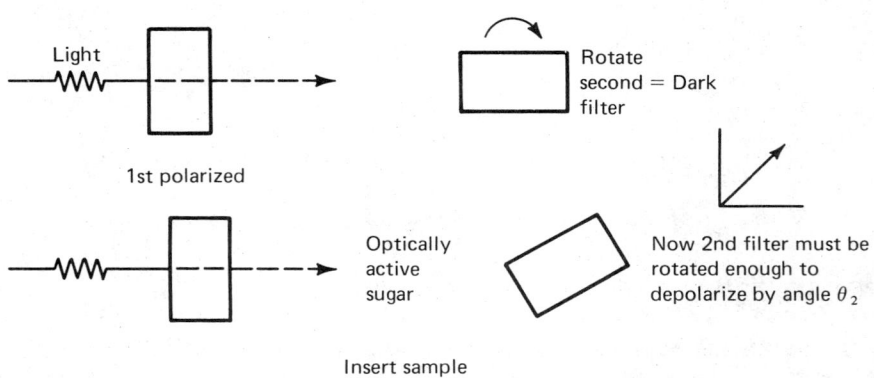

Fig. 7-21. Angle θ_2 used to find concentration of optically active sugar.

These respective orientations of objects and radiation then can apply equally well to molecules and to polarized or nonpolarized light whose vectors for the electric and magnetic fields offer convenient orienting directions, and whose waves may be manipulated by polarizing filters, changed by constructive and destructive interference, and read out as informative visual patterns meaningful to a skilled observer just as they were to the insects. Various instruments, such as polarized light and interference microscopes aid the observer. Oriented biopoles will be described in terms of x-y-z coordinates, EM field component vectors, azimuthal angles, and their mutually parallel versus perpendicular orientations. Then the polarization properties, for example those of protein absorption bands, will be related to atomic bending, stretching, and combination motions, so that known structures go with specific absorption bands.

Polarized infrared informs about certain protein and nucleic acid groupings. Visible and ultraviolet light may be polarized and the vibrations then made parallel to the plane of an atomic planetary structure, such as is more or less the case of the heme group which is like a resonant ring structure. One can then determine when the polarized ultraviolet is parallel because the absorption will peak and additional background on internal molecular groupings permits one to draw conclusions as to their relative planes and parallelism or perpendicularity to molecular axes, that is, their orientation.

Polarized radiation is described as "vibrating" at some angle, for example to the y-direction and the molecules on which it impinges are in some characteristic, but often initially unknown, orientation. If the radiation interacts and divides then into perpendicular and parallel components, the refractive index will be different for the vibration in each direction, and the polarized material is birefringent. Most polarization studies begin with a polarizer that filters out all but the light which vibrates in the desired direction (Fig. 7-22). Then comes the interaction with oriented molecules or structures, resulting in a splitting of the polarized light. Finally, an analyzer transmits radiation parallel to another axis (Fig. 7-23). Analysis of this transmission by its field directions reconstructs what happened in the molecules and informs on their orientation.

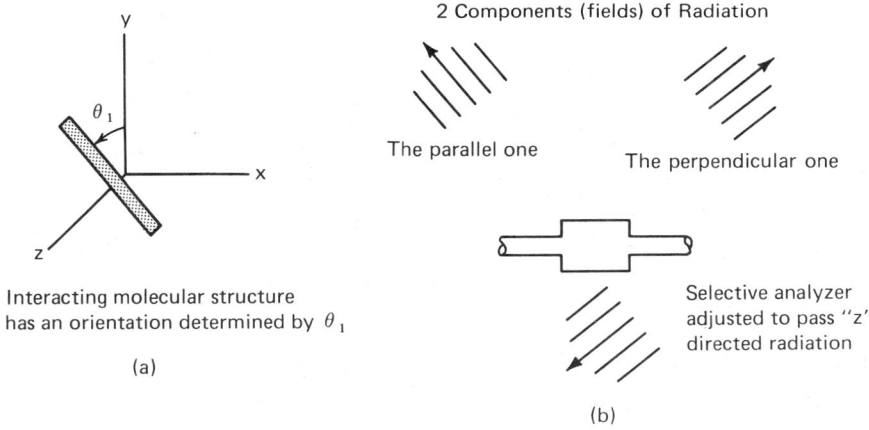

Fig. 7-22. When θ_1 is not zero, radiation divides into two components and the analyzer sees some radiation that informs by "reconstructing" the interaction.

The action in the molecules will be the lagging or leading of the two components, and these phase differences are expressed in degrees or as a fraction of a period of oscillation, giving a resultant E field which is detected by the analyzer. Isotropic substance makes the components equal in velocity and no radiation is transmitted by the analyzer. Anisotropic substance may retard one component by 180°, or one-half period to give a maximum transmission. When the polarizer and analyzer settings are perpendicular to each other, the observation is said to be by "crossed" polarizers.

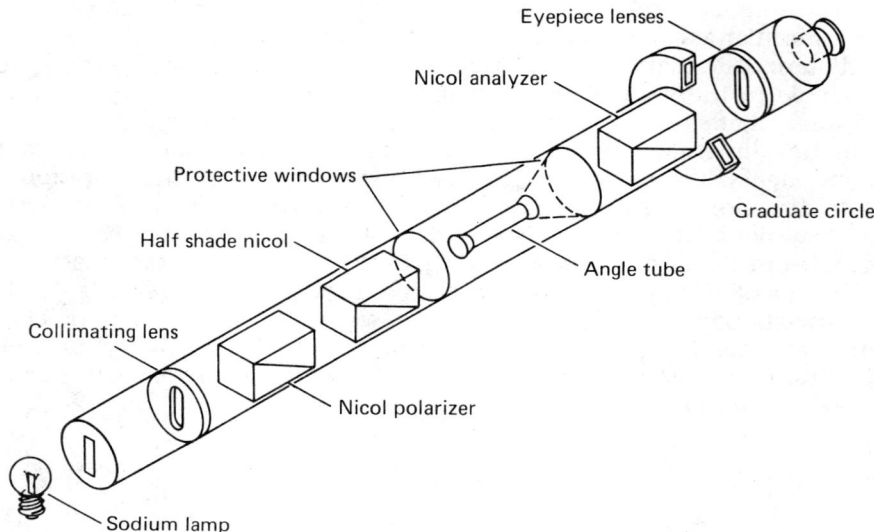

Fig. 7-23. Shows a polarizing analyzer with a choice of intensity comfortably between dark and light for best eye sensitivity. The analyzer prisms permit modifications of the polarization angle for polarized light coming from the first prism. Automatic and semiautomatic vision use approximately the same system.

Polarized rays in the interference microscope divide into such components, one through the specimen and the other around it. The optics then merge these components so that reinforcement or cancellation can occur and be interpreted as visual detail concerning contrasting color or density. The polarizer gives the necessary two components, one of which is made birefringent in a crystal. The component that interacts with the specimen and the other that passes around it are rotated by 90° to each other at that point. The interaction produces information on absorption, radiation polarization, and phase lag or lead in the wave front as it meets details within the interface. When the two components are merged, it is the interference pattern that is meaningful to the observer. The pattern read out is in terms of color or intensity contrasts, both of which are heightened in proportion to the amount of interacting material in the specimen.

There is an interesting social polarization in psychology called the *Shacter experiment* in which a group is oriented by fear, which is the independent variable. This polarization is then seen to be by conditioning. The dependent variable in the case of one group made fearful, and the other reassured of no danger, is affiliation. Then the polarized group is found to be motivated to seek the company of others in what looks like misery loving company.

Another phenomenon, called the *Crabtree test*, uses light deprivation on cats as they are maturing. This radiant, non-radiant polarization makes light deprivation the independent variable. The animal is conditioned by being deprived of the information normally carried by light or of most of it. Lacking this input, the dependent variable (which is the process of pacing or seeking exits in a maze of moderate difficulty in order to gain a reward) is varied over wide limits. The visual development is retarded by lack of light information, resulting in measurably more pacing.

PROBLEMS

1. What do the limits in frequency of the electromagnetic spectrum connote? Can the spectrum be constructed as a circular diagram?
2. What are the functions of the electromagnetic spectrum?
3. What is an electromagnetic wave?
4. What are the principal responses of the subspectra?
5. The frequency of an EM wave is 1 kHz. Determine the wavelength, velocity, and energy.
6. When an excited hydrogen atom relaxes with radiation emission, are all emissions of the same frequency?
7. What media are needed for propagated waves?
8. What is meant by "waves in phase?"
9. What are the principal subspectra or bands?
10. What are the colors in angstroms and meters?
11. Is fluorescence visible or ultraviolet light?
12. How do you refer normally to radio, microwave, visible power, and telephony?
13. Are X and gamma rays identical?
14. Are cosmic rays hazardous?
15. Express Planck's constant in terms of frequency, mass, and the velocity of electromagnetic radiations.
16. What is the mass in kg of the quantum at a frequency of 10^{10} Hz?
17. How can X-ray generation be considered the reverse of photoelectron production?
18. What light characteristic is used by the honeybee?

BIBLIOGRAPHY

Bode, J.D. and J.P. Carrico, "Remote Optical Sensing for Air Pollution Monitoring," *Bendix Tech. Journal*, Vol. 4, 3, pp. 14-28, 1971.

Chapman, R.F., *The Insects Structure and Function*, American Elsevier Pub. Co., New York.

Colwell, R.N., "Some Practical Applications of Multiband Spectral Reconnaissance," *American Scientist*, Vol. 49, 1, March, 1971.

Conrod, A.C. and R. Blythe, "Infrared Imagery of Polluted Water. The Need for Multispectral Scanners," *Bendix Tech. Journal*, Vol. 4, 3, 1971.
Ficchi, R.F., Ed., *Practical Design for Electromagnetic Compatability*, Hayden Book Co., New York, 1971.
Francon, M., *Holography*, Academic Press, New York, 1974.
Giese, A.C., Ed., *Photophysiology*, Academic Press, New York, 1971.
Hills, Kondra, Hamid, *Canadian Journal of Animal Science*, Vol. 54, p. 573, 1974.
Lackchaura, B.D. and J. Jagger, *Radiation Research*, Vol. 49, 631-646, 1972.
Maxwell, J.C., *A Treatise on Electricity and Magnetism*, Oxford Univ. Press, New York, 2 vols., 3rd Ed., 1904, 1873.
Michewer, C.D., *The Social Behavior of Bees*, Harvard Univ. Press, Cambridge, Mass., pp. 16-18, 1974.
Schmitt, F.O., Ed., *Functional Linkage in Biomolecular Systems*, Raven Press, New York, 1975.
Smith, R.A., *The Detection and Measurement of Infrared Radiation*, Oxford Press, 1968, pp. 456, 457.
Sudd, J.H., *An Introduction to the Behavior of Ants*, E. Arnold Ltd., London.
vonFrisch, Karl, *The Dance Language and Orientation of Bees*, Harvard University Press, pp. 380-400, 1967.
vonFrisch, K., *Bees, Their Vision, Chemical Senses, and Language*, Cornell Univ. Press, Ithaca, New York, 1950.
Wolbarsht, M.L., Ed., *Laser Applications in Medicine and Biology*, Plenum Press, New York, 1971.

Film

Return to Bikini (1966) 23½ min., Color.
Produced for the AEC by the Laboratory of Radiation Biology, University of Washington. For sale by the Motion Picture Service, U.S. Department of Agriculture, Washington, D.C., 20250 at $97.00 per print. Available for loan (free) from USAEC headquarters and field libraries. The study of bioaftereffects of nuclear test at the mid-Pacific atolls intermittently since 1946.

8

The Radiation Biology of Ionizing Radiations

> Beware lest neutrons with finality,
> Let men in ashes, gain equality.
>   D. A. C.

AGONIZING RADIATIONS

The author overflew Nagasaki and passed through Hiroshima, Japan within a few months of the devastation caused by the atomic bomb explosions in August 1945. The areas were essentially empty and casualties approaching one-half million were, in many cases, still in hospitals—usually elsewhere because these cities had lost most of their medical facilities and many medical people.

The scene in hospitals after the blast was pitiful; patients did not know what had hit them but they could see that their hair was falling out by the handful, that their burns were not healing, they had no appetite and were weak and subject to strange feelings including fear and despair. The doctors diligently sought explanations in treatment and autopsies but they were handicapped by the novel symptoms; for example, many suspected a bacterial kind of contamination of the patients and their belongings with the attendant possibility of epidemics. Continual reminders of these first impressions during years of teaching radiation have suggested a more apt description of ionizing radiations as "agonizing" radiations.

During the intervening years, in the shadow of the bomb, biologists have been trying to sort out the effects on living systems and, in the interdisciplinary field which forms our background, remarkable progress has been made in turning around the forces of destruction into precise instrumentation and procedures for study, diagnosis, and therapy. At the same time these forces that seem so large, yet are actually only nuclear in dimensions, now appear as mankind's principal security against a total eclipse of energy sources in the future.

TYPES OF RADIATION

The X and gamma radiations and cosmic rays are continuations of the electromagnetic spectrum beyond the ultraviolet toward smaller and smaller wavelengths. Cosmic rays of the high energy associated with these smallest waves can cause the flashes of light experienced by astronauts every few minutes in outer space. The particulate radiations include beta rays or electrons, alpha particles or helium nuclei, deuterons, carbon nuclei, protons, neutrons, and others. These are ionizing particles with various energies (usually in keV or MeV) which produce a proportionate penetration into materials known as their range. Neutrons may be high energy above 10 MeV, then fast, slow, and thermal with correspondingly less energy.

These radiations are interrelated due to the nature of their reactions; thus a common source of betas is the ejection of electrons from atoms by X-rays or radioactive beta decay in which an electron and a neutrino are produced. Protons are commonly produced when fast neutrons eject them from the nucleus. Finally, alphas come from the alpha-type of radioactive decay; for example, that of radon, or the disintegration of boron by slow neutrons.

The neutrino is a particle, apparently insignificant biologically, that is emitted from elements like ^{234}Th, the daughter product in the decay scheme of ^{238}U. The positron is the positive beta particle. These are accompanied by neutrinos and anti-neutrinos, also probably unimportant biologically.

The particulate radiations like alphas are certainly ionizing even if not rigidly electromagnetic in classification. They are compared with X-rays (which *are* a bit more readily seen as electromagnetic radiations) as to biological effectiveness. Moving particles have wave-like properties according to the dual nature of radiation as a function of their velocity, mass, and accelerating potential. With

$$eV = \tfrac{1}{2}mv^2 \qquad (145)$$

one can say, for example, that among very small dose effects on male sex cells (spermatogonia), neutrons lead. This is to say that while three rads or less of ^{60}Co gamma rays are needed to produce detectable change, only two rads of 2.5 MeV neutrons are needed. While they are called ionizing, these radiations may produce excitations and, in certain cases, this is the main route for energy absorption and biological effect.

Half-Life

The physical half-life was developed in Chapter 5 in connection with exponential processes and the decay constant

$$\alpha = 0.693/T_{1/2} \qquad (146)$$

and here, the exponential decay of the radioactivity is seen to combine with the clearance time or elimination rate in an organism to give a third half-life that is a function of the other two and is called the effective half-life T_{eff}.

$$1/T_{eff} = (1/T_{1/2}) + (1/T_{biol}) \qquad (147)$$

Then,

$$T_{eff} = (T_{1/2} \times T_{biol})/(T_{1/2} + T_{biol}) \qquad (148)$$

One or the other or both will be important depending on the relative order of magnitude and T_{eff} will be smaller than both but nearer to the lesser one. A very short physical half-life $T_{1/2}$ (in minutes) will weigh the effective half-life in favor of $T_{1/2}$. A rapid biological clearance time, on the other hand, will give a small T_{biol} and *it* will accordingly determine the radioactive staytime. Bone phosphorous has a biological half-life of about three and one-third years while its physical radioactivity half-life is only 14.3 days. About 20% of the amount of it that is ingested will be found in the skeleton. Liver phosphorous has a T_{biol} of a few days due to the different metabolism there. Potassium with a T_{biol} of about one month and a $T_{1/2}$ equal to 1.3×10^9 years, passes mainly to the muscle system. Carbon-14 radioisotope has a T_{biol} about equal to that of potassium and a long $T_{1/2}$ of 5,730 years. About half of it will be in fatty substance. We have also to consider those elements in the dynamic state such as nitrogen versus lock-in elements like some bone calcium ($T_{biol} = \infty$) and intermediate ones like iron in the hemoglobin of the red blood cells ($T_{biol} = 65$ days). Radioiodine will be concentrated to at least about 25% in the thyroid gland ($T_{biol} = 180$ days).

The actual levels reached around a radioactive source capable of producing these radioisotopes will depend on the housekeeping efficiency and thoroughness in protecting the ambient. This protection can be effective enough to eliminate all hazards unless there is an accident. On the other hand, the chief interest within a framework of maximum permissible concentrations lies in the marginal cases that is, children, the aged, and the ill. Some individuals will be exposed to much more radiation than others due to a pattern of work related to radiation. The dynamic state of body constituents works to turn over elements and thus modify long staytimes even in bone. These and other factors make the T_{phys} much more reliable as a figure than T_{biol}.

The effective half-life provides a logic for determining the MPB or maximum permissible burden of a radionuclide either in a characteristic locus (critical organ) or in the whole body. The critical organ will be an essential one that is most radiosensitive to the nuclide and the one that is the most active in concentrating it. Several nuclides meet these criteria for bone: radiophosphorous, radiocalcium, radium, and radiostrontium. If the radionuclide is uniformly dispersed, the whole body becomes the critical organ.

Table 8-1 lists cells in order according to their radiosensitivity and in Table 8-2 the radiosensitivity of activities and components is listed with comments on the results of the sensitivity.

TABLE 8-1. Radiosensitivity of Cells

Radiosensitivity		Cell Type
1st	Most Sensitive	Lymphocytes, major class of circulating WBC
		Erythroblasts, RBC precursors
		Spermatogonia, primitive level in spermatogenesis—and their mitosis
2nd		Ovum surrounding granulosa cells in ovarian follicles
		Myelocytes (bone marrow)
		Intestinal crypt cells
		Germinal cells of epidermis
3rd		Endothelium lining and gastric gland cells
4th	Moderately Sensitive	Osteoblasts (bone-forming)
		Osteoclasts (bone-resorbing)
		Chondroblasts (cartilage former precursors)
		Spermatocytes, spermatids
5th		Granulocytes, WBC
		Osteocytes, bone cells
		Sperm
		GI surface cells

THE PROBLEM OF DOSE INTEGRATION FROM VARIOUS TYPES OF RADIOACTIVITY, UNITS AND EXPOSURE

The problem of dose integration for radiations is closely related to that of bringing all radiation absorption mechanisms under one general and comprehensive relationship. While highly desirable for unifying radiation mechanisms at the top of the analytical pyramid, such a single relation would be much less meaningful for integrating the biological effect. There are too many ways in which physiological systems can be insulted or benefited by all the radiations that exist to unify them, except at levels where radiations are brought together to emphasize the associations that do exist such as transmission, reflection, absorption, possibly polarization, and exponential attenuation in absorbing tissue according to coefficients of absorption.

Instead of integrating the dose, the immediate or long term exposure is described in terms of the dose description most applicable to the radiation of interest. *Thus body burdens of ingested, inhaled, absorbed, or injected isotopes are given in curies* (Ci) *or microcuries* (μCi) *for each isotope* and these are compared with maximum levels published in sources such as the U. S. National Bureau of Standards Handbooks 62

TABLE 8-2. Radiosensitivity of Activities and Components and the Results

Component	Sensitivity	Result
Amino acid	Side Chain	Functional loss in protein
Protein	Amino acid, side chains, secondary, tertiary, etc. structure	Loss of functions unfolding distortions, exposure of special group
Enzymes	In solution without substrate	Inactivation
DNA	Deamination or loss of a base; Hydrogen bond break between chain; Single and double-strand breaks; Crosslinks to protein, other DNA or own structure	Loss of DNA information
Lipids	Doubly or more saturated lipids, linoleic acid	Peroxides, RO_2, then free radical *chain* reactions
Carbohydrates	Chain integrity, hyaluronic acid-protein complex in polysaccharides	Permeability

and 69. *Beta ray exposure* may be counted in units of rads (100 ergs per gram of absorbing material). This may be converted into absorbed MeV units of energy per gram by multiplying by the factor 6.24×10^7 (see below). To describe the absorption further, a certain activity may be taken for a beta emitter in curies. The linear absorption coefficient (cm^{-1}) may be taken from readily available data for the energy of the betas emitted by this isotope. This will give the absorption ($Ci \times cm^{-1}$) according to the thickness of tissue and handbook data may be consulted for converting to rads/min as a function of the concentration of beta activity, for example, curies/cm^3 and the beta energy in MeV.

The dose received from betas is significant for the outer dermal layers of tissue for external sources, for distances (about a centimeter) from the isotope in encapsulated and implanted form, and for numbers of small organisms (within this range of beta ray effect) in the case of cultures. These conversion relationships are determined in dosimetry studies. The dose for one second in rads for microcurie beta sources may be obtained by converting the number of disintegrations per unit of time.

$$\text{rads} = \frac{3.7 \times 10^{10} \text{ dps/curie} \times \text{microcuries} \times \text{average beta energy in MeV}}{10^6 \times 6.24 \times 10^7 \text{ MeV absorbed per } gram} \quad (149)$$

This conversion is possible due to the definition of rads in terms of ergs absorbed per gram of tissue or other material. The MeV equivalent 6.24×10^7 of 100 ergs absorbed per gram of tissue or other material may then be readily given because the microcurie activity in disintegration per second (dps) is converted to MeV by multiplying the average beta energy per disintegration by the number of disintegrations per second.

In analyzing these dimensions, with C the concentration of the beta emitting isotope in μCi/g, 3.7×10^4 MeV would be the radiation energy passing into each gram of tissue each second. The 3.7×10^4 dps from one microcurie with an average E of 1 MeV produce

$$\frac{(\text{dps}/\mu\text{Ci}) \times (\text{MeV}/d) \times (\mu\text{Ci}/g)}{6.24 \times 10^7 \text{ MeV absorption}/g/\text{rad}} \qquad (150)$$

the dose in rads in one second of disintegration.

It will be noted that the conversion specifies the rate of emission of *betas* of a certain energy. *Gamma* rays or other radiation may also result from nuclear disintegration according to the path of decay for the isotope. Thus, Co^{60} (which is called a gamma source also) emits a beta of 0.094 MeV average energy and gammas of both 1.332 and 1.173 MeV. It is the *actual quanta emitted* that determine the activity and thus the dose.

The gamma ray conversion may use a symbol for intensity Γ. It has units of roentgens per millicurie hour at one centimeter from the point source of gammas. The conversion factor is 6.77×10^4 MeV per cubic centimeter per roentgen. Thus, the distribution of the dose within a ball of radius one cm around the point source may be examined. For gammas with a linear absorption coefficient μ

$$\frac{\dfrac{3.7 \times 10^7 \text{ dps}}{\text{mCi}} \times \dfrac{3600 \text{ sec}}{\text{hr}} \times \dfrac{\mu \times 1 \text{ cm}}{\text{cm}} \times \dfrac{\text{MeV}}{d} \times \text{gamma fraction of atoms disintegrating}}{4\pi r^2/\text{cm}^3 \times 6.77 \times 10^4 \text{ MeV}/\text{cm}^3/\text{roentgen}} \qquad (151)$$

$= $ roentgens/mc hr at 1 cm

A useful and abbreviated dose calculation may be made for an external gamma dosage R in milliroentgens/hr at a distance of one foot from C millicuries with a gamma energy E per disintegration in MeV

$$R = 6CiE \qquad (152)$$

It is then necessary to *reduce* to the recommended maximum permissible (handbook 69) concentration MPC for radiation personnel so that combined external plus internal exposures remain below the maximum for a sensitive organ (National Council on Radiation Protection and

Measurements 1954, 1971, Basic Radiation Protection Criteria, Report No. 39, Washington, D.C.).

Within experimental error, the rads from betas may be added to gamma roentgens to produce the total. In tissue, one roentgen (R) of high energy quanta produces an energy deposition essentially equal to 1 rad. Gammas differ greatly however in their range and ordinary distances in tissue are much less likely to absorb all of them than in the case of betas.

The roentgen equivalent man or rem is any radiation which is equivalent to the absorption of 1 rad of 200 kvp (kilovolts potential) X-rays. To enumerate rem, the value of the RBE or relative biological effect must be known.

$$\text{rem} = \text{rads} \times \text{RBE} \tag{153}$$

If it *is*, only the type of ionizing radiations needs to be specified and the rems will be equal for any type.

For example, the rem unit may be associated with the lethal dose in the case involving high level exposure. The lethal dose LD_{50} thus means lethal for half the population exposed within 30 days of the event. As applied to mammals and X-rays, this dose is about 600 rem but it probably increases in adulthood and then decreases with advanced age. If another ten times more effective radiation is used, e.g. neutrons, then rems would still be 600 but rads would be only 60. Topping the sensitivity series for animals would be all the mammals followed by the tortoise, the newt, and snails with the most resistant organisms represented by yeasts which show at least five times the LD_{50} of mammals.

Some bacteria are resistant, notably *Micrococcus radiodurans* which repairs radiation damage due to 4 microjoules per mm^2 of ultraviolet as well as 1.2 krad of gamma radiation within one minute. As for survival, this organism tolerates about 0.4 millijoules per mm^2 of ultraviolet and 120 krad of gammas.

Radiation sensitivity in plants varies from a high sensitivity in plants like Gillyflower of the *Cruciferae* which can tolerate only 40,000 R, and sensitive cereals of leguminous crops which die after 10kR, to resistant plants like *Arena, Lupinus, Lotus, Melilotus, Trifolium,* and *Linus*. In wrinkled peas, sensitivity is associated with the recessive gene for the wrinkling and smooth peas are resistant.

If the RBE for neutrons is two, it means that the neutrons produce some biological effect to a degree twice as great as the same rads of X-rays. Thus the RBE takes radiosensitivity into account and it is fair to say that most components of organisms vary in their capability to resist the effects of ionizing radiations.

$$\text{Sensitivity Factor} = \text{RBE} = \frac{\text{Rads to effect the biochange with 250 kV (therapy X-rays)}}{\text{Rads to equal this change via another ionizing radiation}} \tag{154}$$

Table 8-1 lists typical cells of radiation interest in a series beginning with the most sensitive. Lymphocytes lead in radiation sensitivity and spermatocytes are well represented. Table 8-2 is the radiation sensitivity of certain important biochemical components, the basis, and result of radiosensitivity. Functional information is an important casualty.

Integrating the dose also requires a consideration of inverse square law effects where the intensity of divergent radiation that is in the process of scattering and interacting equals the initial intensity I_o divided by the square of the distance.

$$I = I_o/d^2 \qquad (155)$$

Therefore, doubling the distance will reduce the intensity to one-fourth (Fig. 8-1). Exponential absorption will modify this result which will

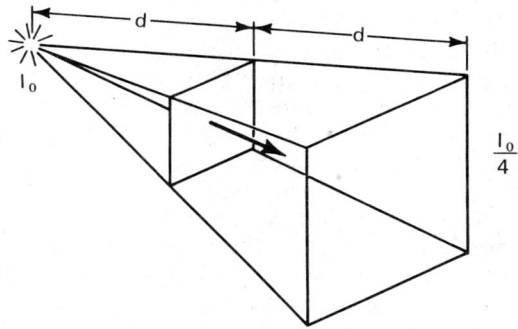

Fig. 8-1. Diverging radiation from scatter effects gives the inverse square law. Let $d = 1$. Double d. What is the intensity in watts/cm^2?

$$I = I_o/(2)^2 = I_o/4$$

then become a combination of the two processes. Other factors are the incomplete or complete decay of the isotope, effect of dose from positrons and the effective half life which is an important modifier for internal isotopes. While ^{14}C is widely and safely used in radioactivity experiments, and has a short *effective* half-life, it may become concentrated merely by virtue of its popularity. Its rapid clearance is due to excretion and turnover, not to its radioactive half-lifetime which is 5.57×10^3 years. It may form more permanent compounds however, and is thus sometimes hazardous. There is a special information deleting problem with an element like ^{14}C beyond the fact that it is a dangerous internally-emitting isotope. It can garble the DNA message as well as prevent the function of an operating form of protein or enzyme. It does this as it decays to nitrogen in an atomic site where interacting molecules had expected to find a carbon. This is then an instance of transmuted atomic information which, in the case of DNA, has genetic consequences.

The relative internal hazard of various beta and gamma-emitting isotopes is shown for three groups and three activity ranges in Table 8-3 as specified by the National Bureau of Standards Handbook 42.

TABLE 8-3. Relative Internal Hazard of Various Beta and Gamma Emitting Isotopes

Radiotoxicity Group	Activity Limits		
	Low	Intermediate	High
Slight Hazard Na^{24}, K^{42}, Cu^{64}, Mn^{52} As^{76}, As^{77}, Kr^{85}, Hg^{197}	Up to 1 mc	1-10 mc	Above 10 mc
Moderately Dangerous H^3, C^{14}, Na^{22}, S^{35}, Cl^{36} Mn^{54}, Fe^{59}, Co^{60}, Sr^{89} Nb^{95}, Ru^{103}, P^{32}, Te^{127} Te^{129}, I^{131}, Cs^{137}, Ba^{140} La^{140}, Ce^{141}, Pr^{143}, Nd^{147} Au^{198}, Au^{199}, Hg^{203}	Up to 0.1 mc	0.1-1 mc	Above 1 mc
Very Dangerous Ca^{45}, Fe^{55}, Sr^{90}, Y^{91}, Zr^{95} Ce^{144}, Pm^{147}, Bi^{210}	Up to 0.01 mc	0.01-0.1 mc	Above 0.1 mc

ENVIRONMENTAL IONIZING RADIATION

Introduction

The natural unenriched environmental radiation from cosmic rays (as well as the tritium, ^{14}C, ^{40}K and nuclide species they produce), the earth materials, atmospheric particles of gases, and absorbed radioisotopes all are in the range of a fraction of a rad per year. In enriched areas, for example, uranium-rich granite terrain with heavy exposure, the level may reach one rad or more. Probably the dose absorbed by the sensitive gonads, bone, and bone marrow cells is the most meaningful as an indicator.

Due to a large concentration of trapped protons in the radiation belt in the outer atmosphere of the earth, astronauts during trans luna and trans earth injection, have taken particular care in avoiding those rays.

Both electromagnetic and particulate ionizing radiations occur naturally from radioactive decay of elements. Enormous increases occur in the region of air explosions of nuclear weapons. Natural radioactivity may increase locally due to manufacturing or building practices such as the use of beautifully-colored radioactive clays in ceramics. Similarly, certain foods have a tendency to accumulate radioactivity, although not usually to the danger level. Mayneord (1961) analyzed many of them and found a maximum of 1690×10^{-12} curies per 100 g in brazil nuts. He concluded that an adequate Western diet will have about 5×10^{-12} curies of alpha activity per day at least and this could vary by 1000 if the slightly more radioactive foods like cereals were selected.

Radiation From Fallout

The testing of nuclear bombs in the atmosphere causes observable changes in the radioactivity of the earth environment. The problems created have led to some significant studies on the incorporation of radioactivity into the various land and sea food and respiration chains. The known mutation consequences of ionizing radiations and other noise input into physiological processes come under special study. Usually the mutations are deleterious, changing the instructions in a program that developed only through long evolution.

The radioactive fallout studies have shown that some merit can be attached to treating the environment radioactivity in terms of pairs of elements that are normally associated with each other. Thus the $^{90}Sr/^{40}Ca$, $^{137}Cs/^{39}K$ ratios have been studied to estimate biological dangers from the radioactive elements ^{90}Sr and ^{137}Cs. These results, especially for the $^{90}Sr/^{40}Ca$ ratio, have been quoted as a unit of exposure hazard for the population and the ratio has been followed through time peaks of activity as well as by regional distribution. The appropriate strontium unit is picocuries (1 curie = 3.7×10^{10} dis/sec) per gram, but the ratio $^{90}Sr/^{40}Ca$ is probably a better description.

The main problems are connected with the incomplete knowledge of the exact normal behavior of these elements and their ratios. For example, cell walls or shells may be made up of many combinations such as cellulose, chitin, calcium carbonate and glucose polymers, but some radiolaria make their shells of strontium sulfate ($SrSO_4$). Perhaps they have an advantage in the nuclear age in that they can leave their shells, relocate, and reshell. In consideration of all the variables, it is essential to use a representative and meaningful sampling technique with due regard for the previous history of the material such as plant and animal uptake, metabolism, and composition.

Before elements like ^{90}Sr become locked into organisms, they may be separated, to varying degrees, from ingested or inhaled material by chemical and physical means (ion exchange, chemical interchange, and filtration). After they are ingested, isotope characteristics, the nutritional state, and body clearance capability become crucial. Dialysis and drug therapy may be useful as well. For example, knowledge that a dangerous isotope is present and may enter some metabolic cycle may suggest a drug that will uncouple that cycle. Short-lived radioisotopes may be left to decay to a safe level if this is permitted by the nature of the food, beverage or other element in the food chain. It is simple and effective for a general reduction in radioactivity.

The air and water around radioactive sources can accumulate radioisotopes by processes like activation or neutron reactions with stable elements. The behavior of ^{131}I, ^{32}P, and ^{65}Zn in this connection are of particular interest and their maximum permissible concentrations are prescribed by Handbook 69 of the U. S. National Bureau of Standards (1959) and its revisions. In μCi/cc, 2×10^{-5} of ^{32}P, 10^{-4} of ^{65}Zn are set for water and 3×10^{-10} μCi/cc is set for ^{131}I in air.

The concentrating action of organisms vastly increases their radioactivity over that of the environment in which they live. Fish may accumulate ^{65}Zn in eyes, ^{24}Na in bone, and ^{51}Cr in blood. Hair may accumulate ^{65}Zn, thyroid glands and vegetation take up large amounts of ^{131}I, and bone stores several isotopes such as ^{45}Ca. The reinforcement of body burdens by the growing use of nuclear reactions means that more radionuclides have become available to the concentrating action of the food chain (especially meat and milk) and to the human organism itself. Thus if one is attempting to take a gamma spectra of whole bodies, an annoying peak will occur at 0.662 MeV which is new in the history of radiation contamination. ^{137}Cs, is a component of radioactive rain after nuclear explosions. These explosions pump many isotopes into the atmosphere, some of which are probably not being characterized as yet due to measuring limitations.

Each period of bomb testing creates a peak in these products and then some like ^{137}Cs can disappear during periods when this testing is underground or limited. Probably some disappear from organisms into the world-wide radioactive contamination pool in which they then work out their half-lives.

RADIATION FROM OTHER SOURCES

The Extreme Case of Chronic Radiation Exposure

A very extreme case of chronic radiation exposure may be illustrated if the habits and life style of a radiation worker, perhaps a not too well-nourished painter of radium-coated dials on instruments, is studied. She could be living in the air currents downwind from an active nuclear explosive site with significant $^{14}CO_2$ therefore in her respiration. If such a worker also becomes pregnant, the fetus could accumulate ^{131}I in the thyroid along with the original burden. It may be that she had an inordinate fondness for foods which are especially rich in natural radioactivity. Milk and milk products such as ice cream would contribute ^{137}Cs and ^{90}Sr.

Her home may be in a place like Joliet, Illinois which has had rather high radium levels in its water system and this element would "mimic" calcium in the skeleton, emitting alphas directly as ossifying points according to its half-life of 1,620 years. Nonazite, rich in radioactive thorium may be the main soil component. She might also be living in a brick house with radioactive uranium clays used for the bricks and use stained glass and dishes with similar colored ceramics. The winter may be long and the in-house time lengthened.

If, to this, were added proximity to a nuclear power or fuel station of large proportions and mediocre nuclear plant housekeeping, some necessary dental X-rays and medical fluoroscopy, the radiation absorbed will undoubtedly be very dangerous.

One would expect at least the development of radiation erythema or skin carcinoma and tumors in various tissues and organs, possibly leukemia, bone necrosis, genetic effects (which are less dose-rate sensitive), radiation damage to sensitive biological cells and a shortening of the life. At the limit, of course, would be the development of a radiation syndrome—generally appearing as a possibly fatal radiation sickness. An acute version of this extreme case would be the large radiation exposure from accident or intention. This extreme case of the woman illustrates the two broad classes of radiation damage; external, such as the X-rays from machines in need of adjustment, and internal, from the ingestion of radioisotopes.

A summary has been made of the results of a survey of 361 people with skeletal burdens of Radium 228 (mesothorium) and Radium 226. Approximately two thirds of these were radium dial painters, the remainder were contaminated as a result of laboratory work, from injection of radium solutions or from ingestion of radium-mesothorium mixtures. The latency period for the appearance of bone tumors was on the order of 20 to 30 years.

Radiotherapy. When radiotherapy is used for certain rheumatic diseases and enlarged thymus in children, more cases of leukemia or thyroid carcinoma occur. Spondilitis rheumatic disease is treated by radiation and excess deaths occur from leukemia in those treated.

Very characteristic, large, lobulated or multiple fibroblasts may be seen in the brain blood vessels after accelerated overexposure of head skin tumors to radium. Small vessels become necrotic, the epithelium swells; later connective tissue becomes thickened, or hyalinotic, not unlike an arteriosclerosis. Blindness, paraplegia and epileptic attacks due to gross atrophy of occipital lobes was the pattern seen in a case where the scalp was irradiated for skin disease but the latency period lasted five years.

Boron compounds provide an example where therapy can be enhanced by injecting them into the bloodstream where they can *collect* in "difficult" tumor roots. Then the neutron activation of the boron makes it emit alphas into the tumor that has captured it. If it is known that a radionuclide will localize in a molecule which, in turn, selects a specific kind of tissue, then therapy could be called on to add intrinsic directivity. For example, there are radioimmunoreactions in which antibodies have the radiation and can *recognize* a medical target in treatment such as a tumor, then attack it with ordinary, plus radiation, weapons held directly on the target by the antibody-antigen reaction.

Radiation as a Diagnostic or Biological Tool. In medical diagnosis, biophysical objectives are to develop indirect viewing methods and better imaging (image intensifier) to reduce hazards. A chest X-ray gives from 10 to 2,000 milliroentgens depending on the operation of the equipment, so that possibly two and one-half chest X-rays in a year would equal the maximum occupational exposure. Thus massive X-ray screening programs against tuberculosis may not be worth the risk. A spinal X-ray procedure may result in an exposure of up to five roentgens.

Between 1928 and 1945 thorium dioxide in a colloidal suspension (thorotrast) was widely used in diagnostic radiology as a contrast medium in injections for the visualization of body cavities and certain organs. Numerous cases of malignant tumors have been reported (liver and biliary ducts) with a mean latent period for appearance of about 18 years. The alpha radiation dose to liver and spleen of these individuals has been estimated to be as high as hundreds of rads per year. These instances are related to those seen in the watch dial painters as well as to those of uranium ore miners and of patients who have received radiotherapy, especially with radium. Radon gas in the uranium mines contributes to lung cancer, which is often fatal after years of exposure to hundreds of rads per year. Table 8-4 shows several radioisotopes and their use as diagnostic or biological tools.

TABLE 8-4. Radioisotopes and Their Use as Diagnostic or Biological Tools

Radioisotope	Use
Radioiodine ^{131}I	Diagnosis of hyperthyroidism
Carbon ^{14}C	Photosynthesis, steps in utilization of C
Iodine ^{131}I	Labeled rose bengal in hepatic disease
Cobalt ^{60}Co	Pernicious anemia and B_{12}

Tomography with computer-assisted readout adds scanning and sequential contrasting to the three-dimensional diagnostic picture of tissue presented to the radiologist. Technetium-99M as a short-lived, injected radioisotope (six hours) makes brain (Mark 3, rectilinear, transverse section) tumor scans in 10 minutes comfortably. Currently, parent-daughter generators permit continuous laboratory separation of ^{99M}Tc from its parent longer-lived ^{99}Mo (67 hours) by ion exchange columns. The column is eluted daily for fresh ^{99M}Tc. There is also a liquid extract generator in which ^{99}Mo, NaOH solvent, and methyethylketone extraction yield ^{99M}Tc by gravity separation. MEK has the desired isotope which is recovered by evaporation of the solvent. Technetium lacks hazards from primary beta particle radiation of tissue and permits ten times more radioactivity for superior resolution in scans without radiosensivity problems in bone marrow, and is easily detected. As the pertechnetate it is selected for thyroid, salivary gland, and stomach scans. Chemically manipulated into serum albumen, it is used for lung films and the colloidal form is for the liver and spleen. In chelate form, the nuclide is selected for analysis of the kidney for, as in the other organs, analysis of function and structure.

THE ENTRANCE OF RADIOISOTOPES

In nuclear radioisotope pollution of the environment, the evaluation is based on weapons tests, reactor experience, and animal and plant uptake studies—that is the food and respiration chains. Important isotopes and compounds include ^{40}K, ^{14}C, ^{137}Ce, ^{90}Sr, ^{131}I, ^{137}Ba, ^{65}Zn (food) and ^{239}PuO$_2$, ^{144}CeO$_2$, ^{106}RuO$_2$, and ^{95}ZrO$_2$ (inhalation) where the characteristics are shown in Table 8-3.

The behavior of organisms with respect to these is very important. ^{137}Cs is evaluated as a co-element of potassium just as ^{90}Sr is associated with calcium and ^{131}I tends to find its way into the thyroid gland. Ruminants like the cow and sheep discriminate against strontium and, so, act as a *filter* against this radioisotope even while these animals are making their overall accumulation of radionuclides in meat and milk. Strontium does not move bodyward against a concentration gradient (it diffuses) but calcium *does*. This is a real service in that ^{90}Sr, like radium, tends to be locked into the skeleton.

Nuclide therapy can take a form such as that used for persons exposed to plutonium poisoning. As this nuclide becomes as common in the world as breeder reactors will make it, human internal exposures beyond the 0.04 μcurie limit will increase. This bone-seeking nuclide concentrates its alpha emission in the range of 50 μm and its half-life is 24,000 years, so it can infiltrate the liver, lymph, bone marrow, and spleen. Then its weakening of cell walls can let it pass into the blood system. The chelating agent (chela = Greek for claw) diethylenetriaminepentaacetic acid binds plutonium in an insoluble complex which may be cleared through excretion. It must be administered before the actinide gets to bone and soft tissue for best results since its action is extracellular. Glucan "ejectors" like pyran copolymers may work to stimulate plutonium-disrupting-polysaccharides *within* the cell for a cooperative action.

Nutritional states are always significant. Thus, with good calcium nutrition before having to cope with ^{90}Sr, and the use of removal agents after exposure, some control may be possible. Some nuclides like tritiated water vapor, will pass through the skin as well as through the lungs and, to suppress such entry, the partial vapor pressure of T_2O or HTO should be low, and the skin cooled (to depress the spread by circulation). If radionuclides can be rendered insoluble—like the oxides and phytates in some cases—they will not penetrate well, which would suggest reactive coatings or phytates. However, this mode of body burden building is not significant, unless the exposure is typically through the skin. It is not with fallout, but it may be in industry or research. For an extreme example, the author has seen immersion tests for percutaneous absorption of a labeled detergent where the experimenter used his own skin. Radioactivity is not the only testing hazard of course, because the toxicity of innumerable elements like arsenic, mercury, and insecticides are simultaneous hazards over and above any radioactivity, and may thus constitute an associated secondary insult; that is, the radioactive insult combined with the physiological one from the same source.

A nuclear explosion will expose earth areas around the blast to various immediate doses which may be shown on a chart by contour lines of equal intensity (Cohn et al, 1960). Organisms will receive an initial dose, which will depend on the kind of bomb, hydrogen, or atomic, and on the blast (air, underwater, etc.) and will decrease quickly with the decay of short-lived isotopes (that is, those with a half-life in minutes or days). It will be very strong in the top soil and will gradually enter the food chain via plant and animal uptake. Particulate radioactivity can settle in water, leading to large concentrations in bottom feeding animals such as clams, snails, and crabs. Animals within the explosion area (a hundred miles or much more depending on the strength of the blast) accumulate radioactivity according to their *stay-time, the normal route of the element in metabolism, the half-life clearance time, and solubility of the radioactivity*. A large initial beta activity can be counted from ^{90}Sr, ^{140}Ba, and the rare earths. It is concentrated in the livers and large intestines. Most of the activity by far is in the skeleton and epiphyseal regions of bones, thus reflecting the new growth that occurs there.

About 3 millicuries appeared in the gastrointestinal tract (of total radioactivity) of persons near Bikini after the nuclear explosion tests that took place there. This was one day after their accidental exposure (when they failed to leave the test area). On the average in the whole area, one found about 12 microcuries accumulated in each person (body burden). The detection of those burdens depends on the available techniques and instrumentation and the figures are likely to be smaller than actual. As for dramatic physiological injury (permanent pathology), none had been reported for these persons five years later. With the exception of strontium concentrators like coconut crabs on detonation sites, gradual return to low radioactivity levels and slow return of animal and plant populations seems to be the pattern. Ocean water diluting and tidal cleaning functions are important factors in this process.

The gross beta activity in tropical plants, two years after a fission detonation, varies with the kind of plant (even on one small area). The fission products are rich in large pieces of CaO, $CaCO_3$, and NaCl and the fallout is soil soluble and absorbable via roots and leaves. The data indicate that levels are low (betas of the order of 10^{-5} Ci/m/kg) but more interesting, the leaves tend to be most radioactive. Certain species, for example, *portulaca*, are likely to concentrate the activity. Ripe coconut milk tends to discriminate against the activity, but by a small factor of less than ten. Uptake depends on dose concentration as to its distribution.

Naturally resistant barriers to absorption may exist due to horny epithelium layers in various parts of the body such as the palms and the bottom of the foot. The water or oil solubility of the isotopic compound as well as the natural functions of structures such as sweat glands, absorption mechanisms, and the interplay of absorption and nutrition will influence radioactivity transfer through the natural barriers. According to the teaching of Stannard (1959), a filter for removing inhalable and absorbable contaminated particles would be extremely fine (0.2 to 0.3 microns in diameter). Several nutrient isotopes, for example, ^{131}I, ^{59}Fe, and ^{45}Ca will be absorbed according to the physiological demand. Perversely, a poor nutritional interrelationship may occasionally dampen absorption, for example, poor vitamin D and amino acid levels will lower the rate of ^{45}Ca absorption. The diarrhea symptom may be a measure which hastens the exit of absorbable isotopes and decreases the residence time of radioactivity sources. As body burdens become of more general interest with greater distribution of radiation, it may become necessary to label food products and building materials with their radiation emissions. A common gas lantern for example, (to brighten the light) used a radioactive mantle cloth whose thorium is easily dispelled into a room, especially when first used. For emergency lighting, the lamps are often used in multiple units in closed spaces where they can emit alphas to be absorbed by respiration, either in a captive or transient group.

Information and Antigen Activity, The Immunoresponse of Radiation

The suppression of antibodies by doses of radiation such as 125 to 175 roentgens of X-rays or more is a characteristic radiobiological response. As a result, the resistance to new infections and parasites is decreased. This suppression may parallel the depression in general protein synthesis, for a special example, properdine, a globulin in the complement system is suppressed. This is also an antibacterial factor and so the risk of infection is doubly increased. There is also an effect from ultraviolet in which the radiation suppresses the immunoreaction to cancer, leaving the tissues more susceptible to the disease.

Makinodan and Gengozian (1960) describe the mechanisms of antibody response as the phenomenon of self and non-self recognition in which there is an acceptance-of-information phase. Thus certain hematopoietic cells are primed by their resident information to produce a certain immunoresponse, a gamma globulin antibody that recognizes strange (non-self) material. Radiation damages this ability, or eliminates it completely for a period, with the effect being proportional to the dose (Fig. 8-2).

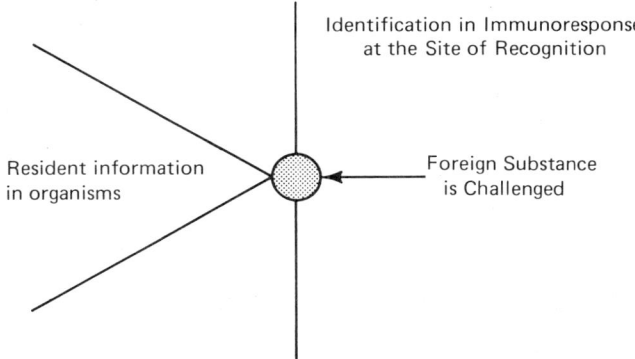

Fig. 8-2. Immunoresponse. There is a radiation-borne capacity to eliminate or alter this response and it is proportional to dose.

The internal membranes, gastrointestinal tract, and the oropharyngeal region are of special interest. They are involved in the pathology of infection as well as in absorption and fluid control. They are weakened by radiation and may develop lesions. The epithelium is sloughed off in the gastrointestinal tract and one possible reason would be that an immunomechanism resurging after the period of antibody suppression due to irradiation, may regard the organism's own epithelium as foreign (Fig. 8-3). This rejection-of-self mechanism is seen as a serious error in biological communication. To speculate further, it may also be true that a "safe" radiomimetic agent like chlorambucil might prolong the

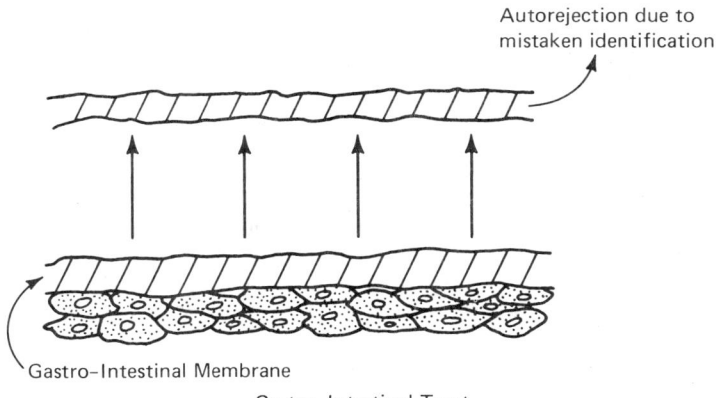

Fig. 8-3. Error is induced into the information status of the immunosystem and when the reparation is complete, the regenerated system may no longer *recognize* its own membranes when they appear.

antibody suppression beyond such a critical period. The infection may be treated with aureomycin or terramycin, so it is much less significant than the loss of gastrointestinal tract integrity which probably determines the four to six day gastrointestinal death. The initial immunoresponse is succeeded by recovery of antibody production. Bone marrow cells are prescribed for transfusion in accidents. These tend to take advantage of the fact that the body at this time has a depressed or absent immunoresponse. The term secondary disease is associated with the attack by the new antibodies against transfused cells or by the antibodies from the transfused cells on the organism.

Capillaries and Irradiated Limbs. The integrity of the capillary endothelium is threatened by radiation. Edema follows in an irradiated limb, labeled isotopes penetrate the membranes, and red cells go into the lymph. The shielding of a limb from ionizing radiation exposure has some special results. There is also a gross effect on the limb. The one that is irradiated will be stunted in growth compared with the control limb. Similarly, one limb of an animal may be irradiated and the other serve as a control. If the control limb is properly protected, its hematopoietic system remains intact to establish a regeneration and repair process.

Oxygen Effects. Anaerobic conditions are protective against radiation injury. The reasons are connected with the action of certain highly reactive free radicals that can be found in the presence of oxygen (see Chapter 10). These perpetuate the reacting schemes that are deleterious to tissue, forming ordinary perhydroxyl radicals and organic perhydroxyl to do so. Some organisms have genic capabilities such as catalase production which gives them unique protection against hydrogen peroxide, an oxidizing agent which can be produced in the radiolysis of water. Oxygen itself, as an isotopic contaminant, appears to be insignificant. After nuclear explosions in which it is produced, the half-life of the main isotopic form is measured in seconds.

THE RADIATION BIOLOGY OF THE RADIATION SYNDROME

Man appears to be one of the most radiosensitive species. The evidence from human radiation studies is that those who are most exposed occupationally are tending to accept less and less intentional exposure. This is due to two things in particular. The early experience of radiologists showed a selective shortening of life over other type of medical personnel, and a careful reduction in exposure has already eliminated this difference. The other is the growing conviction that no threshold dose of whole body radiation exists. In other words, there is no low, low level of irradiation below which the dose can be said to be harmless on the life span.

There is a decrease in injury with decreasing dose. Beagles have been exposed in dog-lifetimes (13 years) experiments which illustrate this well. It is easier to associate these results with man than those of mouse experiments, for example. The mouse has a mean life expectancy of one-sixth that of the beagle, which, in turn, lives about one-sixth the lifespan of man. A dose exposure of 300 roentgens per day is destructive of bone marrow and lethal in two weeks. At lower dose rates of 35 R/day, anemia becomes the primary cause of death. At the level of 5 to 17 R/day, bone marrow disorders, particularly myelogenous leukemia cause death. Fatal diseases are connected with the dose rate rather than with the total dose, showing the influence of repair. Possibly, somewhere below 5 R/day, ordinary diseases may be more important than the radiation connected ones. Sterility has been seen in unborn pups at dose rates between 5 and 17 R/day, but otherwise all was normal.

In a global environment, the gradual introduction of intentional and accidental radioactivity into the biosphere from military and peaceful applications will very likely raise the background against which organisms must survive. Certain consecutive thresholds will then be reached for radiation-generated disease. The normal disease rate in the historical equilibrium for radioactivity in the environment will at some point be doubled by the dose to which populations are then exposed. If the radiation exposure is acute, the doubling dose will be smaller by about one-fourth than that for chronic exposure which may be about 100 rads for man. This doubling dose may be the global crisis level at which planet habitability begins to be seriously threatened.

Sensitive tissues are those such as the bone marrow, the gastrointestinal tract, the epithelium, and the seminiferous epithelium. A localized exposure (for example, to a few millimeters of the spinal cord of an animal) can be destructive of the function of that part and doses of the order of 4 kilorads result in paralysis and death. The basis for cellular sensitivity is found in the reproductive activity, capacity for regeneration and repair, and the degree of differentiation and primitivity.

To generalize, radiosensitivity depends on the proportion of undifferentiated cells, mitotic activity, and the number of cell divisions the cell needs to develop its function. As in the rationale for radiotherapy against cancer, the more active cellular structure is more sensitive (during divisions). Also, the cellular radiosensitivity is inversely proportional to the degree of differentiation away from the primitive state, the latter being hyposensitive. Thus, highly specialized non-dividing muscle and nerve cells are more radioresistant.

An examination of the effects of a steadily increasing dose (Tables 8-5, 8-6, 8-7) shows that less than 300 rads will not bring sudden death. However, between 0.3 and 0.9 kilorads, the dosage is sufficient to cause death in 10 to 15 days. It is likely to be the hematopoietic syndrome, based on destruction of the regeneration system for blood corpuscles, that is seen (Fig. 8-4). The time for death after a dose of 0.9 to 10 kilorads is shortened to 3 to 5 days. The syndrome is that of the gastrointestinal tract which suffers a great loss of sodium and fluid through diarrhea and vomiting. Both hematopoietic and gastrointestinal tissue may be in rapid division and differentiation and the dividing time is the radiosensitive period. There is a loss of epithelial tissue as well as of the mechanism for its regeneration. If this tissue loss is progressive, the shorter survival is likely. Actually, the portal of entry for absorbed substances (nutrients) is closed while the barrier to bacterial invasion is opened. The bacteria can pass through the

TABLE 8-5. The Radiation Syndrome

	Dose	Response	Comment
Bone Marrow Hematopoietic Syndrome (acute near minimal dose)	0.5 kilorads	Transient nausea, vomiting, diarrhea; 2-3 weeks for effect of bone marrow destruction. Chill, fatigue, petechia. Oropharyngial ulcers. Epilation, internal ulceration.	Recovery
The Gastrointestinal Nausea-Vomiting-Diarrhea Syndrome (NVD) (acute lower dose)	0.5 - 2 kilorads	Loss of appetite, listlessness, nausea, vomiting, diarrhea (especially at higher doses), fluid losses, electrolyte losses, abdominal edema and pain, loss of peristalsis, dehydration and hemoconcentration.	Death in 1 week. Loss of intestinal epithelium may be related to pulmonary death.
Acute Circulatory Change Syndrome - Intermediate	8.8 kilorads whole body	Headache, speech and vision problems. Vomiting, cramps, diarrhea, edema, decreased blood pressure	Death-49 hours. (circulation) Also-in others, initial depression, convulsions and ataxia
Central Nervous System Syndrome (acute large dose)	10 kilorads whole body	Shock, lymphocytopenia in 6 hrs, then slight recovery. Intestinal cramps at 30 hours, then cyanosis, coma. Nausea, vomiting and depression are CNS in origin. In monkeys, the same dose produces meningitis, choriodplexitis, vaxculitis, pyknos duces meningitis, choroidplexitis, vasculitis, pyknosis of cerebellum granule cells.	Death-39 hours

TABLE 8-6. Chronic, Lower Dose Effects

Dose	Effect	Comment
A few neutrons per day for a year to rats, rabbits, monkeys and dogs.	Lymphopenia in a broad depression of weeks, anorexia, loss of weight, weakness, corneal opacities, bilateral cataracts	Species variability, peripheral blood effect more common in dogs than in rabbits at few R/d.
A few roentgens per day for two years	Mucoid conjunctivitis, keratoconjunctivitis, pulmonary infection, paratyphoid infection, various neoplastic changes, especially leukemoid. Hypoplastic hematopoeitic organs, gonal atrophy, hemorrhage in lymph nodes, heart, stomach, small bowel and kidneys. Slow neutropenia and even slower erythrocytosis. Constant platelets, and reticulocytes until roentgens increase to a few per day. Erythrocytes relatively stable to chronic peripheral blood dosage. Neutrophiles more stable than lymphocytes but neutropenia occurs.	Chronic varies from acute, especially in absence of severe reticulocytosis.
(Threshold for ultimate chronic effect is fractional roentgens per day)		Bone growth is inhibited even at levels too low to give neutropenia. Bone degeneration may lead to osteosarcoma, and necrotic fractures.

TABLE 8-7. Acute, Critical, Large Dose Effects

Dose	Effect	Comment
Super doses, many kiloroentgens	Acute, CNS death	
1 to 5 kR	Acute, intestinal death in mice, lung and hepatic cell damage in rats. Persistant capillary dilation of skin.	IgM, IgA, IgG, and IG_z decrease in serum and this less defensive state may persist for years. Lung is rather resistant.
Large dose 550 R; 750 rads—man and experimental *animals*.	Quick damage to lymph nodes, spleen, thymus, bone marrow, spermatogonia, epithelial mucosa. LYMPHOCYTOPENIA, neutrophilia then neutropenia, *later* loss of mobility in sperm, anemia in two weeks, RBC precursors in bone marrow decrease, then myeloid types follow, leucopenia is fast, then leucocytosis and another wave of leucopenia. (All against background of mitotic delay.)	Effect requires irradiation of the intact, whole animal or human. Subjects with fewer immunoglobins and granulocytes show more deaths from infection. Repair within days or weeks. Faster in spleen and nodes than in circulating lymphocytes. Lymphatic tissue is resilient. Mature neutrophiles surge to explain neutrophilia but are not replaced so neutropenia occurs. Thymus regenerates faster than bone marrow.
Whole body radiotherapy 100-350 R at 1.5 R/m to 1.5 R/h also produces some of the effects.	Platelets decrease with marrow megakaryocytes.	Hemorrhage and purpura occur.
	Sertoli cells in testes resistant but otherwise testes depopulated.	Repair with restoration of megakaryocytes. Partial or complete sterility after 2 weeks of fertility.
	Stomach and colon more resistant than small intestine. Small bowel extensively damaged.	Mechanism in effect is suppression of mitosis, pychosis.
	Permanent epilation, radiation osteitis is delayed, degeneration of bone after large dose.	Destruction of all germ stem cells insures sterility. Fertility functions prematurely age.

thin intestinal barrier and hasten death. The appetite suffers and a decrease in granulocytes and lymphocytes means a loss of their protection against disease. Along with the decreased regeneration of tissue, there is an increase in glucocorticoids and more adrenal mineralcorticoids appear. Thus the loss of sodium may diminish. However, the capillaries become more permeable but with fewer platelets, clotting is reduced. This sequence of events leads to hemorrhage which brings on anemia and anoxia. The highly radioresistant brain and kidneys finally give in, damaged by the after effects of the dose.

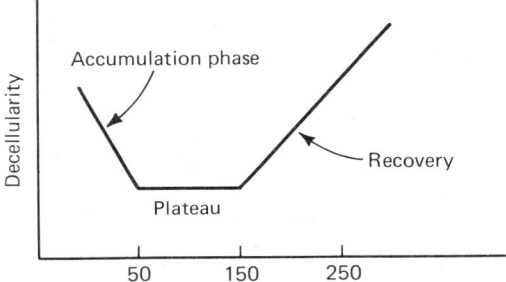

Fig. 8-4. Chronic irradiation leads to plateau in response followed by recovery if dose is not great.

Chronic ionizing radiation is an accumulation phenomenon (Table 8-7). An early sign of change is loss of cells in sensitive organs such as gonads. This loss increases with time. A stability occurs (plateau) when repair balances the degeneration, and the relative rates of damage and repair determine the position of the plateau (Fig. 8-5). The size of the individual doses will be determining rather than the accumulated dose

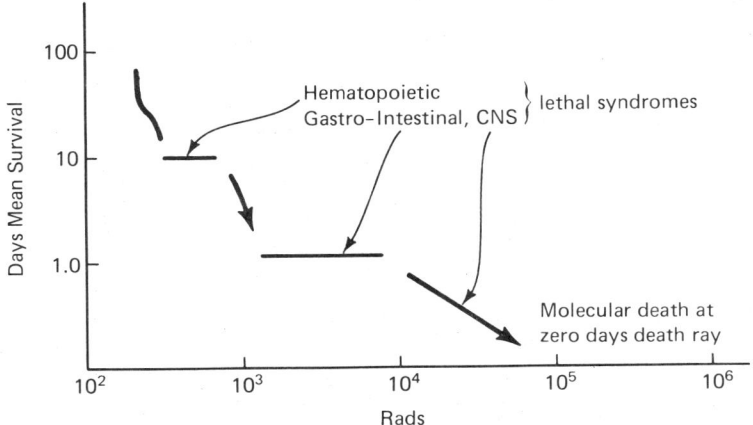

Fig. 8-5. Arena's concept of syndromes as a matter of survivors after increasing doses (log-log) gives a general relationship for most mammals. A plateau represents short (hematopoietic) and longer regions where effect is independent of increasing dose. *After* Arena, V., *Ionizing Radiation and Life*, C. V. Mosby Co., St. Louis 1971.

alone. At low individual doses, the plateau will be like the control, and at high doses, the appearance will be like a peak effect. The greater the level of dose, the greater, of course, will be the accumulated dose and so the chronic exposure can be maintained until death, which will then be associated with the intensity of the chronic exposure or regular dose.

THE COURSE OF HISTOLOGICAL CHANGES IN THE GASTROINTESTINAL TRACT AFTER ACUTE EXPOSURE

Striking changes occur in the proximal gastrointestinal tract (small intestine). Karyorrhexic and karyolysic degeneration in cells and enlarged nuclei become prominant one hour after acute irradiation. The bases of the villi are involved in the degeneration more than the tips. Significant change that occurs throughout the gastrointestinal tract is the cessation of mitotic divisions. This persists for approximately eight hours in the small intestine and for as long as twenty hours in the large bowel.

At four hours after irradiation, injury to the basal cells of the small intestine is more evident. Numerous nuclei undergo fragmentation. Mononuclear phagocytes are present in large numbers, many of them filled with cellular debris. Along the sides of the villi, pycnotic nuclei appear, and mitoses are at a minimum.

The degenerative changes in the bowel are no longer apparent thirty hours after irradiation, but macrophages are still present in the lamina propia. Mitotic figures reappear in moderate numbers at this time, but the number observed is still less than that seen in the intestinal epithelium of the controls. Between twenty and sixty hours after exposure, the most prominent feature of the intestinal mucosa is the presence of the large epithelial nuclei. These cells produce a distorted epithelial pattern in the intestinal crypts. After 60 hours, the appearance of the intestinal mucosa returns to normal.

With a dose of about 10 kilorads or more, death is a matter of 24 to 48 hours away. There is ataxia, then pulmonary and respiratory problems develop with severe lesions and stupor, the characteristics of the central nervous system syndrome in acute radiation sickness. The symptoms include headache, dizziness, malaise, abnormalities in the sense of taste and smell, nausea, vomiting, diarrhea, and a decrease in blood pressure. There is irritability, and insomnia. The white blood cells and platelets both decrease in number.

The development of the radiation syndromes is also indicated by showing the days of mean survival against the dose in rads as in Fig. 8-5. This is Arena's concept of a general relationship for all mammals and a plateau means that more radiation fails to produce more damage for a short period as in hematopoietic death, or for a fairly long period with death from gastrointestinal effects.

THE NERVOUS SYSTEM

In cats, medium doses of 200 to 400 R cause increased brain electrical activity as a function of whole-body or brain-only exposure.

Hippocampal activity is observed and it peaks 3 to 7 hours after the exposure. The animal is made more alert by such doses. To get these changes on a chronic basis seems to require an accumulation of up to 300 R. The hypothalamus group seems to be affected rather promptly after an exposure as shown by observing behavior under exposure. After brain tumor therapy, brain edema develops within one or two days. The tolerance limit for the human brain is probably 150% of the radiotherapy dose. In chronic exposure, it is necessary to accumulate 3 or 4 kR before changes are apparent.

IMPAIRED EMBRYONIC DEVELOPMENT

Radiation causes chromosome changes in the fertilized egg, and the embryo is very sensitive to the time or phase of exposure. For one thing, mosaics occur from the changes that are made *after* the fertilized egg has begun divisions. If the change happens to be at the first division, half of the body of *Drosophila* will be at variance with the rest. Later exposure, when the cleavage divisions have been made, will mean that the mosaics are seen in a smaller part of the body.

Embryos of mice are radiosensitive and develop brain hernias (exencephalia) after 200 R of X-rays, but if the earlier embryo (2-cell) is exposed at 1.5 days after conception, only 15 R will produce these changes. These levels are within certain diagnostic ranges and may be transmitted to subsequent generations (both genetic and congenital). Mammalian embryos react to 5 R delivered before cleavage within hours after conception with increases in lethality, delay in cleavage, resorption, and disintegration. The nuclei are changed with coloration, swelling, and sticky chromosomes.

BLOOD SYSTEM

Ionizing radiations profoundly change both the peripheral blood and the hematopoietic system. This is most evident in the case of whole body exposure and most dramatic at acute dose levels. The changes are discussed in the next sections and include those of blood forming, maintenance, and circulatory system. The changes are fairly consistent among species and data on humans continues to be obtained from atomic bomb survivors and from patients receiving total body tumor therapy. The blood changes follow predictable patterns based on the nature of the exposure so that they can be used as diagnostic aids and important factors in personal medical history as respects radiation. Among the other radiation effects in atomic bomb survivors is the long latency sequel—aplastic anemia which is related to the blood changes, increased incidence of cancers and other diseases. Now, 35 years later, the positive association of clinical findings with the bombs is proving more and more difficult, so that genetic effects with successive generations in the long term, may be the most important lessons that can now be learned.

The importance of radiation medical histories, especially for radiation personnel, is clear. Also, a system of obligatory reporting of accidents *promptly* should be universal so that higher authorities can institute remedies. These nuclear accidents, according to the author's own experience, are due to the failure of people, especially of health physics supervisors, and to the failure of materials. The principal means for avoiding them is adequate planning and continual supervision with the possibility of personal, legal responsibility for negligence.

Accidental exposures often occur in experimental situations where the victims are aware of the radiation bioeffects. Anxiety therefore plays a role as it no doubt did in the Nuclear Center accident in Mayaguez in 1963. Consultation with the attending physician (McCandless 1963) who was simultaneously in the author's radiation biology course, tended to emphasize the surprise of the victims when they realized that they were being irradiated by premature positioning of a strong source of gammas by a crane operator. Subsequently it was estimated that the doses to seven persons ranged from 4 to 100 roentgens to the whole body and there have been apparently no serious aftereffects. The exposure could have been the full 5kR-50kR requested for the experiment which was to irradiate and test the plant cuttings the seven were arranging.

Diagnosis, Dosimetry and Prognosis Based on Blood Analysis

One of the most reliable guides to the occurrence of a radiation exposure is thus a composite blood count. The counts of various cells change within minutes after acute exposure and the timing and extent of these changes function in the manner of a biological clock which begins to run when the radiation is absorbed in the hematopoietic and circulatory system. The leucocyte count, for example, is an excellent personal guide or clinical test for exposure. In the case of radiation accident, the counts and changes in blood cells provide evidence of the time the accident occurred as well as the approximate dose. Sometimes these facts are difficult to obtain otherwise. The basic change in the pattern of blood cells may be illustrated by a family of curves which indicate the changes after time zero and for the short term following an acute exposure (Fig. 8-6). The significant early changes for diagnosis are summarized in the vector diagram (Fig. 8-7). Platelets show an early resistance to decrease and then fall off.

The significance of these blood changes may be pointed out in some detail since it is thought that they constitute a readily available alarm system as well as indicators for the prognosis. The leucocyte count, for example, if followed, reveals whether or not the exposed person will survive and does so fairly soon after the questionable dose has accumulated.

The notation for the pattern analysis follows the definitions of blood cell components and their physiological function. Reticulocytes (usually numbering 1 to 2% of total cells), by their very presence in the blood after exposure, indicate that enough bone marrow or blood forming organ has been spared from quick destruction to allow red blood cell regeneration. The red blood cells have a volume of about 100 μ^3, number about 5 million/mm^3 and live for 120 to 130 days, then die in

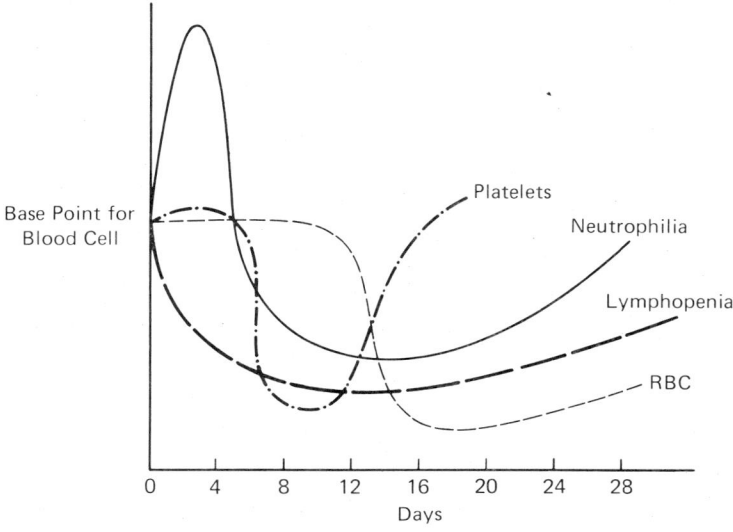

Fig. 8-6. Biological dosimeter. Blood changes, lymphocytopenia severity is $\sim$ dose.

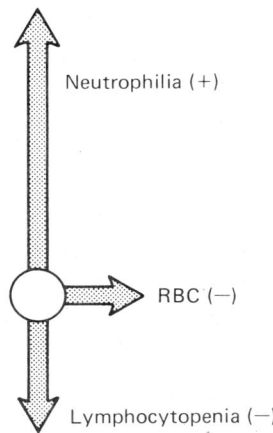

Fig. 8-7. Vectors for diagnostic, early blood changes.

the spleen and liver, contributing to the color of excreted waste products. Their precursors in the hematopoietic system, erythroblasts, myelocytes, and megakaryocytes have varying sensitivity but all decrease after moderate exposure, the erythroblasts within a few hours. The total white blood cells usually number between six and seven thousand per mm^3 in humans. The granulocytes, white blood cells, or polynucleated cells have a much shorter biological life of a few days. They have an affinity for granules giving them their name and may be

Neutrophiles — large and most abundant (to 60-70%).
Eosinophiles or Acidophiles — amounting to 2-4% and normally called up in cases of allergy and anaphylaxis.
Basophiles — numbering 0-1%, often having the two-lobed nucleus. Granulocytes are radioresistant and the time elapsed for their rapid decline in numbers increases with the series rats, rabbits, dogs, and monkeys. The neutrophile component is the one that shows the sudden rise immediately after the exposure giving the neutrophilia. Lymphocytes in lymphoid tissue, are mainly nucleus. Their sudden disappearance (within 15 minutes) shows that a person has been exposed to even a few roentgens. The amount of the change shows, within limits, how much exposure there has been (Fig. 8-8). Thus, in man, lymphopenia seems to be the best indicator of the progress of acute radiation sickness.

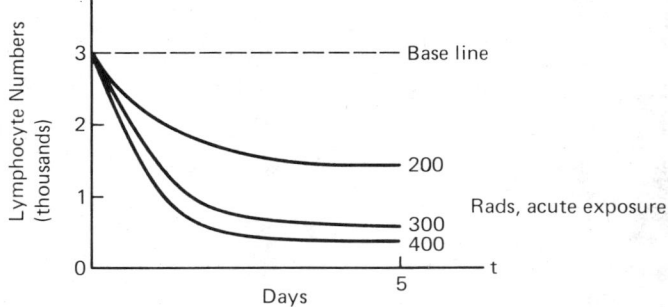

Fig. 8-8. Lymphocyte dosimeter. The initial lymphopenia is directly proportional to dose.

The platelet count is usually 300,000 to 500,000/mm^3. Hemorrhage is a significant irradiation result as is the formation of petechia by loss of red blood cells into the "burst" blood vessel. Platelets are the cell fragments which form from the megakaryocytes and, as normal blood entities agglutinate in allergy, or rush to the repair of damage to blood vessels. A lack of them means a lengthened clotting time which is actually greatly increased after exposure. Just as bone marrow injection is an important treatment for radiation sickness, platelet injections ease the clotting problem. The platelets may increase slightly immediately after the exposure, but then the downward trend sets in. The response is less regular and less proportional to the dose.

LONG TERM EFFECTS — ATOMIC BOMB CASUALTIES

There is a definite link between radiation and the induction of cancer. After the atomic bomb explosions in Nagasaki and Hiroshima in 1945, the survival of exposed persons has been closely followed.

Examination After Twenty Years

Twenty years later, the earlier predominance of leukemia cancer had been replaced by more of the other forms of neoplasms and recently, the excess mortality due to these was much more than before. This is the extra number of deaths due to the radiation in the over 200 rads dose group. Of course, some of this change is due to the fact that survivors were then much older.

Thirty Years Later

Thirty years later, many radiation-related clinical disorders and abnormalities were observed:

1. More lenticular opacities
2. Thyroid tumors
3. Leukemia
4. Chromosome aberrations in the peripheral blood lymphocytes
5. Slight impairment of growth and development of those exposed early in life
6. Microcephaly and mental retardation in those exposed *in utero*, especially if exposure occurred during early period of gestation
7. Heavily irradiated survivors have more solid tumors, especially of the breast and lung but extending to the stomach and elsewhere.

Children

In the new-born children, there is no evidence of *hereditary* abnormality due to exposure of the parents in the early years. There is no loss of fertility and no acceleration of aging. The minimal lenticular opacities do not appear to progress further. There are no rare specific clinical disorders that can be related to the bombs. As yet there are no genetic effects. The increased risk of cancer is a persistent trend.

Some 4,400 pregnant women have delivered after bomb exposures and microcephaly has been seen in 33 offspring, with 15 of these showing mental retardation. Many show mild visual disabilities. Some 205 children who were exposed during the first half of intrauterine life, showed developmental defects associated with the central nervous system when the mothers were unshielded and exposed only 1200 meters from the hypocenter.

Aplastic Anemia

Blood disarrangement and bone marrow damage to the hematopoietic system resulted in cases of aplastic anemia in survivors. During the first 20 years, 156 cases were noted and 40 came from the survivor group, some of whom certainly had the anemia and some who probably had it. Possibly three had received over 100 rads, 13 being exposed to 1 rad or more, and 13 to less then 1 rad.

A Nagasaki resident was pinned in bomb debris, and died later at 51 years. He was estimated to have absorbed 381 rads (377 of gamma, and 4 neutron radiation). He suffered general malaise and hospital findings showed the blood pattern of aplastic anemia which was shortly fatal.

A male working in a factory was exposed to about 330 gamma rads and 8 neutron rads. Like the other casualty, he had lost consciousness, developed gingival bleeding and epilation. In 1954 he began to show oliguria, lower edema and proteinuria. Then in 1955 he had general malaise, epitaxis, and subcutaneous hemorrhage, which were fatal at that age (54). It is possible that renal insufficiency was complicating an aplastic anemia.

In a female patient who died in 1963, her exposure to 283 rads (167 gamma, 116 neutron) at Hiroshima was followed by the fatal aplastic anemia. She also had observed gingival bleeding as an acute symptom.

SURVIVING EXPOSURE

A surprising source of initial injury in nuclear explosions is the first wave, the infrared radiation which causes intense burning (Fig. 8-9). Another is the slower pressure wave which blasts structures into pieces, filling the air with missiles including glass pieces that can kill and maim. The radiation illnesses only come later with the seriousness depending very strongly upon the nearness to the source of the radiation and the body geometry that is exposed and the physical protection of shielding available.

While it may seem unlikely that anyone needs advice on how to survive a nuclear weapon blast, it may be instructive against the back-

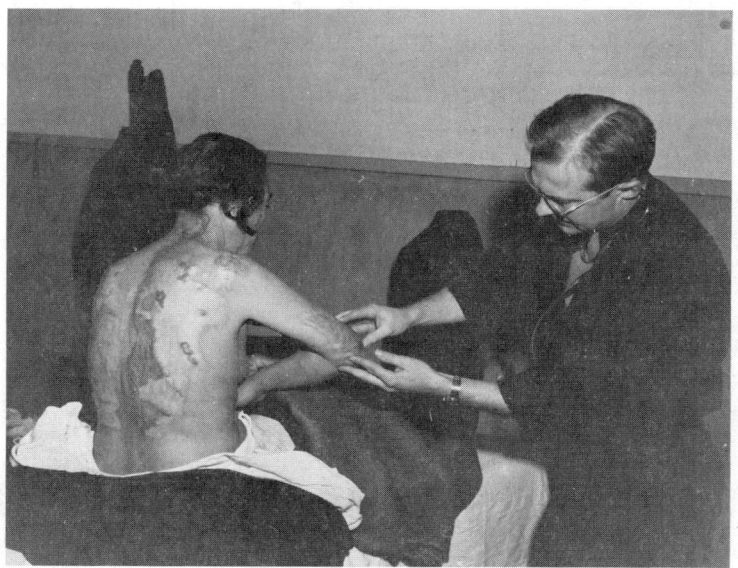

Fig. 8-9. Tama Yamada sustained burns when about 1 km. from the ground center at Nagasaki, Japan. The burns healed followed by extensive formation of heavy elevated scar tissue. She also lost all of scalp and axillary hair, scalp hair now having returned but not axillary hair. Courtesy, U.S. Army.

Radiation Biology of Ionizing Radiations

ground of the study of radiation biology. The nearer the blast, the greater the danger, but for an air blast caused by bombs like those dropped in wartime anger, it would seem that a foxhole or underwater position is a priceless advantage and the well-organized shelter, if in a cave underground, would be enormously better. The foxhole lets the infrared wave pass you by and those incapacitating burns must be avoided at all costs if the total body's mechanisms against injury and infection are to be mobilized against radiation insult. The exposure of even the top of the head is dangerous so that the hole could have to be deep enough for complete safety in this first of the three waves (Fig. 8-10).

More clothing gives a better chance against the whole disastrous event, especially a good head covering down around the neck and face. The simple difference between openly facing the blast or an axial orientation parallel to it may be death or life, in that order, and a patient in a complete plaster cast might make out far better than a short-sleeved orderly in the hospital.

The next requirement is for shock wave protection and here, the same enemy exists—*surprise*—because if you have any warning, you can be out from under the structures that will surely collapse and away from the flying glass and missiles that make serious lacerations. 75% of the buildings are erased in a whole city like Hiroshima. These two blasts

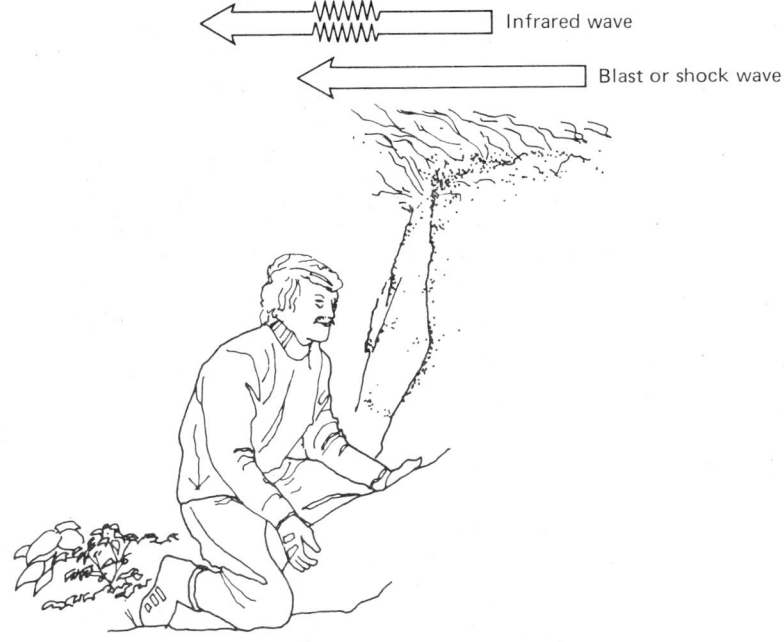

Fig. 8-10. A small ground cavity or depression which is *exactly* oriented to scatter the infrared and deflect the shock wave, is life-saving and permits the survivor to at least consider his chances against the ionizing radiation phases.

are revealing in many ways. At Hiroshima a soldier instinctively dropped to the ground with the blast. The infrared came and left burns on the uncovered part of his head below his cap. Then, in order, the air blast blew off the hat, leaving him unprotected but already saved at least from upper head burns. The real destruction of living cells comes with the rain of radioactive death, the third wave encountered. From a 20 kiloton fission bomb, 194,000 casualties in a city like Hiroshima can be expected and the weapon could be smaller or larger.

The foxhole is no good now but it did let you survive direct rays, move out of the disaster area, remove all that clothing, and start to wash off radioactive contamination with effective chemical scavenger soaps and nonradioactive water. Not eating or drinking anything that has been activated will be extremely wise. An organized community shelter may even have air reprocessing facilities to filter radioactive dust and safe water and food stores, but those above ground lack blast protection and are traps to whole populations.

Survival in the radioactive catastrophe, or after it, has many facets. First there is the analysis of the environmental results from nuclear contamination and the formation of food, water and air chains that lead the radiation to man. Many of the evaluations depend on a knowledge of the ordinary biological behavior of the elements on which must be superimposed the special isotopic effects. It is of special significance to observe the patterns of radiation illness or syndromes for recognition and diagnostic purposes. As mentioned previously, the changes in blood patterns for red and white cells, platelets, neutrophils, granulocytes and other elements are a valuable record of the dose and date of the exposure. These and the syndromes follow fairly specific levels for acute injury due to hematopoietic, gastrointestinal, and central nervous system radiation damage.

For the survivors, all the hazards are not yet past, for the longevity effects and genetic damage create permanent uncertainty. Radiologists who were exposed very early to the biological effects of ionizing radiations had abnormal offspring in numbers that might possibly be statistically significant.

There are some promising lines for protection. Vitamin E is involved in the stability of fatty acids but further research is needed to show that it protects against radiation oxidation of vital body fats. Vitamin C may be of similar potential for aqueous systems for stabilizing against radiation damage. Some agents will exacerbate the injuries by intensifying the absorption and they belong in radiotherapy against cancer where tissue destruction is the objective. Chloramphenicol, an antibiotic, for example, holds down the repair after a first exposure. Then a second one may be more serious because damaged systems, chromosome breaks in this case, can add to those from the first dose. Opportunities for repair include: 1) damage message, for example dimer formation is recognized; 2) incision, local degradation and replacement; 3) rejection of damaged area; 4) repair and rejoining with necessary regeneration of missing information. This kind of result brings about a physical dose rate dependence in which the opportunity for repair between doses is sometimes important in how a given total dose of radiation affects the organism.

Most biological injuries will depend on the dose rate so that it is important to space out the necessary exposures. A more than adequate nutritional status is obviously necessary because all the body's defensive mechanisms are immediately placed in overtime activity. Much more detailed care is then required to *manage* the metabolic interrelations so that metabolic and repair demands are not rendered incompatible by time and phase relationships that are only beginning to be understood. Information management would require that some metabolic stress activity would never be applied at a moment when a local molecular system is mobilized for a different task. When the laser beam hits the retina, for example, in detached retina operations, its effects may alter the local oxygen requirements in a manner leading to the success of the procedure.

Of all the measures that can be considered in therapy after a serious exposure, the bone *marrow* transfusion is currently the most likely. It recognizes the radiosensitivity of the hematopoietic system and the importance of the products of this system if the victim is to survive.

PROBLEMS

1. The fundamental units of ionizing radiation dose are ergs per gram and rads, which equal 100 ergs per gram. Express this unit in joules and watt-hours.
2. Give the approximate ionizing radiation from a 60-Cobalt source.
3. How can a graph of percent radioactivity against time be used to find the activity A_t at time t?
4. How does cell radiosensitivity relate to differentiation, reproduction, and primitivity?
5. What constitutes low level sensitivity to ionizing radiations?
6. What is the most convincing etiology for radiation sicknesses such as those experienced in Hiroshima and Nagasaki?
7. What is the sequence of life endangering waves from a nuclear explosion?
8. What treatment is currently most useful in cases of serious exposure to ionizing radiation?
9. What is the ionizing radiation sensitivity of plants?
10. If a person can stand 5 feet from a particular radioactive source for 10 minutes, how long can he stand 10 feet from the source according to the inverse square law?

BIBLIOGRAPHY

Bacq, Z.M. and P. Alexander, *Fundamentals of Radiobiology*, Pergamon Press, New York, 2nd Ed., 1961.
Bazin, H. and G. Doria,"The Metabolism of Different Immunoglobulin Classes in Irradiated Mice. III. Effects of supralethal doses of X-rays. *Int. Journal of Radiation Biology*, Vol. 17, 4, pp. 359-365, 1970.
Blair, H.A., Ed., *Biological Effects of External Radiation*, McGraw-Hill Book Co., New York, 1st Ed., 1954.
Cassarett, A.P., *Radiation Biology*, Prentice Hall, Englewood Cliffs, N.J., 1968.
Cohn, S.H., J.S. Robertson and R.A. Conrad, *Radioisotopes in the Biosphere*, R.S. Caldecott and L.A. Snyder, Eds., Univ. of Minn. Press, 1960.
Davies, B.R., "The New Low Level Radiation Studies Begin," *Argonne News*, Vol. 26, 4, April
Edmondson, P.W. and A.L. Batchelor, "Acute Lethal Responses of Goats and Sheep to Bilateral or Unilateral Whole-Body Irradiation by Gamma-Rays and Fission Neutrons," *Inter. Journal of Radiation Biology*, Vol. 20, 3, 269-290, 1971.
Edrich, J. and C.J. Smyth, "Millimeter Wave Thermographs as Subcutaneous Indicator of Joint Inflammation," *Proc. 7th European Microwave Conference*, Sept. 1977, pp. 713-717.
Field, S.B. and D.K. Bewley, "Effects of Dose-Rate on the Radiation Response of Rat Skin," *Inter. Journal of Radiation Biology*, Vol. 26, 3, pp. 259-267.
Fitzgerald, J.J., *Applied Radiation Protection and Control*, Vols. I and II, Gordon Breach, Science Book Co., New York, 1969.
Foresberg, A., *Advances in Radiation Biology*, L.G. Augenstein and R. Mason, Eds., Academic Press, New York, 1964.
Goldin, E.M. and R.D. Neff, "Lymphocyte Depletion in Peripheral Blood of Acutely Gamma-Irradiated Rats," *Inter. Journal of Radiation Biology*, Vol. 27, 4, pp. 337-342, 1975.
Grover, R.J., *Transfer of Radioactive Material from Terrestrial Environment to Man*, CRC Press, Cleveland, Ohio, 1972.
Hopewell, J.W. and E.A. Wright, "The Effects of Dose and Field Size on Late Radiation Damage to the Rat Spinal Cord," *Inter. Journal of Radiation Biology*, Vol. 28, 4, 325-333, 1975.
International Atomic Energy Agency, *Effects of Ionizing Radiation of the Nervous System*, 171-185, 207-224, 1962.
Ichimaru, M. and T. Ishimaru, *Radiation Research*, Vol. 49, 461-472, 563-588, 1972.
Kramer, M.W. and S. M. Michaelson, *Radiation Research*, Vol. 49, 461-472, 563-588, 1972.
Kelong, E.W., R.C. Thompson, and H.A. Kornberg, *American Journal of Roentgenology*, Vol. 71, 1038-1045, 1960.
Keyeux, A.D., et al, "Late Functional and Circulatory Changes in Rats After Local Irradiation," *Inter. Journal of Radiation Biology*, Vol. 20, 1, 7-25, 1971.
Makinodan R. and N. Gengozian, *Effect of Radiation on Antibody Formation in Radiation Protection and Recovery*, A. Hollaender, Ed., Pergamon Press, New York, 1960.
Mayneord, W.V., *Progress in Biophysics and Biophysical Chemistry*, Vol. II, J. Butler, B. Katz, and R.E. Zirkle, Eds., Pergamon Press, New York, 1961.
McCandless, J.B., *Accidental Acute Whole Body Gamma Irradiation of Human Beings A Report of Seven Cases*, Masters Thesis, Univ. of Puerto Rico, Mayaguez.
Morse, B, D. Oiuliani and E.R. Giuliani, *Radiation Research*, Vol. 60, 307-313, 1974.
National Academy of Sciences, National Research Council, *Report of the Pathologic Effects of Atomic Radiation*, Publ. 452, Washington, 1956.
Nickson, J.J., *Symposium of Radiobiology. The Basic Aspects of Radiation Effects on Living Systems*, John Wiley, New York, Chapters 21 and 23, 1952.
Pinson, E.A. and W.H. Langham, *Journal of Applied Physiology*, Vol. 10, pp. 108-126, 1957.

Quastler, H., "The Nature of Intestinal Radiation," *Radiation Research*, Vol. 4, 303-320, 1956.
Rossi, H.H. and R.H. Ellis, "Radium Therapy and Nuclear Medicine," *American Journal of Roentgenology*, Vol. 68, 980-988, 1952.
Schimid, E., *et al, Intern. Journal of Radiation Biology*, Vol. 26, 31-34, July 1974.
Silman, V., *The Basic Physics of Radiation Therapy*, C.C. Thomas Publishers, Springfield, Illinois, 1960.
Stuart, C., G.C. Kingdon and E. Balish, *Radiation Research*, Vol. 62, 145-158, 1975.
Stannard, J.N., *Progress in Nuclear Energy Series VII*, Pergamon Press, New York, 70-82, 1959.
U.S. Department of Commerce, *Safe Handling of Radioactive Isotopes*, National Bureau of Standards Handbook no. 42, Washington, D.C., 30 pp.

Films

Return to Bikini, see Chapter 7.

9

Unraveling the Information Routes of the Ionizing Radiations

> Let a full fair tide
> Bear information on . . .
> Those solutions are known
> Whose problems abide.
>
> D.A.C.

ACTIVATION

Ionizing radiations are both characteristically emitted *by* radioactive substances and capable of making other substances radioactive. The word activation used in this sense means the process of making elements radioactive. The term activation analysis is applied to studies which may be made once the elements have been rendered radioactive, but activation itself is a much broader term of more primary nature than analyses which may subsequently be carried out at any time. Conversely, a delayed activation of Napoleon's hair in 1961 showed lethal concentrations of arsenic. Activation involves nuclear reactions as a result of which radioactive decay and the emission of radiation occurs. The atoms are said to disintegrate at a rate of a certain number of disintegrations per second. These emissions are detectable with transducers such as Geiger-Muller counters or scintillating crystals which, with the help of readout systems, give the rate of decay in terms of counts per unit time (Fig. 9-1). Then, a curie of radioactivity is 3.7×10^{10} disintegrations per second and, if 100 per cent of these emissions would be collected and counted in some very special counter, then this same number of counts would be obtained in a second. The same emissions of radiation can be considered as incoming to some other target. There, if they have the proper characteristics including energy, they will activate it and cause very important transformations in its atoms.

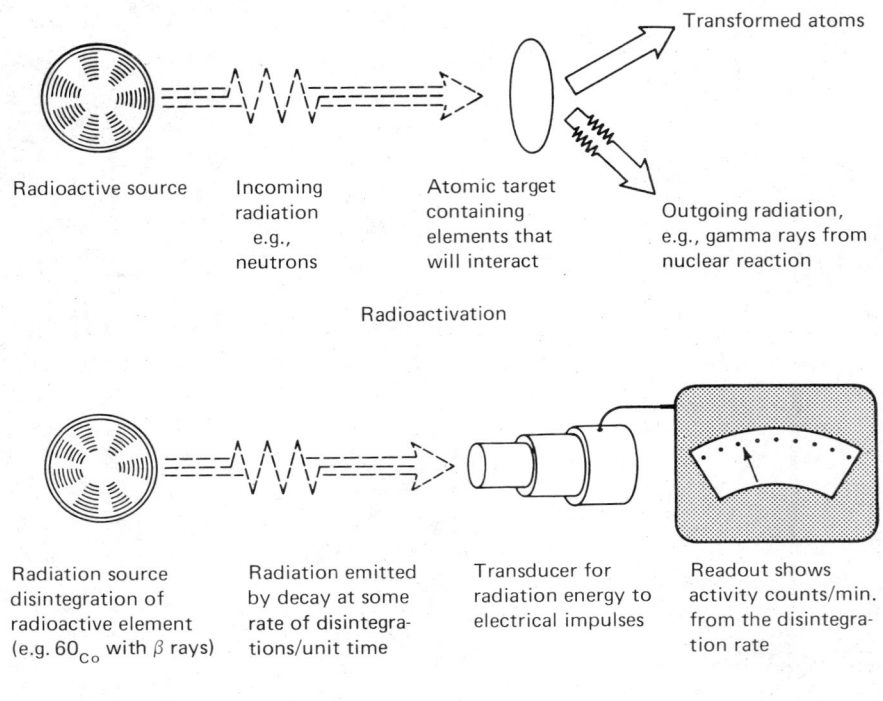

Fig. 9-1. Radioactivation and counting. Radioactive substance has a decay rate measurable in disintegrations per unit time or counts per minute (familiar audible signals click-click-click-click-click, furnished by GM counter. Source of count rate is decaying element and its disintegrations per unit time.

COLLISIONS

Looking at the interaction from the standpoint of an incident radiation particle or beam, it is seen that when this particle or beam hits the nuclear target, it scatters and undergoes a change in energy while the nucleus transforms. It turns out that the nature of this collision is of great significance because it determines the energy distribution and, therefore, the information that is carried to another site. Biological materials respond primarily to this energy—carried information—in what may be a life or death manner.

Elastic collisions, which are by far the most common, leave the energy unchanged, that is, the total kinetic energy remains constant as shown in Fig. 9-2 where the two billiard balls rebound elastically after collision. This type of collision can be readily examined because the relative velocity and moments are the substantial quantities. The collision may only change the directions while leaving the velocity the same,

Information Routes of Ionizing Radiations 361

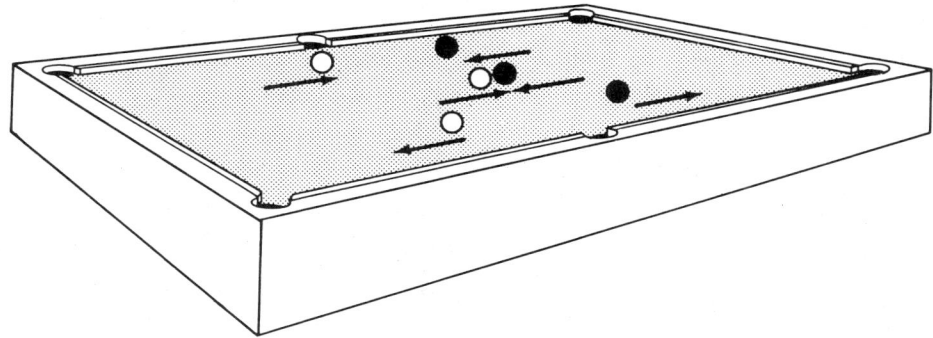

Fig. 9-2. Elastic (billiard ball) collision.

that is, the separation velocity equals the incoming velocity. While these collisions do not produce transformed nuclei, they are very useful. In a reactor, elastic collisions between moderator nuclei and neutrons produce moderated or slower neutrons capable of causing nuclear reactions such as the activation of sodium-23 (Fig. 9-3).

This sequence of events can be visualized and recorded in a bubble chamber which bears a fanciful relationship to a glass of beer which is subjected to a linear disturbance such as withdrawal of a wire, the "equivalent" of intercepting radiation. The actual test equipment is shown in Fig. 9-5. The entering neutron or proton sees a concentration of liquid hydrogen atoms in its pathway and establishes a series of collisions with them before it exhausts its initial energy. The route followed by the particle may be like that shown in Fig. 9-4.

In such collisions, some of the incoming energy may be interchanged, making an inelastic collision (Fig. 9-6). This is illustrated by the dart striking and entering the soft ball (such as plastic putty) and the two continuing on together. The masses involved in these events have commensurate physical values. The collisions between noncommensurate objects may be far less *revealing*, for example, when a tree is struck by a hammer. The redistributed energy must be accounted for and the larger or more energetic the incoming particle or radiation is, the greater the energy to be accounted for.

While the discussion is about ionizing radiations, we may compare a nonionizing radiation, infrared, as it impinges on a molecule. The internal energy changes inelastically via rotational or vibrational kinetic energy. Thus the energy imparted by the impinging radiation excites the molecule, and the excitation energy will be measured by that lost from the incoming radiation. Since such energy changes are marked by corresponding frequency changes, in this case a lower frequency, radiation is *scattered*. An excited molecule may lose excitation energy to incoming radiation and the scatter will be of higher frequency. This is the type of change seen in the Raman Effect when the incoming radiation may be a laser beam. Thus the energy hf, and the frequency,

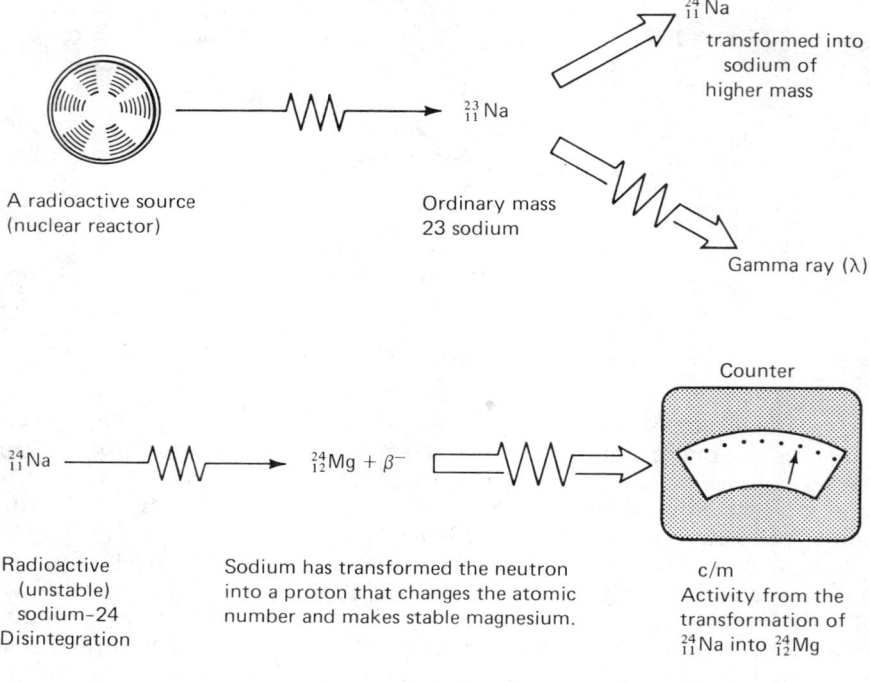
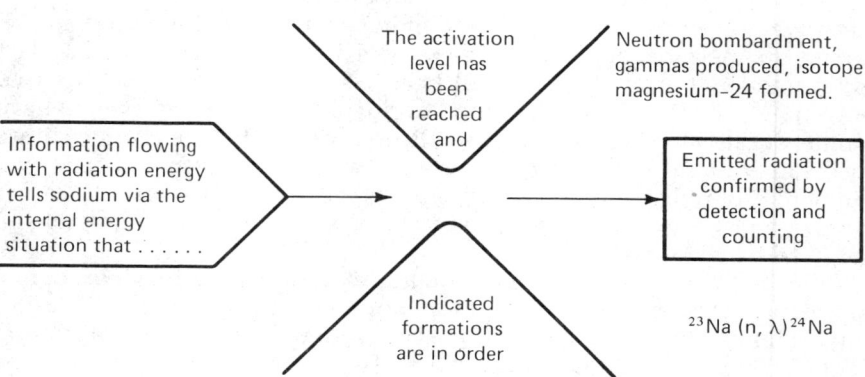

Fig. 9-3. Information and nuclear reactions. Information flows with the radiation energy, and the energy level and identities determine interactions.

Information Routes of Ionizing Radiations 363

Fig. 9-4. Photo of bubble chamber showing instrumentation. Right, a place where bubbles are made in a somewhat similar manner.
72-Inch Liquid Hydrogen Bubble Chamber at the Lawrence Radiation Laboratory, Berkeley Campus, Univ. of California

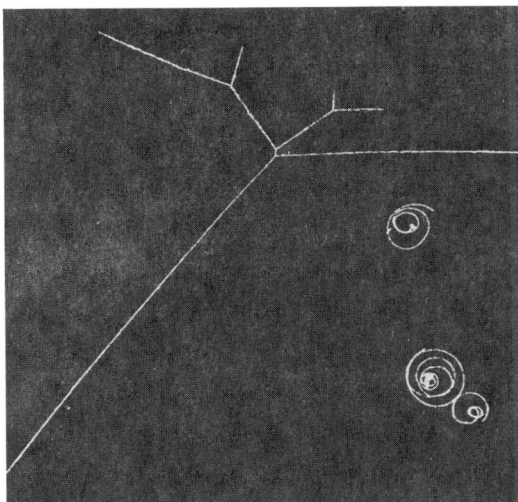

Fig. 9-5. Multiple proton scattering in the bubble chamber. A proton enters the chamber (filled with liquid hydrogen) and bounces off successive protons very much like billiards. Courtesy, Lawrence Berkeley Laboratory.

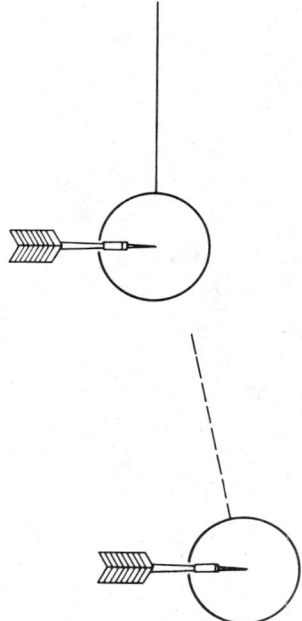

Fig. 9-6. Inelastic (dart and ball) type collision. In the inelastic collision, the combined objects take on the sum of the moments which each one had separately and they stick together. If a beta particle transfers its kinetic energy to an atom, the energy may be transformed into thermal or photo energy.

E/h, are closely related, and a frequency change may be described as going from hf^1 to hf^2.

Super ionizing radiations, greater than about 8 MeV from betatrons and other high energy generating machines, can activate an entire animal so that it is easily seen to be radioactive. Sometimes the large 25 MeV betatron is used in cancer radiotherapy and about 5% of the dose to the patient comes from his own activated carbon 11 made radioactive from carbon 12 after the x-radiation. Slow neutrons will activate metal in an exposed person, and in one case, the radiation made a gold tooth so radioactive that it caused lesions of the gums.

Particles like neutrons may thus interact with a material by elastic or inelastic collision or they may fragment the nuclei or be captured there. These opportunities lead to a sort of target analysis type of reasoning. If the area of nuclear material in the path of incoming particles or radiation is S cm^2 and it has a thickness $\triangle x$ and V nuclei per cm^3, then the target areas exposed to the incoming particles are given by the cross section per nucleus σ in cm^2 and the other values according to the product

$$\sigma V S \triangle x \qquad (156)$$

In this product, σ is the constant of proportionality for the radiation.

If the radiation particles are neutrons, it will be the sum of cross sections for elastic, inelastic, and absorptive interactions.

The cross section is also spoken of as a probability.

$$P = \sigma \, V \, \triangle x$$

When this quantity, evaluating the interacting target, is divided by the whole target area S, the probability *is* expressed that a hit will occur.

If the nuclear reaction yield is to be evaluated, then the cross section σ is measured in terms of the probability that some nuclei will be transformed. Alternatively, the absorption of incoming radiation particles may be the feature of interest. The particle concentration or flux is then measured as it changes by interaction.

Then an atomic mass M, density rho, and Avogadro's number give the nuclei per cm^3 that are exposed. The flux is measured by appropriate neutron dosimetry. With charged particles on the other hand, the flow constitutes an accurately measurable current. Considerable space in the nuclide blocks on the Chart of the Nuclides is devoted to σ with various subscripts and units. Following the same target concept, the probability of neutron interaction with nuclei is expressed. Earlier, in fission studies, the unit was given the whimsical name *barns* per atom, referring to the ability to hit something. This is a cross sectional area of only 10^{-24} cm^2 and even smaller targets may be shown as mb, milli or μb, microbarns which are 10^{-27} and 10^{-30} cm^2/atom respectively. One probable reaction is neutron capture followed by energetic gamma emission designated σ_r and berylium-9 has this mode with neutron cross section 9 mb and Resonance Integral 4 mb. This is an integral of resonant, i.e. particular, neutron energies in eV where interaction is exceptionally probable. These neutron energies are fast, intermediate, and thermal (about 0.025 eV) and are produced in the moderator by the atomic collisions. The flux in the moderator when there is no absorption is inversely proportional to the energy which is then called $1/E$ dependance for neutrons. Velocity enters into the kinetics also so that with absorption and a room temperature moderator, the neutrons move at 2200 meters/sec. Often a target nucleus will have some corresponding cross section and the greater the velocity, the less the absorption cross section so the dependence is 1/vel.

NUCLEAR REACTIONS

Nuclear reactions came on the scene after considerable experience had been gained with ordinary chemical reactions. The difference is that the former deal with the nucleus while in the latter, interactions between the electron structures are involved. The isotopes of elements are called nuclides, which are arrangements of atoms that are stable long enough to be identified. More specifically, Truman P. Kohman gave this term to atoms characterized by their neutrons and protons, so there are both stable and radioactive ones.

There are isomer nuclides with radioactive properties that differ within long lived states, that is, have different radioactive properties in different energy states of long lifetime. There are 264 stable natural elements, 548 isomers and 1838 nuclides known. Some 66 nuclides are natural, mainly among the heavier elements. The Chart of the Nuclides *is* the current status of discoveries in this area of research. Although isotopes can have variable mass they have the same chemical features since the latter are based on the electron structure. In nuclear notation the isotope is then defined with its mass number A above and its atomic number Z as a subscript, both to the left of the symbol. Examples are:

$$_Z^A H \quad _1^1 H \quad _1^2 H \quad _1^3 H \quad _6^{12} C \quad _6^{14} C \quad _7^{14} N \quad _{19}^{39} K \quad _{15}^{32} P \quad _{20}^{40} Ca \quad _{20}^{45} Ca \quad (158)$$

Thus Z for N means that there are seven protons in the nucleus. The notation A then means that nitrogen has fourteen main nuclear particles consisting of seven neutrons and seven protons. These heavy particles determine the mass so the nitrogen atom has a mass number of fourteen. Thus

$$A = Z + N \qquad (159)$$

where N is the number of neutrons. The mass number is the familiar number for the element from chemistry such as 12 for oxygen, and 23 for sodium. The Z value of particles is often encountered in work when the ionization track is photographed. In the film, a heavy bold track indicates a particle having a larger Z, perhaps up to 10 or even 20.

Nuclear reactions permit changes in these numbers and thus transmutations from one kind of atom to another. Neutrons, protons, mesons, deuterons, alpha particles, and many nuclei may have obtained the necessary energy by acceleration or otherwise to cause, on collision, a nuclear reaction. With a collision where a neutron, e.g. from a reactor, has appropriate energy, the neutron may be captured in a hydrogen nucleus. Now it contains two mass units and is an isotope of hydrogen, deuterium. Its nucleus alone without the orbiting electron is a convenient particle for acceleration and directing into nuclear collisions, the deuteron (d). A balance of particles given for this reaction must account for the change in the energy of the incoming neutron. This is done by showing that a gamma ray is given off. An unusual set of *radiation particles* and a ray are involved.

$$n + p \Rightarrow d + \gamma \qquad (160)$$

If a hydrogen nucleus or proton collides with $_3^7 Li$, the nuclear reaction is:

$$_3^7 Li + _1^1 H \Rightarrow _4^8 Be + \gamma \qquad (161)$$

or if it collides with neutrons instead,

$$_3^7 Li + _0^1 n \Rightarrow _1^3 H + _2^4 He \qquad (162)$$

which liberates nuclear energy in the form of an alpha particle, these being bare helium nuclei. Water is the most common biological material and so, the isotopic forms of water are often used in tracer experiments. One isotope appears in tritiated water which is composed of two tritium atoms $^{3}_{1}H$ and oxygen and is

$$^{3}H_2O \text{--}\mathcal{W}\text{--} \beta^{-} \tag{163}$$

radioactive, giving off soft (low energy) beta radiation which can be detected by liquid scintillation counting. The tritium nucleus has two neutrons and a proton. Deuterated water has two deuterium $^{2}_{1}H$ atoms instead and is called heavy water. This form of water ($^{2}H_2O$ or D_2O) with stable deuterium has been useful since the earliest experiments with isotopes and it is detected by virtue of its mass difference from ordinary water. Thus it is often called heavy water and may show differences, the mass or isotope effect, in its biochemical behavior, especially in the kinetics of reactions in which it is used. It is possible to replace almost all the water in a cell with heavy water.

The energized helium nucleus from the reaction above of the lithium with a proton acquires enough energy to proceed as an alpha particle or ray for a distance of a few centimeters. However, the particles are stopped via absorption by inserting even a sheet of paper in their path. While this penetration seems to be very limiting, it must be remembered that the alpha radiation energy is, as a consequence, deposited in a very concentrated (densely ionizing) way or there is a high linear energy transfer (LET). In the information sense, the presence of these high LET rays is sufficient to completely alter the nuclear information and pave the way for massive redistributions or changes in genetic data by recombinations and crossing over. There will be chromosome responses after the disruption, as ends join and heal. Tritium can be formed by neutron irradiation of lithium which in shorthand is

$$^{3}Li(n, \alpha)\ ^{3}H \tag{164}$$

with production of an alpha particle.

$$^{6}_{3}Li + ^{1}_{0}n \Rightarrow ^{3}_{1}H + ^{4}_{2}He \tag{165}$$

If the alpha particle is able to collide with a nitrogen atom, transmutation occurs and a proton particle is released

$$^{4}He + ^{14}N \Rightarrow ^{17}O + ^{1}H \tag{166}$$

This is the nuclear reaction

$$^{14}N(\alpha, p)\ ^{17}O \tag{167}$$

The former reaction illustrates artificially made tritium. Deuterium, on the other hand, may be separated according to its natural abundance from ordinary water by electrolytic enrichment. A substitution reaction on tritium by accelerated deuterons can change it to helium.

$$^{3}H(d, n)\,^{4}He \tag{168}$$

$$^{3}_{1}H + ^{2}_{1}H \Rightarrow ^{4}_{2}He + ^{1}_{0}n \tag{169}$$

THE CHART OF THE NUCLIDES

Armed with these preliminary guideposts, it is possible to use the map of the isotopes, or, more properly, the Chart of the Nuclides. (A 29" x 49½" Chart of the Nuclides is available from Educational Relations, General Electric Company, Schenectady, New York 12345 for $4.00.) This is a display of the isotopic form of all the elements and their relationships by atomic mass, atomic number and the number of neutrons. Thus the nuclides are given a block with the necessary data and all the blocks form a wide upward flowing diagonal, which must be long enough to accommodate some 1838 nuclides, so that it is separated into three diagonals. These, in turn, may be further separated into convenient subsections for closer examination. Table 9-1 begins the study of the chart by listing all 103 of the atomic elements. The nuclides have their neighbors chosen in such a way that their orientation, up, down, left, right, follows the nuclear process and prescribes the product as shown in Fig. 9-8. The chart shows the displacements caused by neutron activation processes on the original nucleus. Thus lithium-6 shown above is displaced one block left and three down to become hydrogen-3 in the (n, alpha) nuclear reaction (Fig. 9-9). A stable isotope is indicated by shading, deuterium being one example (Fig. 9-11), and a stable fission product appears with shading and a small black triangle as in Fig. 9-12. The map uses a bold black outline around the block to indicate what is the ordinary nuclide

Table 9-1. The 103 Atomic Elements by Atomic Number Z, the Protons in the Nucleus, Chemical Symbol and Name

Atomic Number	Symbol	Name	Atomic Number	Symbol	Name
0	n	neutron	6	C	carbon
1	H	hydrogen	7	N	nitrogen
2	He	helium	8	O	oxygen
3	Li	lithium	9	F	fluorine
4	Be	beryllium	10	Ne	neon
5	B	boron	11	Na	sodium

Atomic Number	Symbol	Name	Atomic Number	Symbol	Name
12	Mg	magnesium	58	Ce	cerium
13	Al	aluminum	59	Pr	praseodymium
14	Si	silicon	60	Nd	neodymium
15	P	phosphorus	61	Pm	promethium
16	S	sulfur	62	Sm	samarium
17	Cl	chlorine	63	Eu	europium
18	Ar	argon	64	Gd	gadolinium
19	K	potassium	65	Tb	terbium
20	Ca	calcium	66	Dy	dysprosium
21	Sc	scandium	67	Ho	holmium
22	Ti	titanium	68	Er	erbium
23	V	vanadium	69	Tm	thulium
24	Cr	chromium	70	Yb	ytterbium
25	Mn	manganese	71	Lu	lutetium
26	Fe	iron	72	Hf	hafnium
27	Co	cobalt	73	Ta	tantalum
28	Ni	nickel	74	W	tungsten
29	Cu	copper	75	Re	rhenium
30	Zn	zinc	76	Os	osmium
31	Ga	gallium	77	Ir	iridium
32	Ge	germanium	78	Pt	platinum
33	As	arsenic	79	Au	gold
34	Se	selenium	80	Hg	mercury
35	Br	bromine	81	Tl	thallium
36	Kr	krypton	82	Pb	lead
37	Rb	rubidium	83	Bi	bismuth
38	Sr	strontium	84	Po	polonium
39	Y	yttrium	85	At	astatine
40	Zr	zirconium	86	Rn	radon
41	Nb	niobium	87	Fr	francium
42	Mo	molybdenum	88	Ra	radium
43	Tc	technetium	89	Ac	actinium
44	Ru	ruthenium	90	Th	thorium
45	Rh	rhodium	91	Pa	protactinium
46	Pd	palladium	92	U	uranium
47	Ag	silver	93	Np	neptunium
48	Cd	cadmium	94	Pu	plutonium
49	In	indium	95	Am	americium
50	Sn	tin	96	Cm	curium
51	Sb	antimony	97	Bk	berkelium
52	Te	tellurium	98	Cf	californium
53	I	iodine	99	Es	einsteinium
54	Xe	xenon	100	Fm	fermium
55	Cs	cesium	101	Md	mendelevium
56	Ba	barium	102	No	nobelium
57	La	lanthanum	103	Lr	lawrencium

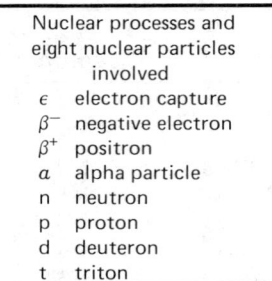

```
Nuclear processes and
eight nuclear particles
       involved
  ε    electron capture
  β⁻   negative electron
  β⁺   positron
  α    alpha particle
  n    neutron
  p    proton
  d    deuteron
  t    triton
```

Fig. 9-7. Nuclear particles.

		³He in	α in
β⁻ out	p in	d in	t in
n out	Original nucleus	n in	
t out	d out	p out	$β^+_ε$ out
α out	³He out		

Fig. 9-8. Legend for neighboring nuclides showing how various nuclear processes give products in specific chart locations relative to the original nucleus.

such as hydrogen-1 or carbon-12 as in Fig. 9-10. An artificially radioactive one is set in a plain block with radioactivity data such as its half life as seen in Fig. 9-13. If it is naturally occurring and radioactive, it will have a solid bar of black spanning the top of the block, for example vanadium-50 in Fig. 9-14. Polonium-218 in Fig. 9-15 has a second short bar which indicates its presence in the decay scheme of the nuclide shown in the little bar. If there are daughters in the scheme, the block is divided and appropriate data given as in Fig. 9-17. Cobalt-60, for example, has the two isomeric radioactive states given in that block. The symbols used in charting the nuclides are summarized in Table 9-2.

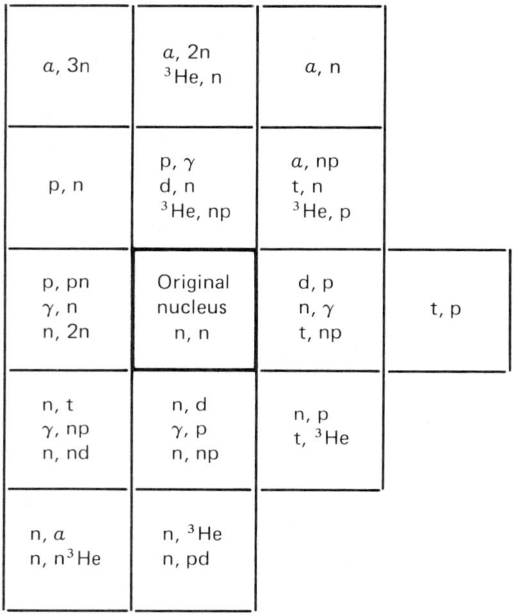

Information Routes of Ionizing Radiations

	a, 2n		
a, 3n	^{3}He, n	a, n	
p, n	p, γ d, n ^{3}He, np	a, np t, n ^{3}He, p	
p, pn γ, n n, 2n	Original nucleus n, n	d, p n, γ t, np	t, p
n, t γ, np n, nd	n, d γ, p n, np	n, p t, ^{3}He	
n, a n, n^{3}He	n, ^{3}He n, pd		

Fig. 9-9. Guide to neighboring nuclides according to the nuclear reactions involved. Nuclear bombardment processes cause the original nuclear condition to be displaced according to the particular process.

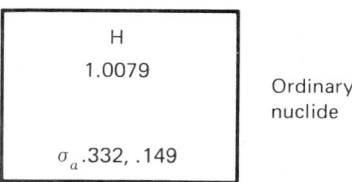

Fig. 9-10. Ordinary nuclide.

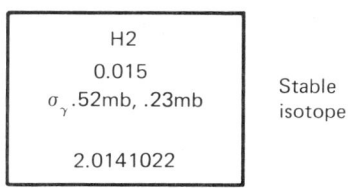

Fig. 9-11. Stable isotope.

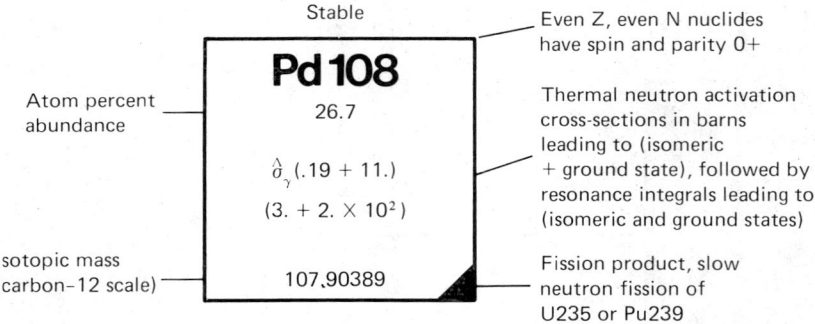

Fig. 9-12. Stable fission product.

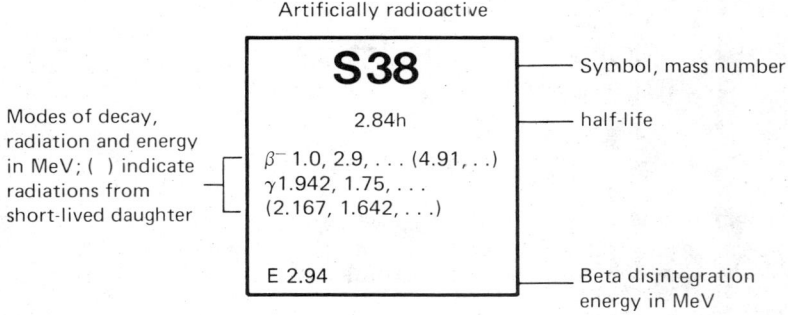

Fig. 9-13. Artificially radioactive.

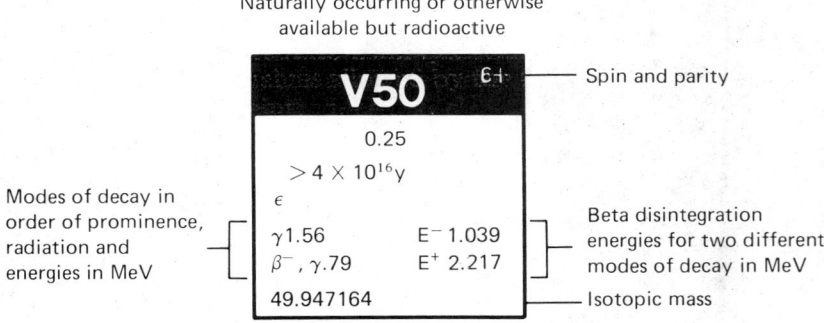

Fig. 9-14. Radioactive.

Information Routes of Ionizing Radiations 373

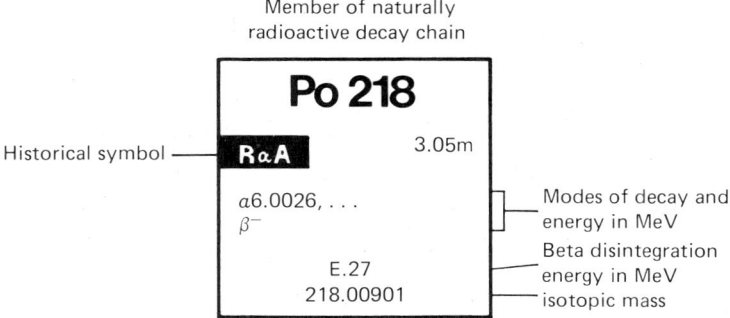

Fig. 9-15. Decay Chain Isotope, RaA.

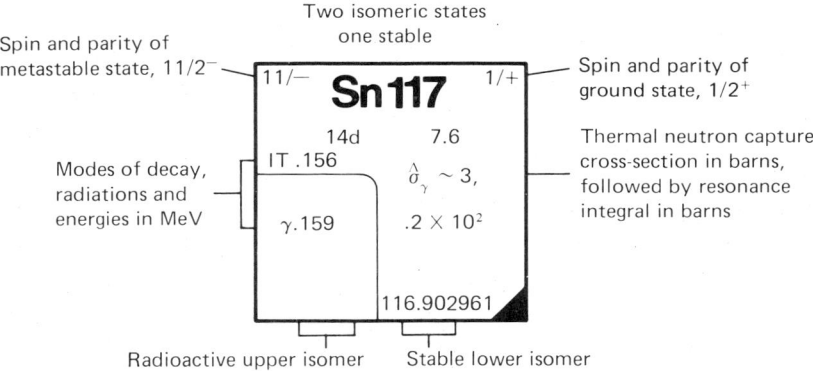

Fig. 9-16. Two state isotope, one stable isomer.

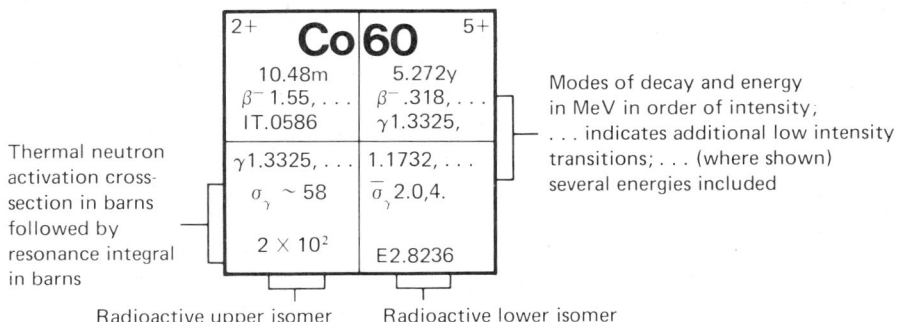

Fig. 9-17. Two state isotope, both radioactive isomers.

Table 9-2. Symbols Used for Charting the Nuclides

α	alpha particle
β⁻	negative beta particle
β⁺	positive beta particle
γ	gamma ray
SF	spontaneous fission
n	neutron
p	proton
e⁻	internal conversion electron receives nuclear energy, is ejected
ε	electron capture from K shell
IT	isomeric transition
D	radiation delayed
E	disintegration energy
μs	microsecond
ms	millisecond
s	second
m	minute
h	hour
d	day
y	year

The chart uses four colors to add to the visual display of information. In the lower half of the block, a warm color, orange, is associated with strong neutron absorption by the nuclide while, in other cases, a cooler absorption is indicated by blue.

Color Code	Neutron Absorption—Barns
Orange	over 1000
Yellow	500 to 1000
Green	100 to 500
Blue	10 to 100

The upper half uses a lighter shade of the colors to relate half lives and activity with highest radioactivity shown by orange which would consequently have the shortest half-life.

Color Code	Half-lives
Light orange	1 to 10 days
Yellow	10 to 100 days
Green	100 days to 10 years
Blue	10 years to 10,000 years

The symbolism in the blocks is necessarily guided in the direction of brevity so that some explanation is necessary. There is, for example, the question of the isomeric states. Among these will be both long and short lived ones but in nuclear terminology, long means microseconds or longer. The ground state is the probably stable, lower energy state in contrast to the isomeric higher state. Parentheses are used to separate parent and daughter radioactivity. The nitrogen-17 parent with a half life of 4.16 sec. as an example produces, via negative beta emission, its accompanying daughter, short-lived oxygen-17 which is a neutron emitter so these neutrons are delayed and in parentheses.

One popular isotope for standardization of some counters is cesium-137 with half-life 30.1 years. Its short lived daughter is 2.55 min barium-137. The cesium decays by negative betas but it is the barium isomer that emits the 0.66164 MeV gammas that are often of special interest in the laboratory and this situation is indicated in the block by a "D."

Some nuclide blocks represent the ground *state* and show the *Spin* and *Parity* of the corresponding energy level. Others in isomeric *states* have these numbers in an upper corner. In a way similar to the electron, neutrons and protons have intrinsic angular momenta of ½ in units of $h/2\pi$ where h is Planck's constant, and this in combination with their orbital angular momenta give the *Nuclear Spin*, the resultant angular momentum. The orbital angular momentum is zero or an integral multiple of $h/2\pi$ so the nuclear spin is integer or half odd integer, according to whether the nucleus has an even or odd number of nucleonic particles. The *state* may give the atomic system an even (+) or an odd (−) parity. In these numbers, if the spin is ½, the blocks show only 1/ for brevity and the sign, that is, 1/+. If it is in the ground state and even, the data omit spin and parity 0+. In a practical way, the energy states and radiation emission are illustrated by the production of gammas. This emission requires a sizeable change in angular momentum to connect the states, energywise, or if they connect with a small energy difference, it means a long lifetime, or metastable state.

In these numbers, there is an assist from wave mechanics for the classical Bohr atom which is the system that can be visualized. The wave concepts are capable of defining permissible angular momenta for the atomic particles. On the wave side, it helps to approximate the picture by referring to a standing wave pattern and realizing that here too, there is a resonant system. The concepts of say, electrons in certain orbits according to wave mechanics and that of standing waves in a rope held at both ends and showing various nodes and maxima (wavelengths), can be used to explain the even multiple of $h/2\pi$ requirement for the angular momentum of an electron in a permitted orbit. In the traditional manner, the angular momentum of the particle is simply the product of its linear momentum times the distance r from the axis of rotation. The linear momentum is the mass times the tangential velocity.

From the emphasis on neutron absorption cross section, it may be surmised that neutron induced reactions which cause various particle or radiation emissions are very common. Another neutron reaction is resonance capture

$$A_x(n,r)A_{1x} \tag{170}$$

The fission mentioned previously will result in fragmentation in size

$$A_X \genfrac{}{}{0pt}{}{\rightarrow A_{1Y}}{\rightarrow A_{2Z}} \tag{171}$$

in which the fragments are equal, Y = Z, or more probably they will be asymmetrical. Fission is a thoroughly variable process with many products. Uranium-235 fissions by thermal neutrons to produce products from zinc with atomic number (protons) 30 to dysprosium, with 66.

The nuclear reactions are then determined by the incident energy, the nature of the particle, and the identity and features such as cross section of the target. Particle reactions are indicated with parenthetical abbreviations such as

$$(n, \alpha), (n,p), (p, \beta^+), \text{ and } (p, \alpha) \tag{172}$$

The growth in internal energy when particles enter results in excitation and emission. In reactions caused by deuterons, they can lose a particle in the collision. There are also heavy ion particles, energetic gamma, and alpha reactions, such as (α, p) and (α, n), and proton, neutron (p, n) in which a proton is incoming and neutrons are produced.

THE DECAY SCHEME

The decay mode permits the unstable nucleus to accommodate itself to the required change. For example, if there are too many protons in the nucleus compared with normal isobars (units of the same mass) *positron emission* occurs and this is the mode of disintegration. This decay involves the formation of a positive beta particle (β^+) which is capable of removing the proton charge when an excess nuclear proton changes into a neutron (Z is reduced by one). If instead, the negative beta particle (β^-) is emitted, a neutron has changed into a proton. The (n, γ) reaction is an instance where the nucleus has been able to accommodate a new neutron by emitting a gamma ray. Positrons can also enter into the annihilation process. At the end of its path of ionization a β^+ meets a β^- and the result is merely two quanta.

$$E = mc^2 \tag{173}$$

$$2 \text{ quanta} = 2\beta \text{ masses} \times c^2 \tag{174}$$

$$E = 0.51 \text{ MeV} \tag{175}$$

The decay scheme of a radioactive nuclide results from such accommodations in the interest of achieving a stable nucleus. The phosphorus-32 isotope decays by negative beta emission but with no gammas, and has a half-life of 14.2 days (Fig. 9-18a).

Information Routes of Ionizing Radiations

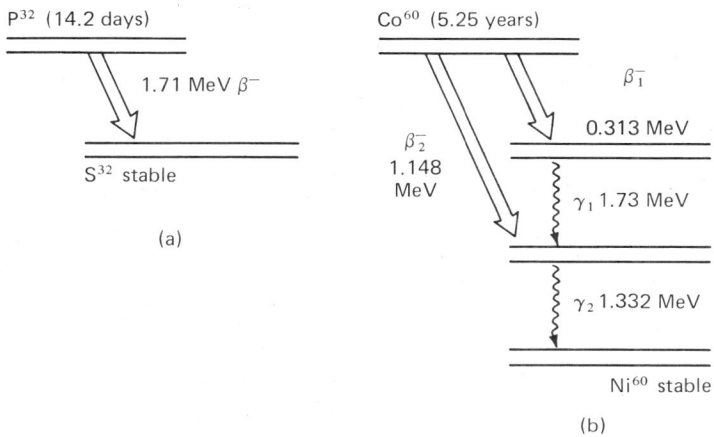

Fig. 9-18. Decay schemes with single and alternate pathways.

The cobalt-60 isotope (Fig. 9-18b) can decay by either of two beta and gamma routes with a more or less constant energy release to the stable daughter element, nickel-60.

The nuclide chart is the source for all of this kind of information. The schemes for nuclear processes and products may be readily associated with the decay chains for elements such as parent uranium-238. Then it can be followed through a natural sequence involving alpha emission to its daughter, thorium 234 *down to the left*. When this emits its negative electron, it increases its atomic number as the neutron becomes a proton so the change is *up one and left one* space to protactinium-234 with its isomer states. From here it is on via negative beta emission *to the left and up* to uranium-234. This decays by alpha emission to thorium-230 and the emission is repeated to yield radium-226. Three more alpha emissions lead to radon-222, then polonium-218, to lead-214. This is an unstable lead which gives off negative electrons to become bismuth-214 after which alpha decay leads on to lead-210. Alternatively, bismuth-214 goes to lead-210 via alphas by way of thallium-210 and the beta process is repeated to polonium-210 after which alpha emission produces stable lead-206.

THE CHART OF THE NUCLIDES AS AN ATOMIC AGE CONTRIBUTION

A careful study of the nuclide chart is extremely profitable, and taken in conjunction with the Periodic System of the Elements, it can serve to unify and concentrate a very large body of atomic age information indeed. If, to these two, is added the body of information represented by the Electromagnetic Spectrum, the reader will have acquired a working knowledge of both energy and materials suitable for progress even in the direction of biomolecular engineering. This reasoning has led to the discussion of all three in this text.

As it looks today, the nuclide chart represents the accumulated work of an army of scientists and is certainly one of the greatest of scholarly concepts. The data in this remarkable chart came from the large body of data needed for atomic fission and subsequent discoveries in nuclide technology and was prepared by G. Friedlander and M. Perlman. It was revised by F. William Walker, G. J. Kirouac and F. M. Rourke from the version of D. T. Goldman and the work of the General Electric Company for the United States Atomic Energy Commission. It is sectionalized in a 29-page booklet available from the General Electric Company, 175 Curtner Avenue, Mail Code 685, San Jose, Claifornia 95125 for $4.00.

Radioisotope work in biology requires the kind of information provided, for example, for half-lives and radiation energy. The chart presents most of the data needed in planning a given experiment for estimating the required amount and type of radioactivity and evaluating the biological hazard. Frequently, the biologist will be involved in radioactive dating as with carbon isotopes or in studies of primeval energy sources, the primitive reactors, origin of the planet and formation of the solar system. It will be useful in nuclear environmental studies, radiotherapy, medical uses of isotopes, activation analysis and theoretical work on stellar synthesis of elements.

An enormous amount of radiochemistry and mass spectra work went into the little blocks with the small black triangles which distinguish fission products. All of the fission modes such as asymmetrical, binary, and ternary have their significance in reactor design and there the final decision is biological as there are many elements of regulation. The chart has the keys to disaster too in the data on nuclides involved in nuclear explosions.

ATOMIC AND THERMONUCLEAR EXPLOSIONS

A nuclear explosion is a nuclear reaction with a very large energy difference. Controllable ones could produce a reservoir of energy. The atomic detonation is the fission of heavy uranium nuclei into the smaller fragments like krypton and barium with the release of enormous energy. This tremendous energy can be used to momentarily create fusion conditions for highly energetic but lightweight hydrogen atoms to produce helium as they do inside the sun where temperatures of about 10 million degrees centigrade exist. The energy then released is familiar as that of the hydrogen or thermonuclear bomb. The atomic explosion looks like that in Fig. 9-19 taken of the Baker test at Bikini atoll in the South Pacific. In this test a 20 kiloton bomb was exploded 200 feet below the surface of the water. Earlier pictures show a rather large ship on the side of the column of water evidently behaving like a rocket bound for the moon. The plume carried about a million tons of water skyward. Its walls were 300 feet thick, it rose about 8,000 feet and served as a stack for venting the radioactive, underwater gas bubble. Around the top of the plume was the broad cap or mushroom. These tests were made underwater, in the air, and on the ground. The bombs dropped on Nagasaki and Hiroshima were air blasts and fission-type weapons. Since then, many of the tests have been of hydrogen bombs, that is, thermonuclear or fusion explosions.

The hydrogen bomb is euphemistically described as clean or having little fallout but the trigger explosion of a fission device caused the largest radioactive rain ever seen when the multimegaton test occurred at Marshall Island's Bikini atoll (March 1, 1954). It is possible to design atomic weapons for *all degrees* of destruction and violence. The neutron bomb or enhanced radiation weapon exploits the sudden massive release of neutrons in the fusion of light atoms. The selection of a "mini" H bomb of about a kiloton enhances the information aspect or at least the directivity of the fusion weapon, because it discriminates between targets. The neutrons will activate, be captured by, or be elastically or inelastically scattered from *a given atom* according to the neutron absorption cross section for capture and activation and its energy. Thermal neutrons are captured; the intermediate or resonance energy neutrons are elastically scattered or captured, fast ones are elastically scattered, and relativistic neutrons over 10 MeV interact by spallation. Carbon moderates neutrons by inelastic resonance scattering, emitting photons in the process. Hydrogen which is of a mass about equal to the deuterium-tritium of fusion devices moderates neutrons by elastic scatter.

It is easily seen that neutron reactions can saturate a small planetary area of impact. These reactions will easily destroy all life and leave property radioactive but otherwise intact if air blast is used. This extermination effect is extremely attractive to countries that feel menaced by potential enemies who could muster thousands of conventional attacking units on land or sea. In the U. S., for example, the defensive use of the large yield H bombs would devastate whole sections, for an unattractive, sacrificial kind of victory. The enhanced radiation weapon is a more selective alternative, since it would send penetrating neutrons in particular through the armor plating or bunkers of attacking units to neutralize whole groups by unseen rays and spare the battle area outside of 140 yards from the blast, infrared damage, and radioactive rain of large yield H bombs.

Radiation death would come soon from an airblast at 300 feet for individuals within 1000 yards and radiation sickness would develop in persons out to 1¼ mile. This calculated effect could come from an eight inch shell (according to results from the Nevada underground tests in 1963) and the shell or rocket could be fired from 10 to 75 miles distance. Like the other great nuclear sources, the neutron weapon would have gamma and particle radiation, gradients of intensity around the burst, high and low exposure zones, and corresponding biological effects. The enhancement of neutrons is at the expense of the other

380   *Informational Bioelectromagnetics*

Fig. 9-19. The atomic bomb. a) First atomic bomb test at Bikini Lagoon, 1 July, 1946. Looking like a giant cauliflower head suspended to an ever-stretching neck, Bikini's billowing cloud of smoke and flame was caught in its various stages of formation by a Navy patrol bomber flying just beyond range of the deadly explosion. These pictures were taken within several minutes of the detonation

Information Routes of Ionizing Radiations 381

and represent the first series of aerial views to be flown to the United States for publication (U.S. Army photograph). b) Second atomic bomb test at Bikini Lagoon, 24 July, 1946, showing column of water as it began to fall. Photographed by an automatically operated camera on a nearby island (U.S. Army photograph).

components by having 10 to 20 fewer byproducts. Another component that is reduced is environmental ^{14}C which is so characteristic of the atmosphere and all living things. More tests should have increased ^{14}C in this chain, but only 50% as much is produced by the neutron device. This nuclide with its 5,750 year half-life comes from activation of nitrogen and it used to be that the source was cosmic rays.

The neutron weapon deals with light atoms and does not have the massive atomic fragments of fission, but fission gives five times as much energy. More to the point, the fusion reaction yields many more neutrons than fission, especially high energy ones. The high LET of the neutrons makes them selective for tissues which are sensitive to such saturation. On the other hand, the neutral electrical status lets them penetrate through orbiting atomic electrons and pass coulombic atmospheres near nuclei. They collide (elastically) with light atomic nuclei because they can be packed in higher concentrations then heavier atoms could be.

No doubt, H bombs will now be considered as more useful and employable thanks to the flexibility introduced and deterrence will be more effective, but not as effective as first appears. As a matter of scale, weaponry requires both the mini and super megaton versions, so that Western Europe and other regions will feel relieved to have some defense against waves of battle tanks. Biologically, it will be difficult to build complete, instantaneous deterrence into the weapon without blast or intense heat and irradiated warriors, while doomed, may well survive to reach their objectives and, knowing they are doomed may give them a kamikasi attitude. At saturation levels, information transfer is obliterated by noise and the neutron is unconcerned whether it encounters a target atom in living or nonliving substance. Nevertheless, as will be seen, there are selective elements involved, and information continues to flow in parallel with the energy, ready to surface when the noise is reduced.

A number of accidents (Table 9-3) illustrate what happens to lethally irradiated persons and instantaneous death does not occur even when the heart is overwhelmed with 12 krads directly to it. There may not even be any discomfort for several hours except for shocked feelings and apprehension.

Other devices besides the atomic detonation may be used to create the necessary temperatures. Plasma reactions (as in the stellerator), magnetohydrodynamics (to contain the seething gases), and masers or lasers of high output (for excitation), are some areas being investigated. The laser may be used, for example on injected frozen pellets of deuterium and hydrogen in the void inside swirling, molten lithium to create fusion of the hydrogen nuclides with lithium to moderate the neutrons produced down to the thermal level. Then the energy might be drawn off in the lithium for a controlled thermonuclear reaction terminating via turbines and a generator as electricity. The lithium provides a reacting "vessel" which is also the reactant so that the problem of containerizing this reaction has a possible solution as does the problem of heat loss to container walls (Fig. 9-20).

In the mirror machines or magnetohydrodynamic containers, the plasma is injected into a straight vacuum tube which is surrounded by magnetic fields which concentrate at the ends. Bending the tube into a doughnut or toroid eliminates the loss problem at the ends. A magnetic field reflects plasma particles as a mirror does light. The deuterium, tritium nuclides must attain a plasma density over 10^{14} ions/cm^3 for one second at 100 million °C or otherwise meet Lawson criteria for fusion.

Whether or not this particular approach will be the one that leads to the age of fusion for mankind in the next century is the question. If it does, there will be an enormous amount of pollution-free electricity, and other forms of energy made available for the environmental support

Information Routes of Ionizing Radiations

Table 9-3. Reported Accidents

Place	Device	Exposure Level	Comment
Bikini Atoll Marshall Islands Pacific 3/1/54	Thermonuclear	Radioactive rain rather than burst effects	Wind shaft deflected fallout. Largest group of casualities. No deaths. Heaviest fallout. Best reported.
Oak Ridge, Tenn. 6/16/58	^{235}U enrichment plant	300-500 rems neutrons and gammas	Critical mass by accident in drainage drum, no deaths. Five victims.
Vinca, Yugoslavia, 10/15/58	Uranium fueled reactor	1 krem	Reactor accidently made critical, six victims. 1 died later.
Los Alamos, New Mexico 12/30/58	Plutonium recovery. Operator may have varied the procedure	1.5×10^{17} fissions, 12 krad to heart neutrons and gammas	3 kg sequestered, Pu made critical conditions in recovery process. Operator dazed and in 5 minutes semiconscious. ^{24}Na abundant in discharges, lived 25 hours.
Lockport, New York, 3/8/60	X-rays from unshielded klystron	1.5 kiloroentgens to head, pulsed 300 R to trunk	No deaths, two injured.
Mexico City, Mexico April to July 1962	Lost ^{60}Co source for industrial radiography	3.0 to 4.7 kR or less gammas	4 deaths after dose accumulated from device used as a home plaything.
Woods River Junction, Rhode Island 7/24/64	^{235}U chemistry work in tank with sodium carbonate	2.2 krads neutrons, 6.6 krads gammas extremely high	Made critical conditions. Victim lived 49 hours. Concentrated ^{24}Na in body, urine retention, edema
Pittsburgh, Pa. 10/14/67	Gulf R&D 3 MeV Van de Graff accidently "on"	6 kR to hands 600 R to body 300 R and 100 R	No deaths, amputation Mildly ill, returned to work in few weeks

of life systems with the hydrogen nuclides being drawn out of the waters of the sea where they constitute the ultimate energy resource.

Because the mass and energy are related by the expression $E = mc^2$ the energy and mass values for a proposed nuclear reaction may be examined to predict whether or not the reaction will proceed. In neutron-induced reactions, the only ones possible are those where there is a quantum fit for the energetic neutrons and energy states in the proposed nucleus.

MASS DEFECT

Some of the redistribution of energy that occurs in an inelastic or absorptive collision goes into that required for the mass defect. This is the binding energy or mass difference for the neutrons, protons, and electrons in the transforming atom. It represents the difference between the mass of these atomic particles and the measurable mass. The factors

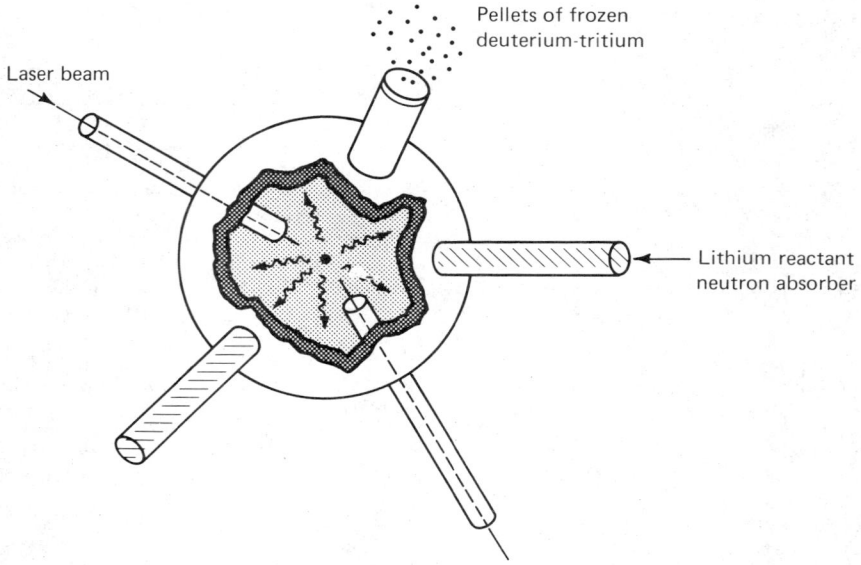

Fig. 9-20. Implosion of D-T plasma by laser ignition method for thermonuclear device.

involved are the space states in which particles exist and their probabilities and wave functions. The energy used in binding nuclear particles is incredibly large since it must produce densities of the order of 100 megatons. To be captured, a particle would have to be accommodated by an equivalent change in these enormous forces and this involves the liberation of radiation equal in energy to the binding energy. To exchange their intranuclear forces in rearrangements, nucleons can interchange mesons which are accordingly described as adhesive particles in the nucleus. They are not so limited, however because, as free particles, they are part of the cosmic radiation.

AVAILABILITY OF ISOTOPES

Not all of the isotopes are available for use. Some, like those in Table 9-4, have been used for many years or have been implicated in studies of radioisotopic contamination of the biosphere. Tables 9-5 and 9-6 show the characteristics and applications for other isotopes which might be made available in the future. The sources for these isotopes are accelerators and nuclear reactors, and their production and distribution constitutes a new technical industry as well as an important governmental activity.

TABLE 9-4. Characteristics of Some Important Isotopes in Biophysics

Name	Symbol	Mass No.	*Half Life	Energy in MeV of β^-	γ	α	Artificial	Natural
Tritium	H	3	12.26 y	0.018	—	—		√
Carbon	C	14	5730 y	0.156	—	—	√	
Sulfur	S	35	87 d	0.167	—	—	√	
Calcium	Ca	45	163 d	0.25	—	—	√	
Iron	Fe	59	45 d	0.46	1.10			
					0.27	1.29		
					1.57	0.19		
Phosphorus	P	32	14.3 d	1.71	—	—	√	
Cobalt	Co	60 (Isomers)	A 10.5 m	1.54	1.33	—	√	
			B 5.24 y	0.31	1.33	—	√	
					1.17			
Strontium	Sr	90	28.8 y	0.54	—	—	√	
Iodine	I	131	8.05 d	0.61	0.36	—	√	
				0.25	0.72			
				0.81	0.08			
Cesium	Cs	137	30 y	0.51	0.662	—	√	
				0.57				
Bismuth (one member of natural decay chain of radium RaE)	Bi	210	A 2.5 × 10⁶ y	—	0.26	4.95		√
			B 5 d	1.16	—	4.7		
Polonium	Po	210	138 d	—	0.8	5.3	√	

*m=minutes; h=hours; d=days; y=years

TABLE 9-5. Characteristics of Potentially Useful Isotopes Which Can Be Made Available

Name	Symbol	Mass No.	*Half Life	Energy in MeV of β^-	γ	α
Iodine	I₃	123	11 to 13 h	—	0.159	—
Xenon	Xe (gas)	127				
		(Isomer A)	36 to 38 d	—	0.20	—
					0.17	
					0.15	
					0.06	
		(Isomer B)	75 s	—	0.125	—
Silicon	Si	32	700 y	0.1	—	—
Gadolinium	Gd	148	85 y	—	—	3.18
Polonium	Po	208	2.9 y	—	0.6	5.1
			0.28			

*s=seconds; h=hours; d=days; y=years

TABLE 9-6. Possible Applications for Other Isotopes if They Were Available

Isotope	Application
^{208}Po	Source of alphas and of secondary neutrons from Li and Be
^{148}Gd	Alpha source and secondary neutrons — for detectors (uranium) and — for activation analysis
^{127}Xe	Relatively good half-life for local stocks
^{72}Zn	Diagnosis of prostate cancer
^{82}Sr	Diagnosis of heart and arterial disease
^{26}Al	Metals research
^{178}Ta	Low energy X-rays
^{123}I	Effective half-life is reduced so patients receive less dose in clinical uptake tests than from I-131
^{32}Fe ^{85M}Sr ^{135}Ba ^{113}Sn ^{15}O ^{13}N ^{11}C	Enriched radioisotopes from cyclotron are like ^{99}Mo, which generates ^{99M}Tc for scans.

DOSIMETERS

The usual environmental problem as far as nuclear reactions are concerned is how to avoid undesired or accidental activation. It is undesirable to have any significant amount of radioactivity resident in the body. Normally, the use of personal dosimeters is the best of several safeguards designed to prevent a person who must be near radiation from becoming "his own dosimeter" via the radioactivity of his tissues. A dosimeter measures the radiation incoming to the body when worn on an exposed portion such as the chest (pocket), belt, or finger. It may be a pencil-sized electroscope that discharges a preset charge or accumulates a charge due to the ionization of the radiation. The result may be read, on an integral scale or on one provided, as the radiation exposure during some period of time. Ordinary photo film in a pocket badge (Fig. 9-21) makes a good dosimeter because the film darkens when exposed to all radiations except weak betas and alphas, so that the exposure is estimated when the film is developed. Various special films, windows, and shields built into the film holder refine the meaning of the darkening by localizing, or by reacting to fast neutrons, thermal neutrons, beta, gamma, and X-ray exposure. Other dosimeters can

Information Routes of Ionizing Radiations 387

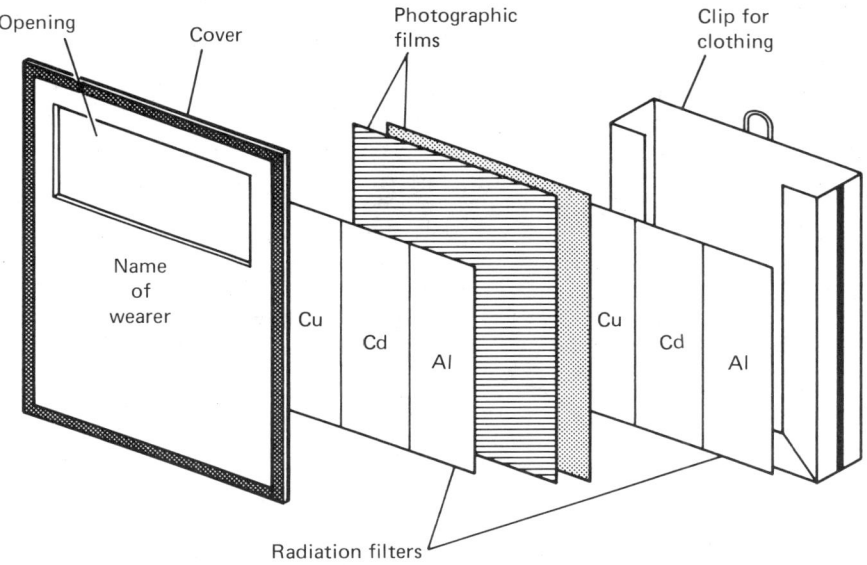

Fig. 9-21. Film badge construction.

accumulate exposure evidence by color changes in appropriate chemicals and may be worn as a medallion, implanted or used in the form of film. The thermoluminescent (LiF, CaF_2, BaF_2) phosphors store a "radiation received" signal which may be obtained by heating the dosimeter and measuring the emitted light. Some dosimeters give audio/visual alarms.

INFORMATIONAL LINKS, ACTIVATION ANALYSIS

Nevertheless, the fact is that many atoms normally present in the body can be made radioactive in order to study some function. This extremely useful application in organic and inorganic material is known as activation analysis. In vitro analysis of blood, urine, and hair are used in criminology to associate evidence found at the scene of the crime with a suspect. Only a very small sample is required but a fairly sophisticated kind of detective work is needed to establish the evidence firmly.

When the atoms in a substance are activated, the radiation begins to transmit information about them. With appropriate receivers, such information as the radiation energy and type, half-life, and rate of decay become available. Not only are unsuspected substances dramatically revealed by their own transmitted information but associated facts may be combined chemically, physically, and biologically to produce a prior history and classification of the atoms and compounds of interest. Activation results also include information on the kind of nuclear reactions and the dose of radiation an organism has received. *In vivo* sensing of the concentration of activation or of tracers in various tissues is possible with needle, silicon, solid state detectors, and scans.

For success, this process depends on a basic knowledge of the normal constituents of the organism, especially its chemical composition and distribution of the elements. Then if dose analysis is involved, the relative biological effect (RBE) becomes important for various radiations as well as the cross sections for activation which indicate the interactions that may be anticipated. It is convenient to use a neutron flux, as from a reactor set-up for irradiation, for these experiments (also see Table 11-4). After activation, the kind of radiation emitted and the counting geometry are typical considerations. Then comparisons are made with known isotopes and their curves for important parameters such as decay time. This will positively identify the substance or at least tell a great deal about it (Fig. 9-22). Some of the procedures can be automated and many of them can be made routine.

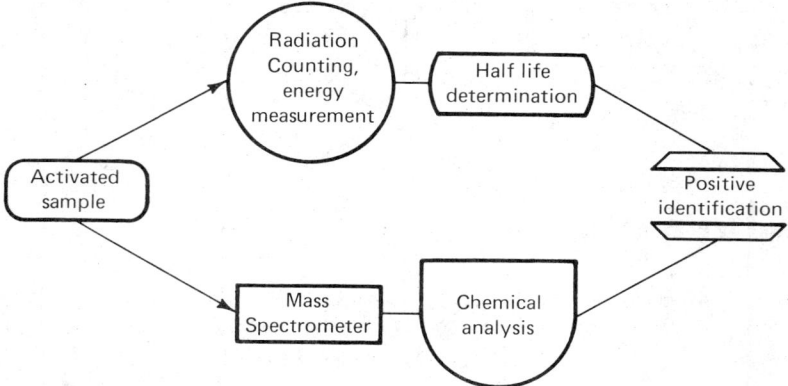

Fig. 9-22. Activation analysis to reveal identity of unknown substances. Activated sample has observed characteristics that narrow the possibilities, (for example, a foil, a colored substance, and other clues associated with its discovery). Mass spectrometer analysis is for further narrowing the possible identity. Positive identification of the substance is made from the accumulated physical and chemical properties and comparisons with known data.

AUTORADIOGRAPHS

Autoradiographs represent filmed images made by an organism or a specimen by virtue of the radiation which it has absorbed. It is not necessary to activate the preparation by irradiation. All that is needed is a way for radioactive isotopes to enter the molecular systems in the preparation, for example, by normal uptake of phosphorus-32 into the leaves of a plant (or tritium into cell nucleus processes). The radiations then emitted by incorporated isotopes can expose a film placed in close contact with the preparation. An exposure time is chosen such that the amount of radiation emitted and the number of photographic emulsion grains affected are sufficient to produce a contrasting pattern similar to an X-ray (Fig. 9-23). The density of the darkening is interpreted for any location, for example, phosphate activity in DNA in the nucleus to

Information Routes of Ionizing Radiations 389

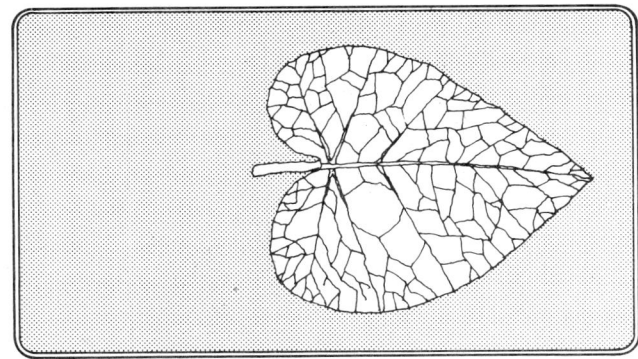

Fig. 9-23. Autoradiograph preparation involves: 1) radioactive isotope added to soil; 2) tissue uptake; 3) time exposure film is exposed to self-absorbed activation; 4) finished autoradiograph to be analyzed for areas of activity.

show division activity and growth or more subtle differences with other isotopes in metabolism, abnormal absorption, toxicity, and synthesis.

TRACERS

While radioactive isotopes are valuable as tracers used to ascertain the behavior of compounds in biology, the techniques used and the supporting biochemistry are of critical importance even though they often begin with well-educated "guesstimates." In nearest neighbor analysis, for example, by tagging the correct atoms, the informational macromolecule DNA can be caused to reveal the coding present in the sequence of bases. This is done by a sort of musical chairs game in which the players are tagged bases and when the enzymes act, the bases can separate. After they separate, it is clear that the next base must have been at the site of the tag (Fig. 9-24).

Since there are millions of atoms involved in these analyses, elaborate care in the form of information analysis of the code structure is required before the messages present in the informational molecules can be read. After sufficient radioactive tracer and chemical evidence has been obtained on the macromolecular system, the power of mathematics can be brought to bear in the form of frequency analysis, probability determinations, and decoding.

ISOTOPE DILUTION

In isotope dilution, a little nonradioactive unknown which must be studied and identified is mixed with a negligible bit of a compound that is thought to be the same but has radioactive tracer. The mixture is allowed to come to equilibrium (for example, in crystallization), then separated and counted. The total counts per minute are now distributed throughout the unknown for which the weight is already known. The count of original radioactivity is also recorded. If the isotope has in fact been diluted through the crystals, rather than

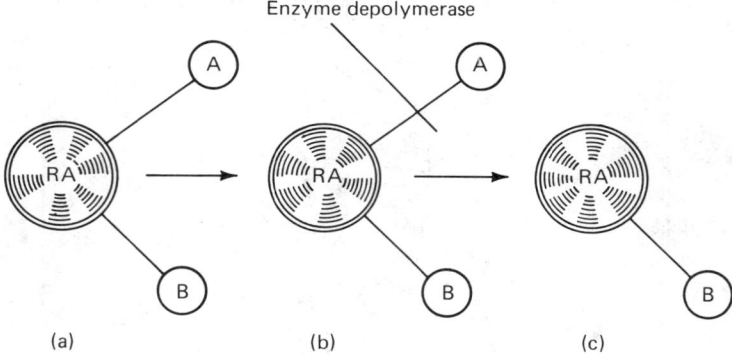

Fig. 9-24. Nearest neighbor analysis. 1) Radioactive tracer (RA) is placed on compound bound to A and the usual synthesis proceeds with the addition of B. 2) Enzyme depolymerase breaks off base A leaving the tracer still tied to B. 3) Examination shows the tracer still tied to B and conclude B used to be nearest neighbor of A where the tag was originally placed.

rejected in the crystallization as foreign, then any tiny fraction taken as a sample will have a constant value in counts per minute of radioactivity. It can then be concluded that the unknown was actually what it was thought to be, the proof coming from the "like" isotope material added.

Probably the recent progress in the understanding of energy pathways in organisms, aerobic and anaerobic fermentation routes, glycolysis, phosphorylation, respiration, oxidation and reduction, electron transport, acceptors and donors of electrons, turnover rates and alternate pathways would not have been nearly as rapid without the help of radioisotopes and analyses such as those that have been described.

PROBLEMS

1. Compare age measurement in trees by counting annual rings and radiocarbon dating with the basic assumption that the latter means respiration stopped when the tree was cut down.
2. To study vitamin B_{12} accumulation in the liver, three possible isotopes of cobalt would be used: ^{57}Co, ^{58}Co, and ^{60}Co. Of these three, the last has the largest LET and a total radiation due to the liver of 13 rem/microcurie. Which would you use? ^{57}Co emits no betas but only low energy gamma radiation during decay. ^{58}Co can be counted with geiger and scintillation counters.
3. Hemoglobin in human red blood cells is labeled with ^{15}N in glycine which is introduced in a relatively short period (3 days), as a "pulse." The radioactivity in the cells rises continuously for 20 to 40 days, but at 120 days, the *count*, as atom % ^{15}N in excess in hemin drops rapidly. What conclusions may be drawn?

4. From the Chart of the Nuclides and its explanations, what activity is associated with the coolest color? With the greatest half-life values?
5. What does a circle around the nuclide mean? Which are circled?
6. Given a source of ionizing radiation with a half-life of 12 hours and strength of 1 curie, what is the strength after one week?
7. Express the activity of a radioactive source in terms of the decay constant and half-life.
8. Express the number of atoms decaying in a radioactive sample as a differential equation.
9. Give a radiochemical equation for formation of tritium by neutrons with alpha particle emission.

BIBLIOGRAPHY

Becker, K., *Solid State Dosimetry*, CRC Press, Cleveland, Ohio, 1973.
Chase, G.D. and J.L. Rabinowitz, *Radioisotope Methodology*, Burgess Co., Minn., 1962.
Danielle, J.F., *General Cytochemical Methods*, Academic Press, New York, 1958.
Kamen, M.D., *Isotopic Tracers in Biology*, Academic Press, New York, 1957.
Moore D.H. and J. Sacks, Eds., *Physical Techniques in Biological Research*, 2nd Ed., Academic Press, New York, 1966.
Oak Ridge National Labs, *Radioisotopes Catalog*, Oak Ridge, Tenn., 1976.
Roy, R.R., and R.D. Reed, *Interactions of Photons and Leptons with Matter*, Academic Press, New York, 1968.
Spencer, R.P., *Radionuclide Studies of the Spleen*, CRC Press, Cleveland, Ohio, 1975.
Snell, A.H., Ed., *Nuclear Instruments and Their Uses*, John Wiley, New York, 1962.
Taylor, L.S., *Radiation Protection Standards*, CRC Press, Cleveland, 1971.
Wang, Y., *CRC Handbook of Radioactive Nuclides*, Cleveland, Ohio, 1969.
Wolf, J., *Isotopes in Biology*, Academic Press, New York, 1964.

Films

Atoms in Agriculture, 1969. 26 minutes. Color.
 For sale by WRS Motion Picture Lab. 210 Sample Street, Pillsburg, PA, 15213 ($83.09). FOB Pillsburg. Available for loan (free) from AEC Headquarters and Field Libraries. This motion picture explores the beneficial applications of atomic energy in the field of agriculture.
Extracorporeal Irradiation of Blood and Lymph, 1966. 7½ minutes. Color.
 For sale by B&O Film Specialists, 619 West 54th Street, New York, NY 10001, ($45.63) FOB New York City. Available for loan (free) from AEC Headquarters and Field Libraries. This film describes a new tool available for medical research.
Radioisotope Scanning in Medicine, 1965, 16 minutes. Color.
 For sale by Handel Film Corp. 6926 Melrose Ave., Hollywood Calif. 90038. ($180) Available for loan from AEC Headquarters and Field Libraries. This film describes the use of a new technique in medical diagnosis.

10

Informing Processes, Energy Transport and Reaction Direction with Hydrated Electrons and Free Radicals and Superoxide Dismutase

Hingehen und gucken

OPERATING RADICALS

The question of where the reactive free radicals belong in the information pathway is clearly related to all the things that have been said about radiations, as well as about free radicals in resonance spectroscopy. In this pathway lie all the receptive molecules and materials that give informational content to the radiation and provide for transduction into operating energy at the organismic and molecular level.

Except for the informing roles of radicals such as in spectroscopy, at the organismic and molecular level, the emphasis to date may have to be on the deleterious effects of the reactive free radicals as described in this section. Ahead of us, however, are all the quite peaceful roles of free radicals, hydrated electrons, and charge carriers which will come to be understood in the area of organic semiconductivity. These concepts are suggested by the heading *Electrons and Holes*.

Found here also is more evidence of the widespread phenomenon of *polarization* as seen in the context of ionizing radiations. The principal reason why photolysis and radiolysis and the related topics must be examined, however, is that these events occur in the dominant biological substance—water.

Radical operations in metabolism include participation in oxidative respiration, electron transport, toxification and detoxification of superoxide and the most interesting enzyme group known as the superoxide dismutases.

THE UNPAIRED ELECTRON AND REACTIVITY, PHOTOLYSIS AND RADIOLYSIS OF WATER

The free radical in radiolysis is a powerful biological hazard as well as a tool, and it comes into being at the very beginning of the train of events after the ionization of water when exciting and ionizing radiation energy is absorbed. It has already been discussed in connection with its paramagnetic effect which acts as an indicator in spectral absorption analysis. There it was shown that it is the unpaired electron in the electronic configuration of the orbiting electrons of the radical that confers the characteristic instability and extreme reactivity. If these radicals collide, their reaction is certain. They exist only for microseconds under normal circumstances. Their lifetime depends on the density of ionization and their concentration. Thus free radical reactions, at least the primary ones, are immediate and they occur within distances which depend on the amount and nature of solutes; this distance being perhaps five nanometers in a yeast cell.

Free radicals such as the hydrogen atom, the hydroxyl and perhydroxyl radicals, and the hydrated electron in the radiolysis and photolysis of water are nevertheless very simple structures which are significant in the study of the nature of water itself. The hydrated electron, for example, is the simplest kind of reducing agent available. It is therefore probably the ideal agent for the newer studies in biophysics, studies that deal with reduction and oxidation involving electron transfer, functioning charge carriers in biotransfer electronics with macromolecules, conductivity, and semiconduction behavior in biology, specifically that of electrons and holes. Although simple in structure, the free radicals become directly involved as elements in radiobiological theories such as those which attempt to explain direct and indirect action of ionizing radiation, polarization, protection, bound water, and frequency effects.

The Nature of Water

A biologist is normally impressed with the multifaceted nature of water but what he sees in radiolysis surpasses even the most dramatic of the molecule's other characteristics. One of the most surprising things is the large number of variations that are possible in the fragmentation of the water molecule (Table 10-1). The ionization model for radiolysis presents a very large number of combinations of the hydrogen and oxygen atoms which show how truly versatile the molecule is. Table 10-2 gives the time scale for radiolysis. Fortunately, the "capture" of free radicals in crystalline structures such as ice at very low temperatures (ice radiolysis) as well as other "irradiate and observe" methods allow these exotic substances to be examined by biophysical methods.

The Hydrated Electron. As might be concluded from the nature of the radiolysis products, the oxidation and reduction of sensitive biomolecular groups is an important cause of damage in tissue when the chemical action renders functional sites inoperative. Tissues usually contain between 65 and 99% water with an average of about 80%. Thus radiolysis, leading to the production of these reducing and oxidizing agents, is a

TABLE 10-1. Some of the Symbols for Free Radicals, Water Fragments, and Molecules that Have Been Demonstrated or Suggested for Ionized and Excited Water and Radiolysis

e^-_{aq}	Aqueous electron ⎫
e^-_{sol}	solvated electron ⎬ alternate
e^-_{hyd}	hydrated electron ⎭ names
$(H_2O)^-$	negative polaron
$\overset{\otimes}{H}$	hydrogen atom or atomic hydrogen also H'
$\overset{\otimes}{OH}$	hydroxyl radical
$\overset{\otimes}{H}O_2$	perhydroxyl radical; peroxyl; hydroperoxy
O_2^-	negative oxygen, also a form of the perhydroxyl radical at high pH, superoxide anion
H_2	hydrogen gas
H_2O_2	hydrogen peroxide
H_3O^+	positive polaron and electron scavenger
H_2O^*	excited water
H_2^+	acidic form of hydrogen
OH^-	hydroxide ion
H^-	negative hydrogen
H^+	hydrogen ion
O^-	oxygen radical ion

significant event. The most important of the reducing species is the hydrated electron, and it is also a primary product in many photochemical processes. Both the formation of this special electron and its subsequent influence in the medium involve polarization. The hydrated electron itself is a polarization product made by the orientation of the water dipole under the influence of the electronic charge (Fig. 10-1).

TABLE 10-2. Time Scale for Radiolysis

Picosecond events ($\approx$ to 10^{-13} s)	These are the more or less instantaneous excitations that occur as the radiation transfers its energy to the molecular matrix (see Fig. 10-2). Photoelectric, Compton, and pair production effects are characteristic.
Nanosecond events ($\approx$ to 10^{-11} s)	During this time, the radiation energy undergoes transduction to chemical and thermal forms. Ionizations, stable molecules, free radicals and e^-_{aq} are features.
Microsecond events ($\approx$ to 10^{-8} s)	Combination chemical species and reaction products are formed in the longer term. The results depend on the LET of the radiation or its specific ionization, and the presence of other components, notably oxygen.

$$H_2O$$
$$H_2O \quad e^-_{aq} \quad H_2O$$
$$H_2O$$

Fig. 10-1. Polar grouping for e^-_{aq}. Some number of water molecules form a positive ion sphere around the charge of the hydrated electron. High concentrations of other solvents may compress the grouping.

The electron comes to be trapped in liquid water by its own polarization field after being ejected from the parent (now ionized) water by radiation energy. Under suitable conditions it has a blue color and its hydration energy is two electron volts. The electron is called by other names, such as the *solvated electron*, which is a good description considering its polarization interaction with the solvent. In view of the polarization, the electron may be called a *polaron*, or sometimes it is referred to as an *excess electron*, a *bound electron*, a *secondary electron*, or as $(H_2O)^-$. The primary electron could have been the radiation which carried the energy into the medium. Solvated electron is a more general term by which all possible solvents are implied. In any case, it is the hydration or solvation energy of these reactive ions that binds them in their medium. This energy is distributed over several solvent molecules which form the cavity or space occupied by the electron. Perhaps the best way to study electron transfer which lies at the base of many, if not most, biological functions is with an electron. Thus one result of the electron's uniqueness is that it permits electron transfer to be analyzed as well as electron bonding. Such "direct" observation has been difficult since a radioactive tracer cannot be placed on an electron. Watching the hydrated electron means watching the most simplistic reducing agent in action, the most refined electron transfer reactions, or the smallest current.

The behavior of the solvated electron in ammonia brings interesting metal relationships into focus. There they may be called ammoniated electrons and are formed with metal ions such as sodium dissolved in ammonia. This leads to special phenomena in areas such as conductance, electrolysis, volume minimum with increased concentration, and the metallic nature of the solution.

Speed and Reactivity. The dipolar forces binding the hydrated electron are fairly weak, thus permitting it to combine with electron acceptors in keeping with its role as an electron "donor" and reducing agent. Yet its behavior will be different from that of the electrons that are present in a liberated or thermalized state. For one thing, a higher activation energy for a given reaction may be demanded with the ordinary electron. Reactions may derive forward proceeding energy from many sources, such as the reservoir of thermal energy in the reacting medium. With the free radicals, the energy comes from that transported into the medium by radiation. This difference means that the radicals are themselves energetic and contain the energy necessary for the radical reactions. Thus these reactions are favored over those with ordinary thermalized radicals because of a much lower activation energy requirement and the presence of the energy in the reacting radical itself.

It takes a fraction of a nanosecond for the hydration of the electron and in this time, it associates its sphere of polarized groups. After this, it behaves in its ultrareactivity according to factors such as those listed in Table 10-3 and endures for a period of only about 3×10^{-5} seconds. The end result is that reaction is many orders of magnitude faster than with ordinary reducing agents combining with reducible ions or molecules.

Radiolysis produces the solvated electron in neutral solution and the hydrogen atom radical in acid pH. This reducing radical is weaker than the hydrated electron and reacts differently. It takes hydrogen out of compounds to form hydrogen gas whereas the electron does not withdraw hydrogen in this manner. Production of the hydrogen atom is an alternative and delayed fate for the hydrated electron should it fail to meet another type of solute,

$$e^-_{aq} + H_2O \Rightarrow \overset{\otimes}{H} + OH^- \tag{176}$$

so that under such conditions, it is the ultimate reactivity and products of the free *hydrogen atoms* that are studied.

The birth and fate of the hydrated electron are observed by monitoring its absorption spectrum, that is, the absorption of visible light between 5000 and 7000 angstroms by the solution. The decay of the spectrum, that is, the leveling of the absorption maximum, is followed

TABLE 10-3. Factors in Reactivity of Free Radicals

Temperature
Pressure
Solid, liquid, or gas phase
Bond energies and biogroup sensitivity
Polarity and polar electron groups
pH
Electron pairs and bond behavior
Electron transfer
Presence of oxygen
Activity of enzymes to disrupt products
Lower activation energy
Free radical carries reaction energy
Hyperconjugation
High and low density electron regions
Neutralization and recombinations

to show changes that occur in the hydrated electron in various solutions after its production and their kinetics (Fig. 10-5).

Similarly, the reactivity of a hydrated electron with a compound can be examined to tell about the compound's electronic structure. In this sense it functions experimentally as an indicator of orbital vacancies and as a tracer in reactions involving electronic rearrangements. In company with radicals formed in the radiolysis of water, it illuminates features of the water molecule not previously appreciated. From wave mechanics, the hydrated electron appears to have a wave function related to the local solvent molecules and responding to their polarization. Thus the behavior of dipoles, their fields, and relaxations once again become of special interest along with the dielectric properties of the solvent. In the behavior of hydrated electrons following radiolysis, indications can be read concerning what reactions can occur. This behavior involves diffusion from the radiation track or spur with corresponding process time constants, half-lives, and reaction times (Fig. 10-2). This allows their reactivity to be followed.

Solvent molecules react to the presence of the hydrated electron by orienting their valence and other electrons at a distance (the radius of a sphere) around the hydrated electron. Thus the "cage" is formed by the positive centers of charge of the solvent molecular dipoles. The electron is orbiting at a high frequency so that it constitutes a region of charge rather than a spot charge. The polarization of the solvent is hindered by constraints offered by the existing bonds and the thermalization. The

Informing Processes 399

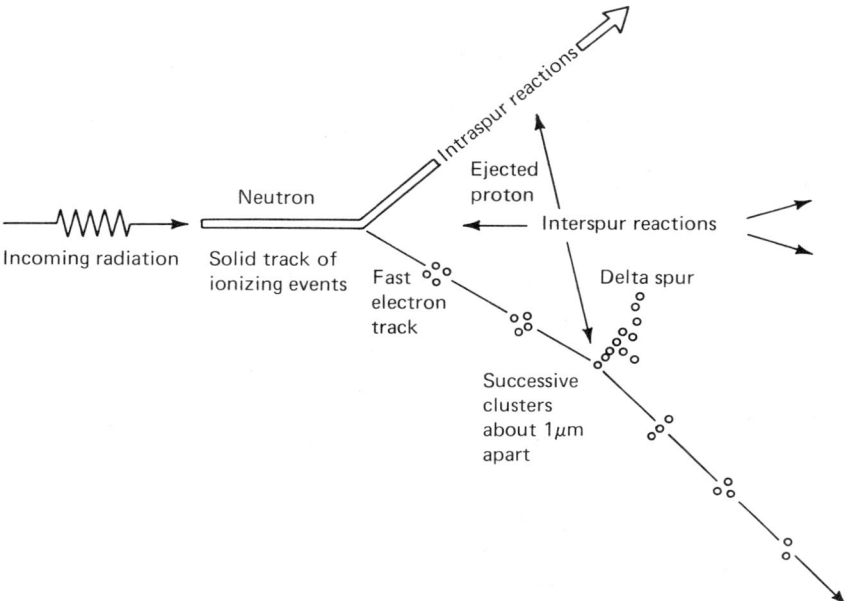

Fig. 10-2. Different ionization tracks, spur events diagrammatically.

interaction is one of distances, that is, if the electron is a "permitting distance" away, it can polarize the solvent molecules coming under its influence. Thus the relaxation time of the molecules interacts with the time-motion of the electron. One result of the energy situation in the cage is that the configuration permits light emission or chemiluminescence.

ELECTRONS AND HOLES

When radiolysis occurs in ice, the electrons and holes may recombine unless attracted solute species absorb them before they have the opportunity. If the solute does not provide suitable acceptors for the electrons and holes, they can exist in the crystalline or aqueous structure as coupled hole and electron *polarons* (Fig. 10-3). This stable existence is an alternative to their disappearance by annihilation. The use of the symbols $(H_2O)^-$ and $(H_2O)^+$ is sometimes preferred to emphasize how the polaron charges are distributed through the solvent; that is, they influence several solvent molecules. The charge or force acting on the ice polaron may arise from either the low frequency or high frequency dielectric constant of ice. The relaxation time τ for water will be much longer in ice than in liquid as shown in Chapter 11. It follows that in water, the dipole orients even up to high frequencies which it cannot do in ice. Thus the high frequency dielectric constant is governing. Therefore, the absorptive polarization in the microwave interaction has features in common with the solvent polarization by the hydrated

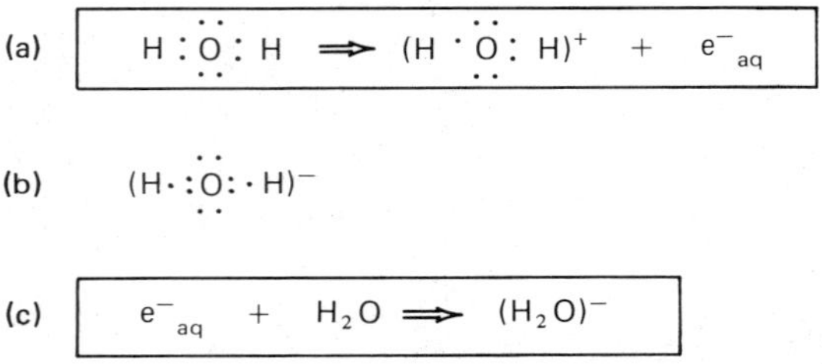

(a) $\quad H:\overset{..}{O}:H \implies (H\cdot\overset{..}{O}:H)^+ + e^-_{aq}$

(b) $\quad (H\cdot:\overset{..}{O}:\cdot H)^-$

(c) $\quad e^-_{aq} + H_2O \implies (H_2O)^-$

Fig. 10-3. a) The polarons come from an electron unpairing in the molecule. The radiation energy is the primary source of this electron removal and water ionization. This hydrated electron is now available for its polarization behavior, for its formation of intermediates, and for reducing action in a diffusion pathway. b) Negatively ionizing water will have a surplus of electrons in the valence configuration and will arise when the hydrated electron disappears into the valence configuration of another water molecule (c).

electron. Hindrance by the solid state affects both. The polarons in the present case, while still formed, remain in an "interinfluencing" or bound state and dipole orientations are constrained. The word *exciton* is used to describe this kind of binding and these concepts may be followed to their natural conclusions—that the radiolysis of ice has semiconductor aspects. Thus the ionization looks like an electron moved into the conduction band with the corresponding positive hole being left behind in the valence band. They attract each other to give potentials in a series of stable bound states in the energy gap between the two bands.

The polarization, dielectric, radical and hydrated electron activities merge with a microwave interaction when electron spin resonance is used to observe the electron, dipole, and microwave frequency effects of the microwaves used in electron spin resonance. These include the g-factors of the free radicals produced, linewidths, and the microwave power saturation effect.

In crystals, electrons trapped in lattice defects make fluorescence possible. Similarly, color centers in glassware and polymers absorb radiation to turn colors such as purple and brown, the color being a function of the composition. If the stored electrons require heating in order to release energy in this manner, the process is known as thermoluminescence. Alpha rays can destroy a lattice (reversible with heat) and produce a storage of about 26 calories per gram. In the graphite moderator material, essential in some nuclear reactors, heat can violently release the stored up energy.

When oxygen is unavailable, the hydrated electron becomes a possible reducing agent but, under *aerobic* conditions, it is a source of the perhydroxyl radical. The important reducing agent had been thought to be the hydrogen atom free radical, until pulse radiolysis made hydrated electrons available for observation by monitoring the light absorption. Now it is known that radiolysis produces the hydrated electron at neutral pH, and the hydrogen atom when the reaction is acidic.

FREQUENCY EFFECTS

What may be described as a frequency effect (Figs. 10-4, 10-5) is observed with radiolysis. This yields such terms as gamma and ultraviolet radiolysis or photolysis, sometimes with different products. Some of the difference is due to the manner in which the radiation energy is deposited, that is, in a sudden versus a distributed way. The hydrated electron is observed in ultraviolet flash photolysis of water and inorganic salts, e.g. chlorides and aromatic amino acids such as phenylalanine

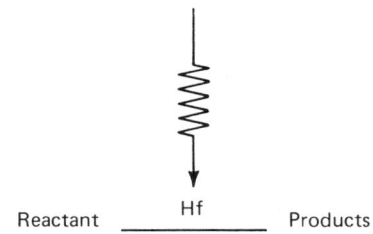

Gamma, Ultraviolet radiolysis, Photolysis, and Pulse radiolysis

Fig. 10-4. Frequency effect. Radiolysis products depend on the frequency.

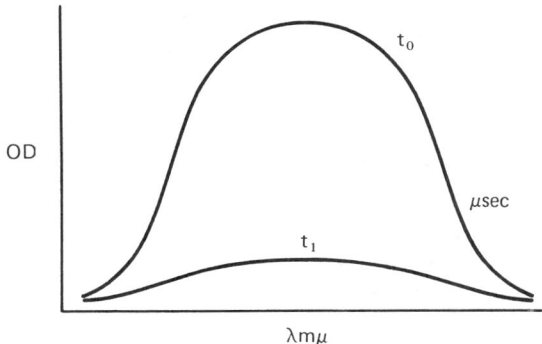

Fig. 10-5. Absorption due to e^-_{aq} and method for observing the e^-_{aq}. Direct observation of e^-_{aq} after pulse radiolysis by monitoring the visible absorption spectrum, t_0 initial time, t_1 a later time. The disappearance of the absorption is due to the fact that the e^-_{aq} is reacting with some solute molecules. Similarly, electrical conductivity is used to follow the radical effects.

and tyrosine in neutral solution. Gamma radiolysis produces only a small yield of hydrogen peroxide from aerated solutions and actually does not change water appreciably if it is oxygen free.

A microwave input to hydrogen gas causes a gaseous discharge or plasma in which the hydrogen atoms are generated. This is an important source of the free radical which can then be harvested by bubbling through water or reacted directly with materials to cause unusual products.

ORIGIN OF THE FREE RADICALS

The hydrated electron will transform into the hydrogen atom radical if it reacts with acids. Reaction with the hydroxide base can turn the hydrogen atom back into the hydrated electron.

Taking some liberties for illustrative purposes, the production of the hydrated electron may be visualized as coming from the unpairing of the water valence configuration (Fig. 10-3).

It is this negative water ion that can decompose in water to produce the hydrogen atom, a reducing free radical, and the ordinary hydroxide ion.

$$(H_2O)^- \Rightarrow \overset{\otimes}{H} + OH^- \tag{177}$$

The hydroxyl free radical, the most important of the oxidizing radicals, is obtained from the positive water ion, along with the usual hydrogen ion.

$$(H_2O)^+ \Rightarrow \overset{\otimes}{OH} + H^+ \tag{178}$$

The crossed circle over the free radical here (and in Table 10-1) means that it has the unpaired electron and is, therefore, one of the main species under discussion here. If oxygen is introduced, the hydrogen atom yields another important and strong oxidizing free radical.

$$O_2 + \overset{\otimes}{H} \Rightarrow \overset{\otimes}{HO_2} \tag{179}$$

This is the perhydroxyl radical which can also be produced by reaction of the hydroxyl radical with hydrogen peroxide, as well as from other combinations. This gathering of hydrogen atom radicals by oxygen is called scavenging and oxygen is particularly efficient at it. Thus the yield of hydrogen peroxide in radiolysis, expressed by such quantities as the G value (G = number of molecules changed/100 electron volts absorbed) is modified under aerobic conditions because of the reaction

$$\overset{\otimes}{HO_2} + \overset{\otimes}{HO_2} \Rightarrow O_2 + H_2O_2 \tag{180}$$

and because the hydrogen atom is being scavenged by oxygen so that it can decompose the peroxide. The latter also results easily from union of two hydroxyl radicals.

Informing Processes 403

$$\overset{\otimes}{OH} + \overset{\otimes}{OH} \Rightarrow H_2O_2 \quad (181)$$

These are two cases where two free radical products react to produce a molecular product and, if this molecule is not decomposed enzymatically or by free radical events, it can readily oxidize tissue components. A free radical chain reaction can produce the peroxide as well.

$$\overset{\otimes}{HO_2} + e^-_{aq} \Rightarrow HO_2^- \quad (182)$$

perhydroxyl accepts an e^-

$$HO_2^- + H^+ \Rightarrow H_2O_2 \quad (183)$$

The H_2O_2 anion accepts a hydrogen ion

The other molecular product of radiolysis is hydrogen gas as would normally be expected in the breakdown of water. A radium solution will cause radiolysis by alpha particles.

$$H_2O \xrightarrow{\alpha \text{ rays from radium}} H_2 + \tfrac{1}{2}O_2 \quad (184)$$

Via free radicals, there are at least two routes to this molecular product.

$$\overset{\otimes}{H} + \overset{\otimes}{H} \Rightarrow H_2 \quad (185)$$

$$2e^-_{aq} \Rightarrow H_2 + 2OH^- \quad (186)$$

These are both radical-radical reactions which represent cases where high radical concentrations predispose the reaction toward the interradical combination rather than toward their outward reaction or toward the formation of intermediates in free radical reactions.

Polarization has provided a natural explanation for so many phenomena that it comes as no surprise to find the track of ionizing radiations making a free radical distribution that looks very much polarized (Fig. 10-6).

SUPEROXIDE DISMUTASE

With the momentary separation of electrons in charge redistribution in metabolism and energetics, it should not be surprising to find radicals present. It does come as a surprise, however, that *oxygen metabolism* should involve a reactive species like superoxide O_2^- which can be very dangerous if not disposed of rather quickly. This has been true all along for H_2O_2 and it turns out that the fate of these products is coupled together in several ways.

The superoxide radical is involved in dismutations which include autoxidations of many kinds, such as that of catechol amines, ferri-

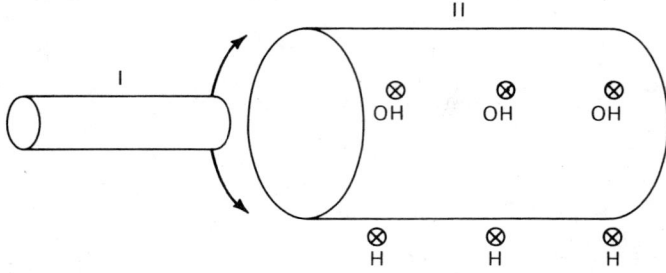

Fig. 10-6. Polarization around the ionization track. I-first, electrons are ejected from water forming a cylindrical pathway. Then the e^-_{aq} gives $\overset{\otimes}{H}$ and the positive water ions yield $\overset{\otimes}{OH}$ in a larger reaction zone II around the original pathway. Then the unreacted e^-_{aq} radicals can polarize spheres of solvent molecules in a cage effect as an additional polarization.

doxins, hydroxyquinones, hemoproteins, leukoflavins, rubredoxins, tetrahydropteridines, thiols, and reduced dyes. Next it is related to enzyme *catalysis* including that of xanthine oxidase. It is anionic and made in chloroplasts in the process of photosynthesis, with acceptance of electrons by oxygen, cytochromes and NAD.

The informational pathway runs through the functional linkages of the radicals in metabolism. It is especially clear in the relation of the superoxide to metalloenzymes. These enzymes include ATPase, carbonic anhydrase, alpha amylase, and aminopeptidase which have diphosphorylating and hydrolytic action. They contain Mg^{++}, Zn^{++}, Ca^{++}, and Mn^{++} respectively. Their bioelectrical modulations are illustrated by Mg^{++} which can modulate the charge of phosphate groups.

The radical is produced because when oxygen is reduced, it utilizes a univalent route, this choice being based on a spin restriction. Then it is a key intermediate, electron transfer (ET) substance. Its dismutation involves reactions such as

$$O_2^- + O_2^- + 2H^+ \rightarrow H_2O_2 + O_2 \qquad (187)$$

which, along with other reactions, prevents its accumulation. A perhydroxyl radical, for example, may combine with it,

$$\overset{\otimes}{HO_2} + O_2^- + H^+ \rightarrow H_2O_2 + O_2 \qquad (188)$$

These two oxygen products, H_2O_2 and O_2^-, would then be extremely dangerous if left to react with oxidizable biosubstances. Instead, catalases, peroxidases, and dismutases are on hand to prevent the damage. Superoxide dismutase accelerates the breakdown of O_2^- since, fortunately, it is readily available in a cell so that the superoxide can easily meet and react with it to prevent superoxide poisoning. Under photochemical conditions, the superoxide can form a radiochemical flux against which the dismutase is seen as protective. This suggests a role for it in primitive radioactive conditions when the earth's atmosphere changed from reducing to oxidizing. Then the enzyme could have guarded early life by acting as a "getter" for the superoxide. Since lipids play an informational role in membranes, it may have been this communicative link that was protected, since the dismutases acted against the oxidation of lipids.

There is an amplification of the superoxide action if it can form the hydroxyl radical

$$O_2^- + H_2O_2 \rightarrow O_2 + \overset{\otimes}{OH} + OH^- \tag{189}$$

which extends the scope of free radical action in oxygen metabolism. Since these events would then be extremely commonplace rather than caused by radiation only, there is obviously a great dependence on instant detoxification by peroxidases, and on catalases against H_2O_2. Then the dismutases work with these enzymes to manage the superoxide. Therefore, they would be recovered in anaerobes, sulfate-reducing, and other bacteria. Their presence in phagocytes and bacteriocidal cells has led to drug development (orgotein and erythrocuprein and others) which are used in medicine against inflammation.

What may well prove to be a significant role for radiation is the "activation" of common atmospheric, unreactive, triplet oxygen $^3O_2$ to singlet oxygen $^1O_2$. This lets radiant energy exert a coupling action on oxygen as it matches electron spins from parallel to antiparallel, the form many of its reactants will have. Visible light has the necessary coupling energy. Alternatively, the transition is catalyzed by lactoperoxidase and myeloperoxidase in phagocytes to kill target cells via the action of $^1O_2$. This is a product of the hypochlorite ion, OCl^- combining with H_2O_2.

Production of O_2 or H_2O is a common enzyme detoxification. Catalase combines two hydrogen peroxides into water and oxygen. Peroxidase makes NAD and water from the peroxide plus NADH and H^+, and superoxide dismutase also makes O_2. The ET system has a $4e^-$ transfer process in which O_2 goes to water while making and removing in the four steps O_2^-, H_2O_2, and $\overset{\otimes}{OH}$.

The transient radicals of oxygen linger sufficiently to modulate aerobiosis. *Spirillum volutans*, in nutrient broth, under nitrogen, tolerates only 1 to 5% by volume of O_2. The sensitivity of O_2 seen in anaerobes is clearly related to their low efficiency in the detoxification of transient oxygen radicals.

LINEAR ENERGY DISSIPATION

Bubble chamber and cloud chamber studies show how the ionizations form in tracks. Then the radicals are known to form on branches out from the ionization point or line in the irradiated material (Figs. 10-2, 10-7). If they are well separated, as would be the case in the path of a radiation with a small linear rate of energy transfer, or LET (Fig. 10-8), then they may not have an opportunity to combine to form the molecular products, hydrogen and hydrogen peroxide, and they will linger long enough to interact in adjacent regions of the medium. Conversely, with high LET (Fig. 10-9), the concentrations of radicals are such that

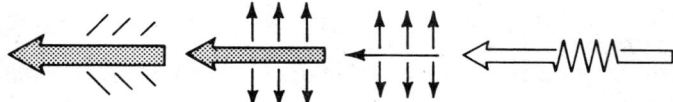

Fig. 10-7. Spur diffusion. Spur *fattens* after first interactions occur by diffusion of radicals outward. Diffusion is related to the availability of a reactant needing the odd electron in the free radical to form a pair. The electron transfer mode is used to eliminate the unpaired e^-. A reactant is needed that can radicalize toward stability. The radicals will initially distribute along track spurs in the path of the radiation energy. The spur of radicals expands itself to reach susceptible biogroups.

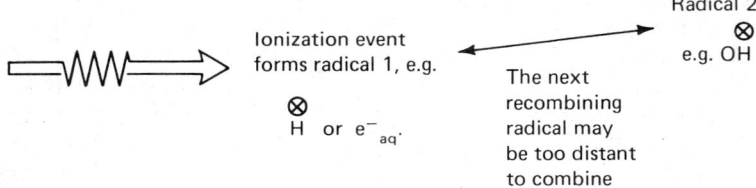

Fig. 10-8. The low LET situation (sparsely ionizing radiation).

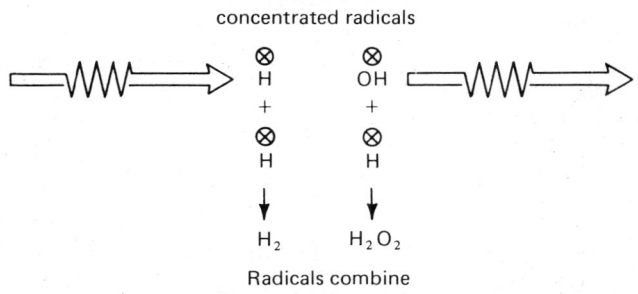

Fig. 10-9. The high LET situation (densely ionizing radiations): Radical-radical recombinations are favored along the spur with high LET and are decreased by the diffusion process.

the hydrogen atoms are likely to rejoin hydroxyl radicals to form water, and the hydrated electrons may produce hydrogen gas and hydroxide ions. The cylindrical distributions around the line of ionization may expand by diffusion and, as the diffusion distance widens, the radicals threaten solute molecules rather than yielding to self-destruction and related interradical products. When a molecular or radical product such as hydrogen peroxide is made, it extends the effect of the original radical. The linear energy dissipation from beta rays will be less than that for deuterons or alpha particles, and from gammas less than betas. The units will be in molecular distances and bond energies, giving a LET, for example, in keV per micrometer or angstrom. The difference in the radical concentrations produces the concept of "dense", for example, alpha particles, versus "sparse", such as gamma rays in ionizing radiations. In the first case, a biological substance suffers a sudden massive disturbance on a molecular scale while the second is a distributed chain of damaging events.

The phase during which the solvent is irradiated also influences the result. There will be more collisions in the condensed phase and reactants are more concentrated. In the radiolysis induced in solutions of radium salts for example, the oxygen molecules may react to produce hydrogen peroxide and this will leave the hydrogen gas in excess. The ionic yield will be ten times as great in liquid water as in ice, and for water vapor, it will be only one-tenth as much as in ice.

The formation of the perhydroxyl radical from the hydrogen atom is an example of the *oxygen effect*. This is the extension of the free radical damage in tissue if oxygen is present, for then the effect of the radiation persists through formation of perhydroxyl radicals. Susceptible organic molecules can also continue the chain of events forming organic perhydroxyl radicals in a similar manner to extend the oxidizing effects.

Due to its multiple sources, hydrogen peroxide should be fairly abundant, yet many times it is difficult to demonstrate. This is because of the tendency for reactions between free radicals and molecular products to produce or reproduce other species. The hydrogen peroxide formed in radiolysis may be destroyed by the hydrogen atom

$$\overset{\otimes}{H} + H_2O_2 \Rightarrow \overset{\otimes}{OH} + H_2O \tag{190}$$

and the molecular hydrogen scavenged by the hydroxyl radical

$$\overset{\otimes}{OH} + H_2 \Rightarrow \overset{\otimes}{H} + H_2O \tag{191}$$

The end result is a combination of the hydrogen atom and hydroxyl radical into water and a disappearance of the peroxide. Thus competition establishes conditions in which the molecular products are eliminated. If the hydrogen peroxide can remain, however, it is a powerful enough oxidizing agent to override any tendency toward the production of a reducing situation by radicals such as the hydrogen atom.

DIRECT AND INDIRECT ACTION

Based on an understanding of free radicals, electronically excited states, and formation of intermediates, it is possible to examine biological effects. Some biomolecules are sensitive while others are resistant to the attack of these reactive radicals. Purines and pyrimidines, for example, are much more reactive than other molecules. Sulfhydryl containing compounds such as glutathione may have a built-in resistance to the attack (function protectively).

Some of the molecules will be acted upon *directly* by the radiation, becoming ionized where the radiation strikes. As opposed to this kind of direct action, there is the type which results *indirectly* from the free radicals generated. Every time a DNA molecule is hit by an alpha particle with 600 electron volts of energy available for breaking the two chains, the chain will be broken by direct action if the breaks are within five nucleotides (Fig. 10-10).

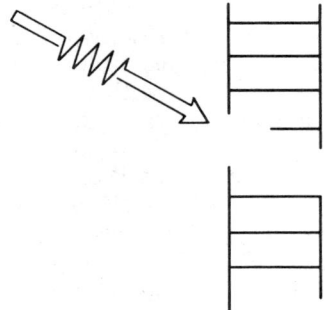

Fig. 10-10. The DNA molecule is separated into fragments by an α hit in Direct Action.

Indirect action follows the polarization and track diffusion route, and is caused by the oxidizing and reducing action of free radicals in forming reaction intermediates in the dispersion pathway. Thus there may be an action distance involved that is not characteristic of direct action. The target theory is used to explain the biomolecular basis of direct action (Fig. 10-11). It visualizes the radiation depositing its energy within a target so that the molecular zone around the ionization is functionally destroyed. This function may be the replicative or coding one, viral, hormonal, enzymatic, or some other sensitive type. The target or sensitive region is directly proportional to the molecular weight in molecular targets so that this value may be quite well estimated. What is estimated is the inactivation cross section and the data are interpreted in terms of the molecular shape as well. For example, ribonuclease is shown to be a spherical molecule with a radius of 2.1×10^{-7} cm and a molecular weight of 27,000. The cross section

Informing Processes

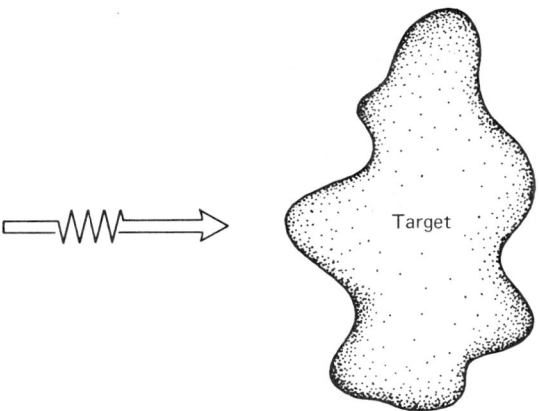

Fig. 10-11. *Direct Action* affects large targets such as biomolecular systems and the Target Theory is used to analyze the results.

may be different for two different bioeffects, that is, the functional organization may be in different molecular regions. If so, the data will not be consistent for determinations by alternate methods, but may illuminate the activation cross section differences.

Target analysis may make it possible to measure the yield of changed molecules in terms of the number per quantum absorbed. This number varies between 1 molecule and one thousandth of a molecule. The G level of 100 electron volts is justified as approximately the cost of the ejection of the electron which begins ionizing events. If it is a question of targets, it should be easier to hit a large target than a small one and this is found to be true; the amount of radiation necessary to inactivate the target to some standard level, such as the 37% one, is inversely proportional to the target size.

The necessity for separating target and indirect effects or even to discuss them as theories, requires that the indirect effects be suppressed to allow analysis of direct effects. This may be done by some experimental device, such as dehydration (if the biological structure is not already dry), or by freezing, which has similar features. These measures reduce the liquid water content and, thus, radiolysis and the production of diffusible free radicals for reaction.

Since the indirect effect is constant for a given amount of radiation energy deposited (Fig. 10-12), it is possible to divert it to protective molecules which harmlessly absorb the radiation effects. Such substances would be energy or free radical acceptors or scavengers. Typical ones are cysteine and glutathione (R—SH or sulfhydryl containing compounds). A cysteamine injection just prior to irradiation and at the site of the exposure, shows that radiation changes can be prevented at the time and place of the injection.

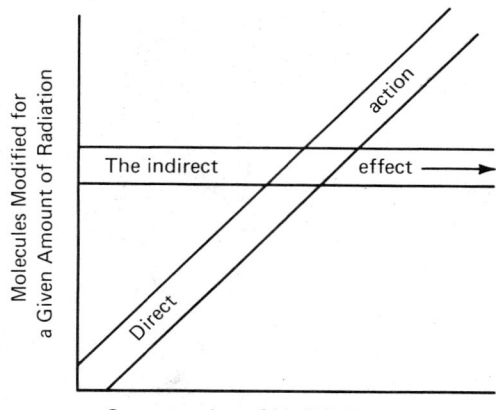

Fig. 10-12. *Indirect Action versus Direct* proceeds via the production of free radicals which interact to form intermediates and ultimately inactivated or modified molecules. Thus the energy available has an equivalent yield of changes. The curve is level in a suitable range of concentrations. Direct action is very sensitive to concentration and changed molecules are directly proportional to concentration. The direct effect thus can be diluted out of existence at very low concentrations.

The hydrated electron is very reactive with neutral sulfhydryl and disulfide groups in biopolymers. The —SS— group is especially sensitive to the electron in nucleic acids and protein. Thiourea is a free radical acceptor (FRA) for the hydrated electron and for the hydroxyl radical, thus protecting biopolymers. The resulting "cooled" free radicals (oxidized and reduced) then combine to produce a repaired polymer.

The direct effect varies with the concentrations of the target molecules since the more of them there are in a volume placed in the path of the radiation, the more the probability of hits and danger to the molecules. Thus this effect can be "diluted out" by reducing the concentration of sensitive molecules.

These basic actions will lie at the root of all the biological effects of ionizing radiations (Table 10-4) and will be involved in the use of radiations in radiotherapy such as in cancer treatment or immunosuppression for organ transplants, and in the development of protective compounds and treatment for radiation damage.

Sometimes the study is aimed at compounds used for enhancing the local destruction of abnormal tissue while protecting normal tissue as in the case of tumors, or for increasing the yield of desirable mutations. Protection is favored also by all of the measures that depress the ionization and production or diffusion of radicals. The oxygen effect is notable and anaerobiosis is protective.

TABLE 10-4. Partial List of Radiobiological Effects

Mitosis interference and delay
Mutation
Lymphocytosis
Carcinogenesis
Genic damage
Fragmentation and abnormal chromosomes
Cytoplasmic damage
Enzyme inhibition
Nuclear damage
Bone marrow degeneration
Intestinal mucosa damage
Spermatogenetic damage
Hair greying and alopecia
Immunosuppression
Selective tumor inhibition
Neutrophilia
Destruction of radiosensitive molecules such as purines

Another example is found in specie differences wherein some species possess greater protection than others. The microorganism *E. coli* has different genic strains; some produce catalase while others do not synthesize it. Those that are catalase negative are helpless to break down hydrogen peroxide when it is formed after a radiation exposure. The catalase positive cells survive while the negative ones continuously undergo inactivation. Many other features contribute to resistance including a high membrane integrity which would protect against the mixing of enzymes and of normally separated substrates.

BOUND WATER

An analysis of the target theory reveals that no such single theory is sufficient to explain the effects of ionization. A suitable compromise would put the target molecule inside a sheath of water molecules, thus enlarging the target size. However, it is extended in a rather special way. The water, so bound and actually a part of the molecule, is receptive to radiolysis so that the parent molecule effectively carries a radiation energy sink in which free radical chain events may be initiated. The zero diffusion distance that is now involved makes the bound water different from the rest of the solvent structure. In essence this means that the region of influence of the target theory becomes indistinct at the boundaries and indirect action can be presumed to add to the biological result.

Bound water is difficult to dislodge in dehydration without causing a loss of function, so that this fraction of the solvent may be presumed to remain in those substances in which the water content is reduced by natural or artificial means. The indirect effect may persist in freezing also. Usually tissue freezes at just below $0°C$, but freezing out the effect of free radicals in bound water may require much lower temperatures. Thus this effect would persist in the experiments designed to test the validity of the target theory.

COMPARTMENTATION

It has been seen that the compartmentation existing at various levels works against diffusion. The hydrated electron, well down in this hierarchy, is bound for a time in its solvent cage and in turn, along with free radicals from radiolysis, constitutes a receptive organization which is divided into cell compartments, the mitochondria, ribosomes, and organelles which limit the receptive ranges. The indirect action that occurs within the cell is more significant than the intercellular indirect action (e.g. with suspensions of cells), because the cell contents are not diluted and the cells are merely dispersed. Similarly, macromolecules display levels of organization into primary, secondary, tertiary, and higher configurations. It has been seen how the inactivation cross section may not be the same as the molecular cross section. Sensitivity may be related to the different organization levels and to the periods of transition between levels during which the molecular exposure and the exposure of sensitive functional sites are varied.

Overall, radiolysis, as described here, produces six radiation products from water.

1. e^-_{aq}
2. $\overset{\otimes}{H}$
3. $\overset{\otimes}{OH}$
4. $\overset{\otimes}{HO_2}$

5. H_2O_2
6. H_2

The last two molecular products predominate at high linear energy transfer rates while at small LET, the radical yields are greater. This shows that molecular products are formed from primary products occurring within convenient combination distances. On the other hand, dispersed radicals widen the zone of reaction and increase the probability of finding appropriate solutes for reaction.

The study of the molecular biophysics also involves the oxygen effect, in which the production of perhydroxyl radicals from water and organic molecules increases the total reactivity. There are also the forward reactions and recombinations, the effect of densely versus sparsely ionizing radiations, the effect of protecting versus hazard-increasing solutes, and the differences due to high and low rates of applying the radiation energy.

The resonance effects in electron spin resonance and chemical reactivity in the free radicals are related to their unpaired electron. The radicals have an odd number of electrons, one being unpaired (not associated with another). This makes the free radicals such as the hydrogen atom or hydroxyl radical much more reactive than their ionic counterparts. Usually an atom has its pairs formed and must find a suitable partner which lacks an electron to react with it or make a chemical bond. Hence the free radical, with its unpaired electron, may react easily. Similarly, due to the unpaired electron, the odd electron radical has a paramagnetic effect. This means that it can resonate when a microwave field is applied. Microwaves and radiolysis are seen to be related when it is observed that polarization of molecular systems continues as a basic response to radiation perturbation, and both microwaves and radiolysis have a special absorptive relationship to water.

The presence of highly reactive free radicals in biological systems presents an immediate physiological challenge to oxidizable or reducable organic units and can easily degrade biomacromolecules.

PROBLEMS

1. a. Of what is the hydrated electron the simplest example?
   b. What is the polarization product from orientation of the water dipole?
   c. What is the product resulting from the interaction of the solvent with an electron ejected by radiation?
   d. To what is the motion of the hydrated electron equivalent?

2. How do hydrated electrons lead to production of the hydrogen atom? Give the free radical equation.

3. Use an equation to show how the hydroxyl free radical is formed from the positive water ion.

4. What are the main results of perhydroxyl radical formation?

5. What primitive biological role is suggested for superoxide dismutase?

6. What is the result of an alpha particle's traversing a DNA molecule?

7. What is the overall result of radiolysis of water?

BIBLIOGRAPHY

Bacq, Z.M., and P. Alexander, *Fundamentals of Radiobiology*, Pergamon Press, London, 1961.
Berliner, L.J., Ed., *Theory and Application of Spin Labeling*, Academic Press, New York, 1975.
Casarett, A.P., *Radiation Biology*, Prentice Hall, Englewood Cliffs, New Jersey, 1968.
Dole, M. *The Radiation Chemistry of Macromolecules*, Vols. I and II, Academic Press, New York, 1972, 1973.
Fridovich, I., "Superoxide Dismutase," *Annual Review of Biochemistry*, Vol. 44, Annual Reviews, Inc., Palo Alto, Calif., F.M. Snell et al, Eds., 1975.

Hart, E.J., Ed., "Solvated Electrons," *Amer. Chem. Soc. Advances in Chemistry Series*, Vol. 50, 1965.
Huyser, E.S., *Free-Radical Chain Reactions*, John Wiley, New York, 1970.
Lea, D.E., *Actions of Radiations on Living Cells*, Cambridge University Press, Cambridge, England, 1946.
Pizzarello. D.J. and R.L. Witcofski, *Basic Radiation Biology*, Lea and Febiger, Phila., 1967.
Platzmann, R., "Physical and Chemical Aspects of Basic Mechanisms in Radiobiology," *Conference of U.S. National Resource Council*, Publ. no. 305, 34, 1953.
Pryor, W.A., *Free Radicals in Biology*, Vols, I, II, and III, Academic Press, New York, 1975, 1976.
Schwarz, H.A., *Advances in Radiation Biology*, L.G. Augenstein, R. Mason, and H. Quastler, Eds., Academic Press, New York, 1964.
Spinks, S.W.T. and R.J. Woods, *An Introduction to Radiation Chemistry*, John Wiley, New York, 1964.
Wang, S.Y., Ed., *Photochemistry and Photobiology of Nucleic Acids*, Vol. I-Chemistry, Vol. II-Biology, M.H. Patrick, Ed., Academic Press, New York, 1976.

11

Impedance and Membranes, Core Conceptualizations in Biology, Dielectric Analogies and Information Propagation in Tissues

> "The time has come," the Walrus said,
> "to talk of many things:
> Of shoes—and ships—and sealing wax—
> Of cabbages and kings—
> And why the sea is boiling hot—
> And whether pigs have wings."
>
> Lewis Carrol

COMMUNICATING AND COMPARTMENTING MEMBRANES

When, early in evolution, organisms learned to form molecular sheets, they gained at once the advantages of compartmentation, communicating, mobilizing, organizing structures, partitions, as well as the status of an isolated, possibly independent working unit. This seems to rest on polarizations, such as the natural hydrophilic and hydrophobic action illustrated in Fig. 11-1, where a membranous sheet can form from micelles by the natural inclinations of associating molecules. Examples are also seen in the interfacial membranes in blood clotting, the surface skin of milk custard, or that of heated milk. With compartmentation came a greater role for the communicating molecules in separated regions. Electrical phenomena provide opportunities for cell, membrane, and tissue components to become information sinks and links and so balance their isolation. These and related matters are taken up in this and following chapters.

When a glass slide is dipped into fatty acids, the molecules can form a monomolecular layer on the slide as it is carefully withdrawn. Thus fatty acids seem to have special surfactant abilities and they are found in the phospholipid portion of cell membranes. These cases furnish examples of molecular ordering based on the information provided by water or lipid, and a negative or positive charge preference at one end of a polarizable molecule.

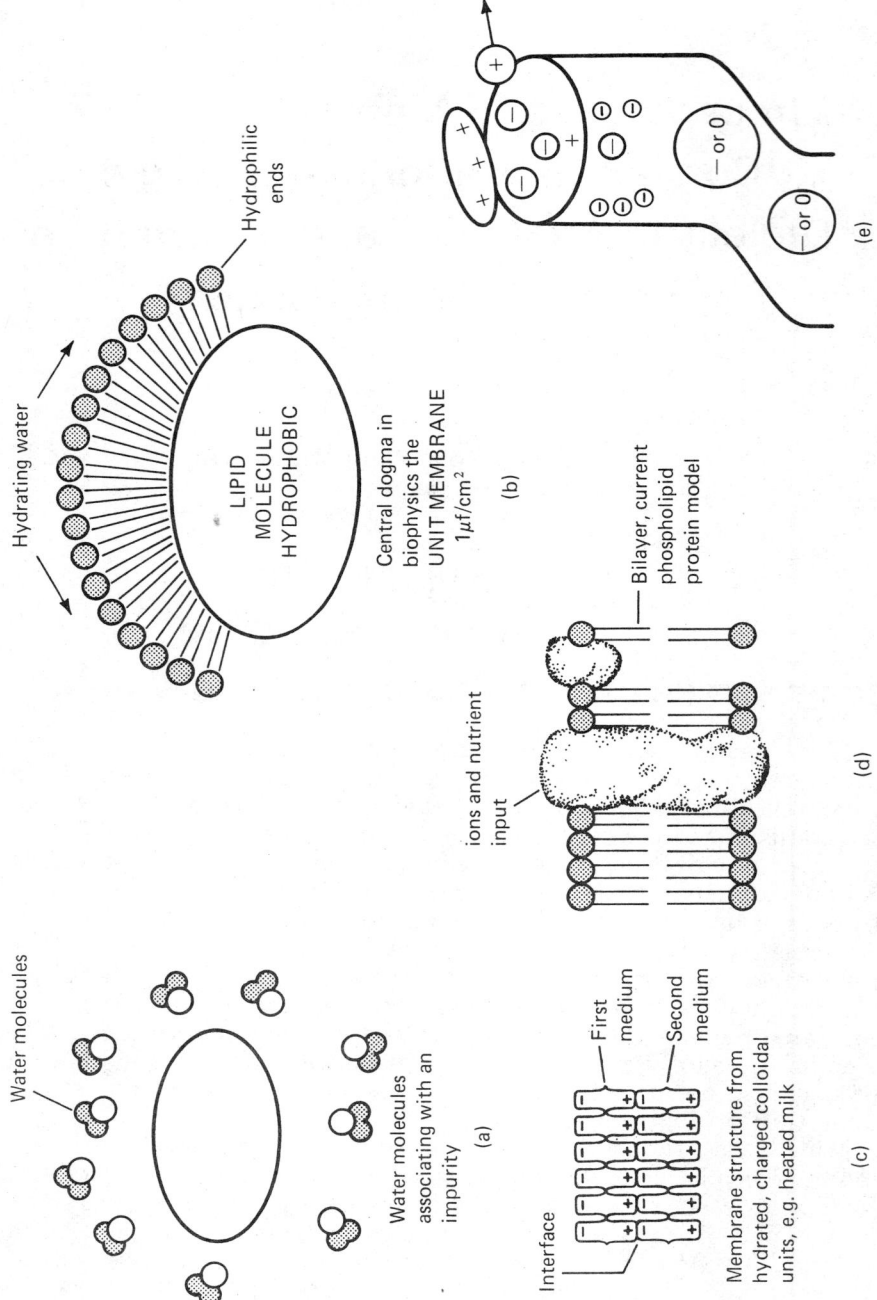

DISCOVERY OF THE MEMBRANE IMPEDANCE CHANGE AND THE UNIT MEMBRANE CONCEPT IN BIOLOGY

The biologist has an enormous field for his observations and he must try to balance the requirements of molecular biology, cytology, morphology, genetics, embryology, physiology and the numerous subareas of his field. In guiding his experimental interests, an expertise in the choice of experimental material is invaluable and may permit special relations with the available theory to enrich the investigation. The physicist resembles the biologist in also having working systems to deal with, but his have physical attributes that usually relate easily to classical or modern theories.

It is quite an event then, when these two intersect in their investigations. So it was when Hugo Fricke in Cleveland, Ohio showed Kenneth Cole his physical techniques for studying biological material. The result was some fifty years of outstanding biophysical investigation by this converted physicist (Kenneth Cole) which led to the nerve propagation and membrane excitation models that are probably the most durable and successful examples of delineation of structure and function in biology. In 1924 Fricke was obtaining electrical measurements on red blood cells and on milk fat globules in suspensions. Not only were his measurements meaningful, they set in motion those fifty years of brilliant work combining electromagnetics with biology.

Any of the persons in this bioelectric history would quickly credit other giants of the nineteenth century, and even earlier observers such as Ben Franklin and Galvani. However, Dr. Eric Ponder, von Helmholtz, Dubois-Reymond, and Kohlrausch introduced bioelectrical investigation. J. C. Maxwell and his *Treatise on Electricity and Magnetism* and Ohms law, which came in 1927, provided the physical attributes. In this century Bernstein offered a very provocative membrane theory based on semipermeability, with the internal side negative, and almost jumped fifty years ahead by stating that the active membrane switched to permeable to allow ions to cross which, in turn, cancelled the membrane potential.

Earlier, the Maxwell side of the study was reinforced by the Clausius Mosotti definition of the dielectric constant which gives a basis for the dielectric nature of membranes. On that same side, Lorenz-Lorentz provided the index of refraction as a fundamental concept while Poisson gave equations for a two-phase system.

Fig. 11-1. Membrane developments: a) Particle contaminates asymmetric water molecules; b) Polarization or orientation of membrane-forming particles on a substance with their water preferring termination outward and hydrophobic ones inward; c) Charge preference directing orientation and electrical distribution leads to a structure; d) Current phospholipid-protein model contains giratory proteins and mosaic distributions of fixed and fluid regimes; and e) Charge distribution in pores.

Fricke's work is probably the earliest use of the impedance concept in biology. In 1924 he showed how conductivity and capacitance could inform on the shape of suspended biological units; their dimensions, volume, and status of colloidal systems. He studied the change of resistivity with frequency from 1.6 to 200 kc. The giant step was the fusion of his results with those of Gorter and Grendel's lipid analysis of red cell membranes in 1925, which couldn't have happened without Langmuir's brilliant techniques with monolayers. *Out of this came the core concept of a biological unit membrane with dimensions.* This is a uniform membrane capacity of 1 $\mu f/cm^2$ with structural descriptions since provided by electron micrographs. The capacity persists even when accompanied by large changes in membrane conductance during operation and is independent of function.

From Peter Debye, Cole obtained other physical tools very much in evidence as this chapter uses Debye's theoretical contributions to give a basis for experiment and analysis. The trend set by Cole continued as several outstanding scientists, both physical and biological, moved about between each other's laboratories, merging techniques, and adding to the story.

Hodgkin and Huxley produced the Na and K ion model needed to explain permeability changes during the propagation of impulses in excitable membranes. Their nerve equations had shown an inductive reactance and were of no small comfort to Cole. However, Alan Hodgkin saw Cole working on the giant squid or cuttlefish axon and this inspired his most significant contributions. Of course squid deserve credit too and no doubt they paid a price as populations were heavily depleted near marine laboratories at Woods Hole, Massachusetts, Plymouth, England, and Misaki, Japan during the height of this work. The squid axon may be as large as 1 mm in diameter and so, lends itself to such remarkable techniques as the *voltage clamp* which requires five or more electrodes, two in, and three outside the membrane. Other organisms such as *Drosophila* would obviously not be suitable. Without this experience in the choice of experimental material as provided by trained biologists, little could be accomplished. Biologists learn that sometimes tissue culture is appropriate and at other times, as in oxidative respiration, the mitochondria of the thoracic muscles of bees are suitable. Metabolic cycles, DNA, and RNA tests might well require the services of *Escherichia coli* or its extracted fiber of DNA; embryologists will often pick out *Arbacia punctulata*, the sea urchin, for its eggs which develop more or less in plain sight. Alternate choices might include salamanders, frog, or chicken eggs for differentiation and development. The biologist knows of little gems like the sand dollar embryo's being nearly transparent so that its gastrulation can be viewed directly instead of inside a uterus.

At various times Cole did work on *Nitella*, a plant cell similar to the squid axon in size, which had an interesting alternate ion (instead of sodium, it included calcium), *Arbacia* eggs, squid, and muscle. He and others also used lobster axon. The renowned biologist, J. Z. Young, first explained the features of the giant squid axon to Cole. Young, a firm believer in the critical nature of information pathways, appreciated the initials his parents provided and so do we, for they allow no uncertainty as to which Young chose the giant axons for study. At first he mistook these thread-like structures for veins but was soon showing their conduction of nerve signals. Knowledge of the 0.1 V potential difference between the interior and exterior, the explosive character of the charge propagation, the refractory period, the fuse-burning effect in which one explosion triggers the next, conduction without decrement and without losing information, and the volley of synaptic transmitters, all came in time.

The Bernard Katz work, for example, started in 1941 and, along with the collaboration of José del Castillo, Paul Fatt and Ricardo Miledi yielded the Nobel Prize in 1970 for clarification of the quantal (energetically stepwise) release of packets of neurotransmitters. These, such as acetycholine, form volleys as miniature end plate potentials rise at neurojunctions.

When Curtis and Cole found an ac impedance change coincident with the rising of the nerve action potential, the ionic permeability shifts in membrane became understood. Then it was José del Castillo and J. W. Moore who showed an almost infinite velocity of propagation for the excitation and synchronous potential change along the whole axon (something startling when it is considered that here nature comes close to the effect of using wires).

ROLE OF FREQUENCY AND OTHER ELECTRICAL PARAMETERS

Again, in the nerve impulse work, the emphasis was on the problems of getting ions inside the cell with charged membranes on them rather than toward a frequency-directed change in the electrical impedance, the opposite line of thinking. As a result, of course, the monumental discoveries were made with the squid axon, showing that membrane permeability changed for the nerve pulse, allowing the entrance of ions during a peak of conductance. Then with an internal space clamp, plus external electrodes, del Castillo and Moore studied the super speed of the polarization impulse, and progress was made toward showing the threshold for exciting the pulse and the use of membrane "capacity" to send an ionic current across when directed. Space clamping, current clamping, or voltage clamping refer to the separation of a nerve axon section by electrical isolation in order to apply a feedback controlled electrical input to it and observe results. This is usually explained as being like a special length of gunpowder fuse. Adding subthreshold heat does not maintain burning to the nearly adiabatic flash point which is like the membrane *current clamp*. Cooling away the heat of combustion makes an even burn isothermally analogous to a *voltage* clamp. The process provides an electrical stability for measurements so that ordinary differential equations can be used to describe the data, making mathematical analysis feasible. An overshooting of the action potential and negative resistance produce a complex active mechanism.

The burden of responding to membrane potential information is placed on the Na and K ions in a *nerve axon*. If the potential suddenly increases, it means that the membrane must increase its conductivity to Na quickly, and then let the conductivity to Na rest while K conductivity surges up to a steady value. There is a "shock" value to the stimulating directive when it hits quickly so there is rate information in the stimulus in the sense of a reinforcing positive feedback with respect to the conductivity to Na. Opportunities for information management include: resting value, rectification, quantal behavior, frequency dependency, potential at rest and at critical points, anomalous *inductive and capacitive* reactances (therefore impedance), ion fluxes, duty cycles, strength of signal and its form for threshold action, also accommodation action inherent in the utilization time and limited gradient, refractive and recovery phases, repetition for continuing currents and form and velocity of propagation. Had Cole added frequency effects directly instead of in the frequency dependent impedance, his reports would have had aspects of electromagnetic communication. Instead, frequency had to be stabilized so that other measurements could surface. Yet some of the conductances and impedances required a careful choice of frequency or they would have been quite different. The serious student of these matters sees that the electrical functions in tissue can be associated with both classical physics, with which he is familiar, and with communications engineering (especially with cables). Thus this discussion must have a combined physics and engineering tone to reflect its true nature.

The communication is two way, locally at least, giving a feedback control which is recognized as always being based on information. A subthreshold potential starts Na conductivity inward across the membrane, but this is quickly countered by outward K conductivity to return the potential toward the resting value. A threshold potential can direct a sustained Na ion current up to the desired Na potential where it then decelerates as K turns on. The task of relaxing the system to rest values is taken up by K with longer time constants.

Fricke made calculations on the basis of a dielectric constant of 3, similar to lipid. Then the capacitance and electrostatics suggested a thickness of about 33 angstroms, similar to the lipid size. The information content specifies about 100 angstroms for membrane thickness with the protein and double phospholipid layers and indicates that it is not loose and wandering but solid, at least where such order is necessary, and this structure is continually replaced. Then, proteins can have rotating bases for active transfer of components as shown by other studies. The membrane capacity and its *relative* independence, suggests that this protein portion, with its active transfer function, should be a small fraction just as seems to be the case with the giratory proteins.

Electrically, the temptation is to look at the membrane as a semi-crystalline structure rather than as a liquid or a gaseous one.

The membrane potential changes (from inside) allowing ions to enter. Logically, nutrients should have such access just as products must have egress. When the ions enter, the potential disappears and the opportunity is provided.

It will certainly be significant in studying the model for the influence of frequency that 200 monovalent ions permeate in 1 msec, giving a current of 200 ma/cm^2 across a hexagonal surface of 2×10^4 A^2, and they are fed by an ionic bombardment of the external hydrophilic faces which amounts to about 10^9 ions per msec; and that this current is directly related to the field applied and the conductance, where the latter depends on the mobility and concentration of charges and on a high field strength acting on relatively few ions. It is also significant on surfaces that can be electrically charged or surface-active with fatty acids present. The giratory units are also notable in that they can orient, relax, and reorient in active transfer giving possibilities for biology that are unlikely in inorganic semiconductors, and such activity is carried out "without motors and alternators."

In conversations at the laboratory of José del Castillo, the author found that he modestly defers to the discoveries of others—especially his one time collaborator, Bernard Katz (Nobel Laureate in 1970)—and does not think of his work on velocity as unique. Instead, he is happy to have continued patient developments in micro-electrophysiology, carefully following such models as the cardiac and fresh water shrimp muscle cell physiology with voltage clamp and multi-electrode techniques. Painstaking work such as his proceeds amoeba-like, with scientific pseudopods, based on a long sequence of manipulated variables. Addition of some biochemical substance is made and the model is then searched for significant information. There is a close coupling *between* the biophysicists, with loops formed by their publications and lectures. Katz received the Nobel prize for work that spanned thirty years and explained elements of mechanisms in the release of transmitter substances from nerve terminals. Paul F. Fatt aided him in visualizing the miniature end-plate potential and José del Castillo collaborated on the idea of quantum transmitter release. Acetycholine is ejected from the nerve endings in sizeable molecular groups which infiltrate the opposite side as voltage changes signal the opening of ionic channels in muscle membranes. During successive stimulations, none, one, two, or three quanta are released according to the stepwise potential changes in the end-plate. The smallest step matched the miniature potential in size and form and one such quantum would mean a volley of about 104 acetycholine molecules. The laureates like Katz, the Swiss scientist Werner Alber, Hamilton O. Smith and Peter Simon of Johns Hopkins who shared honors for discovering DNA "restriction enzymes," and F. Lipmann, the great biochemist, inspired admiration and respect in their research associates who willingly followed the pathways suggested. They were grateful for the perception and acute sensitivity of their teachers who showed, again and again, the most promising of the many routes open to investigation.

In a paradox of convergence and divergence, the area of investigation is ever widened and enriched as the specialization becomes more intensive, so that these inspired people enjoy a panoramic life of disclosure, essentially denied to others.

There will have to be many more of these brilliant scientists before it is known how liquid, electrical junctions operate in membranes and electrolytes, the precise field configurations to be expected, ionic and colloidal behavior, combined hydrophilic and hydrophobic effects with osmotic and electrical mobilities of charge carriers as well as how all the other parts of the biophysical puzzle fit together. Thanks to the progress made by these and other dedicated investigators, it may not take the 50 to 100 years that membrane work required to bring us the polarization, Na, and K model for membranes.

HOW THE MEMBRANE, MOLECULE, INDIVIDUAL CELL OR TISSUE IS INVOLVED IN INFORMATION AND ENERGY TRANSMISSION

The interacting radiations described in Chapters 8, 9, and 10 all belong to the short wavelengths side of the electromagnetic spectrum. Their quanta contained significantly larger amounts of fixed energy vibrating at super frequencies. For the remainder of this book we will be concerned mainly with interactions involving the long wavelength side, that is, the electric, radio and microwave frequencies. These give broad-band absorptions in tissues and their electromagnetic waves are usually *distributed* rather than concentrated in small volumes.

The region concerned is essentially that from 10 to 10^{10} Hz, and the emphasis is on the sinusoidal nature of the currents produced. That is to say that, ordinarily, as in the soothing effect of medical diathermy or gross absorption experiments, the waves affect a commensurate or larger-sized region and the wavelength in the medium is less than the wavelength in space. In order to concentrate the energy, various devices are used, outstanding ones being the maser and the laser or the helical director as used by the author (Copson, 1975).

In the spectroscopies used in this large region, associated instrumentation can help to produce the sharp spectral bands required, as described in Chapter 6. The emphasis is now on the electromagnetic wave features and on the forcing functions of the energy on the susceptible particles and on how this energy allows them to do work against fields or against viscous resistance in the tissue medium. Such media then come to be described in terms of their impedance which is another of the significant conceptualizations in biology. The impedance Z is frequency dependent (Table 11-1) as are the important properties of the tissues and cells, ϵ, the dielectric constant, μ, the magnetic permeability and σ, the conductivity. Thanks to this double duty for these

Impedance and Membranes 423

TABLE 11-1. Relation Between Components of Loss and Frequency

σ	=	ω	ϵ'	tan δ
conductivity		frequency dependent	dielectric constant	loss tangent
↓				
salt solution		ω = 2 π f	dipole response absorptive polarization	phase relations in the interacting systems
↓				
$I^2 R$ loss				

properties (the electrical and tissue functions) they are used to help describe biological structure and function at several levels beginning especially with the molecular one. Thus the molecule, individual cell, or tissue becomes an information and energy sink in which the deposition can accomplish a communication, informing, or energy delivery function. One of the most significant structures is the membrane, and some of the concepts developed in this chapter, in the hands of the aforementioned biophysicists, provide the most successful analysis of biological structure and function ever made in the characterization of membranes and nerve impulse transmission. In order to enter these areas resolutely, the background that follows in terms of polarization phenomena is extremely useful.

RELAXATION DEFINED

Most informational processes of this kind which are related to radiations are linear, and the response of the informed system is proportional to the stimulating information under conditions of equilibrium. Electromagnetically, the stimulated systems do not respond immediately to the directives for changing and this delay is known as a *relaxation* process. As seen in biology, a very large number of systems can be described as dielectric or semiconductor ones, electrical conductors being reasonably few. Semiconductors are imperfect dielectrics or insulators, and these broad classes of materials have also been discussed in Chapter 6 under dielectromagnetics. With the exception of electric organs as found in eels or in the fish *Anathonemus petersii* and *Gymnarchus niloticus*, biological materials have small or reasonable conductivity. There is organic semiconductor-like local electron activity in special systems under certain conditions and it is true that organisms are permeated by fields, waves, and local conductance of this nature. Radiations, as stimulating forces, polarize the organic materials which respond in this way to the electromagnetic fields. The delay in response is an equilibrium-seeking *mechanism* by which the radiant energy can be transformed into other forms of energy notably, but not exclusively, heat. In equilibrium-seeking processes such as those of biomolecular systems after being hit by a forcing fuction, molecular reversibility is assumed. This means that any process and its reverse will occur, on average, at the same rate. This condition is satisfied when an intersystem is observed which includes responding units such as dipoles at one end and stimulating oscillators at the other.

In this chapter, the consequences of the time delay are examined mathematically and interpreted for the reader. Our search for informational pathways is again rewarded because the systems are informed as they are polarized and the process of relaxation has an informational content that could be measured in a separate study. Once the process has been identified as a relaxation, it is a generic one, applicable to innumerable models especially in biochemistry and enzymology. The radiation biologist sees the process as one informing about the structure and function of unknown areas, using radiations for probing and measuring the output or response. The behavior of the stimulated elements is described in terms such as impedance, resistance, capacitance, currents, and voltages and a biological model is derived which has these elements as equivalent or analogous concepts.

The relaxation encounters resistance or forces that would dampen it, and it is the input of radiant energy that allows it to do work to overcome it and let the system respond or polarize. The resistance may be viscous drag and temperature dependent. The relaxation is due to a withdrawal or reversing of the radiant field and is aided by thermal motion as well as a natural tendency to return to the favored position *after* the forcing function has been withdrawn or changed. Force can be exerted against relaxing systems by a simple, local flow of electrons or by highly complicated procedures in complex proteins. Relaxation is a mechanism or process in a circuit or equivalent circuit and is not a characteristic of the circuit itself. It is a process that may happen when elements of the circuit are arrayed in a suitable manner.

The significance of the biological model is that it suggests an equivalent circuit that, in turn, suggests a given electromathematical treatment. Thus, no single analysis serves for all circuits with respect to relaxation or impedance. Solving this problem requires a simplification of the circuit by rearranging the elements in an equivalent, manageable way. If they are equivalent, they will respond to applied voltages in a similar manner. Figure 11-2 shows such a change made in the interest of simplicity or in recognition of the fact that measurements could not tell one from the other. Fricke saw correctly that the equivalent circuit for the cell should be parallel capacitance and resistance with an added resistance in series with the capacitance. The latter represents cyto-

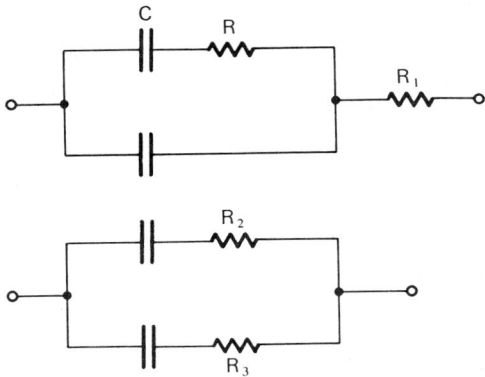

Fig. 11-2. Representation of R_1 in an equivalent circuit by R_2 and R_3. Circuits with several resistances and a capacity can be classified as equivalent to either of the two small circuits shown. Another way of saying this is that electrical measurements at the terminals do not distinguish one from the other as to the number, position, and size of the elements. Thus a given set of data can be used for a number of circuits.

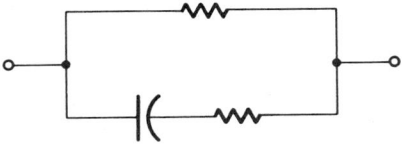

Fig. 11-3. Equivalent circuit for cell.

plasm while the parallel elements stand for the membrane (Fig. 11-3).

Relaxations are broader than electrical effects, for example, with their solutions, they often apply to analogs such as acoustical cases. In biology, where molecules, cells, and tissues may be the substances that determine capacitance in place of air in the capacitor, the dielectric analogy is important. An isothermal expansion of air in pulmonary passages is a relaxation and the behavior of ultrasound on tissues shows ultrasonic relaxation. An example of a curve of relaxations is given in Fig. 11-4 where the frequencies of relaxation of whole tissues are indicated. In our overall theme, these frequencies suggest pathways by which electromagnetic radiations influence tissues and organisms.

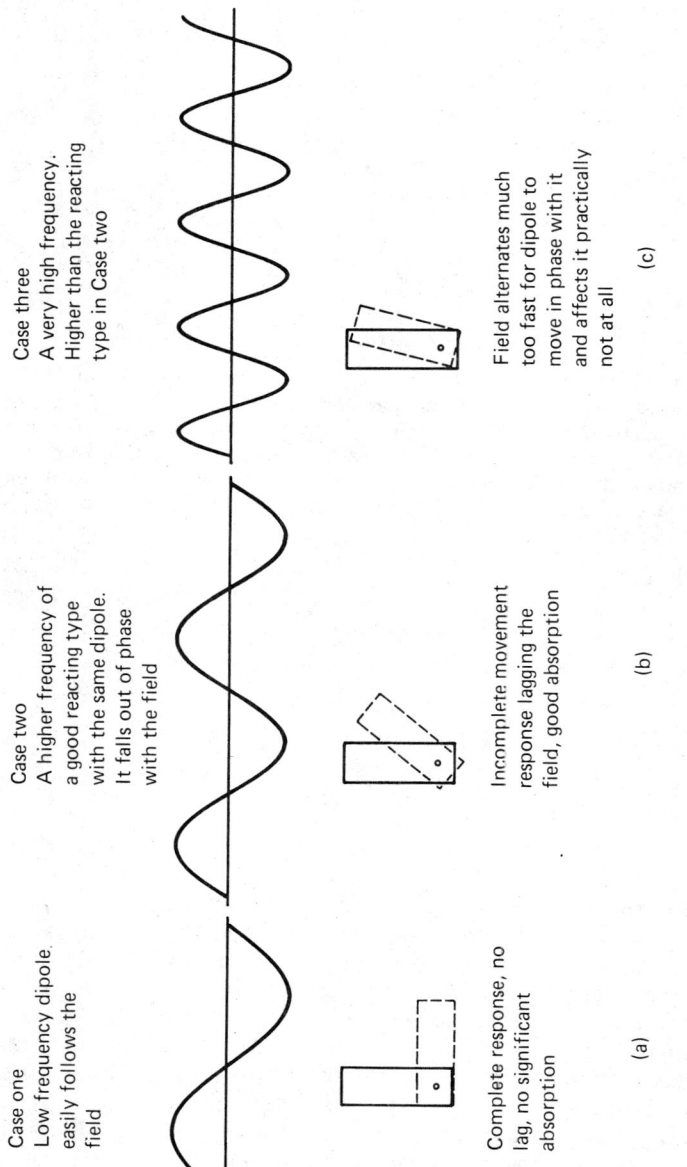

Fig. 11-4. Frequency response. Between 0.1 and 10 times the relaxation frequency, there is a good dipole response to the fields in terms of reacting frequency, the phase of the field, and the phase of the responding dipole, and the absorption of radiation energy.

HOW THE MOLECULE, SINGLE CELL OR TISSUE BECOMES AN INFORMATION AND ENERGY SINK

If the displacements due to radiations that occur in exposed media or in samples are interpreted as polarizations, then the latter become a very common response. Thus, resonance of the NH_3 molecule at its maximum frequency f_m in the microwave region can be regarded as a *reversing polarization* and many other reversing displacements have features in common with the NH_3 resonance. Informationally, resonance is a molecular recognition process. Thus each kind of chlorophyll has its own wavelength which absorbs solar energy (Fig. 7-11). While these are molecular responses, the single cell or aggregate of cells may also be the responding system in the interactions described in this section and dimensional resonances, especially with electromagnetic fields, are equally important.

The reversing nature of the forcing function, the electromagnetic wave, and the relaxation mechanism are the paramount considerations in the absorption process. Then the events at the molecular level progress to a widespread disturbance which is observable at the level of the whole medium or organism. The molecules act as energy transducers for the force exerted on them, gaining potential energy in the process. This may be further degraded into thermal energy as the molecules respond to their energy status against opposition in the medium. While the information running in parallel is heading for the same type of fate, with the aid of appropriate analysis and interpretation, it also communicates considerable molecular data. Once the degradation to the thermal status occurs, the informing power is drastically reduced since it depends on subtle interactions at energy levels far below saturation, but above thermal agitation levels. This said, the analysis proceeds in depth for those interested in gaining a command of the energy sink situation.

Polarization absorption processes follow the type of change where the dipole orients according to the cyclic forcing function of an alternating field as in microwaves. The wave shows changes in phase and amplitude, the molecules orient and, at high intensities, measurable temperature changes occur. One approach is to follow the decay in this orientation of position. These motions absorb energy as dipoles attempt to overcome potential energy barriers ("an opposing rotation") and, consequently, absorption occurs in the medium. A great deal of information about the dipole, its charge, shape, size, mass and bonds is revealed by the time constant, so this is given the special name of relaxation time τ which equals the reciprocal of alpha α, the decay constant. In particular, the dipole may have a short or long relaxation time and dipoles accordingly respond to the frequency. τ is studied as a function both of molecules and frequency.

THE FREQUENCY RESPONSE

At some special frequencies, an unusual response may occur, for example

$$\tau = 1/2\pi f_m = 1/\omega_m \tag{193}$$

where ω is the angular frequency, and f_m is the relaxation frequency or point of maximum response of the dipole to the radiation. The electromagnetic energy passes into the absorber most readily at ω_m, the frequency which is specific for a dipole, its temperature, and the viscosity of the medium in which it exists. In microwave absorption the whole interest may be in these characteristics of the medium, especially in the relaxation time of dielectrics. If the value of τ is known in seconds, the frequency response may be predicted. For example, if τ_{H_2O} equals 10^{-10} sec, then

$$f_{relaxation} = 1/2\pi 10^{-10} = 1.6 \times 10^{11} \text{ Hz} = f_m \tag{194}$$

In the opposite sense, the value of τ may be found from the relaxation frequency. A protein may have

$$f_{relaxation} = 1.6 \times 10^6 \text{ Hz} \tag{195}$$

$$1/\tau = 2\pi (1.6 \times 10^6) \tag{196}$$

$$\tau \approx 10^{-7} \text{ sec} \tag{197}$$

In comparing these two dipoles, protein will have a longer relaxation time than water, the time being related to its size, and its relaxation frequency is a radio frequency, while that for water is microwave. Observations also indicate that a good frequency response (i.e., absorption of the energy) is obtained if the frequency of interest lies between 0.1 and 10 times the relaxation frequency. For radiations alternating at less than 0.1 times the $f_{relaxation}$, dipole orientation can keep pace; between 0.1 and 10 times $f_{relaxation}$, they lag the field or are out of *phase*, and the orientation involves a struggle against opposition. Beyond 10 times $f_{relaxation}$, they can change practically not at all (Fig. 11-4). Thus in the above example, the no-response region begins at $10 \times 1.6 \times 10^{11} = 16 \times 10^{11}$ Hz for H_2O.

There are many examples of reacting systems that exist both in and out of phase to be found in biology. The diurnal tide cycles function on a lunar basis and physiological, behavioral periodic phenomena based on circadian cycles (24 hours) day-night, light-dark and biological clocks follow this general concept, e.g. bone marrow cell mitoses have AM and PM phases. Perhaps a more direct example is the relative phase situation for current and voltage in a capacitive electrical circuit. There the current, as shown by a waveform on an oscilloscope, leads the voltage by $90°$ or is out of phase by this amount as a characteristic of the

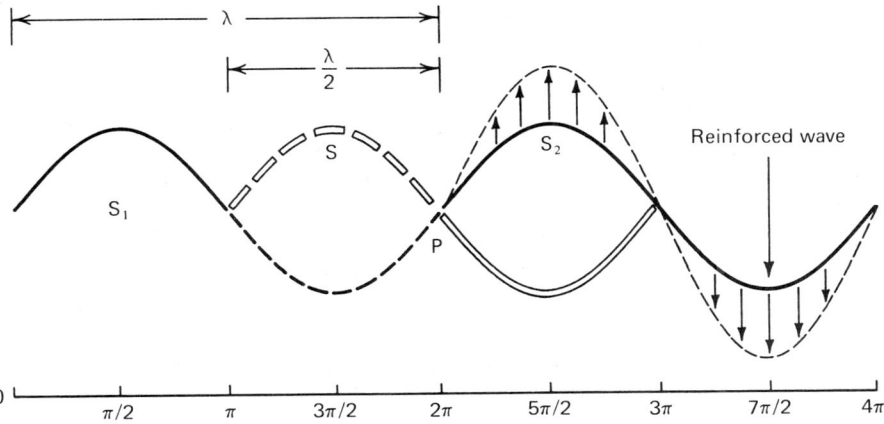

Fig. 11-5. In and out of phase wavelengths exert quite different effects on the quantities that the waveform represents. Thus wave S starting at a point $\lambda/2$ distant from point P will be fully *out* of phase with the dark wave and its effect will be cancellation. Wave S_1 begins a distance λ away and will be in phase. Its effect will be reinforcement of S_2.

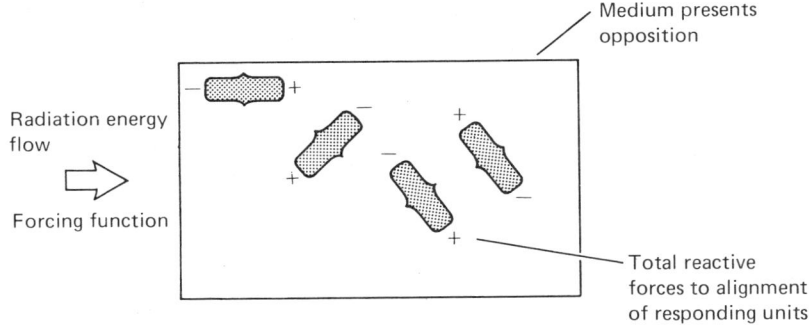

Fig. 11-6. The forcing function and the system reaction to it create an energy sink. Ensemble of dipoles under forcing function *fights* to orient in the crowd of restrictions—neighboring charges, molecules, and the viscosity in general. This struggle extracts the energy out of the wave letting it dissipate to zero. The description of the dipolar response involves the relaxation frequency and time, and the dissipation involves the loss factor and loss tangent.

circuit. The periodic, sinusoidal nature of these phenomena permit the mathematical power of sine functions and wave notation to be brought to bear on the problems that need to be solved (Fig. 11-5).

Polarization is of interest for units ranging along the size scale from electrons to cells, organisms, and crystals. The time constant of instruments used in measuring or responding to processes also involves the same considerations, that is, the instrument errors may be larger if its response period is not commensurate with the time constant of the process being measured. The ubiquitous water molecule is a microwave absorbing dipole that is often fastened to other larger molecules as bound water. Such units or terminal groups may give a dipole function near their characteristic relaxation frequency even if that of the main molecule is different—or even if it is not polar.

ORIENTATIONS—THE RELAXATION RESPONSE EXTENDED TO INCLUDE SYSTEMS

The relaxation mechanism is wider in meaning and application than first appears. It includes most responses of systems exposed to an external influence if there is an associated delay time in the response. With polar molecules, the *orientation* response is the one observed and it is the variable which sets various materials apart into groups according to their absorptive polarization. Relaxation of polarization is at a maximum when the dipole or polar particle is exposed to an alternating electromagnetic field with such a cycling time that the particle can relax to a new position before being called upon to reorient. The particle characteristics as well as the frequency applied constitute the system.

Organization Levels for Polarization

At each level of organization from electron to organism, some frequency will be in harmony as far as time constants and the responding system are concerned. Electrons may displace in atomic electron clouds in about 10^{-15} seconds, the corresponding frequency in cycles per second being in the near ultraviolet spectrum, 10^{15} Hz. Many atoms, bound in molecules, may distort toward positive and negative centers of charge and the time required, about 10^{-14} seconds, would put the process in the infrared at 10^{14} Hz. For crystalline structures, ion displacement in the lattice would be similar in time, taking about 10^{-12} seconds for a far infrared process. Gases often show similar polarization times. The kind of isolated displacement illustrated by these effects seems to be a response to a transient radiation pulse with the relaxation immediately following it. Then the intrinsic relaxation is a characteristic recovery from the disturbance according to system time constants.

Electronic polarization can be resonant and it responds in an insignificant time at these radio and microwave frequencies. Thus it is described by an infinite polarization and by the optical dielectric constant ϵ_∞. Electronic polarization gives the optical refractive index n. The region between 10^{11} and 10^{14} GHz is one of transition from microwave to light frequencies. Resonances begin at about 10^{11} GHz and this is about where microwaves end. However, for most intents and purposes,

Impedance and Membranes 431

$$\epsilon_\infty = n^2 \qquad (198)$$

for measurements of n in this spectral region.

If the relaxations in two systems are compared, then the effect of various components, buffers, experimental conditions and even the functions may be studied with respect to process time constants and their meanings in terms of structure, function, communication and responsiveness. This is seen in the case of nuclear magnetic resonance relaxation from which fundamental information may be gained. Here interest centers not so much on absorption of energy as upon the relaxation time *per se*.

Energy Flow

In the usual case, the process continues and radiation energy flows into the system in response to the total reactive forces opposing the free motion of the reacting units (Fig. 11-6). Sometimes it is the transient relaxing system response that is of interest. At other times it is the behavior of the medium such as its activation or the production of heat or other radiation from it. Polarization decay and dielectric relaxation are equivalent expressions in dielectric media while dielectric absorption (sometimes called dielectric loss) is absorptive polarization, arising from the energy transfer process associated with the dielectric relaxation. The model is the capacitor and the loss is in the dielectric when it is inserted between the plates.

In the polarization absorption process—a negative exponential change for the energy—the relaxation time would allow a relaxation of 1/eth of the orientation. This is

$$E = E_o\, e^{-1} \qquad (199)$$

$$E/E_o = 1/e = 1/2.718 = 36.8\% \qquad (200)$$

also the 36.8% value and 63.2% of the orientation of dipoles is lost.

PHASE RELATIONS, THE CAPACITOR MODEL AND THE DIPOLE ENSEMBLE

It is also necessary to consider the phase relations of the quantities involved in the system. Here the intention is to describe the polarization as lagging behind the alternating field by some phase angle. The radiation energy or wave is alternating from the generator or source toward the absorbing medium and back to the source according to the time relations established by the frequency (Fig. 11-7). The medium may vary from a thin slab of tissue to the infinite regions of space (Fig. 11-8). There will be a temporal relationship between the dipoles and the period during which the energy is going in one direction. During this period they will have an opportunity to respond before the influence

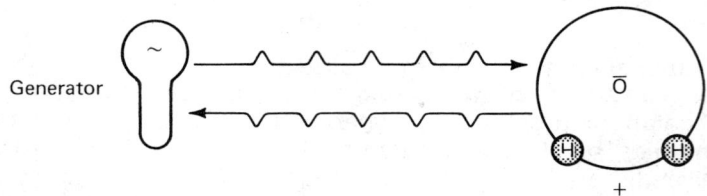

Fig. 11-7. Energy travels to and from the generator at a rate of f cycles per second.

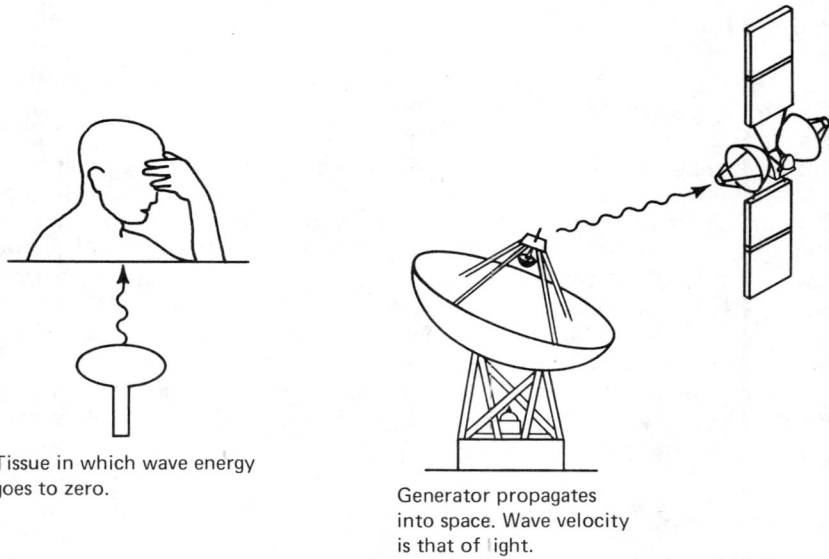

Tissue in which wave energy goes to zero.

Generator propagates into space. Wave velocity is that of light.

Fig. 11-8. Variations in a medium may extend from a thin slab of tissue to vast distances in space.

shifts in the other direction. The alternating field will be sinusoidal with f cycles per second. The dielectric dipolar response in tissue to such a time varying process could be described by how much the orientation *lags* the applied field. There is a similar situation in a capacitor where the voltage lags the current in a capacitive circuit by 90° (one quarter of a cycle). This particular phase angle is 90°. A very convenient model is available for dielectric absorption based on this result. With it, one can proceed to describe current and voltage relations in a condenser with a dielectric having some loss between the plates. Then the amount of loss can be logically associated with the phase angles involved, that is, those for orientation of dipoles and for current leading the voltage.

Impedance and Membranes 433

In comparison with dipole behavior and relaxation spectroscopy this, in effect, associates some basic electrical engineering concepts in dealing with power loss in transmission circuits. The shift is also from the dipole organization level to the dielectric material level which is essentially one in which there is an ensemble of dipoles. Capacitance is the basic electrical property contributed in a circuit by a condenser or capacitor. With C in farads, the RC product has units of time.

The function of the dielectric material in a capacitor is immediately applicable to the study of dielectric loss in a biosystem under the influence of frequency, temperature, and variations in its compositions and the biosystem becomes part of this circuit.

If voltage is applied across the circuit terminals of a capacitive circuit, the capacitance does not follow the voltage instantaneously. There is a charging time required for full charging of the capacitor plates. Thus there is a curve (Fig. 11-9) in which the voltage across the plates rises with time toward the maximum voltage. The greater the circuit resistance, the longer the time required for the capacitor to reach V_{max} because of the opposition to the flow of charging current. The time required for the capacitor to become fully charged depends on the product of the resistance and capacitance. In engineering, this RC product is the capacitive circuit time constant and analogous biological processes give relaxation times (Fig. 11-10). The RC constant gives the seconds required for the capacitor to reach 63.2% V_{max}. Also, for a discharging capacitor, the RC time constant equals the time required for it to lose 63.2% of its full initial charge. The plates offer more and more resistance to the flow of current as they become charged. In an ac circuit, current flows continuously "across" the capacitor as its charges vary between the maximum and zero. In the dc circuit, the capacitor produces an "open" circuit with zero current. As the frequency increases, a complete decay of the voltage is prevented.

Capacitive and Resistive Circuits

The relation $P = IE$ measures, in watts, the potential power that exists in the circuit. However, it does not become energy until some use for it is made or it does some work. Thus a line may supply power at 115 V and 15 A or 1650 watts. If a resistance heater is plugged in, it will begin to use this power and the power times the time during which it is used is the energy in watt hours or other time unit. For a half hour, for example, the energy used may amount to 825 watt hours, which is considerable heat. The electrical energy may be used for heating and the 825 watt hours may be converted into several other units, for example, kcal, Btu, ergs, and eV, which may happen to be in mechanical or chemical energy instead of electrical. Thus these are interconvertible.

Power is dissipated in electrical circuits which, as a result, need to be cooled. Most bioelectrical power is used quite efficiently (e.g., in neurons) and the heat that must be removed is largely due to metabolic processes.

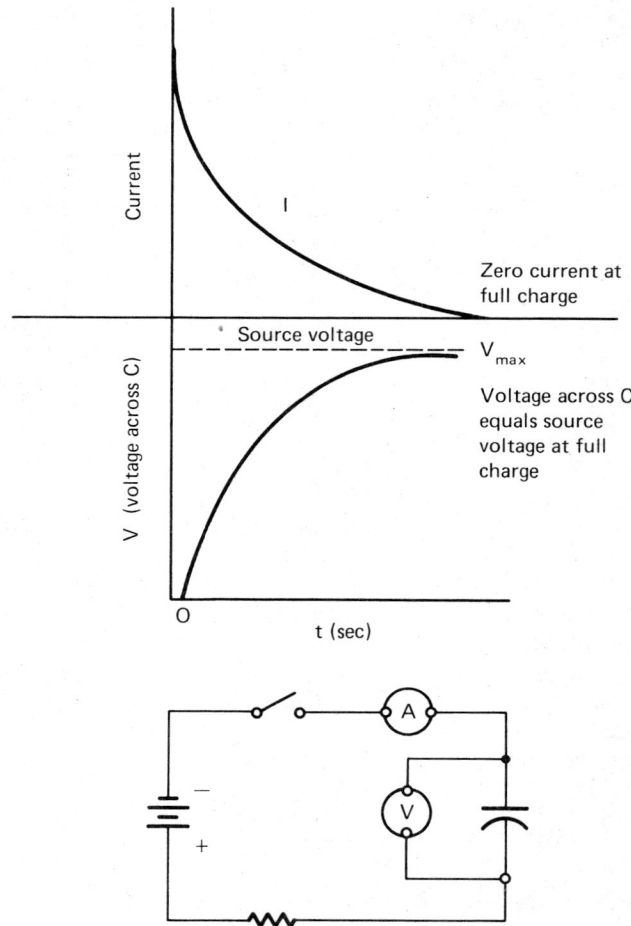

Fig. 11-9. Voltage on capacitor plates. Rates of change in resistance capacitance series circuit.

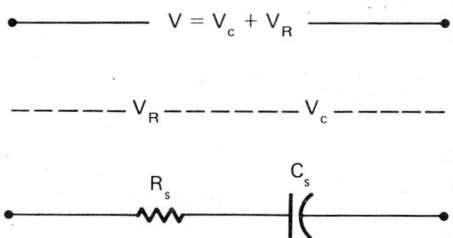

Fig. 11-10. RC time constant. RC seconds = R ohms × C farads = time constant.

Voltmeters and ammeters, the instruments used to measure the quantities of interest in this process are concerned with the relative magnitudes of the phase angle between the actual and indicated magnitude, for example, that between the actual current and voltage or the angle θ in the capacitive circuit. The current wave leads the voltage by 90° because the current must flow to charge the plates before voltage can exist across the capacitance (Fig. 11-11). As the capacitor charges,

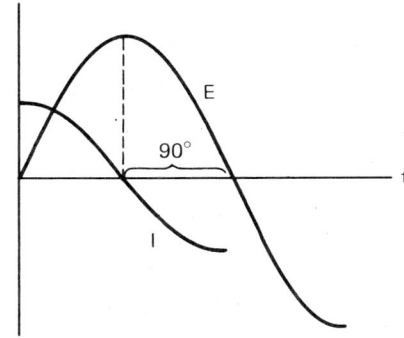

Fig. 11-11. Current sine curve (I) crosses zero 90° before the voltage curve E.

the current drops toward zero as the voltage rises to its maximum. The actual or true power used will be less than the apparent power. The power is $IE \cos \theta$ and, with $\theta = 90°$, in the capacitive (or inductive) circuits, $\cos 90° = 0$ and no power is absorbed. This circuit element can only store energy or redeliver it. The resistance decreases the phase angle, putting the current more in phase with the applied voltage and increasing the true power so that $P = IE$ in a pure resistor.

It is convenient to define ϵ_r, the relative dielectric constant, as the measure of the capacitance fraction C/C_o for a dielectric material C as opposed to free space C_o in the capacitor. Then, ϵ_r will equal ϵ'/ϵ_o where ϵ' is the handbook dielectric constant which is about 72 for water. At times this constant is simply ϵ, while in engineering the symbol κ (kappa) is used and the name is relative permittivity. This "constant" will be found to decrease with increasing frequency for a given material so that its constance is with respect to time, not to frequency. The appropriate frequency response of the principal dipolar substance will be seen macroscopically in the behavior of the capacitance and shown on instruments which measure ordinary electrical quantities. In the biological model for explaining dielectric absorption, the material of interest between the plates of the condenser is biosubstance rather than the usual air, plastic, paper, mica, glass, etc.

The *relaxation frequency* is a time of sharp change in the dielectric response, that is, it depends on the dielectric constant and the resistivity of the substance. Both are less at higher frequencies. The range during which phase relations and process time constant predict an interaction will be from 0.1 to 10 times the relaxation frequency.

It is therefore convenient to define a loss angle δ arising from the phase relations between current and voltage in the model. It is the angle complementary to θ, the phase angle and together they establish the $90°$ phase difference between current and alternating potential E. The charging current leads the voltage by this amount and the current has a loss component in phase with the voltage (Fig. 11-12). Then

$$\tan \delta = \frac{\text{Loss I}}{\text{Charging I}} = \epsilon''/\epsilon' = \frac{\text{Loss Factor}}{\text{Dielectric Constant}} \qquad (201)$$

With this relationship, the energy sink is seen to be associated with the phase angles and loss factors, the link being communication while the meaning is information on the status of a biosystem.

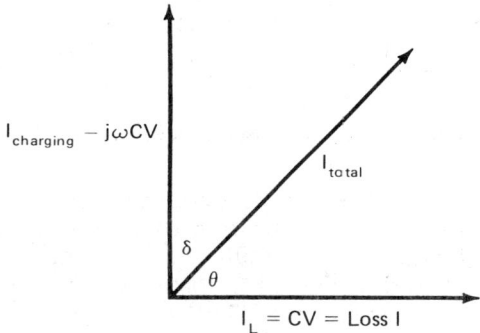

Fig. 11-12. Model for absorptive polarization. Relations between currents expanded with complex notation.
$I_{charging} = dQ/dt = -j\omega CV$
$-j$ shows a $90°$ lead
ω is the angular frequency
C is the capacitance
V is the voltage
I_L = conduction current in phase with voltage
$\tan \delta = I_L/I_{charging} = \epsilon''/\epsilon'$

Complex Dielectric Constant

The larger $\tan \delta$ is, the greater the loss current and the loss factor. In connection with dielectric loss, some form of complex notation is

Impedance and Membranes

required to account for the polarization absorption in the presence of phase relations between the fields. The complex number is of the form $a \pm jb$ where vectors or values of a being real and jb imaginary, describe positions in the complex plane. Thus a complex number indicates the sum or difference of a real number and an imaginary one (Fig. 11-13).

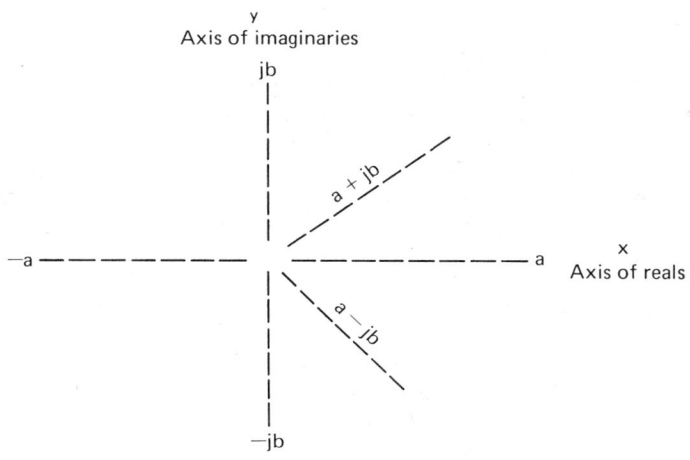

Fig. 11-13. The complex plane and rotation clockwise.

Thus j is an operator equal to $\sqrt{-1}$ which is simply an addition to the real number system. The following angular positions are defined:

$$
\begin{array}{ll}
\textit{Rotation of} & \textit{Shown by} \\
90° & j \\
180° & j^2 = -1 \\
270° & j^3 = j^2 j = -1j = -j \\
360° & j^4 = +1 \\
\text{also} & j^{-1} = j/j^2 = 1/-1 = -j
\end{array}
\qquad (202)
$$

Thus multiplying a real number n by -1 is equivalent to rotating the line that represents the number n about the origin through 180° to a new position for n in the opposite direction and n units from 0.

$$ j^2 n \qquad -n \longleftarrow 0 \longrightarrow n \qquad -1 = (\sqrt{-1})^2 = j^2 \qquad (203) $$

In the process of absorption, the EMR traverses the material, and loses its energy while causing changes in molecular orientations. At high intensities of microwave energy certain changes in the temperature would be observed. The wave itself experiences changes in both phase and amplitude. The dielectric constant is a measure of these changes in terms of the complex interaction *permitted* by the material with the wave energy,

$$\epsilon^* = \epsilon' - j\epsilon'' \qquad (204)$$

The complex form ϵ^* consists of the real part ϵ' which measures the phase changes in the wave, and the imaginary part, ϵ'', which measures the transfer of energy as the dipoles feel the forcing function of the wave of polarization from the applied field and tend to lag behind it. In this sense, the behavior of the molecules contributes to absorptive polarization and so, ϵ'' is known as the loss factor. It varies greatly as a strong function of frequency and material in the microwave spectrum. The real part ϵ', however, is fairly constant up to the dispersion region.

In the special conditions of an absorptive cell in an appropriate instrument, the resonant absorption and relaxation are measurable at specific frequencies. The results may be applied to a proposed transformation process for the material using this energy, or to an elucidation of the material's structure and function, in the sense of the physics of molecular biology as applied to both single cells or to their aggregation in tissue.

The capacitance model is seen to emphasize the vector values for current and voltage during a period (or a unit frequency) of change in the alternating electromagnetic field. The changes in the phases of currents and voltages and the effects of interplate, dielectric material properties are familiar from capacitance theory.

The capacitor electrodes are seen to be charged regions whose mutual influence depends on the charge, area, and amount of separation (distance). For example, as the distance between them decreases, because of the added effect of closer charges, the capacitance increases (Fig. 11-14). The angle between the current I through the dielectric and $I_{charging}$, the current that would flow through an ideal dielectric 90° out of phase with the voltage V, is determined by the loss angle δ (Fig. 11-12). θ is the angle between the actual current through the dielectric as measured and the voltage causing this current.

$$\tan \delta = 1/Quality\ Factor$$

where (205)
$QF = 2\pi \times$ energy stored per half cycle/energy dissipated per half cycle
In terms of this ratio, microwave cavities are described as being of either high QF or low QF. This factor is thus seen to be a figure of merit as far as what the cavity does with admitted energy is concerned. The denominator quantity can be dissipated as heat (irreversible). An actual capacitor C ──┤├── having losses which give rise to a dissipating factor = $\tan \delta \neq 0$ may be shown as the equivalent circuit with an ideal capacitor C_o in parallel with R (Fig. 11-14a). Therefore

Impedance and Membranes 439

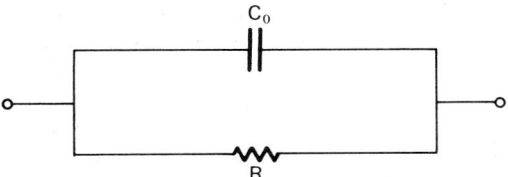

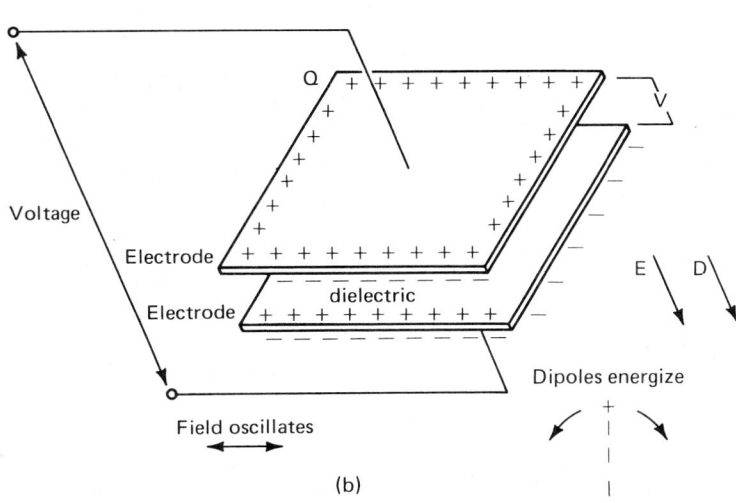

Fig. 11-14. Parallel plate capacitor as it appears when an applied field affects special dielectric between the plates. Reacting (displacement) dipoles may be such as the linear, cylindrical, or spherical forms within the dielectric between the plates. The dielectric may be microscopic charged units consisting of *cells* with *tissue* or other substance, and these increase the capacitance by polarization and holding charges on the plates. E-electric field and direction *with vacuum*; D-displacement of reacting dipoles; V-potential difference; Q-plate charge. Part of the plate charge is compensated for by a displacement of the charges in the dielectric and then the field E between the plates does not exist in vacuum as it would with matter between the electrodes.

$$R = 1/\sigma \qquad (206)$$

and

$$\tan \delta = 1/\omega R C_o \qquad (207)$$

relating the voltage and charge via the implication of tan δ.

Fields and Forces

Displacements occur in the direction of E, the applied field, and they equal the sum of $\epsilon_o E$ and P, the polarization, where ϵ_o is the dielectric constant of free space.

With polarization, dipoles orient with their positive charges closer to the negative plate (inside fields). This tends to increase the flow of current in the circuit (outside fields) because the positive poles attract more negative charges in the plates (Fig. 11-14b). Then a surplus of electrons on this plate repels electrons from the opposite plate leaving it positively charged. Q, the capacitor charge $= C_o V$ where C_o is capacitance with no dielectric (free space).

$$\epsilon_r = \epsilon'/\epsilon_o \qquad (208)$$

The charge neutralized by dielectric polarization at the plate interface is

$$Q[1 - (1/\epsilon_r)] \qquad (209)$$

and the remaining charge that contributes to the field outside is

$$Q/\epsilon_r \qquad (210)$$

These electric forces are exerted according to fundamental charges acted upon by fields

$$E = F/q \qquad (211)$$

Here a field intensity E is produced when an electric field F acts on a charge q. Between two charges in newtons, meters, and coulombs,

$$F = q_1 q_2 / 4\pi \epsilon_o r^2 \qquad (212)$$

where ϵ_o is the dielectric constant or the permittivity of free space and r is the distance between the charges

$$E = q/4\pi \epsilon_o r^2 \qquad (213)$$

An electric field polarizes the dielectric and the displacement $D = \epsilon E$ depends on the dielectric constant of the material. The dielectric constant ratio

$$\epsilon_o/\epsilon' \qquad (214)$$

is a reducing factor which describes how the force on a charge in the material is reduced by the presence of the dielectric. It will be reduced by this factor. When the plates become charged by applying a voltage, the charge is

$$Q = CV \qquad (215)$$

and it also depends on the capacitance C. In turn, this quantity is directly proportional to the dielectric constant ϵ and the area of the plate and inversely proportional to the separation between them.

$$C = \epsilon' A / \ell \tag{216}$$

The capacitance will therefore increase when any material with a dielectric constant greater than that of free space ($\epsilon' > \epsilon_o$) is inserted between the plates. The insulator or dielectric material between the plates permits absorptive polarization. If the value of C increases, the potential difference required for a certain charge is reduced.

The ratio of the dielectric constants is the relative dielectric constant

$$\epsilon_r = \epsilon' / \epsilon_o \tag{217}$$

which in turn influences the capacitance

$$C/C_o = \epsilon_r \tag{218}$$

Dielectric polarization produces a field P which affects both the capacitance and the electric fields.

$$P = (C_o - C)V = [(\epsilon_o/\ell) - (\epsilon/\ell)] V \tag{219}$$

$$P = \epsilon_o (E_o - E) \tag{220}$$

where E_o is the electric field in a vacuum

$$P = (\epsilon_r - 1)\epsilon_o E \tag{221}$$

$$\epsilon_r = 1 + (P/\epsilon_o E) \tag{222}$$

Another field arising from partial mobility of displaced units in the dielectric under the field influence is the displacement D.

$$D = \epsilon E = \epsilon_o \epsilon_r E \tag{223}$$

In the absence of a dielectric, no ϵ exist and $D = E$

$$D = P + \epsilon_o E \tag{224}$$

$$P = D - \epsilon_o E \tag{225}$$

$$\epsilon_r - 1 = P/\epsilon_o E \tag{226}$$

The last, which is derived from

$$\epsilon_r = 1 + (P/\epsilon_o E) \tag{227}$$

is the dielectric susceptibility χ, which informs on the ratio of bound to free charge carriers.

$$P/\epsilon_o E = \chi \tag{228}$$

and

$$D = E(1 + \chi) \tag{229}$$

so that

$$\epsilon_r = 1 + \chi \tag{230}$$

provides a definition for the dielectric constant of a material.
There are dipole moment contributions

<-------- E Field -------->

from electronic   + ←— e⁻ —→ −,   (231)

ionic   + ←— ion —→ −,   (232)

and permanent dipoles.

TYPES OF DIPOLES

Permanent dipoles are all diatomic molecules having two different, unequal atoms separated by the distance established by the chemical bond between them, and the many asymmetrical structures and ionic configurations. The elasticity in these structures persists through the many states and variations in which they are found. Examples are water, hydrochloric acid, carbon monoxide and complex molecules. The water molecule is of prime importance as the foundation of the microwave effect in biology. It retains its dipole nature permanently as it forms a symmetrical triangle with the angle between the two O—H bonds being about 105°. The electron of the hydrogen is attracted to the oxygen nucleus, or better, electron pairs make a firm polar connection to oxygen. Oxygen, with its own electrons, makes a negative pole or center while the hydrogen nucleus is the center for the positive charge. On the other hand, dioxane, hydrogen, oxygen, benzene, cyclohexane, carbon bisulfide, paraffin wax, carbon tetrachloride and tetracholoroethylene, are all nonpolar and do not have moments. The polarization P or the polarity in debyes is the sum of all the dipole moments whose polarizabilities α may be evaluated,

$$P = \bar{\mu}N = N(\alpha)E \tag{233}$$

For example, the permanent dipoles have a total moment

$$N = (\mu^2/3kT) \tag{234}$$

where T is the temperature (°K), and μ is the dipole moment.
    The dipole moment is in response to the force exerted by the field and it helps to describe molecules in terms of the distribution of charge

and shape. When a molecule like carbon dioxide is found to lack a permanent dipole moment, it leads to a specification of a three-atom-in-line form with the carbon halfway between the oxygen atoms (Frohlich 1958). Some molecules like carbon tetrachloride contain polar groups but the individual group effects are balanced by their symmetrical arrangement. The total moment is seen to be dependent on the temperature because ordinary thermalization works to randomize rather than to align the dipoles. The higher the temperature, the less the moment or polarization. The alignment may be retarded or instantaneous. In the former case, a potential energy barrier to rotation must be overcome. Changes in temperature affect the height of this barrier and, thus, the energy needed to activate the dipole. The resonance frequencies for this behavior occur in the microwave and infrared regions.

On the other hand, the *electronic polarizability* or displacement is instantaneous and tends toward dominance in the ultraviolet and visible spectrum at about 10^{15} cycles. For ionic polarizability with natural frequencies near 10^{13} cycles, the infrared is the spectral action region.

From a material standpoint, some substances will have only one, two, or all of the possible polarizabilities. Their dielectric constants will be due to the components that apply. Some very symmetrical crystals will be limited to electronic polarizability. In some salts, there is considerable ionic polarization (sometimes called atomic polarization) in which the anion and cation portions are displaced by the applied field. Sodium chloride is an important example. With each pole held in a crystal, the molecule can displace but not rotate. The Na ions displace with the field, and Cl ions against it, with a displacement proportional to the local field and important in the infrared region. From a frequency point of view, most of the polarizabilities will be observed at low frequencies while in the visible range, electronic polarizability is especially significant.

The polarization is the product of the dielectric susceptibility, the local field, and the dipole moment.

$$P = NxE_{loc} \qquad (235)$$

This quantity carries information on the dipole moment per unit volume of the dielectric when the concentration tends to produce local interaction. An atom in the local field is influenced by

E_1, the field due to charge density on the plates and free charge carriers in the external field E.

E_2, the field due to dipole polarization and induced charges on two sides of the dielectric opposite the plates, and

E_3, the adjacent dipole effect.

Then,

$$E_{loc} = E_1 + E_2 + E_3 \qquad (236)$$

The Lorentz force is obtained from a spherical model for an atom in a cavity

$$F = q(P/3\epsilon_o) \qquad (237)$$

and for the field with dielectric ϵ_r

$$E_2 = P/3\epsilon_o \qquad (238)$$

$$E_{loc} = E + (P/3\epsilon_o) = E + (4\pi P/3) \qquad (239a)$$

Thus the local field near the atom is greater than the applied field E by an amount directly dependent on the polarization density. Various corrections for non-spherical shapes will modify these results. The behavior may be, for example, more nearly ellipsoidal with several axes, especially when actively polar groups are similar to distortions on the vaguely spherical character of the dipole.

The Clausius-Mosotti equation is obtained from an examination of the atomic polarizability α as it makes up the dielectric polarization in the local field (E_{loc}) from the number N_i of atoms. The result is

$$(\epsilon_r - 1)/(\epsilon_r + 1) = 1/3\epsilon_o \sum_i^n N_i \alpha_i \qquad (239b)$$

which shows that the dielectric constant with a certain α is dependent on the number of atoms in the volume. Their relaxations, polarizations and displacements lie behind the complex interaction as susceptible units respond to the microwave which, in turn, experiences phase and amplitude changes.

ENERGY CONVERSION

The interaction between dipoles and field will ultimately result in the dissipation or transfer of the wave energy as kinetic, potential, or thermal energy depending on the energy levels reached, that is, how much power, how many dipoles, and the duration of the interaction. Before heat appears, the dipoles will have polarized with the field using field energy according to their dipole moment in so doing. Almost at once, however, depending on the thermal randomization, the temperature of the material has an effect which opposes the dipoles and forms an activation level or potential energy barrier to alignment. The factors used, such as τ and phase changes, describe the passage of wave energy to the molecules making up the medium. This is because they are a measure of the hindrance and interference encountered by the forcing function. *When these forces operate, they always involve the equivalent energy of accelerating masses.* Thus there is a basis for the transfer of wave energy. The material state (solid, liquid, gas) also determines the degree of freedom for the dipoles. It is a factor in the interference due to the state of the medium in which energy is transferred. The response of a *free* dipole will be greatest at low temperatures, that is, full alignment is possible. Then maximum polarization occurs in the system. This is also attained with very large field gradients (volts/unit distance).

For some applications such as microwave freeze-drying, the persistence of loss components in materials in the frozen state is significant. Without this loss, the application could not succeed. It is seen that the ice state hinders the absorptive polarization by apparent immobilization of the water dipole but does not completely prevent it. Dielectric constant values decrease from about 72 to 4 as the change of state to ice occurs. Below freezing, the loss factor decreases as the temperature falls further. As a factor, polarization tends to vanish at temperatures well below freezing and at the high microwave frequencies. Other loss components come into play, particularly the ohmic loss due to change in the conductivity. These other loss components are related to supercooled regions and salt migrations in frozen substances (Fig. 11-15). If

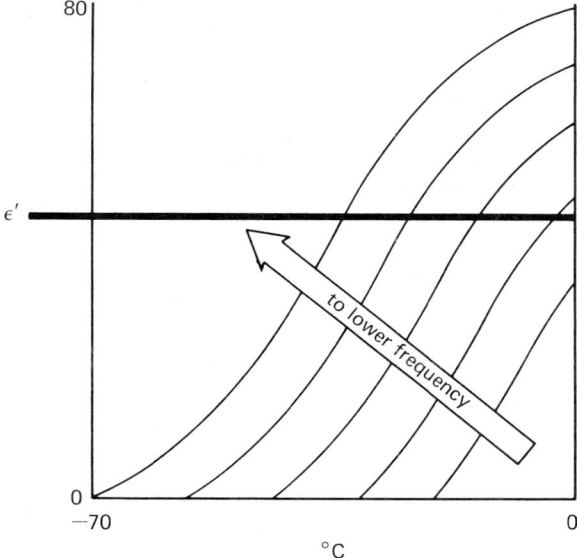

Fig. 11-15. To maintain the value of ϵ', the frequency must be reduced in the kHz region below 0°C.

in ice, ϵ' is to be held near the high value (near 80) that it has as in liquid water, then the frequency must be reduced. Otherwise, ϵ' will decrease to nearly 0 between 0°C and −40°C in the kilocycle range.

The temperature dependence of dielectric loss varies with the frequency. As the values from von Hippel's laboratory show, water has a negative temperature coefficient for a given frequency below three GHz and a positive coefficient above ten GHz for the dielectric constant.

From 1.5 to 95°C between 3×10^8 Hz and 2.5×10^{10} Hz including tests at 3×10^9 and 1×10^{10}, tan δ almost always decreases as the temperature rises, and ϵ' decreases except for an increase above 1×10^{10} Hz. In the same temperature range, between 10^5 and 10^6 Hz, tan δ increases with temperature and ϵ' decreases. The highest two frequencies in general show much larger tan δ and lower ϵ'. The temperature-viscosity relationships of water are thus important at the higher frequencies with the factors contributing to tan δ thus being more prominent at about 10^{10} Hz with less effect at higher temperatures (data from 1.5 to 95°C).

Around the threshold for polarization the question is whether the field force and relative freedom of the dipole to rotate will overcome the intrinsic thermalization forces. The alignment and restoration forces can prevail to produce microwave-type heating. It turns out that in most media, rotation will be opposed by the thermal agitation, Brownian movement, collisions and mobility *dampening* factors giving a phase relationship or lagging between the force and the response of the dipoles (Debye 1929).

Complex Time Variation

When an alternating field is applied, its intensity has a complex time variation factor. The intensity is

$$E_o e^{j\omega t} \tag{240}$$

where ω is the angular frequency. $j\omega$ replaces a time differential operator. In $e^{j\omega t}$, ωt is any real number of radians

$$e^{j\omega t} = \cos \omega t + j \sin \omega t \tag{241}$$

and obeys exponent laws. Also

$$a + jb = r(\cos \omega t + j \sin \omega t) = r e^{j\omega t} \tag{242}$$

gives the exponential form of the complex number. For example

$$3 + j4 = 5(\cos 53.13 + j \sin 53.13) = 5\, e^{j53.13} \tag{243}$$

Also,

$$10^{j0.37} = 10(\cos 0.37 + j \sin 0.37) \tag{244}$$

$$= 9.32 + j\, 3.62$$

if

$$\theta^z = e^{x+jy} = e^x(e^{jy}) = e^x(\cos y + j \sin y) \tag{245}$$

where y is radians of angle and may be used in defining $\cos y$ and $\sin y$.

Impedance and Membranes 447

Frequency Response (Tan δ)

The field E is a function of time

$$E = E_o \cos \omega t \qquad (246)$$

and the dielectric displacement is likely to be time separated from it by an angle ϕ

$$D = D_o \cos(\omega t - \phi) \qquad (247)$$

These vector relations are shown as follows. In terms of polar coordinates, the representation is clock-like (in this example counterclockwise) (Fig. 11-16).

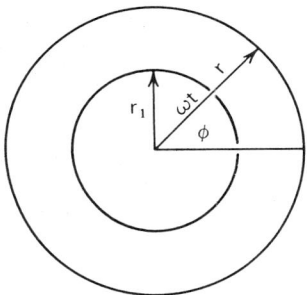

Fig. 11-16. Vector relations for two quantities r and r_1 in polar coordinates.

Then, in the sine wave representation, an amplitude y varies with the angle (Fig. 11-17). The phase between two sinusoidally varying quantities y and y' is given by the angular difference between the vectors (Fig. 11-18).

$$D = D_o \cos \phi \cdot \cos \omega t + \sin \phi \sin \omega t \qquad (248)$$

Since

$$\cos x - y = \cos x \cos y + \sin x \sin y \qquad (249)$$

$$D = D_1 \cos \omega t + D_2 \sin \omega t \qquad (250)$$

$$D = \epsilon^* \epsilon_o E_o e^{j\omega t} \qquad (251)$$

from

$$\epsilon^* = \epsilon' - j\epsilon'' \qquad (252)$$

By separating the real and imaginary portions which behave differently according to frequency,

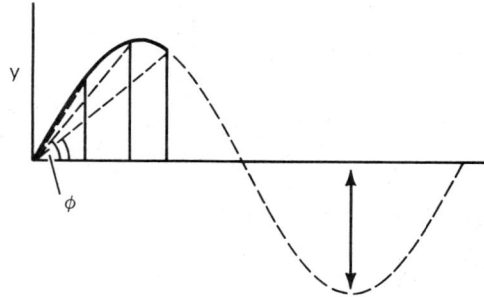

Fig. 11-17. y is the amplitude of the sine wave and behaves as a vector that alternately lengthens and contracts as the wave frequency prescribes. Since this line is the opposite side for the angle ϕ, a sinusoidal variation in the value of sin ϕ is obtained. A tracing of the length of y as a function of time produces the familiar sinuous track.

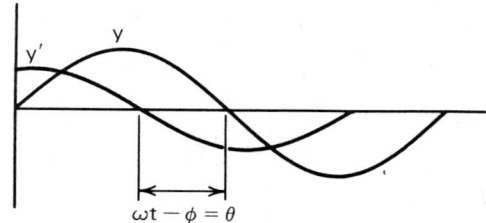

Fig. 11-18. Dual sine wave representation for the phase between two quantities. $y = r \sin \omega t$. y lags y' by the phase angle θ.

$$D_1 = \epsilon' E_o \quad (253)$$

$$D_2 = \epsilon'' E_o \quad (254)$$

then, from the fact that the field $E = E_o \cos \omega t$ and also from

$$D = D_1 \cos \omega t + D_2 \sin \omega t \quad (255)$$

a relation for loss factor tan δ is obtained

$$\tan \delta = \epsilon''/\epsilon' \quad (256)$$

All three of these quantities are regularly obtained by investigators and may be found in handbooks.

Tan δ distinguishes lossy, absorbing substances from the nonloss types. It separates the polar molecules which show a significant tangent value (such as 0.169 for water at 3 GHz and room temperature) from the nonpolar kind which have a value of less than 0.0001 for tan δ. If the nonpolar materials were truly passive they should give no effect. The existence of a slight response, which is also a function of frequency and which increases as the frequency does, indicates a displacement within the molecules rather than a full dipolar response. The displacement will have a moment and be similar in nature but much less in effect than the dipolar orientation (Fig. 11-19).

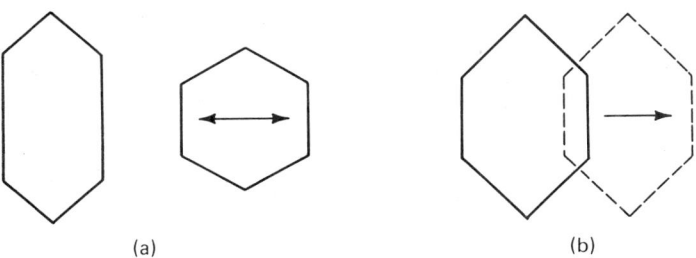

(a)   (b)

Fig. 11-19. A distortion A or displacement B of nonpolar molecules such as benzene may account for the residual value of the loss tangent tan δ in these molecules. Its small value increases noticeably with the frequency.

Dielectric Conductivity

With the loss factor ϵ'' defined in the relation for the complex dielectric constant,

$$\epsilon^* = \epsilon' - j\epsilon'' \tag{257}$$

its equivalence to the product $\epsilon' \tan \delta$ can be noted. Then the dielectric conductivity σ in reciprocal ohms/meter is

$$\sigma = \omega \epsilon' \tan \delta = \omega \epsilon'' \tag{258}$$

This equation relates real conduction, dielectric constant, and absorptive polarization effects with frequency (Table 11-1).

High frequency dielectric conductivity is theoretically less important as a source of loss than that at lower frequencies (from 10 to 1000 MHz). In the latter range, conductivity is a significant absorptive component. This is applicable to water mixtures and large concentrations of ions or great surface conductivity will alter the frequency response, predisposing the absorption in favor of the $I^2 R$ loss.

The conductivity influences the wave propagation in the material. The free space value of the wavelength will decrease by several hundred times in passing into a medium having a conductivity of a few mho/m.

At very high conductivities as in conducting metal, the propagation is effectively limited to a small region at the surface which is known as the skin depth d. This depth may be more generally related to the attenuation and would then be the distance inside a material where the surface fields have decreased to the 1/e or 37 per cent value (e = natural logarithm constant = 2.718). In meter-kilogram-second units,

$$d = (1/2\pi)\sqrt{(4\pi\rho\lambda/\mu c)} \qquad (259)$$

where d is in meters, ρ is the reciprocal conductivity or resistivity in ohm meters, μ is the permeability in henrys per second, λ is in meters, and c is 3×10^8 meters per second. If d is 10 cm at 30 MHz, it will increase to 52 cm as the frequency decreases to 3Mc. The attenuation is α or one-half the inverse distance to the 37 per cent level ($\alpha = 1/2d$).

If the absorption is attributed to the total conductivity, it would be seen to be derived from ionic and dielectric or dipole motion components and these will vary with the frequency. As the dielectric constant decreases with increasing frequency, the conductivity does the opposite (Fig. 11-20). The crossover point is at the relaxation frequency where

$$\sigma = [(\sigma_0 + \sigma_\infty)/2] \qquad (260)$$

and

$$\epsilon' = [(\epsilon_0 + \epsilon_\infty)/2] \qquad (261)$$

or these properties equal one-half the sum of their low and high frequency values. The increase in conductivity (equivalent parallel circuit ⎯⎯⎯) means that thermal degradation of the applied field energy recurs at "infinite" frequency.

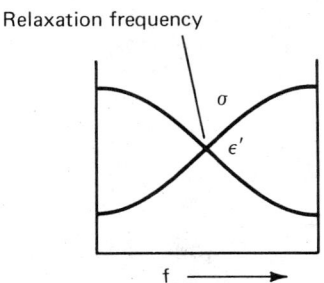

Fig. 11-20. The crossover point for the decreasing dielectric constant and the increasing conductivity occurs at the relaxation frequency.

Cavities, Loads, Resonance and Temperature

The microwave energy may be applied to an absorbing load in a closed resonating cavity. This model emphasizes the presence of the load as a part of the circuit and one which immediately influences the operation (and well-being) of the generator. In the cavity, the electromagnetic fields oscillate in some specific modes or resonances according to the design. A microwave oven is an easily adapted extension of a waveguide transmission pathway in which these oscillations continue until the wave fully attenuates to zero in the load. A large cavity size and a perturbation device such as a mode stirrer are called upon to increase the resonances and the oscillations by a modulation equal to the stirrer's rotational speed and resonances and thus, evenly distribute the energy. The most difficult case is probably the very lightly loaded cavity in which the load *damping* factor modifies the circuit resonance. At the other extreme, the cavity may be saturated with the load which, accordingly, acts as a transmission medium with corresponding attenuation (Fig. 11-21). In such a case, the zero limit for the field with 100

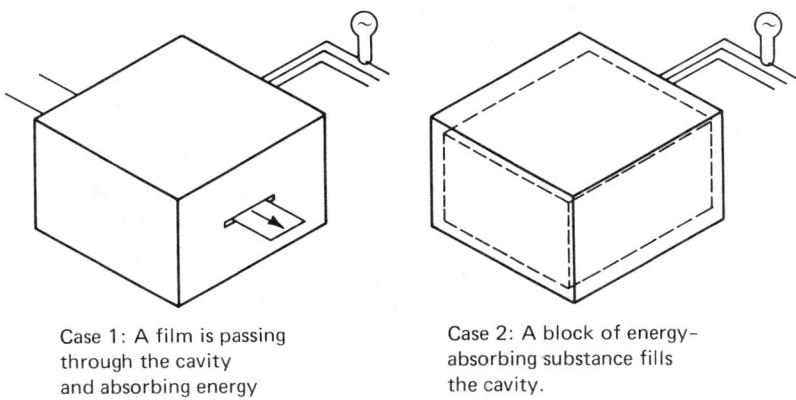

Case 1: A film is passing through the cavity and absorbing energy

Case 2: A block of energy-absorbing substance fills the cavity.

Fig. 11-21. Extreme cases for the loading of a cavity. Case 1-Film passing through the cavity and absorbing energy. Case 2-A block of energy-absorbing substance fills the cavity.

percent absorption is easily visualized. The cavity will not usually be filled and the damping situation exists. Then various tuning procedures may be used to maintain the best oscillation. When the loss factors change in the load, as they will with a change in its temperature, the attenuation will also be temperature dependent and sensitive therefore to cavity tuning. The resonant cavity can then come to require a different frequency—the new resonant frequency with the load at the new temperature. However, the generator usually has a fixed frequency.

Thus it will then be less tightly coupled to the load which will result in a difference in its energy output. Less energy will be seen in the load and probably more in the flow of coolant from the generator. In the design, provision is made to reduce this sensitivity to a reasonable level, so that the output is more constant. If a still more precise output is required, a feedback device is considered to retune the cavity (Feiker and Gittinger 1959).

Gas Excitation

Vacuum is commonly introduced into the microwave instrumental system for various reasons. There are two situations as far as the resulting absorption is concerned. A body of material is placed in the vacuum microwave cavity or line may be the only absorbing medium present or it may have some company in the sense that a vacuum drawn on the cavity may cause the molecules of the residual gases to become excitable. Under suitable conditions as to low pressure and high energy field strength, these gas molecules may become sufficiently excited to form a reacting, characteristically colored, pseudoplasma. The same air molecules that are originally decoupled at high concentration, become coupled at low pressure.

The gas discharge, while colorful, may often be undesirable if one is not interested in plasma production or gas breakdown phenomena. In microwave freeze-drying of biological material for example, the glowing may damage the material being dried with the aid of microwave energy.

Even at low pressures, the residual gas may be maintained in an uncoupled condition by reducing the field strength around the body of material. The energy is then usefully employed in the body where it accomplishes sublimation as in freeze-drying or some other result. The question is how to go about handicapping the residual gases in the competition for microwave energy with the body. In one sense it is a geometrical question which asks how and where the energy can be concentrated in the body instead of in the free space. Such concentration may be logically related to the *distribution* of field amplitudes in the cavity. The existence of a specific transmission mode already establishes a particular configuration for these amplitudes, leading to the possibility that the body can be moved around until its position coincides with the field maximum. The essential features would be the direction of the electric vector of the field, the dielectric constant of the materials, the dimensions of the cavity, the frequency, and the mode excited. If the mode presents two maxima, then two bodies positioned as in Fig. 11-22 would act as energy sinks and deprive the free space of excitation energy.

A simpler mode with a single maximum (Fig. 11-23) would require only one body. A further refinement may be visualized if it is determined that the location L (Fig. 11-22) coincides with some construction such as a seam or discontinuity in the metal cavity wall. Since the field is minimal at L, the probability that local corona effects may trigger gas

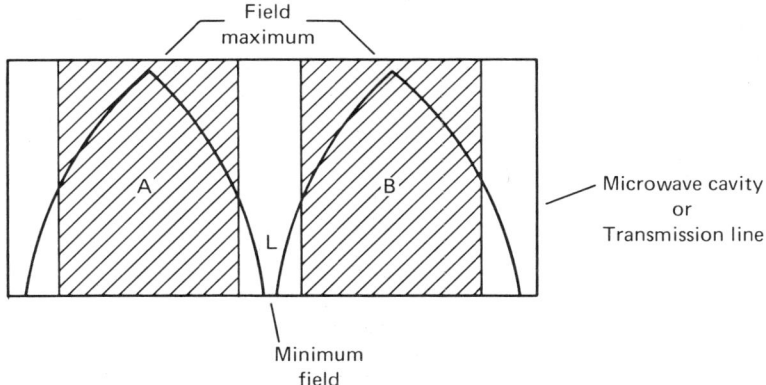

Fig. 11-22. Absorbing bodies A and B compete for free space cavity energy by virtue of special positions at the field maxima.

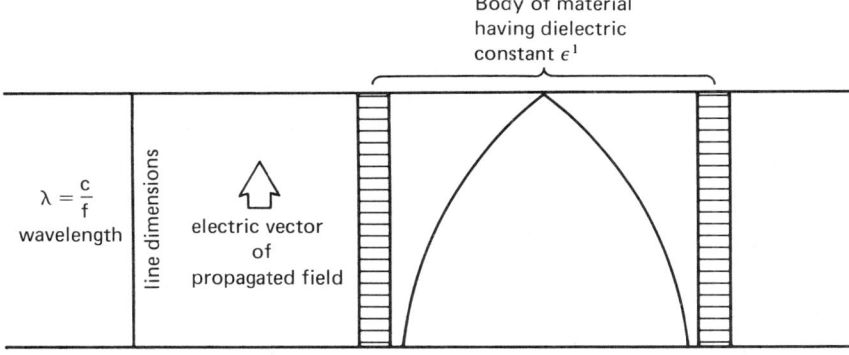

Fig. 11-23. Essential features for decoupling residual gases in a microwave vacuum system having a wavelength λ propagated in a specific transmission mode.

excitation and breakdown is also minimal. To achieve these conditions, it is only necessary to analyze the transmission mode and adjust the dielectric as to its constant and position.

THE IMPEDANCE CONCEPTUALIZATION

Combining Absorption and Geometry Factors with Impedance

Impedance Matching. The interplay of geometry, impedance and dielectric aspects of microwave interactions are illustrated in another situation which has been described by Parker (1968) following on the work of Copson (1962) in connection with the fundamentals of microwave freeze-drying in a vacuum.

Taking a large body of only slightly absorbent material and using it to intercept a wave traveling down a vacuumized transmission line will produce a reflected wave or a voltage standing wave (Fig. 11-24). The

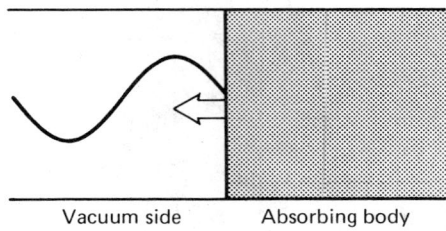

Fig. 11-24. Reflected wave at dielectric interface.

voltage standing wave ratio $VSWR$ will be

$$VSWR = \text{impedance in vacuum/impedance in body} = \sqrt{\epsilon'} \quad (262)$$

This ratio is the measure of the relative voltage or field strength which, in turn, measures the energy available for transfer into the body substance. It measures it from two aspects: firstly, the energy is very strongly a function of the field since it varies as the square of this value, and secondly, any reflected energy cannot contribute. The geometry of the situation specifies that the VSW maximum be located as a quarter wavelength distance from the interface where the wavelength is the same as that in the space (Fig. 11-25). Inside the body, the wave will

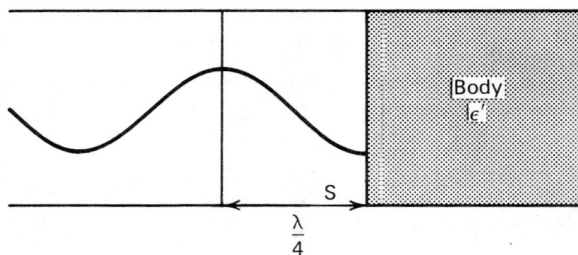

Fig. 11-25. Wave maximum at a distance S = one quarter wavelength from a substance having dielectric constant ϵ'.

continue at a reduced level due to the $VSWR$ and resulting loss of the reflected part of the energy. There is also the low value of the dielectric loss which directs that energy affecting the body will not be readily

Impedance and Membranes

absorbed, or that there will be only gradual attenuation. Thus, inversely, the penetration will be great. The *body* may also be pictured as having a special thickness, such as that of a quarter wave. This condition makes it a quarter-wave transformer and the behavior at the second interface is then of particular interest (Fig. 11-26). There is an impedance ratio at

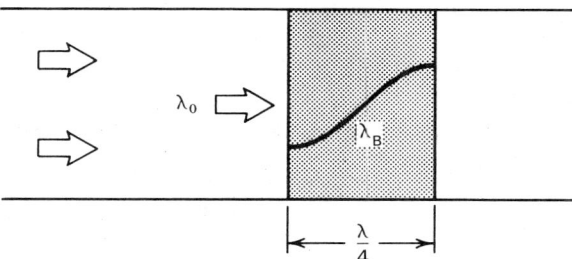

Fig. 11-26. Impedance of $\lambda/4$ transformer for wave in space λ_0 and in a body λ_B.

the far side

$$\text{impedance at 2nd interface} = \frac{\text{impedance in body}}{\text{impedance in the space}} \quad (263)$$

which quite effectively reflects wave energy back into the body and establishes a wave minimum at the near side of the quarter wave thickness. There will be quite a range of energy densities within the body, that is, all the way from the maximum to the minimum value. If uniformity of energy density is desired, then enough other waves would have to be introduced or this one varied to equalize the distribution of energy in the body.

The impedance at the second interface equals

$$Z = \sqrt{\epsilon'} \quad (264a)$$

and the impedance depends on the dielectric constant. As an impedance transformer, the quarter-wave pad would connect with some other substance such as tissue and its dielectric constant would then be engineered to *facilitate* entry of energy into the tissue.

Still other pads or laminates can dampen reflection, which is important for anechoic chamber walls. Then the outer laminate can cancel entering waves caused by reflections from the back laminate. Assuming the back laminate to be conductive, two waves will have a cancellation frequency if the pad is a quarter-wave or a multiple of it in thickness. This action depends on magnetic permeability and magnetic loss in the outer laminate such as might be contributed by plastic sheets filled with iron powder for frequencies from 1 to 16 GHz with commensurate thicknesses in this range from 0.04 to 0.2 inches or 0.1 to 0.5 cm. The damped reflectivity could then be described as a negative dB at resonant frequencies.

The response to transient, sinusoidal forcing functions from electromagnetic waves can cause a current in conductive substances, including cells. In these situations there is an alternation of current I or alternating densities and concentrations in acoustics, with the densities and concentrations forming a wave,

$$I = A \cos(\omega t + \phi) \quad (264b)$$

where A and ϕ are constants, or as the real part of the complex current

$$I = I_o e^{j\omega t} \quad (265)$$

where the *complex current amplitude* is related to the angle function relationship by

$$I_o{}^* = A e^{j\phi} \quad (266)$$

In terms of voltage and current again

$$Z = V/I = V_o/I_o \quad (267)$$

or

$$Z = R + jX \quad (268)$$

Z is the impedance in sinusoidal currents and potentials or in their Fourier transforms. Z is expressed in complex form with R and X (resistance and reactance) because of the current sine wave effect of the phase relations. Resistance R gives the ratio of the portion of V in phase with I while X gives the ratio of the portion of V 90° out of phase with I

$$Z = R + jX \; \underbrace{\frac{V \text{ in phase with } I}{I}} \quad \overset{\displaystyle \frac{V \text{ 90° out of phase}}{I}}{} \quad (269)$$

R and X are real if $X = 0$ and Z is real and I and V are in phase. If R is 0 and X is not 0, V and I will be 90° out of phase.

It is this use of Z as a coefficient for the flow I subject to a pressure V that leads to so many biological analogies and simplifications. Perhaps the most well known are in acoustics where a speaker is matched to a circuit, or in telecommunications where oscillators are matched to antennas for propagation, or in sound-powered systems where behavior at junctions is critical. An example of the latter would be the stethoscope and an example of the former is telemetering in bioelectronics.

When the magnetic permeability (μ) of the material equals that of free space (μ_0), R and X can replace σ and ϵ in cells and tissues. Both are used here to explain the arc relationship but interest returns to σ and ϵ in other discussions of dielectrics. The difference between μ and μ_0 is less than 0.01 so for these studies, μ is taken equal to μ_0. At low frequencies, cell membranes are insulators and since an applied current circles the cell, this makes its pathway longer than if it had gone through. With the cell in a suspension therefore, R will be proportionally greater in the composition of Z and it both depends on and measures the volume of the cell. It turns out that the low frequency impedance is similar for skeletal muscle, liver, and cardiac muscle. The resistivity ρ is about equal to 900 ohm cm and it decreases as the frequency increases. Blood has relatively greater conductivity in accordance with the fact that its cell "structure" is much looser while fat and bone have less conductivity. Thus, from 1 to 100 kHz, Z falls rapidly, and in this region a good cellular model is that of a spherical homogeneous conductor with a lipid insulating layer. Therefore a cell is a capacitor and the interior of the cell is a resistance.

At low frequency

$$Z_{cell\ capacitor} = j/\omega C \tag{270}$$

is <u>large</u> *and the frequency is in the denominator. Thus the cytoplasm does not conduct and the cell acts as an insulator.*

At high frequency

$$Z\ is\ very\ small \tag{271}$$

the cytoplasm is a conductor, and the impedance of the suspension is also small. While Z may be small at radio and microwave frequencies, it is certainly meaningful.

The membrane may be regarded as a "leaky" capacitor which means that it resembles a large resistance in parallel with a capacitance (Fig. 11-27). The membrane capacity is in $\mu f/cm^2$, the resistance is in ohm cm^2, and Z of the cytoplasm is in resistivity ohm cm.

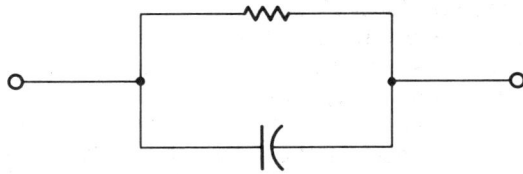

Fig. 11-27. Membrane as a "leaky" capacitor resembles a large resistance in parallel with a capacitance.

The concept of leakage becomes important in the description of the stimulus and disturbance in an excitable membrane. This disturbance travels like a wave, that is, in a constant manner along axons. On a cable it would die out quickly. To communicate then, some axon technique had to provide for maintaining the wave propagation. As the impulse passes, the membrane impedance first decreases and then recovers. This is the variation in the membrane that is needed to explain a period of different characteristics associated with the operation or function and is, thus, full of meaning. The leakage changes but the capacity does not and thus there is a change in conductance or in resistivity. We can thus write a variable resistance into the membrane model (Fig. 11-28) and associate the change in leakage with permeability, with excitation, and with changing potential, obtaining an overshoot in the case of the axon potential (Fig. 11-29).

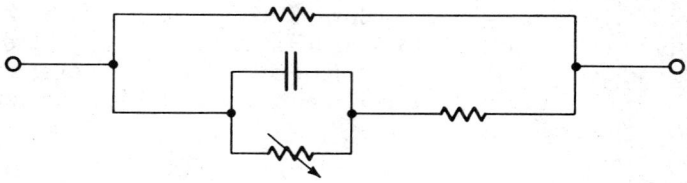

Fig. 11-28. Variable resistance and membrane model.

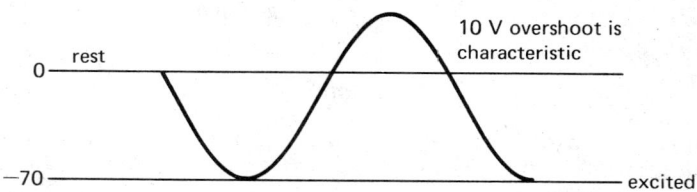

Fig. 11-29. Action potential overshoot seen in excitation.

At low frequency, the current *around* the cell in suspension is important; at high frequency, the internal conductance is important because the current conducts *through*. This "around" versus "through" behavior is what is meant by the frequency effect as an information communication link. As applied to dielectrics, the cell has a high frequency dielectric conductivity that also changes the situation in a meaningful way for communication between or with living systems.

The basis for our analysis of the behavior of dielectrics in the capacitance model could be called an analog for Z. A complex capacitance is then illustrated in the dielectric analogy

$$C^* = C_s/(1 + j\omega\tau) \qquad (272)$$

where capacitance is a function of frequency ω and is in series, C_s. This means that complex voltage, charges, etc. are to be presented as complex plane values where the

$$Z^* = x + jy \qquad (273)$$

Impedance and Membranes

and the coordinates are x and jy. Voltages V^* rotate here counter-clockwise as a function of ω

$$V^*(t) = V_o e^{j\omega t} = V_o(\cos \omega t + j \sin \omega t) \qquad (274)$$

and the charge and current are represented by a vector with an angular rotor, or circular motion.

In conventional circuits,

$$Z^* = 1/j\omega C^*(\omega) \qquad (275)$$

where relations between $Z^*(\omega)$ and $V^*(\omega, t)$, and $I^*(\omega, t)$ give the complex impedance or in polar coordinates

$$Z^* = Z e^{j\phi} \qquad (276)$$

Here Z at the right is real and is the ordinary circuit impedance. A series circuit

for example, gives

$$Z = [1/(\omega C_s)]/(1 + \omega^2 \tau^2) \qquad (277)$$

In another analogy, there is the resistive-compliant trachea-lung system where the resistive element (friction in trachea) and capacitive part (compliance in the lung) are also in series (Fig. 11-30). These can then

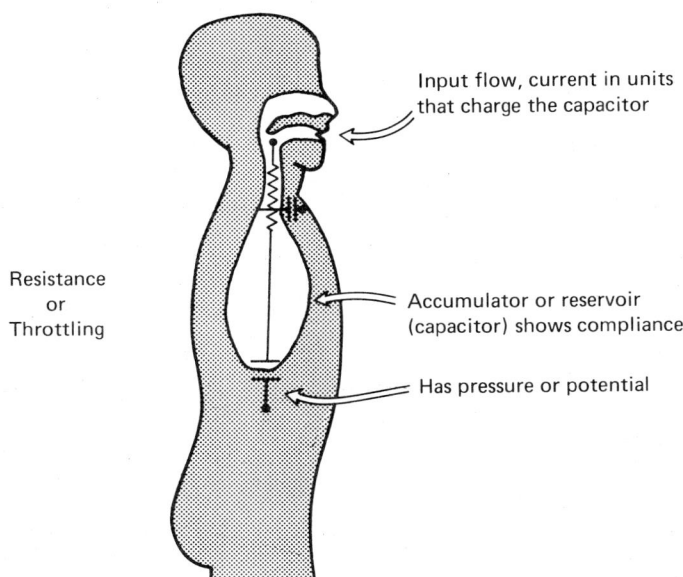

Fig. 11-30. The lung-trachea system.

be modeled by the RC analog using Kirchhoff's current law for finding the system parameters and the biological components are treated as a network. The lung is a lossy storage element somewhat like an elastic, reactive balloon, primarily capacitive but with some resistance. Then the *system* has both a loss part and reactance which give the impedance. A capacitance added sideways to the circuit in the trachea (Fig. 11-30) can represent the flexibility of that tubular section. Flexing in vessels is thus seen as a manifestation of capacitance and friction as resistance and both are considered in the total impedance.

For these same electrical elements in parallel, the circuit would have a system analogy wherever its impedance matches the transfer function of a system (Fig. 11-31). The system would need to have a resistance and a storage element, but arranged in parallel. A colloidal or cellular suspension is like a parallel resistance and capacitance and its conductivity will depend on the resistance.

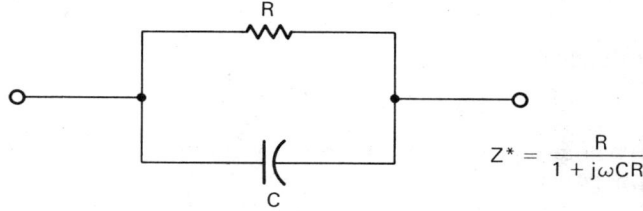

Fig. 11-31. Parallel RC circuit.

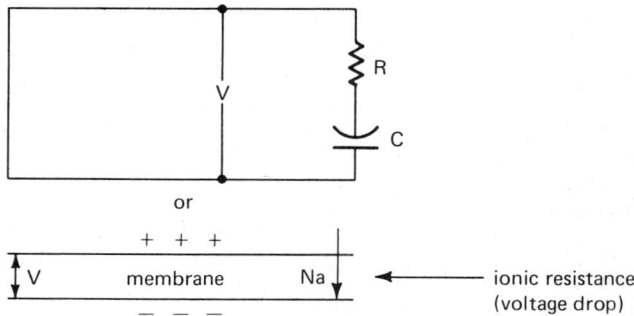

Fig. 11-32. Membrane measurements.

Informationally, the circuit has its most useful analog in a membrane (Fig. 11-32) conveniently treated as a "leaky" capacitor with the capacity in $\mu f/cm^2$, and a resistivity plus a cytoplasmic impedance. The leakage resistance of the membrane is 25 to 10,000 ohm cm^2 or more. On a linear basis (with distance) the protoplasm is about 10 to 30,000 ohm cm. The bioelectrical properties are very similar for a wide variety of cells and give useful values for such calculations as those being discussed here. Thus protoplasm has a capacity, on an area basis, of 0.1 to 3 $\mu f/cm^2$, or an "average range" from 0.6 to 1.2 with the very common value of 1 $\mu f/cm^2$ discovered by Cole and others.

Nerve membrane leakage would suggest that ions were crossing. If these are known, then the effect of a stimulus is predictable. Cables have similar features—capacitive membranes with leakage, core resistance and membrane resistance. When resistances are measured, it is found that as distance increases along the cable or nerve, resistance increases greatly. In fact, if we double the distance, we double the resistance.

In the sense that frequency is a factor in the quantities observed, radiations probe biological structures, especially membranes, and the quantities observed are Z, ϵ, C, R, V, current, etc. since these depend on the frequency. It is then found that the membrane depolarizes, that it transmits ions, and that the nerve transmits a pulse. Biological materials contain important capacitances like those in membranes with a high lipid content. These show dielectric behavior similar to that of a material with insignificant electrical conductivity. Dielectric relaxation occurs, polarization is based on ϵ, and the stimulus is an electric field. Other materials like bone, electrolytic fluids and tissues can behave as semiconductors in the sense of imperfect conductors and this behavior is based on σ. Radiations then influence both types according to the two parameters. The radiations force systems to respond due to the sinusoidal nature of the electric fields with voltages varying with time. The interactions, taken in this way, are informing, using sensors in the instrument sense, that is, impedance bridges or the equivalent. Within narrow limits of the information content, they tell just what the interaction is.

What is being observed is analogous to a natural communicative process within the organism which gains simultaneously meaningful electrical parameters.

The impedance as shown in the following equation is a function of the shape of particles, that is of the form factor γ in a colloidal suspension, the volume concentration ρ, and the membrane characteristic impedance Z_m:

$$Z_{suspension} = r_1 \frac{(1-\rho)r_1 + (\gamma+\rho)(r_2 + Z_m/a)}{(1-\gamma\rho)r_1 + \gamma(1-\rho)(r_2 + Z_m/a)} \qquad (278)$$

where r_1 and r_2 measure resistances of inner and outer spheres in the concentric sphere model of the suspended units, and a is the radius of the outer shell corresponding to the membrane. Thus, as the impedance responds to frequency, it in turn informs on these dimensions. The accuracy is within fractions of a microfarad with respect to one element, the membrane capacity, which measures (by bridge methods) about 1 μf/cm². "Unfortunately" the groups working on this aspect were hopeful of generalizations like all cells having membranes with this capacitance and all cytoplasm appearing with about 100 ohm cm resistivity. The variation with frequency was a problem which had to be circumvented by choosing instruments that tended to cancel out the frequency dependence. Herman Schwann may have been an exception with his efforts to interpret frequency phenomenon by repeating the experiments at many frequencies, but this tended to confound his listeners somewhat where it probably should have intrigued them! The most interesting offshoot from the work is the discovery that the dielectric properties of membranes have an information content, and that a dielectric loss process occurs as f (frequency), where the membranes look like solid dielectrics and semiconductors.

Absorption in Tissues. I saw Schwann's laboratory at work on these problems and it was clear that he saw the frequency dependence as a highly informative factor in biological electromagnetics. Schwann is an electrophysicist and is probably the most informed scientist on frequency effects in tissue. He found that tissues tended to be similar to blood in regard to the frequency dependence of resistivity, and dielectric constant. When the log of ϵ/ϵ_o, which is the effective dielectric constant, is compared with the log of frequency, a very interesting curve results with three regions (Fig. 11-33). Region number I indicates the dielectric response for large bodies compared to protein or low frequency effects at higher organization levels. In region II, the conductance shifts from *around* to *through* cells in tissues with the cells varying greatly in structure as suggested by the curve. In region III, it is the relaxation effect in molecules that is characteristic, e.g., the relaxation of water is at 10^{10} Hz.

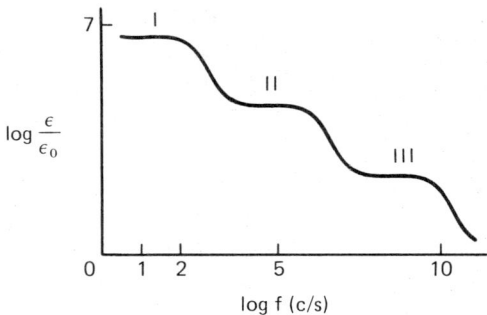

Fig. 11-33. Frequency dependence and relaxation regions in whole tissue dielectrics (muscle).

The dielectric constant suggests the capacitive portion of the impedance because $\epsilon = C/C_o$ and, if the resisitivity followed the same curve (which it does), then it might well be possible to say that these two factors Z and ρ are lumped for these considerations. The relaxation frequency is seen as a breakpoint which signals a rapid decrease in the lumped values with still higher frequencies. These values are similar for muscle, liver, spleen, pancreas, lung and kidney at frequencies over 10^5 Hz suggesting a commonality like water content. Bone, with its concentration of calcium phosphate, has an impedance that is much greater than that of soft tissue at low frequency which suggests that it is lucky indeed that EEG traces can be seen outside the cranium which imposes such hindrance to the flow of fields. The frequency response for fat is like that of muscle but the values are displaced to indicate much higher values of resistivity and lower ϵ.

Brain tissue is similar to watery tissues as far as ϵ goes, but when it comes to resistivity, it has a step change. It is like fat at low frequencies with high resistivity, but from 10^6 to 10^7 Hz the resistivity drops, and brain tissue then acts like nonfatty tissue at high frequency with less resistivity. We study the dielectric loss because membranes have a loss component which is interestingly similar to that of solid dielectrics and brings in the studies of semiconduction and the solid state. Here it is seen how the megahertz range of frequency introduces information not available at low frequency and relating the dielectric constant and resistivity to the biological composition and vice versa are equally true.

Impedance, a Core Concept

Thus impedance matching is fundamentally a process of balanced couples designed to assure a desired flow of energy in a system. Then one could arrange for any current across two points S and T (Fig. 11-34) by varying the resistance R. Then ST could represent any system

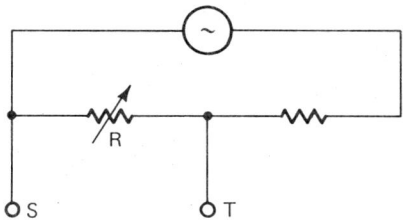

Fig. 11-34. Power source matched to attached system.

being served by the power supply with its internal resistances and power source. In the more general sense, it is the impedances that should be matched so that voltage and current will be in correct phase, or so that resonances in the power supply or added system will be maintained in proper relation. Devices like the one-fourth wave pad are called impedance transformers because the result is often like matching the impedance in a primary with the impedance in the secondary. The reason for this is that the main circuit cannot function unless it sees an acceptable impedance. If they differ, the impedance transformer has an effect like a turns ratio

$$N_{primary}/N_{secondary} \tag{279}$$

which is usually provided by the manufacturer. So by varying elements such as transformer windings, the impedances may be made to match between primary and secondary. In this sense,

$$Z_{prim} = E_{prim}/I_{prim} = (E_{sec}/I_{sec}) \times (N_{prim}/N_{sec})^2$$

$$= Z_{sec} \times (N_{prim}/N_{sec})^2 \tag{280}$$

or the impedance ratio is the square of the turns ratio.

TRANSFER FUNCTIONS AND MODELS IN BIOLOGY

In modeling, impedance matching serves the purpose of transferring an attribute from one black box to another without loss of effect. Z may also represent the transfer function that transforms the input into the output. Known rules for impedances in series and parallel allow the transfers and connections to be made. The transfer function could be one involving *storage*, or the latter may arise out of an operating function in a system such as the transfer function in a white box. Some informing function may be assigned to cyclical storage in which the organism wishes to store rather than lose *information*, for possible retrieval later. This causes impedance to enter into the capacitor model for absorptive polarization. Resistance is not sufficient because, while it is also measured in ohms, it stresses the loss of energy associated with friction against the flow of energy. Obviously, an ideal capacitor would have none of this property, but if ohms are also connected to the storage of energy, then the capacitor can be described in terms of this function. Storage is an effective way of impeding flow in the sense of a buffer. There may be some dissipative energy loss as well and the total is impedance.

In practice, each part of the line that is to be matched has an impedance symbol. A dielectric may be called Z_{DK} for example, with the input impedance Z_e; for free space, it is Z_o; the characteristic impedance of the transmission line is Z_H; and Z_L is that of a coaxial type of line.

If the flow of electromagnetic energy is subject to impedance, this is essentially a manipulation of the information that flows in parallel and can, therefore, serve the ends of communication. The signal flows in response to characteristics of the force or pressure. Each part of the flow route presents its own characteristic impedance. Then, as impedance is equally applicable to other flow regimes, analogs would be found for liquids, solids, and gases *and the evidence of mismatch is reflection of energy at an interface which is both informative and measurable.*

Impedance and Membranes

It is easy to see how the information pathway can be marked because the interfaces or junctions all have the mark of Z, the intercepted characteristic impedance. Then field and flow will be governed by the path of best impedance. In chapter 13 the uses of this information in bioelectronics will be indicated.

One can design a series impedance for the circuit (Fig. 11-35) including C and R but not a series resistance which is R alone. Then Z is the voltage drop across the series impedance divided by the current that flows at any time to C.

$$Z_{RC} = V_{SZ}/I_{CS} \qquad (281)$$

Fig. 11-35. Designing a series impedance for the circuit.

Biologically, the series impedance can be visualized in gas and liquids flowing against resistive-compliant elements in a flow system. The advantage of Z is that it can be closely connected to a transfer function for such a system or the mathematical expression for the operation in a black box. Then this valuable option is available along with expression of the system as a differential equation or as a model in the sense of an analog of the circuit, all the options giving the same information but with different degrees of convenience in an experiment.

Detailed understanding of the structure may suggest the *analog*, whereas lack of the understanding may suggest the system transfer function, which has virtue through its use of lumped linear relations. Alternative models are Laplace transforms, $P_{(s)}$, a mathematical method that works well both here and with differential equations, time functions, or tables for the s function that is involved, or equivalent differential equations. The easy interchange from model to s function requires the use of the impedance concept for the elements of the circuit.

For example, if we take a transfer function relating the capacitor current (the output) to the generated voltage $\ominus$ in a series-resistance circuit (the input)

$$I/V = 1/[R + (1/s\,C)] \qquad (282)$$

Here I/V is a conductive property in reciprocal ohms (mhos) and may be taken as a transfer function. Looking then at the denominator on the right, R is in ohms and $1/s\,C$ has to be in ohms too, so the transfer function is one of conductance, 1/ohms. The way to describe this property in $1/s\,C$ where it has units of ohms but belongs to capacitance, which does not have such units, is as an impedance. The latter may be of this type (series RC circuit) or sinusoidal.

$$Z = R + (1/sC) \qquad (283)$$

Since this equals V/I, the impedance of the network is the transfer function. The transfer function implies an information process on the black box with input and output information which must be related to storage as the capacitive function. The input is the stimulating voltage, the output is current response and the transfer function is the variable V/I which is expressed operationally as $1/sC$ and is directly seen as Z.

Various flow regimes may be defined for communication in tissues, vessels, and membranes such as: 1) Molecular; 2) Diffusive; or 3) Pressure (Copson 1975). These may be used for freeze drying tissue with microwaves in which the impetus is due to collision effects, concentrations, and pressure drops respectively. For example, with a concentration K forcing function, the ions or nutrients N will flow from high to low concentration as current does under voltage. Then with the impedance defined as

$$K/N = Z \qquad (284)$$

progress may be made toward transfer functions, analog models and differential equations based on Z.

Z has already been discussed in relation to membranes and an information role has been attributed to it. It has also been seen as regulating energy transmission between interfaces other than the membrane one, and above all it has been seen as a very convenient frequency responsive signal whenever there is a combination of materials with impedances and fields that apply some forcing or informing function. In Figs. 11-12 and 11-13 the real and imaginary parts of a quantity are explained; for example, in 11-12 as $I_{charging} = j\omega CV$ for the current that leads by 90° while the loss current I_L is in phase with the voltage. The impedance Z is the complex ratio of E/I and it has the complex notation in which reactance and resistance coexist (Fig. 11-36). The resistor represents the real part while the reactive part of Z is considered imaginary and, by mathematical convention, has a "j" in it.

The capacitive reactance is $-j/2\pi fC$ and the inductive reactance is $j2\pi fL$. Thus the Z vector is described by a phase angle (Fig. 11-37).

From Ohm's law, ac impedance by analogy to resistance is $Z = E/I$ but Z has both resistive and reactive parts. The resistance is 90° out of phase with the reactance, a problem for which these special mathematical solutions are available. The phase angle between the impedance and resistance components is the angle of the impedance, often called θ (Fig. 11-36). Impedance is universal with resistance having inherent capacitance and inductance, capacitance having inherent resistance and inductance (lossy capacitor), and inductance having inherent resistance and capacitance (lossy inductor). A *resonant* circuit (to which we attribute special meaning and information processes) lacks reactance and the resonances are functions of frequency. At resonance the impedance is zero, giving a maximum current in a series-resonant

Impedance and Membranes

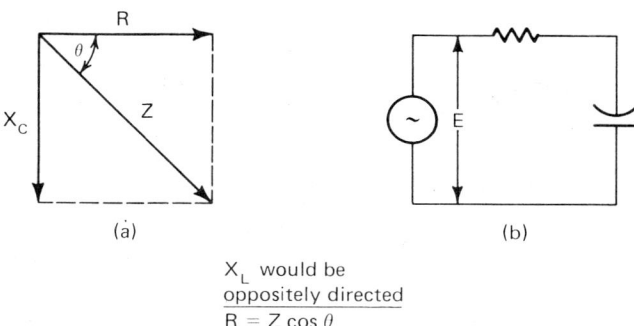

Fig. 11-36. Z diagram for resistance and capacitance Z is vector sum of resistance vector R and reactance vector X_C. The phase angle of the resulting impedance is the angle θ between R and Z vectors.

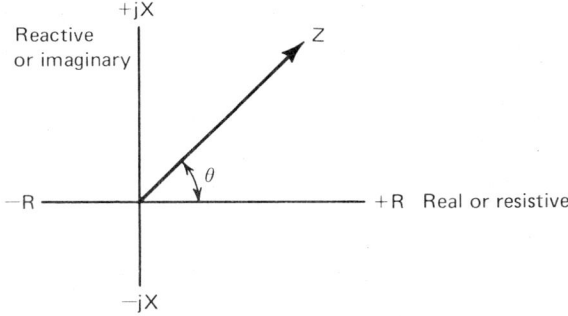

Fig. 11-37. The Z vector is described by a length and phase angle.

circuit. In a parallel-resonant circuit the impedance is infinite at resonance and the voltage is at a maximum. Z is thus a core conceptualization or an element in the pervasive *capacitive model* which has now given not only an analog for Z but a host of other relevant insights including those of:

Compliance in biosystems as represented by reactance which combines with resistance in storage functions.
First order changes in system status for A → B.
Complex time relations.
Bioprocesses, their time constants, and their spectra of relaxation times.
Penetration of EMR and skin depth.
Polarization in bioelectrical responding systems and membranes.
Conformation and cooperation.
Instrumental response.

Microwave heating and diathermy.
Forward, positive exponential changes such as growth and
Backward, negative exponential changes such as decay or degradation with their associated coefficients.
Instant relaxations and delayed effects.
Guides such as the 1/e level for intersystem comparisons.
Resonances, special relaxation frequencies and the facilitated entrance of energy and information.

MEASUREMENTS AND THE TEMPERATURE EFFECT WITH MICROWAVE PARAMETERS

The capacitance model is actually a measurement or working model because the polarization parameters are observable in it. Capacitance measurements and other tests at low and microwave frequencies give dielectric constants or permittivity (either static with dc ϵ_o or other ϵ) and loss factors such as tangent δ and dielectric conductivity σ. Relaxation times are studied by frequency measurements under controlled temperature. Careful temperature control is essential for most loss measurements. Tan δ for example, increases with temperature at frequencies where the refractive index is measured as shown by

$$\tan \delta = T^n \, e^{-\alpha/T} \qquad (285)$$

where α is a constant and n is the refractive index and T is the Kelvin temperature. Since optical and microwave properties overlap, the refractive index is a useful parameter in many measurements. The relaxation time τ is inversely proportional to the temperature. If W is the potential energy required for overcoming the alignment barrier or the activation energy for the dipole relaxation, the exponential change in τ is

$$\tau = \tau_0 \, e^{-W/kT} \qquad (286)$$

Or τ may be related to the alignment probability where suitable process time constants will produce alignment of a susceptible body in the alternating field

$$\tau = (A/T) e^{W/kT} \qquad (287)$$

and W is the height of the potential energy barrier, A is a constant, k equals Boltzmann constant and T is Kelvin temperature. Also, as a function of reactionary forces of viscosity in the medium,

$$\tau = \text{the frictional constant for the medium}/2kT \qquad (288a)$$

The dipole relaxation time is seen to decrease as the temperature rises, a result that in turn displaces the maximum frequency toward the higher frequencies.

The various dielectric loss parameters are different mathematical approaches to the identical end effect—an irreversible dissipation of electromagnetic energy in a material interaction.

ANOMALOUS DISPERSION

There are two special dielectric loss conventions which are pertinent in the observation of the main parameters. Both have to do with the relaxation frequency

$$\omega = 1/\tau \tag{288b}$$

One is called the *frequency dependence* or anomalous dispersion (large attenuation) of the dielectric constant which relates the dielectric constant, loss factor ϵ'' and the frequency. The other forms an interesting geometrical relation, the Cole-Cole arc plot, for the dielectric constant as a function of the loss factor and frequencies from 0 to ∞.

The meaning of the real and imaginary parts of the complex dielectric constant ϵ^* are illustrated rather strikingly by plotting them according to their variation with frequency in the region near the relaxation frequency or natural frequency of the dielectric material (Fig. 11-38). At this frequency, the relaxation time τ is seen as a special

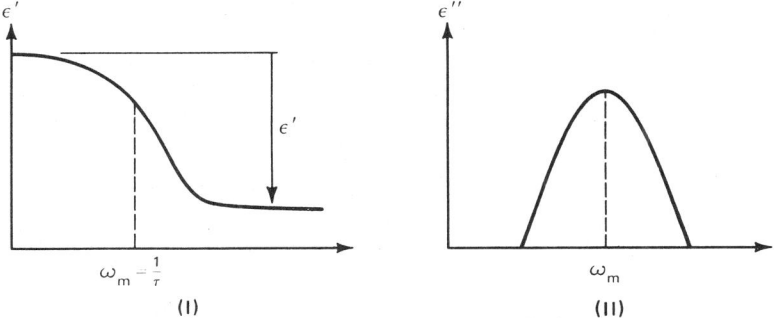

Fig. 11-38. Behavior of dielectric constant and loss factor in the region of the natural frequency. I-ϵ', the real part of ϵ^* versus $\log_{10} \omega$. II-ϵ'', the imaginary part of ϵ^*, versus $\log_{10} \omega$.

absorption time and an exponential function related to the natural frequency of the polarizable substance at which it can absorb energy via resonance absorption. Strictly, resonance is for behavior like the NH_3 inversion which does not continuously decrease as ϵ' does here, but instead is like a small attenuation with normal dispersion and loss only at the resonant frequency. Over the entire absorption spectrum for dielectrics, ϵ' goes from the static permittivity ϵ'_o to the value at the high frequency end ϵ'_∞. The maximum value of ϵ'', the loss, is at ω_m.

The electrical susceptibility χ is of general interest for electronic processes in materials. It is the susceptibility to oscillation in connection with the dipolar response. It may be examined for its behavior in the complex plane and for its frequency dependence in which it shows a dispersion of the real part and a maximum of the imaginary part in the vicinity of the natural frequency. The susceptibility may be related (since it is directly proportional to the polarization) to the various components of the polarization such as dipolar, electronic and ionic, with only the dipolar one being temperature dependent. In addition to this electric susceptibility, materials may also have ferroelectric, diamagnetic, paramagnetic, and static susceptibilities.

THE CIRCULAR FREQUENCY GRAPH

Kenneth Cole, using a neat method that he developed with his brother Robert, transformed the reactance, resistance, impedance, volume concentration and radius of a spherical unit in suspension, a model (cells, colloids) with an interfacial surface layer or membrane, into an equation with both sides squared. He used the limiting cases, infinite and zero resistance and specific reactance per unit area and the absolute value of the impedance.

Then he let the internal resistance be infinite (no conductivity, as at low frequency) and introduced the frequency-dependent specific impedance per unit area for the spherical particle, still regarding the frequency as that "incidentally" associated with the measuring *current*. In any event this meant that when frequency goes to 0, the impedance became infinite and when frequency goes to infinity, Z became 0. The change in Z with frequency is similar to that of ϵ' (Fig. 11-38). The specific resistance is seen as

$$R_3 = mX_3 \qquad (289)$$

where it varied by a constant from the specific reactance per unit area. Thus Cole produced the circular frequency graph known as the Cole arc for suspensions and showed that he was then seeing the frequency as the *informing influence* which points out the advantages in the circular presentation to its own originator.

The circularity is due to the way that this reactance changed from 0 to ∞ with the locus of the arrow on the impedance vector becoming the arc of a circle in the complex plane. Thus it is a circular presentation of the impedance loss. This fact is indicated in Fig. 11-39. One can then evaluate distances such as a and b to see that their ratio varies directly as the absolute impedance and inversely as terms for the resistance and volume concentration arranged in such a way as to form a constant. Similarly, the depression of the center below the R axis is significant and geometrical considerations produce the indicated chord distances. Also when m is 0, the circle is centered on the R axis. The depression of the center below the axis also measures the distribution of relaxation times as opposed to a single relaxation. If the distribution is zero, the center is on the ϵ' axis. If it is below, then measuring the angle of how

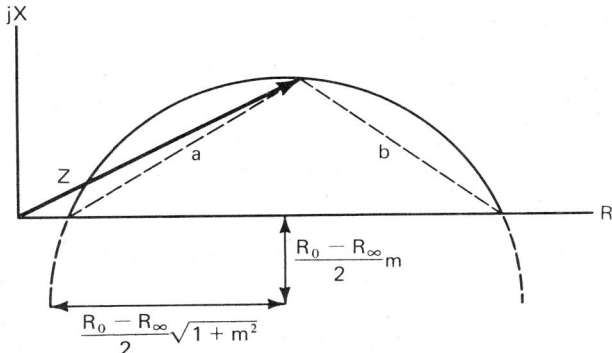

Fig. 11-39. Cole-Cole arc for Z, applied to R and X for spheres in suspension. Interior nonreactive, surface is a thin layer with R and X.

much it is below will estimate the distribution. In water, the relaxation mechanism shows only slight spreading of this kind. *For red cells*, the data of Fricke give an application of the circular plot as shown in Fig. 11-40 after Cole (1928).

These arcs mean that R and X behave in such a way with R being dissipative and giving rise to heat and X being conserved and later returned to the system, that the descriptive equation is a semicircle from complex variable (jX) theory with the semicircle having a depressed center as in Fig. 11-40.

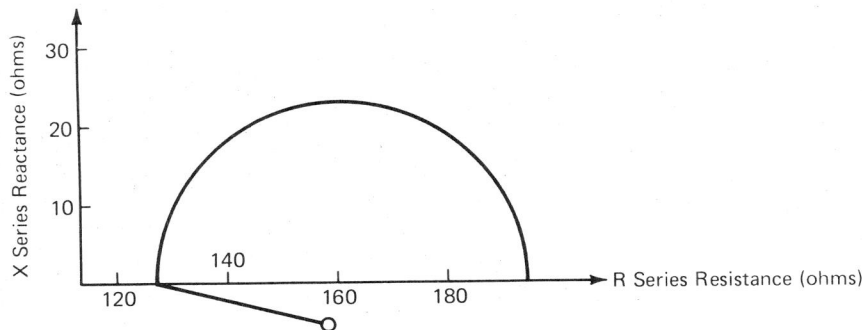

Fig. 11-40. Complex impedance locus red cell suspension where membrane capacity varies with frequency.

The impedance locus is a rather elegant way to relate:
1. frequency dependence of ϵ', ϵ''
2. ϵ_o, ϵ_∞
3. Debye equations

4. relaxation times and frequency
5. stored or conserved and returned energy
6. dissipative energy

In addition, the several areas given show how data could be given in terms of R and X or in terms of ϵ' and ϵ''. Such a compilation of data is anything but simple, but biologists have long been resigned to the complexities that biology-in-depth tends to present. In addition to the meanings given, the diagram can be expected to give some degree of management over dielectric parameters and so be useful in applications involving high frequency energy. These range from medical diathermy and analysis of biohazards to domestic and industrial use of the radiations (Copson 1975).

If there is a composite dielectric, the dispersions are sometimes separable and are temperature sensitive. At 40°C two individual dielectric dispersions (ethanol, cyclohexanol) are separated 1 and 2 in Fig. 11-41; in the liquid state when supercooled, the two are "not" resolved 3.

When data from several different materials are plotted, their different losses and dielectric constants, but common mechanism, may be illustrated by areas which fit inside one another (Fig. 11-42). In the

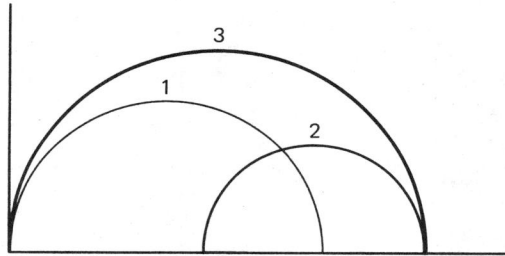

Fig. 11-41. Resolution of components.

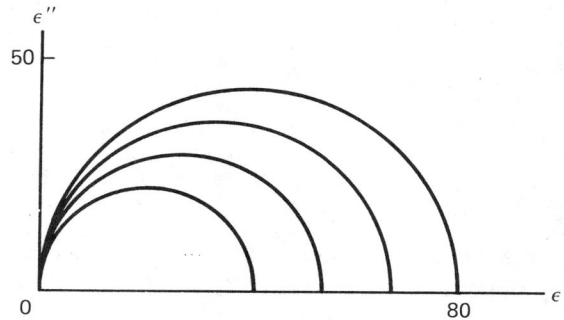

Fig. 11-42. Several arcs with a common mechanism.

example of de Loor, the materials are water, 4% agar, potato, and starch (consisting of 2% agar and 25% starch) in order of decreasing radius.

The complex dielectric constant may be analyzed in much the same way to produce the Cole arc with respect to frequency (Fig. 11-43).

$$\epsilon^* = \epsilon' - j\epsilon'' \qquad (290)$$

$$\epsilon^* = \epsilon_\infty + [(\epsilon_o - \epsilon_\infty)/(1 + j\omega t)] \qquad (291)$$

The Debye relations give

$$\epsilon' = \epsilon_\infty + [(\epsilon_o - \epsilon_\infty)/(1 + \omega_m^2 \tau^2)] \qquad (292)$$

τ is the relaxation time of the dielectric

$$\epsilon'' = [(\epsilon_o - \epsilon_\infty)\omega_m \tau]/(1 + \omega_m^2 \tau^2) = \epsilon' \tan \delta \qquad (293)$$

The right side of (293) allows ϵ'' to rise to a maximum from zero, then go to zero over the right frequencies as a relaxation mechanism, and the loss tangent is

$$\tan \delta = [(\epsilon_o - \epsilon_\infty)\omega_m \tau]/(\epsilon_o - \epsilon_\infty \omega_m^2 \tau^2) \qquad (294)$$

Then ϵ' and ϵ'' change near the natural frequency according to Fig. 11-38. In curve I, the approximate middle of the dispersion and the maximum in curve II are points where

$$\omega_m \tau = 1 \qquad (295)$$

where

$$\tau 2\pi f = 1 \text{ cycle} \qquad (296)$$

and

$$\epsilon'' = (\epsilon_o - \epsilon_\infty)/2 \qquad (297)$$

and

$$\epsilon' = (\epsilon_o + \epsilon_\infty)/2 \qquad (298)$$

Manipulation of (298) gives

$$[\epsilon' - \frac{\epsilon_o - \epsilon_\infty}{2}]^2 + \epsilon''^2 = [\frac{\epsilon_o - \epsilon_\infty}{2}]^2 \qquad (299)$$

the equation of a circle which relates ϵ'' and ϵ' in the complex plane. With all positive values, this semicircular plot ϵ'' versus ϵ' is called the Cole-Cole arc (Fig. 11-43) and is a useful distribution diagram for relaxation times which may be compared for various media. As the figure illustrates, it is a distribution of relaxation times giving a maximum value for the loss factor ϵ'' and an associated change for the dielectric constant ϵ' in a range of frequency ω. When ϵ'' is on the y axis and ϵ' is on the x axis, the equation for a circle that relates them appears, in turn, as an arc which meets the ϵ' axis at the limits for ϵ' which are ϵ'_o and ϵ'_∞. Then

$$(\epsilon_o + \epsilon_\infty)/2 \tag{300}$$

is at the center.

Observations reveal whether the values to be compared fall on such an arc. If so, conclusions of Debye are applicable and the dipolar mechanism is justified (Fig. 11-43).

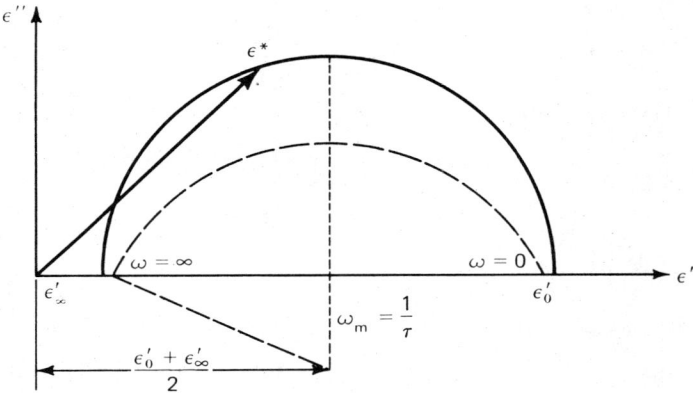

Fig. 11-43. Cole-Cole arc for the behavior of the complex dielectric constant ϵ' in a complex plane also using the values for the frequency ω and loss factor ϵ''. The arc results from the resulting equation for a circle, $a^2 + b^2 = c^2$ or in the variables used

$$[\epsilon' - (\epsilon_0 - \epsilon_\infty/2)]^2 + \epsilon''^2 = [(\epsilon_0 - \epsilon_\infty)/2]^2$$

The arcs mean that, on the complex plane, ϵ'' for dilute polar molecules rotating against kinetic agitation and viscosity, is a circle with a rather small time constant.

The values of ϵ'' may fall below the abscissa so, on an arc representing the distribution and intersecting at ϵ'_o and ϵ'_∞, the center of the circle is below the ϵ' axis and the diameter through the center from ϵ'_∞ makes an angle with the abscissa (see dotted lines).

BEHAVIOR OF CONDENSED MIXTURES COMPARED WITH THE NH_3 RESONANCE

Two rather opposite extremes may be considered in this connection. At one extreme, one effect of concentration is to produce a large interaction field on neighboring dipoles. The dielectric relaxation is broadened over the situation with uncoupled dipoles (Figs. 11-44 and 11-45).

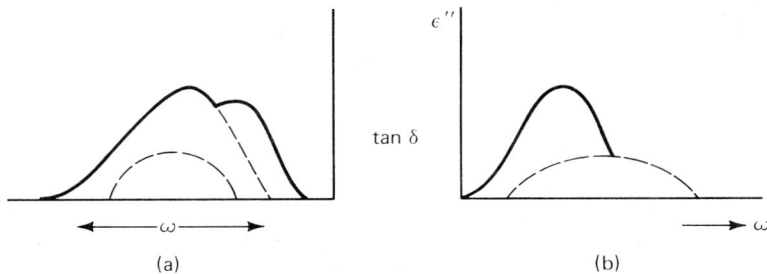

Fig. 11-44. Effect of diluting of mixture. Individual components interacting. Coupling of dipoles.

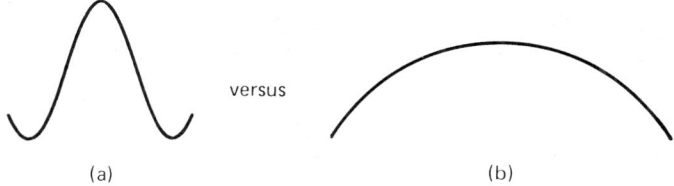

Fig. 11-45. Coupling of dipole species may smooth out the absorption band character of the dielectric absorption. If the trace (a) is taken to be that of ϵ'' as a $f(\omega)$, then most biodielectrics would give a curve like (b) in which the Debye peak has nearly vanished.

In Fig. 11-44a, the curve for frequency for a maximum tan δ has two peaks reflecting the mixture of two components. The loss factor ϵ'' curve flattens out as the individual components interact. The effect of the different dipole species is sometimes additive and sometimes not. If it is, it is probably due to the adding to the dampening factor but smoothing the absorption "band" character of the response (widening it) (Fig. 11-45). Several absorption times τ_m may arise in crowded fields, and with more relaxation times, the dispersion range is widened by dipole coupling. In this sense resolution is reduced. Functions $\epsilon'(\omega)$ and $\epsilon''(\omega)$ correspond to the ϵ' distribution and ϵ'' widening.

The situation with systems with many relaxation times is the common one in biology. To obtain the ideal Debye condition we would need large polar substances detected in a nonpolar solvent. Many simple solutes in nonpolar solvents exemplify *modified* Debye dielectrics. When the difference extends to multicomponent ordinary biodielectrics —fatty tissue for example—then the Debye peak characteristics for ϵ'' (Figs. 11-44, 11-45) become almost unidentifiable. For a typical gas, quite the opposite of a condensed mixture, the relaxation time increases as the pressure does, until it gets to be like a liquid beyond a still critical pressure and then the relaxation time decreases. The mechanism is relaxation for withdrawing radiation energy but the band is broad and the process difficult to see and estimates may be necessary. (If dielectric polarization and absorption follow ultrasonic relaxation as they probably do, then water is classed as an associated liquid like methyl alcohol or glycerol, with a negative temperature coefficient for the process.) It is interesting that both the water molecule and the ammonia molecule are notable for absorptions in the microwave region, one as a liquid, the other as a gas; and both have symmetrical pyramidal models.

In crystalline solids, the common form of dipole reorientation is molecular rotation and its dampening would be called hindered rotation. In this, the dipole can behave like a cylinder or globule. Many plastic crystals, although solids, can have molecular orientation, and changes from rotation to nonrotation or intermediate changes accompany changes in phase from normal-to-plastic-to-liquid and gas.

If the relaxation time and its activation energy for a liquid and a solid are the same, this informs that the motion responsible for relaxation is the same. Polymers have relaxation that depends on the type and the order. Smooth, spherical, molecular dispositions (globular and cylindrical) lead to large changes of orientation in ideal crystals. Disorder increases the chances for molecular movement. A polymer like polyethylene is not permanently dipolar, has low dielectric loss (for polymers) that is more or less constant with frequency and is elastic. One like polymethylmetacrylate has permanent dipoles that are mobile, intermediate loss and is rigid. One like polychlorisoprene has high loss and is soft and rubbery with characteristic *disordering* and mobile dipoles. Cellulose is of intermediate loss but its characteristics differ due to considerable hydrogen bonding.

The NH_3 molecule, a gas, vibrating or oscillating between two positions is an example of the other extreme in extracting microwave energy from the applied field (Fig. 11-46). N can be in either position

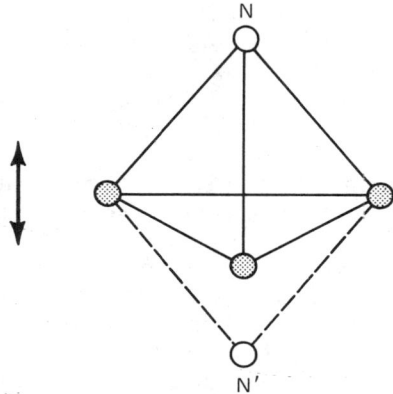

Fig. 11-46. Microwave resonance in NH_3 molecule.

Impedance and Membranes 477

N or N' impartially if thermalized. An electric field can bias the molecule in favor of a position on one side of the energy barrier. Absorption is only in the vicinity of the resonant frequency. In a population of NH_3 molecules, the densities of the two occupied states is N and N'. The natural frequency is one of transition between the occupied states

$$\omega_m \rightarrow \omega_{NN'} \tag{301}$$

The quantum transition

$$\Delta\omega_{NN'} = hf \tag{302}$$

requires the emission* or absorption of a quantum depending on the direction. If N' is the higher energy state and also the less occupied level, quanta are absorbed in a quantized transition to unoccupied energy states (Fig. 11-47). The potential energy barrier is greater in an

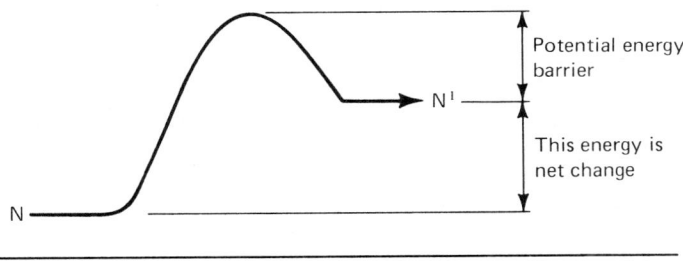

Fig. 11-47. Transition to higher energy state with absorption of energy.

ordered structure due to less steric and dipole interaction when the dipoles separate in the ordered structure and make a transition. The value of the relaxation time is proportional to the activation energy and increases with order.

In optical absorptions the foregoing (see above asterisk) may stimulate the emission of coherent light in the reversed, equilibrium-seeking transition in an appropriate medium. A laser results if the transition is arranged to include amplification.

In the orientations, the rotation means that dipoles gain the necessary activation from the field. Passing the energy barrier in rotation may be quantum-mechanical tunneling. Wave mechanics provide for the transition by tunneling or penetration of the potential barrier in NH_3, in other molecules, or in groups attached to larger molecules. Possible positions for the atoms are separated by energy peaks and the atoms, or groups must tunnel through to reach alternate energy positions.

MOLECULAR EXCITATION AND THE QUANTUM PUMP PRINCIPLE OF THE MASER

The resonant frequency of principal absorption ω_m is the transition frequency at which molecules, according to quantum mechanics, are allowed to absorb or emit the energy

$$E_3 = E_2 - E_1 \tag{303}$$

This is the energy difference between the ground and excited states.

$$E_3/h = \omega_m \tag{304}$$

The transition is possible when absorptive polarization extracts the required energy from an electromagnetic wave. Some molecules will normally be in the excited state but more will be in the ground state. If the tendency is toward excitation. there will be absorption; otherwise emission will occur. It depends on the quanta of energy available, the population density of molecules in each state, and the available states ready for occupancy. Absorption may then be followed by emission as the population works toward equilibrium. Under special conditions, this emission may be reinforced or encouraged to obtain the special source for coherent radiation. The device which provides the special conditions is called a maser for microwaves or a laser for light, these being the acronyms for microwave or light amplification by the stimulated emission of radiation. This source will produce a radiation of the same wavelength and phase as the absorbed radiation under control by an applied electromagnetic field (Fig. 11-48).

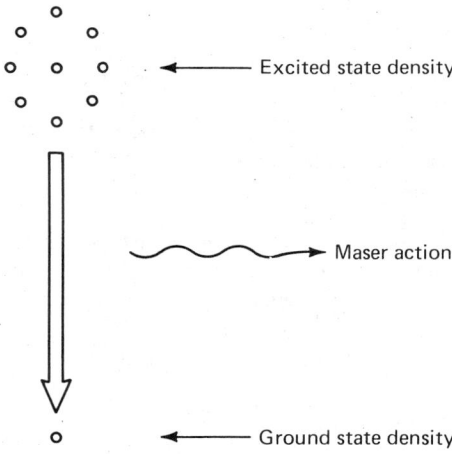

Fig. 11-48. Coherent emission results from molecular transitions and internal selection.

Impedance and Membranes

Microwave Molecular Resonance

The maser is thus related to dielectric absorptive polarization because it is this absorption which enriches the population of molecules in the excited state. This process, as taught by its originators, Gordon, Zeiger, and Townes (1955) required four principal parts: 1) a source of susceptible molecules such as NH_3; 2) an enrichment section in which those in the excited state (higher inversion, vibration, or rotation levels) are selected for reaction; 3) an input of microwaves at the resonance frequency; and 4) an output for the coherent radiation produced (Fig. 11-49).

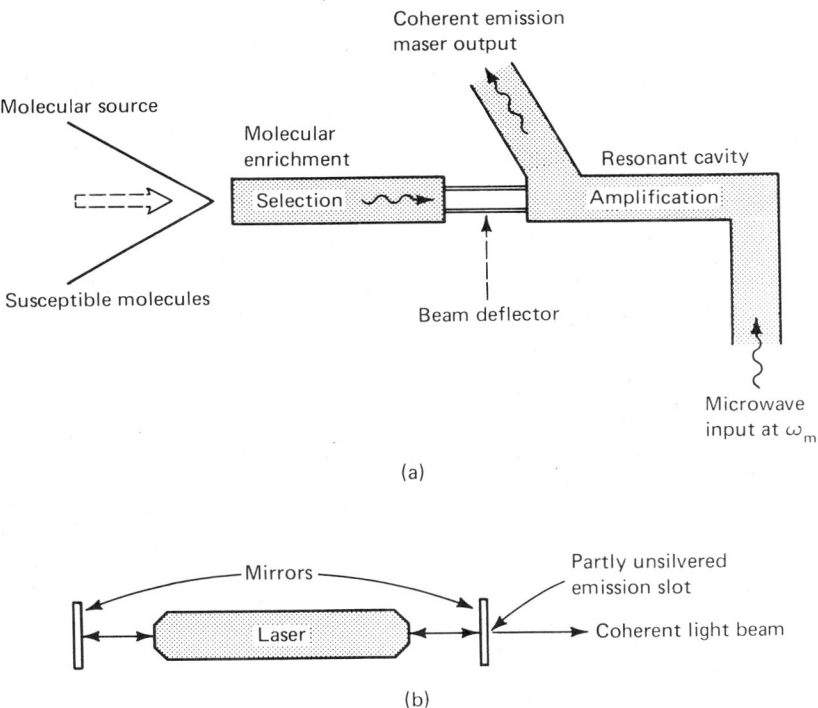

Fig. 11-49. Maser diagram (a), Laser (b). The maser is an oscillator and molecular beam emission instrument which uses molecular resonance of some molecule such as ammonia and a microwave input which extracts excitation energy from the molecules and delivers it via the increased output. In the beam deflection, the upper energy state molecules are focused toward the axis and into the resonant frequency cavity. This process is equivalent to an enrichment of the energetic molecule population in the beam.

The concept has been extended to visible light waves, and

several other wavelengths which are usually identified as lasers. In the laser, only parallel photons (to the system axis) produce the magnified light beam. Beams off the axis are multireflected and absorbed at the sides.

Coherence results when two or more wave motions at the same frequency maintain constant phase relations between themselves. The wave motion is then a steady, continuous wave train oscillating at right angles to the direction of propagation. Most excited molecular generators of radiation naturally emit at random with sudden and discontinuous phase changes. The laser enriches the beam in the desired emission. This property makes the lasers and masers important in instruments that need a stable wave action, for example, interference devices, holography, and communications.

The action of the beam deflector results in a selection of excited molecules for the narrow beam entering the resonant cavity. The input at frequency ω_m need only be sufficient to stimulate the molecules to lose their excitation energy at resonance so that this energy can be added to the maser output line. The device is then a power amplifier, a molecular resonator, a high-resolution spectrometer, and a super stable microwave oscillating source. The input of microwaves controls the molecular resonance as a frequency response over the likely molecular resonant frequencies. If ω_m is varied and the output observed, the stimulated output gives a signal as molecular action spectra are reached (molecular spectroscopy). The stimulated emission thus works like a quantum pump and has been extended to many molecular sources, solids, and semiconductors. The resulting devices find application in about every field of science, where coherence, monochromaticity, and directionality are required. The usefulness of a laser beam that one can focus is greatly increased. Then it can be directed onto retinal detachments, obliteration of tumors, cauterizing and surgery in a nearly bloodless field. Extremely large potentials on the order of 10 billion volts per cm can be obtained at a focus across preparations that transmit the beam and this tends to accentuate the electrical effects of the radiation. Figure 11-50 shows a laser spectrum with the wavelengths indicated in nanometers.

PROBLEMS

1. Give an equivalent circuit for
2. If the relaxation time is 1 microsecond, what is the relaxation frequency?
3. A wave starts traveling at a distance of $\lambda/2$ from another wave. If the waves can mutually interact, what effect will they have on each other?

Impedance and Membranes 481

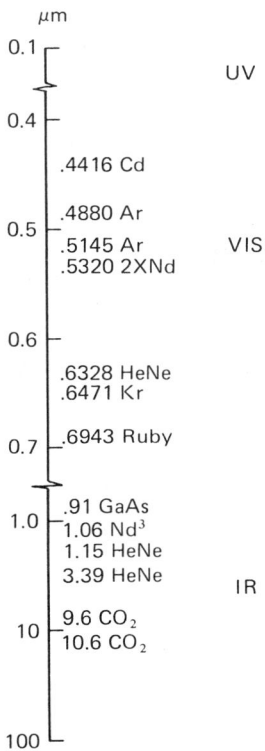

Fig. 11-50. Spectrum of lasers.

4a. What is the relaxation time for forcing functions? What are the corresponding time constants for displacements of atomic electrons? What is the band?

b. How many seconds will it take for atoms bound in molecular structures to orient in an applied field? What would the band be?

5. The dielectric conductivity in a biophysical model is 10,000 reciprocal ohms/m. If ϵ' is 74 and tan δ = 0.001, what is the angular frequency? What will the relaxation time, τ, be if ω is the relaxation frequency?

6. An equivalent series circuit for a tissue has a resistance of 100 ohms and a capacitance of 10 microfarads. What is the time constant?

7. A particular waveform has a maximum at 90°. Express this position in terms of time, sine, and radians.

8. Express $E = E_o e^{-\alpha x}$ in natural logs.

9. One absorber of radiation has a density thickness of 10 g/cm^2 and a density of 16. Another has a density of 4. What density thickness of the second will absorb equal radiation?
10. In a tissue absorbing EMR exponentially, the product of the absorption coefficient and a measured distance is 0.5. The impinging radiation at the surface was 10 watts/cm^2. What is the strength at the distance measured?
11. If there is 100% transmission in a blank in a colorimeter and calibration shows that 12 g of myoglobin/100 ml transmits 30% of the passing light, find the myoglobin concentration when the optical density is 0.70. Assume that the optical density of different concentrations will give a straight line on a graph and that 100% transmission gives the position of one point.

BIBLIOGRAPHY

Ackerman, E. *Biophysical Science*, Prentice Hall, Englewood Cliffs, N.J., 1962.
Baba, K, T. Fujimua, and K. Kamiyoshi, "The Dielectric Relaxation of Binary Liquid Mixtures," *Journal of Physical Chemistry*, Vol. 73, 4, April, 1969.
Blesser, W.B., *A Systems Approach to Biomedicine*, McGraw-Hill Book Co., New York, 1969.
Boisson, S.D., *Mem. Acad. Sci. Inst., France*, Special Report 5, 488, 1826.
Bora, B., *4th State of Matter, Plasmas*, St. Martin Press, New York, 1972.
Burns, D.M., and S.G.G. MacDonald, *Physics for Biology and Pre-Med Students*, Addison-Wesley, Reading Mass., 1970.
Cole, K.S., "An Analysis of the Membrane Potential Along a Clamped Squid Axon," *Biophysical Journal*, Vol. 1, 401-417, 1961.
Cole, K.S., "Electric Impedance of Suspensions of Spheres," *Journal of General Physiology*, Vol. 12, 29-36, 1928.
Cole, K.S., "Electric Impedance of Suspensions of Arbacia Eggs," *Journal of General Physiology*, Vol. 12, 37-54, 1928.
Cole, K.S., "Ionic Current Measurement in the Squid Giant Axon Membrane," *Journal of General Physiology*, Vol. 44, 123-167, 1960.
Cole, K.S. and R.H. Cole, *Jour. Chem. Phys.*, Vol. 9, 341, 1941.
Cole, R.H., *Jour. Chem. Phys.*, Vol. 6, 385, 1938.
Copson, D.A., *Microwave Heating*, Avi Pub. Co., Westport Conn., 1975.
Copson, D.A., *Microwave Power Engineering*, E.D. Okress, Ed., Academic Press, New York, 1968.
Daniel, V., *Dielectric Relaxation*, Academic Press, New York, 1967.
Debye, P., *Polar Molecules*, Chemical Catalog Co., New York, 1929.
deLoor, G.P., "Dielectric Properties of Heterogeneous Mixtures Containing Water," *J. of Microwave Power*, Vol. 3, 2, 1969.
Feiker, G.E. and N.C. Gittinger, "Rapid Heating of Dielectric Materials at 915 mc," *Elec. Eng.*, 78, 11, 1089, 1959.
Fitzhugh, R., "Thresholds and Plateaus in the Hodgkin-Huxley Nerve Equations," *Journal of General Physiology*, Vol. 43, 5, 867, 1960.
Fitzhugh, R. "Impulses and Physiological States in Theoretical Models of Nerve Membrane," *Biophysical Journal*, Vol. 1, 6, 445, 1961.
Fricke, H., "A Mathematical Treatment of the Electric Conductivity and Capacity of Disperse Systems," *Physiological Review*, Vol. 40, 575-587, 1924.
Fricke, H. and S. Morse, "An Experimental Study of the Electrical Conductivity of Disperse Systems," *Physiological Review*, Vol. 25, 361-367, 1925.

Frolich, H. *Theory of Dielectrics*, Oxford University Press, London, 1958.
Hodgkin, A.L., A.F. Huxley, and B. Katz, "Ionic Currents Underlying Activity in the Giant Axon of the Squid," *Arch. Sci. Physiology*, Vol. 3, 129-163, 1949.
Hodgkin, A.L. and A.F. Huxley, "A Quantitative Description of Membrane Current and its Application to Conduction and Excitation in Nerve," *Journal of Physiology*, Vol. 117, 500-544, 1952.
Hodgkin, A.L., *The Conduction of the Nervous Impulse*, Liverpool University Press, England, 1967.
Huxley, H.E., "The Contraction of Muscle," *Scientific American*, Vol. 119, 5, 67, 1958.
Katz, B., "How Cells Communicate," *Scientific American*, Vol. 201, 5, 209, 1961.
Keynes, R.D., "The Nerve Impulse and the Squid," *Scientific American*, Vol. 199, 6, 83, 1958.
Maxwell, J.C., *Treatise on Electricity and Magnetism*, Clarendon Press, Oxford, Vol. 2, 247-262, 1892.
Melia, T.P., *An Introduction to Masers and Lasers*, Chapman & Hall, London, 1967.
Orton, J.W., D.H. Paxman, and J.C. Walling, *The Solid State Maser*, Pergamon Press, New York, 1970.
Parker, W.N., "Freeze Drying," *Microwave Power Engineering*, E.C. Okress, Ed., Academic Press, New York, 1968.
Ready, J.F., *Effects of High Power Laser Radiation*, Chap. 7, Academic Press, New York, 1971.
Schupp, P.O., "Zur Physik der killektrischen Verleste," *Wiss. Veröffintl Siemens*, Vol. 17, 149, 1938.
Schwann, H.P. and C.F. Kay, "Conductivity of Living Tissue," *Annals N.Y. Academy of Science*, Vol. 65, 1007, 1957.
Schwartz, M., "The Theory of Impedance in Biological Systems," *Journal of Biological Physics*, Vol. 1, 123-142, 1973.
Siegman, A.E., *An Introduction to Lasers and Masers*, McGraw-Hill Book Co., New York, 1971.
Smith, C.P. and C.S. Hitchcock, *J. Am. Chem. Soc.*, Vol. 54, 4631, 1932.
Townes, C.H., "The Maser—New Type of Microwave Amplifier, Frequency Standard and Spectrometer," *Phys. Rev.*, Vol. 99, 282L, Thesis, Columbia University, 1955.
Von Helmholtz, H.L.F. *Die Lehre von der Tonempfindungen als Physiologishe Grundlage für die theorie der Musik*, Vieweg, Brounschweig, 1863.
Von Helmholtz, H.L.F., *Handbuch der Physiologischen Optik*, Voss, Leipzig, 1867.
von Hippel, A. *Dielectric Materials and Applications*, Technology Press, M.I.T., Cambridge, Mass. 1954.
Waterman, T.H. and H.J. Morowitz, Eds., *Theoretical and Mathematical Biology*, Academic Press, New York, 1965.

12

Dielectric Loss in Mixtures, Solids, and Complex Materials

> Being is, Non-Being is not.
> Parmenides

ABSORPTION MECHANISMS

Frozen Body

Turning to the effect of molecular loss mechanisms such as absorptive polarization on larger scale systems, the water system in particular, it is apparent that the status of the system at the time that the energy is applied should affect the results. Molecular responses will depend greatly on whether or not the system is in a liquid, solid, or gaseous state, the conductivity, and whether the dipole which responds is free to orient. The dipole may also be at an interface or bound to a macromolecule. The "freezing out" of the dielectric loss is actually a counter dampening action in which the dipolar mechanism is gradually eliminated in the crystalline structure. There are important informational consequences in these mechanisms which are miniformed in Table 12-1 and taken up in this section.

As the temperature of water systems is decreased, the system crystallizes in a pattern which depends on the velocity of the freezing process and on the presence and nature of inclusions. Slow solidification is a vastly different process than the instantaneous formation of a solid water system. The last materials to crystallize in the mixture in the slow process will be forced into a core which will alter the uniformity of the material as it is exposed to the radiation energy (Fig. 12-1). The concentration gradients of solutes in and near the core will make that region very susceptible to other absorption mechanisms including I^2R or ohmic loss. The presence of super-cooled free water will also have important effects on processes in which the electromagnetic energy and polarization absorption are special features such as in the freeze-drying of biological substances.

TABLE 12-1. Frequency Information-Meaningful Interactions, Including Go (where there is an arrow), and No Go (with no arrow), Emphasizing Water — the Biosolvent, and Dispersions, Relaxations, and Partial Resonances

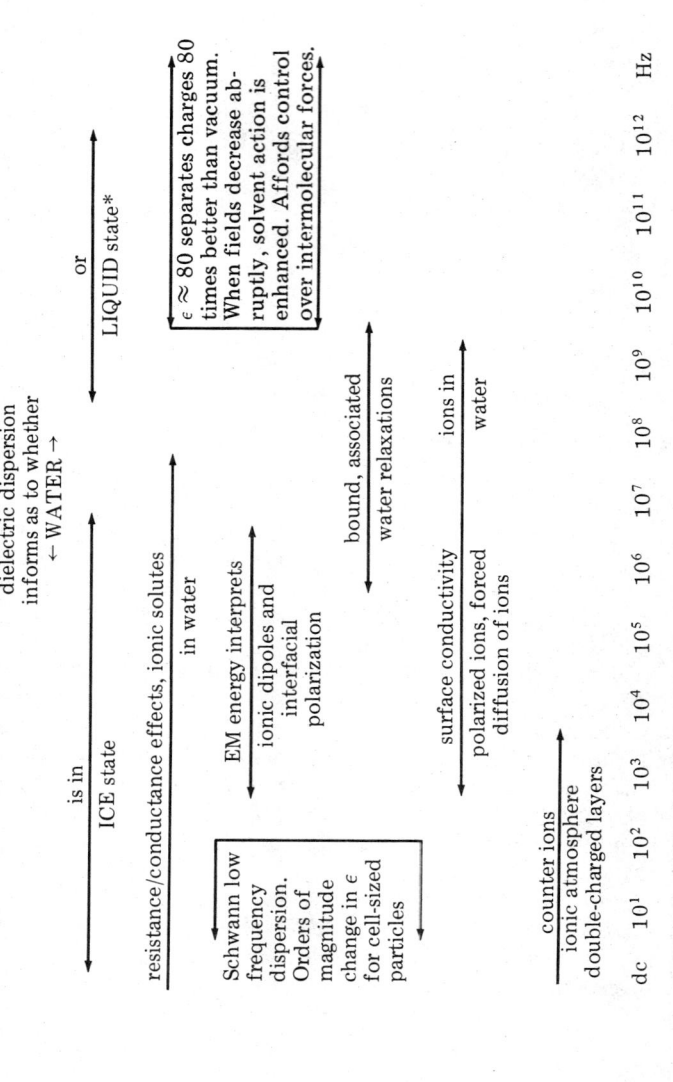

Dielectric Loss

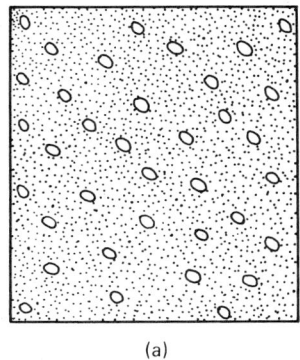 

(a)   (b)

Fig. 12-1. The rapid freezing of water systems. A produces a higher dielectric constant than slow solidification B. The two solids will have a different composition due to migration of impurities and the formation of gradients and cores.

The dielectric constants measured on the systems during freezing or after they are frozen will show the effect of freezing velocity, and will be higher in the case of fast freezing and solidification. Thus the factors which are influenced by the freezing velocity may be investigated by means of dielectric constant measurements.

The supercooling before solidification of the water may endure to very low temperatures even to $-50°C$ and may greatly affect the dampening action on radiation in a material. As the temperature of the ice decreases, it requires a lower frequency to maintain a high dielectric constant in the low frequency range, that is, a frequency on the order of kilocycles/sec, (deLoor 1964).

Loss Data for Water

In the low frequency dielectric dispersion region for water-in-oil emulsions, the losses vary around the end of the supercooling range. With concentrations from 10 to 30%, $-37°C$ is the breakpoint. Above this point, there is no loss when the frequency is low, that is, less than 100 kHz. Below $-37°C$, the loss is seen to increase as the frequency does while the dielectric constant decreases and this behavior is true down to 3 kHz. If the emulsion is warmed, the loss maximum vanishes along with the loss at around the normal freezing point of $0°C$. If the frequency is held at about 100 kHz as the temperature is changed, the dielectric constant falls away at $-37°C$.

Table 12-2 shows how the loss parameters for water behave with temperature and frequency. As the wavelength decreases, the loss factor increases and the dielectric constant decreases. In the variation from 0 to $70°C$ the dielectric constant decreases in about the same proportion whether the wavelength is 17.2 or 52 cm. The loss factor however, changes much more at 17.2 cm than it does at 52 cm. *The relaxation wavelength which is 3.33 cm at $0°C$ steadily decreases until it is equal to 0.73 at $60°C$.*

TABLE 12-2. Loss Values for Water at Various Wavelengths and Temperatures

Water at 20°C
Wavelength, Dielectric Constant and Loss Factor
(Water has only small distribution α with microwaves)

λcm 20°C	1.26	3.225	8.22	10 by three methods	17.2 by two methods	52
ϵ'	30.8	61.5	76.3	17.2	79.3	80.3
ϵ''	35.2	31.4	15.6	13.1	7.9	2.75

Ionic conductivity corrections applied but never greater than 1%

Water at 0 and 70°C
Wavelengths at 17.2 and 52 cm,
Dielectric Constant and Loss Factor

	0°	70°
As T increases at 17.2 cm λ_0	ϵ' goes from 85	to 64
As T increases at 52.0 cm λ_0	ϵ' goes from 88	to 67
As T increases at 17.2 cm λ_0	ϵ'' goes from 17	to 2.5
As T increases at 52.0 cm λ_0	ϵ'' goes from 6	to 1

Relaxation, Wavelength and Temperature
(λ_m = Relaxation Wavelength, cm)

°C	0°	10°	20°	30°	40°	50°	60°
$\lambda_m \Rightarrow$	3.33	2.38	1.74	1.36	1.09	0.89	0.73

Note: After Grant, Buchanan, and Cook, 1957

Distribution of Relaxations

In general, the Debye equations are appropriate for a single relaxation time. That is, the distribution of relaxation times, commonly called alpha, is equal to 0. It is known that water has only a small distribution in the microwave region. It is interesting to see what difference it makes in the loss factor ϵ'' when there is a small distribution at 20°C and wavelength of 10 cm. ϵ'' calculated from the Debye equations with a single relaxation time equals 4.18. When measured, it is only slightly larger or 4.44, with the intrinsic distribution operating.

In the circular arc for ϵ'', the depression of the center below the axis is a measure of the distribution or spread of the relaxation time and if it is a single relaxation with alpha = 0 then the center is on the axis. For reference purposes both ϵ'' and ϵ' can be taken from the values shown on the arcs.

The dielectric loss of a mixture will increase with the amount and dielectric constant of the energy dampening inclusions. If the ratio of components is held constant, and the dampening by doping (adding loss) is varied, then the dielectric constant of the mixture will vary along with the lossiness of the additions.

In mixtures in the condensed state, the relaxation process is strongly dependent on both the temperature and the viscosity because the orientation of dipoles involves interactions with neighboring dipoles. The presence of many different dipole shapes tends to widen the dispersion band of frequencies producing a distribution at ω_m. The coupling that results involves displacements and reacting motions between the molecules and the dipolar masses. When the temperature increases, the dispersion frequency (ω_m) of a dipolar substance increases. A mixture with two components will have a composite ϵ'_{mix} that favors the concentrated component. In a solid, a molecule will have varied potential energy barriers in the way of its rotation due simply to the several initial angular positions that it can have. The natural frequency for a solid will usually be lower than that for a liquid. Glasses have a strong ionic polarizability which is their loss mechanism at low temperatures. At high temperatures however, it is the $I^2 R$ loss that assumes first importance. Plastics and rubber have a wide dispersion band due to their composition. In a water mixture the ϵ'_{mix} dielectric constant and ϵ''_{mix}, the loss factor, will be generally proportional to the concentration of any lossy additions to the water.

Bound Water

A change in state for water means that a change will occur in the dispersion frequency. This dispersion is due to polarization changes and for ice is in the frequency region from about 100 to 1000 c/s. Free water relaxations, on the other hand, are found at microwave frequencies. The bound water relaxes at intermediate frequencies (10^7 to 10^9 c/s) depending on the degree of binding. This range is a very large one and is apparently necessary to cover all of the states in which water can be found. When the total amount of water present is known, the fraction of free water present may be found by microwave determinations and then the bound water fraction calculated (deLoor, 1968).

Bound water = known water content − free water (determined)

Similarly, the degree of binding may be studied as a function of the relaxation frequency to reveal at least the main divisions, such as, tightly and loosely held water molecules.

Dominant Component

The question of how many relaxation times there will be in a mixture depends on the degree of association of the components in a liquid and on whether they are high or low in dielectric lossiness. There is a dilution effect in which a single loss peak for ϵ''_{mix} may resolve into component peaks when the dipoles are separated by dilution. However, the relaxation of the mixture and the loss tangent maximum at ω_m may be due to mechanisms other than a blending of the individual component actions. It is a question of whether the dipole coupling and changes in the potential energy barriers between equilibrium positions are significant in the interrelation between components. The molecular explanation is favored over a regional one in the mixture (Smyth 1955). That is, the molecular loss mechanism characteristic of the dominant component is stronger than local interactions. If the mixture has a large water fraction (as in many biological substances), it will have a single relaxation time and this will be a water relaxation (Bottcher, 1952). deLoor has observed that the relaxation frequency of a mixture is always the same or higher than that of the relaxing component (deLoor, 1968).

Events at Interfaces

When interfacial polarization or the formation of charged double layers comes to be important as at frequencies lower than microwave, the dielectric loss is increased. These interfaces form in colloids. The applied field forms charged layers on inclusions too, which cause the dielectric constant to increase well over that of water (at frequencies from a few cycles per second to 10^5 c/s. At microwave frequencies, absorbing inclusions in a mixture may become semiconducting due to their surface charges (O'Konski 1960).

COMBINING THE ABSORPTION EFFECTS

In Chapter 11, in many cases, it was the fine structure dielectric that was examined and such differences as the current pathway through the cell at high frequency versus *around* it in a longer pathway at low frequency were of prime interest. Here we have examined mixtures like tissues of a heterogeneous nature in which combined dielectric effects of various components give a gross response. In biology the component of principal interest is water, but water exists in several states that change the results. The dielectric constant will differ for dispersed substances as opposed to the continuous phase.

At low frequencies ice would be lossy in the dipolar sense and liquid water is lossy at a microwave frequency. I^2R losses will be both gross and local and will be related to O'Konski's polyelectrolytes, Debye-

Falkenhagen interionic effects, and Maxwell-Wagner interface effects (Maxwell-Wagner conductivity). These many contributions cause the broadening of the absorption maximum (or ϵ'' is more uniform in a given frequency region). A relaxation due to one component will be at the same ω_m or at a lower frequency $\omega_m{'}$ than that of the mixture. Interfacial absorption increases ϵ due to charges on counter ions at such boundaries. These easily exceed the ϵ for water.

Surface conductivity of "dispersed" substances is important at microwave frequencies and surface conductivity is often masked by bulk conductivity in water or in ice. The Maxwell-Wagner two-layer condenser model (for interfacial polarization) is important in heterogeneous systems or for one-layer suspended in another (sphere model) as well as in the familiar water in oil suspension.

The charged carriers migrate with the applied field through the suspensions and congregate at the interface, stimulating equal or opposite charges on the other side of the bound. The charge displacements in water-in-oil may be protons rather than electrons. Thus we have an accumulation of little dipoles with resulting moments in the ac field. The Debye equation for ϵ'' then needs a term for the component dielectrics and for their wavelengths. The ohmic conductivity is added to the polar relaxation and the new term will cause ϵ'' to increase, going to infinity as ω goes to zero. Therefore ϵ'' will be great at low frequency and, after decreasing with increasing frequency, will reach its maximum at *intermediate* frequency and then fall to zero as ω goes to infinity. This means a large loss at low frequencies *instead* of Debye orientational loss which is the large loss at higher frequencies.

SENSIBLE HEAT PRODUCTION

A given volume of dielectric material will absorb microwave energy according to the loss factors. This dissipation will be determined on a gross material level by the square of the voltage applied in the material, the frequency of the microwaves and the loss factors. The latter include the relative dielectric constant often written ϵ_r', the loss tangent, tan δ of the material, and the dielectric constant of free space ϵ_o,

$$\text{Power in watts/m}^3 = E^2 \epsilon_r' \tan \delta \; 2\pi f \epsilon_o \tag{305}$$

The relative dielectric constant ϵ_r' has a value of about 68 to 76 for water and it, along with the loss tan δ, are obtained from handbooks or by experiment. If the frequency $2\pi f$ (which is ω) is combined with the value of ϵ_o, the dielectric constant of free space = 8.85×10^{-12} farad/meter (MKS system),

$$2\pi f \epsilon_o = 2\pi f \times (10^{-9}/36\pi) \text{ farad/meter} \tag{306}$$

or $0.0556 \times 10^{-9} f$ farads/meter or $5.56 \times 10^{-11} f$ farads/meter as a lumped value when the linear dimension is in meters. Then with 1 farad = 1 amp-sec/volt and f = 2.45 × 10⁹ Hz and Hertz with units of sec⁻¹,

$$2\pi f \epsilon_o = 0.0556 \times 10^{-9} \times 2.45 \times 10^9 \qquad (307)$$

$$= 0.136 \text{ amp/volt-meter}$$

Then by dimensional analysis at 2.45 × 10⁹ c/s,

$$\frac{1}{\text{sec}} \frac{\text{farad}}{\text{meter}} \times \frac{\text{amp sec}}{\text{farad-volt}}$$

and (308)

$$0.136 = \frac{2.45}{18} \frac{\text{amp}}{\text{volt-meter}}$$

For the energy density let

$$E = 500 \text{ volts/m, } \tan \delta = 0.15 \qquad (309)$$

$$\epsilon' = 72 \qquad (310)$$

$$E^2 \epsilon' \tan\delta = \frac{500 \times 500 \times (\text{volts})^2}{\text{meter}^2} \; 72 \times 0.136 \; \frac{\text{amp}}{\text{volt-meter}} \times 0.15 \; (311)$$

$$= 3.68 \times 10^5 \text{ amp-volt/meter}^3 = \text{watts/meter}^3$$

$$= 3.68 \times 10^2 \text{ kw/meter}^3$$

or multiply both sides by m³ to get power dissipated,
or multiply by time to get energy dissipated,
or in terms of a smaller specific volume,

$$3.68 \times 10^{-1} \text{ watts/cm}^3 = 0.368 \text{ watts/cm}^3 \qquad (312)$$

PENETRATION

The energy being absorbed in a dielectric material is described as being attenuated in the propagation direction. The attenuation coefficient is

$$\alpha = \pi \epsilon'' / \lambda_o \sqrt{\epsilon'} \qquad (313)$$

The distance to which the energy will reach inside the material before it falls off to 1/eth (0.368) of the initial strength is reciprocally related

$$D = 1/2\alpha = \lambda_o / 2\pi \sqrt{\epsilon' \tan \delta} \qquad (314)$$

Other special penetration depths such as the half power level may be obtained by suitable conversion factors.

Dielectric Loss

Net Absorption

The nonreflected microwaves will be either absorbed or transmitted through the material.

Absorbed = Incident − (Reflected + Transmitted)

The reflection at the surface is determined by the impedance of the media involved. Thus, with the impedance of the first media Z_1, and that of the second Z_2,

$$\text{Reflection} = (Z_2 - Z_1)/(Z_2 + Z_1) \tag{315}$$

In this case the energy involved is that which arrives at the load to be either reflected or to continue. Instrumentation in the propagation line can measure the net power flow in one direction by taking the difference between the power going in both directions. If the reflection = 0, the impedance is said to be matched and a smooth flow of energy into the material occurs. If not, there are standing waves and some energy is reflected.

MICROWAVE HEATING

The discussion of volume power absorption in the dielectric material leads naturally to applications where a certain mass is required to absorb enough energy to change its temperature or state. This is a matter of satisfying its specific heat requirements and objective temperatures. However, certain form factors influence the absorption of energy and are peculiar to microwaves. The density of the mass or the load is a typical factor as exemplified by rapid heating of a leavened dough product or frozen bread. The polarization behind the effect is inversely proportional to the density.

The loss mechanisms at the molecular level as they have been presented are appropriate for the microwave when it is propagated into a space, from any propagating device. This may be a radar antenna, a diathermy director, or from within a metal enclosure from which the energy is not allowed to escape. In such a confined space, it can reflect from metal walls until, from single or multiple transits through the load, it attenuates to zero. The capacitor model persists here if the metal walls are regarded as being like parallel plates having the electric field between them. Actually the generator provides the field which is transmitted by wave guides, coaxial cables, or directly injected into the cavity which is designed to accommodate the load. Thus the alternator is one end and the place in the load where the load goes to zero is the other. These terminations resemble the capacitor plates which become charged and impress a field force on the interplate material. If this analogy failed, the logic of the model and, therefore, the experimental data from it would be in doubt. Thus the metal walls are good reflectors and function geometrically and electrically with respect to the wavelength and the load.

Once the transmission and confinement have been arranged, the material-energy interaction is the main interest. This absorption will be in proportion to the dielectric loss. Very great concentrations of power are possible so that, with a reasonable choice of frequency and load form factors, dramatic absorption results occur. This is the basis for the growth of the industrial, medical, and scientific applications which have translated the results into selectivity, noncontact, high efficiency, large capacity, and throughput or other appealing factors.

Microwave applications are wider than this however, encompassing radar detection and ranging, missile guidance, and satellite and earth communications or power relay links. In all these, accidental absorptive loss is usually avoided at all costs because the objective is to retrieve as much as possible of the power or intelligence resident in the echoing or communicating signal. A basic incompatibility would, in fact, exist if international regulations did not establish ground rules for the mutual guidance of both the "loss and non-loss" applications or schools of thought. The communications and radar application engineers develop procedures in minute detail for managing their operations with adequate rejection or discrimination against noise or extraneous signals. The microwave power engineers on the other side concentrate on the load loss and limit their operations to regulated ISM frequencies (industrial, scientific and medical). The Federal Communication Commission has allocated four such frequencies which conform to the International Radio Regulations adopted at Geneva in 1959. These are:

$$915 \text{ MHz} \pm 25 \text{ MHz}$$
$$2{,}450 \text{ MHz} \pm 50 \text{ MHz}$$
$$5{,}800 \text{ MHz} \pm 75 \text{ MHz}$$
$$22{,}125 \text{ MHz} \pm 125 \text{ MHz} \quad \text{undeveloped}$$

(433.93 and 896 MHz are permitted in parts of Europe.)

Other frequencies may be used but elaborate precautions would then be required to restrict illegal radiations from the equipment. It turns out that leakage of microwaves is intolerable, even from devices using the assigned frequencies, because of the indiscriminate exposure with possible hazard to persons in the vicinity. Thus a field of microwave biophysics comes into being which is involved with the collected interests of both sides, that is, all of the microwave industry. In this area of study, the biological effects of the microwave radiation come into focus and an objective is to find new biological uses as well as to establish safe operating levels or conditions for those who work with the energy. It is of considerable interest to note that one scientific field, that of microwave spectroscopy, has been operating for as long as generators have been available. In this field the scientist is on both the "loss" and "no loss" sides because he uses the dielectric interaction mechanisms to establish the structure and function of dielectrics.

SEMICONDUCTION

It develops that *all* material substance is divisible into three categories, conductors, semiconductors, and dielectrics. All that occurs electronically in all substances may thus be classified in terms of three processes associated with the three basic types of substances. Biology, because of the importance it places on the water dipole, is clearly associated very closely with dielectric processes. Microwave interactions are unique methods for studying dielectric substances and so biology comes to be more and more concerned with microwave energy. The presence of physiological salts and semiconductors in all tissues tends to place biology in the front of the line eagerly awaiting announcements of parallel progress concerning the other two types of substances as well. Microwave energy, it turns out, has loss components in the conduction and semiconduction processes, that is, nondielectric loss interactions. The influence of microwave energy on biology thus appears to begin at the molecular level. However, even this barrier needs bridging because, through developments in nuclear magnetic resonance and electron spin resonance, microwaves are commonly used to study *sub*molecular structure and function in biological and nonbiological spectroscopy. Finally, tissues themselves are recognized as microwave generators, producing signal information which may be managed by suitable recorders and analyzers in keeping with the μwatt power available.

HISTORICAL LANDMARKS

Until about 1960, the main wellsprings of information for this overview of EMR came from scientists such as Peter Debye whose book *Polar Molecules* published in 1929 described the fundamental theory for the behavior of dipolar substances in the microwave field. He later earned the Nobel Prize while working at Cornell University. The comprehensive book of C. P. Smith (1955), called *Dielectric Behavior and Structure*, along with subsequent investigations from his laboratory at Princeton, provide the principal background for the widest possible view of the microwave interaction with material.

Another book, written from the standpoint of a theoretical physicist, is that of H. Frohlich (1958). Another treatment of the subject comes from the Laboratory for Insulation Research at Massachusetts Institute of Technology and is edited by its head of many years, A. von Hippel. This work (1954) is published in several volumes under the title *Dielectric Materials and Applications*. The scope of this writing and thinking is by far the deepest excursion for the microwave-material interaction. In later meetings, von Hippel managed to extend the concept of dielectrics and their processes into the broader idea of molecular engineering which is the engineering of substances by matching their properties to the desired specifications. The motivation for this work came from the field of communication cable theory, electric power distribution and military insulation research. According to something of a pattern that seems to happen with microwave engineering, these studies have been extended to semiconductors, ceramics, plastics, and a very wide variety of materials. The excursion into the field of molecular engineering was a natural development of the work with the electrophysics of the dielectric materials. At one far out end, it represents what may well be the ultimate in interdisciplinary bridges, that between materials science and engineering. For biology, seemingly the only requirement for participation appears to be the prefix bio- for biomolecular engineering. Information is an ideal bridge since some aspects of nerve nets require cable theory for analysis, and so many elements of communication are common to human information processing.

This philosophy is particularly appropriate in the realm of microwave heating applications and continuous material processing due to the special characteristics of the energy and the unusual opportunities for matching them to special material or process requirements. The design of equipment, selection of materials, and the management of the process are all based on the frequency response and molecular engineering. The possibility of highly selective interactions and unique solutions is sufficient motivation for considerable effort in this area.

Merely covering a radar antenna against the weather is an exercise in matching characteristics of radar energy of a certain frequency with the lossiness of potential materials. Such a radome must transmit the full signal without absorption. The plastic, glass, or ceramic which will be selected will have resin, filler, and finish all highly selected for least opposition and least absorption of microwaves, that is, low dielectric loss.

Developments in these areas, mainly from his continuous microwave experiences, were reported by D. A. Copson in the book *Microwave Heating* in 1962 (revised 1975), a portion of which is devoted to the shape of things which where thought to be coming in the radiation biology of microwaves. In the same year, A. F. Harvey at the Royal Radar Establishment in England was preparing his epic work called *Microwave Engineering*. This book presents in three pages what must be the ultimate in a concise treatment of the microwave as an interacting electromagnetic radiation and then goes on for 1313 pages to fill in the details. *Heating with Microwaves* came in 1964 from H. Puschner from his work in Europe and elsewhere. This book was described as naturally complementary to Copson's book, not only with respect to titles but for content as well. Where Copson dealt with the theory and applications, Puschner described the generators and microwave engineering. The momentum of development leading to the popularity and relevance of microwave today is seen in the two volumes *Microwave Power Engineering*, edited by E. Okress (1968).

The importance of the influence of developments in microwave spectroscopy on the overall subject of microwaves is recognized by the fairly extensive treatment given here. The origins of these influences are seen in the book by C. H. Townes and A. L. Schawlow, *Microwave Spectroscopy* which appeared in 1955. A natural momentum was generated for these concepts using the power of spectroscopic methods to understand the microwave-materials interaction. This is shown in the work of the Hungarians P. Hedvig, and G. Zentai called *Microwave Study of Chemical Structures and Reactions* (translated into English by E. D. Morgan).

Finally, the forward-looking experience of the Czechoslovakians and Eastern countries in the area of microwaves as they constitute a public health concern has been reported in English by K. Marha, Jan Musil, and Hana Tuha in a book entitled *Electromagnetic Fields and the Life Environment* (1971). Thus the current surge of apprehension over unwanted inclusions in our ambience has been logically extended to include the ever increasing concentrations of electromagnetic radiations in our biosphere—a matter which has been taken up in this book in the context of the electromagnetic radiation compatibility of lifeforms.

THE WIDENING OF THE SCOPE FOR MICROWAVES

In addition to altering the biological or material status on the gross level, microwaves are most certainly a means for information transfer and the transport of energy. There is a possibility that non-degradative coupling will have storage and communicative features in organisms and that, alternatively, high power densities will produce the obliteration of information on demand. The information is resident in the coupling sequences of energy to substance, in the transducer mechanisms, in the modulation of the carrier by pulse-amplitude, pulse-code, pulse-count, pulse-duration, pulse-position, and in the modulation by pulse-repetition frequency, pulse-repetition rate, pulse-time and pulse-width. The basic frequency itself carries considerable information as does the power. Given an interaction which influences organisms to begin with, the final development of this interaction may well tend to equal the development of microwaves in communications for the transmittal of information to the nervous systems, control loops, and responsive components of the human body.

SOUNDING BOUNDS

Almost the same procedures are seen in acoustic science which furnishes many direct examples of information transfer, since this energy is so often specifically meant to communicate. There is the matter of reflections which probe bounds as a function of acoustical impedance, a quantity which depends on the density and velocity. A cathode ray oscilloscope shows the reflections at the penetrating distance so the energy "sounds" its bounds. Doppler sound effects in which there is an informing frequency shift associated with moving objects gives movement information. Then the intensity may be associated with the frequency, area, density and volume or average sound pressure p, as follows;

$$I = \tfrac{1}{2}\omega^2 A^2 (\rho V) = p^2 / 2\rho V \qquad (316)$$

Signals transmit through tissue with a velocity of about 1.5 to 1.6 km/sec to distances 10 to 20 cm deep, returning echoes within a certain number per second where each lasts perhaps a microsecond and they may have an associated power in W/cm^2. All of these informing factors may be combined in echo-locating of tumors or in external sensing of fluid flows in the body.

Impedance values are used in plethysmography, an analysis in which the volume of a region changes with its mode of operation, examples being blood vessels, lung, and heart activity.

These developments in advanced technology need:

1. Administrative freedom in the advanced uses of the electromagnetic spectrum to guarantee the march of progress.
2. Recognition of an electromagnetic age in biology with a consequent heavy demand for trained paramedical personnel, continuing education, and voluntary certification.
3. Encouragement of remote imaging methods to separate the operators from hazards inherent in diagnostic work with emphasis on ultrasound and thermography.
4. Sophistication in management of the information content in electromagnetic radiations so that less energy is needed in devices for an equivalent return. Improved directivity and selectivity will reduce the associated biohazards. Alternatively expressed, if the U.S.A. had relied on higher power than necessary for radio location and microwave military uses, the U.S.S.R. may have gained an advantage in terms of learning to operate at lower signal strength. In terms of the power flux unnecessarily emitted, the difference by a factor of 1000 in population exposures permitted by the two nations would then have worked to the disadvantage of the U.S.A.

FRANK INJURY

Any housewife would know that exposures at the multiwatt level are not thresholds, acute, or chronic, they are simply "cooking" and injury is merely a question of time. At 400 watts an object or load with an area of 154 cm^2 or 14 cm (5.5 in) across would receive about 2.5 W/cm^2. For a biological substance of 4 ounces or 115 grams, the exposure should boil its water content in about 2 minutes. This type of frank injury applies to genitalia or similar presentations. At 40 watts it would mean 250 mW/cm^2 or about 20 minutes for the result, and here the MPL_{10} of 10 mW/cm^2 is smaller by a factor of only 25. The results would be worse if the surface were nonuniform so that certain places received concentrated energy. The situation would be favored by plane waves and zero reflection, or attenuation to zero in a cavity; otherwise the MPL_{10} and instant denaturation would diverge by a suitable factor.

Persons near microwave generators could easily expose themselves to energy levels near this amount—that is, within 25 times the safety factor. At a minimum, fertility is then in danger but the skin of the closest body surfaces would be irritated and show lesions if the practice is continued. There would be burning of the eyes, and rash with inflammation of the dermis and subdermis. Due to the external presentation and interference with thermoregulation by humidity and clothing, the penis and scrotum should show hardening, suggesting the possibility of interference with spermatogenesis, confirmable by biopsy and a sperm count approaching zero from normal levels in the millions.

The hyperthermia may produce inflammation but, contrary to that due to bacterial infection, this reaction is expected to *follow* the denaturation of tissue proteins. Such results may require temperatures near 64°C if the process is to be irreversible. The most important tissue properties would then be vascularity and thermoregulatability. The susceptible tissues have been defined in numerous studies as the lumen of the gastrointestinal tract, gall and urinary bladders and ocular interfaces as well as the externally exposed genitalia. Hollow spaces are vulnerable so that epigastric distress is a typical complaint after moderate exposures while warming is noticed in casual ones. Finally, the walls of the gastrointestinal tract may be inflamed and this may lead to petechia. The blood pressure should decrease along with nausea and vomiting and death may occur due to shock or burns. There is some possibility of reduced resistance to invasion by ever-present bacteria, as they colonize the necrotic regions. Thus, one might expect acute appendicitis and peritonitis. Small blood vessels are sensitive to such treatment leading to hemorrhage. The size of the adrenals may suggest a negative shift of hormones and changed platelet activity may produce abnormal blood clot management.

In demographic studies, important sources of data on electromagnetic radiations in populations would be found in higher incidence of birth defects. The size of the embryonic cerebrum may make it susceptible to herniation and congenital development disorders such as clubfoot, heart disease, cleft palate, defects of the genital organs, and the possibility of a higher fetal death rate for an equivalent reporting basis may be worth investigating.

PROBLEMS

1. In a tissue, what factors would you expect to give meaningful interactions with respect to frequency information?
2. Express the EMR power lost in a tissue volume as a function of field, frequency, and loss factors.
3. What are the units for the dielectric constant of free space in MKS?
4. How deeply will 0.1 m, EMR penetrate before it attenuates to the 1/eth level when $\epsilon' \tan \delta = 0.7$?

5. Fill in the blanks:
   a. In an ideal _____, current leads voltage by 90°.
   b. The current lead is true for all _____.
   c. The impedance _____ with increasing frequency.
6. What is the relation between impedance and resistance in an ideal resistor?
7. What is the impedance at an interface between two media?
8. What is meant by frank injury from EMR?
9. The radiation energy in ergs impinging on tissue per second through 1 cm^2 of area and the energy specify the quanta passing. What information from the relation $E = hf$ completes the specification?

BIBLIOGRAPHY

Baba, K., T. Fujimura, and E. Kamiyoshi, "The Dielectric Relaxation of Binary Liquid Mixtures," *J. Phys. Chem.*, Vol. 73, 2, 1146, April 1969.
Bardos, I., et al, "Microwave Oxidation of Silicon," *J. Microwave Power*, Vol. 11, 2, 184, 1976.
Bottcher, C.J.F., *Theory of Electric Polarization*, Elsevier, Amsterdam, 1952.
Bucci, O.M., A. de Bonitatibus and C. Savareses, "Time Domain Spectroscopy in Open Structures," *J. Microwave Power*, Vol. 11, 2, 209, 1976.
Chapman, I.D., "The Effect of Temperature on Dielectric Properties of Water-in-Oil Emulsions," *J. Phys. Chem.*, Vol. 72, 1, January 1968.
Cope, F.W., "A Review of the Applications of Solid State Physics Concepts to Biological Systems," *J. Biol. Physics*, Vol. 3, 1-41, 1975.
Cope, F.W., "Biological Sensitivity to Weak Magnetic Fields Due to Superconductive Josephson Functions," *Physiol. Chem. and Physics*, Vol. 5, 173-176, 1973.
Copson, D.A., *Microwave Heating*, Avi Publishing Co., Westport, Conn., 1962.
Daels, J., "Microwave Heating of the Uterine Wall during Parturition," *J. Microwave Power*, Vol. 11, 2, 166-168, 1976.
deLoor, G.P., *Appl. Sci. Res.*, B11, 310, 1964.
deLoor, G.P., *Journal of Microwave Power*, Vol. 3, 2, July 1968.
deLoor, G.P., "Dielectric Properties of Heterogeneous Mixtures Containing Water," *J. of Microwave Power*, Vol. 3, 2, 1969.
Debye, P. *Polar Molecules*, Chemical Catalog Company, New York, 1929.
Debye, P. and H. Falkenhagen, *Physik Z.*, Vol. 29, 121, 1928.
Fletcher, N.H., *Chemical Physics of Ice*, Cambridge U. Press, England, 1970.
Froelich, H., *History of Dielectrics*, Oxford U. Press, London, 1958.
Grant, E.H, T.J. Buchanan, and H.F. Cook, "Dielectric Behavior of Water at Microwave Frequencies," *J. Chem. Phys.*, Vol. 26, 1, January 1957.
Harvey, A.F., *Microwave Engineering*, Academic Press, New York, 1963.
Hasted, J.B. and M.A. Shah, "Progress in Dielectrics," *Brit. J. of Applied Physics*, Vol. 15, 825, 1964.
Hedvig, P. and G. Zental, *Microwave Study of Chemical Structures and Reactions*, Chemical Rubber Co. Press, Cleveland, Ohio, E.D. Morgan, Ed., English version, 1969.
Kretzschmar, J.G., "Waveguide and Cavity Systems for Rapid Heating of Continuous Products with Small Cross-Section," *J. Microwave Power*, Vol. 11, 2, 188, 1976.
Mackay, A., W.R. Tinga and W.A.G. Voll, "Excitation of Microwave Cavities with Frequency-Agile Source," *J. Microwave Power*, Vol. 11, 2, 186-187, 1976.
Marha, K., J. Musil, and H. Tuha, *Electromagnetic Fields and the Life Environment*, San Francisco Press, San Francisco, Calif., 1971.
McNiven, D.R. and D.J. Wyper, "Microwave Therapy and Muscle Blood Flow in Man," *J. Microwave Power*, Vol. 11, 2, 168-170, 1976.

Mearns, A.M. and E. Ekinci, "Hydrogen Dissociation in a Microwave Discharge," *Journal of Microwave Power*, Vol. 11, 2, 183, 1976.

Ohlsson, T., "Temperature Distribution in Microwave Oven Heating Experiments and Computer Simulations," *J. Microwave Power*, Vol. 11, 2, 178-179, 1976.

Okonski, *J. Phys. Chem.*, Vol. 64, 602, 1960.

Okress, E., Ed., *Microwave Power Engineering*, Academic Press, New York, 1968.

Puschner, H., *Heating with Microwaves*, Philips Technical Library, Springer-Verlag, New York, 1964.

Risman, P.O., "Microwave Energy Control by Humidity Sensing," *J. Microwave Power*, Vol. 11, 2, 177-178, 1976.

Schwan, H.P., G. Schwarz, J. Maczuk, and H. Pauly, *J. Phys. Chem.*, Vol. 66, 2626, 1962.

Schwarz, G., "A Theory of the Low-Frequency Dielectric Dispersion of Colloidal Particles in Electrolyte Solution," *J. Phys. Chem.*, Vol. 66, 1636, December 1962.

Smyth, C.P., *Dielectric Behavior and Structure*, McGraw-Hill Book Co., New York, 1955.

Stevens, A. and F. Peluso, "Temperature Changes Generated by Microwave Therapy in the Thighs of Human Subjects," *J. Microwave Power*, Vol. 11, 2, 170-171, 1976.

Townes, C.H. and A.L. Schawlow, *Microwave Spectroscopy*, McGraw-Hill Book Co., New York, 1955.

Wolf, A.A. and E.H. Halpern, "On a Class of Organic Superconductors," *Proc. IEEE*, Vol. 64, 357-359, 1976.

Wagner, K.W., *Arc. Electrotech.*, Vol. 2, 371, 1914.

13

Biology and Electromagnetic Radiations — Nonionizing

> The qualities of wholeness suffer surely,
> With tiny systems isolated insecurely.
> D.A.C.

CRITICAL MASS

The biology of electromagnetic radiations is an extremely large field blessed with many recent innovations. It is so large that when the dangers of these radiations began to be studied with urgency as the potentially great environmental hazard of the century, a task force in the Office of the President of the United States was deemed appropriate. This organization, as it spread out among disciplines and agencies, became a sort of biological Manhattan Project such as the one which had led in earlier times to the atomic bomb rather than to safety measures. Senior people in the electromagnetic radiation work repeatedly referred to a "critical mass" for the enlarged electromagnetic radiation research effort by which was meant that, at some point, there would be enough energy input to cause a significant step change in understanding or an information explosion.

Ordering Forces

If the reader has covered the great conceptualizations that shape modern biology as described in Chapter 1, he will appreciate the position of information pathways in the ultimate objectives of such a significant program. For good or bad, these must include the idea of *ordering forces at the disposal of information managing elements.* These connect atoms and molecules with their control and regulation systems as in kinds of bonds and operations such as photosynthesis, phototaxis, radiation activation, inactivation, inhibition, metabolism, division, repair and periodicity. The relation is more urgent and exciting as sense modes and field forces are now seen to directly interact in these fascinating studies. One can, with complete confidence, use the background in information, communication theory, and regulation and control developed in Chapters 2 and 3 as well as the receptive biological systems involving DNA, and organization for development, growth, and longevity covered in Chapter 4, to understand the interactions with nonionizing radiations and their field forces. The evidence is convincing that the "critical mass" will be measured in this format as the effort intensifies to understand intersystem reactions and direct effects on organisms. This structure is appealing because it phrases new input of information positively in terms of partial solutions, instead of busily proliferating questions. Anything that avoids "gap" problems between generations of serious students is welcome and there is no doubt that the bad habit of ending dissertations with a well-posed problem instead of a partial answer has given us gap trouble. In this spirit which asks that several hypotheses be given their fair day, some answers may be offered.

Certainly the first premise is that the electromagnetic age has come to biology. These studies with nonionizing radiations can no longer be put aside as minor trivialities and most major countries in the scientific scene have programs as has the United Nations in its WHO organization which sponsored activities in Poland with the help of blocked currency. URSI, the International Union of Radio Science, has reported on the biological effects of RF. The earlier and even some current work often produces statements like "convincing evidence for nonthermic effects for electromagnetic radiations has not been demonstrated to date" which was probably correct because under the experimental conditions, they could not be seen. The polemics or exacerbated semantics of the thermic versus athermic argument for explaining biological effects of microwaves were set out in a discussion (Copson, 1975 p457) where it was shown that spectroscopic grade preparations were more likely to yield success and that athermic effects *always* occur but may well be masked by saturation effects in the fields or systems. To this should be added the effect of a very rapid rise in temperature. Although thermic, the associated responses may be nonlinear and dependent on the velocity at which the status is altered.

TERRORWATTS

After Heinrich Hertz pointed out electromagnetic fields in the environment in 1888 and biological effects due to such fields were observed by D'Arsonval and Tesla, the first noted instance of concern was when the White House, where the President of the United States lives, was electrified in 1891. No electricity was permitted in close proximity to the President's person. This was 60 Hz electromagnetic radiation and we have only recently become aroused at the hazard to the rest of the population. Interestingly enough it was a microwave device, the microwave oven, that brought out this official interest.

In the meantime, of course, the fields observed by Hertz have become enormously more powerful and very much more common and the dreams of Tesla about EM alterations of the climate seem to be materializing in some ways. There has been a progression of higher and higher power in the transmitters so that we used to be exposed to *milli*, then *kilo*, and now *mega*watts. Terawatts are obviously coming and by then the spelling may have to be revised to suggest the terrible possibilities in such super transmissions.

Nations are therefore bracing for a storm in this area. It is a natural field for the sensationalists. Parade Magazine of the Washington D.C. Post on October 8, 1972 related how Italian workers blamed their anti-sex machines for their personal problems. They were using high frequency EMR generators to assist in the assembly of furniture. A columnist, Jack Anderson, described the perils of the microwave oven in the New York Post on December 23, 1971. J. Randall told of the Veteran's Administration shunning a dying serviceman who had been exposed to electromagnetic radiations (*Washington Star*, January 27, 1972) and there is the real exposé by P. Brodeur, "The Zapping of America, Microwaves, Their Deadly Risk, and the Cover-Up" which was published by W.W. Norton in 1977. His story goes up to, but misses by a matter of days, the highly significant fire that destroyed the U.S. Embassy in Moscow in 1977. It obviously took a great deal of effort to bring all of his material to the surface but it is difficult to put out a book of that kind without seeming to have an axe to grind. With our present text having been in preparation for over a decade, it was a pleasant surprise to have Brodeur illuminating, in an entertaining way, the sections here on biohazards with electromagnetic radiations. While it may be heartless to call such alarming facts entertaining, they are so from the viewpoint of a scientist who cannot take many liberties in his reports. The section on Moscow, page 291 for example, is absorbing where a journalist, William Beecher, tells his story of personal microwave exposure while in the Moscow embassy with positive and negative episodes, repetition, and all the scientific aspects except biomedical tests. Beecher eliminated artifacts, such as a malfunctioning air conditioner, and described mental and physical discomfort in four out of eight interviews in the Tchaikovsky street embassy.

Hazardous and Useful Modes

In this present study, it is easy to see perils in every experiment but our approach is that the radiations, as such, are not the whole matter. What counts is the interaction with organisms. This too is not the beginning and end because it comes down to special interactions involving specific communication, molecularly or organismically within a hazardous or useful mode. Thus humans can see *light* with pulsed ultrahigh frequency (UHF) which is ordinarily invisible (Frey and Messinger 1973) and it is rare indeed when a substitute mechanism for normal vision, even one as crude as this, is observed. On the other hand, the fact that insect pupae develop abnormally after microwave exposure also merits a close look when reported by Carpenter and Livingston (1971).

Scaling Factors, Sense Modes. We are not comforted however, that reported connections between eye lens membrane effects from electromagnetic radiations and arterial heart disease could not *be* because these organs arise from different embryonic layers (ectoderm and mesoderm, respectively) (Tyler, 1975). Far more projected extrapolations are necessary and routine in these same studies. With electromagnetic radiations, this resembles scaling when taken up from microorganisms or small animals to humans. A 2450 MHz exposure at 10 mW/cm^2 is assumed tolerable for a human but is a serious if not lethal matter for a rat or mouse.

Electromagnetic pulses superimposed on cardiac evoked potentials, and synchronized, can vary and stop the heart beat in frogs as shown by Frey (1961, 1962). At a minimum one would expect cardiac irregularities and associated respiratory effects. Cardiovascular symptoms are prominent in the chronic, long-term effects and range from slower heartbeat and reduced blood pressure, less ventricular capacity to stabbing heart pains moving to shoulder blades and arms. These are, nevertheless, not always serious and are controlled by rotation of workers. They form a syndrome with other complaints which may be called microwave sickness and are known largely from Czech surveys made by the Institute of Industrial Hygiene and Occupational Diseases in Prague in the Department of High Frequency, and reports of M. Sadcikova and S. V. Gordon (from the Soviet equivalent) in the Proceedings of the International Symposium, Warsaw, Poland in 1973.

Low level, chronic exposure to electromagnetic radiation has prompted a long list of complaints, probably much too long. The U.S.S.R. 10μW/cm^2 occupational level comes from observing these complaints at a 0.1 mW/cm^2 and applying a safety factor of 10. The global symptoms may have included, in addition to those mentioned:

1. Dizziness and vertigo after long standing period in field.
2. Headache and eye pain with skin tightening on forehead and head.
3. Restlessness and sleeplessness, or less profound sleeping.
4. Irritability and changes of mood.
5. Hypochondria.
6. Fear.
7. Nervous tension, depression, under-performance intellectually with some emphasis on memory.
8. Behavioral changes in monkeys (operantly conditioned) and similarly prepared rats which refused cues or hesitated after microwave pulses, or chose routes other than ones having 0.2 mW/cm^2 microwave flux.
9. Stomachache or epigastric pain.
10. Muscular pain.
11. Pain in bone regions of limbs near electromagnetic sources.

12. Changes in bone marrow cell chromosomes and lymphocytes more numerous, with blastic transformations.
13. Alopecia—loss of hair.
14. Increased incidence of appendicitis.
15. Lymphatic cancer and leukemia (in heavily exposed, highly susceptible mice).
16. Brain tissue alterations in cytogenetics and histology.
17. In humans (but not in monkeys), elevated serum triglycerides and loss of mental functions with ELF.
18. Embryogenesis defects, miscarriages, fewer births, more females, sterility in offspring. Down's syndrome when fathers are exposed to nonionizing radiations.

The symptomology is strange due to the unusual connections and lack of positive or favorable effects at these low levels, particularly on circulation, and the possibility of malingerers in the surveys must be considered.

Microwave diathermy has been popular in the U.S. for over forty years and it may give an exposure of 100 mW/cm^2 for 20 minutes, more or less. No acute ill-effects are on record but the adequacy of the follow-up is in question. In the light of what we now know, this application of EMR seems somewhat unsophisticated and dangerous in unskilled hands. Further general understanding of the chronic effects awaits the declassification of all associated studies.

Acute microwave exposure from powerful sources may endanger sight and hearing, joint health, integrity of periodontal capillaries, capillaries of the eyes and ears, the gastrointestinal tract, and may give irregularities in cardiovascular performance and outright pain in the chest. The warning may be fuzzy vision, burning around eyelids, fibrillation of the heart, dizziness, plus associated non-bioeffects such as heating of metal jewelry, interference with EM receivers, decreased performance in hearing aids, remote radio controllers, recorders, and communication devices. The fingertips may change color, merely itch, or tingle. The exposure direction may alternatively focus the energy on the stomach, producing epigastric distress, possibly when the flux is from 5 to 10 mW/cm^2 and usually from 8 to 12 GHz.

Even though the emphasis in Russia produced a long list of chronic ailments, there is a trend toward neural observations, or effects on the nervous system and behavior. Russian investigators seem to consider bioresponses as more appropriate and generalized than bioeffects or biohazards and the U.S. philosophy was built with electromagnetic diathermy. Future plans suggest that the U.S. and Western Europe will be probing lower exposure levels more sensitively but one of the most interesting possibilities is that neural responses after EMR, especially pulses, will show persisting or lingering effects after exposure to the flux levels chosen as maximum permissible level by the super powers. The thalamus changes in chronic exposure for more than a month. At 5 to 7 mW/cm^2, an exposure for 3 to 4 months in pulses will alter the chemistry and function of the central nervous system.

Shock Waves. Nonionizing radiation effects have already been outlined in the "minispectrum", Table 12-1, but the electromagnetic conditions must be examined in more detail. The actual frequencies of interest go to the extremely low region of the spectrum, that is, ELF, which from 45 to 80 Hz, find use in the U.S. Navy's Seafarer communication system and the ubiquitous 60 Hz of modern power lines.

The acronym Seafarer arises from surface ELF antenna for addressing remotely deployed receivers. This project, which has been closely identified with the EM environment and its control or abuse, much to the discomfort of senior people in the U.S. Navy, has an even wider significance. It is no doubt related to the massive Russian electromagnetic radiation emissions in 1977 which could also communicate across enormous distances by simply using brute EM force. Another significant result is that these large scale emissions actually change the atmosphere via electrical permeants which relate back to Nikoli Tesla's work in which he properly felt that even relatively small modifications to the bioelectrosphere might result in significant planetary abnormalities.

Not insignificant, at least in terms of bird behavior, would be the fact that statistically significant deviations in heading are shown by birds responding to the electromagnetic fields and bird watchers uncovered or undercover could locate ELF or high pulse microwave sources merely by taking note of the orientation of birds in flight.

The conditions range from the more common electromagnetics of 60 Hz to the informationally-rich high peak power, HPP pulses in nanoseconds, and pulses with extremely rapid rise time which can introduce shock wave conditions in biological tissues. The subject of exposure conditions is a topic in itself, as is the direct effect of EMR on organisms via some specially receptive system. The question of dimensions is vital because of the geometrical relationships between wavelengths, frequencies, sizes and distances. Two very specific sense modes with which EMR quite effectively interact are sight and hearing. There is a developing awareness of specific interactions with blood and lymph from which certain patterns are discernible. Effects on metabolism and on into growth and development comprise an extremely fascinating field of effects in which broadband specialists can discourse knowingly on molecular biology.

Multiple EMR Parameters and Biological Levels. Multiple EMR parameters include those guides which are at least one level higher in complexity and add the joint effects from more than one frequency in the spectrum. Thus there are:
1. Waveforms and commensurately-sized targets to deal with.
2. Resonances; different, biologically significant levels of power density.
3. Thresholds for action.
4. Continuous cw, versus pulse, and associated peak power.
5. Extended exposures of months and years.

6. Polarization as an applied field pattern or orientation and as a target presentation.
7. External and internal fields and field force effects.
8. Distribution of energy.
9. Form factors of shape as well as size, homogeneity.
10. Differences between species and those due to age and other circumstances in the individual.
11. Simultaneous effects from multifrequencies.
12. The nutrition in a cell or microorganism.

With the present intensity of inquiry, there arises a shortage of intellectual frameworks on which to build. There are signs of strain at some points. Why is the whole equal to more than the sum of the parts? When related to experiments on tissue cultures versus intact organisms, the answer seems to be more than organization and rather in transformations. Thus the nervous system is more apparent as a receptor system in the *intact* organism. Every molecular interaction with EMR has its input, output and transfer function. When these systems, which are very numerous, are multiplied by the hierarchies of systems, there are at least the following levels to manage:

1. minimolecular
2. macromolecular
3. subcellular
4. cellular
5. supracellular or suborganismic
6. superorganismic or interorganismic
7. intersystematic
8. operational

The last seems a more complete term than behavior in our informational sense. At the least, new concepts and laws are needed for each major area of the spectrum and some are beginning to be visible.

Managing the Biohazard. There is also a pragmatic twist to the management of biohazards. It is not an absolute matter of zero hazard or perfect safety but instead, the risk must be weighed against the benefits just as has become the habit for the operation of vehicles. One healthy sign is a desire to upgrade the approach and to understand the essential nature of the EMR. This avoids the pitfalls in playing down the biohazards and using inadequate criteria for biological evaluations. The management is now convinced that these problems cannot be "wished" away. At the same time the sophisticated networks in the informational pathways clearly strain the available understanding in the disciplines. In biology this is no more apparent than in the case of the audio effect of microwaves or "hearing" EMR yet at the same time, investigators are adding new basic understanding of the hearing process within the mission to evaluate the hazard.

THE ELECTROMAGNETIC ENVIRONMENT

Obviously it is too much to expect to fully analyze the electromagnetic environment when it can have frequencies from 0 to 300 GHz in unpredictable combinations superimposed at any location. The other extreme, a fairly well-defined hazard, would be in the main beam of a powerful transmitter. Thus restrictions are introduced, the most likely being frequency ones. Sometimes many frequencies can be integrated when there is no harm in losing frequency information. Even 60 Hz is important because of the high voltage, 345 to 765 kV, used in overland lines. These create high fields, questions of coupling, corona shock, perception, and annoyance thresholds, not to mention lighting all the fluorescent lamps in the vicinity. In a large city like Boston in 1975, from 54 to 890 MHz, one may find environment levels to be less than 2 $\mu W/cm^2$ and usually in the range of 0.1 to 0.5 $\mu W/cm^2$. Some guides to the significance of these numbers are:

10 mW/cm^2 = Thermoregulateable (OSHA—U.S. Occupational Safety and Health)
5 mW/cm^2 = *Used* microwave oven limit BRH ⎫ U.S. Bureau Radio-
1 mW/cm^2 = *New* microwave oven limit BRH ⎭ logical Health
1 $\mu W/cm^2$ = Russia, occupational limit

Near the Post Office at Mt. Wilson, California, a transmitter site, 1-44 mW/cm^2 were calculated and 4.8 mW/cm^2 measured. At the Post Office, 0.2 to 2 mW/cm^2 outside densities and 0.06 to 0.6 inside were found. The main tower was at >10 mW/cm^2 with its several antennas working. Beneath a 765 kV line, a field of 7.5 to 8 kV/m will exist or about 5 kV/m at 46 feet. Such levels can induce body currents from 150 to 250 μamp. When hospital technicians observe their instruments showing strange fluctuations, they should immediately suspect electromagnetic interference. This is often from 88-108 MHz and may be at field strengths from 1.5 V/m or more, but anything greater than 1 V/m is probably intolerable in this interference mode. In one case a hospital was located 3,200 feet from a transmitting tower that was 977 feet high but the hospital was of the high rise type. The tower had eight TV and four FM transmitters at about 777 feet or higher and was fitted with emergency transmitters at the much lower level of 180 feet. The latter were activated during repairs to the primary antennas but, being lower, they gave the most trouble.

In this situation it turned out that there were several danger areas in the hospital. As might be expected, one was the solarium, another was the infant intensive care and cardiac care sections as well as the pediatric units, where readings of 1 V/m or more were obtainable.

Extremely Low Frequencies

The effects of EMR in the ELF range might affect mitosis, embryogenesis and circadian rhythms. When combined with another stress such as an infection, the multiple stresses are considered in the appropriate frequency range. In ELF experiments the field strength, V/m, is established and any necessary modulation is included. It is also necessary to set the polarization, for example vertical in the E field, or horizontal in the H field. The modulation may be MSK, meaning multiple shift key, and it constitutes informational input. This is where meaningful signals may be passed into the organism and such modulation does in fact consequently modulate mitotic events to produce mitotic delay. Normal division in the cell is completely eliminated in strong magnetic fields but this is a biological endpoint that is strongly dependent on selection of the correct time for sampling.

TV and AM

Most transmitters create very low exposure levels, for example $\geqslant 1\mu W/cm^2$ but powerful UHF TV transmitters or their clustered or colocated combinations may present a total emitted power significantly greater than a megawatt. This kind of emission will then cause exposure levels of 10 $\mu W/cm^2$ at points of access for persons at the transmitter site such as the workers at the maintenance stations, while at a distance of three miles away, exposures above 1 $\mu W/cm^2$ are unlikely. Within the equipment room at a TV station with the visual power at 10 kW, measurements show 1 $\mu W/m^2$. From microwave ovens with properly closed doors, 0.32 mW/cm^2 and a median 0.25 mW/cm^2 have been measured at two inches at the sides of the door. For AM radio in the Washington and Baltimore region, 0.2% of the population would be exposed to fields of over 2 V/m and this would seem to be a common, city, nationwide status. A station with an output of 50 kW should have a maximum of 13 V/m. The height of the antenna is determining, as is seen in the hospital case. If UHF TV at 1 megawatt mentioned above is taken as the source, then exposures above 1 $\mu W/cm^2$ are received by 13 million North Americans, but if antenna height is factored in, then this reduces to four million North Americans. The exposure of this population seems to be usually at less than 10 mW/cm^2 and generally is below 1 mW/cm^2. A level of 10 mW/cm^2, which is considered thermoregulateable for humans, is normally very difficult to exceed. A deliberate effort is required at current transmitter output levels. Climbing the antenna at a site, looking into the waveguide or standing in a beam will usually produce the exposure. However chronic, long-term athermic exposure conditions are easier to achieve at very low levels below the 10 to 30 mW/cm^2 (low level) range of power density.

Again, besides the levels for acute or chronic exposure, there are a number of intrinsic hazards due to biological coupled EM wave modulations. Perhaps the best appreciated one is peak power and the associated rapid rise time in absorbed energy.

In this range, various bioeffects are possible depending on the exposure type and including changes in the operations of an animal or its behavior. Specifically, effects on learned behavior, perception of EMR, and nervous system activity changes occur that may be observed physiologically.

Shipboard, Civil Aviation, Diathermy.

Shipboard, Civil Aviation, Diathermy. Tactical air navigation radars at 1205 MHz and emitting 10 kW peak power or an average of 200 W will measure 2 mW/cm^2 at one foot from the antenna on a tower. The ASR-7 Radar at 2820 MHz peaks at 425 kW and averages 425 watts. To find 10 mW/cm^2 from this operating radar one needs to insert the probe between the feedhorn and antenna and there is no detectible exposure in the control room. Other radars show a *few* mW/cm^2 in the immediate vicinity of the antenna. On a heliocarrier deck, 1.5 mW/cm^2 are locally measurable but usually less than 1 mW/cm^2 is found. At an airport it is easy to find 1.0 to 8.0 mW/cm^2 at the end of the runway and high frequency contributions are added, for example, 18 to 70 V/m near the airport buildings.

In the area of civil aviation, localizer arrays radiate 200 watts at 110 MHz with 11 to 15 elements and to get 100 V/m, the subject must be 0.3 m or less from the elements. Aircraft radars give 10 mW/cm^2 out to 3 to 10 m along the boresight line of the pilot and an EM shield protects the cockpit when the radar reverses. If the safety measures fail, they may give a sensation producing 20 V/m at the feet of the pilot.

A medical diathermy set with its output at 10 watts which is perhaps low, gives a field power density of 500 mW/cm^2 for the equivalent plane wave but it is important for the operator to know that 10 mW/cm^2 of scattered radiation will occur around the applicator if the measurements are made in simulated tissues and taken as a function of the applicator design and the absorber. The transverse component of these plane waves seems responsible for the heating. The diathermy causes some 300% stimulation of the blood flow which is a remarkable result at these power levels not duplicated by short wave, massage or infrared. The diathermy after-effect is also real and advantageous. This means that after 15 minutes beyond the treatment there is a residual temperature increase in muscle, but not every animal and every occasion will give equal stimulation.

Laser and maser sources would give a welcome selectivity and directivity to ranging, directing and detection electronics and could be safer to service. TACAN transmitters which power Tactical Airborne Navigation Equipment may be located at the end of runways and use a powerful klystron generating 1000 MHz or 30 cm microwaves with an input voltage of 24 kV. Such powerful incoherent radars are difficult to service unless running and the serviceman is heavily exposed. Coherent beam devices or instrument landing systems with radio may be much less hazardous. If the big "roasters" are used, it is obvious that remote controls should be provided for servicing the high powered tubes. Sensitive groups or pacemaker users should not approach closer than 2000 feet to medium and high-powered radars if they wish to avoid precipitating a crisis or, in the case of pacemakers, if they wish to continue having the benefit of at least 80% of their life and death signals. Airport transmitter towers may need servicing at times and the repairman will be exposed to 60 to 80 watts according to estimates quoted by Brodeur (1977) if the 300 MHz VHF is running. These are control tower-to-plane communicating radios.

The over-the-horizon backscatter radars OTH-B, which are installed in Maine and in the Oregon-Washington region, have a frequency of 3 to 30 MHz. They emit megawatts of power and are hazardous for tissues up to one-half mile away so that the Armed Services will provide shielding and signs to delineate the dangers and no access will be allowed at least to 1000 feet. These had better be observed until more coherent and selective devices are developed. Radars such as the PAVE/PAWS, SLBM (AN/FPS-85) and Cobra Dane are of special biological interest. The PAVE/PAWS unit is a massive, completely-housed unit with multiple frequencies that scan 240° with 3600 radiators. At 425 to 450 MHz, it can be expected to have optimum whole body energy transfer to man. The range is an effective 2500 nautical miles and, with 3° elevation, the units can see submarine-launched missiles which are the perfect offensive nuclear weapons.

The SLBM has a 700 kW maximum and 140 kW average output of which some 40% is dedicated to ICBM detection and 60% to detection of space objects. Probably a flux of 10 mW/cm^2 would be found out to 3000 feet in the scan mode. Private planes with nonmetal skins would be expected to be hazardous for "pacemaker passengers" out to about two miles. Even the BMEW type (early warning) radars for ballistic missiles have over 1 mW/cm^2 flux densities out to a third of a mile in side lobes.

In their bursts, nuclear weapons generate many electromagnetic radiations. The gammas and particle radiations produce secondary electron emissions which may find ideal conditions for generating electromagnetic radiations or for exciting atoms and molecules to emit electromagnetic radiations. Their range can be much larger than that of the immediate destruction. A high burst has the earth's magnetic field to provide a crossed field effect on its EMR emissions and, under such conditions, a very large generator can be imagined with effective distances for electromagnetic contamination of many miles.

Unless the exposure is intentional (as in diathermy), it is evident that the electromagnetic transmitters should be treated like weapons and angled in a harmless direction. Antennas will have to be kept high and life sectors carefully controlled in operation by the use of stops which prevent accidental exposures. Certainly it would be folly to remain in front of boat radars. If these should stop rotating they will give 1 mW/cm^2 average power density and therefore, they should be secured in populated areas.

ELECTROMAGNETIC STATUS, CHARACTERIZATION OF THE EXPOSURE

An electromagnetic status is definable for any given situation, which means that the exposure to electromagnetic radiations falls into one of several categories or types of exposure. Every exposure is measured in units which should depend on factors such as the frequency of the electromagnetic radiation. Highly specialized measurement equipment is utilized according to the frequency, strength, and the units of interest. These units and the exposure characteristics or EM status then lead to specification of safety levels, the maximum permissible level or exposure limits. When exposure is extreme, various syndromes are recognizable which may be a prelude to death for the organism.

TYPES OF EXPOSURE

The exposure can be characterized according to the many criteria that have been developed under previous topics. It is becoming clear that periodicities play an important role, referring to the phase of the system as related to the exposure, its cycles, and time constants. Cardiac exposure for demonstrating brachycardia, tachycardia, or no change depends on the phase of the exposure with respect to heart waves, for example, at the peak of a rising R wave or elsewhere. It is also apparent that coupling is the significant factor, for example, again in cardiac exposure, whether or not the excised heart is immersed in Ringer's solution. Anesthesia is clearly a factor in experiments and this relates to narcotics and other drugs, such as curare. They may give cardiac stability against arhythmic effects *in situ*. Field patterns, for example, from a coaxial probe versus a feedhorn, and concentrations, standing waves, and resonant or nonlinear effects all play a part. Of course the absorbed energy is critical in an exposure and this depends on the power density which is, in turn, a function of generated power. This is then modified by the rate of energy absorption.

A major distinction may be drawn between energy that is directly communicated and has informational significance and that which is scattered. The first uses a receptive system, such as a nerve net or an antenna-like structure in an organism while the second reveals particulars by analysis of energy scattered from a structure. Thermic effects include near threshold or local hyperthermic ones. Then athermic ones will include force field effects which must struggle with Brownian motion and thermal agitation to surface and polarization influences which depend on the orientations of subjects and fields.

In a 70 kg hypothetical man exposed to 20 MHz, the ratio of magnetic to electric field will favor the former almost 4,000 to 1 and the H field power absorption is the important one. Thus for HF between 3 and 30 MHz or VHF from 30 to 300 MHz, the exposure must be specified in both V/m and A/m or G/m (volts, amperes, gauss per meter). The ANSI (American National Standards Institute) standard specifies 200 V/m RMS and 0.5 A/m as a maximum. This organization also uses 10 mW/cm^2 as a standard limit.

Yet it is not so simple, and tentative maximum exposures that can be permitted relate more closely to the frequency. Thus,

at 10 MHz $\quad E_{max} = 274.6$ V/m
$\qquad\qquad\quad H_{max} = 0.73$ A/m

at 40 MHz $\quad E_{max} = 97.1$ V/m
$\qquad\qquad\quad H_{max} = 0.26$ A/m
(318)

Another distinction may be drawn between exposed subjects in the free space *mode* and the immobilized or constrained one. The free space one is where the field is free to influence various aspects of the subject as in an animal moving about in an exposure facility rather than a subject sitting at controls or an animal held in a harness.

ACCEPTABLE UNITS

While the power density unit of mW/cm^2 is quoted here because of its widespread use, the definition of this unit should be intensity and the measurements are made with intensity meters now misnamed densiometers. They read the EM flux impinging on an area of one cm^2. Then the investigators who extend the unit to energy by taking an exposure time are doing us all a service. This produces mW hr/cm^2 which is an EM exposure and still not an energy density. Yet it has the flexibility of an exposure period, one step beyond a performance standard that merely has to describe the power intensity in a space of interest from a safety standpoint. With corrections for reflection and transmission, the exposure becomes an absorbed energy, in units of mW hr/g for safety studies. Then since a joule/sec is a watt, this may be expressed as joules/unit mass, usually g or kg.

The intensity level 10 mW/cm^2 is a calculated value that is considered to be normally capable of thermoregulation in humans. The great variability in this among persons working in EM fields suggests that it should be a range rather than one number, perhaps 8 to 12 mW/cm^2. Then persons whose working environment is cold and balancing would use 12 and those in warm and exacerbating ambients would use 8 or less.

The microwave oven performance standard prompted the first departure and the first law on maximum leakage permitted with the limits of 1 mW/cm^2 and 5 mW/cm^2 at 5 cm from the door for new and used ovens respectively in the early seventies. Another field test for exposure from "used ovens" specifies a maximum space level of 0.01 mW/cm^2 at 2 feet or arm's length. This level for repairmen would put them in the same flux as the U.S.S.R. permissible exposure to 10

$\mu W/cm^2$ for eight hours and unfix the notion that the two countries are 1000 times apart in all cases.

IMPINGING EMR

According to the usage above it is clear that the power density (mainly in mW/cm^2 but also in $\mu W/cm^2$) is the most favored description of EMR and it represents the intensity, flux density or impinging radiation in the microwave region without specifying the absorption, reflection, or transmission or the factors that modify them. This is logical in a program designed to establish safe exposures.

Poynting's vector or its summations describe the electromagnetic field vectors through which the electromagnetic energy flows and with time, the vectors describe the amount of flowing energy according to the x-y or x-y-z coordinates used to describe the space. Insertion of an absorbing subject or material in this space then requires that the exposure be taken as the impinging energy modified by numerous factors especially the reflection, transmission, and distortion coefficients. Then the absorbed part may be further modified by its actual interactions with targets of interest down to atomic levels.

Up to the time that an interaction, response, or dissipation occurs, and even afterward, there is no mention in this section of dose or dosage, terms reserved for ionizing radiations or medicinal use. Usually the energy may be substituted for the dose, and exposure for its meaning with advantage. The word dose is too vague to connote all of the direct effects of communication that can be attributed to EMR. Very low power densities would be measured in $\mu W/cm^2$; low would then be about 1-15 mW/cm^2 and high above this level.

ABSORBED ENERGY

The next requirement is for a unit of energy absorbed per unit of absorber and this is expressed as joules per gram or kilogram. Thus, in a study of the average absorbed energy to death, the exposure rate of microwaves to the time of death and the total absorbed energy in an animal is measured. For a tiny animal like a mouse, the average absorbed energy to death increases as the rate of absorption decreases. Here it is seen that high rates of microwave heating produce more effective thermoregulation (Rugh et al, 1976). The total energy absorbed in J/g divided by the exposure time and the weight gives the average rate of energy absorption. For example,

Millijoules energy absorbed/[time (sec) × weight] = mW/g (319)

Sometimes the absorbed power is related to that moving either forward or away from the EM source. As the forward power increases, the absorbed energy does also, death is faster and the average absorbed lethal energy decreases. The power in watts for the tests cited ranged from 4.82 to 8.56 while the mW/g of absorbed energy rate went from 50 to 125, and the absorbed energy in J/g went from 182 to 56. This range of energy units absorbed coincided with a decrease in death time from over 60 to 7.6 minutes. At the lowest rates of exposure survival was possible for a matter of weeks and females were more durable than males.

Other modifiers besides sex and rate of exposure would include all those other factors that go to make up the exposure characteristics. *Polarization* would be an orientation factor, *delayed* effects would not demonstrate themselves for a longer period, possibly being subject to *cumulative* damage via chronic exposure as opposed to acute. Development and growth observations would require consideration of embryonic and fetal sensitivity to EMR. A metabolic effect may require appropriate units of exposure and in fortunate circumstances such as in the exposure of *Photobacter fischeri*, the luminescence under microwave stimulation serves as a unique biological, living photointensity meter (photodensitometer).

Specific Absorption Rates

In certain standard cases, it may be advantageous to observe that a certain energy density produces a fixed rate of energy absorption such as,

$$7 \text{ mW/g} \equiv 20 \text{ mW/cm}^2 \tag{320}$$

Specific absorption rates are rates of energy absorption that can be specifically related to a biological endpoint. The specific absorption rate (SAR) for cataracts in rabbits is 100 W/kg in 140 minutes.

When the power density is related to joules per unit volume of space, it can measure an intensity. Then if a certain dB change is specified, there is a range of intensity expressed. For example,

$$0.3 \text{ nJ/m}^3 \approx 2 \text{ mW/cm}^2 \tag{321}$$
$$3.0 \text{ }\mu\text{J/m}^3 \approx 200 \text{ mW/cm}^2 \tag{322}$$

and the dB change is 50.

The radiation exerts a pressure on the receptive system which may mean that deformations are "felt" by cells just as they are by large aggregations. In the solar system this pressure is sufficient to "steer" objects entering the spaces commanded by the sun. When a comet comes into this sphere of influence, the radiation is sufficient to push on the tail particles, polarizing the tail away from the source of radiation thus pointing the head. The average EM power impinging on particles is the Poynting vector $\overline{E}\overline{H}$ which is in watts per unit of area such as mW/cm². Then,

$$W = pc$$

where p is the radiation pressure and c is the velocity of light. If the power is reflected, then pressure is increased,

$$p = 2W/c$$

At about 100 mW/cm^2, $p = 0.67 \times 10^{-4}$ dynes/cm^2 which is near threshold levels for human detection by pressure sensors including audio ones. Actually the hearing threshold is 0.2×10^{-3} dynes/cm^2.

INSTRUMENTATION

The power density measurement requires a probe or detector with a sensitivity appropriate to the field of application, for example, in the *near* as opposed to the *far* field, of a source or antenna, absorption transduction is most important. Most power density meters do not show magnetic field power and multi-loop antenna probes or the equivalent are needed.

In sensitive measurements, disturbances are inherent in the presence of the probe itself so that the emergence of new optical fiber, liquid crystal, or light emitting diodes with microwave transparent semiconductor leads is very significant. Schottky diode detectors for X-band for 1 μW/cm^2 to 10 mW/cm^2 and 8.5 to 9.6 GHz have been developed. The U.S. Air Force has an energy meter that covers from 3 to 300 MHz and takes only 0.5 m sec to give 90% response. This is a frequency range in which there is not enough biological or electrical information for standards. Their thermister is nonperturbing on the field and probes temperatures from 0-100°C. The Air Force magnetic field probe goes from 5 to 50 A/m from 100 to 300 MHz.

Liquid crystal optical fibers were used as temperature probes to measure exposures to a few mW/g in organs separated for study (Olsen et al, 1977) in tests by the Navy Aerospace Medical Research Laboratory in Pensacola, Florida.

An example of exposure instrumentation is the Raytheon 4J31 magnetron with its power supply and the Manson Labs pulse tube modular, Model 275, for example with a pulse repetition frequency of 500 Hz and a pulse width of 1 μsec. There would be an S-band horn Model 299 for cw and pulse power. The forward power would be monitored in such a facility with a 30 dB waveguide, directional coupler and a power meter such as hp Model 432. Then the probe is calibrated with forward power measurements of power density plotted linearly against forward power so that densities greater than 100 mW/cm^2 are attainable by extrapolations. The Narda 8305 density meter with the 100 mW/cm^2 probe (No. 8323) gives the high exposure measurements. Scanning infrared thermography (SIT) has been useful for measuring EM energy patterns in any plane of the subject under exposure.

Most of the armed services have some kind of exposure facility. HEW/BRH (Bureau of Radiological Health in the Department of Health, Education and Welfare) has a microwave power density calibration facility using 2450 MHz and other frequencies to evaluate prototype antennas and probes. NIOSH (the National Institute of Safety and Health) has an exposure and calibration facility for 10 to 40 MHz with up to 5000 V/m and 50 A/m.

Spectrum analyzers may read RMS power (versus peak). Then RMS × 0.14 = peak. Rod and loop antennas give values for E and H fields respectively, for example the Stoddart NM-25T (HF). In the microwave region omnidirectional units (isotropic mode) and horns or dipoles for oriented sensing are available devices.

The Environmental Protection Agency (EPA) has spectrum analyzers and minicomputers for EM environmental levels with coaxial transmission lines of rectangular cross section, and Crawford cells for small animal and *in vitro* work. Their small cells go from 500 to 1100 MHz with amplitude modulation. The large types are for 100 to 450 MHz and are used for exposure of rats and mice in order to study absorption under temperature control. For example, there is the maximum frequency of absorption which gives a special peak in the spectrum with *in vitro* analysis. An interesting double line microwave spectrometer provides for both sample and a standard or reference exposures simultaneously. A time domain reflectometer can be used for the dielectric constant up to 9 GHz. Sensors such as discones which begin saturation at 200 V/m or monopoles (2V/m) from Antenna Research Associates read from 1 kHz to 1 GHz and may be used in EM surveys as in the case of the hospital near the transmitter site where the field strength, for example, was decreased from 13.2 to 6.5 dB by shielding the roof. Instruments for Industry's EPS-1 reads from 140 kHz to 10 MHz.

Microwaves and EM processes in general are fast phenomena along with enzymatic processes and many biochemical activities. Fast transients can be repeated in the CRO display with a persistent trace and generators to give a fast rise time of a few picoseconds. If a tissue bound is being investigated, for example, the thickness of a tissue layer such as the dermis, a square wave pulse may be used and the sounding done by time domain reflectometry (TDR). Then the rise time of the impulse will be small compared with the delay in the events in the time domain.

Perfect transmission would produce no echo but, on the other hand, mismatches return reflections giving standing waves. The time base then shows the *range* of the interface along the transmission line. The location could be indicated by a probe, as well, when it is inserted *on the* target and its echoes coincide with those of the specimen. The procedure is:

1. Start sweep of CRO and initiate pulse from generator with velocity v.
2. Pulse hits bound and at $2d/v$ sec, echo reflects to be seen within $2d_o/v$ sec.
3. Resolution by fast response, sampling pulse of CRO within 30 picosecond allows reflection to be ranged within 500 μm or 0.5 mm. The nature of the bound is suggested by its resistivity, impedance, or admittance by

$$\rho = (Z - Z_o)/(Z + Z_o) = (Y_o - Y)/(Y_o + Y)$$

Here Z and Y refer to the load and Z_o and Y_o to the line.

4. Let v = line velocity, d_o/v = time for impulse to be sensed at probe. Then $2d/v$ = time from probe to first interface and back to probe.
5. With no mismatch, the pulse generator will see the reflection in phase and couple it in. If it is out of phase, it will reflect to the CRO and take $2d_o/v$ seconds to reach the screen.
6. The $2d_o/v$ seconds is the "good" observation time in which the CRO can see exactly what happened at the interface before noise hits it.

The CRO has a dynamic range, for example, the limits of reflected to incident wave at interfaces detectable may be 1/10,000. This would be the resolution in terms of the mismatches recognized or resolved. To expand the concept somewhat, it may be observed that the frequency domain may be selected instead of the time domain. This type of measurement may be superior in some cases and is given by standing wave ratios in the line.

MAXIMUM PERMISSIBLE LEVELS

Exposure levels below 10 mW/cm^2 are much more common than levels above this density and most occupational MPLs use this figure in some combination of exposure time. Lower frequencies like 13.56, 27.12, and 40.68 are ISM (Industrial, Scientific, and Medical) assigned frequencies which are very important in the textile, lumber, and plastics industries for volume heating. The guide for MPLs is about 200 V/m and 0.5 A/m for electrical and magnetic field power respectively. These two ways of expressing MPLs are related in many ways. For example, the NATO MPL of 1000 V/m $\equiv$ 266 mW/cm^2 from 1 to 10 MHz and 500 MHz and 500 V/m $\equiv$ 66 mW/cm^2 at 10 to 30 MHz (more lax than OSHA). NATO also uses 10 mW/cm^2 for frequencies above 30 MHz. In the USAF, 50 mW/cm^2 is the average power density for exposures longer than 6 minutes and 3,600 mW/cm^2 for less than 6 minutes, both at frequencies under 10 MHz.

For safety in high peak power densities, on an average over an exposure period, the status may be described in mW hr/cm^2 or an energy density due to the peak power acting over a period of time. It is now felt that the MPL is 1 mW hr/cm^2 for 6 minutes or 3.6 J/cm^2 but lower levels are probably advisable, perhaps 0.02 mW hr/cm^2 or even 0.01 mW hr/cm^2 for pulses coupling into a sensory mode such as the audio one.

PROBLEMS

1. What is meant by ordering forces at the disposal of information managing elements?

2. Do athermic effects always occur with biological absorption of EMR?
3. Why does the long list of chronic effects from exposure to EMR appear to have unexplained aspects?
4. What is the importance of diathermy with EMR in assessing the hazards?
5. What is the difference between bioeffects and biohazards of EMR?
6. List some informationally rich EMR conditions.
7. What is meant by the thermoregulateable level of EMR power intensity?

BIBLIOGRAPHY

Allen, S.J., "Measurement of Power Absorption by Human Phantoms Immersed in Radiofrequency Fields," *Annals of the New York Academy of Sciences*, Vol. 247, 494-498, 1975.

Barron, C.I., A.A. Love, and A. Baraff, "Physical Evaluation of Personnel Exposed to Microwave Emanation," *J. Aviation Medicine*, Vol. 26, 442-452, December 1955.

Bassen, H. and M. Swicord, "A Miniature Broad-Band Electric Field Probe," *Annals of the New York Academy of Sciences*, Vol. 247, 481-493, 1975.

Bassett, H.L., et al, "New Techniques for Implementing Microwave Biological Exposure Systems," *IEEE Trans. on Microwave Theory*, MTT-19, 197-204, 1971.

Bassett, H.L., G.K. Huddleston and B.W. Nolte, *Study of Microwave Dosimetry*, Techn. Rpt. no 1, Georgia Inst. of Technology, Atlanta, 1972.

Bassett, H.L., G.K. Huddleston and J.B. Langley, *Study of Microwave Dosimetry*, Techn. Rpt. no 2, Georgia Inst. of Technology, Atlanta, 1973.

Beischer, D.E., and V.R. Reno, "Microwave Energy Distribution Measurements in Proximity to Man and Their Practical Application," *Annals of the New York Academy of Sciences*, Vo. 247, 473-478, 1975.

Bernstein, J., "Physicist—Part II (I.I. Rabi)," *The New Yorker*, October 20, 1975.

Bierman, W. and M.M. Schwarzschild, *The Medical Applications of the Short Wave Current*, William and Wilkins, Baltimore, 1942.

Brody, S.I., "Military Aspects of Biological Effect of Microwave Radiation," *IRE Trans. on Medical Electronics*, Vol. ME-3, 8-9, February 1956.

Cope, F.W., "Superconductivity—A Possible Mechanism for Nonthermal Biological Effects of Microwaves," *J. Microwave Power*, Vol. 11, 3, 267-269, 1976.

Daily, L.E., "A Clinical Study of the Results of Exposure of Laboratory Personnel to Radar and High Frequency Radio," *U.S. Naval Medical Bulletin*, Vol. 41, 1052-1056, July 1943.

Frey, A.H. and R. Messinger, Jr., "Human Perception of Illumination Pulsed UHF Electromagnetic Energy," *Science*, Vol. 181, 365-368, 1973.

Gandhi, O.P., "Strong Depedence of Whole Animal Absorption on Polarization and Frequency of Radio-Frequency Energy," *Annals of the New York Academy of Sciences*, Vol. 247, 532-538, 1975.

Griffith, J., and R. Ballantine, *Silent Slaughter*, Henry Regnery, Chicago, 1972.

Guy, A.W., "Electromagnetic Fields and Relative Heating Patterns Due to a Rectangular Aperture Source in Direct Contact with Bilayered Biological Tissue," *IEEE Trans. on Microwave Theory Tech.*, MTT-19, 2, 214-223, 1971.

Guy, A.W., "Future Research Directions and Needs in Biologic Electromagnetic Radiation Research," *Annals of the New York Academy of Sciences*, Vol. 247, 539-545, 1975.

Hirsch, F.G., "Microwave Cataracts—A Case Report Reevaluated," *Electronic Product Radiation and the Health Physicist*, Proc. of the 4th Ann. Symposium of the Health Physics Society, Louisville, Ky., January 1970, HEW pub. BRH/DEP 70-26.

Ho, H.S., et al, "Microwave Heating of Simulated Human Limbs by Aperture Sources," *IEEE Trans. Microwave Theory Tech.*, MTT-19, 2, 224-231, 1971.

Huddleston, G.K. and D.I. McRee, "A Pyroelectric Probe for Measurement of Microwave Power Density Under Far-Field Conditions," *Annals of the New York Academy of Sciences*, Vol. 247, 510-525, 1975.

Hunt, E.L. and R.D. Phillips, "Absolute Dosimetry for Whole Animal Experiments," *Joint Army/Georgia Institute of Technology Microwave Dosimetry Workshop*, 74-77, Walter Reed Army Institute of Research, Washington, D.C., 1972.

Johnson, C.C., et al, "Fiberoptic Liquid Crystal Probe for Absorbed Radio-Frequency Power and Temperature Measurement in Tissue During Irradiation," *Annals of the New York Academy of Sciences*, Vol. 247, 527-531, 1975.

King, R.W.P. and T.T. Wu, *The Scattering and Diffraction of Waves*, Harvard University Press, Cambridge, Mass. 1959.

Lepage, W.R., *Complex Variables and the Laplace Transform for Engineers*, McGraw-Hill Book Co., New York, 1961.

Lidman, B.I. and C. Cohn, *Air Surgeons Bulletin*, Vol. 2, 448-449, 1945.

Lin, J.C., A.W. Guy and G.H. Kraft, "Microwave Selective Brain Heating," *J. Microwave Power*, Vol. 8, 275-286, 1974.

Liu, L.M., F.J. Rosenbaum and W.F. Pickard, "The Insensitivity of Frog Heart Rate to Pulse Modulation Microwave Energy," *J. Microwave Power*, Vol. 11, 3, 1976.

Matelsky, I., "The Nonionizing Radiations," *Industrial Hygiene Highlights*, Vol. 1, 140-178, 1968.

Mumford, W.W., "Some Technical Aspects of Microwave Radiation Hazards," *Proc. of IRE*, Vo. 49, 427-447, February 1961.

O'Neil, J.J., *Prodigal Genius: The Life of Nikola Tesla*, I. Washburn, New York, 1944.

Patrusky, B., *The Laser*, Dodd Mead, New York, 1966.

Phillips, R.D. and E.L. Hunt, "Problems of Physical and Biological Dosimetry of Microwave Irradiation," *J. Microwave Power*, Vol. 6, 90, 1971.

Phillips, R.D, E.L. Hunt and N.W King, "Field Measurements Absorbed Dose, and Biologic Dosimetry of Microwaves," *Annals of the New York Academy of Sciences*, Vol. 247, 499-509, 1975.

Presman, A.S., *Electromagnetic Fields and Life*, Plenum Press, New York, 1970.

Reno, V.R., "Microwave Reflection, Diffraction and Transmission Studies of Man," *Technical Report, NAMRL-1199*, Naval Aerospace Medical Reserach Laboratory, Pensacola, Fla., 1974.

Rugh, R., H. Ho, and M. McManaway, "Relation of Dose Rate of Microwave Radiation to the Time of Death and Total Absorbed Dose in Mouse," *Journal of Microwave Power*, Vol. 11, 3, 1976.

Shapiro, A.R., R.F. Lutomirski and H.T. Yura, "Induced Fields and Heating Within a Central Structure Irradiated by an Electromagnetic Plane Wave," *IEEE Trans. Microwave Theory Tech.*, MTT-19, 2, 187-196, 1971.

Shereschewsky, J.W., *Radiology*, 20:246, 1933.

Smith, S.W. and D.G. Brown, "Radiofrequency and Microwave Radiation Levels Resulting from Manmade Sources in the Washington, D.C. Area," *Radiation Data Report 13*, 335-345, 1972.

Tyler, P.E., "Biological Effects of Nonionizing Radiations," *Annals of New York Academy of Sciences*, Vol. 247, February 1975.

Wacker, P.F. and R.R. Bowman, "Quantifying Hazardous Electromagnetic Fields, Scientific Basis and Practical Considerations," *IEEE Trans. Microwave Theory Techn.*, MTT-19, 178-187, 1971.

Walker, C.M.B., W.A.G. Voss and W.R Tinga, "Dielectric Properties of Dimethyl Sulfoxide and Their Importance in Cryobiology," *J. Microwave Heating*, Vol. 11, 2, 206, 1976.

14

Direct Information Transfer from the Electromagnetic Environment

> "Sensory ideas are only sets of physical occurrences in our nervous system. To say we see yellow is explained as a result of certain light waves stimulating the optic nerve which then sets certain brain patterns in motion."
>
> *Thomas Hobbes answering Descartes*

POSTULATIONS AND EVIDENCE FOR DIRECT INFORMATION TRANSFER

In nonionizing radiobiology, the reader comes to appreciate what is meant by the electromagnetic environment, and becomes concerned about his compatibility with it. The appreciation is formed by an understanding of electromagnetic waves, the spectrum of frequencies and their interactions with biomaterial. The units, dimensions, nature of radiation, and polarization are initial elements in this understanding. Microwaves illustrate these elements well and seem to offer the added incentive of a rapidly growing importance in most sections of human activity from molecular biology through industrial, medical, commercial, and domestic application. One can then anticipate that some collisions might occur between momentum factors in this technological growth and in biological constraints. This in fact occurred amid great international interest in questions of environmental compatability. The background is found in the refinement of new radiation generators and devices, innovations and introductions, as well as extension of older applications. In the biophysics at least, there is considerable contention between the eastern and western nations over the matter of compatibility. At this time the U.S.S.R. seems to have taken a much more safety-conscious attitude than the U.S. Whether this is actually true or justified is still being debated. Some U.S.S.R. radiation safety levels are disturbingly close to microwave tissue emissions and natural earth levels so their meaning is in doubt. In a962 Soviet counterintelligence activities gave the world the first documented instance of the intentional application of radiation (microwave) to influence people and equipment over a long period of time. Efforts to understand what might happen to persons so exposed have led to the analysis of compatibility which is made here.

In addition, many investigators such as Hjeresen (1980) have found that EMR transfer information directly to animals as seen by changes in their behavior.

THE MICROWAVE COLD WAR

It is of keen interest that an East-West conflict should surface over electromagnetic matters. This is smoke around the flame and the fire itself may seldom be discussed because it is much too confidential to the powers concerned. The first smoke signals were seen on the technical-biological side. They are now disappearing amid a growing cooperation at the intergovernmental level even as new incidents are revealed. The crux of the technical matter is the shape of a curve—the one for tolerance limits for permissible exposures. This kind of curve defines the time during which microwaves, at some intensity, flow through a working space. Maximum permitted levels are 1.0 mW/cm^2 for 20 minutes, 0.10 mW/cm^2 for 2 hours and 0.01 mW/cm^2 for 24 hours in a day during which a person is immersed in such fields in the U.S.S.R. In the U.S., there was the 10 mW/cm^2 level in which humans can thermoregulate and the 1 mW/cm^2 emission maximum for new microwave ovens and 5 mW/cm^2 for used ones. In general however, the U.S.S.R. seemed to be talking of current safety levels of 10 $\mu W/cm^2$ while the U.S. was holding at the 10 mW/cm^2 which separates the East and West by an enormous 1000 times on a matter of safety, with the U.S.S.R. being that much more concerned about its citizens. Levels below 1 $\mu W/cm^2$ are of interest probably only in spectroscopy. If the U.S.S.R. data are properly interpreted, for example, by taking an 8 hour exposure to 0.01 mW/cm^2, this *is close to* the used oven exposure or field test in the U.S. There we have a maximum of 0.01 mW/cm^2 at two feet or arm's length.

Exposure Standards — East vs. West

It is hard to say how genuine the U.S.S.R. standards are, but they would prohibit the use of microwave ovens which pass the U.S. tests. For such a violation, a person would need to cook with his whole body two inches from an oven emitting 1 mW/cm^2 for more than 20 minutes. Thus it becomes clear that the U.S. exposure standard does not apply to a whole body because a person would have to be wound up into a small ball and placed almost up against the oven to receive the emitted microwaves. Thus the standard applies where only a portion of the body is intercepting at a time.

Most of the material in these chapters on the electromagnetic environment is concerned with these biological and safety matters but it is very difficult for a specialist to separate what is biological and what is communication in this area of conflict. If the superpowers diverge greatly on permitted levels, one of them could choose whichever level was suited to its purposes in offensive operations. This turned out to be what Russia did in the embassy case where the beam of EMR was well below the U.S. standards, but dangerous by theirs.

On the EM Frontier. The U.S.S.R. probably believes that its strength lies, as in the case of the Red Armed Forces, with superior strength and a massive capability. Communication matters are so information-rich that they are more subtle in their manifestations. Early in 1976, it came to light that strange, relatively massive, swept-frequency EMR was being directed at the American Embassy in Moscow and had been for 22 years, which means that the children growing up in this environment, as well as adults working in it, had a daily exposure from across Tchaikovsky Street. Microwaves at 5 to 15 microwatts/cm^2 were beamed at the west facade from 1953 to 1975 and at the south facade from 1975 to 1976, mostly at the upper floor. The Russians were using electronic countermeasures against the communication-intelligence functions and this was an intelligence-security problem with innocent victims by the score. It certainly shows a weakness that the U.S.S.R. can exploit all over the world, confident that the U.S. will not or cannot always counter the ploy. The domestic problems seem so severe that anyone who leaves the shores of the U.S. is doubly on his own against any hostilities created by the mother country, as well as hazards produced by foreign countries. American ambassadors, their staffs and families, as in the case of the Moscow Embassy, seem to be a particularly valiant, if unsung group of heroes. When not being exposed with their families to EMR, they are likely to be kidnapped or simply murdered in cold blood. Evidently, the famous Henry Kissinger during the Nixon and Ford administrations looked at them as sort of shock troops. From 1973 to 1976, the American ambassadors to Lebanon, Cyprus and the Sudan were cut down in cold blood with the killers left unpunished. A prepared counteraction, taken swiftly and with determination is necessary in cases where other states are powerless to help. This seems to be the least that can be expected of a superpower. The observation cannot be resisted that, in our context, the U.S. is much more of a super(microwave oven)-power than the U.S.S.R., with one in ten households having an oven but the real power is in Japan which leads the world in the use of the oven.

Interestingly enough, interference with electronics (for example, receivers or instruments being used for other purposes) is likely to be the first warning of an environmental hazard unless people learn to sense it. Unless this happens, the population will normally be unaware of any danger.

Electromagnetic Blitz. In 1976, the U.S.S.R. began a program of electromagnetic disturbance that effectively disrupted some world communications for months. This was, in effect, a ticking sound that came on when maritime, aeronautical, telecommunications, and ham radio operations were attempted. The sources were three powerful stations in the U.S.S.R. that caused interference as far away as Australia and the U.S. After complaints poured in from the numerous countries affected, the transmissions stopped about November 10, 1976, and the protests were officially acknowledged by the U.S.S.R. without explanation about December 15, 1976. While it can be presumed that the Russians were simply taking a major new communications step to maintain contact with their far flung surface ships, submarines, and agents, busier and busier all over the earth, other possibilities exist. Since 1950, they and their Eastern European laboratories have conducted a steady research program on EMR. They emphasized effects that are based on nonspecific functional changes, headaches and sexual irregularities usually of chronic origin and subject to disappearance. There was a tendency to look at organismic responses and effects mediated by the central nervous system while this connection was actually difficult to interpret (Copson, 1975 p.572 and Chapter 21 and 22). Then, with the communication blitz in 1976, plus the Russian penchant for doing things in a big way, it was tempting to speculate that these supertransmissions had a sinister motivation behind them in the sense of science fiction and terror watts, to paraphrase the prefix connoting power levels of 10^{12} watts.

Then an intentional intersystem biohazard from EMR could be created by beaming at populations in an unseen electromagnetic war. Selected targets, for example, cities already saturated or at MPL, could be caused to have levels greater than this limit. Then marginal members of the population with special sensitivities would be disabled, electronics would suffer from interference and information systems put in disarray.

Cooperation to Defuse Situation. This scenario is less reasonable in the light of cooperative efforts in this area of research into EMR bioeffects which began with an agreement in 1975 for a joint research program. One can only hope it is not like people on both sides of the law studying electronics with obviously different motives.

Ordinary cooperation in scientific meetings between East and West has existed for many years especially since 1950 but U.S.S.R scientific reports in EM research have had this strange bias toward the electronarcosis and direct effects on the nervous systems. In addition, there have been suspicions generated by the distance between scientists, the language barrier, and the fact that the Russian research had a different philosophy than publishing openly and in detail. However, the same problems develop in any scientific exchange and are by no means limited to the U.S.S.R. The U.S.-U.S.S.R. cooperation is prepared by HEW via its concerned agencies and, on the U.S.S.R. side, the Institute of Municipal and Communal Hygiene at Kiev is coordinator. Areas of stated interest are the central nervous system, behavior, blood, and biochemistry. The frequencies are 500 and 2300-2450 MHz at 5 to 500 μW/cm^2 while in the U.S. this level goes up to 10 mW/cm^2. At the lower frequencies, the U.S. has already found that less than 2% of its population would suffer exposure to 2 V/m but it should be noted that the MPL in the U.S.S.R. is 20 V/m for the AM broadcast levels.

Characteristic Transients. No doubt the most significant findings in the Soviet studies is that, in their low-level exposures, *practically all animal tissues show responses*. The point that has been missed in trying to duplicate the Soviet work is that these are often evanescent and *transient* changes whereas in the U.S. the search has been for biological *"endpoints,"* a word which sometimes connotes a more permanent response such as an irreversible cataract. Thus we may be on firmer ground in following, even at this early stage, the hematopoietic patterns which are strictly variable with time and above all, the concept of direct EM effects.

Decimeter, centimeter, and millimeter waves are least ambiguous for referring to the frequencies from 300 MHz to 300 GHz used in Soviet microwave biophysics because they designate super high frequency (SHF) to cover the UHF, SHF, and EHF bands. Their interest in microwatt exposures came after 1957 as a shift in emphasis from the milliwatt range. If their data are surveyed from high to low exposure in the range from 1 watt to 20 microwatts/cm^2, the most notable observations are:

1. Thermoregulatability at 10 mW/cm^2 is ignored as a significant guide.
2. 5 mW/cm^2 is a significant exposure for testicular and blood pressure change, transient leukopenia and erythropenia, and darkening of crystalline lens. This exposure is a common choice in their tests.
3. Sensation of pain occurs at about 600 mW/cm^2.
4. Cataractogenesis is at the high levels of 1 W/cm^2 for 3-5 hours but bilateral cataracts were also seen at 100 mW/cm^2. "Normal age" cataractogenesis has to be distinguished from microwave disease and this may be troublesome. Retinal change and color discrimination may be more meaningful observations.
5. Blood pressure depression is very common over a very large range except for transient increases followed by a decrease at the single high level of 100 mW/cm^2.
6. At 20 microwatts/cm^2 skin temperature of conditioned persons (exposed earlier) clearly increases.
7. Color discrimination increased for red, green, and blue, especially red at a fraction of 1 mW/cm^2. Workers became transiently more sensitive to light.
8. Headache, fatigue, drowsiness, and loss of memory are to be expected among EM generator service personnel exposed up to 1 mW/cm^2 for a few *years*. Their leukocytes decreased, lymphocytes increased, and certain neutrophils decreased along with reticulocytes and thrombocytes. Cellular and humoral immunoresponses were depressed.

9. Microwaves act on coenzymes producing vitamin deficiencies in chronic exposure of rats to 570 microwatts/cm^2 for 15 days at 2000 MHz, cw.
10. Change in electron transystems were transient.
11. In brain neurons of rabbits at 460 MHz, cw, 2 and 5 mW/cm^2, with bodies protected, activity in dorsal hippocampus, specific and nonspecific nuclei of thalamus, *hypothalamus*, and reticular formation of midbrain showed greatest response in the central brain and in neurons of *hippocampus*, and least in reticular formation. There may have been deinhibition by the cortex of larger hemispheres on subcortical structures. An EEG activation appears and adoption of light flashing rhythms is alleviated at 2 mW/cm^2 and 13 to 25 Hz. Transients in brain catecholamine metabolism included a peak in epinephrine on the 20th day of exposure.
12. The cardiovascular response in rabbit to 2375 MHz for 60 days was:

    3 to 6 V/m about 10 microwatts/cm^2 increased the heart rate, then decreased it. EKG spikes were modified.
13. At 42 days, blood serum cholinesterase decreased (1 hr/d, 3000 MHz, pulses of 3 microseconds, PRR 350 Hz, average power 10 microwatts/cm^2) neutrophil cell metabolism showed transient increases in alkaline phosphatase, and glycogen at low microwatt levels, and at 500 microwatts/cm^2, glycolysis was activated in these cells at 28 days.
14. Pulse microwaves decreased, transiently, hematocrit, leucocyte, and lymphocyte counts.
15. Neutrophil phagocytosis is altered, as well as lysozyme and complement titers (nonspecific immunoreactivity). The phagocytes were less active with 6.5 mm waves, 1 mW/cm^2, 15 min/d for 20 d., and more active with 2375 MHz and 30 d., at lower power density. A power density of 1 microwatt/cm^2 increased their activity most, but these are transients, returning to normal afterward.
16. The chromosomal aberrations, polyploids, aneuploids, deletions, acentric fragments and chromatid gaps increased with 12 cm waves and 2 weeks at 50 microwatts/cm^2.
17. Tissue culture cells survive less, develop morphological and functional disturbances at 1 mW/cm^2 (6.5 mm microwaves) exposure generally impaired cell monolayer. Virostasis occurred with waves at 6.5 cm, 2 h, with 1 mW/cm^2.
18. Significantly, the most sophisticated of transients, direct EM communication purposefully directed, is not reported in this survey for the post-1976 period by McRee (1980). That experimental design is challenging even for the potent basic research capability of the U.S.S.R. academies, or the results are being kept secret.

19. Transients in membranes need spectrographic grade preparations for showing instabilities. Changes are not expected to be thermal but membrane surface ones through losses by conductivity and internally through dielectric losses. Permanent depolarization and reduced membrane potential are expected (stomach, waves of 8 mm, power 1 mW) when sodium ion permeability increases and active transport decreases.
20. EEG shows brain is stimulated at 10 to 50 microwatts/cm^2 2375 MHz, cw. but inhibited at 500 microwatts/cm^2 with reduced work efficiency. Duplicate experiments in the U.S. have now shown reduced SH activity and blood cholinesterase, animals suffering aldosteronism, behavior studies, blood chemistry and pathology, supporting Soviet results 500 microwatts/cm^2. Above all, the joint U.S. and U.S.S.R. or cooperative efforts reveal the increased sophistication of microwave biophysics especially in the elucidation of transient and evanescent phenomena, whether directed purposefully or not, at time and space processes of cells. It would be foolhardy, if not outright dangerous to assume that the Soviets are not near this remarkable breakthrough in informational bioelectromagnetics.

WAVES AND SYSTEMS

Microwaves in particular, polarization in general, and specifically dielectric phenomena have interrelationships that we have already seen with biosystems. Inputting a direct information transfer on this basis alone would be hazardous. It is the commonality that is advanced however, and we would like to examine, against this background, some of the growing evidence involving organic semiconductors, "pearl chain" formation and, more generally, field forcing and informing functions in organisms that seem appropriate in extending the direct interaction concept. It must be advised, however, that in this area, the studies suffer from a surplus of interest and lack of analysis.

Commensalism

First, Table 14-1 summarizes ten selected features suggesting the mutuality of microwaves as part of the environment and biological systems which we have taken up.

This is probably the first time that these mutual features have been presented and the student will appreciate them after studying the preceding sections. With this evidence, and the results that have been reported in this area of study, it appears justified to search for extension of the concept of direct information transfer, and to decide whether conclusions on the organismic reception and interaction are possible at this point in time.

TABLE 14-1. The Mutual or Interacting Features of Microwaves and Biosystems

1. Microwave spectrum as an array of forcing and informing functions.
2. Polarization as in the microwave interaction model is a remarkably general phenomenon in biology.
3. Relaxation mechanisms apply to biosystems and are common as well to the microwave interaction model.
4. There is a commonality in the microwave absorption parameters of dielectric constant and loss factors with physical and chemical factors of structure and functions in their solvents and solutes.
5. Influence of water so important in biology and specific for absorptive polarization with microwaves.
6. Microwaves interact directly with biostructures in electron spin resonance.
7. Microwaves as a plasma and free radical source.
8. Microwaves as a chosen carrier for information in circuits for communication between humans.
9. Membranes, cells and tissues as an information and energy sink.
10. Electromagnetic radiations with their meaningful space and time constants in informing processes relate to similar constants having to do with phases of metabolic, reproductive and other biological processes.
11. Microwave generation in cells uniquely represents a source of ordering forces commensurate with organismic dimensions and at the disposal of information management elements.
12. In the direction of decreasing frequency, for the nonionizing radiations, microwaves have the first significant penetration into biosubstance. Thus, thin layer techniques are not as essential and the under layers are accessible to the forcing function.

EMR Forcing and Informing Functions

The concept of an electromagnetic environment is useful for visualizing an organism immersed in a field of electric and magnetic forces capable of interacting with it (Fig. 14-1). It is difficult to see how a human being could fail to interact and this recognition leads us to adopt him as an element in the circuit. In radio frequency and microwave circuits, theory permits this adoption and creates a need for the conclusions that must follow. There is a special significance to the human organism with his well developed and measurable fields emitted by systemic activity such as those seen in electrocardiography and encephalography and many other evoked and stimulated potentials, his

Information Transfer 531

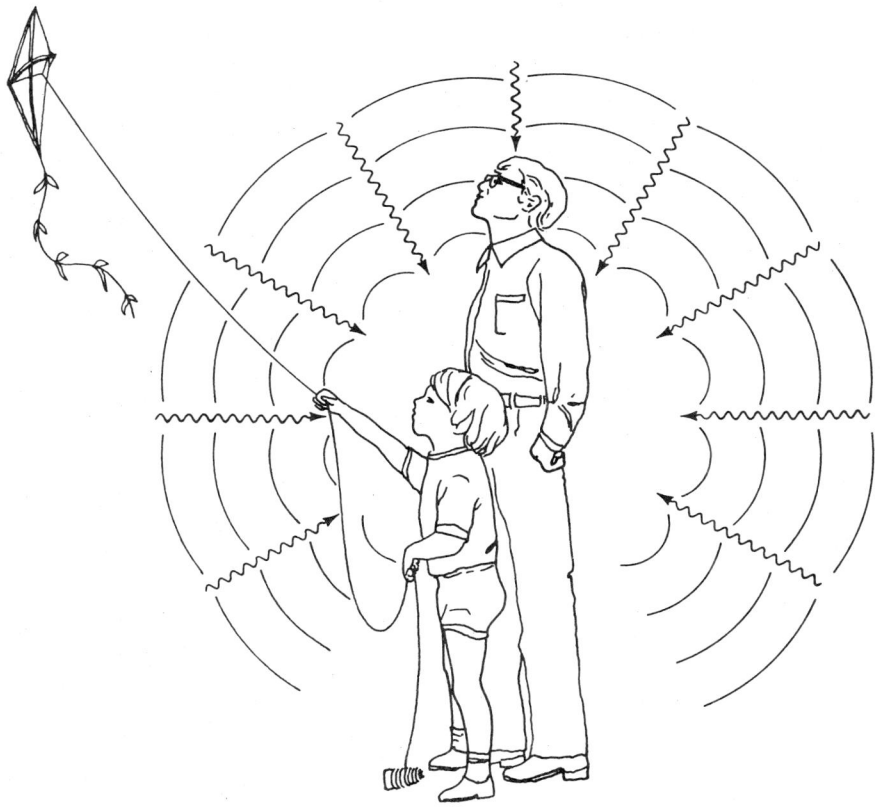

Fig. 14-1. Environmental fields interact with those generated by human activity and direct reception of microwaves by the nervous system also occurs.

neuroelectrical network, and information factors. If the special interaction is admitted, as it must be, due to the evidence that is presented shortly for microwave-induced signals in the nervous system and direct effects on organisms, then we must be ready to accept that microwaves can alter the status of biological systems. The organism is certainly coupled into the electromagnetic field and it is of interest to know how the signal behaves. If it retains a great deal of information rather than being simply degraded, e.g., in heating, then a response may be expected or the information may be stored. In some manner there is a threshold for the survival of signal information, and systems with high power may have modes which obliterate information through saturation effects. An exception may exist where a fast rise at the edge of the higher power input leads to "organized" disturbances. On the other hand, coupling mechanisms, choice of frequency, and modulation of the carrier appear as factors for transmitting information that can influence the organism and reveal nervous system control loops.

Electromagnetic radiations are known to interact with the sensory system. Although soundless, they produce the sensation of sound and although well out of the visible spectrum, they can produce light sensation. The absorption and crossing of EM fields and their interaction around and within an animal can be shown with properly positioned electrodes on the head with the assistance of a recorder for the responses. The normal response or graphical trace is interrupted when the energy is applied giving an action peak which is followed by a return to normal. Specifically, the potential of the hypothalamus or its electrical output in cats has been so modified as has the function of the brain stem or *medula oblongata*. Skin receptors "feel" VLF energy and in some unexplained way, the EM energy occasionally produces headache. While far simpler than other systems to be considered, these indicate the reception of electromagnetic energy.

Communication with an organism via this kind of energy should, if positive, indicate an extraordinary capability. If the radio-frequency energy is pulsed and these pulses synchronized with the R wave of the ECG potential, then there is tachycardia, arhythmia, and cardiac interruption in an isolated frog heart preparation. However, synchronizing with earlier ECG portions gives a negative result. Synchronizing with the T wave (later) also gives tachycardia but less than with the R wave. In vivo, this effect is also found, i.e., tachycardia and arhythmia, but to a lesser degree. This result is essentially successful time synchronization with the electrical field around an organism (for example, communicating information by radiation)(Frey 1969)(Fig. 14-2).

INFORMATION AND NOISE IN THE COMMUNICATION

In this EM linkage the information transfer possible in mere microwatts of energy flow is significant. Telecommunication systems frequently operate over long line-of-sight distances using something on the order of a watt of microwave power, passing most forms of communicated intelligence between population centers in this manner. The first bit is the presence or absence of information followed by modulations of many types. It is attractive to examine the word information in this context and to relate it to noise and biophysical systems. While it may have to be accepted with some faith, noise can be visualized in a randomizing role, that is, in a *positive* one, forming and informing. It would be somewhat like shaking a pinball machine to let balls enter the holes. An obliterating role can, on the other hand, make noise deformational. Finally these are on the input side so that the other side of a system may be classed as outformational (Quarton, 1967). Those correcting measures the biosystem takes to neutralize the disturbing aspects of noise are logically antinoise, and work in the sense of a biological squelch.

Information Transfer 533

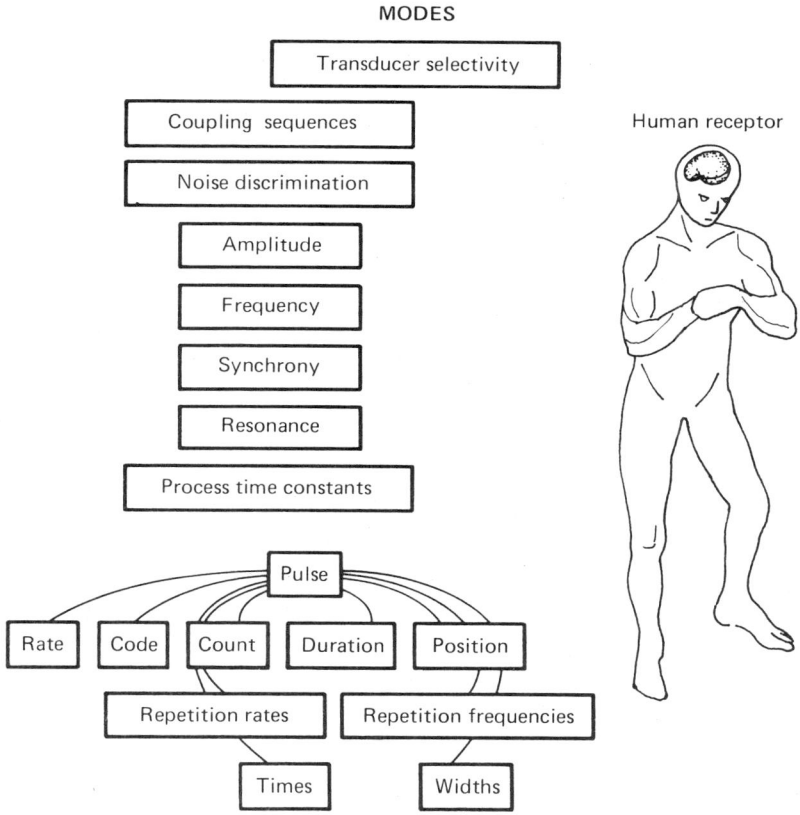

Fig. 14-2. Information input via the electromagnetic fields in the environment where Watts (average power) = Energy per pulse (Joules) × Pulse repetition Rate (Hz); and energy per pulse = Watts (maximum power) × pulse duration (sec.).

In the teaching of Professor von Foerster, noise is upgraded into undirected energy input to a system. In one of his illustrations, magnetically-faced cubes are designed with their magnetic fields pointing in different directions. This means that they will assume certain specific orientations decided by the fields and the partners they meet. However, the meetings can be influenced by shaking up the cubes in a special container with ball bearing surfaces. Using this motion as an undirected energy input or noise, the system rejects a hapless pile formation in favor of a highly ordered structure. It resembles the kind of thing that might be put together from blocks in a stick-together toy game but it is self-ordering with the help of the noise input (Fig. 14-3). Perhaps it is possible to see in this exercise that the cubes might be molecules or cells with favored positions and, under suitable forces, they combine

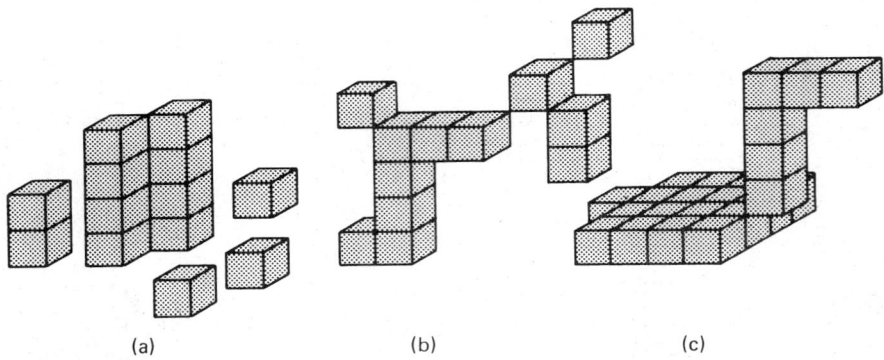

Fig. 14-3. a) Mass of blocks; b) with undirected energy input (noise); c) with addition of more information (directed energy).

with the aid of noise input—the noise being readily available in the chosen environment. Professor Foerster's suggestion is that while an organism feeds on negative entropy or order, it also consumes noise in the bargain (von Foerster, 1960). The entropy decreases and there is more order in the system as the self-organizing system organizes. Then the entropy in the ambient that supports this system, that is, in the universe, will increase according to the Second Law of Thermodynamics. As seen in Chapter 1, this law says that the total amount of disorder or entropy in the universe can increase but it cannot decrease.

POWER DENSITY AND EXPOSURE

Practically all organisms are exposed to electromagnetic radiation. With humans, an analysis is urgent as to the biological hazard associated with the exposure. Either intentionally by experimental design or by accident, animals will be exposed to some known or unknown energy input and the results may often be observed. The organism is said to be "illuminated" by the particular frequency using a term either borrowed from the visible spectrum or taken from presentations such as the radarscope where the profile of objects is actually delineated by microwave echoes passed to a sensitive screen. Sometimes the term "immersion in the field" is used or simply "exposure" to the energy. In no case does there seem to be a need for the word *dose* except when comparing ionizing radiations and nonionizing radiations.

First we must specify the amount of exposure at the particular frequency or band of interest. The power density, usually in units of watts per square centimeter, tells a lot about the conditions. The maximum permissible density exposure (MPD) of 10 mW per square centimeter happens to come from calculating the heat input that a human can absorb more or less harmlessly or for which he can thermoregulate. Although it is an empirical figure, it has shown remarkable persistence. For one aspect, the absorption coefficient used for the impinging energy is unity which, while not strictly realistic, seems to have served the need via some sort of averaging effect.

ABSORBED ENERGY VERSUS EXPOSURE — PHILOSOPHY

An absorbed exposure standard has virtues that appeal to many radiation biologists. Much of this appeal comes from recognizing the improvements that occurred in studying the biological effects of ionizing radiations when the rad was introduced. This is the radiation-absorbed-dose or 100 ergs per gram of tissue. Against the absorbed dose concept is the fact that depth of penetration into the material is not considered, only energy per unit of mass. The rad tends to relate effects that are not always related. Any special effects such as dimensional, coupling, resonance or high amplitude (modulation) effects, tend to complicate an absorbed exposure standard that might otherwise be proposed for nonionizing radiations. Watt-seconds per gram or joules per unit weight would be typical of unit possibilities. To the extent that an absorbed energy unit is equally implicated in various effects, it is valuable. Even though inconsistent, i.e. the unit does not have equal relative effectiveness in various experiments, it would seem that speaking of absorption is better than speaking of mere exposure. Possibly this is true in low intensity exposure studies, and the validity of an absorbed exposure unit would be more marked. Also, for frequencies where all of the energy is absorbed, or where the coefficient of absorption is unity, it is attractive to simply take the power integrated over the exposure time, that is, the energy, and divide this by the weight of the exposed material. Another place where an absorbed exposure unit is advantageous is in the events related to electromagnetic energy leaks from a device or inadvertent beaming of radiation at personnel. In the evaluation of the related hazard it would be good to relate the amount of energy escaping or beamed to the amount absorbed. Since polar substances react to absorbed quanta or to the actual energy density in the tissue, the hazard from dielectric polarization absorption could only be related to the energy density. It may not be obvious, but deliberate rejection of a unit of absorption may be better since it forces attention toward the total picture of both the energy and the component coefficients in the interaction scheme, even if it is awkward to be left uncertain about a "ready" absorption figure. Thus the electrical matters will be stressed for special field effects, the molecular or material factor in the molecular level of interaction, and the behavioral for that level etc. Perhaps only in the health area where interested users could be less expected to bring a multidisciplinary approach to bear would the unit of absorbed exposure be helpful. Yet interestingly enough, this is precisely where the most recent acceptance has been gained for the power density unit. It is, in fact, almost world wide in acceptance. Transformation into energy density from watts per unit of area is very direct. This is illustrated in calorimetry, which is of great value as in absorption tests, and uses energy absorption per unit volume or energy density in standard material such as water.

Perhaps the greatest reluctance to use an absorption unit for general practice arises from the problem of perturbation caused by insertion of the measuring device and assumptions concerning the uniformity of the field. Even for a power density measurement, a probe is often used to search a space for the maximum reading on the meter or recorder. Then this is compared with the allowable power density to evaluate the hazard to personnel of operating in the space.

EXPOSURE AND ABSORPTION MODELS

It is essential to produce the absorbed exposure where absorption itself or calorimetry *per se* is studied, but the first step is specification of the power density or illumination. Then the presence of the preparation is specified and the uniformity of the illumination is accounted for or observed. Good standardization usually demands observations in the undisturbed field and then observations for the changes under exposure conditions. When these perturbations can be determined, it can be quite possible to duplicate the conditions at least up to the point where uniformity of illumination on, or effect in, the material is the problem. Some experimenters avoid the natural model as being too dangerous or individualistic. This tends to explain the popularity of standard simulating models or phantoms in the reports.

Calorimetry carries much of the measurement burden for determining the thermal characteristics of electromagnetic systems. Factors such as the amount of energy generated by the source are determined in this manner. If a container is chosen with little or no energy receptivity, then the contents will absorb the energy exclusively. The rise in temperature, exposure time, and the specific heat are the only quantities needed for a fixed amount of absorber material. Then this energy may be expressed in watts of output or other units. A very convenient material in the microwave region is water which may often be distilled, 1% saline, or Ringer's solution, all of which have nearly equivalent absorption under conditions commonly found and within the usual errors in temperature measurement. An obvious refinement is to employ a tubulated system with circulating fluid. Then the configuration and rate of flow may be adjusted and a power meter used to read the output in watts directly. The flexibility of this system permits the sensor to be installed in small waveguides or in large experimental structures to provide additional information as to the distribution of power.

SPECIAL MICROWAVE EXCITATION AND BOND BREAKING

A brief scanning of the power densities and the responses suggests the scope of the electromagnetic events. It is partly intuitive although supported by evidence that in reaction kinetics, a pulse of dense radio frequency or tightly beamed microwave energy can trigger a special excitation and/or a chemical change. The step change or burst of energy input (super rise time) is followed to observe the time constants of the response in the manner of nonequilibrium thermodynamics rather than following the system via the mechanics of collisions. Then, at extremely high power concentrations, a microwave resembles ionizing radiations and is capable of stripping off electrons from molecules (Illinger, 1969) (Fig. 14-4).

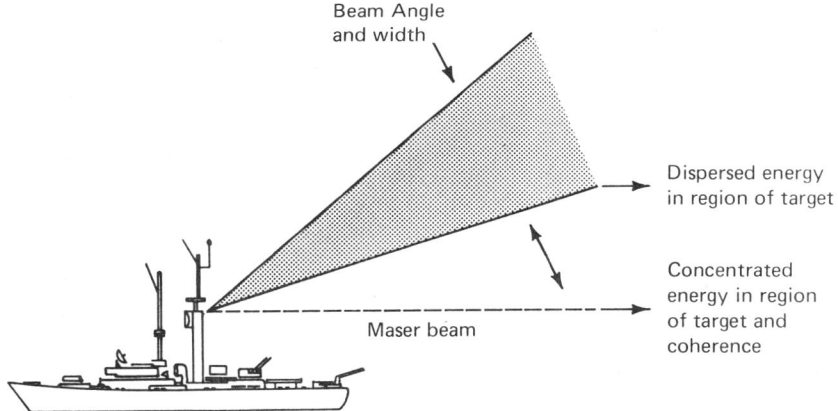

Fig. 14-4. Maser-type input forms a new dimension for bioeffects of microwaves. Concentrated "spot" of microwaves tends toward ionizing-type of reaction even if quanta are inadequate. Area $(cm^2) = 7.854 \times 10^{-9} (\mu m \text{ diam.})^2$.

At lesser power densities, the quanta available are absorbed, thus rendering a molecule more reactive. In the sense that reactions of these activated molecules occur as in exchange or other reactions which involve the breaking of chemical bonds, the quanta even from subionizing radiations may be regarded as capable of breaking bonds.

One dramatic experiment uses a dense microwave field obtained with 5 kilowatts of input into a suitable cavity such as one of about one cubic meter (2.45 gigacycles) to vaporize a solid iron wire. At one fifth this level or even less, photographic flash bulbs with metal "meshworks" are ignited and steel wool is caused to "burn." Still less power is needed to cause glow discharge and neon bulb lighting. At very low power, the other extreme, there is thermal equilibrium during application of power and molecular rotations or similar responses are promoted. Most evidence from theory and experiments with animals and people place the thermal rise density or thermogenic threshold just above maximum power density or between 10 and 15 mW/cm^2. Then the microwatt region below 10 mW is nonthermogenic.

CONTINUUM OF BIOLOGICAL EFFECTS

The microwatt cannot be dismissed as nonhazardous. It is established prominently in the U.S.S.R. and Czech maximum allowable power density. Within this region, and down to the molecular response level, the analysis focuses upon possible athermic effects or effects that are both thermic and athermic in origin. It is valuable to recognize this continuum of effects down the power density gradient rather than to see the responses as cut off at some arbitrary level. It may be tempting to make this cutoff because experiments at the lower density levels seem more difficult and the gradient, in an experimental sense, crosses a number of disciplinary lines. Dismissing the low energy studies as a Russian hoax does not seem reasonable in view of the scope and organization of their work, but interpretation should certainly be made with care.

LOW AND HIGH POWER DENSITIES

It requires about 600 mW/cm^2 to give the sensation of pain in the skin but hearing sensitivity is affected at 1 mW/cm^2. Pulses lasting 50 μsec at 40 μJ/cm^2 may introduce the pseudoaudio sensation. In Czechoslovakia, Marha's threshold (1971) for bioeffects is 0.1 mW/cm^2 and, by applying a safety factor of 10 to this in the conservative direction, the U.S.S.R. and Czech maximum power density of 10 μW/cm^2 is obtained.

Usually in biological hazards work, a field described as of high relative intensity means one above 50 mW/cm^2, while "low" means 1 mW/cm^2 and below. In the U.S.S.R., hundredths to tenths of mW/cm^2 would describe a weak field, tenths of mW/cm^2, a moderate one, and 3 to 4 mW/cm^2 periodically is an intense field. Thus, the U.S. embassy events involved weak to moderate fields. At frequencies lower than microwave, the convention shifts from power density to voltage gradient to describe the field intensity.

THE ELECTRIC AND MAGNETIC COMPONENTS

At the same time, the electromagnetic field may be split into its component forces and each one studied separately. The voltage gradient equivalent of the pure electric field component at a power density of

10 mW/cm² is 195 V/m. In this case it is a matter of describing the field strength in terms of voltage per unit length, for example, V/m at some distance from a source and this field may persist for some period to give an exposure. This description is applied to such matters as interference measurement between radiation oscillators and receivers, field gradients between charged bodies, and to biologically effective fields at frequencies lower than microwaves.

With the composition of the acting field as a point of interest, the magnetic component may be expressed in amperes per meter, A/m. Experiments applicable to animals grazing in a field and exposed to the magnetic field, show that 1 to 10 A/m may be tolerated for up to six minutes. For the electric field, the tolerance is 0.4 to 4 kV/m for this period of time in the 6 to 14 Mc. high frequency band (HF = 3 to 30 Mc.) It has been observed that rats will tolerate such a field, that is, up to 8 kV/m for between 5 and 15 minutes.

RUSSIAN WORK AND THRESHOLD LEVELS GIVE INTERPRETATION PROBLEMS

At least in the U.S.S.R. and Czechoslovakia, industrial health rules specify a maximum exposure to 10 V/m fields. This is based upon observations that, at 100 V/m or higher, biological effects occur and a safety factor of 10 puts the maximum level at 10 V/m. These eastern countries also use a time factor in prescribing safety levels which will be discussed more fully along with later observations on safety levels.

Unless there is a strong frequency dependent factor at the low and medium frequencies, the 10 V/m safety level is rather difficult to understand. The natural voltage around the earth (earth's electrical field) is about 100 V/m so that the Czech and Russian scientists are

TABLE 14-2. Radio and Microwave Spectral Bands

	Frequency	*Frequency Limits (per second)*
ELF (or VLF)	Extremely Low Frequency	10 to 30 kilocycles kc
LF	Low Frequency	30 to 300 kilocycles kc
MF	Medium Frequency	300 to 3000 kilocycles kc
HF	High Frequency	3 to 30 megacycles Mc
VHF	Very High Frequency	30 to 300 megacycles Mc
UHF	Ultra High Frequency	300 to 3000 megacycles Mc
SHF	Super High Frequency	3 to 30 gigacycles Gc

Below 300 Mc may be called "radio frequencies"
300 Mc and above are called "microwave"
Cycle/second = Hz = Hertz

very close to natural levels. There are obviously some measurement problems involved. However, enough atmospheric electricity exists to power tiny motors and to ring bells.

Extra control over work space is applied by the U. S. Army and Air Force Departments when the power density increases into the 10 to 100 mW/cm² zone as of 1965. Above this higher density, no occupancy is allowed. Corrective measures for reducing the power density in an area may be directed in terms of decibels (db). This is the logarithm to the base 10 of the ratio of the two power intensities before and after correction, and thus a logarithmic change in quantity.

$$db = 10 \log_{10} (Power_1/Power_2) \qquad (325)$$

Suitable tables may be consulted for decibels for any power ratio of interest.

Some spectral bands in which nonionizing biological effects occur are shown in Table 14-2. The band is a convenient set of definite frequency limits, starting above the lower limit and including the upper one. Biological responses, of course, continue beyond those seven Federal Communication Commission bands into millimeter waves at 300 gigacycles and further.

An interesting point arises when taking up safety in the electromagnetic environment. Solar radiation is a rich source of these frequencies which have been under discussion. Yet the combined illumination from solar, mixed, and electromagnetic radiations at the earth's surface is about six times as much as the maximum power density in use in the U.S.A. for purely microwaves which, in turn, is 1000 times as much as the Russian maximum power density. The latter is the more conservative. Considerable refinement occurs when time factors for duty periods at these densities are applied. Nevertheless, a *natural* earth illumination of 60 mW/cm² (up to 90 mW/cm²) is enough to explain the respect that is taught for the sun's effects even if we have no formal regulatory scheme for this natural electromagnetic environment. Thus, the fact that a high natural electromagnetic environment exists does not mean that the radiation itself is absorbed. Many forms of protection exist and are naturally utilized. As for solar radiation, it is extremely variable and discontinuous due to rotation of the earth. Clothing alone is enough to screen the infrared portion of the mixed radiation, reflecting if it is light colored, and absorbing it if dark.

In the U.S.A. there will be a tendency for cities and states to set their own population exposure limits according to the best available bioelectromagnetic data. The proposed limit for the New York metropolitan area is 50 μW/cm² which is $1/200$ of the legendary 10 mW/cm² but so far, no information has come through as to whether the city, and other areas which are sure to follow this trend, will include modulation and carrier frequency guidance so as to refine the limits.

One cannot escape the observation that the W/cm² flux given as a maximum permissible level depends, as in Table 14-3, on whole body exposure in terms of energy and time because in an organism, the

TABLE 14-3. The International View of Biohazards from Microwaves and Probable Permissible Levels

United States	Exposure should be to less than 10 mW/cm^2. Theoretically, any input (*not subject to thermoregulation*) or losses, could accumulate to the tissue denaturation level.
U.S.S.R.	Population should be exposed to no more than 10 μW/cm^2
Metropolitan Area	Proposed 50μW/cm^2 —NYC
Spectroscopic levels	Usually below few μW/cm^2
Probable maximum permissible levels according to total experience and best judgment to date	Exposure for few minutes only at 10 mW/cm^2 to be permitted. Part of the day may be spent in exposure to few μW/cm^2. Persons with electrical, cardiac, or neurophysiological problems to be excluded. Soviet permissible levels (as they are) probably applicable to microwave ovens.
For comparison: 1) Brain Power	Sensory information processing including memory, association, and emotion about 10—30 watts.
2) Natural Planetary Biosphere Illumination	60—90 mW/cm^2 solar EMR

geometrical cross section has all of the cm^2's connected. A meter reading about 10 mW/cm^2 concerns only that particular area and, to be meaningful for the whole body, that body would have to be reduced to a tiny ball.

For 1.5 to 300 GHz a radiofrequency protection guide of 5 mW/cm^2 deposits over the "average body", about 144 J/kg in 6 minutes. This allows an SAR average in terms of space and time for the whole body to be given as 0.40 W/kg.

PROBLEMS

1. Does noise always have a negative connotation?
2. What are the advantages and disadvantages of the power density or intensity units for EMR?
3. Can microwaves be ionizing?

4. How can the difference in maximum permissible exposure in the U.S. and U.S.S.R. be reconciled?
5. What is the most significant interacting feature of microwaves and biosystems?
6. What is meant by the unique position of microwaves in the dimensional spectrum?
7. How do the power densities or intensities in the "hazard" range compare with biological power levels?
8. A sound has a power of one erg/cm^2/sec. A biological medium decreases the sound at 10 dB/cm. After passing 2.5 cm, what is the power?

BIBLIOGRAPHY

Aebi, O. and H. Aebi, *The Art of Adventure of Beekeeping*, Unity Press, Santa Cruz, Cal., 1975.

Appleton, B., "Results of Clinical Surveys for Microwave Ocular Effects," *Selected Papers from the Division of Biological Effects Lecture Series*, Bureau of Radiological Health, DHEW Publ. no. (FDA) 73-8031, BRH/DBE 73-3, February 1973.

Appleton, B., "Microwave Cataracts," *J. Amer. Medical Assoc.*, Vol. 229, 4, 407-408, July 1974.

Appleton, B, et al, "Microwave Lens Effects and Humans,"*Archives of Ophthalmology*, Vol. 93, 257-258, April 1975.

Asanova, T.P. and A.I. Ravok, *The State of Health of Persons Working in the Electric Field of Outdoor 400 and 500 KV Switchyards*, translated by G. Guy Knickerbocker, Electrical Safety and Life Sciences Subcommittee, Power Engineering Society, Inst. of Electrical and Electronics Engineers, Inc.

Barniska, S. and Czerski, P., *Biological Effects of Microwaves*, Dowden, Hutchinson & Ross, Stroudsburg, Pa., 125-136, 168-169, 1976.

Baranski, H. et al, "Influence of Microwaves on Genetic Processes of *Aspergillus nidulans*," *J. Microwave Power*, Vol. 11, 2, 146-147, 1976.

Bureau of Radiological Health, BRH/OBD 70-3, *1969 Annual Report to the Congress on the Administration of the Radiation Control for Health and Safety Act of 1968*, April 1970.

Catravas, G.N., et al, "Biochemical Changes in the Brain of Rats Exposed to Microwaves of Low Power Density," *J. Microwave Power*, Vol. 11, 2, 147-148, 1976.

Cleary, S.F., Ed., *Proc. of the Symposium on the Biological Effects and Health Implications of Microwave Radiation*, Richmond, Va., September 1969, U.S. Department of HEW, BRH/DBE 70-2, June 1970.

Czerski, P., M.L. Shore, Eds., "Biologic Effects and Health Hazards of Microwave Radiation," *Proc. of Inter. Symp. at Warsaw*, Poland October 1973, Polish Medical Publishers, Warsaw, 1974.

Frey, A.H., "Human System Response to Radio Frequency Energy," *Aerospace Medicine*, Vol. 32, 1140-42, December 1961.

Frey, A.H., "Human Auditory System Response to Modulated Electromagnetic Energy," *J. Applied Physiology*, Vol. 17, 4, 689-92, July 1962.

Frey, A.H., "Biological Function is Influenced by Low Power Modulated RF Energy," *IEEE Trans. Microwave Theory Techn.*, MTT-19, 2, 153-164, 1971.

Frey, A.H., S.R. Feld, and B. Frey, "Neural Function and Behavior: Defining the Relationship," *Annals of the New York Academy of Sciences*, Vol. 247, 433-439, 1975.

Frey, A.H., "Effects of Microwave and RF Energy on the Central Nervous System," *Biological Effects and Health Implications of Microwave Radiation Symposium Proceedings*, BRH/DBE/70-2, U.S. Dept. HEW, September 1969.

Galloway, W.D., "Microwave Dose-Response Relationships on Two Behavioral Tasks," *Annals of the New York Academy of Sciences*, Vol. 247, 410-416, 1975.

Gandhi, O.P., "Polarization and Frequency Effects on Whole Animal Absorption of RF Energy," *Proc. IEEE*, Vol. 62, 1166-1168, 1974.

Grundy, R.D., S.S. Epstein, Eds., *Consumers Health and Product Hazards, Vol. I*, MIT Press, Cambridge, 173-257, 1974.

Hirsch, F.G. and B.A. Bruner, *Proc. of the Technical Coordination Conf. on EMP Biological Effects*, sponsored by the Lovelace Foundation for Medical Education and Research, Albuquerque, N.M., July 1970.

Hunt, E.L., N.W. King adn R.D. Phillips, "Behavioral Effects of Pulsed Microwave Radiation," *Annals of the New York Academy of Sciences*, Vol. 247, 440-453, 1975.

Illinger, K.H., "Molecular Mechanisms for Microwave Absorption in Biological Systems," *Biological Effects and Health Implications of Microwave Radiation Symposium Proceedings*, BRH/DBE/70-2, U.S. Dept. HEW, 112-258, 1969.

Johnson, C.C. and A.W. Guy, "Nonionizing Electromagnetic Wave Effects in Biological Materials and Systems," *Proc. IEEE*, Vol. 60, 6, 692-718, June 1972.

Justesen, D.R., "Microwaves and Behavior," *American Psychologist*, Vol. 30, no. 3, March 1975.

Kalant, H. "Microwave Radiation Hazards," *Canadian Medical Association Journal*, Vol. 81, October 1959, 575-582.

King, N.W., D.R. Justesen, and R.L. Clarke, "Behavioral Sensitivity to Microwave Irradiation," *Science*, Vol. 172, April 1971.

Marha, K., J. Musil, and H. Tuha, *Electromagnetic Fields and Life Environment*, San Francisco Press, Inc., San Francisco, Calif. 1971.

McLaughlin, J.T., "Tissue Destruction and Death from Microwave Radiation (Radar)," *California Medicine*, Vol. 86, 5, May 1957.

McLaughlin, J.T., "Health Hazards from Microwave Radiation," *Western Medicine*, Vol. 3, April 1962.

McRee, D. T., "Soviet and Eastern European Research on Biological Effects of Microwave Radiation," *IEEE 68*, 1 January, 1980.

Pattishall, E.G., Ed., *Proc. of Tri-Service Conf. on Biological Hazards of Microwave Radiation*, George Washington University, July, 1957.

Pattishall, E.G. and F.W. Banghard, Eds., *Proc. of the Second Tri-Service Conf. on Biological Effect of Microwave Energy*, University of Virginia, July 1958.

Peyton, M.F., Ed., *Proc. of the Fourth Annual Tri-Service Conference on the Biological Effects of Microwave Radiation*, Aug. 1960, Plenum Press, 1961.

Prausnitz, S., and C. Susskind, "Effects of Chronic Microwave Irradiation on Mice," *IRE Transactions on Bio-Medical Electronics*, Vol. BME-9, no. 2, April 1962.

Püshner, H., *Heating with Microwaves*, Springer Verlag, New York, 1966.

Quarton, G.C., T. Melnechuk and F.P. Schmidt, Eds., *The Neurosciences*, Rockefeller U. Press, 75-79, (L. Onsager), New York, 1967.

Reno, V.R., "Microwave Reflection, Diffraction and Transmission Studies of Man," *Naval Aerospace Med. Res. Lab. Tech. Report 1199*, Pensacola, Fla., 1974.

Roberti, B., G.H. Heebels, J.C.M. Hendricx, A.H.A.M. de Greef, and O.L. Wolthuis, "Preliminary Investigation of the Effects of Low-Level Microwave Radiation on Spontaneous Motor Activity in Rats," *Annals of the New York Academy of Sciences*, Vol. 247, 417-424, 1975.

Rose, V.E., et al, "Evaluation and Control of Exposures in Repairing Microwave Ovens," *American Industrial Hygiene Association Journal*, Vol. 30, 137-142, March-April 1969.

Rosenthal, D.S. and S.C. Beering, "Hypergonadism after Microwave Radiation," *Journal American Medical Association*, Vol. 205, 4, 105-8, July 1968.

Schwan, H.P., "Radiation Biology," *Microwave Power Engineering*, E. Okress, Ed., Academic Press, New York, 1968.

Servantie, B., et al, "Comparative Study of the Action of Three Types of Microwave Fields Upon the Behavior of the White Rat," *J. Microwave Power*, Vol. 11, 2, 145-146, 1976.

Sigler, A.T. et al, "Radiation Exposure in Parents of Children with Mongolism (Down's Syndrome)," *Bulletin of the John Hopkins Hospital*, Vol. 117, 374-99, 1965.

Susskind, C., Ed., *Proc. of the Third Annual Tri-Service Conference on Biological Effect of Microwave Radiating Equipment*, Aug. 1959, Univ. of California.

Thomas, J.R., et al, "Effects of Low Level Microwave Radiation on Behavioral Baselines," *Annals of the New York Academy of Sciences,* Vol. 247, 425-432, 1975.

Tyler, P.E., Ed., "Biologic Effects of Nonionizing Radiation," Conf. held by the New York Academy of Sciences, New York City, February 1974, *Annals of the New York Academy of Sciences,* Vol. 247, 1975.

U.S. Government Printing Office, *Hearings before the Committee on Commerce,* United States Senate, Ninetieth Congress, May 6,8,9,13 and 15, Washington, D.C., 1968. Serial No. 90-49.

U.S. Government Printing Office, *Radiation Control for Health and Safety,* Hearings before the Committee on Commerce, United States Senate, Ninety-third Congress, March 8,9,12 1973. Washington, D.C., Serial No. 93-24.

vonFoerster, H., "Self-Organizing Systems," *Proceedings of an Interdisciplinary Conf.,* M.C. Yovits and S. Cameron, Eds., May 5 and 6, Pergamon Press, New York, 1959.

Wolf, A.A. and E.H. Halpern, "Diamagnetic Levitation in the Fractionally Superconducting Bile Cholates," *Chem. and Physics,* Vol. 8, 34-36, 1976.

Zaret, M. "Electronic Smog as a Potentiating Factor in Cardiovascular Disease," *Medical Research Engineering,* Vol. 12, 3.

Zaret, M. et al, *Occurrence of Lenticular Imperfections in the Eyes of Microwave Workers and Their Association with Environmental Factors,* Progress Report, Rome Air Development Center, TN-61-226, 1961.

Zaret, M., "An Experimental Study of the Cataractogenic Effects of Microwave Radiation," *RADC Tech. Documentary Rpt.* N. 64-273, October 1964.

Zaret, M., "A Study of Lenticular Imperfections in the Eyes of a Sample of Microwave Workers and a Control Population," *Final Report, RADC-TDR-63-125,* 1963.

15

Intersystem Theory — Direct Electromagnetic Effects

"The event that exists irrespective of our perception is an intersystem interaction between radiation and materials, the result of which is a sensory note if someone happens to be focusing on that interaction. Thus the appearances we experience do not have to be the very objects themselves as Bishop Berkeley contended because those objects have materialized by the interaction."

D.A.C.

INFORMING FORCES FOR COUPLING AND UNCOUPLING

The theory that emerges from our examination of informational pathways as related to the electromagnetic environment is of great value in that it depicts the microwave, for example, as both connecting systems and as informing a receptive system that it must change in some logical manner. We see that in nonionizing radiation it is apparant from the nature of the electromagnetics of the wave and of its receptive systems or destinations that both the wave and the receptor feature symbolic logic in the sense of communicated information. This approach motivates an investigator to look for signs of direct stimulation or inhibition of a biosystem that has an associated logic such as systems with phase or cyclic characteristics and changes, e.g., membranes in the brain controlling the blood barrier and permeability. Similarly the search involves seeking out communicative-type cells which have strong intrinsic bioelectric fields wherein direct information transfer may be feasible, depending on the exposure characteristics, especially a matching of important features or coupling.

If the theory is sound, it is to be expected that EMR can inform the system and not merely expose it. It would be very tempting indeed then to visualize the information as being used in inhibiting unwanted biological processes, for example, in microwave cancer therapy, or in stimulating normal or desired processes such as healing. Further use of the information theme would be in cell identification and diagnosis, interrogating distant samples, or the exertion of influence on development, replication, or in metabolism and the biochemistry of nutrition.

In regulation, inhibition is seen to be more than enzymatic. If an EM forcing function is attenuated, then the substrates already present block *motions* that would normally activate enzymes leading to the production of those substrates. This is another manifestation of the presence of a metabolite inhibiting the enzymes used to synthesize it, but now EMR and cellular streaming are involved. The connection is geometrical and has to be developed in terms of space-time constants along with the nature of the forcing function, so that some step leading to the enzyme is not forced. Essentially though, the loss of an intersystem signal means that an influence is felt on metabolic events. The details concern how this might come about.

Chloroplasts and some other organelles that have radiation sensitive membranes have small, protruding, mushroom-like coupling factors which undergo conformational change. These are essential for photophosphorylation but not for electron transport. ATPases also couple the Na^+ and K^+ fluxes across membranes. Still other coupling factors involve mitochondria, and bacteria enzymatically couple phosphorylation. These relate to intersystem theory when, for example, the EMR impinging on the photosensitive membranes are informing and communicative in that they force a coupling of ATP reactions that lead to phosphorylation. As enzymatic, the coupling is very information-specific and as conformational, the changes can be in the membrane-coupling factors. Illumination can be the *go* or *no go* agent in such systems which can normally be *go* if used regularly and *no go* for irregular action. It is more efficient to apply the inhibitor only rarely for the first case, and to activate the second on demand via informing or messenger deinhibitors.

We are on the threshold of understanding these matters and already are experimenting with the ordering forces and observing the effects. Considerable sophistication, many times more than in ordinary interdisciplinary effort, is required to show intersystem direct information transfer and the quality is easily that of the best Nobel Prize work. The intersystem communication mode has been detailed against contemporary studies in the first four chapters but it now will be examined as a direct effect of electromagnetic radiation.

THE DIRECT MODE

Possibly some idea of the demands made on specialists in this area has been given by reference to the need for new laws for each spectrum and the strain placed on basic biology of sensory systems but, in addition, there are new requirements in instrumentation so that when the effort is brought together, it can appear sizeable. Success comes about when biological responses are demonstrated and an explanatory model is fabricated. Some of the considerations are:

1. Exposure conditions—characterization of the exposure-frequency. Power and input of energy and precise control of all these conditions. Frequency series and multiple frequency responses.
2. Type of cell, the temperature in relation to metabolism.
3. The growth stage and mechanism for division and development.
4. Previous nutrition, energy needs and type of supporting metabolism. Adequacy of nutrients, sources of carbon and nitrogen.
5. Periodicity in time and space, process constants, phases of activity with relation to exposure to EMR.
6. Cellular composition, especially informational molecules DNA and RNA, and their orders under which the biosystems are operating.
7. Morphology, pathology, and transformations.
8. Water status in relation to dielectric processes, bound, partially bound and free water.
9. Inhibition and stimulation of receptive systems.
10. Conformation in terms of morphology at multiple levels of organization.
11. Communication channels for molecular systems for the intersystem informational network needed to guide metabolic events.
12. Vibrational, electronic, and rotational states of excited molecules generate coherent intersystem EMR for communication.
13. Water systems function as receivers on macromolecules and the reception is simultaneously a protective device, both functions being lost in dehydration.
14. Quantum mechanical analysis is applicable to the study of the excited states.
15. Raman-shift frequency lines inform of specific metabolic activity according to the status of the cell as to log, lag, or germinating phase and water activity.
16. Synchronous growth is a special status that is important in the interaction of space and time constants governing processes and revealed by communicating EMR in the intersystem as well as spectroscopic probes.
17. Input or output, frequency-dependent communications constitute spectral signatures accompanying bioevents and are specific for the genus of cell.
18. A modulation is discernible in the communication based on constants that separate meaningful, interacting frequencies by a series of characteristic integers. Integral multiples of 7 GHz and 5 GHz for two sets of such frequencies suggested for example, quantum energy transitions with different J values for one rotational process and rotational constants B for each such action could be expressed as about 2.5 and 3.5 GHz.

19. Theoretically, coherent EMR from vibratory and rotational generators from 10^{11} to 10^{12} Hz are capable of forcing functions in intermolecular system networks.

DIRECT COMMUNICATION, A FUNCTION OF PENETRATION

Several scientists have shown that the *E. coli* system, which is so well-characterized in terms of many of the necessary operations, obviously enjoys enhanced cell growth under microwave influence (Blackman 1975, Webb 1968). Whether the studies of Webb and his coworkers, Dobbs, Booth and others are inspired or suspect due to difficulties in their duplication by "lesser" teams is now quite clear. It is clear that Webb is a bold innovator who is very able in the advanced game of orienting and reorienting the physical and biological elements in new and stimulating ways. When he exposed samples in a spectroscopic manner he was immediately on the right track but this restraint also meant that he was studying thin layer effects only, with no suggestion of the usual penetration common to microwaves. Thus Blackman, attempting to include penetration as a characteristic, could not duplicate Webb's results. Interestingly, he saw 28.5 dB/mm at 140 GHz and 24.2 dB/mm at 70 GHz decrease in power in samples 1.5 mm deep. To avoid this apparent lack of control, it is necessary to lower the frequency until sufficient penetration is always assured to make a constant, uniform exposure. Yet this inherently voids this frequency-dependent response. Thin layer techniques circumvent this difficulty, and avoid the frequency or penetration effect, but the technology of thin layer growth is crucial. This is the place where matters of safety part from basic science. Penetration is a vital matter of safety in exposure to electromagnetic radiation. Thus, since ordinary clothes shield a body from infrared and ultraviolet, those electromagnetic radiations are much safer. However, in thin layers of exposed cell preparations, any part can be reached by the signal information and the matter of penetration is ruled out as a variable just as it is *in vivo* for appropriate distances.

Assuming that rotational transitions have a place in the interaction mode, it becomes a question of which ones, for example, J-0 or J-1 may not be populated at manageable frequencies. At the easier, lower frequencies in microwave, the transitions would be 2-3 or 3-4. Similarly, the transitions would be related to resonant, narrow bands while the line widths observed are a few GHz wide in this range.

Waves of 10^{11}, 10^{12} Hz are excited in metabolism (Fröhlich, 1968) and bacteria show these emissions at 3×10^{12} GHz by Raman effect spectroscopic analysis. Thus, electromagnetic radiations are consequences of tissue bioactivity and have the usual quantum sources, rules, and the associated informational content which must be presumed to be associated with communication, regulation and control between cells, tissues, and organs. Evidence of enhancement, stabilization or interruption of growth rate by *exposure* to such frequencies would be presumptive clues to the actual existence of such information management. Thus the report by Webb and Booth (1969) that such evidence had been obtained with *E. coli* was most significant. Then, J. Berteaud, M. Dardalhon, N. Rebeyrotte, and D. Averbeck (1975) specifically confirmed the response of growth depression in *E. coli* at 71 and 73 GHz as very frequency-selective or dependent. When the external EMR is superimposed on an existing quantum source of the same frequency in the tissue, the result is resonance which means an easing-in of the energy and its associated message. Russian confirmation of the responses using coherent 4-6 $\times$ 10^{10} Hz (Devyatkov 1974) was interpreted as meager in experimental detail by Grundler, Keilmann and Fröhlich (1977) who set about to duplicate the work with precise methodology. At 1.1 to 2.7 mW/cm^2 and heating of only 0.4°C in a volume of 2.5 ml between 41.83 and 41.96 GHz their resonances were observable and as narrow as 0.01 GHz bandwidth.

Microwave generation *from* tissues, when made into a record, are clearly like EEG and ECG in significance, giving thermographs that diagnose subcutaneous inflammation or other information as used by Edrich and Smyth in 1977. The effective distance for the signals is greater than with infrared but it is the proliferation of these emissions through the EM spectrum that gives the concept of organisms as wide spectral sources. Then their absorption or scattering of incoming radiant information completes the loop of intersystem interaction. In diagnosis or therapy terms, it is the capability for *in vivo* impedance, reflectance, absorption and interference methods that is of interest along with the use of multiple frequencies in various modulations (Magin and Burns, 1971).

Similarly the proliferation of nuclear particles adds the informing properties of particles such as the neutrino to spectral capabilities. This particle with zero or negligible mass and charge, but with finite momentum and energy, accompanies the emission of betas and adjustment of the nuclear energy from one energy level to another, so that in disintegration, their combined energies equal the disintegration energy. The two kinds of neutrinos have discrete forms of antiparticles. An electrically neutral, small body would carry energy through matter and over vast distances. Since information flows with the radiant energy, neutrino messages can provide vital communication for compatible receivers through planetary material or across interplanetary distances and across eons of time to bridge both time and distance.

RADIANT AND DNA INFORMATION INTERACT

The intersystem direct effects of ultraviolet, infrared, and visible radiations are very numerous, frequency specific, and often associated with spectrographic grade preparations or thin layers as opposed to large organisms. In addition to the frequency for photosynthesis, phototaxis, photoinactivation of microorganisms and photoreactivation, there are bands for absorption for practically every chemical known. As an ultraviolet direct effect at 2537 Å, mutations depend upon the nutritional status of the organism, as well as upon its ability to alter the status. If a certain metabolite is present in the culture medium, then the information controlling the synthesis of that same metabolite is changed by the ultraviolet. DNA contains the pertinent information, so more of it is needed when the complexity shifts from substrate to organism (as succinate is more complex than glucose and amino acids more advanced than NH_4). The greater the amount of DNA, the more likely the organism's protection against mutation or lethality at 2537 Å. In this frequency region, proteins and nucleic acids receive electromagnetic energy easily and directly but above 3200 Å, the oxidative respiration machinery (e.g., flavenoids and quinones) function as receptors. Opportunities for directing the mutation are then presented from 3200 to 4000 Å, for example, by semidessication of the cultures, controlling the period allowed for incubation, or management of other growth factors. Then the absorption shifts to flavenoids and quinones from protein and nucleic acids may be observed.

If Webb's assumptions are taken, then the *order* in which the genes mutate is the same as the one observed in conjugation (*E. coli*) for transfer. The slowing, increase, or stopping of DNA operations and protein synthesis in dividing cells can then be related to EM polarization of water molecules bound on or near macromolecules. Once again these molecules carry their EM receptor units with them as they move.

TIME AND SPACE ACTION SPECTRA

The selection of a guide to the growth response to EMR in cells is important. Labeled thymine (^{14}C) or protein synthesis by monitoring the elaboration of key enzymes such as β-galactosidase or intermediates are typical indicators but the selection must accommodate time and space constants for processes in cell biochemistry. The ultraviolet mutations depend on the source of carbon and the frequency for influencing protein and DNA synthesis with microwave is related to the source used for carbon and nitrogen.

The Chronology

Most processes are time-dependent and cellular events are clocklike in moving toward division or the repair of damage. The EM response is frequently keyed to such periodicity and out of phase exposures may be deceptively negative for the biological response. In particular, any subprocesses such as genetic transfers in conjugation are periodic with scheduled stockpiling of materials, and elaboration of the apparatus; then transfer and time constants characterize them. The pattern of time intervals between gene transfers varies with the type of gene; being short, 0 to 7 minutes, medium, 16 to 20 minutes, or long, 32 to 40 minutes. Intervals for transfer for genes governing serene, thymine, and nicotinic acid are 32, 34 and 30 minutes, respectively. The PABA gene is passed after 35 minute separations. Tryptophane S, the regulatory gene, is separated from ordinary tryptophane by 39 minutes. Thus the genetic map is a realization of space constants. Instead of a map, it is a *time* program for passing genes into recipient cells during conjugation. The time constants, in turn, establish direction changes. Looking now at the DNA as a dynamic, folding, macromolecule, the direction changes or steering information are space events where, as in handling a rope, a turn is made at time t establishing a coil of time constant t characteristics. However, the fold or turn is chemical and spatial and due to enzyme-substrate-DNA contacts with the coenzyme and the respiratory chain as energy source. The evidence for such periodicity is in the repetitions in time of nutritional mutants under ultraviolet radiation. When ultraviolet mutations happen, they look like repeated mutations, repeated at specific times. Webb connects these loci with an inducer, as characteristic in genetics for operon and the like, and lets it move to induce (space control) according to a periodicity (time control) and this, reflecting the sensibility to EMR, is called a chromatophore. Then the master control is based on information (specifically conformation) as to the *carbon source*, which tells the DNA how or where to change in mutation and informs the chromatophore where it should start (usually the locus for arginine synthesis).

A Metabolism-Frequency Model

A series of microwave frequencies will inhibit or stimulate, that is, exert *go* or *no go* control over metabolic action where the ability to absorb at a given frequency and the space between frequencies (separation) interact with the nutrition source. Thus, the microwave-direct-effect depends on metabolism as the cell senses its medium. If it is glucose with prepared amino acids, then f_1, a set of multiples of 7 GHz influences the result; if it is glucose with NH_4, the set of pertinent frequencies is spaced by 5 GHz. This constitutes a directing signal definable in frequency or corresponding wavelength, the dimensional constant, and in energy, the power being from 10 to 50 mW/cm^2. The action can depend on J quantum values for the frequencies and rotational constants B associated with these movements when the latter equals 2.5 and 3.5 GHz or one half the frequency separation. If the separation action spectrum is 2B, then the N source controls. This corresponds to an action spectrum for protein, DNA, and RNA synthesis and cell growth. The action spectrum depends on the frequency rotation forcing function *on* a molecule that is susceptible by virtue of its *dimensions* and those of its associated water molecules which change with hydration, and on *where*, for example on N or PO_4 etc. that the H_2O binds (conformation). This depends directly on the type of N used as a source. NH_4, as an example, gives a larger diameter to DNA than it does to amino acids.

The mechanism for metabolic control would then be intracellular motion in division which segregates the synthesized proteins in the daughter cell, and sets up time and distance features on operations such as the reading of the genetic map by a chromatophore. Some features of the mechanism are the retarding of an enzyme action time from its synthesis time, and the components interacting with certain quanta only in a specific microwave frequency series.

In the amino acid and succinate medium, the action frequencies space by 6 and 7 GHz, giving two action spectra distributions. In the NH_4 and succinate one, the separations are by 6 GHz in a positive series which give a *no-go* signal to the metabolism. Separations in the frequency spaces by 5 GHz give the negative series or *go* information which is recognized by the key enzyme β-galactosidase.

Patterns Vary with Abnormal Cells to Provide a Forcing Function

In a simple extension of this sensitive series logic, the signals may be applied comparatively to normal and tumor cells. Then the normal cells demonstrate a different series with action spectra separated by 2 and 2.5 GHz for protein and DNA synthesis while the tumor cells have 1.9 to 2 and 2.3 and 7 GHz or probably three series for this activity. These signals could be lost to the cells, for example by attenuation, so that the information would not be received and the energy not absorbed. When this happened there was a shift away from tumorgenesis by the cell strain and 2.3 GHz as a *directing* frequency is antitumorgenic. Under these circumstances of suppressed tumorgenesis, the variable cells (BHK, Ehrlich ascites, normal mammalian cells at 50-90 GHz) appear as normals and give the 2.5 GHz-series-response.

Processes involving movement of cytoplasm and nuclear material involve space relationships and space constants. EM exposure and mutation studies show sensitive times for certain loci in *lag* phase growth of *E. coli*. These times are spaced by a "constant" t that relates to *motion* of the absorber over the DNA with a time period such that 1 or 2 overpasses occur in the period of the *lag* phase. The absorber sequences itself over the genes within 40 minutes in the life of the cell and makes contact with DNA during duplication of some genes as well as when mRNA is transcribing. In this situation, any EM forcing function such as microwave rotation of water and small ancillary substances constitutes a functional receiver appended to nucleic acids and protein. the medium, e.g. succinate which is a key metabolic intermediate, sets the beginning of a cycle of movement of the chromatophore. At this starting point, the succinate commanding loci are located.

Mechanics and Periodicity with DNA

What is needed next is a periodicity in the information molecules so that their data "turn up" at the right place in the conformational scheme. Spiraling of the coil of DNA creates periodic nearness of related enzymes and their genes even though they are linearly well separated in genetic terms. The operations are then permitted in the sense of space coordinates. Coiling a rope would similarly allow every turn to be marked at the same distance, for example, with a meaningful code. Then all sorts of tricks might be done with the rope. Depending on the stiffness and the "lay" or tension in the turns, it can be made to move in a downward or outward direction through a matrix by pushing one end (by new growth). The motion can be very specific for example, this kind would produce a spiral turn to the whole complex with DNA rotating in it.

The connotation of metabolic bridges then becomes apparent as amphibolic enzymes form such connections between metabolic routes. This motion is also a basis for replication of sequences, because the purpose of the spiral conformation is to make a copy of the parental configuration. This it does by the DNA position. With respiratory chains furnishing energy via $NADH_2$, FAD, and cytochromes, plus coenzyme availability and enzyme substrate positioning, the conformation assumes an operating status to which suitably chosen EM signals may be effectively coupled.

CONFORMATION AND MORPHOLOGY

Suitable structures are essential as bases for these events and cells are uniquely equipped to fabricate these as necessary. Then, in the fashion of an oscillation of a receptive surface, microwaves can impart a motion to the membrane and to associated substrates. Thus displacement of membranes and their attachments are forcing functions which can trigger a reaction by collision of reactants where the influence is exerted over molecular distances.

The small molecules H_2O and NH_4, as well as the carbon source, create a transmission mode with the help of dielectric constants, dimensions, and EMR coupling features. Space conformability is provided by the motions, absorbing molecules and their turns, and the pitch of the spiral relating meeting places on DNA with enzymatic energy, and substrate factors.

This provides a morphology to which tRNA and mRNA can conform, following the spatial directions set up for amino acid production, so that they operate in *response* to, rather than as a cause of, the events occurring in metabolism. The governing regime is geometrical, leading to shapes and planes and growth pathways established by the periodicity and form factors inherent in the spiraling turns. When their active areas at a given moment are all adjacent, then the enzyme action is in a configuration. The spiraling motion is superimposed, which gives a basis for amplification, propagation, and a dynamic status.

The membrane responses allow DNA growth in preferred directions as DNA rotates either *by* the membrane or *with* it, and both of them form and reform according to the dynamic status. EMR impinges on a resonant or at least absorbing structure, e.g. a chromatophore as in the electron transfer system which is in "go" status and has a cell membrane and DNA connection. The EMR receiver can then travel, the motion being contributed by new synthesis and the receiver can then rotate in the manner of planetary gears with the DNA motion. Daughter cells can result from this set of events. Webb saw this pattern in observing that the daughter, the only one from the first division, is fitted out with the new (mutated) metabolic scheme and from time factors (retarded enzyme reactions) that favor a spiral, repeated geometry and UV mutation sequences.

Microwave absorption is a coupling of periodic frequencies with constant time intervals. This coupling displaces or interferes with DNA and protein synthesis. Numerous motions, rolls, twists and lack of them, both position genes and impart the information needed to proceed with metabolism based on the *already* formed metabolite and varying from the other possible configurations in terms of the amino acid products. Forcing functions are of the order of 10 μW to 150 mW as a function of the frequency. In the correct configuration (which obviously implies considerable experimental artistry) these provide an EM signature recognizable within the operating metabolism so that it can respond to the time and space information.

NEURAL DIRECT COMMUNICATION

Cell membranes generate electric fields so that external EM fields should naturally interact with the membrane potential or with the fields generated by the membrane. This phenomenon has been known as long as the stimulus of a neuron by a dc pulse, and experiments have been extended to skin, bladders, healing or repair of tissue and to ac fields. There is a positive promotion of growth at fracture sites in an imposed electric field with directivity observable in the polarization of fields and direction of new growth. A pacemaker is an example of a relatively simple EM communication. It serves as a communicator with cardiac muscle and is designed to substitute for normal electrical communication within the tissue. Pacemakers as communicators may be incompatible with EMR, and to qualify for allowing their wearers unrestricted access, they must not be affected themselves by square waves at 400 MHz of field strength 200 V/m.

A carrier wave, amplitude modulated by EMR which are numerically commensurate with an EEG, gives an EM coupling to the EEG and reinforces the brain waves. Sensing of such information can depend on electrochemical designs in receptive gradients and dimensions that can discriminate against the carrier in favor of the modulation information.

An example of this is an Adey EMR communicating model which uses 147 MHz at 1.0 mW/cm^2 amplitude modulated at 0.5 to 30 Hz.

Then, with antenna-like fibers grouped 0.5 μm wide, at a synapse poised across the synaptic space toward a similar receptive array with about 2.5 μV of synaptic potential between them, conditions exist for EM interaction. Fields must be delicate to be sensitive to ambiental signals which are of the same magnitude so that such weak fields probably need *amplification* to influence membrane transmitter substances. Such a system is very frequency-dependent and requires many bands.

Coupling requires a matching of signal characteristics in receptive networks or the existence of transducer mechanisms which normalize the signal. The average incident power will be important when the time constants make it so. Thus, in the audio effect of microwaves, cw power rather than pulse is effective in some designs tested. All the receptive systems are also sensitive to poor signal-to-noise ratios. The interactions, while often delicate, must exceed background noise in order to give modulated neural operations or have the benefit of filtration. Information is present in the arrival time of the signals at the synapse. This time is subject to amplitude modulation, then *global* modulation for radiation signals in *networks* by velocity modulation, producing a coordinating kind of control among neurons. This is phase information and besides the time of arrival, the frequency of arrival and the spatial distribution of arrival of action potentials at multineuronal locations are involved.

Direct Effects on the Brain

Substances like protein are normally excluded from brain tissue by a membrane barrier that is lowered by EMR. Normally water, glucose and the like are passed while high molecular weight proteins over 5000 are discriminated against, at least in quantity. The EMR required is low, that is, below 10 mW/cm^2 in rats; pulsed waves are more effective, and the effect is greatest in the medulla, cerebellum and hypothalamus. A calcium efflux which is the exodus of Ca from brains after ELF or modulated VLF waves at 6 to 20 Hz is observed in chick brains. In humans, headache is a chronic effect from EMR and the author is inclined to accept this from personal observation. As an information pathway, the membrane permeability can be directly affected. The effect is like a nerve impulse in reverse, which is to say that the wave *causes* the permeability change or influx.

A use of microwaves is to rapidly inactivate the neurotransmitter substances actually present, and in brains these are found to be in much greater concentration than was thought to be the case. With this capability which is like "freezing" a status for examination purposes, EMR can then be applied in exposure tests in specific brain areas (no effect at 19 MHz in mice). High fields, for example 8 kV/m with 40 A/m, show positive changes in catecholamine and acetylcholine neurotransmitters in the brain, while 3 kV/m with 8 A/m is insufficient to show the change. These fields also give a colonic fever of 2°C in mice which is associated with the neurotransmitter changes at 1600 MHz which are thermal and sensitive to the heat distribution.

With microwaves, intracranial heating is at a maximum near the brain's center and falls off toward the periphery. The thalamus is near the site of peak heating which may produce saturation in the hypothalamus and subthalamus which are regularly seen to become vacuolized and chromatolyzed. There is a possible peak effect about 40 to 200 W/kg and 25 to 50 mW/cm^2 at 2450 MHz applied in the chronic pattern for days or for 30 minutes to cat heads which reduces the level of protein synthesis. With these high levels of microwaves or any (equivalent) thermic input, the functioning of the spinal cord is altered. When used to balance cooling, microwaves restore normalcy. Cat spinal cords show an increased reflex response and synaptic transmission even at 10 mW/cm^2.

In electrocardiograms, for example, localized for the occipital or frontal region, a microwave exposure can alter the record taken after exposure. There is the suggestion here and elsewhere that animals sense a shock wave which, among other things, they convert to modulation of their record. Longer microwaves are more effective than short and pulses are more effective than cw. The vacuolization and chromatolysis are thermal neuron effects. In the long term, say 3 to 4 months, exposure produces abnormal metachromatic substances around the blood vessels and on the myelin sheath in the brain and induces a loss of *active* acetylcholinesterase centers. In the 1-30 Hz range for modulating EEG records, only 1 mW/cm^2 is necessary and the behavior is like that of a resonant frequency.

In the thermic tests, the cooling curve is a reversing of the heating curve. When the heating curve cannot be followed, for example, due to instrumentation difficulties, one can use the thermodynamics of exposed preparations as equilibration occurs and relate the absorbed electromagnetic energy to equal losses to the surroundings. In direct observations of the heating curve one may utilize liquid crystal instrumentation and infrared microscopy.

EXPOSURE, DIMENSIONS, NETWORKS, AND THERMIA

Both exposure levels and the subject's dimensions are substantive elements in interactions. Athermic and thermic levels have roles to play and the nervous system is a receptive network for organism-informing purposes.

Intersystem communication is most direct and dramatic when EM radiation affects the nervous system. This occurs regularly in many ways, one of the easiest to visualize being a nervous system's sensitivity to local thermal gradients, including an effect on subdermal nerve processes with consequent effects on muscular and other operations. Many experiments are done on rats and mice in which a special metabolic or physiologic thermoeffect occurs. These animals show a colonic temperature rise of 1 to 3°C even in weak and intermediate exposures. This effect might therefore be described as metabolic rather than as thermic or athermic. These colonic temperatures at low levels of EM radiations are due to dimensional and related conditions. In the rat, a special diversion of blood in this area is known as a shunt. This is thus a nongeneralized heating and is independent, for example, of cerebral gradients, since the brain is able to rise in temperature without the colonic rise and vice versa. These effects resemble Q effects with gradients of about 0.1°C which are normally seen in activity or body operations with the environment. Thus they relate to, or suggest communication with, internal systems having the requisite functional linkage or the same kind of operating information borne by the electromagnetic radiation. Information-rich or sensing systems are most directly of interest, especially the nerve networks which closely resemble electrical nets. These are mediated by the controlling structures of the intact organism and thus are less likely to respond in experiments when the "whole" is absent as in isolated neurons or sections. Also, when each dendrite makes a tiny input, its contribution may be missed unless integrated with that from thousands of its neighbors. These EM radiations may be at subneural strength, a few volts per centimeter or even practically dc at low frequencies. Properly applied they change the system operations, and can be observed by watching an appropriate record (e.g. the EEG). Inevitably the results are seen in operational events of the recipient that are tightly coupled to information transfer operations that rest on periodicity, or time-based phenomena, for example, cycles and the firing of synapses, pulsing of nerves and inter-dendrite-neuron-neuroglia transreception of weak local signals. Separately and combined, these mean electric and magnetic fields, plus electrochemical and neurochemical changes.

Adey's Trigger Condition

Adey's trigger conditions for such events involve coupling into a brain. The monkey is one model, with its 10 cm^2 cross section for conduction and brain tissue impedance of 300 ohm cm, producing an electrical gradient of 0.1 to 0.01 microvolt when the length of 7 cm is on an E field axis and the strength is 10 V/m from an EM radiation input. On the neurochemical side, the communicating substrate for interneural EM activity includes negative polyelectrolyte nerve membranes with macromolecules and their bound water (glycoprotein, glycocalyces structure as interstitially active biochemicals). The electronegativity attracts acidic units. The cellular Z pushes electrical signals around such membranes and around the cells (extracellular) and conductance follows the brain operations, probably participating via membranes in the access to and retrieval of brain information. The membrane structure seems to be accordingly thickened and modified for such information storage and sensing, using information communicating substances such as hormones, effectors, ions, and neurotransmitter chemicals like gamma-amino butyric acid, which accordingly receive space in the expanded membrane. To obtain responses requires the molecular information plus ionic currents, particularly that of Ca, switching with materials like prostaglandins, and energy step changes through enzyme activity detection and transduction. The energy amounts are small, and their interception or interdiction with EM radiations involves very small fields. Both *propagation* and *amplification* are key conditions, the response to input of information requiring sudden energy for execution of the command.

Elements in the Communication

Polyelectrolytes and counterion layers are receptors for EM radiation with Ca ion fluxes, a uniquely satisfying medium of exchange for binding and releasing selectively in cooperation with the EM fields in long waves. The information molecules are uniquely selforganizing and status-sensitive and can reorganize both zonally and conformationally on receipt of ionic signals. The molecules, membranes, currents, fields, and ions associate in a propagating, amplifying, switching network of a communicating system. Amplification, for example near dc, can arise from enormous dielectric constants in a sudden frequency response to influence the orientations and the movement of cations. Electrically, the fields can be sorted out in a manner meaningful to the organism when long waves are received around or between, and faster pulses are received *in* the neurons. Applied fields of even low power density produce opportunities for resonance especially when they persist and the electrical interaction is self-amplifying. Space and time constants and phase permitting, a very small input could then alter the muscle, gland, and nerve cell output. Specified patterns of excitation in overlapping dendritic fields in palisades of cells, and the development of rhythmic electrical waves, affect the ability of the tissues to change their excitability as the tissues "remember" such patterns as having existed. Communication will be via a frequency-dependent attenuation of high-frequency components that depends on Z (extracellular). A 10 mV neuronal wave will force about 10 μV of signal strength out of the cell into the volume-conducted wave.

The EEG is a record of both a behavioral state and of a volume-conducted forcing function for receptive extracellular substances. In particular it may participate in membrane amplification and sensing operations and basically communicate signals to brain substances signaling the neuron and then being signaled as a weaving in and out of brain data. In the brain, communication takes place in the presence of much larger synaptic operations. The brain operation is complex and the small divisions often studied are too simple to serve as models.

The alpha (8 to 13 pps) waves are intermediate, beta (13 to 30 pps) fast, theta (4 to 7 pps) slow, and delta (0.5 to 3 pps) are very slow in frequency in reflection of these activities from the sleeping to super alert status.

The sialic acids are a common family of 9-C sugars occurring in glycolipids, mucins, secretions and glycoprotein. Those like N-acetyl, O- and N-diacetyl and N-glycolyl are N-acylated derivatives of neuraminic acid. Sialic acid terminals in the brain are calcium ion carrier loci and morphologically these appear in a matted arrangement of "mobile" fibers associated with plasma membrane protein and terminating in the ion site. With a preferred direction for these charges being toward the membrane, a diode type rectifier is formed by charge migration. The applied EM radiation and locally generated fields, in activity, will interact with these "huge" ions, especially at low resonant frequencies such as the intrinsic ones and applied ones near dc, acting to select appropriate frequencies.

As in the case of the precautions in applying the skin effect to EM radiations and biology, antenna theory gives many useful lessons if interpreted correctly. Resonance, or maximum absorption occurs when the body's long axis is parallel to the plane of the E field polarization. With free field exposure, possible resonances occur when the body has a length of about 0.4 times the wavelength, or at 68 MHz for a 1.75 m man. The resonant frequency is halved if the object is grounded because it then forms a monopole. In terms of safety the neck is very sensitive in free field exposures and the ankles or legs are especially sensitive in grounded exposures.

SPATIAL THERMIC CONDITIONS AND DIMENSIONS

To an animal the size of a rat, a 1 to $3°C$ colonic rise due to EM radiation fields gives a signal even from small stimuli which distracts from tasks or gains the animal's attention. This amounts to redirection or modified behavior just as with any other sensed signal, thermal or otherwise. It is a core effect in a dimensional sense with added significance due to the colonic blood shunt in the rat. In exposure to 500 MHz, rats become supersensitive to orientation, that is they become bipoles with their long axis parallel to the plane of the E field polarization, probably because of the core heating.

The dog has countercurrent brain cooling but man, monkey, and rabbits do not show this limitation on the brain temperature rise.

The extrapolation of animal data to man is constrained mainly by dimensional relations. Global data, e.g. effects on molecular and cellular events like information transfer or division with common mechanisms, are ideal but dimensional effects are probably the only other basis for extrapolation. The principal group of such effects is the one related to size as commensurate with wavelength or its fractions.

Thus the animal or subject is being affected by EM radiations of given dimensions and the greater the wavelength, the greater the size of the animal that should be used in the same studies, other factors being equal (which they rarely are). An exposure of 10 mW/cm^2 is a serious matter for a rat but not for man. This is only to point out the inherent problems of extrapolation and to aid in interpretation. The same thinking applied to molecular levels would change the frequency to fit the molecular size, usually with the rotational transitions to higher energy levels in mind, but here it is necessary to sort out bound water, and other effects due to associated small molecules and molecular groupings. Even in thermic effects, with strong fields and general heating as for rapidly inactivating enzymes, there is dielectric modulation by matter and dimensional effects from *form*, absorption cross section, layers and interfaces.

The internal EM fields depend on and accompany all operations and when they interrelate with applied fields, the imposed fields depend on the source configuration, frequency, shape, size, composition, and orientation of the exposed subject.

Just as intact organisms have special reception capabilities, some effects are seen in cultured cells that do not appear in the whole organism when exposed (and regional effects occur on exposed parts of a body as opposed to the whole organism or its cells. The intact organism retains its regulatory, compensatory and homeostatic functions in an experiment while the isolated cell or tissue is more limited. Still, the latter are valuable for modeling, designing and testing in relation to mechanisms. There is probably an overlap in the dominance of the E or H field. The E or H field can influence as a function of frequency, where at upper MHz or GHz, the E field is determining and in KHz or lower MHz, the H field is.

The rapid rise time (RRT) effects due to the great step changes in applied fields are modulations of great significance depending on time constants. This is probably more true in athermic than in thermic events and pulse modulation in general is more significant in biological than in physical processes. Then pulse repetition frequency (PRR), pulse frequency, and the peak power contribute more information to the signal. In the near field of EM radiation, the absorbed energy (the significant portion) is likely to be greater than in the far field and thus, from a safety standpoint, near fields are more dangerous.

Effects of relations between phonons and phonons, and phonons and protein are usually unknown but they could potentially affect conformation and membranes. Also dipole-dipole waves in tight coupling situations slow to suitable interaction speeds for effects. All of the internally generated waves, *EEGons* etc., will create possible interactions and couplings as will important charge carrier migrations, particularly at their sites of action.

As EM radiations strike muscle, bone, and fat layers, hollow chambers, liquid pools, reflective surfaces, wave guide-like structures, antenna-like dimensions, special conductors etc. within a body, it would be expected that the distribution of energy would be very nonuniform. This unevenness might well begin at weak power levels and build into hot spots with strong fields as a function of time or higher applied power densities. Thus the tissues show average intensities and maxima. The contribution of E and H field components will add to the variable distribution as will the dimensions and orientation (Table 15-1). The

Intersystem Theory

TABLE 15-1. "Critical Mass" in an Effectively Polarized Head

Subject	Head (cm)	Frequency (kHz)
rat	1.0	73.0
guinea pig	1.2	60.8
cat	3.0	24.3
Rhesus monkey	3.5	20.9
humans	7.0	10.4
	8.0	9.1
	9.0	8.1
	10.0	7.3

maximum power disposition for fields polarized along the long dimension of the body is at frequencies such that the major length or axis of the subject is equal to about $\lambda/4$ for radiation. Within the limits of all these variables, a "critical mass" may be suggested by a region of the body such as the head. The effective absorption area is from 2.5 to 3.5 times the geometrical cross section, so that effectively a subject presents a target quite a bit larger than its shadow.

If a container is used for a preparation then its axes bear a relation to the wavelength, with the flask and its contents behaving like an antenna. Currents are induced by EM fields, and dimensions such as 1.5 wavelengths create the possibility of a modified resonant absorption of power. To avoid this, the container should be greater than 1.5 times the wavelength. A flask exposed at 3 GHz interacts with a 10 cm wavelength. Its dimensions may be 14.5 cm high, 4.5 cm wide and 3 cm deep. The largest dimension, the height at 14.5 cm, is less than 1.5 × 10 cm and therefore it introduces the resonance possibility.

SHOCK WAVE

No end points have been observed that can be clearly associated with shock waves unless the audio effect is such a result. Yet conditions seem suitable for the development of such waves, especially in local regions. This is a mode of interaction that is both thermic and athermic since it is a suddenly applied temperature gradient with "no place to go" except into a wave form that it induces, and yet the results are more related to the wave than to the heat. The expression "rapid rise time" with reference to the pulse modulation expresses the nature of applied force, and it is known that the temperature increase per pulse in tissue can be $10^{6}\,°C$ but it is applied in a few microseconds versus a common pulse rise time of 100 microseconds. The effect is not unlike the Ludwig-Soret effect where a thermally-induced electric field is set up by the ionic liquid and depends on the *thermal gradient* and not on the difference in temperature alone. This gradient is enormous, perhaps

$10^5\,°C/cm$ in tissues, and creates a drastic physiological result if ultrasound offers any criterion. In membranes, even small temperature gradients can depolarize or cause Ca fluxes. The shock wave effect is a more distributed result that can communicate with other organs, especially sensory end organs, to produce sensations and shock effects. Embryos at least may be killed by shock waves in their medium. Such microwaves can produce shock wave amplitudes of about ⅓ of an atmosphere which is not great as a biological threat. *Laser* beams have a much greater inherent shock effect at time zero because of the sudden rise time in affected structures and their concentrated energy ray. They are easily damaging.

ARRANGING ATHERMIC CONDITIONS

Much of the difficulty in demonstrating athermic effects is related to the discipline required to think in spectrographic terms, but there is also the problem of the large amounts of electromagnetically-reactive water in biological preparations. The use of refrigerated cultures and preparations has its counterpart in hypothermia in animal experiments, for example with rabbits and dogs. Dampening the temperature rise while using strong fields in the exposure is, however, self-defeating. Tissue cultures, e.g. a whole lens in a culture medium, make preparations that moderate the conditions. Genuine thermic events, though, are found in pulse modulation, the effect *per se* of rapid rise time, and delicate communication via phase and periodicity with due respect for system space and time constants. Then, in pure information transfers, the thermic distinction loses its significance. Pulsing, in part, contributes its effect based on the argument that cw offers less dissipation or less repair and, therefore, more heating or that peak power has more dissipation. If an effect is thermic, then cw versus pulse modulation makes less difference in the result. Sensory modes may give athermal responses, for example, audio ones at a temperature increase of only $10^{-6}\,°C$ in the tissue. Yet one must conclude that EM radiations always have an athermic effect which may or may not be observable and that effects are always both thermic and athermic because the ultimate fate of EM radiation is always heat. The heat may be commanding or unimportant in the result depending on the exposure conditions.

The importance of information transfer is very clear in sensory effects, normally at low levels of power. Biochemical effects including inactivations are prominent at high levels using an increase in temperature in order to speed reactions, denature protein or the like. Then with biomolecules in highly organized, operating states with phase, periodicity, space and time sensitivities, low level exposures can again communicate with biochemical elements. Here, the signature of the EM field and its recognition provide an information-rich mode of interaction. The configurations become significant as do dimensional derivatives based on the interacting field patterns and the dielectric, colloidal properties of the medium. Then frequency shifts and excitations are possible in the manner of raman spectra.

MODIFICATION OF GROWTH AND DEVELOPMENT

Biochemical Component

The level of serum triglycerides increases at 2450 MHz at or below 10 mW/cm^2 in mice. The threshold is at about 1.5 mW/cm^2 and this endpoint means that there will be more formed fat in blood and possibly most will occur at this frequency in the ultrahigh frequency range. This fact will mean a new method for modifying lipid metabolism and possibly is related to cholesterol levels and connected problems. The special significance is that this is a nondiet method of changing lipid levels, which could be related to stroke.

Enzymes. Functional macromolecules such as enzymes, when specific targets for inactivation, are thermally inactivated by next-door collisions in the vibratory mode by EM radiation at 3 MHz. This is near the natural frequency for some whole enzymes. The intramolecular groups have a higher frequency. The absorption in the solvent is important because the energy is transferred to the enzyme and some solvents, e.g. phosphate buffers, are very lossy. For many enzymes, lipozymes and trypsin as instances, microwaves provide an alternative method of heat inactivation. A serious interest in the enzymes *per se* brings methods of handling the substrate into the picture. Modification of the pressure, combined foaming and inactivation by microwaves, yields specially rapid inactivation. These changes are impossible in *in situ* experiments such as the inactivation of those enzymes active in changing brain neurotransmitters. There, the natural absorption of the brain tissue is sufficient. A helical applicator can "focus" microwaves in a counter-flowing liquid to inactivate the enzymes and sterilize fluids (Copson, 1976).

Protein. For whole protein, the dispersion region is 1 to 10 MHz but for water, the relaxation frequency is 19 GHz. The dielectric constant ϵ' for protein starts at about 1000 Hz in the region of 85, drops a bit at 10^5, disperses downward at 10^9 to 10^{11} Hz and levels off at about 5 from 10^{11} to 10^{13} Hz. Meanwhile water is dispersing between 10^9 and 10^{11}, bound water from 10^7 to 10^9, and amino acids or small peptides from 10^8 to 10^{10} Hz. Thus the dispersion is below 2.5 GHz for groups and bound water, or about 300 MHz. Organization of protein *in vivo* allows various order effects in keeping with the conformation. These multiple effects are like dipole on top of strong dipole effects and may be quadrupole, octopole and higher order results related to the signatures of the EM fields. These may appear as force field effects or derivatives of the field acting on components.

Organisms and Bone Healing

The intrinsic electrical properties of molecules and cells interact in growth and development, being especially sensitive to interference before maturation. Thus, on the question of EM radiation intervening, growth and development may be seen as the expression of DNA information while the organism is interacting with EM radiation. The EM radiation disturbs circadian rhythms of bone marrow cell mitosis in guinea pigs as a function of time (of day). In healing bone fractures in EM fields, new collagen, fibrocyte activity and deposition of bone are oriented by the applied field and they may be promoted or retarded as a function of the field factors.

Bone substance undergoes a much more rapid turnover than one would expect. According to Woolf's law, a stressing force on bone causes reshaping. For example, deposits may occur on one side of the longitudinal axis and losses on the opposite side. In this sense, weightlessness is a negative stress and, in some cases, astronauts experienced an enormous 30 per cent loss of bone weight. These stresses produce electrical charges with positive and negative surfaces. Thus bone is piezoelectric due to its collagen. Other components that are involved are long chain polymers, DNA, and polyelectrolytes but the collagen is a large proportion of the total. A downward force on a small sample of bone will produce a positive charge and electrons can be made to flow from the sample. In a case of information flowing and directing the operation, the forcing function serves to reform damaged portions. With this situation known, it is possible to direct an electrical impulse around a defect to influence the repair. This can be demonstrated by creating a defect in a bone preparation, inserting platinum electrodes tightly on each side of the defect, and applying a dc source of the hearing aid battery type. Other possible applications involve the directed regeneration of lost bone and repair of pseudoarthritis. In an action that resembles communicating with the defect, an electrical input on the order of 10 microamperes at 0.5 V stimulates inactive defect cells. Cytological events during the electrostimulation involve: 1) secretion through numerous vacuoles of a dense material of about 1000 Å, activity of bone cells with hydroxyapatite polysaccharides, collagen and its precursors; 2) ribosome-lined vacuoles; 3) mitochondria with a dense lucid material. Pulsed microwaves at low frequencies have given positive results in healing bone with low power in an athermic mode.

A channel for stimulating DNA synthesis in cartilage cells is provided by an electric field of 1166 V/cm at 5 Hz. This field probably generates a Na^+ and Ca^{++} flux which stimulates the synthesis and may account for therapeutic effects in bone healing. The electromechanical shock of a field as strong as this provides a temporary channel through actual disruption or dielectric breakdown of the membrane as a function of the rise time of the field in terms of microseconds or with pulses of lesser intensity.

Embryos. Japanese quail embryos in the form of fertilized eggs react to 5 mW/cm^2 of EM radiation during development (related to 2450 MHz and safety as in much of the current work). The same level of radiation makes hemoglobin increase and monocytes decrease in two day old chicks. When such embryonically exposed males mature and mate, they have 20 per cent less fertility, that is, more of the eggs are nonfertile. The exposure also diminishes the male livers and develops an asymmetry with one lobe larger. In the female it enlarges the spleen and the bursa which is the source of B cell lymphocytes.

In the teratology of fetal mice and rats, 100 minutes of exposure daily to 14 and 28 mW/cm^2 at 2450 MHz, free field, inhibited growth and encephaloceles were noted. At the upper exposure, adult males were 50 per cent less fertile after four weeks when tested in immediate breeding.

Electromitotic and electrometabolic inhibition may be involved in chick embryos. The magnetic field of 1, 5 or 8 gauss or the E field at 1 or 10 V/m inhibits at 60 Hz while the E field at 1 or 10 V/m accelerates when the frequency is 75 Hz.

Zebra Fish. One can tell the difference between thermic and athermic damage in zebra fish (*Brachydania rerio*) by the appearance of the fish blastula at death. The exposures were in 1 microsecond pulses at 0.5 peak megawatts and 2760 MHz with a PRR of 20, 50, and 270 per second. Lethal exposures were only slightly less, starting in a room temperature bath, as opposed to starting at 20° below room temperature showing that a lethal temperature was not operating.

Chromosome Aberrations. An exposure of 15 mW/g at 2450 MHz on kangaroo rat cell cultures showed chromosome aberrations. The cells were subcultured in many passages after each five to seven days growth. In each transplantation, 25,000 cells were moved to each new flask. The culture was exposed for 320 days and incubated normally. After 20 passages or 120 days minimum, it was found that the RH5 cell line *lost* a chromosome to give a chromosome complement of only 10. Then this variation became more common. As the time of exposure increased, the variation accumulated. Chromosome aberrations (but not necessarily deletions) are rather common as a result of physiological insult, but their careful analysis is something else. As far as EM radiation is concerned, the probability of changing the autosomal genetic complement or expression at the level of DNA, DNA transcriptase, or repressor loci is low in stable chromosomes.

The significance of a chromosome break is that the acentric fragment is mitotically inactive, depriving the cell of possibly important information in the portion deleted. Induced chromosome breakage results from irradiation with ionizing radiation, or from radiomimetic agents such as the mustard gases, as well as from exposure of cells to sudden drastic changes in temperature. Alternatively, spontaneous chromosome breaks are less likely and both types of breaks may be repaired with possible rearrangement or shift in a segment to another location.

Viruses. Microwaves depress *Herpes* virus after infection but not before, and make virostats more effective (Cytosar of Upjohn Company and ARA-C) as shown by lesions on mice.

If interferon is the virostat and the virus is *Sarcoma 180*, then microwaves acting to produce hyperthermia enhance the effect about 200% at 42-44°C. This is then a bright hope for cancer therapy or at least for viral tumor inhibition. The therapy shows that two types of treatments are more effective and more flexible than one. Other frequencies are involved. In oxidative phosphorylation with mitochondria, low frequency EMR can uncouple the respiration. At 3000 MHz an effective exposure for the inhibition of cell growth is 20 mW/cm^2, but at 5 mw/cm^2, there is an increase in the use of glucose and more protein synthesis using the myxovirus, virulent Parainfluenza 3. 3000 MHz may be shown to *stimulate* these cells' metabolism at 3 mW/cm^2 in exposures of 30 minutes, while at 20 mW/cm^2, the membrane permeability is increased and there is less succinic dehydrogenase activity (as a measurement of respiration). An endpoint is inhibition of the myxovirus. In this kind of effect near a threshold for stimulation versus inhibition, one can stimulate the survivors selectively while the weaker cells are damaged, so that for the *culture* as a whole, the influence is reversible, but certainly not for the nonsurvivors. To demonstrate the effect, it is necessary to expose, then infect the cells that are then cultured and tested.

There tends to be a heavy bias in favor of positive results which is somewhat supported by the skill necessary to manipulate the fields and forces in EM radiation as well as the biological preparations. This is especially true when field effects are tested on rapidly dividing cells where measurements may be made on growth rate, pregnancy ratio, and number of progeny.

Insects. In tests on insects, pupation may be analyzed after exposure. Measurable abnormalities occur in the Darkling Beetle, *Tenebrio militor* at 900 MHz with day-old pupa exposed in the wave guide in styrofoam supports at 100 mW or about 8.6 mW/cm^2 for 1 to 2 hours (1 mW $\approx$ 0.86 mW/cm^2 at the center of the wave guide). Such experimental results as anomalies in adults show a dependence on exposure, power density, and time, but apparently no effect is seen from pulses as opposed to cw, from parallel versus perpendicular orientation with the E field, or in terms of variation in the day of development. If the exposure is reduced, so are the effects between 20 and 40 mW-hr but the energy is controlling; that is, the power level is changeable as long as the time-power is constant and such energy gives constant anomalies. A trained person can recognize morphological changes and observe that the pupation time and count of anomalies increases exponentially with the power level. Thus at 2 mW one finds 33% anomalies in adults but at

20 mW, this increases to 55%. Ordinary oven heating did not produce these effects.

Microplasmodia. *In studies on the chronic effect of low EM radiation exposure, there is seen to be a delay in the cycle of mitosis and cell division.* At 2.0 gauss and 0.7 V/m, weak ELF exposures of the microplasmodia of slime mold *Physarum polycephalum* cause a time effect; delay in the mitotic cycle of 30 to 150 minutes is demonstrable. This is normally a cycle that occurs regularly at 15 hours. The frequency need be only 75 Hz and the exposure is 1100 days but the E field is sufficient and the cells return to normal afterward. During the period of the effect, less oxygen is consumed and the period of protoplasmic streaming is longer.

TUMOR CELLS — ABNORMAL DEVELOPMENT

In Webb's sensitive frequency series, as seen by laser-raman spectroscopy at 50-200 GHz, one can separate normal from abnormal cells. Cells have natural absorptive frequencies that may, as the reader has seen, form a frequency series and ω_{m1} differs from ω_{m2} by a constant frequency. This constant or frequency constant separates the absorbed frequencies in one or more of the series of frequencies absorbed by different cell types. Then there are cell type frequency series I and cell type frequency series II and the difference is a frequency constant. Some *in vivo* absorption mechanisms that may account for the differences have been discussed. For example, CTFS I has: a) the input energy distributed differently than CTFS II, and b) for some reason, energy transition states in I are different from those in II.

One can look at b) with raman spectra using laser spectroscopy. The cells have to be in a synchronous population which can be obtained in various ways, for example, by filtering out the small, growing cells. Raman shifts are periodic and based on the metabolic time clock or life cycle in a given type. At one moment in the cycle there will be a certain metabolic phase. One can now study the effect of the exposure to microwaves (of some frequency) on the metabolic phase that exists and get the raman shift lines associated with the metabolic phase. Then 2-4 GHz speeds up the metabolic cycle just as heat does and >50 GHz leaves the metabolic cycle constant. These effects occur as *frequency series specific changes* and are visible as either the presence or absence of raman shifts. Absence means that no such energy state exists whereas such states usually do exist.

We can then assume that certain metabolic cycle events (that show up as raman shifts because excited molecular states are present in the cycles present) have been inhibited by microwave and that these events are otherwise necessary to give the biological endpoint. The effect occurs at frequencies >50 GHz which alter the metabolic phase while normal cells do not absorb these same frequencies.

Selectivity of EMR

There is then the possibility that microwaves could selectively absorb in tumor cells, leaving normal cells unchanged if the right "cell type frequency" is used. Thus the potential for nonionizing EM radiations in the diagnosis and therapy of cancer depends on how well normal cells tolerate the exposure.

Diagnosis. For diagnosis, cells are seen to have their living microwave spectra consisting of energy interactions and energy state transitions or absorption bands. The history of a cell during a period may be displayed as a spectrum that records its status. Frequencies of 2 to 4 GHz speed up cell cycles in the manner of a thermogenic effect. Microwaves also *delete* some bands. Thus, by logical inference, there is an intrinsic change in their *source metabolism*, i.e. cell events are interrupted. Using microwaves against tumor cells selectively is a case of exploiting the differences in absorption between normal and tumor cells.

Radiotherapy. For radiotherapy, all cells are sensitive to combined hyperthermia and X-radiation. Microwaves, if applied uniformly to a tumor, $\pm 1°C$, can produce the radiosensitivity desired but special applicators are needed (obviously probe devices). Heat treatment alone at about $45°C$ is carcinotherapeutic. A selective destruction is present over normal cells and a selective thermoresistivity favors the survival of normal cells.

Pulsed 27.12 MHz applied for several days shrinks Lewis lung carcinoma in mice when the exposure is after tumor infection with up to 60 to 90% tumor inhibition. This is like a direct command to the tumor cells to stop growing but the inhibitory effect is lost for some reason. Even when the same cells are exposed *in vitro*, there is a selective antitumor action which is dependent on the power, the peak power, and the average power. Still this effect is not considered hyperthermic.

The known effects of EM radiation on blood-forming systems, the immunosystems, and the endocrine system may be important in cancer studies at the level of basic human physiology. There is no effect at 2.45, 3, or 3.4 GHz and 41 W/kg on mitochondria or oxidative phosphorylation.

Experiments at some stage of development could communicate conflicts such as might produce monsters when the whole organism relates the input to the development activity at the particular time. Exposure of rat embryos to very high frequency, for example on day 13 or 14, produces a curled up tail anomaly. The phase of development at the time of exposure is determining. DNA synthesis is sensitive to VHF where the wavelength in air is 11 m, frequency 27.12 MHz and a quantum is only 8×10^{-7} eV. This effect is sufficient to terminate tumors.

MICROWAVE HYPERTHERMIA AND TUMOR THERMOTHERAPY

Specialized Directors and Impedance Transformers

If the thermal effects of microwaves are frankly accepted and any selectivity taken with gratitude (even though not completely understood), then it is possible to get down to the business of talking about the clinical application. Thus a fairly sizeable medical technology has developed out of the earlier diathermy procedures but now directed against cancer. This physiotherapy uses 2450 MHz with generators of very low output which are distinguished by having very specialized directors. These can be related to the size of the tumor and, in the future, there will be more variation in the frequency to provide physiotherapy for deeper tumors at low frequencies, probably even to 400 MHz in shielded situations to take care of FCC requirements where necessary.

There is also a tendency to use the impedance transformer pad or dielectric-loaded transmission line termination to match the electrical properties of the tissue to the transmission lines. This assures propagation into the tumor without reflections, at least from the air-to-skin interface.

Comparison with other Treatment

The microwave physiotherapy is selected over hot lamps, hot water baths, parallel plate diathermy machines and infrared because of the convenience. The rationale of the treatment does not seem to be applicable to these alternative methods with the exception of high frequency diathermy although sometimes superficial tumors are susceptible to management by the simpler methods. It is almost always better to use microwaves and adjust the directors and to try and extend the frequency range to take care of deeper lying neoplasms. The problem of the uniformity of the heating within the tumor is much easier to manage with microwaves than with perfusion methods or hot water immersion because at least the thermal energy is deposited in approximately the volume of interest.

Metastasis. When it comes to metastasis, a systemic hyperthermia of $41.5°C$ for several hours in three or more treatments may be necessary but it may be difficult to apply microwaves and much better to use perfusion hyperthermia. These procedures arise from evidence from inadvertent treatment of tumors where high fevers were present from concurrent infection.

The effect of microwaves at the cellular level in hyperthermia suggests that a given temperature of about $42°C$, if maintained for a sufficient time, destroys the tumor cells but inflicts only reversible damage on normal cells. This result is accompanied by a lower oxygen utilization in the tumor cells. Amino acids, thymidine and uridine, are severely depressed in terms of their synthesis into protein and nucleic acid in hepatoma cells. This was shown by isotope studies on cells incubated for two hours at 43°C and could not be demonstrated on regenerating rat liver cells.

Cell Physiology and Thermia. Lysosomes are sensitive to hyperthermia. These bodies in phagocytic and other cells secrete lysozymes which can digest foreign substances but are normally isolated behind security membranes to avoid their dangerous effects on vital cell structures. Chinese hamster ovarian cells *in vitro* lose proliferative activity under hyperthermia. Cyclic AMP-dependent protein kinases are thermally sensitive at 43°C. The enzymes adenosine monophosphate kinase and ornithine decarboxylase are involved in growth and development in cells with the latter being involved in the synthesis of polyamines. The ornithine decarboxylase is suppressed at 47° rather than at 43°C, but the lower temperature did inhibit protein synthesis temporarily. With the emphasis on the fact that the observations are made on a culture of tissue, hyperthermia shows spherulation, bubbling of the surface, enlargement of the nucleolus and breakdown of polyribosomes in growing Hela and normal cells. However, if the normal cells are not growing, then these changes are not observed. The basic fact seems to be that, in an active state, cells are synthesizing enzymes for metabolism and this makes them labile as compared to the more thermal resistant, dormant or nongrowing status.

Like most physiological insults, hyperthermia and microwaves can, both together or individually, cause chromosomal aberrations and the cell membranes have often been observed to be sensitive to hyperthermia. The cellular effects are most obvious when combined with the importance of cytotoxic agents as used in chemotherapy. Here the hyperthermia may prevent the tumor cells from repairing themselves and if the same cells are being medicated, then rejection of this treatment is less likely. This leads to a synergistic result when hyperthermia is combined with radiotherapy with ionizing radiations. Thus the multiradiotherapy treatment is possible in which microwaves can enhance the direct and cytotoxic effects of ionizing radiations.

Effect of Oxygen

There is a great deal of variation in the vascularization in a solid tumor. As a consequence, the tumor cells vary greatly even from aerobic to hypoxic in terms of oxygenation as well as in nutrition. The aerobic cells are more sensitive to the ionizing radiations. The hypoxic cells are more separated from vascular supplies and, incidently, more protected from chemotherapeutic agents when these are involved. Therefore, for these two reasons, the hypoxic cells may resist conventional therapy.

Triple Threat with Chemotherapy. This is where the microwaves are useful so that the multiple radiations combined with chemotherapy pose a triple threat to the cancer cell population. Resistant tumor cells in the depths of the neoplasm would see themselves as hypoxic and thermally isolated with poor thermal regulation and so most hyperthermic, while at the same time being better protected from chemotherapy which depends on vascularization. In this scene, the microwaves are selectively directed in a concentrated manner at the most resistant sector of the population.

The triple threat approach to the tumor cells may reduce the doses of radio and chemotherapy necessary. The microwave hyperthermia makes the normally protected parts of the tumor sensitive and accessible with less danger of their recovery from radiotherapy which, in turn, must be limited by the tolerance of the adjacent normal tissue. While hyperthermia at 41.5 to 42°C is rather well tolerated by the normal cells, an even higher temperature would be desired for rapid tumor destruction. The upper level might possibly be raised to 43°C for animals.

The threshold seems closer to 42°C for thermotherapy of malignant tumors with an operating range as close to 43°C or 43.5°C as possible without side effects. At the maximum, side effects may come from nonuniformity and temperature peaks that should not occur. These temperatures may be obtained with an input of 20 W or less, well-matched into the region. Then penetration is "doubled" to about 6 cm. at 2.45 GHz by applying to both sides of an accessible region such as a breast, muscle, or neck.

The reports of Mendecki, Friedenthal and Botstein at the Albert Einstein College of Medicine in New York stimulated a great deal of interest in tumor eradication with microwave-induced hyperthermia (1976). Except for anesthetization with Nembutal, the only treatment used on the mammary adenocarcinomas in C3H mice was microwave hyperthermia at 43°C in four treatments of 45 minutes each. The results were 100% in favor of the physiotherapy. It may be concluded:

That microwaves are ideal for many solid-tumor masses;
That elevated temperatures greatly enhance the cytotoxicity of chemotherapeutic agents such as nitrogen mustard and Bleomycin;
That the duration of the hyperthermia determines the maximum temperature reached in tumors;
That the surviving fraction of cells increases with intertreatment intervals within the range of one to four hours; and
That the frequency of chromosomal aberrations depends upon the stage of the cell cycle and is greatest in the growth stage, or after mitosis but before DNA synthesis.

Support for combined hyperthermia and radiotherapy in far-advanced but previously untreated neck and gynecological malignancies is provided by clinical trials reported by Hornback et al (1977). These trials involved 433.92 MHz microwaves and radiotherapy on 70 patients. In 21 patients who completed the course of therapy, 90% had complete, and 10% partial relief of symptoms. There were 16 cases of complete clearing in the 21 patients and 9 of the complete responders stayed clear of neoplasms from 9 to 14 months, the survey period. The patients had been termed *probably refractory* to further medical treatment, but then showed regression of symptoms.

Rabbit metastases were considered accessible to microwaves in experiments by Goldman and Dreffer (1976). They considered that suspensions of iron compounds as illustrated by ferrofluids are non-toxic and, directed by applied magnetic fields, could be mobilized in the neoplasm. Whether directed or deposited, they significantly increased the temperature experienced in the target tissue and, for laser therapy in superficial neoplasms, the darker colored iron compounds added selective absorption. Laser energy could be directed into deeper tissues by wave guide probes alone.

STIMULATING NATURAL IMMUNORESPONSES

Metastatic cancer represents an invasive amplification via circulatory routes or other channels. Treatment by the methods now to be discussed is able to use some of these same routes to distribute a curative agent. Thus, when combined with other procedures such as *ex vivo* absorption of IgG using *Staphylococcus aureus*, metastases can be managed through changing the natural antitumor substances in the blood in favor of the host. The extracorporeal perfusion technique which is commonly associated with the names of Bansal and Terman and their teams means that populations of T and B lymphocytes are altered, serum blocking factors and immune complexes decrease, complement-dependent serum cytoxicity appears transiently, and antitumor antibodies rise as shown by more immune complexes after the process. Then these complexes may become too large to block cellular immunity. The idea is to *stimulate* the immunoresponse, while saving essential plasma components when *S. aureus* selectively absorbs IgG, the immunoglobulin that constitutes 80% of these proteins. In the human trials described by Bansal et al (1978), E rosetting lymphocytes decreased, IgM, C3 and C4 levels increased, and Ig-bearing lymphocytes increased causing a cytotoxic effect on the target which was a colon carcinoma. In the *ex vivo* perfusion, hyperthermia is apparently self-organizing. In the case reported, the tumor decreased in size while healing and improved condition accompanied a disappearance of generalized intraperitoneal metastasis.

The organism is supposed to release its own tumor cell identifying immunoglobulins and to prepare antibodies to follow with suppressive and lytic attacks. Somehow, the swarm of noise caused by marker proteins blinds the antibodies; so to fight the tumor, masking IgG must be separated out. This task is assigned to *S. aureus*. Immunocomplexes of antibodies and antigenic globulins also mask lymphocytes with specific capabilities against the tumor cells, but if these are *separated* on the coccus, the attack of antibodies is probably permitted. On dogs, heat-killed, *S. aureus*, Cowans I in the membrane filtration system for the plasma component causes an acute tumoricidal response. The key agents as suggested by Terman (1980) are tumor-specific antibodies that surge for 48 hours after the treatment, infiltrate the tumors, and cause lysis but spare normal tissue. The perfusion, somewhat like an immobilized enzyme treatment, but over *S. aureus* of this particular strain, generates or activates the tumoricidal antibodies and seems both logical and effective in canine breast carcinoma. The cited team observed but discounted hyperthermia in the results since some of the dogs showed only minor temperature elevation. It has been observed also that serum C3 component, by inflorescent analysis, is attracted to tumor cells and chemotaxis may occur by macromolecules and their signals so that leukocytes pass into the tumor.

Regimen

Computerized application to tumorous tissue from an antenna array at the optimum frequency for desired penetration is an appealing regimen for certain neoplasms. While 2450 MHz is in use, 915 MHz is an attractive alternate when, as in prostate cancer, the result is more heat and uniformity. This particular site is accessible for a rectal probe as shown in dog tests. While the antenna is sending microwaves into the neoplasm from alongside, X-radiotherapy may be added from outside. Internal applicators have been prepared for 2450 MHz based on a stripped coaxial, quarter-wave termination, a design that will become larger at lower frequencies. There is a rationale developing around the concept of delivering the nonionizing radiation in the S phase of cell growth when the sensitivity to microwaves is greatest and resistance to ionizing radiations is also greatest. This would presuppose a synchronous population of tumor cells gained by some strategem. The idea is to prevent cytological repair of damage from ionizing radiation. The S phase is the period devoted to synthesis in the cell division cycle of a eucaryote with the G_1 growth period coming before, G_2 after, and D the final mitotic and cytokinetic phase. Various procedures such as nutritional control, irradiation, and heat shock or chemical agents are able to synchronize a population of cells.

The neoplasms that are susceptible to additional selective treatment due to their type and origin may be treated by cytotoxic and antiviral agents, such as interferon along with the multifrequency radiotherapy. The function of the computer is to calculate exposures from suitable sensors and to manage the applicator topography and radiation energy accordingly. Unfortunately, thermocouples even as thin as tens of micrometers pick up stray fields, act as antennas or heat sinks, and distort the therapy field, with these disadvantages becoming progressively worse with lower radiofrequencies. Thus power meters and reflectometers are preferred as controller elements for computer input. For tissue hyperthermia, the matching transformer becomes topographical with the dielectric loading being adjustable to body contours. Then matching and this contour-following complement each other. Mendecki and his team (1978) used the RCA 4×4 dipole array with space transmission at 10 GHz. It is shifted to 2.45 GHz by making a special dielectric antenna out of the printed dipoles, with support from the microwave cavity, and the loading element (bag of contact, dielectric powder). Ecoflex HiK, K12 dielectric was the frequency-shifting element. In the array, computer control can program the activation of the dipoles to smooth the thermia.

Interferon, as a host-specific antiviral agent interferes with viral takeover and counters the transformation of the host cell, Both DNA and RNA viruses are sensitive and interferon or its inducers block mouse viral leukemias and sarcomas in concentrations of about 30 units per ml (interferon stock activity of 3000 units per ml). The viral functions that suffer are also replication and specification of the T-antigen production. Viruses may block the natural synthesis of interferon as a ploy in their attack.

It has been known since Bart and Kopf, 1969, that tumor cytotoxins such as the double strand synthetic RNA, homopolymer, polyinosinic-polycitidylic acid inhibits malignant mouse melanomas. The mechanism is not an antiviral interferon one but probably enhancement of immunoresponses.

Cancer seems apt as a candidate for the list of special information diseases since it functions directly with DNA master molecules as they lose organizational control over proliferation of cell masses. Treatment is an effort to destroy this mass of cells selectively by an agency that recognizes them or induces better information procedures in the immunosystem so that it carries out its role of recognition of this foreign cell line and prepares suppressive and lytic agents against it. We can foresee a happy coincidence of these early applications of hyperthermia and a far more sophisticated communication of signals directly to the cancer system (Chapter 14). The current efforts are producing applicators and controllers which always precede a surge of applications (Taylor 1980). Concurrently, the analytical thinking toward the direct interaction goal has shifted toward Germany with S. J. Webb at the Max Planck Institute in Stuttgart where he can collaborate with D. Gericke, F. Dietzel and others who have used clostridial hyperthermia and X-radiation at Liebig and Frankfurt. The organism used by the latter is *C. oncolyticum, S. butyricum* (ATCC 13732) which shows the tumor environment is anaerobic and the effort is to impose a differential stress not bearable by the tumor cells by high-frequency hyperthermia and local X-radiation. First came the X-radiation at 2 kilorads, then the hyperthermia at 400 MHz, inductive, 150 W to 40-41°C and finally, after 12 hours, the anaerobe was injected (10^8 spores in 0.5 ml isotonic NaCl). The spores germinate, lyse the tumor cells, and do not trouble the injected mice which tolerate repeated applications. Some of the animals needed treatment for four or five relapses.

The concept of using microwave signals to influence the growth of tumors is suggested by the obvious contrast in thermographs of tumors (as opposed to hazardous X-rays) when enhanced by microwave heating. The tumor may naturally be hyperthermic to about 5.5°C. Tumor tissue, as opposed to tumor cells, is more vascular than normal and absorbs microwaves better and the better circulation also thermoregulates to cool the tissue faster, at least for some tumors. Cells on the other hand, may well go along with this, but present the possibility of direct absorption into components other than blood (thermal or athermal cytotoxicity). Tumors may vary in vascularity, cellularity, and in connective tissue. Thompson and his coworkers (1978) found about 2°C greater temperature increase in liver hepatoma of a guinea pig, under microwave (2450 MHz) when this energy was used for image enhancement of thermographs (about 75 s. at *24 microwatts/ cm^2*).

In whole body hyperthermia with microwaves (clinical practice), the exposure of vital organs becomes a consideration because exposure may take over three hours, especially when the objective temperature is as low as 41.5°C. In local treatment, the patient can tolerate temperatures of 42.5 to 44°C and the time for treatment will be short. Cure depends on temperature and tumor sensitivity to ionizing radiation doubles for each degree of microwave hyperthermia, an effect probably due, in part, to inactivation of DNA repair enzymes. The exact timing for optimum predisposition to neoplasmic damage is not clear, since reasonable flexibility in scheduling is allowed. Even a single exposure to microwave thermia with radiotherapy or chemotherapy delays regrowth of a mouse tumor more than radiotherapy or chemotherapy alone. Problems remain for exploiting the penetration possible in LF bands which show excessive heating of certain lossy tissues such as fat as well as problems in directivity. At the frequencies currently in use, standing waves may be set up in tissue layers. Solutions have utilized matching dielectrics, precision hyperthermia, needle antennas, custom fitting of applicators, and adjustments in directors during single or multiple applications of microwaves to accommodate variations in target tumors. With implanted 0.5 mm or larger antenna directors, hyperthermia can be "inserted" into tumor masses the size of a large bean. Multiple antennas will then treat larger or intractable tumors with an input of 8 W or less by puncture or implant techniques. The use of several directors and their motion to alternate targets tends to allow frequencies in the 0.4 to 3 GHz range to perform well, even with limited penetration. "Pinpointing" the treatment and energy coupling procedures at 0.915 or 2.450 GHz are additional techniques.

"Pinpointing" in the hands of Taylor (1980) means the use of a UG/178 BU coaxial line with a 2450 MHz microwave syringe acting as a termination or injector. The coax measures only 1.8 mm and so can follow endoscopy routes. The small size does not mean cut off, as with wave guides used in contact externally. Rather, the termination is in a needle only 0.8 mm, or in a modification of the coax. This produces a "water drop" thermia pattern about 5 cm long. To extend the thermia, the coax termination is made into a communications-type cross switch. The inner and outer conductors of the coax are cross connected, one-half wavelength from the needle radiator. As in other implantations, the arrangement can permit slight retraction of the syringe to leave the energized termination in the tumor. Various positioning devices can be adapted to make the correct placement and the small antenna insures local thermia with fractional centimeter control. For human esophageal squamous cell carcinoma, microwave thermia after surgery is promising. Brain glioblastoma multiforme is being studied with microwaves preceding radiation therapy and surgery. The wave guide contact applicators for external malignancies give a heating pattern where the width is one-half to two-thirds the width of the wave guide.

LYMPHOCYTES AND THE IMMUNORESPONSE

Lymphocytes are convertible to macrophages, plasma cells, reticulocytes, red blood cells, white blood cell precursors, fibroblasts (which make collagen), and small lymphocytes involved in antibody production. The spleen makes T cells for the immunoresponse and the T cells, as spleenic lymphocytes, have an important recognition and information role in the cellular immunoresponses. Also from the spleen come the B cells which carry out the humoral (body fluid) response. Antibodies come from the lymphocytes via the plasma cell *phase* and are connected with the reticuloendothelium system. Monocytes are specifically directed against longer term infections. The mast cells are the wandering macrophages that absorb the debris, e.g. the remains after the fight against bacteria. Thus the lymphocytes originate in the lymph nodes, the spleen, and also in the thymus and are distinguished by these multiple roles and are directly in the information pathway. After exposure, the Japanese quail embryos grew into female birds with larger bursa. Bursa make the B cells or lymphocytes which can transfer to and clone up in the spleen. In the quail, 90% of the antibodies have this spleenic source. Many of these lymphocyte operations are influenced by electromagnetic radiation. The *lymphocyte division endpoint* illustrates the phasic nature of the biophenomenon. Lymphocytes are known to be sensitive to the stimulating agent or mitogen, phytohemagglutinin, (PHA), after a single high frequency exposure. First the lymphocyte activity decreases, then it "explodes" by several orders of magnitude three days after exposure. Within 24 hours in primates, the lymphocyte division response is cancelled and then it reappears and is inductively coupled to radio frequencies. Lymphocytes may undergo blast transformation but then fail to divide in Chinese hamsters. This raises the question of the adequacy of the lymphocyte response to antigen challenge after exposure. When exposed to low levels, less than 30 mW/cm^2 at 2450 MHz of cw energy for 15 minutes per day for 5 days, the Chinese hamsters show the lymphocyte transformation response. This is seen as a direct information transfer in which the message is "undergo transformation". Then more lymphocytes transform to lymphoblasts in a response that is proportional to the exposure. *Thus microwaves are mitogenic* and, at 30 mW/cm^2, there are 4.3 times more transforms. The EM radiation stimulates the division of lymphocytes and their transformation in the intact animal with its information system complete, but does not do this in cells exposed *in vitro*. Thus, if the whole system is not there, the electromagnetics cannot affect the biological endpoint. One modification of this hypothesis may be necessary. The reception by the intact animal is regarded as a primary reaction with the message to the lymphocytes following.

Susceptible Part of a Cycle and Models

The effect on cell divisions in the bone marrow of the guinea pig is time-of-day dependent. There is an immunorhythm of inhibition/stimulation and an oscillation with respect to phagocytic and antibacterial activity. Thus the demonstration of the endpoint depends upon knowledge of the susceptible part of the cycle as has been shown by a growing number of investigators using the technique. The experimental model must not be a tissue culture even though these preparations are ideal for other purposes. Similarly, phantoms or biophysical constructs offer computer advantages but it would be difficult to demonstrate a response that requires the whole animal. Even in phantoms it is sometimes difficult to include the electromagnetic-sensitive region and to conduct the exposure in the appropriate mode (thermic, athermic) to make the test relevant. When the sensitive part of the animal that is receptive is hormonal, for example pituitary or a rapid glucocorticoid response that in turn stimulates lymphocytes, it may be doubly difficult to replace the intact animal.

Glucocorticoids are immunorepressive and they demonstrate the point that there will usually be molecular or subcellular effects before the results are seen in the intact animal. Nevertheless the endpoint is so certain that it seems as if lymphocytes "feel" an exposure, for example, to 2000 to 3400 V/cm at 26 MHz and RNA "feels" 100 MHz at 30 V/cm in the exposure media. Yet some of these effects may be absorption and excitation effects when the H bonds in tRNA are affected at 10 kV/m and 100 MHz, or when the peptide backbone of chymotrypsin is stiffened in an electric field (not magnetic) as shown by changed raman signals.

Transition from Information to Noise

The rapid glucocorticoid response at 2450 MHz comes to be immunosuppressive by glucocorticoids instead of directly from the force of the applied field, while, at the same time, EM radiation can stimulate the division of lymphocytes. Information may be transferred to the endocrine system directly by the thermic content of the EM radiation. This occurs when, in one of the modes discussed, there is selective heating of an element in the endocrine system. This selectivity is informational and it is not a general hyperthermia. There is a suggestion that the radiation represents information at 5 mW/cm^2 but a noise saturation at 20 mW/cm^2.

When lymphocytes are under the influence of the phytohemagglutinin stimulation, 5 mW/cm^2 significantly alters the lymphocytes in their uptake of tritiated thymidine. The susceptibility to this stimulant mitogen is associated with a rhythm so that certain effects must be in phase with the organism. The regulation and control of cancer may be related to this rhythm in view of the emerging pattern on blood-forming, immune and endocrine systems.

In primates, where the results are more easily extrapolated to man, the lymphocyte division response (stimulation) in circulating lymphocytes at high frequency is the same at all pertinent exposures.

The stimulation of hematopoiesis in mice in the bone marrow and spleen is at 100 mW/cm^2 so that the nonionizing radiation (in this case microwaves) can modify the bone marrow hematopoietic syndrome that is so damaging in sublethal doses of X-rays. Once again reparation and the incorporation of iron are functions of the phase of hematopoiesis in which the microwaves are applied.

BLOOD PATTERN IN CHRONIC AND ACUTE EXPOSURE

The blood pattern responses to nonionizing radiation are delicate expressions of the organism's reception of energy and information but are different from the patterns seen with ionizing radiations. In particular, abnormalities are reversible and repair is apparently much more common. Stimulation appears to be a more common response than inhibition at this stage in the study. More ^{59}Fe is incorporated in the spleen, and metabolic activity in granulocytes increases as do acid phosphatases and lysozymes. There are changes in interphase nuclei of erythroblasts with more chromosome aberrations and mitotic abnormalities in these after exposure.

At 14 mW/cm^2 of cw, 2450 MHz, E field exposure along the long axis of Japanese quail eggs held at a constant 37°C, the white blood cells and lymphocytes drop in number while heterophiles increase. The hemoglobin is up slightly, monocytes may increase and the hatch is normal. When the lymphocytes decrease it is natural that the total leucocytes also decrease. But after a few months in rats exposed to 2450 MHz at 24.4 mW/cm^2, these are normalized.

Intersystem Theory

Under low exposure, erythrocytes experience a loss of potassium. There is often an increase in cell membrane permeability and a decrease in phagocytosis so that the host organism is less prepared for a bacterial challenge. An infection with *Staphylococcus aureus*, introduced as a stress to the organism exposed to microwaves, increases the change in granulopoiesis over animals which have only the stress of infection. Rabbits have a lower leukocytosis and granulocytosis and lysozyme activity and the bone marrow reserve pool shrinks more. This response is an example of a low exposure, chronic microwave effect which lowers the resistance to infection. The point is too, that normally granulopoeisis can cope with the microwave exposure but under infection with a pathogenic strain of staphylococcus the animal loses some capacity. Figure 15-1 illustrates these and other changes in a diagrammatic form dividing the blood pattern responses into those that occur soon after exposure and those that occur later.

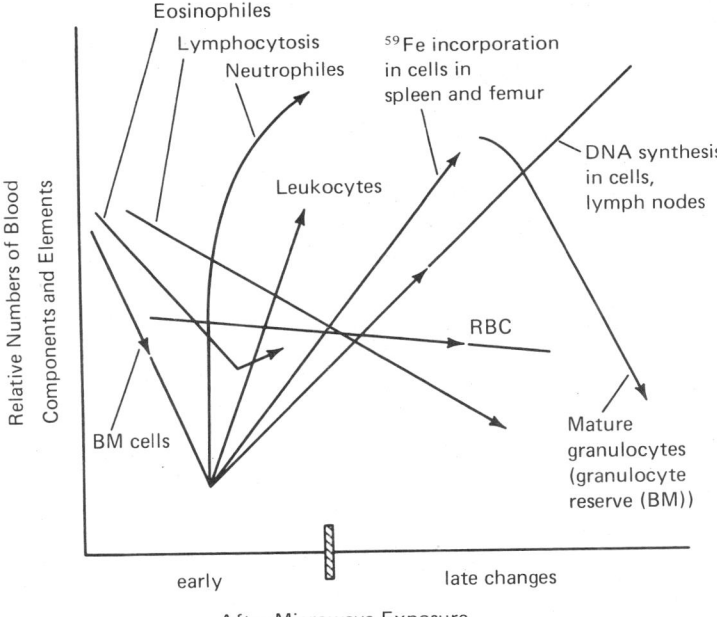

Fig. 15-1. Blood changes from EMR exposure—Diagrammatic

The chromosome aberrations reported by Baranski and Czerski (1976) were only in the erythroblasts and lymphoid cells of exposed bone marrow, but they saw bridges between daughter erythroblasts, deformed chromatin structures in nuclei, and hyperpycnotic metaphase chromosomes. The point here is that routine lymphocyte counts which are significant in diagnosis would not reveal these chromosome molecule changes.

PROBLEMS

1. Name three systems with associated logic for direct electromagnetic stimulation or inhibition.
2. What is meant by cell nutrition and heterotrophic control of growth, and how can these be related to electromagnetic radiations?
3. What is meant by the interaction of systems according to time?
4. Is commensurate interaction the sole basis for intersystem communication in biology?
5. What is the logic of the sensitive series model?
6. What is the Adey communicating model?
7. What type of informing action is assigned to K^+, Na^+, Ca^{++}, and Mg^{++} ions?
8. Would rectal temperatures in rats be representative of the animal's general temperature?
9. Why is a thermal input from EMR equivalent to direct information transfer?
10. How does EMR affect serum triglycerides?
11. Is microwave hyperthermia effective against tumors?
12. What biological endpoint is demonstrated by lymphocytes exposed to electromagnetic radiations?
13. What is the significance of a chromosome break?

BIBLIOGRAPHY

Adey, W.R., R.T. Kado, and J. Didio, "Impedance Measurements in Brain Tissue of Animals Using Microvolt Signals," *Exp. Neurol.*, Vol. 5, 47-66, 1962.

Adey, W.R., R.T. Kado, J. Didio, and W.J. Schindler, "Impedance Changes in Cerebral Tissue Accompanying a Learned Discriminative Performance in the Cat," *Exp. Neurol.*, Vol. 7, 259-281, 1963.

Adey, W.R., et al, "The Role of Neuronal Elements in Regional Cerebral Impedance Changes in Alerting, Orienting and Discriminative Responses," *Exp. Neurol.*, Vol. 15, 490-510, 1966.

Adey, W.R., "Efidence for Cerebrum Membrane Effects of Calcium, Derived from Direct-Current Gradient, Impedance, and Intracellular Records," *Exp. Neurol.*, Vol. 30, 78-102, 1971.

Adey, W.R., "The Influences of Impressed Electrical Fields at EEG Frequencies on Brain and Behavior," *Behavior and Brain Electrical Activity*, H. Eltshuler and N. Burch, Eds., Plenum Publishing Corp., New York, 1974.

Adey, W.R., "Effects of Electromagnetic Radiation on the Nervous System," *Biologic Effects on Nonionizing Radiation, Annals of the New York Academy of Sciences*, Vol. 247, 15-20, 1975.

Adey, W.R., "Evidence for Cooperative Mechanisms in the Susceptibility of Cerebral Tissue to Environmental and Intrinsic Electrical Fields," *Functional Linkage in Biomolecular Systems*, F.O. Schmitt, Ed., Raven Press, New York, 1975.

Albert, E.N. and M. DeSantis, "Do Microwaves Alter Nervous System Structure?" *Annals of the New York Academy of Sciences*, Vol. 247, 87-108, 1975.

Allis, J.W., Discussion paper, "Irradiation of Bovine Serum Albumin with a Crossed-Beam Exposure Detection System," *Annals of the New York Academy of Sciences*, Vol. 247, 312-321, 1975.

Austin, G.N. and S.M. Horvath, "Production of Convulsions in Rats by High Frequency Electrical Currents," *American Journal Phys. Med.*, Vol. 33, 141-149, 1954.

Averbeck, D., M. Dardalhon and A.J. Berteaud, "Microwave Action in Procaryotic and Eucaryotic Cells and a Possible Interaction with X-rays," *J. Microwave Power*, Vol. 11, 2, 143-144, 1976.

Bansal, S. C. and team of physicians, "*Ex vivo* Removal of Serum IgG in a Patient with Colon Carcinoma," *Cancer 42*, July 1978.

Baranski, S., "Histological and Histochemical Effects of Microwave Irradiation on the Central Nervous System of Rabbits and Guina-Pigs," *Amer. J. Phys. Med.*, Vol. 51, 182-191, 1972.

Baranski, S. and S. Szmigielski, "Effect of Microwave Irradiation in Vitro on Cell Membrane Permeability," *Biological Effects and Health Hazards of Microwaves*, P. Czerski, Ed., Polish Medical Publishers, Warsaw, Poland (In press) 1975.

Baranski, S. and Z. Edelwejn, "Experimental Morphologic and Electroencephalographic Studies of Microwave Effects on the Nervous System," *Annals of the New York Academy of Sciences*, Vol. 247, 109-116, 1975.

Bart, R. S., and Kopf, A. W., "Inhibition of the Growth of Murine Malignant Melanosas with Synthetic Double-Stranded Ribonucleic Acid," *Nature*, 224, October 25, 1969.

Barthakur, N., "Stomatal Response to Microwave Induced Thermal Stresses," *J. Microwave Power*, Vol. 11, 3, 247-253, 1976.

Bawin, S.M., R.J. Gavalas-Medici and W.R. Adey, "Effects of Modulated Very High Frequency Fields on Specific Brain Rhythms in Cats," *Brain Res.*, Vol. 58, 365-384, 1973.

Bawin, S.M., *Cat EEG and Behavior in Very High Frequency Electric Fields Amplitude-modulated at Brain Wave Frequencies*, Ph.D. Dissertation, University of California, Los Angeles, 1972.

Bawin, S.M., Kacamarek, L.K. and W.R. Adey, "Effects of Modulated VHF Fields on the Central Nervous System," *Annals of the New York Academy of Sciences*, Vol. 247, 74-81, 1975.

Bawin, S.M., R.J. Gavalas-Medici and W.R. Adey, "Effects of Modulated VHF Fields on Specific Brain Rhythms in Cats," *Brain Res.*, Vol. 58, 365-384, 1973.

Ben-Hur, E., B.V. Bronk and M.M. Elkind, "Thermally Enhanced Radiosensitivity of Cultured Chinese Hamster Cells," *Nature New Biol.*, Vol. 238, 209-211, 1972.

Ben-Hur, E., B.V. Bronk and M.M. Elkind, "Thermally Enhanced Radioresponse of Cultured Chinese Hamster Cells: Inhibition of Repair of Sublethal Damage and Enhancement of Lethal Damage," *Radia. Res.*, Vol. 58, 38-51, 1974.

Bereznitskaya, A.N., "Studies of the Gonadotropic Activity of the Pituitary Gland of Female Mice irradiated with 10-Centimeter and Ultrashort Waves," *Gigiena Truda Prof Zabolev*, Vol. 9, 33-37, 1968.

Bertharion, G, B. Servantie and R. Jolly, "Electrocorticographic Modifications After Exposure to Microwave Field, on the White Rat," *J. Microwave Power*, Vol. 6, 62, 63, 1971.

Blackman, C.F., *et al*, "Effect of Nonionizing Electromagnetic Radiation on Single Cell Biologic Systems," *Annals of the New York Academy of Sciences*, Vol. 247, 352-365, 1975.

Blackman, C.F., *et al*, "Effects on Nonionizing Electromagnetic Radiation on Single Cell Biologic Systems," *Annals of the New York Academy of Sciences*, Vol. 247, 352-365, 1975.

Carpenter, R.L. and E.M. Livston, "Evidence for Nonthermal Effects of Microwave Radiation: Abnormal Development for Irradiated Insect Pupae," *IEEE Trans. Microwave Theory Tech.*, Vol. 19, 173-178, 1971.

Chou, C.K. and A.W. Guy, "Effect of 2450 MHz Microwave Fields on Peripheral Nerves," *IEEE-G-MTT International Microwave Symposium Digest*, S.W. Maley, Ed., 318-320, Boulder, Colo., 1973.

Cleary, S.F., Ed., *Proceedings of the Biological Effect and Health Implications of Microwave Radiation, Richmond, Va.*, 17-19 September, 1969, U.S. Department of Health, Education and Welfare, Bureau of Radiological Health, Rep. 70-2, (PB 193-898), 1970.

Copson, D.A., *Microwave Heating*, 2nd. Ed., 1-9, 12-14, 555-560, Avi Publishing Co., Conn., 1975.

Czerski, P., E. Paprocka-Slonka, M. Siekierzyński and A. Stolarska, In *Biologic Effects and Health Hazards of Microwave Radiation:* 67, Polish Medical Publishers, Warsaw, Poland, 1974.

Czerski, P., "Microwave Effects on the Blood-Forming System with Particular Reference to the Lymphocyte," *Annals of the New York Academy of Sciences*, Vol. 247, 232-242, 1975.

Denisiewicz, R., R. Dzuk and M. Siekierzynski, "Evaluation of Thyroid Function in Persons Occupationally Exposed to Microwave Radiation," *Polsk. Arch. Med. Wewnetrznej*, Vol. 45, 19-25, 1970.

Dickson, J.A. and D.S. Muckle, "Total Body Hypothermia versus Primary Tumors, Hyperthermia in the Treatment of Rabbit VX-2 Carcinoma," *Cancer Res.*, Vol. 32, 1916-1923, 1972.

Dickson, J.A. and H.A. Ellis, "Stimulation of Tumor Cell Dissemination by Raised Temperature ($42°C$) in Rats with Transplanted Tumors," *Nature New Biol.*, Vol. 248, 354-358, 1974.

Dietzel, F., "Effect of Electromagnetic Radiation of Implantation and Intrauterine Development of the Rat," *Annals of the New York Academy of Sciences*, Vol. 247, 367-376, 1975.

Dietzel, F., D. Gericke, L. Schumacher and G. Linhart, "Combination of Radiotherapy Microwave-Hyperthermia and Clostridial Oncolysis on Experimental Mouse Tumors," *Strahlentherapie*, (Sonderbd.) *154*, 1978.

Doury, P., P. Boisselier and J.G. Bernard, "Pathological Effects on Man of the UHF Electromagnetic Radiation of Aircraft Radar," concerning an observation, *Sem. Hop. Paris*, Vol. 46, 2681-2683, 1970.

D'Yachenko, N.A., "Changes in Thyroid Function with Chronic Exposure to Microwave Radiation," *Gig. Truda Prof. Zabolev.*, Vol. 14, 51, 52, 1970.

Edelwein, Z., "Attempt at Evaluation of the Functional State of Brain Synapses in Rabbits Exposed Chronically to the Action of Microwaves," *Acta Physiol. Polon.*, Vol. 19, 791-799, 1968.

Elder, J.A. and J.S. Ali, "The Effect of Microwaves (2450 MHz) on Isolated Rat Liver Mitochondria," *Annals of the New York Academy of Sciences*, Vol. 247, 251-262, 1975.

Elul, R., "Specific Site of Generation of Brain Waves," *Physiologist*, Vol. 7, 125, 1964.

Elul, R., "Use of Nonuniform Electric Fields for Evaluation of the Potential Difference Between Two Phases," *Trans. Faraday Soc.*, Vol. 62, 3484-3492, 1966.

Elul, R., "Fixed Charge in the Cell Membrane," *J. Physiol.*, (London), Vol. 189, 351-365, 1967.

Frey, A.H., "Human Auditory System Response to Modulated Electromagnetic Energy," *J. Appl. Physiol.*, Vol. 17, 689-692, 1962.

Frey, A.H., "Brain Stem Evoked Responses Associated with Low-Intensity Pulsed UHF Energy," *J. Appl. Physiol.*, Vol. 23, 984-988, 1967.

Frey, A.H. and R. Messenger, Jr., "Human Perception of Illumination with Pulsed Ultrahigh-Frequency Electromagnetic Energy," *Science*, Vol. 181, 356-358, 1973.

Frey, A.H. and E. Seifert, "Pulse Modulated UHF Energy Illumination of the Heart Associated with Change in Heart Rate," *Life Science*, Vol. 7, 11, 505-512, 1968.

Frohlich, H., *Proc. Nat. Acad. Sci. U.S.A.*, Vol. 72, 4211, 1975.

Frohlich, H., *Phys. Lett.*, Vol. 51A, 21, 1975.

Frohlich, H., *Nature*, Vol. 228, 1093, 1970.

Frohlich, H., *Intern. J. Quantum Chem.*, Vol. 2, 641, 1968.

Gavalas, R.J., D.O. Walter, J. Hamer and W.R. Adey, "Effect of Low Level, Low-Frequency Electric Fields on EEG and Behavior in *Macaca nemestrina*," *Brain Res.*, Vol. 18, 491-501, 1970.

Gerner, E.W., *et al*, "Biochemical Aspects of Hyperthermic Damage. The Effects of Elevated Temperatures on Two Enzymes Involved in Cell Proliferation," *Radiation Res.*, Vol. 70, 3, 1977.

Gerner, E.W., *et al*, "Hyperthermia-Induced Alterations of Macro-Molecular Synthesis," *Radiation Res.*, Vol. 70, 3, 1977.

Giarola, A.J. and W.F. Krueger, "Continuous Exposure of Chicks and Rats to Electromagnetic Fields," *IEEE Trans. Microwave Theory Tech.*, Vol. MTT-22, 432-437, 1974.

Giese, A.C., *Cell Physiology*, 4th Ed., W.B. Saunders, Phila., Pa., 1973.

Guy, A.W., "Analyses of Electromagnetic Fields Induced in Biological Tissues by Thermograph Studies on Equivalent Phantom Models," *IEEE Trans. Microwave Theory Tech.*, Vol. MTT-19, 205-214, 1971.

Goodman, L.S. and A. Gilman, *The Pharmacological Basis of Therapeutics*, 5th Ed., Macmillan, New York, 1975.

Gordon, Z.V., *Biological Effects of Microwaves in Occupational Hygiene*, N.Kaner, Trans. NASA TTF-633, The Israel Program for Scientific Translations, Jerusalem, Israel, 1970.

Grodsky, I.T., "Possible Physical Substrates for the Interaction of Electromagnetic Fields with Biologic Membranes," *Annals of the New York Academy of Sciences*, Vol. 247, 117-124, 1975.

Guy, A.W., F.A. Harris and H.S. Ho, "Quantization of the Effects of Microwave Radiation on Central Nervous System Function," *Proceedings of the 6th Annual International Microwave Power Symposium*, Monterey, Calif., 1971.

Hahn, O.M., "Metabolic Aspects of the Role of Hyperthermia in Mammalian Cell Inactivation and Their Possible Relevance to Cancer Treatment," *Cancer Res.* Vol. 34, 3117-3123, 1974.

Haidt, S.J. and A.H. McTighe, "The Effect of Chronic, Low-Level Microwave Radiation on the Testicles of Mice," *IEEE-G-MTT Int. Microwave Symposium*, IEEE Cat. No. 73, CHO 736-9 MTT, 324, 325, 1973.

Hall, R.R., R. Schade and J. Swinney, "Effects of Hyperthermia on Bladder Cancer," *Brit. Med. J.*, Vol. 2, 593-594, 1974.

Hamer, J., "Effects of Low Level, Low Frequency Electric Fields on Human Reaction Time," *Commun. Behav. Biol.*, Vol. 2(A), 217-222, 1968.

Hamid, M.A., R.J. Boulange, G.C. Hodgson, P.A. Kondra, K. Smith, and D.B. Bragg, "The Effect of Microwave Radiation on the Growth and Reproduction of Chickens," *J. Microwave Power*, Vol. 4, 253-256, 1968.

Health Effects of Ionizing and Nonionizing Radiation, Working Group Report: 15, Report No. EUR 0-4701, World Health Organization, Copenhagen, Denmark, 1972.

Ho, H.S., E.I. Ginns, and C.L. Christman, "Environmental Controlled Waveguide Irradiation Facility," *IEEE Trans. on Microwave Theory and Techniques*, Vol. MTT-21/(2), 837-840, 1973.

Johnson, C.C. and A.W. Guy, "Nonionizing Electromagnetic Wave Effects in Biological Materials and Systems," *Proc. IEEE*, Vol. 60(6), 692-718, 1972.

Johnson, C.C. and C.M. Durney, "A Theoretical Estimation of Tissue Anisotropy Effects of Electromagnetic Power Deposition 0.1-100 MHz.," *Proc. URSI/GAP Conf.*, 171, 1973.

Kaczmarek, L.K. and W.R. Adey, "Weak Electric Gradients Change Ionic and Transmitter Fluxes in Cortex," *Brain Res.*, Vol. 66, 537-540, 1974.

Kaplan, I.T., et al "Absence of Heart-Rate Effects in Rabbits During Low-Level Microwave Irradiation," *IEEE Trans. Microwave Theory and Techniques*, MTT-19, 168-173, 1971.

Katchalsky, A., "Polyelectrolytes and their Biological Interaction," in *Connective Tissue: Intercellular Macromolecules*, 9-42, Little Brown and Co., Boston, Mass., 1964.

Kirchev, K., P. Eftinov and S. Sivchev, "Some Experimental Data on the Effect of a UHF Electric Field on the Adrenals," *Prob. Physiother. Health Report*, (Moscow), 81-88, 1959.

Kim, J.H. et al "Local Tumor Hyperthermia in Combination with Radiation Therapy. 1-Malignant Cutaneous Lesions," *Cancer*, Vol. 40, 161-169, 1977.

Klainer, S.M., discussion paper, "Raman Spectroscopy of Molecular Species During Exposure to 100 MHz Radio-Frequency Fields," *Annals of the New York Academy of Sciences*, Vol. 247, 323-326, 1975.

Kolesnk, F.A., V.M. Malyshey and BF. Murashov, "Disorders of Endocrine System in Chronic Exposure to a UHF Field," *VoyennoMed. Zh.* Vol. 7, 39-41, 1967.

Korbel, S.F. and H.L. Fine, "Effects of Low Intensity UHF Radio Fields as a Function of Frequency," *Phychonom. Sci.*, Vol. 9, 527, 528, 1967.

Krueger, W.F., A.J. Giarola, J.W. Bradley, "Effects of Electromagnetic Fields on Fecundity in the Chicken," *Annals of the New York Academy of Sciences*, Vol. 247, 391-401, 1975.

Lambert, P.D., R.C. Nealeigh and M. Wilson, "Effects of Microwave Exposure on the Central Nervous System of Beagles," *J. Microwave Power*, Vol. 7, 367-396, 1972.

Lenko, J. et al "Effect of 10-cm Radar Waves on the Level of 17-Ketosteroids and 17-Hydroxycorticosteroids in the Urine of Rabbits," *Przeglad Lekar.*, Vol. 22, 296-299, 1966.

Leytes, F.L. and L.A. Skurikhina, "The Effect of Microwaves on the Hormonal Activity of the Adrenal Cortex," *Byul. Eksp. Biol. Med.*, Vol. 52, 47-50, 1961.

Lin, J.C., A.W. Guy and G.H. Kraft, "Microwave Selective Brain Heating," *J. Microwave Power*, Vol. 8, 275-286, 1973.

Listova, N.M., *Influence of Microwave Radiation on the Organism of Man and Animals*, I.R. Petrov, Ed., Meditsina Press, Leningrad, USSR, (NASA TT F-708), 1963.

Liu, L.M., et al, "Further Experiments Seeking Evidence of Nonthermal Biological Effects of Microwave Radiation," *Digest of Technical Papers, IEEE G-MTT International Microwave Symposium*, 1973.

Martin, J.B., "Plasma Growth Hormone (GH) in Response to Hypothalamic or Extrahypothalamic Electric Stimulation," *Endocrinology*, Vol. 91, 107-115, 1972.

Marton, J.P., "Conjectures on Superconductivity and Cancer," *Physiol. Chem. and Physics*, Vol. 5, 259-270, 1973.

Mayers, C.P. and J.A. Habeshaw, "Depression of Phagocytosis: A Non-Thermal Effect of Microwave Radiation as a Potential Hazard to Health," *Int. J. Radiation Biology*, Vol. 24, 449-462, 1973.

McLees, B.D. and E.D. Finch, "Analysis of Reported Physiologic Effects of Microwave Radiation," in *Advances in Biological and Medical Physics*, J.H. Lawrence and J.W. Gofman, Eds., Vol. 14, 164-220, Academic Press, New York, 1973.

McRee, D.I., "Environmental Aspects of Microwave Radiation," *Environ. Health Persp.*, (2), 41-53, 1972.

McRee, D.I., "Determination of the Absorption of Microwave Radiation by a Biological Specimen in a 2450 MHz Microwave Field," *Eng. Health Physics*, Vol. 26, 385-390, 1974.

McRee, D.I., P.E. Hamrick, and J. Zinkl, "Some Effects of Exposure of the Japanese Quail Embryo to 2.45 GHz Microwave Radiation," *Annals of the New York Academy of Sciences*, Vol. 247, 377-390, 1975.

Michaelson, S.M., R.A.E. Thomson, and J.W. Howland, "Physiologic Aspects of Microwave Irradiation of Mammals," *Amer. J. Physiol.*, Vol. 201, 351-356, 1961.

Michaelson, S.M., et al, "Biochemical and Neuroendocrine Aspects of Exposure to Microwaves," *Annals of the New York Academy of Sciences*, Vol. 247, 1975.

Mikolajczyk, H., "Hormone Reactions and Changes in Endocrine Glands Under Influence of Microwaves," *Med. Lotn.*, Vol. 39, 39-51, 1972.

Milroy, W.C., and S.M. Michaelson, "Thyroid Pathophysiology of Microwave Radiation," *Aerospace Med.*, Vol. 43, 1126-1131, 1972.

Okamoto, Y., and W. Brenner, *Organic Semiconductors*, Vol. 1, Reinhold Publishing Co., New York, 1964.

Ordman, L.J. and T. Gilman, "Studies in the Healing of Cutaneous Wounds. 1. The Healing of Incisions Through the Skin of Pigs," *Arch. Surg.*, Vol. 3, 6, 857-882, 1966.

Overgaard, J. and K. Overgaard, "Investigation of the Possibility of Thermic Tumor Therapy—Shortwave Treatment of a Transplanted Isologous Mouse Mammary Carcinoma," *European J. Cancer*, Vol. 8, 65-78, 1972.

Palzer, R.J. and C. Heidelberger, "Studies on the Quantitative Biology of Hyperthermic Killing of HeLa Cells," *Cancer Res.*, Vol. 33, 422-427, 1973.

Parker, L.N., "Thyroid Suppression and Adrenomedullary Activation by Low Intensity Microwave Radiation," *Amer. J. Physiol.*, Vol. 224, 1388-1390, 1973.

Pasderova-Vejlupkova, J. and Z. Frank, "Influence of Pulsed Microwaves on Haematopoiesis of Adolescent Rats," *J. Microwave Power*, Vol. 11, 2, 1976.

Pease, D.C., "Polysaccharides Associated with the Exterior Surface of Epithelial Cells: Kidney, Intestine, Brain," *J. Ultrastruc. Res.*, Vol. 15, 555-583, 1966.

Petrow, I.R., Ed., *Influence of Microwave Radiation on the Organism of Man and Animals*, NASA TT-F-708, National Technical Information Service, Springfield, Va., 1972.

Petrow, I.R. and V.A. Syngayevskaya, "Endocrine Glands," in *Influence of Microwave Radiation on the Organisms of Man and Animals*, I.R. Petrov, Ed., 31, Meditsina Press, Leningrad, USSR, 1970.

Pettigrew, R.T., "The Effect of Whole Body Hyperthermia in Advanced Cancer," *Brit. J. Cancer*, Vol. 30, 179, 1974.

Presman, A.A., Ed., *Electromagnetic Fields and Life*, Plenum Publishing Corporation, New York, 1970.

Prince, J., K. Mori and J. Frazer, "Cytologic Aspects of RF Radiation in the Monkey," *Aerospace Med.*, Vol. 43, 7, 1972.

Pyle, S.D., D. Nichols, F.S. Barnes, and E. Gamow, "Threshold Effects of Microwave Radiation on Embryo Cell Systems," *Annals of the New York Academy of Sciences*, Vol. 247, 401-407, 1975.

Robinson, J.E., D. McCulloch, and E. Edelsack, "Microwave Heating of Malignant Mouse Tumors and Tissue Equivalent Phantom Systems," *J. Microwave Power*, Vol. 11, 2, 87, 1976.

Robinson, J.E., M.J. Wizenberg and W.A. McCready, "Combined Hyperthermia and Radiation: An Alternative to Heavy Particle Therapy for Reduced Oxygen Enhancement Ratios," *Nature*, Vol. 25, 521-522.

Robinson, J.E., D. McCulloch, and E.A. Edelsack, "Evaluation of Tumor Immersion Technique for Radiobiological Hyperthermia Studies," *Medical Physics*, Vol. 2, 3, 159, Abstract H-8, 1975.

Robinson, J.E., M.J. Wizenberg, and W.A. McCready, "Radiation and Hyperthermal Response of Normal Tissue *in Situ*," *Radiobiology*, Vol. 113, 195-198, 1974.

Romera-Sierra, C., et al, "Electromagnetic Fields and Skin Wound Repair," *J. of Microwave Power*, Vol. 10, 1, March, 1975.

Rotkovská, D. and A. Vacek, "The Effect of Electromagnetic Radiation on the Hematopoietic Stem Cells of Mice," *Annals of the New York Academy of Sciences*, Vol. 247, 243-250, 1975.

Rotkovská, D. and A. Vacek, "Modification of the Repair of Radiation Damage of Haematopoiesis in Mice by Microwaves," *J. Microwave Power*, Vol. 11, 2, 141-143, 1976.

Rugh, R., E.I. Ginns, H.S. Ho, and W.M. Leach, "Are Microwaves Teratogenic?" in *Biological Effects and Health Hazards of Microwave Radiation, Proc. Int. Symp.*, Polish Medical Publishers, Warsaw, Poland, 1973.

Rugh, R., "The Relation of Sex, Age and Weight of Mice to Microwave Radiation Sensitivity," *J. of Microwave Power*, Vol. 11, 2, June 1976.

Ranck, J.B., "Specific Impedance of Rabbit Cerebral Cortex," *Exp. Neurol.*, Vol. 7, 144-152, 1963.

Rupp, T, J. Montet, and J.W. Frazer, "A Comparison of Thermal and Radio-Frequency Exposure Effects on Trace Metal Content of Blood Plasma and Liver Cell Fractions of Rodents," *Annals of the New York Academy of Sciences*, Vol. 247, 282-291, 1975.

Schliephake, E., *Short Wave Therapy. Influence of Microwave Radiation on the Organism of Man and Animals*, I.R. Petrov, Ed., Meditsina Press, Leningrad, USSR, (NASA TT F-708), 1970.

Schwan, H.P. "Interaction of Microwave and Radio Frequency Radiation with Biological Systems," *IEEE Trans. Microwave Theory Tech.* MTT-19, 146-152, 1971.

Schwan, H.P., "Microwave Radiation: Biophysical Considerations and Standards Criteria," *Trans. IEEE-BME*, Vol. 19, 304-312, 1972.

Schwan, H.P. and L.D. Sher, "Alternating-Current Field Induced Forces and Their Biological Implications," *J. Electrochem. Soc.*, Vol. 116, 22C-26C, 1969.

Schwartz, G., "Cooperative Binding in Linear Biopolymers. I. Fundamental States and Dynamic Properties," *Eur. J. Biochem.* Vol. 12, 442-453, 1970.

Servantie, B., A.M. Servantie, and J. Etienne, "Synchronization of Cortical Neurons by a Pulsed Microwave Field as Evidenced by Spectral Analysis of Electrocorticograms from the White Rat," *Annals of the New York Academy of Sciences*, Vol. 247, 82-86, 1975.

Shapiro, A.R., R.F Lutomirski and H.T. Yura, "Induced Fields and Heating Within a Cranial Structure Irradiated by an Electromagnetic Plane Wave," P4458-1 Rand Corp. Santa Monica, Calif., 1971, *IEEE Trans. Microwave Theory Tech.*, MTT-19, 187-196.

Smirnova, M.I., and M.N. Sadchikova "Determination of the Functional Activity of the Thyroid Gland by Means of Radioactive Iodine in Workers with UHF Generators," in *Biological Action of Ultrahigh Frequencies*, A.A. Letavet and Z.V. Gordon, Eds., 50, 51, Moscow, USSR, 1960.

Stehlin, J.S., et al, "Results of Hyperthermic Perfusion for Melanoma of the Extremities," *Surg. Gynecol. Obstet.*, Vol. 140, 338-348, 1975.

Straub, K.D. and P. Carver, "Effects of Electromagnetic Fields on Microsomal ATPase and Mitochondrial Oxidative Phosphorylation," *Annals of the New York Academy of Sciences*, Vol. 247, 292-300, 1975.

Subbota, A.G., "The Effect of a Pulsed Super-High Frequency SHF Electromagnetic Field on the Higher Nervous Activity of Dogs," *Bull. Exp. Med.*, Vol. 46, 1206-1211, 1958.

Subbota, A.G., "Changes in Functions of Various Systems of the Organism," in *Influence of Microwave Radiation on the Organism of Man and Animals*, I.R. Petrov, Ed., 66, Meditsina Press, Leningrad, USSR, 1970.

Suit, H.D. and M. Shwayder, "Hyperthermia: Potential as an Anti-Tumor Agent," *Cancer*, Vol. 34, 122-129, 1974.

Syngayevskaya, V.A., G.F. Pleskena-Sinenko and O.S. Ignatyeva, "The Effects of Microwave Radiation in the Meter and Decimeter Wave Ranges on the Endocrine Regulation of Carbohydrate Metabolism and the Functional State of Adrenal Cortex in Rabbits and Dogs," *Summaries of Reports. Questions of the Biological Effects of SHF-UHF Electromagnetic Field*, 51, 52, Kirov Order of Lenin Military Medical Academy, Leningrad, USSR, 1962.

Szady, J., et al, "Effects of Microwaves on the Twenty-Four Hour Rhythm and Twenty-Four Hour Urinary Excretion of Seventeen Hydroxycorticoids and Seventeen Ketosteriods," *J. Microwave Power*, Vol. 11, 2, 1976.

Szmigielski, S., J. Jeljaszewicz, and M. Wiranowska, "Acute Staphylococcal Infections in Rabbits Irradiated with 3-GHz Microwaves," *Annals of the New York Academy of Sciences*, Vol. 247, 305-311, 1975.

Szmigielski, S., "Effect of 10-cm (3GHz) Electromagnetic Radiation (Microwaves) on Granulocytes *in Vitro*," *Annals of the New York Academy of Sciences*, Vol. 247, 275-281, 1975.

Szmigielski, S., M. Luczak, and M. Wiranowska, "Effect of Microwaves on Cell Function and Virus Replication in Cell Cultures Irradiated *in Vitro*," *Annals of the New York Academy of Sciences*, Vol. 247, 263-274, 1975.

Tanner, J.A., C. Romero-Sierra, and S.J. Davie, "Nonthermal Effects of Microwave Radiation on Birds," *Nature* (London), Vol. 216, 1139, 1967.

Tasaki, I., T. Takewa and S. Yamagishi, "Abrupt Depolarization and Bi-ionic Action Potentials in Internally Perfused Squid Giant Axons," *Amer. J. Physiol.*, Vol. 215, 152-159, 1968.

Taylor, E.M., A.W. Guy, B. Ashleman and J.C. Lin, "Microwave Effects on Central Nervous System Atrributed to Thermal Factors," *IEEE-G-MTT Int. Microwave Symp. IEEE Cat. no. 73 CHO 736-9 MTT*, 316, 317, 1973.

Taylor, E.M. and B.T. Ashleman, "Analysis of Central Nervous System Involvement in Microwave Auditory Effect," *Bran Res.*, Vol. 74, 201-208, 1974.

Taylor, E.M. and B.T. Ashleman, "Some Effects of Electromagnetic Radiation on the Brain and Spinal Cord of Cats," *Annals of the New York Academy of Sciences*, Vol. 247, 63-73, 1975.

Terman, D. S., and team of seven, "Extensive Necrosis of Spontaneous Canine Mammary Adenocarcinoma After Extracorporeal Perfusion over *Staphylococcus aureus*, Cowans I," *Journal of Immunology*, 124, Feb., 1980.

Thompson, J. E., T. L. Simpson and J. B. Caulfield, "Thermographic Tumor Detection Enhancement Using Microwave Heating," *IEEE, MTT-26*, 8, August 1978.

Thrall, D.E., E.L. Gillettee, and C.L. Bauman,"Effect of Heat on the C3H Mouse Mammary Adenocarcinoma Evaluated in Terms of Tumor Growth," *European J. Cancer*, Vol. 9, 871-875, 1973.
Todorovic-Tesic, P., R. Genci and M. Kosanovic, "Influence of Microwaves Upon the Adrenals of the Rat," *Arth. Biol. Nauk* Vol. 17, 121-128, 1965.
Tolgaskaya, M.S. and Z.V. Gordon, "Change in the Neurosecretory Function of the Hypothalamus and the Neuro-Pituitary Body During Chronic Irradiation with Centimeter Waves of Low Intensity," *The Biological Effects of Radio-Frequency Fields Issue*, Vol. 3, 87-97, Institute of Work Hygiene and Occupational Diseases. Armenian SSR, 1968.
Tolgskaya, M.S. and Z.V. Gordon, *Pathological Effects of Radio Waves*, B. Haigh, Transl., Consultants Bureau, New York, 1973.
Traub, R.J., R.J. Vetter, and G.A. Stoetzel, "Microwave Hyperthermia Chemotherapy and Co-60 Radiation in the Treatment of Hamster Melanoma," *J. Microwave Power*, Vol. 12, 1, 40, 1977.
U.S. Standards Institute, New York, "USA Standard Safety Level of Electromagnetic Radiation with Respect to Personnel," *USAS C95.1 1-1966*, 1966.
Valtonen, E.J., "Giant Mast Cells—A Special Degenerative Form Produced by Microwave Radiation," *Exp. Cell Res.*, Vol. 43, 221-226, 1966.
Wallen, C.A., et al, "Microwave Induced Hyperthermia as an Adjuvamt to Cancer Therapy," *Proc. in J. Microwave Power*, Vol. 11, 2, 175-176, 1976.
Webb, S.J., *Bound Water in Biological Integrity*, Charles C. Thomas, Springfield, Ill., 1965.
Webb, S.J. and D.D. Dodds, "Inhibition of *E. coli* by Microwaves," *Nature* (London), Vol. 218, 374-375, 1968.
Webb, S.J., "Genetic Continuity and Metabolic Regulation as Seen by Effects of Various Microwave and Black Light Frequencies on These Phenomena," *Annals of the New York Academy of Sciences*, Vol. 247, 327-351, 1975.
Webb, S.J., "Effects of Microwaves on Normal and Tumor Cells as Seen by Laser Raman Spectroscopy," *J. Microwave Power*, Vol. 11, 2, 1976.
West, B.L. and W. Regelson, "Pulsed Radiowave (27.12 MHz) Effects on *In Vivo* and *In Vitro* Tumor Growth," *J. Microwave Power*, Vol. 11, 2, 176-177, 1976.
Yatvin, M.B., "Influence of Membrane Lipid Composition on Hyperthermic and Radiation Cell Killing," *Radiation Res.*, Vol. 70, 3, 1977.
Yeargers, E.K., et al, "Effects of Microwave Radiation on Enzymes," *Annals of the New York Academy of Sciences*, Vol. 247, 301-304, 1975.
Zaret, M. "Clinical Aspects of Nonionizing Radiation," *IEEE Trans. Biomed. Eng.*, BME-19, 4, 313-316, 1974.

16

Fields, Waves, and Penetration into Biological Preparations — A Beginning Basis for the Hazard Analysis for Various Frequencies

"Fear not the tyrants shall rule forever
Or the priests of the bloody faith.
They stand on the brink of that mighty river
Whose waves they have tainted with death."

Shelley, *Rosalind and Helen*

MODEL IS PART OF THE CIRCUIT

In microwave biology, fields and waves react specifically with the material according to the bioelectrics, that is, radiant, capacitive, inductive and reactive conditions of the situation. There is a matching of the field energy into the receptive biological system which responds according to the special signals or information received and the nature of the impedance seen by these signals.

The transmission and reception of the signals depend on modified antenna concepts, the regions in front of the transmitting antenna having special zones where different microwave-biological conditions will be found. Any microwave exposure will thus originate with a transmission of energy, the units being chosen to describe the biological features of the radiation. The experimental design must be one in which the animal or preparation becomes, by necessity or accident, an integral portion of the microwave circuit. Electronically, this may not yield a precise, well-defined circuit and biologically it puts a premium on an understanding of the interacting electrical parameters.

This is the situation faced by the investigator undertaking work in microwave biophysics and his model may be a special phantom chosen for its experimental value, an organism, or even a spectroscopic preparation. Nevertheless, quite accurate equivalent circuits for tissues and cells may be drawn where the experimental evidence is sufficient.

COMPOSITION AND UNITS FOR THE ENERGY FLOW

The power density in watts per unit area is an average in time of the energy passing and is the product of the velocity of light and the total energy density

$$S = cD \tag{326}$$

When $S = 10$ mW/cm^2, the heating threshold in biological hazards, D will be ⅓ picojoules/cm^3. D is composed of the electrical field component plus the magnetic component on an average time basis

$$D = D_E + D_H \tag{327}$$

Also,

$$D_E/D_H = 1 \tag{328}$$

The relation between D and ϵ_o and μ_o, the dielectric constant and magnetic permeability of free space is,

$$D = \tfrac{1}{2}(\epsilon_o E^2 + \mu_o H^2) \tag{329}$$

Both these characteristics and the energy density may be related to the impedance. Thus, for the characteristic impedance of a vacuum Z_o

$$Z_o = (\mu_o/\epsilon_o)^{1/2} = |\bar{E}|/|\bar{H}| = 377 \text{ ohms} \tag{330}$$

For a given material, the values of the complex dielectric constant and complex magnetic permeability of the material are substituted. Finally, the power density is related to the field components and the impedance.

$$Z_o H^2 = S = E^2/Z_o \tag{331}$$

Among the main factors which determine the power density are the output of the generator, the size of the propagating antenna structure, and the distance at which the measurements are made. The main divisions for distance are termed the *near* field and the *far* field. Not all microwave sources are described in this manner. A typical situation is one in which the radiation is accidentally propagated from the distribution line, that is, a place where energy escapes. While the main factors are still operating, the special situation presents, not a designed set of features, but an arbitrary one.

REFLECTED AND PLANE WAVE EXPERIMENTAL PREPARATIONS

The broad categories for waves operating under experimental conditions are plane and reflected waves. Plane wave experiments involve beaming radiation at a target at a specific point or in a transparent enclosure. Usually this entire preparation, the target (which is perhaps an animal), and the enclosure are in an electrically anechoic (non-echoing) chamber which is carefully lined with absorbent material to prevent reflection of the radiation. This is like having the target remote from all reflecting surfaces. In this arrangement, all of the radiated electromagnetic energy is not used so that for any given power density needed at the target, an excess is supplied which ends up in the absorbent material. Exposure to plane waves is thus a geometrical cross sectional exposure in two dimensions. Therefore it is dependent upon the cross section or profile presented as well as on the characteristically different kinds of energy fields found near, as opposed to far, from the radiator.

The reflected wave method is used in special resonant cavitites or metal boxes into which the electromagnetic energy is delivered directly. These waves will reflect from the metallic walls just as light is reflected from mirrors, and a target material within the cavity will ultimately absorb practically all of the electromagnetic energy. Some will not be absorbed because the fate of a wave advancing through the cavity may be to find the entrance again and reflect back to the generator. Some may be lost in the metallic skin internally penetrating a tiny distance known as the skin depth which varies with the kind of metal. These waves are incoherent in the sense that they vary in temporal and spatial characteristics, in their concentration, and in their angles of incidence. Again they resemble the incidence or illumination from light waves coming from a spot light that is revolved around the absorber centered among mirrors.

Usually the preparation, being the most favorable absorber present, will ultimately dissipate the energy regardless of reflection or transmission coefficients. To do so, it may transmit in part, reflect, and again transmit until the cycle is repeated enough times. Consequently, in such a cavity, practically all amplitudes will be found. It is usually possible to say that an *available* energy density will be the absorbed energy, which is not true for the plane wave exposure.

MODULATION DUE TO ELECTROMAGNETIC FAN

The random nature of the reflections can be cleverly increased by introducing a rotating element shaped like a fan. The effect is to add many surfaces (those of the moving vanes) for infinitely more reflections. This rotating reflector helps to prevent the development of static points of energy density by randomizing the field. "Static" reflections from the enclosure walls are replaced, and uniformity of energy density in an absorber is improved (more diffuse absorption). The dynamic reflector imposes a modulation on the carrier frequency equal to its particular rotational speed. For example, if the RPM is three cycles per second and a four-bladed fan is located so that the blades cross the aperture where the radiation enters, there will be twelve crossings per second. This is a modulation component equal to twelve cycles per second. Due to the angles offered by the metal fan blades to the energy, the preferred direction of reflection will be into the cavity.

EFFECT OF THE MAGNETIC FIELD

While its many advantages suggest the continued use of the power density concept, units other than watts may better express the exposure in certain cases. In any case, the alternate units are usually directly obtainable from the power density. For obtaining the current density, it is convenient to illustrate the conversion by considering the interception of the electromagnetic energy flux by a human target. The resulting dimensional relationships are quite clear. This kind of alternate unit may be used when the magnetic field is of main interest. The rare situation when the target organism is placed near the base of a perpendicular antenna provides the special conditions necessary and only a magnetic field will have an influence (Schwan, 1968).

Current may be the quantity of interest near a point where energy is escaping from a transmission line. In such a case, the nature of the energy field is more difficult to predict and may not be strictly related to a power density safety criterion. If it can be shown that the propagation here follows far field models, then power density would be most appropriate.

For visualization and occasionally by necessity therefore, current density units have direct, meaningful relations to the biological effect. A man has a profile of about 0.56 square meters if the torso is taken facing the energy. As an approximation, a 0.5 square meter person will intercept a flux of 50 watts at a level of exposure of 10 mW/cm^2.

$$(10 \text{ milliwatt/cm}^2)(5 \times 10^3 \text{ cm}^2) = 50 \text{ watts} \tag{332}$$

Converting to energy density requires taking a volume of interest, for example, the all important first centimeter dermal layer. Then again using the almost standard man of 0.5 m^2,

$$50 \text{ watts}/5000 \text{ cm}^3 = 0.01 \text{ watts/cm}^3 = 10 \text{ mW/cm}^3 \tag{333}$$

This calculation illustrates another new virtue of the ubiquitous maximum power density figure. Even for absorbed power it is the same number if the approximate dermal layer is taken as the absorbing volume.

The power loss or heating effect is expressed with the aid of Ohm's law as the $I^2 R$ loss. The resistivity in this volume of tissue is about 100 ohm centimeters from 10^8 to 10^9 cycles per second.

$$I^2 R = 10 \times 10^{-3} \tag{334}$$

$$I = [(10 \times 10^{-3})/10^2]^{1/2} \tag{335}$$

$$I = 10 \text{ mA/cm}^2 \tag{336}$$

The allowable current density is 10 mA/cm² and, for lower frequencies, a lower value would be valid, according the increase in resistivity (Belding, 1955).

At specific gravity one for tissue, this current input for one centimeter of depth is

$$(10 \text{ mA/cm}^2)/(1 \text{ gram/cm}^3) = 10 \text{ mA/gram/cm} \qquad (337)$$

This result is intended to show an exception to the usual conclusion that the magnetic field component is not of great significance in the biological effect of nonionizing radiations. Here currents are caused in a body and current produces I^2R heating as well as related effects. Specific fields may be found, e.g. under a perpendicular antenna base, which are strong magnetic fields and can easily have a biological effect. Based on these calculations using the maximum power density, something on the order of 0.5 amperes per meter may be found to be the allowable current gradient for exposure to the magnetic field.

When hydroelectric power generating stations become very large, significant current gradients exist near them. In the U.S.S.R., fields of 115 to 125 microamperes have been measured and associated with headache and fatigue in the workmen. In oersted units, the magnetic field strengths that produce these effects are said to lie in the 150 to 1500 range.

Microwave Quanta and the Magnetic Component

In terms of quanta, a nonionizing frequency range may be taken such as microwave, and the quanta are calculated to be from 0.024×10^{-6} to 4×10^{-6} eV, greater at the upper frequency for microwave. When such quanta are absorbed, the magnetic field energy is inseparably associated with that of the electric field. Absorption of one requires absorption of the other. These E and H fields are equal for the infinite plane microwave in a vacuum. Their vector ratio E/H varies considerably in the near field of an antenna; it will be mainly H near an electromagnet and almost purely E between the plates of a large capacitor. Biological substance is usually considered to be a poor source of absorbing magnetic material. This leads to essentially a zero interaction between H fields and biological substance. Yet the energy associated with the H field will be indirectly absorbed in the biological substance as shown in the I^2R loss effect. There are no clear grounds either for bioeffects of the athermic type.

CAPACITIVE AND INDUCTIVE COUPLING INTO AN ANIMAL

In the frequency region of a few megacycles, it is common practice to adapt the power amplifier output to applicators of a capacitive or inductive type. Capacitive coupling produces dielectric effects such as heating in material between the electrodes. Inductive coupling can be made to an inductive load placed inside a container constructed of non-energy-absorbent plastic or glass. An air wound solenoid on a 5¼ inch Plexiglas cylinder 10 inches long would provide, for example, a dense magnetic field inside an animal caged in the cylinder at 3 to 30 MHz, the frequency of the generated energy. The dielectric load on the other hand, will absorb an electric field in a fairly pure state. An animal in the inductively coupled system will absorb power equal to $I^2 R$ watts. Here the resistance is "reflected" resistance in series with the solenoid inductance and I is the square root of the root mean square current supplied. In the capacitively linked system it would absorb V^2/R watts where the resistance is the equivalent shunt resistance contributed by the animal's presence and the voltage is the rms voltage across the electrodes. The electric field between is

$$E = V/d = \text{Volts/meter} \tag{338}$$

INADVERTENT RADIATORS

In the process of ferreting out the biological effects of nonionizing radiations it develops that two large classes of events can be examined, the nature of the exposure in the near field and that in the far field. While the proper antenna excites a clearcut division of this kind with a coherent wave front propagating outward, cavities such as microwave ovens and transmission lines emitting inadvertent radiation create sources of a different type. In the first case, the near field may be the one inside the oven or near the point of energy emission.

As the frequency increases, any position near the radiating point will show a greater power density because less power is required to produce a given power density at the higher frequency. In this near field, the length of the field increases exponentially with the diameter of the antenna. Therefore, to maintain a low power density, the diameter of the antenna must be small. At the same time, the power density passes through a maximum value for a particular distance in the near field. For example, at 3 Gc, an antenna of about six meters puts out the maximum power density at 14 meters. In the far field, the power varies inversely as the square of the distance according to the inverse square law and doubling the distance will decrease the power to one fourth.

$$w/\text{cm}^2 = (w_o/\text{cm}^2)/d^2 \tag{339}$$

This is most easily visualized if the first position is taken at 2 units of distance, which is then doubled and the power density remeasured. The new power density will be one fourth.

In the near field close to an inadvertent radiator, the composition of the field reflects on the imperfect nature of the radiator. It is not reasonable with the perturbations present to expect a time average power flow, but instead what is obtained is an instantaneous flux. If measurements are based on expected field values, they may mislead due to reflections from metallic surfaces giving a more diffuse field (from the leaky cavity).

The power density unit implies a flow of power and, in the near field, the power flow may appear to be zero with an instantaneous flux, yet the hazard at points of maximum amplitude in *standing waves* may be significant. In any case, to be hazardous, the power must flow into the absorber, that is, actually be absorbed.

TYPES OF FIELDS FOUND AT VARIOUS DISTANCES FROM THE SOURCE

The two components of the total field then are the radiation one and the reactive one. The former gives a power flow from the source but the reactive component lacks an equivalent loss. Energy flows out and back locally, giving an oscillation without much attenuation. At microwave frequencies, the energy resembles sound waves more than visible or ionizing radiation. There are standing waves, polarizations, resonances, and echoes. In optics, these standing waves form interference patterns, while in acoustics, they give the distance between positions of constructive and destructive interference. Multipath interference is created in a situation where the interference is between direct propagation from a source and reflected or scattered fields from objects illuminated by the source. This may be in the immediate vicinity of a transmission line leak and it may be further complicated by additional leaks nearby.

The meters used to search these fields may actually read in energy density instead of in power. The probe of the instrument can disturb the system and even change the power by its intrinsic effects. Externally, in space measurements, it can scatter the radiations while internally, measurements may suggest that the probe is actually "pulling" the generator or changing the energy withdrawn from the source.

The "Ideal" versus Biological Preparation

In the ideal situation, the magnitudes of the electric and magnetic fields, the energy density, and the power density or energy flow are all directly measurable for monochromatic, linearly-polarized, infinite plane waves. The medium in such a perfect case would be a linear, homogeneous isotropic one and would give fields and densities that are mutually definable. In this sense each actually describes the field except for its phase polarization, and propagation direction. Biological preparations must of necessity introduce considerable arbitrariness into this situation. Similarly, in near fields the polarization is quite arbitrary.

Near and Far Fields

The standing waves or reactive zones that form near the electromagnetic source will have large local fields but no power flow. This type of wave is formed of two plane waves of equal amplitude and opposite direction but the *same* polarization. At certain places, the electric and magnetic fields are twice those of the original waves but the average power density is zero. The energy map for the region near a horn antenna will have peaks that vary by at least a factor of two.

Near cavities with mode stirrers, the phase and amplitude of the fields is given by the modulating effect of the stirrer cycles and its position. Thus what comes out of an opening into free space near such a cavity will depend on the stirrer position at the time the external reading is taken.

If the frequency is greater than one gigacycle per second (30 cm), then the reactive fields are not larger than the radiation fields at distances larger than one quarter wavelength from the source. The significance of this is that the biological effects may be evaluated by the electrical power density in this region. Between this wavelength and down to 2.54 centimeters, the electric field may be deliberately searched for the maximum field of the standing wave. This electrical maximum is likely to coincide with the point of maximum biological effect.

In the ideal conditions possible in the far fields where the distance is greater than a wavelength from the source, the E and H components of a plane wave are in proper time phase and space relation for a straightforward measurement of the power density. Then, on approaching the source, the induction and reactive components emerge more prominently as the radiation ones decrease.

POLARIZATION OF WAVES AND ORIENTATION OF BIOLOGICAL SPECIMEN

The optical explanation for polarization is the most apparent among many possible ones due to the wide use of polarizing sun lenses. This lens medium absorbs the radiation that is polarized in one plane and lets the rest pass as may be demonstrated by superimposing, and then rotating the lenses. The transmitted part is plane polarized. Polarizing slots, either accidentally or intentionally created, will pass a polarized wave for microwaves. For example, a slot about two inches long by $1/8$ inch wide would serve at 2.45 Gc. The characteristics of the plane wave propagation are easier to describe than are those of arbitrary waves or unpolarized, generalized radiation (Born 1964). The vibrations in the plane electromagnetic wave are transverse as opposed to a sound field for example, where they would be longitudinal. Thus the plane wave propagating toward the reader is linearly polarized when its electric field vector is in the vertical page direction. The words *linear* and *circular* come from the special cases of the elliptical path traced by the E or H vector, that is, the polarization ellipse is the locus of points traced by the top of the vector. Whether it is linear or circular depends then on the axial ratio and eccentricity of the ellipse and linearity or circularity represent the limiting situations. When linear, the minor axis has disappeared and when circular, the eccentricity equals zero.

Linear or circular polarization of the radiation will make absorption of the energy dependent on the orientation of an absorber. The energy density will vary by a factor of two depending upon the polarization

Fields and Waves 597

and orientation. Polarization is thus a field factor to be considered in evaluating the electromagnetic hazard.

In the reactive near field, or where there are standing waves, the energy densities are greatest at the field maxima which are λ/4 apart. There is no net power flow and the energy may resonate between the electric and magnetic components. If the power is to be transmitted, then standing waves must be reduced (Fig. 16-1).

Fig. 16-1. Standing wave energy resonates between the E and H forms.

ABSORPTION AS A FUNCTION OF SHAPE, SIZE AND LAYERING

The human body will absorb most of the impinging energy between 3 and 3000 MHz except for the portion reflected.

If a body were tiny with respect to the smallest wavelength to be encountered, then it would partially escape. Resonant effects may come from the field interaction with an object with a dimension comparable to the wavelength. Scattering or reflection from a body will vary a great deal with frequency. Due to the high dielectric constant of tissue, reactive field components may be reflected without absorption as in the near field due to the effect of impedance. The body will be exposed to both peak power densities, modulation, and to average power density.

A spherical shape as approximated by a head may be especially independent of the direction from which the energy comes, although the impinged surface will be first to absorb. Other geometries may be very dependent. The sphere will be sensitive to the electric component orientation and the radius and conductivity will make it frequency sensitive. Sensitivity will be expressed in terms of preferred absorption patterns inside, with more possibility for standing waves, resonant energy transfer, and special internal reflections at high frequency (Kritikos, 1971).

Introducing multiple layers in the body's path of interception creates a strong frequency effect because, at medium and low frequencies, the wavelength in the medium is greater than the whole body size. If the reflected wave situation is taken rather than the plane wave one, multiple path and multidirection reflected waves greatly reduce some of the differences. Internal reflections at boundaries persist however, but the analysis is more difficult.

The radiation absorption cross section S^1 is the *ratio* of absorbed power to power incident on an object's geometrical cross section or shadow. For biological materials it varies around unity in the frequency range from 300 to 10,000 MHz, that is, from 0.5 to 1.25. It is a strong function of the dimensions and shape, the dielectric constant and loss and conductivity. The fact that it gets very small at low frequencies is another way of saying that for large λ, organisms are no longer commensurate.

$$S^1 = w_{abs}/w_{incid} \qquad (340)$$

ABSORPTION IN TISSUE BASED ON EQUIVALENT CIRCUIT PARAMETERS

The equivalent circuit of a body is an approximation that focuses attention on its electrical characteristics. For example, a single cell may be represented by its resistances and capacitances as in Fig. 16-2.

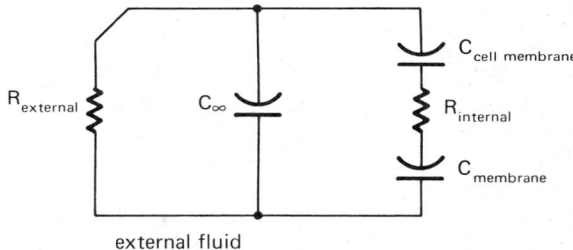

external fluid

Fig. 16-2. Diagrammatic equivalent circuit for a single cell, in which membrane is represented by a capacitance and variable resistance. S.J. Rogers, Radio Hazards Group, Admiralty Surface Weapons Establishment, Potstown, Cosham, Hants England.

From this type of circuit analysis it is concluded that certain well known electric parameters will control the absorption of power in the various tissues. Then, remembering how the dielectric constant decreases from its low frequency value ϵ_o to the high frequency one ϵ_∞, the principal tissues are examined to see how they contribute to this dispersion of values (Figs. 16-3, 16-4).

The fatty tissue, for example, has a fairly flat ϵ which equals 20 over a wide range but blood, with its large water content, is quite the opposite. A similar analysis for σ shows that blood and fat again represent the maximum and minimum effects respectively.

Clarification of Skin Effect and its Relation to Penetration

Microwave absorption is inversely related to the depth to which the microwave energy can penetrate. The overall parameters controlling

Fields and Waves

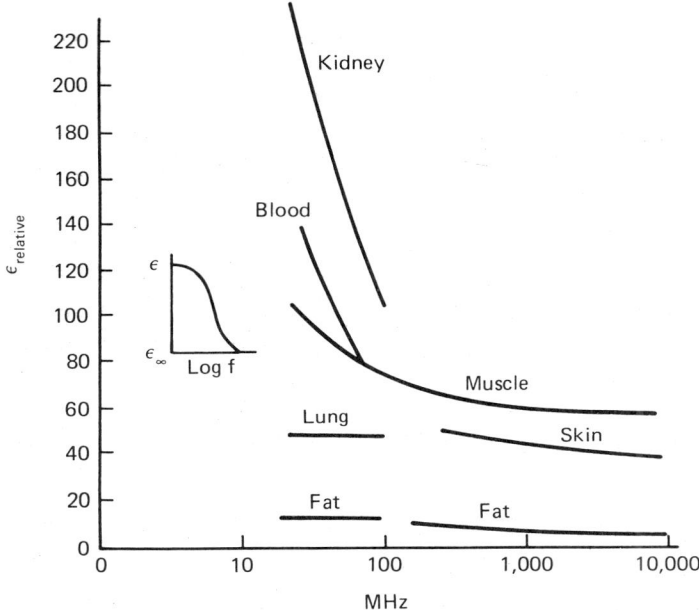

Fig. 16-3. Contribution of various tissues to the change in ϵ from low to high frequency. (*After* S.J. Rogers)

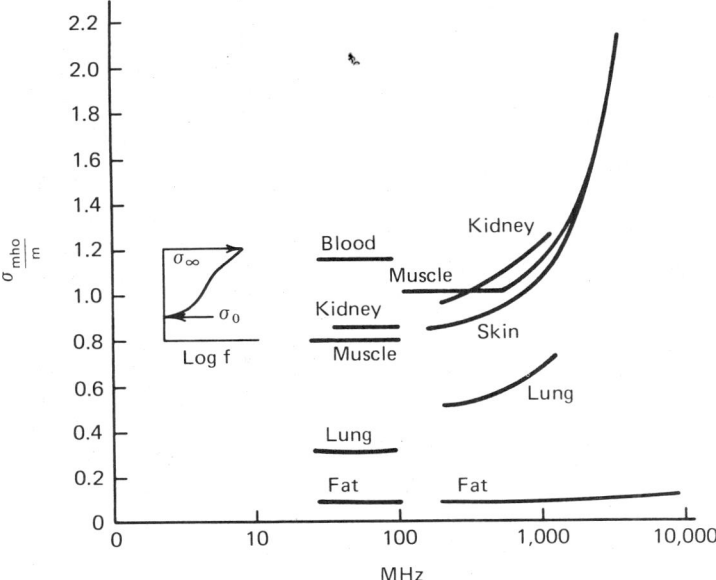

Fig. 16-4. Change in conductivity with frequency for several tissues. (*After* S.J. Rogers)

these factors are the dielectric constant ϵ' and the conductivity σ in reciprocal ohms per centimeter (mho/cm). This is the dielectric or semiconductor conductivity of tissue. If one should emphasize the conductivity factor, an interesting concept develops as to the behavior of the energy according to the conductivity of materials. One could, as Grant does, show that the conductivity of absorbing material is proportional to frequency, increasing quite rapidly in the ultra high frequency range (Figs. 16-4, 16-5).

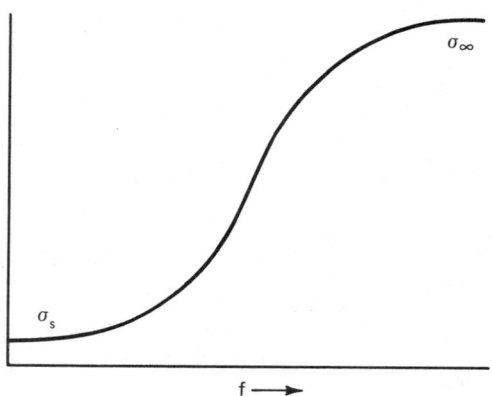

Fig. 16-5. Conductivity increases rapidly in the ultrahigh frequency range.

These concepts, however, may be overextended rather easily. Electrical engineers have an important parameter for evaluating field penetration, known as the skin effect δ, which is inversely proportional to the square root of the conductivity. It is usually necessary to construct the transmission lines in such a way that the cavities and guides that transmit the energy also contain it, so that it will not radiate in an unrestricted manner into space. This transmission begins at the source and terminates in the energy sink—the endpoint, where it is used up or absorbed. Between these limits, the alternating field propagates at the given frequency. The transmission elements such as coaxial conductors, waveguides, and cavities contain the flow of energy on their surfaces or "skins". While low frequency currents travel on the outside of conductors, the microwaves transmit through the inside of metal guides which are designed to prevent losses into space. It is the skin effect or the tendency of microwaves to travel within the superficial portion or skin of the metal which makes this layer effective as a containing wall. Of special interest is the penetration of energy because the thickness of the metal must be sufficient. The skin depth is the $1/e$ (amplitude of the applied field) thickness or the depth at which the energy has diminished exponentially to the 37% value. This and other measures, for example, the half width, decrease with frequency.

In units of meters, kilograms, and seconds it is found that skin depth δ is a function of frequency, magnetic permeability (μ) in henrys per meter

$$\delta = (1/2\pi)/(4\pi\rho\lambda/\mu c) \tag{341}$$

where the resistivity ρ is in ohm meters and c is the velocity of light (3 × 10^8 meters/second). Most good conductors will have a value for δ in the region 1.5 × 10^{-3} millimeters at 3 GHz, so that the design of wall thickness or application of plating to a few thousandths of an inch will give a sufficient pathway for the current to flow along transmission lines.

The result of interest is that depth of penetration or δ can be *reduced* by increasing the conductivity. Only for frequencies that are very high in the microwave region does this reasoning particularly apply to biological materials since they are not used as intentional transmission lines. The conductivity can become very great and one can then speak of skin effect in tissue. Skin may be a rather appropriate word when the epithelium is involved. It is less logical to apply the term "skin effect" to lower microwave frequencies where penetration is the factor of interest and penetration is much greater than the thickness of the epithelium. Many phenomena connected with microwaves in biophysics depend on this depth of penetration which commonly produces an internal energy peak in the subdermal region. Thus the difference in concept is significant and skin depth is an engineering measurement and an afterthought rather than the central parameter.

As an illustration, Vogelman (1961) introduced skin depth but taught about penetration. Taking one special, incident, power density, (the maximum recommended safe value of 10 mW/cm^2) and calculating the penetration, he demonstrated the 1/eth levels for the power as shown in Fig. 16-6.

The maximum effect of this safe power level would be "felt" at one centimeter if the frequency is 8 GHz and this amount is 0.42 milliwatts. The absorption analysis or the inverse effect may, at times, be more informative although penetration and absorption are but two sides of the same coin. Another fundamental point about the skin effect is that it rather vividly connotes a conductive shield effect on tissue in the region of the "overlap" frequencies, that is, in the high microwave region approaching and passing into the infrared.

REFLECTION AND PROTECTION

The dielectric constant measures how much of the incident energy a tissue will absorb or reflect. The reflection coefficient is used to show the amount passing the interface for normal incidence as

$$r = \sqrt{\epsilon} - 1/\sqrt{\epsilon} + 1 \tag{342}$$

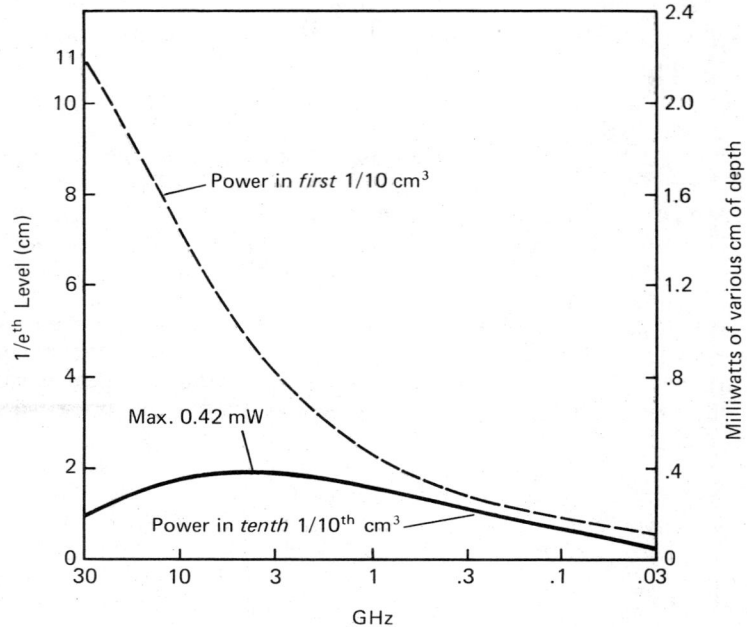

Fig. 16-6. Skin depth analysis gives penetration information.

Thus the dielectric constant makes reflection one of the main protective factors for organisms in electromagnetic fields.

Suppose that a tissue of moderate to high conductivity such as perspiring skin is taken, in which case, a fairly high value of ϵ_r, the relative dielectric constant of living tissue, is obtained. Then the incident power will suffer an immediate reflective rejection at the interface of about 80% (Fig. 16-7).

With metallic conductors, semi-free electrons in the conduction band account for the effect. If the sodium ion is considered to be the significant carrier in biology, it is many thousands of times more sluggish due to the macromolecular complex. The dependence on the nature of the envelope of the cross sections and specific transparencies as a frequency effect, *modifies* the skin effect both at very high and at other frequencies.

PENETRATION STUDIES MAY NOT REVEAL THE INTERNAL PEAKS OF ENERGY

The remaining energy is often said to absorb into tissue exponentially. For the safe level, therefore, 10 mW/cm² is reduced to 2 milliwatts at an interface of the particular area 1 cm² (8 milliwatts lost by reflection). If 10 depth increments equal to 0.1 cm are taken, then the final volume is 1 cm³. At the end of each 0.1 cm, the energy has been reduced by

Fields and Waves

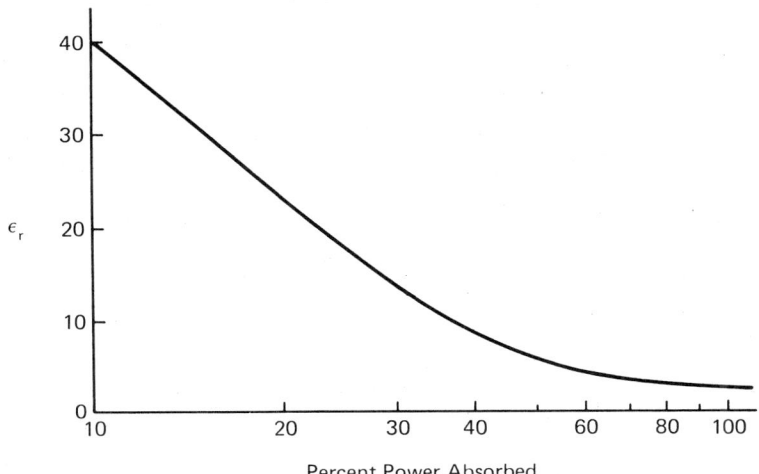

Fig. 16-7. Immediate loss of about 80% by reflection from tissue at dielectric constant about 20.

10% of the part incident on this increment. However, the exponential curve that results can be misleading as compared with a study of actual penetration. Alternatively, absorbed energy may be observed by a tissue response; in particular, calorimetry is the simplest method. Then it will be found that the tissues of interest will absorb characteristically with a shoulder in the curve or a peak region below the surface which is located more deeply with decreasing frequency (first internal peak - Fig. 16-8). The maximum in the dermis is for human tissues, but for embryos or small animals like mice, it is in corresponding organs. *In vivo*, thermoregulation will shift the amplitude and site of this vital peak.

Penetration as a Function of Frequency

Penetration is greatest at low frequencies and also much more constant with varying frequency, changing approximately as the square root of the frequency. At frequencies greater than 300 MHz it varies with frequency inversely as the square of the frequency. This gives about four times as much penetration at 1 Mc than at microwave. Then, at 100 kc, an extremely low frequency, there is still some absorption even though transmission is the main effect and the energy penetrates about three times more than at 1 Mc.

Schwan's useful approximations (1965) for penetration in terms of absorption coefficients, ϵ' and ρ (the resistivity), are divided into the low frequency case

$$D = \sqrt{\lambda\rho}\,/\,17 \simeq f^{-1/2} \tag{343}$$

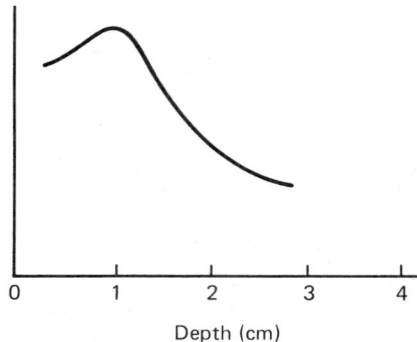

Fig. 16-8. Penetration curve showing first internal peak with frequencies in the region of 3 Gc. Change in penetration as function of frequency.

(for values of "60 times the wavelength" that are greater than $\epsilon'\rho$) in which the square root of the wavelength times the resistivity *product* is determining, and the high frequency case

$$60\lambda < \epsilon'\rho \qquad (344)$$

$$D = \sqrt{\epsilon'\rho}\,/377 \simeq f^{-2} \qquad (345)$$

If absorption is broadly analyzed for tissues with *much water*, and penetration is expressed in terms of ϵ' and ρ, then this kind of frequency dependence emerges. Here the penetration will be a *depth* that decreases rapidly with frequency when the frequency is in the range over 300 MHz, going down to millimeters above 3000 MHz.

HEATING EFFECT, VOLTAGE GRADIENT, AND FREQUENCY

There is much less absorption of energy incident on dielectrics at frequencies below 30 MHz. As to distribution between fatty tissue and muscle, the percentage absorbed in the former becomes greater as the frequency decreases from 100 to 1 MHz. To maintain a given heating rate, for example 1°C per hour, inside the skin depth in fields at the medium to low frequencies, the field strength gradient in volts per meter must be increased over that necessary at higher frequencies. This requirement for more strength for the same absorption will vary with the dielectric loss values and the shape. The most field strength is required with a lossy dielectric slab in the capacitive type of field even though *all* the energy theoretically passes from the electric field through the slab. It is least in a skin depth of a semi-infinite layer of lossy dielectric in a plane wave field and intermediate field strength would be required for a spherical shape of lossy dielectric in a plane wave field. Therefore, the best "heating" frequencies are the high ones in terms of voltage strength needed.

Calorimetry is used to test many of these conditions and phantoms have many advantages. A 0.02 N KCl salt solution will have a dielectric constant of about 81 and a conductivity equal to 6.8×10^{-2} mho/m. It will need from 10^3 to 10^5 V/m to maintain an absorption of 1°C/hr. This requirement will increase inversely with frequency. A fatty tissue may be approximated by KCl, dioxane and water to give a dielectric constant of about 10 and a conductivity of 62.7×10^{-2} mho/m (Schwan 1956). A disadvantage inherent in the rate of rise criterion in phantom tests is that there is not likely to be thermoregulation in the same sense as in the human body.

In the medium and low frequencies where power density is less applicable, the 1°C rise is about that obtained from 100 watts or 10 mW/cm² over one half of the average adult body. Translated into volts per meter near an antenna, the equivalent exposure is 195 V/m for the far field, or beyond 10 meters for frequencies below 30 MHz.

THE HAZARDOUS CONDITIONS AT LOW AND HIGH FREQUENCIES

Damage to biological tissue from radiation will be a function of the exposure, intensity and time factors. In the frequency range from 30 to 300 MHz, the time and field curve for safety (no anticipated injury) will appear as shown in Fig. 16-9. This is a characteristic hyperbolic

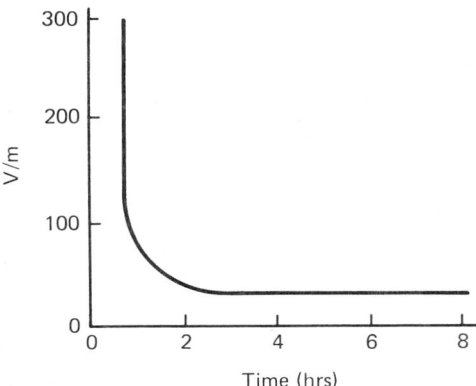

Fig. 16-9. Curve for hazardous conditions at low frequencies.

type of hazard curve tending to endanger more rapidly for field strengths above 50 V/m and exposure times longer than one hour. The power density works into a similar relationship at higher frequencies. How dangerous any prescribed power density actually is will depend on the exposure time. For example, 10 μW/cm² would seem conservative, that is, a high burden in the management of irradiated spaces, but this level may be absorbed safely for eight hours which eases the physical problem. For values above 300 Mc, these relations are shown in Fig. 16-10 with a typical U.S.S.R. maximum permitted level indicated at 100 μW/cm² for two hours. These and related safety levels in the U.S.S.R. and Czechoslovakia are summarized in Tables 16-1 and 16-2.

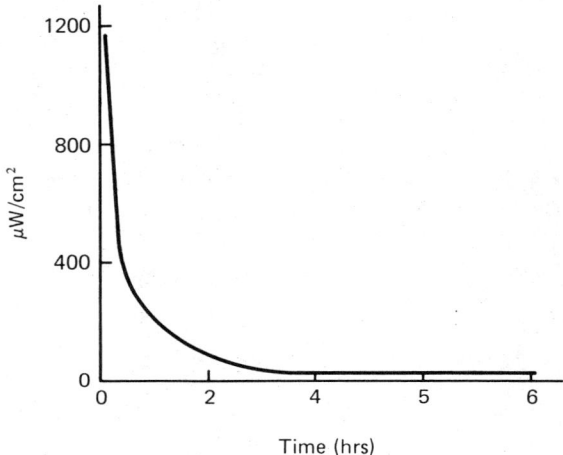

Fig. 16-10. Hazard curve for high frequencies.

EXPERIMENTAL CONDITIONS IN MICROWAVE DEVICES

The fields inside a biological preparation may be designed (within limits) for study. The configuration will depend on the transmission characteristics at the frequency of interest and on the type of material to be observed. The experimental microwave cavity is the most convenient and available device for heating experiments but the frequencies are often limited to 2.45 and 0.915 GHz. In these cavities, the reflected wave condition holds as opposed to the plane wave one. The output can be reduced to low levels by modifying the dc current in electromagnetic windings of the magnetron. An increase in this current decreases the microwave output.

At lower frequencies and microwave, a fringing field can be designed in a coaxial applicator as used in a Stark cell. The center conductor and walls of the coaxial line produce an electric field which fringes between conductor and the side elements. The longest dimension should be considerably less than $\lambda/2$ for single mode propagation. The electric field in a body inside the cell is computed from the input power and the dielectric parameters. This kind of device developed from the need for a uniform field inside a material of uniform dielectric constant. Other fringing applicators are of the interdigital waveguide type.

TABLE 16-1. Safety Levels

Frequency Range	
30 Kc to 30 Mc	400 = V/m × hours For example, 8 hours at 50 V/m for workers who are regularly examined and work near generators on a day shift at the indicated mean value field strength
30 Mc to 300 Mc	30 = V/M × hours For example, 8 hours at 10 V/m
UHF and cw	200 = $\mu W/cm^2$ × hours For example, 8 hours at 25 $\mu W/cm^2$

TABLE 16-2. Population Environment

Mean Calendar Week Value	Mean Field Intensity	Frequency
120	24 hrs. × 5 V/m	30 Kc to 30 Mc
24	24 hrs. × 1 V/m	30 Mc to 300 Mc
60	24 hrs. × 2.5 $\mu W/cm^2$	(mean power density) UHF
24	24 hrs. × 1 $\mu W/cm^2$	(same, pulsed units) UHF

The field varies with the position in the waveguide cross section and, if the material is thick compared to the wavelength, the field also varies with longitudinal position. Some of this variation is reducible if the waveguide is filled with material of the same dielectric constant and magnetic susceptibility as the material being studied. This procedure reduces reflections. A filled waveguide will mean a different propagation medium than air so its dimensions will be based on the wavelength in the filler material. Filling both sides will help to insure single mode transmission. An irregular shape such as an animal would require a liquid that approximates the dielectric constant of the tissue and which produces little or no attenuation.

Another interesting cavity that allows full communication as far as the ambient is concerned is the copper mesh type where the screen forms the walls and the mesh is fine as compared to the wavelength. Waveguide applicators can be made the same way if the largest inside dimension is between one wavelength and one-half wavelength. This construction is capable of providing an isolated electromagnetic environment while permitting free circulation (communication) of the chosen atmosphere around the sample preparation.

There is no doubt that certain biological structures such as horns, feathers, insect antennas and the like play an important role in electromagnetic interactions. Similarly, ceramic, fiber or glass pipes, plastic or glass rods, and plastic water pipes may, intentionally or otherwise, function as waveguides above cutoff. Both conduct energy according to the direction in which it is propagating and make it possible for illumination of whatever absorber happens to terminate the line. Lower frequency dielectric applicators with capacitive plates on either side of the specimen and inductive devices using a coil have already been discussed.

While these examples are not all-inclusive, they are chosen to illustrate some unique aspects of nonionizing radiation transmission and application.

PROBLEMS

1. If the power density of an electromagnetic field measures 10 mW/cm^2, what is the total energy density of the electrical and magnetic components?
2. A man has a profile of about 0.5 m^2 with his torso facing an energy flux. At 10 mW/cm^2, he will intercept 50 watts.
   a. Taking a one cm layer, calculate the energy density.
   b. Taking the resistivity as 100 ohm cm, find the current density for the same flux.
3. How does the dielectric constant vary with increasing frequency?
4. How does the conductivity change with frequency?
5. What is the relation between penetration and absorption?
6. What is the reflection coefficient? How does it affect organisms?
7. What are Schwan's approximations for penetration? Do they allow for internal peaks?

BIBLIOGRAPHY

Agarwal, R.K. *et al*, "A Miniature Portable Microwave Power Monitor," *J. Microwave Power*, Vol. 11, 2, 153-154, 1976.
Aslan, E. "A Low Frequency H-Field Radiation Monitor," *J. Microwave Power*, Vol. 11, 2, 155-156, 1976.
Babij, T.M., "Power Density Measurements in the Near Fields," *J. Microwave Power*, Vol. 11, 157-159, 1976.
Belding, N.A. and T.R. Hatch, *Heating, Piping and Air Conditioning*, p. 129, 1955.
Bielec, M. and S. Samigielski, "Use of Thermography for Quantization of Energy Absorption in Animals Irradiated with Microwaves," *J. Microwave Power*, Vol. 11, 2, 152-153, 1976.
Born and Wolf, *Principles of Optics*, 2nd Ed., Macmillan Co., New York, 1964, 34-36.
Brodin, M.E., "Passive Telemetry of *in Vivo* Measurement of Fields in Biological Materials," *J. Microwave Power*, Vol. 11, 2, 151-152, 1976.
Clarke, J., "Junction Detectors," *Science*, Vol. 184, 1235-1242, 1974.

deVecchis, M., "High Sensitivity Electromagnetic Leakage Monitors,"*J. Microwave Power*, Vol. 11, 2, 154-155, 1976.
Deficis, A. and A. Priou, "Non-Perturbing Microbes for Measurement in Electromagnetic Fields," *J. Microwave Power*, Vol. 11, 2, 148-149, 1976.
Fanslow, G.E. "A Liquid Crystal Calorimeter for Radiation Monitoring," *J. Microwave Power*, Vol. 11, 3, 149-151, 1976.
Gordon, Z.V., "Biological Effect of Microwaves in Occupational Hygiene Izdatel'stvo "Meditsina"." *Leningrad Otdelenie* (TT 70-50087), NASA TT F-633, 1970.
Grant, E.H., "Fundamental Physical Constants Underlying Absorption of Microwave by Biological Material," *Nonionizing Radiation* Sept. 1969, 77-79.
Guy, A.W., "Electromagnetic Fields and Relative Heating Patterns Due to a Rectangular Aperture Source in Direct Contact and Bilayered Biological Tissue," *IEEE Trans. Microwave Theory Tech.*, MTT-19, 214-223, 1971.
Guy, A.W., et al "Study of the Effects of Chronic Low Level Microwave Radiation on Rabbits," *J. of Microwave Power*, Vol. 11, 2, 1976.
Johnson, C.C. and A.W. Guy, "Nonionizing Electromagnetic Wave Effects in Biological Materials and Systems," *Proc. IEEE*, Vol. 60, 692-718, 1972.
Kamal, A., K. Al-Badwailay and E. Hashish, "An Upper Bound on Coefecient of Transmission of Microwave Leakage into Biological Tissues," *J. Microwave Power*, Vol. 11, 2, 159-160, 1976.
Kanton, G. and D.M. Witters, Jr., "A Comparative Performance Study of Spaced Applicators in Microwave Diathermy," *J. Microwave Power*, Vol. 11, 2, 164-165, 1976.
Kaplan, I.T., et al, "Microwave and Infrared Effects on Heart-Rate, Respiration Rate and Subcutaneous Temperature of the Rabbit," *J. of Microwave Power*, Vol. 10, 1, 1975.
Kashyap, S.C., J.Y. Wong and J.G. Dunn, "Microwave Leakage Indication," *J. Microwave Power*, Vol. 11, 2, 156-157, 1976.
Kritikos, H.N. and H.P. Schwan, "The Possibility of Nonuniform Temperature Rise Resulting from Microwave Exposure," *J. Microwave Power*, Vol. 6, 1, 89, 1971.
Kuo-Chun and Chun-Ju Lin, "Cytoxis Effects of Electromagnetic Radiation on Chinese Hamster Cells in Culture," *J. Microwave Power*, Vol. 11, 2, 1976.
Larsen. L.E., R.A. More and J. Acevedo, "A Microwave Decoupled Brain Temperature Transducer," *IEEE Trans. Microwave Theory Tech.*, MTT-2, 438-444, 1974.
Lin, J.C., A.W. Guy and G.H. Kraft, "Microwave Selective Brain Heating," *J. Microwave Power*, Vol. 8, 275-286, 1973.
Luczak, M., et al, "Effect of Microwaves on Virus Multiplication in Mammalian Cells," *J. Microwave Power*, Vol. 11, 2, 173-174, 1976.
McAfee, R.D., "Physiological Effects of Thermode and Microwave Stimulation of Peripheral Nerves," *Amer. J. Physiol.*, Vol. 203, 374-378, 1962.
Michaelson, S.A., "The Cutaneous Perception of Microwaves," *J. Microwave Power*, Vol. 7, 2, 67-73, 1972.
Michaelson, S.M., et al, "Influence of 2450 MHz cw Microwaves on Rats Exposed In-Utero," *J. Microwave Power*, Vol. 11, 2, 165-166, 1976.
Paglione et al, "27 MHz Ridged Wave Guide Applicators for Localized Hyperthermia Treatment of Deep-Seated Malignant Tumors," *Microwave J.* 24, 2, 71, Feb. 1981.
Prucha, R.V., "Human Thermal Loading by Exposure to Emissions from a Microwave Oven," *J. Microwave Power*, Vol. 11, 2, 160, 1976.
Rexford-Weldh, S.C. and I.R. Lindsay, "The Practice of Microwave Radiation Safety," *J. Microwave Power*, Vol. 11, 2, 160-162, 1976.
Rosenthal, S.H., "Alternations in Serum Thyroxine with Cerebral Electrotherapy (CET)," *Arch. Gen. Psychiat.*, Vol. 28, 28-29, 1973.
Rugh, R., H. Ho and M. McManaway, "The Relation of Dose Rate of Microwave Radiation to the Time of Death and Total Absorbed Dose in the Mouse," *J. Microwave Power*, Vol. 11, 3, 279-281, 1976.
Schwan, H.P. and K. Li, *Proc. IRE*, Vol. 41, 1953.
Schwan, H.P. and K. Li, *Proc. IRE*, Vol. 44, 1956.
Schwan, H.P. in *Therapeutic Heat*, 2nd Ed., Eliz. Licht. Publ. New Haven, Conn., Cpt. 3, 63-125, 1965.

Schwan, H.P., in *Microwave Power Engineering*, E. O'Kress, Ed., Academic Press, New York, p19, 1968.

Schwan, H.P. "Microwave Radiation:Biophysical Considerations and Standards Criteria," *IEEE Trans. Bio-Med. Eng.*, BME-19, 4, 304-312, 1972.

Siekierzynaski, N. et al, "Microwave Radiation and Other Harmful Factors of Working Environment in Radiolocation: Method of Determinantion of Microwave Effects," *J. Microwave Power*, Vol. 11, 2, 144-145, 1976.

Skitikos, H.N. and H.P. Schwan, "Hot Spots Generated in Conducting Spheres by Electromagnetic Waves and Biological Implications," *IEEE Trans. Biomed. Eng.*, BME-19, 53-58, 1972.

Taylor, L. S., "Inplantable Radiators for Cancer Therapy by Microwave Hyperthermia," *Proc. IEEE*, 68, 1 Jan., 1980.

Wachtel, H., R. Seaman, and W. Joines, "Effects of Low-Intensity Microwaves on Isolated Neurons," *Annals of the New York Academy of Sciences*, Vol. 247, 46-62, 1975.

Wallen, C.A. and S.M. Michaelson, "Microwave-Induced Hyperthermia as an Adjuvant to Cancer Therapy," *J. Microwave Power*, Vol. 11, 2, 175-176, 1976.

Webb, M.D., A.W. Guy, and J.A. McDougall, "Assessment of the EM Field Coupling of 915 MHz Oven Leakage to Human Subjects by Thermographic Studies on Phantom Models," *J. Microwave Power*, Vol. 11, 2, 162-164, 1976.

Film

The Hazards of Microwave Radiation, ½,¾. VC; C/30 min.; 946, Commonwealth Productions Inc., 21 Ellery St., Cambridge, MA 02138.

17

The Audio Effect

"My cochlea and basilar
Are standing at attention.
And other things are happening
I think I shouldn't mention.
Is someone paging me?
                                V.M.C.

COMPETITIVE ANTAGONISM

Many informational roles for EM radiation have been discussed. It is clear, for example, that microwaves may act as competitive antagonists, usurping the place of the normal EM band but not carrying out its total function. Another role is as white noise or green noise and the like, simply functioning in biocommunications as a sort of electronic jamming, in which case there would be a question of the power level. One sees situations in which excessive foreign fields build up in a region normally the exclusive province of precisely directed fields. Still another would be a modification of antagonism in which the EM energy can actually use the same pathways but in an awkward mode as may well be the case in the audio effect. There is always the possibility that such an effect could be harmful by simply damaging the circuit. In such a case the hearing of microwaves would be associated with "ringing of ears" during fever or chemotherapy, or seeing stars after a concussion on the eye. This would produce an abnormal sensory stimulation that causes a spilling over of excitation into the auditory or optic nerves. Thus infrared and microwaves could antagonize pathways in the eye normally reserved for the visible spectrum, creating noise and causing damage.

In the audio effect however, we have a reducible, reliable, biological response for communicating directly with the organism, if not directly with the associated neural net, that has been confirmed in different laboratories. This endpoint is usually pulse modulated with information from the pulse width communicated indirectly to audio nerves via transduction to pitch elements, but some have emphasized a cw input. As a low level effect it is possibly less hazardous, probably tolerable and likely to be very useful in exploratory and corrective medicine as well as in information processing studies within biophysical psychology and electrophysiology. The sound appears to originate from inside or a little behind the head. The pulses sound like buzz-tick-bip-hiss or knock near the source, and persons who detect this type of sound should regard themselves as living detectors rather than question their sanity. The effect occurs at frequencies from 300 to 3000 MHz and from less than 10 mW/cm^2 to over 30 W/cm^2. It can be demonstrated directly on the cerebrum and hypothalamus.

THE HEARING MECHANISM

The diagrams of Fig. 17-1 show the information pathway for a sound message as it is transduced into impulses in the auditory nerve. The

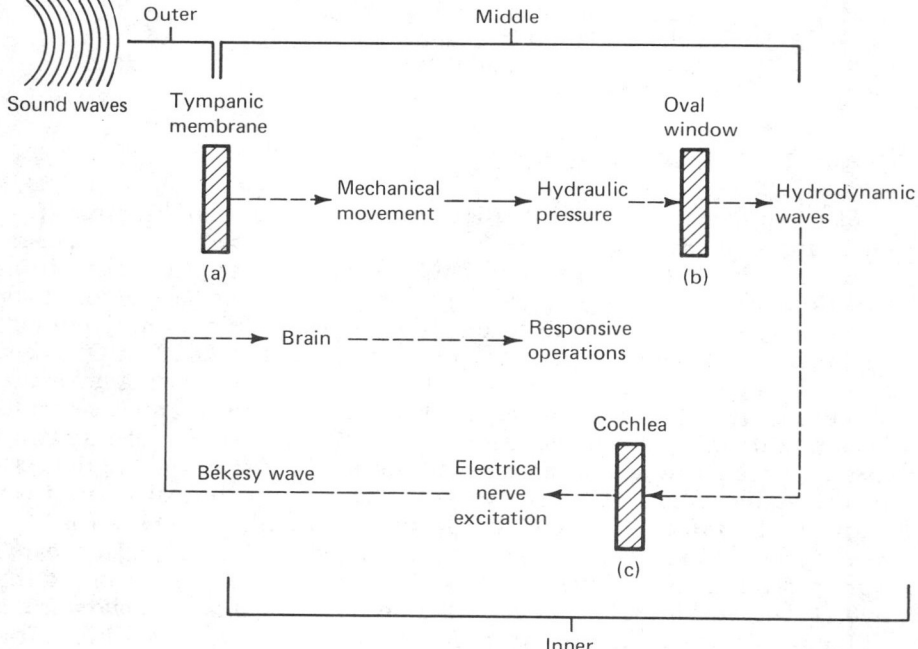

Fig. 17-1. Information pathways for a sound message as it is transduced into impulses in the auditory nerve. Phase and form transducers (a), (b), and (c) represent passage of the energy from sound to mechanical or gas to solid; mechanical to hydraulic or solid to liquid; and liquid to electrical, respectively.

The Audio Effect 613

sound pressure on the tympanic membrane is transferred as shown by the arrows to the ossicular chain and the force of the anvil is passed on to the cochlear window and the basilar membrane. The Eustachian canal equalizes the pressure in the inner ear (Fig. 17-2). Some aspects of

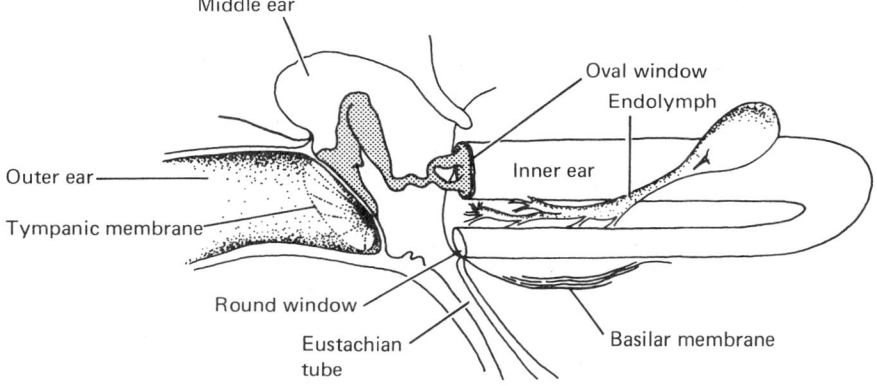

Fig. 17-2. Sound pressure on the tympanic membrane is transferred through the elements of the outer, middle, and inner ear.

the sound information media are:
1. the input to the tympanic region
2. the depth of fluid in the canal
3. the location of the source
4. the length and stiffness of the basilar membrane
5. possibility of shock waves due to high rates of change in amplitude
6. stability functions
7. elasticity

The membrane varies by 100 times from one end to the other in stiffness. (See Fig. 17-5.)

Sound Pressure Pathway

In operation, the cochlear fluid vibrates and the round window vibrates in harmony with it. In Figs. 17-2 and 17-3 the cochlea is "uncoiled" and in Fig. 17-4, it is in the normal, coiled configuration. In operation then, a hydrodynamic vibration responds to a mechanical one.

Regions of Hydroelectric Transduction. The basilar membrane separates two columns of cochlear fluid and contains auditory nerve receptors. In man it is 35 mm and varies in stiffness and width (Fig. 17-5). As a mechano-sensory transducer it is an essential element of the organ of hearing.

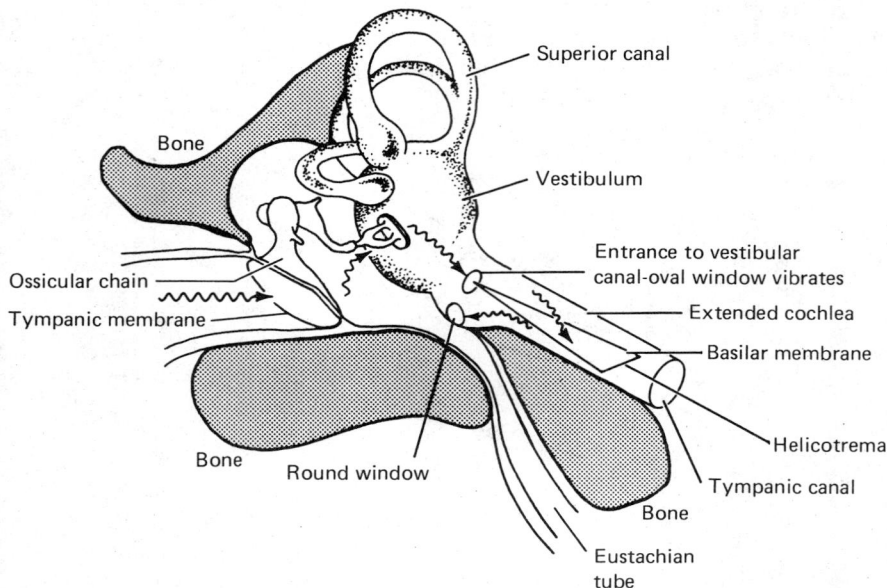

Fig. 17-3. Anatomy of ear with cochlea uncoiled.

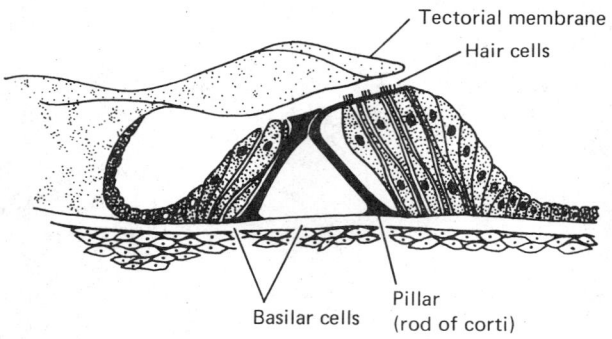

Fig. 17-4. Inner ear details showing cochlea in the coiled configuration and a possible site for the initiation of the nerve pulse. Mechanical displacement of the upper hair cells changes the electrical resistance.

Interfaces. In the cochlea (Fig.17-6) three regions can harmonize,
1. the scala media with its gel, a viscous fluid,
2. the basilar membrane, and
3. Reisner's membrane.

The Audio Effect

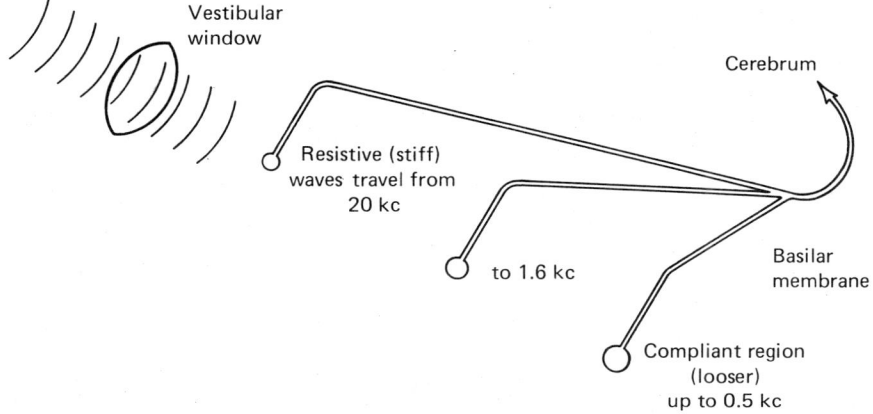

Fig. 17-5. Vestibular window and basilar membrane.

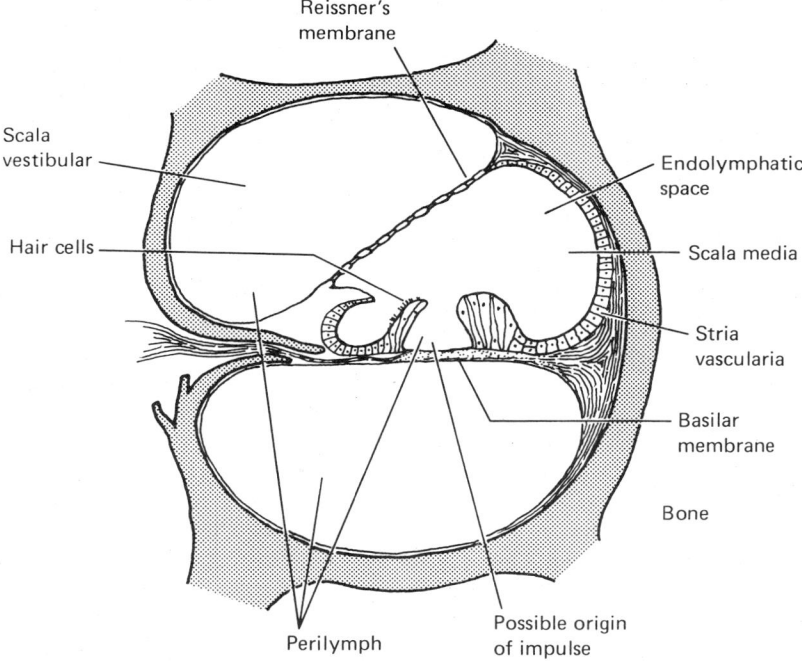

Fig. 17-6. Some of the liquid spaces, membranes and hair cell origins in cochlea.

Waves and Frequency Region. Waves in the cochlear fluid resemble shock waves inducing a traveling wave on the membrane. High frequency audio is near the vestibular window and low frequency audio is near the heliocotrema, and 1600 c/sec would be received near the midpoint of the membrane. In this manner *positional receptor information* is translated into pitch information for the brain. In operation, the air vibrations of low stiffness pass to the ear drum, then to the ossicles, and then to the stiffer fluid of the cochlea. This transmission is perfect at 1000 cps which is thus the frequency of maximum sensitivity. Losses at other frequencies in the range from 400 to 20,000 cps are based on backward reflections.

Species Differences

The guinea pig cochlea is the size of a mere drop of water while the elephant basilar membrane is 10 mm long. The human and mouse discriminate pitch from 400 to 20,000 cps, the elephant from about 30 to 8,000 cps. Above 20,000 is the ultrasound region which can be sensed in other systems. Much higher, there is microwave sound and microwave acoustics. Then the fact emerges that one can sense electromagnetic microwaves with a hydrophone in water receiving strong signals, possibly by a model based on rapid thermic expansion. For hearing ordinary pitch there is probably a series among animals. For capacity to hear the highest pitch this would range from the chicken, to the rat, guinea pig, and finally to the elephant. Electrical stimulation gives a definite hearing effect but the tone is not exactly related, for example 1 kc of EM stimulation gives a white noise but not a pure tone.

STRUCTURES RECEPTIVE TO EMR

The semicircular canals are sensitive inclinometers using the capula and displacements within the hair cells to monitor via certain motions, thermal expansion, convection, wave-mechanical couplings, gravity, radiation pressure, the angular acceleration and inertial guidance information. There is a compression mode, and a vestibular thermic mode where this structure has its nerve terminations as opposed to or in conjunction with cochlear stimulation. This section is also supplied with nerve endings. Thus there is a vestibulocochlear effect and the intralabyrinthine fluids are like a differential dielectric since they are lossy, while within a less lossy, bony container. The intensity of sound is proportional to the thermal expansion coefficient of the fluid as shown by the ability to "freeze out" the sound at 4°C. Microwaves are known to absorb selectively in regions of lower density or at their interfaces where possible standing waves are set up. Standing waves are also produced by mechanical forces and, when they occur, it is possible to perhaps have an enhancement of 10 times in the forces. Looking at the cross section of the cochlear and semicircular canals,

Hair Cell Displacements and Nerve Networks

The cupula and otolith seem to be audio receptor transducers. Here are "calcified pellets" in viscous matrix with motion-sensitive elements (cilia, hair cells) embedded in such a way as to undergo meaningful displacements or polarizations with changes in electrical potential. The threshold is at about 4×10^{-6} dynes. Acting as transducers, combined vestibular and cochlear hair cells give a source for cochlear microphonics and audio sensation. The organ of Corti is completed by the hair cells on the basilar membrane which then has the nerve network of audio nerve terminals. These hair cells vibrate with changes in the membrane, stimulating the terminals positionally for pitch meaning, from the high frequency end near the oval window to the low end at the heliocotrema end. Along this range, the maximum moves up and down as a function of frequency, with 1600 cps being near the midpoint.

Information Switching. Waves can be formed by compression in elastic media in which the units being compressed are alternately bunched and spread out. Alternatively, waves can be formed by dilation as illustrated by a tubular balloon which expands and constricts in an alternating wavelike manner. Waves may also bend or show a shearing effect. Then they can be radial, vertical, or longitudinal in a complex communicating system like hearing and can intercommunicate as information switches, for example, from a diminishing wave to a reinforcing one or vice versa. The diminishing force may be an informing one with an ultimate effect of selectively *damping* certain waves and forcing them to discharge their energy into another train or system. Putting these effects into focus at some transducer region, such as near the stapes, allows information to pass into hair cell waves vibrating with the meaning that came to them via a chain of stimuli. A wave is like an alternating polarization with associated information.

Balancing Microwaves with Sound. Speech-hearing uses a sensory envelope in which information management envelops the effect and, with practice, leads to effective communication. The EM wave in the audio effect can produce a pressure wave into the ossicular chain leading to cochlear hydrodynamics and the transmission of phonics. The reverse should also occur, with the round window being electrically excited to give a chain of events up to the air interface with subsequent communication by an airborne pressure wave. Then tests should reveal if balancing microwave or electrical input information with equivalent airborne sound results in silence. This is true, it does do so.

Polarization and Velocity Modulation. In addition to the alternate polarization in the wave, we may add velocity modulation (VM) or a bunching effect to add descriptive meaning and then the combination of the two can produce complex microphonics from hair cells and cochlea. The nature of wave information in the hair cells is transmitted to the nerve network as the hair cells wave, tilt, accelerate, and decelerate in response to the hydrodynamic stimulus and this is taking place on a membrane which functions as a mechanical oscillator.

Resistive and Compliant Model. A traveling wave explanation would have a disturbance on the basilar membrane sending a wave traveling along it (Fig. 17-7). This could continue a tone, being a vibration that

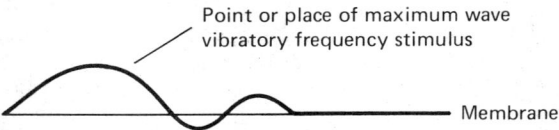

Fig. 17-7. Wave traveling along membrane.

travels. The membrane is relatively stiff on one side and more compliant on the other so that there is an oscillation seen in a thinner and thinner model of a rare traveling wave. These can reflect at the end to give a standing wave. A standing wave pattern on the membrane is information for the distributed nerve endings which inform the brain and produce the meaning (Fig. 17-8). The stimulating frequency makes a maximum

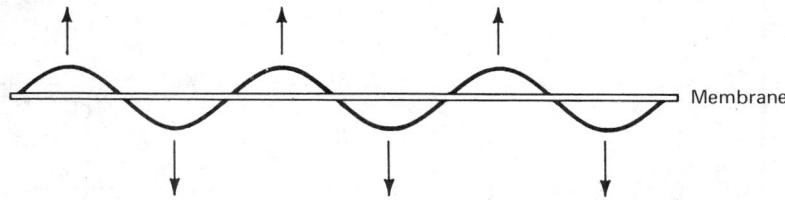

Fig. 17-8. Standing wave pattern on membrane.

displacement at a certain place on the membrane proportional to that frequency. This gives a spatial frequency discrimination and saves the brain the trouble of sorting out this information.

The skull is somewhat like an EM sounding board in that conditions for resonance would exist when the wavelength is commensurate with the skull and the orientation is correct. This would mean some four times the ordinary effect so that in an audio waveform, a power density of 35 mW/cm^2 that is required for sensation would be reduced to 9 in man and to 3 mW/cm^2 in a small monkey. Thus a modified resonance means better than average absorption.

Morphological Modes. Similarly, the canals may be seen as wave guides with certain dimensions, just as they may be in various anatomical parts where the measurements are commensurate with exciting wavelengths. Then in a *morphological mode*, some 10 times the reinforcement of a normal compressive force may be induced with microwaves and standing wave patterns. In a combined thermic mode, the reinforcement may mean hot spots in sensitive regions. In the monkey, 10 mW/cm^2 of incident power density results in about 600 μcal/g sec. The human cerebral heat at 10 mW/cm^2 is about 250 μcal/g sec. These numbers compare with the 1 mW/cm^2 that is considered high for nervous system effects.

Normally, for mechanically resonant processes, the wavelength would be short whereas for thermic ones, it would be longer. Then resonance would be anticipated in commensurate-size situations. The imposition of modulation means that the information is much greater in it than in the carrier frequency or average power. At the threshold, an adequate stimulus is based on the pulse *width*, as opposed to the amplitude or microwave power density, because sensory process time constants appear to be controlling for coupling or admitting versus rejecting the stimulus. Then, the derived sound power intensity and its neural analog come from the incident microwave energy density by some matching mechanism.

Hair Cell as a Pump and the Cyclic Hologram

Some see the utricle and saccule as coupled resonant cavities in which acoustic waves, resembling electromagnetic ones, can propagate and the macules of the utricle could then reflect like mirrors. The hair cells become cylinders with charged particle emission from the open ends. When the cavity pressure falls, this sets up an oscillation as these particles impinge on the gel surface and the oscillation stops when the pressure rises. The wave energy is then passed into sensitive receptors, i.e., the calcified nodules in the gel. Then in the case of direct information transfer, the charge is absorbed and induces hologram records in the gel which are, according to the hologram idea, in-depth-cochlear-images. The record has the characteristics of an image which records information received and has features of memory, pattern recognition, and amplification if the wavelength is greater in the imaging beam. The hologram cycles with the pressure, wave interference, polarization and optical activity (rotation), diffraction, and phase shifts as elements of information. Other features are focusing and a pressure responsive type of image clearing. The result is a set of responses in the eighth nerve, vestibular division. Malcolm (1975) sees this as an engineer's model with the semicircular canals equivalent to an oscillator with elastic walls which resembles a cw laser having two cavities.

Basis for Sensitivity Range. The traveling wave theory needs some such energy pump to amplify the low energy wave and this amplification would be in the organ of Corti. With suitable geometry and adequate energy pumps, the audio information can be transmitted and separated and the information management becomes a question of sound energy detection, amplification and transmission. Then a dipole resonance theory is advanced in which the inner and outer rods of Corti become dipoles and the stapes vibrates instead of simply pushing. The several hundred gradations of pitch are related to the 4,000 rods, 3,500 inner hair cells or 6,000 inner rods in a Vernier measurement. Voice produces a diffraction pattern of hair cells in a transistor action which is resolved to understand the speech.

RAPID RISE IN TEMPERATURE (RRT) AND SHOCK WAVE MECHANISM

Sound waves may be generated in the tissues of the head by a rapid thermal expansion of the tissues when microwave energy is deposited as heat. This is an unusual effect since there is generally very little heat deposited—perhaps a small fraction of a degree rise in temperature. The input of heat is by pulse modulation where the rise time can be *very short*. In any case, the information affects the organism's operations, but the changes in behavior can be attributed to the audio effect rather than to other signals.

Like any sound, the effect of EMR can be a meaningful cue or signal to an animal. The sudden introduction of EM energy into a carefully stabilized system such as the head would be sensed, but the skull has its resonant natural frequency which can result in "eased in" energy and amplification of the response. The skull is known to be a resonant structure and subject to vibration if exposed. A typical possiblity would be the formation of a wave when the sudden rise in temperature represents an input with no other place to go, with respect to equilibration under the applicable time constants.

Thus a wave may be generated to transmit the stimulating energy to the cochlea by expansion and conduction. Then if the subject is otherwise able, he hears the associated pitch of 6 or 10 Kc/s and higher, as the audio effect takes place. This wave appears more like a shock wave coming from the short time constant of the pulse with associated EM and thermic effects of the disturbance, which is a wave that then transmits through the audio system. Subjects deaf to these tones would not receive the EMR audio information.

Dimensionally, antenna theory suggests responsive criteria, and polarization is a significant factor especially in the orientation of the ear canals to the source, or polarization of the incident energy. Peak power in pulses is the key in this mechanism but cw also has an effect on some operations. In rats, the rapid rise of pulses in exposures at 2450 MHz for two weeks gives increases in some activity, decreases in others. Failure of the equivalent average power to give the same effects emphasizes the possible role of the shock wave as a temperature forcing function.

Impedance Transformers. The tympanic membrane and ossicular chain of the midear are impedance transformers, changing the low acoustic impedance of air to the high impedance of the cochlear fluid. The sound power change is up by 30 dB so that without this impedance transformation, a microwave audio effect would have to go via the bone conduction route versus that of sound waves on a membrane. However, a bone mechanism which loses the benefit of this power gain is unlikely.

Time Constants. A compression wave would be one of short time constant, but a temperature difference of even $0.1°C$ does give a possible convective, mechanical force across the semicircular canals and a slow caloric wave with a long TC on the order of seconds that matches the mechanical TC of the labyrinth. This function will be subject to enhancement by resonance and standing waves. In water, microwaves can induce sound in the manner of a liquid transducer action. These systems suggest dependence on both pulse intensity and width so that, by rapid thermal expansion of water (RRT), a shock effect is produced. At $4°C$, there is no microwave effect, because there is no thermal expansion coefficient.

Dielectric. A direct reliance on the known dielectric absorption processes simplistically puts the receptor system in the role of a dielectric crystal. In actual simulation using Eccosorb ANW-77, the material does sense microwaves, emitting audible sound when stimulated by 918 MHz pulses. The Eccosorb was 3 cm thick and 5 cm in diameter and formed into a compressible (*low density*) sponge. Such detectors would be a boon as microwave-powered sound amplifiers for those in danger of exposure to unseen fields.

In dielectrophoresis, the EM fields act like pumps, moving neutral particles along a dielectric gradient toward the greatest E field intensity with nonoscillation, that is, the effect is not alternating even at high frequency. Instead it is constant in direction and has a force proportional to the square of the gradient. The pumping action may be related to parts of the cochlear and vestibular regions with high mechanical sensitivity or small time constants. The displacements depend on the dielectric constants of the particles and their medium; for example to hear microwaves, the cupula would have to be displaced, in the dielectric sense, giving synchrony between stimulating pulses and audio perception.

Dielectric Pump or Siphon

Dielectric mobility or diffusion is observed at very low frequencies when a heterogeneous system, such as layered detergent solutions of varied dielectric constants, reacts with the electric field component to produce a mass movement up the dielectric constant gradient. (That is, matter flows in the direction of increasing dielectric constant.)

POLARIZATION FORCE

A dielectric fluid can move under the influence of a voltage of this kind from one vessel into another in a siphon action. While it is tempting to look at this as a dielectric pump, it is a dielectric dipole orienting system which exerts a *polarization force*. The fluids move up the voltage gradient, for example *between* two "U" shaped electrode walls between reservoirs, and gravity can take them over the "hump" into the other vessel. In the presence of gravity, the dielectric action is of a low order of magnitude but increased pressure on the system can let the polarization become dominant. The system needs an electrode (or the equivalent) at each side to impose the EM fields (about 400 Hz) and runs only with the voltage applied (greater than 19 kV), but as long as it has been primed, and has its voltage, it runs continuously.

The thermic mode with vestibular-cochlear sensing is reminiscent of the infrared sensor in snakes which lets the animal follow its prey and strike. Warm water in the ear is said to give a sense of acceleration as if the endolymph fluid were responding to a thermal gradient. If human reception of EMR of 10 mW/cm^2 is about equal to 250 μcal/g. sec, then sensation in the vestibule should occur at about 35 mW/cm^2. Then as the head gets smaller, the possibility of resonances increases at wavelengths commensurate with skull size. The monkey with its 600 μcal/g. sec, then has a decreased vestibular threshold of 14 mW/cm^2.

Resonant absorption is also directly associated with some contained structures. The vestibule, with a radius between 0.5 and 0.15 cm (axial), has its suitable wavelengths for good penetration, especially at about 1 cm or smaller. "Absorptionally" speaking, the vestibule is some three or four times its actual cross-section. The vertical semicircular canal looks like a finger ring (membranous canal with labyrinth, in bone) with a large stone (the vestibule). The configuration is a toroid overall with density gradients. If the reinforcement of absorption due to dimensions, orientation, resonance, size of canal itself and other factors is 10 times, then this is translated into a lower necessary signal strength, e.g. about 3 mW/cm^2 instead of 30 mW/cm^2.

Mediation by the saccule and utricle could change the dependence from cw with good average power to pulse modulation and explain why both modes are advanced. The mediation would give informing signals as opposed to simpler interaction or, in other words, it would give more than sensation. Then the sensation would be of torque or angular acceleration.

SUMMARY OF AUDIO EFFECT

While at this point the explanations clearly support the existence of EM powered sound in the auditory system, they give more of a fascinating cruise through the EM biophysics of the ear than a hard and fast mechanism. It is logical that a thermally derived shock wave (of the type seen in the effect of explosions, fast aircraft, and in fluid mechanics) can be based on the sudden rise in temperature which then affects the upper sound spectrum, especially from 10 to 20 kc/s, as it propagates as a pressure wave due to expansion of liquid in the vestibular-cochlear medium. The loss mechanism is exerted on the ear liquids which are easily susceptible to EM radiation in the S band of microwaves. It may then be that it is not so much a question of transducer sites as one of a medium that can propagate a wave. The medium is the liquid in the vestibular-cochlear complex and it probably includes a dielectric detector and associated vibrations. For brief pulses, the threshold energy has to come out of the time-limited pulse so that higher energy is probably essential. When pulses are longer, the peak power would be determining. The pulse lead and lag edges can exert a special *boost* in the rapid rise time in liquids. Thus the status of the resting or near resting fluids in the ear is suddenly changed by absorptive polarization, and the shock wave propagates faster than slower caloric events. The ear interprets the shock wave as a harmonic tone whose source is the pulses as they approach and exceed a threshold. The EM audio effect is, of course, alien to the sound receptor system. It can merely cause the system to be placed in "go" or "no go" status and, in this sense, it fails to discriminate tones faithfully. Absorptive polarization and wave propagation can be further processed into inner sensor vibrations which are followed by the microwave neural analog of a true sound powered stimulus. Tone perception is still possible by using pulse trains to get pitch sensations corresponding to the pulse repetition rate.

PROBLEMS

1. What should a person's reaction be when he hears a sound such as a continuous buzz in an EM field?
2. What are the information pathways for a sound message?
3. What is the significance of the variable reaction to waves in the basilar membrane?
4. What force can be exerted in dielectrics?
5. Why is it tempting to see a possible shock wave mechanism in the audio effect?
6. How could an athermic effect surface in a case where there is a masking effect from heating?

BIBLIOGRAPHY

Bell, G., J.N. Davison, and H. Scarborough, *Textbook of Physiology and Biochemistry*, 7th Ed., Williams and Wilkins, Baltimore, 1968.
Candiollo, L., G. Filogamo, and G. Rossi, "The Morphology and Function of Auditory Input Control," *Belton. Inst. For Hearing Res.*, Transl. No. 20, 1967.
Dallos, P., "Dynamics of the Acoustic Reflex," *J. Acoust. Soc. Amer.*, Vol. 36, 2175, 1964.

Davis, A., "Biophysics and Physiology of the Inner Ear," *Physiol. Rev.*, Vol. 37, 1, 1957.
Foster, K.R., and E.D. Finch, "Microwave Hearing: Evidence for Thermoacoustic Auditory Stimulation by Pulsed Microwaves," *Science*, Vol. 185, 256-258, 1974.
Frey, A.H., "Auditory System Response to RF Energy," *Aero. Med.*, Vol. 32, 1140-1142, 1961.
Frey, A.H. and R. Messenger, "Human Perception of Illumination with Pulsed Ultra-High Frequency Electromagnetic Energy," *Science*, Vol. 181, 356-358, 1973.
Guy, A.W., *et al*, "Microwave Induced Acoustic Effects in Mammalian Auditory Systems and Physical Materials," *Annals of the New York Academy of Sciences*, Vol. 247, 182-193, 1975.
Guy, A.W., *et al*, "Microwave Interaction with the Auditory Systems of Humans and Cats," *IEEE G MTT Intern. Microwave Symp. Digest*, Boulder Colorado, S.W. Maley, Ed., IEEE: 321-323, 1973.
Jones, T.B., M.P. Perry, and J.R. Melcher, "Dielectric Syphons," *Science*, Vol. 174, 1971.
Lin, J.C. "Microwave-Induced Hearing: Some Preliminary Theoretical Observations," *J. Microwave Power*, Vol. 11, 3, 295-298, 1976.
Lebovitz, R.M., "Detection of Weak Electromagnetic Radiation by the Mammalian Vestibulocochlear Apparatus," *Annals of the New York Academy of Sciences*, Vol. 247, 182-183, 1975.
Liu. L.M., F.J. Rosenbaum, and W.F. Pickard, "The Insensitivity of Frog Hearing Rate to Pulse Modulated Microwave Energy," *J. Microwave Power*, Vol. 11, 3, 225-232, 1976.
Malcolm, J.E., "Action of Corti's Organ and the Cochlea: A New Theory," *Annals of the New York Academy of Sciences*, Vol. 247, 219-231, 1975.
Najim, M. "Microwave Kerr Effect," *J. Microwave Power*, Vol. 11, 2 210, 1976.
Rasmussen, G. and Windle, W., Eds., *Neural Mechanisms of the Auditory and Vestibular Systems*, Charles C. Thomas, Springfield, Ill., 1960.
Richert, E.S. and A.H. Frey, "Human Auditory System Response to Lower Power Density, Pulse Modulated, Electromagnetic Energy: A Search for Mechanisms," *J. Microwave Power*, Vol. 11, 2, 1976.
Sharp, J.C., H.M. Grove and O.P. Gandhi, "Generation of Acoustic Signals by Pulsed Microwave Energy," *IEEE Trans. Microwave Theory Tech.*, Vol. 22, 5, 583-584, 1974.
Taylor, E.M. and B. Ashleman, "Analysis of Central Nervous System Involvement in Microwave Auditory Effect," *Brain Res.*, Vol. 74, 201-208, 1974.
URSI Symposia, "Biological Effects of Electromagnetic Radiation," 11-15 Oct. 1976, Amherst, MA., *Radio Science 12*, 6, (S), 293 p., Nov-Dec 1977, American Geophysical Union, 1909 K Street, N.W., Washington, D.C., 20006.
Von Bexesy, G., *Experiments in Hearing*, McGraw-Hill Book Co., New York 1960.
Wever, E.G., *Theory of Hearing*, John Wiley, New York, 1949.
Wever, E.G., "Electrical Potential of the Cochlea," *Physiol. Rev.*, Vol. 46, 102, 1966.

18

The Information Contained in Geometrical Relations, Space and Time Constants, and Sense Modes

> Errors like straws, upon the surface flow;
> He who would search for pearls, must dive below.
>
> J. Dryden, *All for Love*, prologue

PULSATIONS AND CONTINUOUS WAVE

The biological result of microwave exposure depends on the material susceptibility, dielectric loss, frequency, waveform, intensity, and energy density as well as on geometrical and dimensional factors (Fig. 18-1).

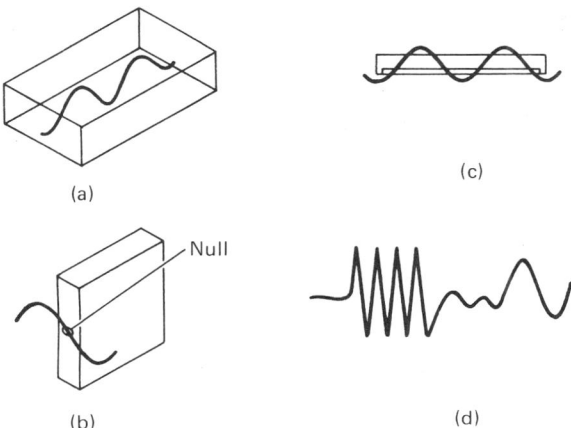

Fig. 18-1. Hypothetical waveforms illustrating some wave and target variables. a) Horizontal, two maxima inside a solid, versus b) vertical, one maximum due to varied orientation; c) Slab avoiding maximal amplitudes; d) Forms

By definition, the microwave is alternating its electrical and magnetic field at the given frequency as it propagates through space at the speed of light. This basic waveform may be modified by interruption as in the case of *pulsed* microwaves used in radar as opposed to *continuous wave* energy used in many industrial, scientific, and medical applications. Pulsed fields are often compared on an equal basis with continuous wave fields by taking the continuous wave power equal to the average power of the pulsed field. This procedure can deemphasize the effect of the pulse, *per se*. This is to say that the pulse amplitude (peak power) may be the determining stimulus for a biological effect. In any case, the pulsations occur at the *pulse frequency* which may be varied from a few to thousands per second. The time during which the pulse operates is the pulse width or duration. The pulse may follow a pattern based on the pulse repetition rate or frequency and the duration. This pattern may consist of very short pulses of high amplitude with long off duty cycles or the reverse. This difference is expected to influence the biological result, even with equal total energy input. The high amplitude situation may be more effective biologically.

The pulse frequency and width may produce a modulation by virtue of the information which it imposes upon the basic frequency. The modulation may then be related to the biological response in terms of whether or not the response is dependent on the energy exposure-rate. This type of relationship may be clarified by reference to the time constants involved. The application of a pulse will be like a stimulus with a time constant. When the first is greater than the second, an interaction in which the response is accentuated becomes possible, and all degrees of response may be related to these interaction times. The significant thing about the pulse is the application of peak power within this spectrum of action (Fig. 18-2).

PULSES—THE DIMENSIONAL RESPONSE AND GEOMETRICAL FIT

Another way to look at the information transfer is to consider the geometrical fit for a frequency and interacting material. That is, the dimensions of the wavelength, compared with those of the target, at least partly determine the interaction. A *dimensional constant* (in wavelength fractions, $\lambda/4$, $\lambda/2$, etc.) results, which predicts the possibility of an electromagnetic wave interaction (Fig. 18-3). This type may be considered a *spatial* constant while forcing functions or field influences brought to bear on a system react according to the *time* constants. These constants predict the interaction as consequential responses of systems for which they are a function. Centimeter and millimeter waves behave with biological materials according to a general tendency predictable from the observation that the materials have dimensions commensurate with the wavelength. The field is much more affected by the presence of such absorbing objects than would be a field with

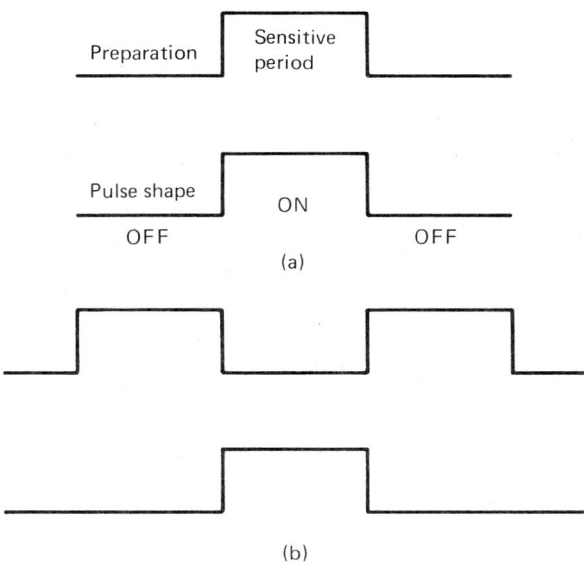

Fig. 18-2. Receptive period of biological preparation perfectly coincident with pulse (a) or anti-coincident (b).

much longer wavelengths such as those of many meters or kilometers (Fig. 18-3).

In tests on animals, some experimenters may not be able to demonstrate a dimensional response, probably because it is not always possible to obtain equipment and to make preparations that are optimized for

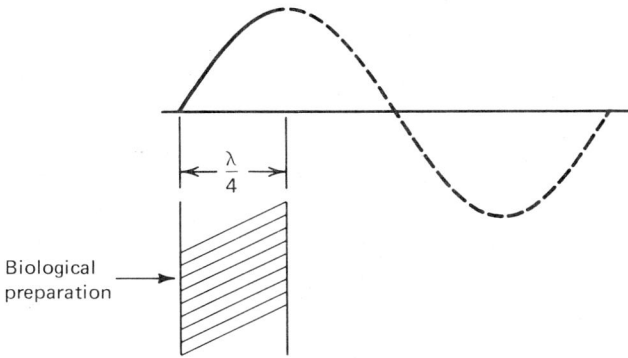

Fig. 18-3. Geometrical fit. Slab of biological preparation matches wavelength. A dimensional or spatial constant.

the study of the effect of size on absorption. For example, Michaelson (1967) teaches that the thermal response in dogs is independent of animal size from 4 to 20 kilograms where the animals are unrestrained in a plastic box at frequencies of 2.8 and 0.2 GHz. Similarly, at 3 to 30 MHz, when 100% of high field strength energy is absorbed, the size or weight of a rat does not influence its survival, only the energy in watt seconds per gram. These animals may not present constant geometry that can be related to the wavelength or they may be too similar to show the difference. The difference may also be masked by saturation and reflection effects. The geometry for absorption includes both penetration distances for the energy and size relations for interception or propagation as a function of body or body part size. The limb of an animal, for example, may be unusually receptive in this relationship. The experiments with rats, rabbits, and optical injury are of special interest.

The rabbit often demonstrates a sensitivity as shown by the injury produced by energy absorption. Rats, on the other hand, may be more resistant. At 2.4 GHz Michaelson (1969) observed that the tissue "focuses" microwaves to a point in the eye lens area, while below 1 GHz, the focus is less sharp and probably within the aqueous humor. A reflecting interface for the injury leading to posterior cortical opacities may be provided by the boundary between the lens cortex and the posterior capsule-to-vitreous body interface. This effect may be missed altogether if the thermal injury is monitored by means of temperatures read more proximally.

TIME CONSTANTS AND THE DELAYED OR FLYWHEEL EFFECT

In general, pulsed or continuous microwave energy is deposited in a manner described by a characteristic curve for its distribution within an absorber in the direction normal to the surface. At the surface, the demonstrable energy is less than that just beneath due to factors which at one time or another may include tissue transformer or matching effects, interface reflections, reradiation or cooling. There is a resulting maximum in the curve or internal peak at a depth which is on the order of centimeters and is a function of the frequency. Then an energy redistribution occurs which is an accommodation particularly of the prominent maximum. It is shown by an equilibrated distribution (leveling of the temperature curve for example) observable after many minutes in objects of a size commensurate with or larger than a wavelength. To show this, reflecting waveforms in cavity tests are preferred since plane forms according to our distinction allow only one passage.

Those skilled in the ways of microwaves have always attached a special significance to the delayed redistribution of the deposited energy which is in the sense of an effect from stored and reacting energy. This delayed effect may appear to be a function only of modulation, that is, due to the pulse repetition rate, or some such change in the carrier. Actually, it is a special result which must be carefully interpreted in terms of homeostatic mechanisms or thermoregulation for in vivo studies. Suppose that a pulse deposits a certain amount of energy inside the insulating dermal layers. With typical skin-fat layers, such a deposit of energy is usually observed. It is a concentration of energy that cannot immediately disperse. Before it does, the delay time may permit local effects due to a lingering activation. This is easiest to demonstrate as a possibility in the thermal response but this latter should be regarded as only one level in the continuum of possible levels of energy and responses. The same amount of energy deposited by a demodulated input, that is, an averaged, non-pulse or continuous input, has clearly more time to dissipate since it can do so while more energy is arriving. If the threshold for an observed change is identified with this difference, the change that occurs will be caused by the pulse modulation and equilibration delay acting together.

In terms of time constants, the biological change will have a reaction time and, if the delay during which activation conditions persist is insufficient, the time constants involved may counterindicate the particular effect. Similarly, the time constants associated with the repair process may be unfavorable for recuperation in the time between repetitions of the insult (Fig. 18-4).

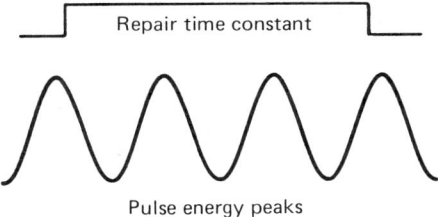

Fig. 18-4. Damage control as determined by the repair process time constant may be impossible in the time allowed between repetitions of the insult.

Examining Michaelson's work (1969) with dogs and other animals in this sense, we see that immediately lethal exposures produced temperature distributions that suggest important redistributions of thermal energy, distinctly different temperatures or susceptibility in different organs, and persistent temperature gradients 15 to 60 minutes post mortem. Taking the rectal temperature as a base after this sort of delay period, that is, well after exposure and death, the *higher* temperature regions were gall bladder, urinary bladder, stomach and lungs. Regions in which lower temperatures were found included the testes and the anterior chamber of the eye.

Thermoregulation and the Microwave Significance of the Dermal Shell

The dermal shell is a primary absorption region and the post absorption equilibration is modified by vasodilation, evaporative cooling and other thermoregulatory activities in the skin and mucous membranes during a continued exposure. The dog apparently has a limit to which he can regulate well and beyond which serious instabilities in his control loops occur. Accordingly, this animal recovers rapidly if at all. The rabbit, on the other hand, goes into violent reaction in a power density of 165 mW/cm^2. Examination reveals that a strong attempt toward control by peripheral engorgement of vessels has been inadequate. This control and its frustration by anesthesia can be illustrated in both the rabbit and the rat. In these animals, the rectal rise on exposure to 2.86 GHz is delayed by anesthesia. With thermoregulation impeded, the animals are hypersensitive to the exposure. Thermoregulation is different for low exposure, chronic conditions than for acute injury because within a month a dog can adapt to microwave thermal exposure. His regulation improves along with his thermal tolerance. The hypophyseal thermoregulating complex may receive chronic stimuli in the manner of a stressor. On the other hand, more acute exposure directly influences heat receptor loops in all the warmed regions which have them. The human sensation of microwaves in the dermal layers has a threshold of about 46 to 47°C at 3 GHz, but this sensitivity is not particularly acute.

Thermoregulation between species is subject to interpretation according to the special physiology involved. For example, during recovery from heating, the rat shows a special rectal temperature that will produce a large error unless the instrument is inserted deeply. This is because of the blood shunt that bypasses the thermometer.

When birds enter fairly low energy fields, their dramatic evasion tactics suggest a particular species effect. The subsequent behavior is also affected by the exposure. The feathers and their insertions which provide a modified dermal and subdermal regional response allow the skin region to rise to an athermic threshold (reflexive temperature) within five seconds in typical radar beams. The special response may have a dielectric mechanism as its basis or it may be a combination of dimensional and dielectric effects.

McAfee's description (1971) of an analeptic response contributes to an understanding of the subdermal reflex based on the neural and hormonal responses to stimulation of peripheral nerves by microwaves. The response is a well-documented microwave effect that is quite reasonably attributed to subdermal peaking of the thermal absorption curve. In this reaction the blood pressure increases and there is a cessation of respiration followed by hyperpnea. These responses disappear if the nerves are cut or if the peak heat region is locally anesthetized both at 10 and 2.45 GHz. Again, secondary effects such as those studied in biopsychology and animal behavior are altered because special effects, particularly thermoregulatory ones, appear.

PROTECTIVE FUNCTIONS OF SKIN

The skin has a variety of tissues (Fig. 18-5) including blood vessels, connective tissue, fat, smooth muscle, nerves and nerve receptor endings, secreting glands and both living and nonliving regions. Aquatic creatures are at the mercy of their rather effective liquid medium for radiation protection while with land forms, several skin mechanisms function to compensate for hazards in the environment. With a sort of shield designed to protect against primitive external insult, it is of interest to find out how the usually well-guarded inner cells fare when the organism is exposed to twentieth century hazards such as strong microwave radiation. The brunt of the defensive task is borne by the epithelial cells or the top layer but many of the nonionizing radiations do not see this barrier, and pass at least to the dermis and subcutaneous fat layers that come next. The dermis is thicker, being millimeters thick, and has the nerve terminal loops, sweat glands, collagen and wound

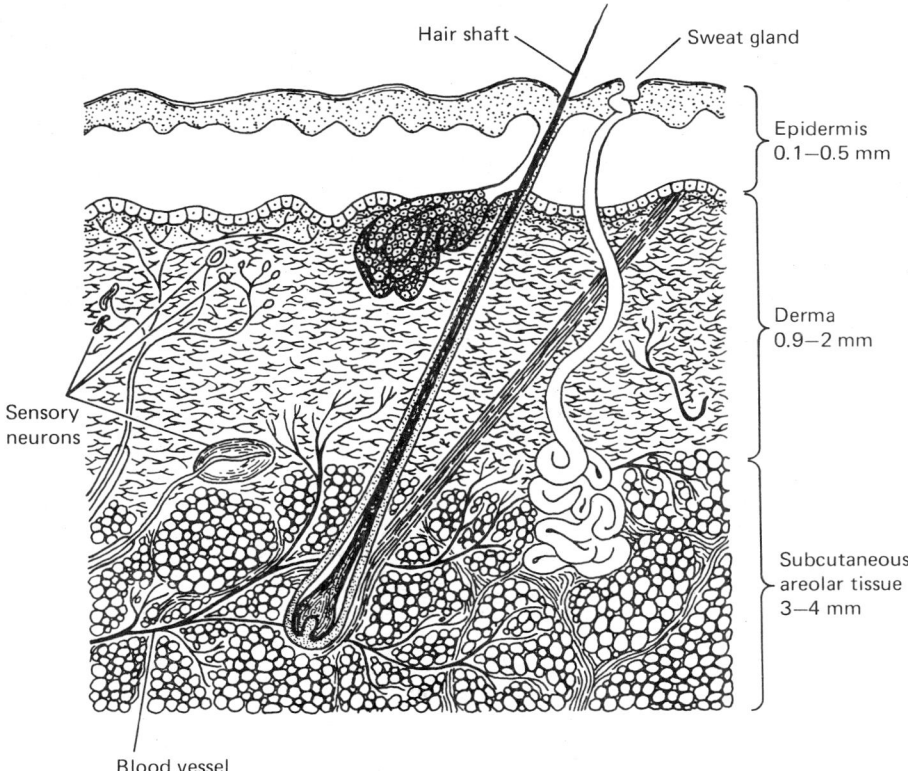

Fig. 18-5. Details of dermal region. Skin portion with hair.

elastic fibers in a gelatinous structure. The thickest region, the subcutaneous fat, is packed with fatty cells and varies in width among individuals, often being on the order of 3 to 4 millimeters. The design of the skin reflects the primitive abrasive hazards to which it was exposed, being both leathery and cushioning. The tough keratin material of the epithelium contributes the former quality while the collagen fibers and fat give protection against impact.

SELECTIVITY ENHANCED BY CHANGED DIELECTRIC LOSS, MICROWAVE THERMAL STRESS

The dermal blood supply is both venous and arterial, combining to provide the vasodilation essential in thermoregulation and detoxification. In the acute responses, these vessels become engorged and inflate the dermal layer which is elastic and displaceable. From a dielectric loss standpoint, the changed character of this flooded tissue is quite important. The sweating process and vasodilation both tend to hydrate the skin, adding to its dielectric loss.

The dermal region sandwiched between the epidermis, which is about 0.1 mm thick, and the thicker subcutaneous fat layer, is overall about 0.5 cm thick, but this dimension actually depends on the region of the body. For example, the skin of the back is twice as thick as that over the stomach, and the flat of the foot and palm of the hand have still thinner skin covering. On the scalp, even with hair follicles present, the skin is only about one and one-third millimeters thick.

The sweating mechanism is the most effective thermoregulatory device in man for reducing excessive internal heat because of the large heat requirement for evaporation. In turn, the mechanism is very sensitive to air velocities. As the fluid to be secreted passes through the outer skin it greatly hydrates the region, changing its dielectric characteristics in favor of dielectric absorption. A high relative humidity (over 90%) causes water to pass into the skin with the same result. Perspiration, on the other hand, is accelerated into a salt solution stronger than 10% and into drier air of relative humidity below 90% (Rushmir et al, 1954). Some differences of interest to radiation studies exist between the skin on different body parts. One such place is the scrotum which does not have the usual sweating control loop and *diffuses* its evaporative cooling water outward as opposed to sweating.

From an information transfer point of view, the ability of electrical signals to pass the skin barrier varies a great deal with the water and impurities present but dry skin will present a definite impedance barrier to the information. The impedance will decrease just inside the outer layer. If the proper sites near the hypothalamus are warmed by microwaves, the thermoregulation responds with a demand for the dissipation of heat. Another skin control mechanism, heat conduction, will be increased when a deposit of heat inside the skin (as with microwaves) creates a gradient in favor of more heat loss from the body and more reradiation of infrared. Extrapolation of results with animals to man is fraught with more than the usual dangers. In addition to the dimensional and geometrical features in the radiation interactions, the dog loses heat due to sweating on foot pads and respiratory panting but the discharge is otherwise inhibited by the fur.

The energy input curve peaks in the skin or subdermal layers for many microwave wavelengths, certainly between 3 and 12.5 centimeters, leaving the heat sink surface at the epithelium much cooler. This peak appears inside the main thermal dissipation region which includes the vasodilation region. This delays its equilibration for a significant biological time period as the thermal insulation layers inhibit the outward flow. Proximally, the peak energy input can equilibrate within the organ, muscle, and bone regions below. The time for equilibration to a flattened curve in these tissues is surprisingly long, on the order of 15 to 30 minutes. The reason that the internal thermal peak is better considered to be a generalized energy peak is because there is no basis for doubting that an athermic energy distribution would follow the same general pattern. This follows from the common absorption mechanism.

DIMENSIONAL RESPONSES

While the variable electrical characteristics of the vasodilating, secreting, and perspiring skin layer must help to shape the energy curve, this layer is not essential for obtaining a peak because a peak occurs quite clearly in phantoms and in tissue *in vitro* without circulation. The consequences of the *insulated* peak are important. The first is the flywheel effect in which the consequences of an energy input can continue to operate well after the exposure has terminated and during the period while this energy curve is leveling. The second is that the underlying structures in the sensory system can experience analeptic responses at 45°C, well below the burning temperature. This response is equivalent in both anesthetized and normal animals. These reflexes are crossed extension cardiovascular and autonomic nervous system activity changes. While the epithelium stays cooler, the nerve pathways are rapidly warmed in the C and Delta fibers.

In a small cylindrical absorber such as an animal body or a human leg, the peak in the energy curve becomes a core maximum, with reinforcing contributions from 360° of direction when the preparation uses reflected waves (Fig. 18-6). This is an additional physiological insult in an animal where the radius of the cylindrical shape and the depth of the maximum are nearly equal distances. These conditions may well lead to a strong frequency dependence. Some mechanism of this nature can operate in the rat and may account for the almost instant convulsions and death on exposure.

BRAIN ENZYME INACTIVATION

To induce microwave thermia in the rat, the E field is arranged to be in parallel with the sagittal axis in plane wave exposures, and the animal

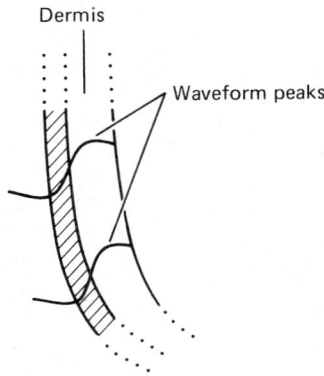

Fig. 18-6. In the reflected wave or cavity-contained preparation, the nerve-rich dermal region of an animal's leg is equivalent to an annular ring wherein interaction is concentrated. One waveform showing peaking in the region.

will probably be held in a polycarbonate wave guide insert. This orientation enhances the hyperthermia. Similarly, to inactivate brain enzymes in neurochemistry, the longer axis of the rat's head is held parallel to the E field. A typical inactivation may use about 5 kW for 0.5 or 1 sec, so it is also a test for massive exposure of the brain. The neurotransmitters, serotonin, dopamine, and norepinephrine decrease with exposure as evidence of hypothalamic thermoregulation which involves a loss of the substances. In the process, it is found that the rat thermoregulates in depth in an exposure of 20 mW/cm^2. The brief exposure freezes the metabolism at a point in time so that the neurochemists can make rational analyses. For a thermolabile enzyme like acetylcholinesterase, activity observed after this fixation decreases as a function of the heating velocity so that its substrate is held in a fixed time status. Concurrently, cyclic nucleotides decrease in the neural tissue and 3,4-diaphorase is more thermoresistant but follows acetylcholinesterase down. After microwave fixation, the tissue can be frozen in liquid air or Freon -12 at -150°C for 30 to 90 seconds to a temperature of 30°C. If the adenylate cyclase and phosphodiesterase enzymes for synthesis and degradation are not inactivated within 1 or 2 seconds, there is a *post mortem* surge of activity. AMP is stable but the others are sensitive to the microwave treatment.

This massive exposure makes the nuclei pycnotic, compacts the nucleolus, inflates the histological appearance and slows the Nissl staining reaction. The metabostat device sold by Gerling-Moore in California is a typical inactivator.

It is very difficult to obtain accurate energy density readings during exposure along the energy distribution curve, even for a thermic distribution. There is an antenna effect in which strong local fields develop around the end of a metal instrument such as a thermistor or thermocouple. In strong fields the metal may melt. Even if the fields are reduced, the metal continues to act as a heat sink and may obliterate an important region such as the maximum or peak by drawing off the energy. The temperatures will either tend to be low according to this error or to be variable from the antenna effect. The usual remedy is to stop the exposure long enough to insert the measuring instrument. Color-indicating, temperature sensitive chemicals may also be used but for the curve these would have to be held in an impregnated suture or the like.

THE INTERNAL PEAK AND THRESHOLDS FOR BIOLOGICAL RESPONSE

The significance of peaking in the energy distribution is extended with respect to thresholds for injury. If a threshold exposure is defined as one of sufficient time at a given power density to cause an injury, then the absorbed energy density would describe the chain of events a step closer to the primary biological effect and this would not be obvious from a given power density exposure unless the maximum is accounted for. This reasoning can be applied to the acute responses ranging from analeptic behavior associated with a thermal but nonburning stimulus and possibly to a "single exposure" eye opacity. A great many other responses develop from chronic exposure. These are *favorable* and *unfavorable* biological effects caused by repeated multiple subthreshold exposures and may have a cumulative nature. These are included so as to give the range of possibilities and, in a tentative manner, to relate their energy requirements to the ubiquitous peaking in the distribution curve.

BIOLOGICAL MOLECULES HIGHLY SUSCEPTIBLE TO ABSORPTION OF ROTATIONAL ENERGY

Radiofrequencies and microwave (nonionizing radiations) are characteristically rotating influences on molecules. The far infrared ones are inverting and ring deforming. In the rotation mode, segmental rotation in biopolymers is of importance at microwave, while overall rotation of these molecules is important at radiofrequencies. In the overlap region between microwave and infrared, the rotation of water molecules is common to both bands, while lattice vibrations, hydrogen bond and ring displacements and the rotation of terminal groups occur mainly in various parts of the infrared spectrum.

Biopolymers are characteristically provided with a large number of degrees of freedom especially when the usual appendages in the shape of small polar groups like water molecules are both present and accounted for. These degrees are in the sense of constraints on motion. An aircraft launching from a catapult has one degree of freedom. At the end of the slide it has at least two and a somersault would represent three.

Rotation and the axes about which polar members of a dipole can rotate both together and independently quickly add up to several degrees of freedom. Other modes such as vibration, inversion, or translation add still further to the freedom. In a protein, the number of amino acids may be 300 and some of this motion may be found in each one. A rotating influence may also be superimposed on an existing motion. Each degree of freedom will mean a partition of the molecule's kinetic energy. Superimposed thermal agitation varies the energy associated with a degree of freedom. As one example, collision kinetics can modify the motion caused by a forcing function or vice versa. With modes of motion as a common basis, these observations are applicable to the study of Brownian movement of particles in a fluid, their thermal agitation and motion as in large molecules, as well as to the study of electromagnetic forcing functions. The mean energy connected with a degree of freedom depends on the gas constant R for a gram molecule of gas and Avogadro's number, N. The ratio of these, for a single molecule, is the universal constant k, or the Boltzmann constant.

$$k = R/N = 8.314 \times 10^7 / 6.023 \times 10^{23}$$
$$= 1.38 \times 10^{-16} \text{ erg per degree}$$
(346)

The Boltzmann constant multiplied by the absolute temperature or kT gives the mean energy for a degree of freedom from the combination of kinetic and potential energy. This is the amount by which degrees of freedom can increase the thermal energy in the system. The introduction of electromagnetic energy to the system, infrared for example, allows the radiation energy, $E = hf$, to be compared with kT. The thermal energy distribution is a strong function of the wavelength.

Photon or collision-induced energy absorptions produce relaxation processes in the high frequency region, particularly in high pressure gases and condensed systems. The susceptible systems will have natural frequencies appropriate to the electromagnetic frequency. Rotational behavior depends on molecular dipoles responding and changing energy states according to the wave function and the angle function between the axes of the molecule and the electromagnetic field. Collision results for polar molecules depend first on proximity and then on the angle function. In liquids they are of similar importance as shown by fine structure interactions, such as when collisions neutralize rotational transitions.

An action spectrum is broadened, bandwise, with pressure in a gas until it is a relaxation in the condensed phase. The liquid has a dampening effect on the otherwise sharp response, e.g. to collisions in a gas. An average potential energy change emerges as a system result. The relations of the frequency dispersion of the dielectric constant (ϵ') and loss factor (ϵ'') distribution about the natural frequency result. The action of collisions is to cause a mixture of basic response motion. They can change rotation into other modes such as vibration and translatory movement.

ENERGY AND INFORMATION TRANSPORT IN PARALLEL

Usually these motions should be biologically harmless except when high power densities and the associated thermal degradations are present. Informational effects however, could be associated with polarizations and concentration gradients or ion distributions in the biomolecular and water systems. These would have their relaxation process and descriptions as functions of the frequency and dielectric polarization.

Since biomacromolecules carry functional information for metabolism, growth, and repair, they are especially vulnerable to sufficiently enduring disturbances. Even ordinary heat destroys hydrogen bonds, letting helices unwind and, according to the amount of heat, setting up a chain of reversible and finally irreversible changes. A pH change of only one will weaken H bonds (Trindle and Illinger, 1967). Microwaves can, by several rotational modes of interactions, impose disturbances on these polymers and even slight changes in any of three levels of structure may cancel the functional bioinformation. A polarizing disturbance which is typical for alternating waves can influence nucleotide base pairs and this kind of disturbance can be carried out within the polymer "cage" formed from secondary, tertiary, or helical organization. In such a case, the effects may occur in a relatively sheltered region free of the outside randomizing effects. If the organization is vulnerable, the energy may rotate properly joined sections of the whole molecule or smaller groups. It is easiest to visualize and demonstrate resulting disorganization but the transfer of radiation information need not be negative. In the low megacycle frequency range, chromosome aberrations including bridges, fragments, and the appearance of micronucleii have been produced and these effects which inevitably involve both good and bad mutations are found, not only in motile protozoa, but in human lymphocytes as well, (Wilkins and Heller, 1963). Some motionally induced regulatory mechanisms on blood components and even the ordered behavior of whole microorganisms are suggested by the results of studies of dielectric mobility (Fig. 18-7).

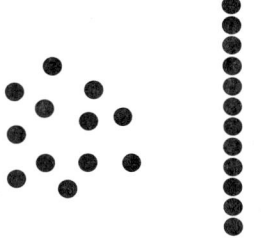

No field   Field applied

Fig. 18-7. Functional information transmitted to susceptible units by applied electromagnetic field. A second order orientation would be a wheeling movement of the aligned particles. In either case, the new position will be a more favorable energy state.

SPECIFIC MOLECULAR AND MOBILITY EFFECTS

The electrophoretic and antigenic behavior of human gamma globulin is affected by frequencies between 10 and 200 MHz, and distinct effects may be due to harmonics of 6 MHz. The action spectra are narrow and within 10 to 20 KHz. A field strength of 25 V/m may cause changes in antigenic reactivity. The activity of alpha amylase is reduced at 12 MHz with a 1 Kw amplifier source (Bach et al, 1961).

Probably the cholinesterase in human erythrocytes suffers a change in activity on exposure in whole blood at 13.56 and 23.75 Mc. The histamine activity decreases at constant cholinesterase levels at the lower frequency, while at 23.75 Mc, the reverse is true with cholinesterase increasing in specific activity.

Evidence is accumulating in favor of a number of electrophoretic type phenomena, ranging from the moving of small particles to the moving of minor organisms in electromagnetic radiation fields. Protozoan experiments for example, suggest that one frequency may be used, (5 to 7 MHz) to cause migration parallel to the electric field. Then, at another frequency (27 to 30 MHz), the organism will change and move perpendicularly to the original orientation. Intracellular organelles are apparently oriented also.

RESONANT EFFECTS

The concept of resonances in nonionizing radiation biology is a broad one that can be somewhat difficult to manage. The term resonant circuit is common in the power supplies used to produce nonionizing radiations where it means a frequency dependent state and the maximum value of one of the circuit parameters. Still in the area of equipment, there are resonant cavities such as those in a magnetron that are coupled to an applicator cavity which is said to be tuned to resonance by adjusting the conditions. The resonant condition may indicate an optimum performance. Similarly, numerous mechanical, electromechanical, dimensional and chemical resonances may be involved at one time or another. On the strictly biological side, a resonant radiation effect in the nonionizing region may, in one sense, be one where there is a narrow band frequency action spectrum that is not involved with an observable change in temperature. This concept is broadened and discussed in some detail owing to the opportunities for confusion in this important area. The idea is to underline any unique results in the interaction such as the easy transfer of energy and any tendencies toward selectivity or special communication of information.

POWER DENSITY LEVELS FOR DIRECT INFORMATION TRANSFER

Interpretation in terms of system time constants, dimensional constants, and energy couplings results in descriptions of resonant-like behavior in appropriate systems. These conditions create a strong demand for fundamental information. A search results for valid explanations and predictions applicable to new energy states, and some striking conclusions emerge. The biologically complex organism is "innundated" in its self-generated fields. These arise from the normal operation of muscular, nervous, and secretory tissue. The most important or well-organized fields may be coupled into measuring devices to record normal or abnormal states of cardiac, cerebral or other essential systems. Examples of these devices are electrocardiographs, electromyographs, and electroencephalographs. The last is unique in that it was developed on fields so strong that they can pass through the impedance presented by the skull. The body's fields are in constant interaction with the natural and artificial electromagnetic fields that constitute this portion of the environment. There is no question that there is an exchange of information but the exact nature of these couplings is far from clear. Biophysical, psychological, and chemical efforts are obviously essential to evaluate the observations because these do not, at first glance, seem probable against accepted rules.

An example is the sensation of hearing attributed to microwaves. The microwaves are received as a sound, but the auditory loop may hear the pulse modulation or repetition or, in a different mode, microwaves may be rectified (perhaps usefully) to sound. There is a good possibility that this abnormal sensing may indicate potential harm to persons who receive the microwave stimulus over a long period and, for the short term at low intensity of 0.5 to 1.5 mW/cm^2, there may be increased auditory sensitivity. Frey (1968) teaches that low level microwaves, simultaneously offered to rats with tone cues, lower the levels of response. This is like a depression of the hearing acuity or the sensing of a distraction. At higher microwave energy levels rats could develop better discriminative efficiency for the presence or absence of a tone cue of 525 cycles. There is clearly a threshold for the auditory response and arousal at about 1 mW/cm^2, but the time of exposure, stimulatory, and depressive aspects are not very firm.

PEARL CHAIN EFFECT, INDEPENDENCE OF PEAK POWER, AND BASIS FOR TRANSFER OF INFORMATION

The components of a biosystem will be subject to forcing functions in terms of applied electromagnetic field forces, and in the consequent energy interaction, the system components will tend to progress toward a lowest potential energy state. Cellular particles in particular may reorient in an applied field in order to adjust to their favorable energy states in a new position (Fig. 18-8). A sort of control over appropriately sized particles is then possible in which the reorientation is according to plan. The observations have been made on many small, natural and artificial particles or globules such as polymeric spheres in suspension and microorganisms. This reorientation is sometimes called the pearl chain effect. It is known to be dependent on the particle dimensions, the frequency, and the strength of the forcing field. Subcellular particles may require a field strength of the order of 0.1 kV/cm. This forcing function can then be evaluated in terms of known-to-be dangerous heating fields such as a level of 10 mW/cm^2 or 200 V/cm. The latter would not only injure the tissue but they would override any subthermal responses. According to evaluations such as those of Schwan and Sher (1969) it would be necessary to find a system in which, for some reason, heating is not masking and yet the required strong fields are present.

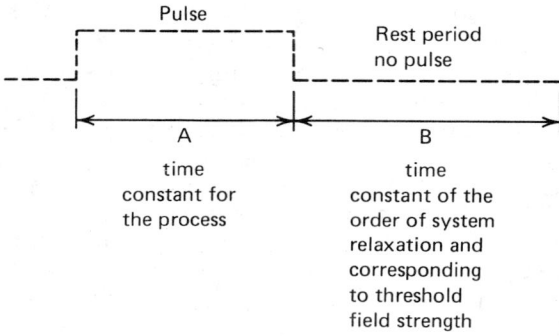

Fig. 18-8. One cycle, time sequence and constants for energy coupling in Pearl Chain effect. A) Time constant for the process; B) Time constant of the order of system relaxation and corresponding to threshold field strength. If pulse time is greater than process time constant, heating occurs.

Schwan (1969) and his school find that description of the total minimum energy in the forcing function as $E^2 T$ results in better agreement for the alignment effect. This is the square of the field strength times the time constant or relaxation time for the process. To obtain the effect, the field threshold must be attained (E_{th}) and then the reaction time constant is inversely dependent on the square of the applied field. The threshold energy allows a stimulated response against the background of noise, Brownian movement, and thermal motion as randomizing influences in the medium. It is then clearly a question of process time constants versus the time during which the energy is applied or the pulse. A fine balance is needed between the optimum time for application of the forcing function, E or E^2 and the particle specifications (for example, time and dimension constants) which allow it to react or relax. Duty cycles T_1 ON and T_2 OFF equal to force, no force time constants will, theoretically, let the process surface in the midst of a potential thermal masking. Beyond the threshold energy, the time constant which depends on field strength and particle characteristics will decrease. The T_1, T_2 modulation required is pulse type in which the *ON* and *OFF* periods are optimized for the react or relax response of the components. The coupling achieved under athermic conditions is when

$$T_2/T_1 = (E_{threshold}/E)^2 \qquad (347)$$

It is the time of application and nonapplication that is determining, and the average value rather than the peak value of E is the causing stimulus. The reacting or chain-forming particle fits the electrical conditions due to its dimensional and other form factors. Then under these "go" conditions, the available energy is absorbed for the reorientation process in a narrow fit of conditions which siphon off the field energy, and in so doing, may exclude a gross temperature rise by a slim margin. The Schwan analysis provides an apt description of the conditions expected for this kind of resonant event (Fig. 18-8). It is worth pointing out in connection with the heating alternative here that all radiations may ultimately degrade to thermal status. However, except for microwave or dielectric heating and the like, this fate may not be the experimental objective.

There is, again, the analogy of pushing a swing at just the moment when the motion will accept energy most effectively (Fig. 18-9). By delaying the pulse application until the reaction is progressing, the new input is directed toward orientation rather than toward random thermal agitation. The cycle energy is averaged over the two time constants, the information passing by way of the time relationships or the effective modulation. With this concept, the peak values of the field are secondary.

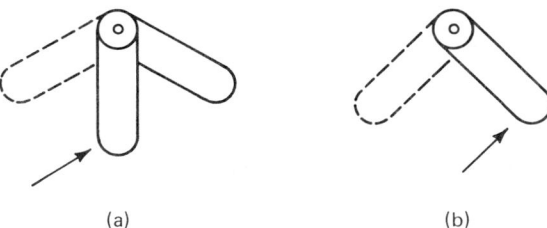

Fig. 18-9. Forcing function (a) and (b). The swing pushing model is applied to a pendulum-like biological preparation. The forcing function (the alternating field) could be applied at any position up to the limits (dotted lines) but at (a), it produces a greater effect than at (b).

Fine Adjustment for Effect to Surface

In attempting to apply these evaluations to observed conditions of electromagnetic energy absorption, it must be noted that the application of fields does not produce a linear response in tissue with depth. Instead, there is a profile which is determined by the tissue constants and the frequency. A low power may be directed into the tissue or, alternatively, attenuation may lower the power. Various particles under differing conditions such as solid, liquid, or gaseous states, and many degrees of freedom are encountered. The variations in energy disposition create opportunities for the spectrum of thresholds to be swept through as long as the masking thermal input is suppressed. When a match occurs between susceptible components and the forcing functions, then the response may well be observed. Considering the diversity of preparations presented, and the delicate adjustment of the window through which the energy may enter nonthermally, it is understandable that few experimenters will have obtained the responses, let alone explained them. Resorting to model systems has been very fruitful and many simple organisms have been caused to align according to a frequency and field strength plan. The potential gradient of interest in such interactions is evidently between 1 and 2 V/cm or more.

RELAXATIONS IN MOLECULAR BIOLOGY

Macromolecules or their components react to the alternating field by a relaxation process which is then their dielectrical property. Measuring the capacitance shows that the effect of capacity decreases with increasing frequency. At some frequency, the large molecules will lose the ability to reorient with the alternating field due to the system time constants and this is the relaxation frequency. It is a resonance with a flattened peak or a degenerate resonance frequency. Counter-ion relaxations occur at low frequencies and molecular groups on macromolecules relax at high frequencies so that the entire structure possesses a wide range of resonances.

If the solvent water could be made inert with respect to its dampening effect on orienting solute molecules, rotational diffusion would be ineffective and there would be intramolecular resonance in the solute. However, the price of canceling the solvent-solute couple would be an uncoupling of the energy input. If the energy is matched into the macromolecules in the medium, there would be no way to avoid mismatch without having both components.

NATURAL PROTECTION

While it would be unwise to underestimate the innovative role in phylogeny and the genetic role in adaptation, it seems that, in the latter part of the twentieth century, nature has left the soft creatures somewhat vulnerable to the negative effects of radiation. At least two mechanisms are important in this situation. One is the dermal vasodilation which, in some configurations, can offer a special protection against many nonionizing radiations. These encounter a variable-sized water matrix with absorbing macromolecules in a system highly responsive to thermic and other stimuli. In large individuals this system produces a significant flood of radiation absorbing substance which may be called upon when needed to protect underlying critical tissues, provided there is the necessary thermic or athermic coupling into feedback control loops. Similarly, bony structures offer protection because they offer a significant impedance to the passage of the fields. The feedback loop which can call up the inundation mechanism is unclear unless it can be either a purely thermic response or an athermic response which can ride on the thermoregulation apparatus.

SENSE MODES FOR INFORMATION INPUT

Van Dam, Tanner, and Romero-Sierra (1969) have studied another mechanism of interest—a directly sensed information input from nonionizing radiation to the feathers, especially tail feathers, of birds. In a plane wave exposure at an average of 50 mW/cm^2 at 9.29 GHz, modulated energy at 416 pulses per second with a 2.35 microsecond width, is detected at once. In 10 to 20 seconds, the Leghorn hen is immediately distracted away from ordinary behavior. It sounds an alarm, defecates, and tries to fly. There is a form of adaptation that can sharpen this response. Reverting to normal, with the power turned off, the hen carried out a body feather fluffing and preening routine. This normalizing behavior is associated with the feathery structure which serves a sensory role in the matching of energy into the bird.

It is tempting, although not conclusive, to see an interaction between the biopotentials and the applied field. There is certainly a visible interaction at the organismic level and the incoming information is a clear signal of possible danger which deforms the prevalent behavior predictably.

Feathers do not absorb appreciable energy directly. Nevertheless they or their associated structures are essential to the effect. The quill section does have electromechanical properties of the piezoelectric type, producing resonant peaks in the 0.1 to 10 kHz band on electrical stressing via a resonant mode. Quill cross-sections and insertions vary greatly from the hummingbird to the California Condor, but waveforms also vary greatly from one end (meters) of the microwave spectrum to the other (millimeters). The dielectric composition is fairly constant and electrical coupling involves both size and the composition of the material.

It follows from the bird's behavior that the elaborate feather control neuromusculature is interacting with the microwave field in a very sensitive manner. That is, the bird tries to "iron out" the after-effects of the response when the field is removed, and he always senses the electromagnetic field as a stressor of some undefined urgency and nature. The urgency of the evasive response is proportional to the energy applied. Experience with other absorbing matrices suggests that the wing, with its low density network structure, may be absorbing energy directly in a selective manner.

THE BIRD AS A BIOPOLE IN THE INFORMATION FIELD

The nature of the natural thermal environment modifies a bird's ability to sense microwaves. The bird acts as a multiple biopole immersed in the applied electromagnetic field containing pulsed-type elements of information for its nervous system. All areas of the bird show changes in potential fields and the result is a modification of motor activity which accumulates into an arousal behavior.

The microwave field interacts with muscle control as shown by electromyograph studies of either the specified contractions involved in spreading the appendages or of those involved in the collapse response. The dislocation in contraction is nerve controlled, with dominant contralateral muscle action that extends to paralysis. Normally the central nervous system would control this weighted response so it is obviously involved. These conclusions are summarized in Figs. 18-10, 11 and 12.

ANIMAL SENSE MODES

In considering the tenuous connection provided by possible sense modes for microwaves, there is the work of Justesen and his group (1969) in biophysical psychology. This showed that rats use electromagnetic energy in cueing to both unpleasant stimuli and to anticipated rewards in the way of sugar. The animals preferred to regard the microwaves as a prediction of an undesirable event much as the birds did. This distinction would be more understandable as a direct information transfer if the mechanism were below the thermal threshold. A thermic sensing mechanism would have to explain how an animal can relate the stimulus to a distinction and prediction on a strictly thermal gradient.

The behavioral effects at low intensity are therefore ones which can probably rule out instantaneous step changes in response and instead focus attention on longer term results. From 300 to 900 MHz, hypoactivity in rats is the only behavioral result of a low intensity 0.5 mW/cm^2 exposure for more than forty days. Korbel's work on this aspect (1969) showed a time response that is indicative of time scales as expected in chronic exposure study. Results may take several months to demonstrate and time differences are measurable in weeks or even months.

It is clear that animals adapt to microwaves, increasing their tolerance with time, but how the adaptation occurs is inconclusive. In humans, an adaptation occurs in three years of occupational exposure, and is somewhat easier to follow because the neurotic symptoms diminish. If the exposure continues for many years, there is a regression to sensitivity.

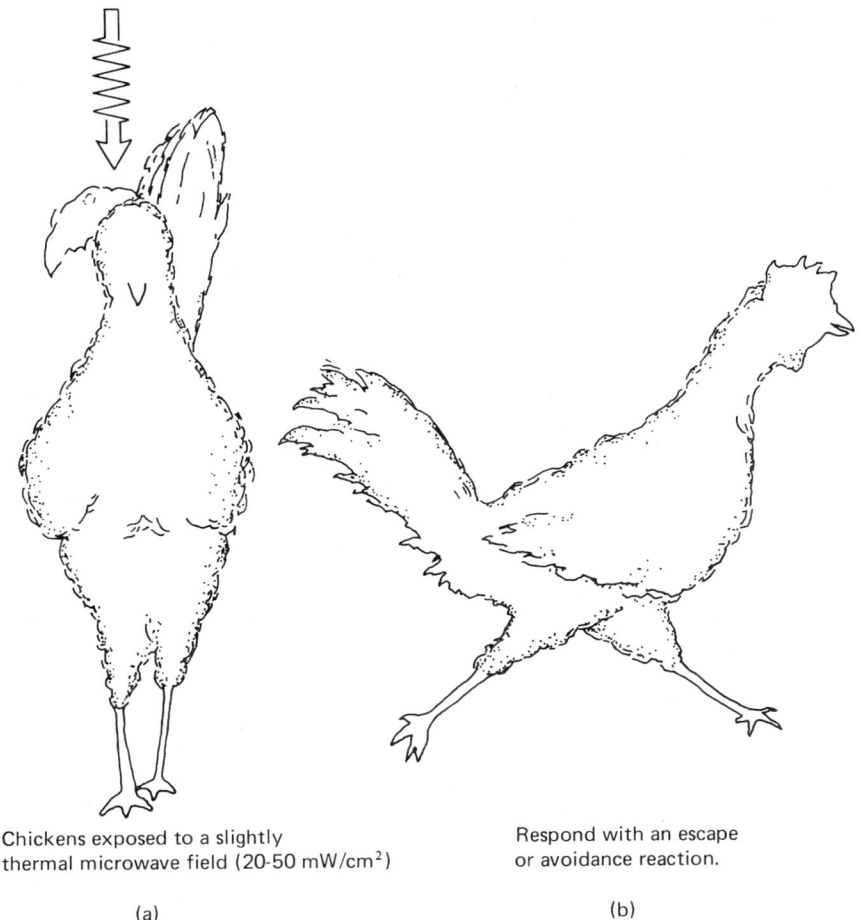

Chickens exposed to a slightly thermal microwave field (20-50 mW/cm^2)

(a)

Respond with an escape or avoidance reaction.

(b)

Fig. 18-10. Microwave avoidance.

The work with cats and the analeptic response tend to implicate the thermo-sensing system. Cats learn to avoid the stimulus that they associate with an apparent peripheral analeptic response. Having previously experienced the threshold at 45°C, they can associate later clues with the feeling and attempt to evade it. The application is most effective on the lower leg where there is a rich supply of peripheral nerves. An animal population can apparently be conditioned and then maintained in a behavior pattern by very small microwave signals.

One of the quick effects is the alerting one. This effect is strong enough to cancel out anesthesia. McAfee (1971) obtained this response by analeptically exposing the feet of animals but it is also obtained

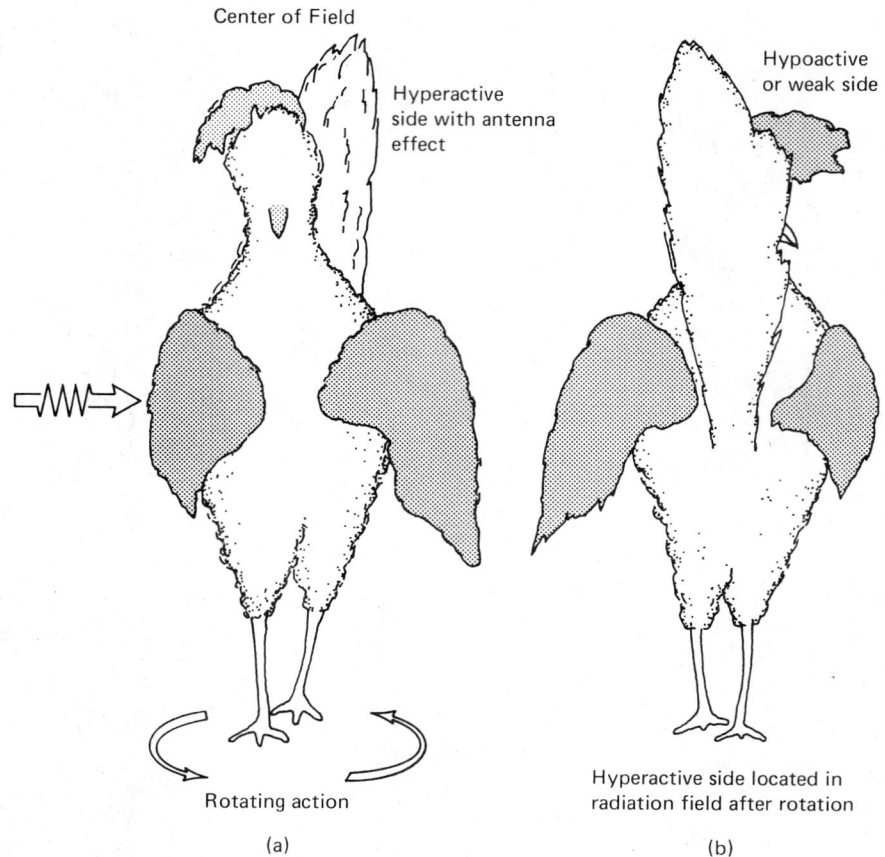

Some animals present a hyperactive side. When radiated with the weak side facing the center of field they rapidly orient, presenting the strong or hyperactive side to the field. Previously described behavior ensues.

Fig. 18-11. Polarization response. Reactions from a bird depend on surface exposed with little reaction if exposure is to ventral surface, supporting the idea of sensors acting like antennas (for example, certain feathers). When exposed from above, distinction is clear between distal dorsal and tail regions of the animal and the head, and the neck to proximal dorsal regions. (*After* Tanner, Romero Sierra and Davie 1967)

from the head region. Smaller animals respond more readily, e.g., rabbits and rats more readily than dogs and cats. The 45°C threshold in the dermal nerve region is a distinct one within ±2°C. If the head is exposed, then the response involves the facial cutaneous nerves and/or scalp nerves. In rats and rabbits, the blood pressure decreases whereas

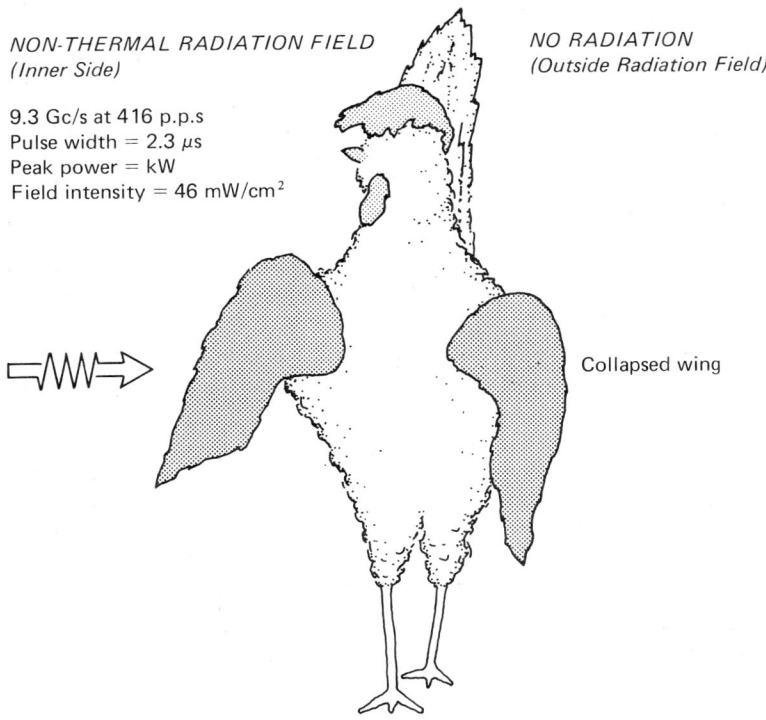

At the onset of radiation the wing outside the field of radiation becomes collapsed and the opposite wing becomes extended. The chicken inclines head with eye closest to field of radiation oriented to the field and the sagittal axis of the head is kept in line with appropriate axis of body.

Fig. 18-12. Pre-collapse. Bird then turns outward from field following axis. In this turning reaction, the outer side of the animal is seen to be paralyzed. On the floor, reaction is increased extensor reaction of inner field side and head turns to face center of field. (*After* Tanner, Romero Sierra and Davie 1967).

it increases in cats and dogs. In humans in high intensity fields the blood pressure drops significantly and may lead to collapse. There are wide individual differences.

The rat is very sensitive and radiation directed to the leg of this animal sees a cylindrical shape and nerve networks inside. Immediate absorption from many radial directions within this cylindrical shape can cause convulsions. These animals also struggle with the swallowing reflex from local absorption in the neck which is another cylindrical region. Even the thermic effect seems to carry special information when combined with appropriate target dimensions and form factors. Conventionally, i.e., aside from radiation, the analeptic response is a well described heat response at 45°C, the heat being applied to an exposed peripheral nerve by a warming device known as a thermode. The unique aspects of the microwave directed response are shown by its being *independent of contact warming*, the special *coincidence* for microwave disposition in the nerve-rich dermal region, and the *delayed* or flywheel effect in which the warming effect is caused to linger in the region.

PERCEPTION AND BEHAVIOR

There are effects in humans and animals in connection with perception. Over the long term there is either a sensitization or the person becomes more perceptive as a cumulative effect depending on the intensity of the power and especially that of the peak power. There is the matter of immediate alerting or arousal in the animals and then a reduced sensory awareness over the period covered by typical behavioral tests. This extends to a stunned effect after an hour of exposure. The animal's ability to decide to respond on cue-information is constant or improves, that is, the discriminative efficiency potentiates. The operant task training for response and conditioning with a Pavlovian stimulus such as a shock or a tone may be combined. The animal then suppresses operant responding when it gets a cue but goes on working. Justesen (1969) shows that microwave power applied for a minute at 600 μW to 6.3 mW per gram gives a cue, that is, information leading the animal to expect a brief unpleasant stimulation but it does not function as a cue to predicting pleasant stimulation. For ionizing radiations, X-rays have been found to carry similar information but only indirectly via the ozone produced by these radiations in air. The ozone can serve as a sensory cue. Exposure to pulses of 10 mW/cm^2 or less leads to a poorer performance in animals and the more complex the task, probably the less the exposure needed.

PROBLEMS

1. How does a wavelength in the power and telephony spectrum affect a human target?
2. What extreme species difference exists for electromagnetic radiation?
3. What is the meaning of redistribution of thermal input from EMR?
4. What is the subdermal reflex?

BIBLIOGRAPHY

Adams. R.L., and R.A. Williams, *Biological Effects of Electromagnetic Radiation (Radiowaves and Microwaves)* Eurasian Communist Countries, Defense Intelligence Agency, U.S. Army Medical Intelligence and Information Agency, Rpt. no. DST-1810S-074-76, March, 1976.
Bach, S.A., A.J. Luzzio, and A.S. Brownell, "Effects of Radio-Frequency Energy on Human Gamma Globulin," in *Biological Effects of Microwave Radiation*, Vol. 1, M.F. Peyton, Ed., Plenum Press, New York, 117-133, 1961.
Bach, S.A., *Fed. Proc.*, Vol. 24, S22, 1965.
Breuer, M.M. and D. Robinson, *Nature*, Vol. 221, 1116, 1969.

Dumanskij, J.D. and M.G. Sandala, "The Biologic and Hygienic Significance of Electromagnetic Fields of Superhigh and Ultrahigh Frequencies in Densely Populated Areas. Biologic Effects and Health Hazards of Microwave Radiation," *Proc. Of an Intern. Symp. at Warsaw, Poland*, Oct. 15-18, (Warsaw, Polish Medical Publishers), 289-293, 1974.

Elden, H.R., Ed., *The Biophysical Properties of the Skin*, John Wiley, New York, 1971.

Frey, A.H., "Auditory System Response to Radio Frequency Energy," *Aerospace Medicine*, Vol. 32, 1140-1142, Dec. 1961.

Frey, A.H., "Human Auditory System Response to Modulated Electromagnetic Energy," *Journal of Applied Physiology*, Vol. 17, 4, 689-692, July 1962.

Frey, A.H., *Physiol. Behav.*, Vol. 3, 363, 1968.

Frome, K.D., "Determination of the Velocity of Short EM Waves by Interferometry," *Proc. Royal Society*, 213A, p. 123, 1952.

Griffin, D.W., "MW Interferometers," *Microwave Journal*, May 1978.

Hearings Before the Board of Public Utility Commissioners, Newark, New Jersey, Docket No. 754-248, Vols. I and II, May 25, July 10, 1975.

Jones, T.B., M.P. Perry and J.R. Melcher, "Dielectric Siphons," *Science*, Vol. 174, 1233, Dec. 1971.

Justesen, D.R. and N.W. King, "Behavioral Effects of Low Level Microwave Irradiation in the Closed-Space Situation," *Symp. on Biological Effects and Health Implications of Microwave Radiation*, S.F. Cleary, Ed., BRH/DBE 70-2, U.S. Dept. HEW, Sept. 17-19,154-179, 1969.

King, N.W., "The Effects of Low Level Microwave Irradiation Upon Reflexive, Operant and Discrimination on Behaviors of the Rat," Ph.D. Dissertation, Univ. of Kansas, 1969.

King, N.W., D.R. Justesen, and A.D. Simpson, *Behavior Res. Methods and Instrumentation* (In Press 1970).

Korbel, S.F., "Behavioral Effects of Low Intensity UHF Radiation," *Symp. on Biological Effects and Health Implications of Microwave Radiation*, S.F. Cleary, Ed., BRH/DBE 70-2, U.S. Dept. HEW, Sept. 17-19, 183, 1969.

Lanzyl, L.H. in *High Energy Radiation Therapy Dosimetry*, J.S. Laughlin, Ed., Annals of the New York Academy of Sciences, New York, 368-370, 1969.

Lin, J.C., "Noninvasive Microwave Measurement of Respiration," *Proc. IEEE*, Vol. 63, p. 1530, October 1975.

McAfee, R., *IEEE MTT-19*, Feb., 1971.

Michaelson, S.M., R.A. Thomson, and J.W. Howland, *Biologic Effects of Microwave Exposure*, Report, (RADC-TR-67-461)AFSC Griffiss AFB, New York, 1967.

Michaelson, S.M., "Biologic Effects of Microwave Exposure," *Symp. on Biological Effects and Health Implications of Microwave Radiation*, S.F. Cleary, Ed., BRH/DBE 70-2, U.S. Dept. HEW, Sept. 171-9, 37, 44, 1969.

Michaelson, S.M., "Thermal Effects of Single and Repeated Exposures to Microwaves — a Review, Biologic Effects and Health Hazards of Microwave Radiation," *Proceedings of an International Symp. at Warsaw, Poland*, Oct. 15-18, (Warsaw: Polish Medical Publishers), 1-14, 1974.

Milroy, W.C., Ed., "Biomedical Aspects of Nonionizing Radiation," *Proc. of a Symp. held at the Naval Weapons Lab., Dahlgren, Virginia*, July 10, 1973.

Onsager, L., in *The Neurosciences*, G.C. Quarton, T. Melnechuk, and F.O. Schmitt, Eds., Rockefeller Univ. Press, New York, 75-79.

Ricketts, L.W., Bridges, J.E. and J. Miletta, *EMP Radiation and Protective Techniques*, John Wiley & Sons, New York, 1976.

Rushmir, R.F. et al, "The Skin," *Science*, Vol. 154, 3747, October 21, 1966.

Schwan, H.P. and Sher, L.D., *J. Electrochem. Soc.* 116, 170, 1969.

Tyler, P.E., Ed., *Biological Effects of Nonionizing Radiation*, Conference held by the New York Academy of Sciences, New York City, Feb. 12-15, 1974, Annals of the New York Academy of Sciences, Vol. 247, Feb. 28, 1974, 10-11.

Tanner, J.A. and C. Romero Sierra, "Response of Leghorn Chickens to Microwaves," *Nature*, Vo. 216, 1139, 1967.

Trindle, C.O. and J. Illinger, *J. Chem. Phys.*, Vol. 46, 3429, 1967.
Van Dam, W., J.A. Tanner, and C. Romero-Sierra, National Research Council, Div. of Mechanical Eng., Control Systems Lab. Report, LTR-CS-19, Dec. 1969.
Wilkins, D.J. and J.H. Heller, *J. Chem. Phys.*, Vol. 39, 3401, 1963.
Zaret, M., "Blindness, Deafness and Vestibular Dysfunction in a Microwave Worker, *The Eye, Ear, Nose and Throat Monthly*, Vol. 54, July 1975, 49-52.

19

Cataract — Noise and Antinoise in Visual Information Flow

"We see things not as they are but as we are."

H. M. Tomlinson, *Out of Soundings*

MULTIRADIATION BIOEFFECT

Radiation cataract is a multiradiation bioeffect in that the cause can be absorbed radiation such as X-rays, gamma ray, neutrons, uV, IR, particle, or microwave. The effect is to produce a defective lens with histological aberrations that disperse light. With the lens altered in this manner, sight may be totally and permanently lost. Normal passage of light is assured only when the cornea, aqueous humor, lens, and vitreous body all remain clear of discontinuities. Then the light-bearing visual information can be successfully carried to the retinal receivers.

The eye segments are optically specialized for refraction, focusing, and transmission under variable conditions of visible light and information presentation. While there must be provision for homeostasis and maintenance of structures in a viable state, this is often accomplished in an avascular mode. The chromatic problems in having a rich blood or nerve supply in the light-transmitting pathway would be great, so in this organ these are not found in the immediate regions involved in radiation cataract (Figs. 19-1, 19-2). The presence of discontinuities in the informational pathway introduces noise. This is usually neutralized by maintenance and repair which thus act in the role of antinoise or biological squelch mechanism.

LENS RECORDS PHYSIOLOGICAL INSULTS

The defects which produce the lesion occur in the lens where both the transmission of light and its focusing on the retina depend on a normal state of the lens fibers. The lens is a biconvex body with one or more layers of epithelial cells between the anterior polar and the equa-

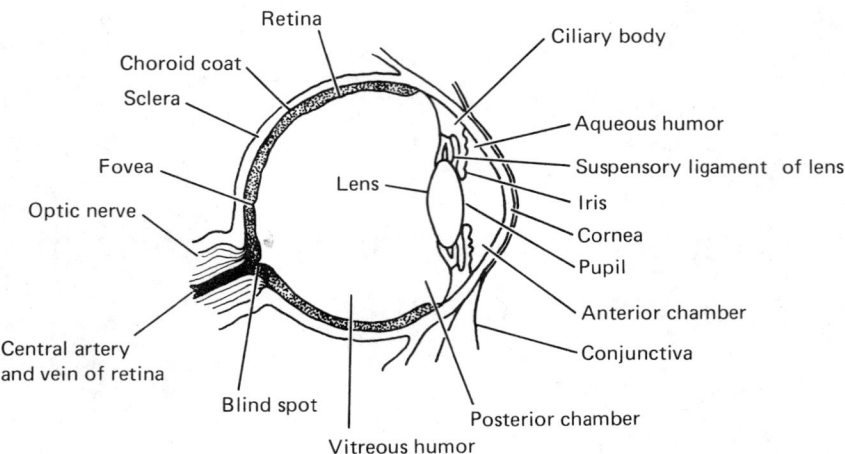

Fig. 19-1. Human eye structures through which visual information passes. The eye reaches its full size early in life.

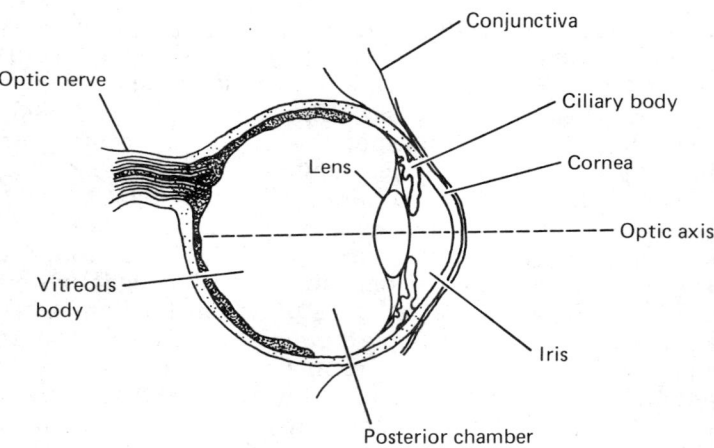

Fig. 19-2. Rabbit eye structures. The rabbit eye is about three-fourths as large as the human eye. Its blood color reflects the density of thin-walled capillaries in the rear segment.

torial zone. These cells are constantly growing and developing into lens fibers that are laid down over the old fibers which are therefore retained and concentrated toward the nucleus along with their lifetime record of physiological insults. This is in contrast to regeneration of cells in other parts of the body. Proper function depends on a smooth capsular surface which is also elastic. The capsule has two poles—the anterior and the posterior. The light pathway is along the polar axis and the equator divides the lens into the anterior and posterior zones. The nucleus is the core of packed older fibers while the developing ones are in the cortex. They join in the *anterior* and *posterior sutures* (Fig. 19-3).

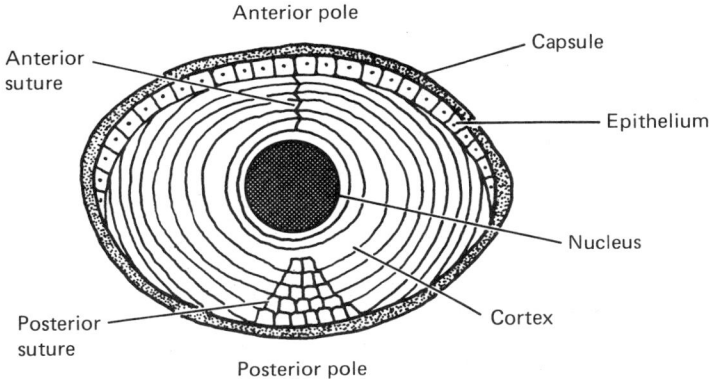

Fig. 19-3. Lenticular structures (rabbits). New fibers form continually but damaged lens fibers do not regenerate. Defects are accumulated.

The synthesis and stability of protein is crucial in lenticular health. Cataract is the coagulation of protein, although opacities may also be associated with the density of crystalline salts like calcium phosphate, cholesterol and cystine. Accordingly, there are radiation cataracts, sugar cataracts (in which merely the optical rotation D or L of a sugar sets off the disease), toxic, nutrition, and genetic cataracts, and hot and cold cataracts.

The maturing lens accumulates chemical and physical defects such as progressive dehydration, concentrations of insoluble protein or albuminoid, decrease in metabolism and ascorbic acid levels, compaction and hardening in the nucleus which interferes with optical accommodation, and degeneration of quality in the epithelium. Maturity nevertheless brings with it a greater resistance in animals.

If the *mitotic index* is involved, as it is in ionizing radiation injury, then it would be greater in immature still-active epithelium and least in the mature lens. This index, the ratio of dividing and nondividing cells, may influence the sensitivity to radiation since actively dividing cells are more sensitive. Young fibers in the cortex are rich in neuraminic acid, which is both polar and hydrophilic, confers electrical characteristics, and affects water permeability (Kuck, 1970).

The lens protein is in the form of a tight gel, rich in more or less bound water. The osmotic strength is 308 to 342 mosmoles which is the osmolarity of 0.9 to 1.0% sodium chloride (Bellows, 1944). The collection and recording of radiation insults occurs in the fibers as they move in a posterior and inward direction. Their disabilities are reflected in reduced transport of metabolic substances and waste products between the humors and deep fibers.

COMPARATIVE RADIATION CATARACTOGENESIS

Radiation cataracts are of two types. This follows from the division of radiations into two kinds, ionizing and nonionizing. The former affects the anterior and equatorial lens and the epithelium and takes months or years to develop while the latter, at least those best understood such as the microwave and infrared cataracts, are in the posterior capsular-subcapsular region. The microwave cataracts require only days or weeks to begin. Very great exposures to X-rays, say 400 rems, may produce cataracts but this exposure is also lethal to many animals. Bachem (1956) teaches that ultraviolet cataracts occur mainly at 297 millimicrons, with the cornea absorbing shorter wavelengths and so protecting the lens. This action spectrum is a broad peak formed by absorbing substances such as lens protein acting with the radiations admitted by the filtering cornea. Other absorbing substances are flavoproteins, nucleotides, and cytochromes. Ultraviolet cataract is probably caused in many cases by solar radiation and no doubt solar eclipses are sometimes involved. Lens sensitivity is very great and cataract occurs after large doses which are not able to bring about other apparent epithelial tissue damage.

Pitt and his colleagues (1977) put the corneal UV cataract in a narrow action peak at 300 to 335 nm. The threshold went from 0.022 $J\,cm^{-2}$ at the smaller wavelength to 10.99 $J\,cm^{-2}$ at the longer side, and twice the threshold energy made irreversible lesions in rabbits. For lens cataract, the range was 295 to 315 nm and only twice the threshold exposure produced permanent injury.

Photosensitization to lenticular cataract is illustrated by many drugs, for example, by methoxsalen, a remedy for vitiligo and a compound used to encourage suntan. This substance can sensitize the lens to long wave ultraviolet and make hazardous the portion of this radiation that passes the cornea. The appearance of photodermatological symptoms elsewhere can serve as a warning of possible lens sensitization and cataract potential.

A few roentgens will be cataractogenic in rabbits and the effect is not strictly dependent on the dose rate. Densely ionizing radiations such as alpha particles are relatively more effective in causing cataract. The 250 MeV oxygen nuclei (heavy-ion radiation) used by Bonney *et al*, (1977) caused ionizing radiation cataracts beginning at 10 months in the posterior capsule with a concurrent increased density in the anterior cortex also. The ring-shaped opacity of other ionizing radiations was absent but not the *granular, frost-like, stria types* and *vacuoles*. These results on monkeys were aimed at clarifying the astronaut "light flash" problem and its significance in planning missions. Chronic exposure to neutrons gives bilateral cataracts in one year at doses of only a few neutrons per day and are thus dangerous. Fractionation of the dose delays and may prevent opacities or cataracts. Over 15 rads of X-rays increases the incidence of opacities in mice and 20 to 45 rads give the estimated range for human cataractogenesis. Membrane permeability which controls hydration is altered in both ionizing and nonionizing cataract and these membrane effects are in turn related to the water-binding and electrical characteristics of neuraminic acid and fucose.

The interspecies differences in molecular structure of the eye are great and exist even between mammals. The bird, for example, is apparently resistant to ionizing radiation cataract and, among other differences, has no gamma crystallin in the nucleus. Instead, this substance is apparently replaced by glycogen. Gamma crystallin is a soluble lenticular protein fraction made up of a mixture of molecules. It is not necessarily related to the species resistance. The laboratory rat on the other hand is resistant. Radiation cataract can occur in a single lens after 200 R, about a hundred times the dose of an X-ray picture exposure, and children are more sensitive than adults.

Infrared

The infrared cataract, like the microwave one, starts as a posterior cortical opacification. Looking at glowing glass dips will endanger the eye of a professional glass blower and looking into a live waveguide or aperture in a microwave system will jeopardize the vision of an electronics engineer. The infrared seems to produce *hot* cataract at the site in the lens nearest the external site of most intense heat. If burns occur on the left side of the iris, for example, the lens below is in most danger. The thermoregulation via aqueous humor or the vitreous currents is poor and pigments abound which increase infrared absorption. The thermoconductivity is about 5.94 in the vitreous humor and 2.13 mW/cm/°C in the lens. With iris, choroidal and ciliary arteries, eye thermoregulation is only about 1.7 cc/min. This circulation is aided by convection and evaporation. Assuming that a conscious person would move out of the heat zone as soon as sensible heat at about 45°C indicated danger, the surface burns would seem to be from almost tolerable temperatures near 45°C. Then cataractogenesis would be associated with temperatures near this level, for example 42 to 44°C.

METABOLIC LESION

There is an instance of enzyme loss in galactose cataract which may be related to metabolic processes common in radiation cataract. Animals fed diets rich in lactose or galactose show a typical central cataract. Dulcitol, a sugar alcohol metabolized from galactose, accumulates as enzymes to attack it seem to be missing. Water moves osmotically inward to correct for the accumulation. Vacuoles form with resultant lens opacification. A similarity to microwave cataract is the sudden drop in glutathione and decrease in the cation pump action. As metabolism is depressed, so is the ATP level.

The maintenance of lenticular health is, to a large extent, the maintenance of lens protein and therefore the synthesis of this substance and availability of amino acids are essential factors. In some forms of cataract, the latter come to be in short supply. Cataract itself as opposed to preliminary opacification can be a rapid event, occurring in a matter of hours with sugar cataract. When potassium does not get replaced inside lenticular membranes and sodium accumulates, the lens hydrates and swells. The cataract is a dramatic result and in the case of galactose, the location is nuclear. These events suggest the term "metabolic lesion" for the sugar cataracts. On the other hand, some amino acids such as methionine may play a role in the protection against cataract. The lens is fairly stable to impairment of oxidative respiration.

Many toxic substances such as ergot produce chemical cataract in various locations, but the radiomimetic chemicals can simulate most ionizing radiation injuries. These chemicals such as myeleran and mimosine are cataractogenic. The latter chemical comes from *Mimosa pudica*, the common tropistic plant, and it proliferates the equatorial epithelium and changes the anterior cortex fibers of the lens (Kuck, 1970).

OTHER OCULAR CHANGES AND AGING CATARACT

The Soviet work reported by McRee (1980) includes a noncataract change for evaluation of microwave exposure. It is the measurement of color discrimination, and tests showed that this function is enhanced by microwaves for discrimination of red, green, and blue, especially red which is improved at 87%. Retinal lesions which impair vision but are not cataractogenic have not been reported because the examiners have not associated them with microwaves. Yet the pathology which can be paramacular or macular is found in workers in related industry and the lesions are in the central part of the fundus.

Another complication is the increased incidence of cataracts in the population in aging. Over age 45, there are always more cataracts and by the ages of 65-94 the incidence is up 50%. Thus epidemiological surveys for microwaves or other etiologies must be aimed at distinguishing between "normal" incidence due to age and that due to exposure. Young populations as in the Armed Forces have a notably low incidence of cataract.

THE EXPOSURE MODEL FOR MICROWAVE CATARACT

Often in the following model, good coupling into the eye must be assumed so that, in animal eyes, the possibility of a hot spot along the direction of the light pathway or beyond is reasonable. The model considers both the energy distribution before equilibrium and following it. As in other exposures, less permanent and unique in effect, that leading to cataract shows an inverse relation between penetration and frequency. Centimeter distances and waves seem logically related in the morphology, and positive interaction is assured but for fractional millimeter waves, the site of action moves to the surface structures. Similarly when the wave approaches one meter, the eye is not commensurate and the whole head or brain will be of interest (Cleary, 1980).

The lowest exposure to microwaves that induces cataracts in man is quoted by Marha (1969) as 10 mW/cm^2 as an occupational exposure over a period of time. Persons have developed cataracts after 15 years of waveguide and transmission testing. A worker in the process of adjusting a generator while observing through an end window in the transmission line may receive 1 watt/cm^2 exposure or much more with high powered generators. Discounting the specific frequency for the moment, since all are in the lower gigacycle range, the exposures capable of causing opacities or cataracts range from 280 mW/cm^2 for five minutes (Carpenter, 1968) to 5000 mW/cm^2 for one minute (Baillie, 1969). The lower power density is about 120 mW/cm^2 for the acute effect and it must be maintained for about 4.5 hours. If this is increased to 600 mW/cm^2, five minutes will be enough (Williams et al, 1956). Other than the very long range Marha estimate, the least accumulated exposure for cataract in terms of power density and repetitions is 280 mW/cm^2 repeated five times for three minutes each time (Carpenter, 1968). The least power density is 80 mW/cm^2, but it must be repeated daily for one hour ten times. A relatively intense exposure of 5000 mW/cm^2 requires seven repetitions, less than a minute each. These results were obtained with rabbits and dogs and excellent intrinsic control is exercised in such experiments when only one eye is exposed.

The cataracts will accumulate in two months with multiexposures at the 5000 mW/cm^2 level but at one-tenth of this power density, they require years, although opacities may appear in months. A permissible exposure curve will have the usual shape for cataractogenesis exposure levels, but the allowable time is on the order of minutes for the power densities referred to in this section (Fig. 19-4).

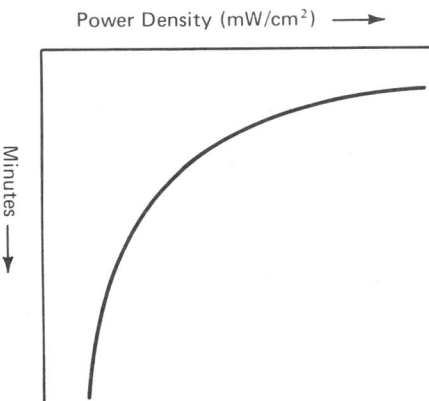

Fig. 19-4. General shape of cataractogenesis curve.

Expressed in terms of absorbed, 2450 MHz microwaves, the minimum cataractogenic level is 100 W/kg maintained for 140 minutes as an acute exposure by a directed slot held 5 cm from a rabbit's eye. A medium term chronic exposure is illustrated by a freely moving rabbit exposed to 10 mW/cm^2 at the same frequency in a cage for six months. There was no cataract. For this animal, 200 mW/cm^2 for 30 minutes in a single exposure is also negative but 300 mW/cm^2 and above for over 15 minutes is positive (2450 MHz). These power densities represent serious injury to the exposed area causing iritis miosis and injection of the cornea but, as observed in the comparison, not necessarily cataract. They also appear like single accidental exposures might appear. If the lower exposure, 200 mW/cm^2, which is subthreshold, is repeated, it may mean cataracts. Here the temperature threshold is taken as 41°C. Absorbed energy in the following combinations is positive for the injury in rabbits at 2450 MHz under a diathermy director:

Time (min)	Power Density (mW/cm^2)	W/kg (4 kg animal)
30	200	184
20	300	276
15	400	368

The probable positive temperature conditions are 41.5° for 20 minutes and 43.5°C for 10 minutes.

There are unique features in microwave cataractogenesis. Conduction heating of the posterior pole of the lens to temperatures above the "threshold temperature" for microwave induction of cataracts *fails* to produce the disorder. Two forms, at least, are recognized:

1. Capsulopathy, an early opacity of the lens which is rare and may often be overlooked.
2. Hydrops, like (1), a thermal cataract.

Opacification develops along the suture lines some 7 to 9 days after the intense exposures cited as cataractogenic. The overall effect seems like a coagulation of lens protein and it is probably a passive event in that the cells are not actively responsive to the injury. However, following exposure, there is cellular (lens epithelium) damage near the lens equator, cells migrate and they *accumulate* debris plus bladder cells. Optically opaque particles lead progressively to the subcapsular defect. The cataract is detected by its iridescence and the presence of vacuoles and opacities. In the closed waveguide exposure where the eye actually abuts the aperture, local burns can be severe but if the injury is more dispersed as it would be in a free field exposure, then posterior lens damage takes place.

FLYWHEEL EFFECT POSTULATED BASED ON IMPORTANCE OF PEAK POWER AND REPAIR TIME CONSTANTS

If the exposure is subthreshold for the acute injury, it will be brought up to the stage of "delayed opacities" by increasing the repetition rate and number of exposures when the preparation is pulsed. The peak power is important for positive acute cataract. For example, 400 mW/cm^2 pulsed for 60 minutes is capable of forming a cataract even though the average power is only 80 mW/cm^2 (Fig. 19-5). This is seen as an

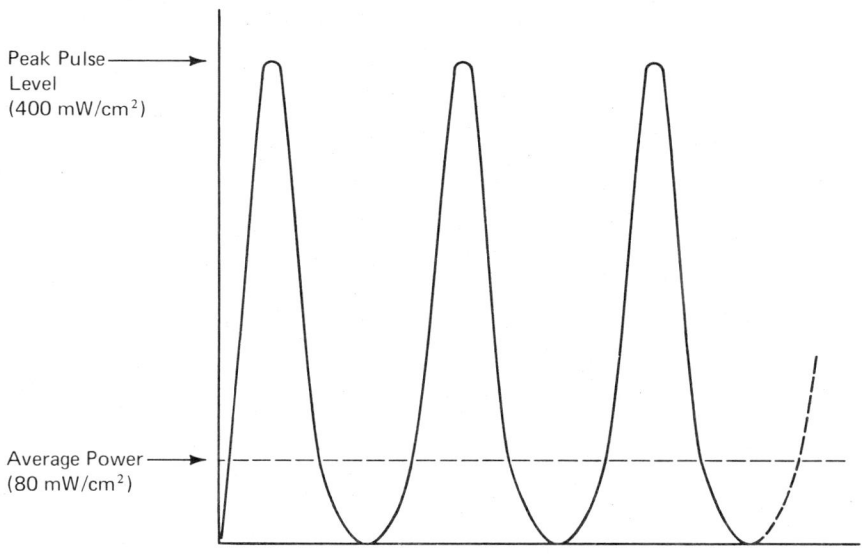

Fig. 19-5. Pulsed power contributes to flywheel effect.

instance of flywheel (delayed) effect. The delay time during which induction conditions (for example, greater than 80 mW/cm^2) persist, permits multiple insults. Another way of expressing this is that the primary lesion is followed by further insult before recovery of the repair process. What the lens feels is a reinjury while its status from the previous exposure remains acute. Then, as the reexposures become more frequent, the acute injury is almost continuous. The influence of repair-time-allowance will persist until exposures are truly reexposures, say with 24 hour separations versus pulsing within the time scale of seconds. However, fast repair processes will be superceded by the slower ones.

Slow recovery processes in the rabbit take up to 8 weeks so that, in order to show the effect of time constants, multiple exposures spaced after recovery would be needed. The situation in dogs is similar in dependence on time constants. It may also be interpreted in thermal terms by showing that higher temperatures are maintained by the pulsed input since energy arriving on an average input basis would have longer to escape. A cumulative rise in temperature is allowed if the time constants for pulse, distribution, and process do not permit relaxation to normal temperature

The result of an exposure of 5000 mW/cm^2 for one minute, one of the cataract threshold conditions, is clearly instant denaturation or thermic destruction. The same power density held for *less* than one minute during seven exposures is also positive unless the subject is rendered hypothermic. This cumulative injury does not bring the lens to the protein coagulation point of 60° and therefore may be related to an altered repair metabolism. Baillie (1971) used hypothermia to protect his rabbit preparations by immersing them to the neck in ice water so that their temperature became 22°C. This prevented lens heating and avoided cumulative cataract. However, the critical temperature for induction of the acute injury is probably less than that for the cumulative one, that is, 45° to 55°C. (The normal human temperature is 1.2°C less than the rabbit's which is 38.7°C.) Nonthermic microwave cataract is suggested by the posterior lenticular positioning until the characteristic curve for energy distribution with its internal peak is considered.

A common model for cataract production is that of the restrained or anesthetized animal which offers a definite geometrical and electrical design. If the animal is not restrained, its freedom of movement is protective against the injury. The animal can and will move about so that the exposure will depend on the ratio of the time during which energy is lens-coupled to the uncoupled time. This dependence is also influenced by the amount of sensible coupling because the animal will have more ability to avoid this portion of the energy.

JUXTAPOSITION OF INTERNAL PEAK AND SITE OF PRIMARY LESION

Considerable significance may be attached to the nature of the microwave energy disposition curve in relation to the site of the radiation cataract. In the lower gigacycle region, the dimensional relations put the peak of energy distribution about one to two centimeters inside the tissue (Fig. 19-6), (Copson, 1954, 1962). The pathway through the cornea and lens to the site of cataract development with microwaves in the posterior capsular region is of the same order of distance, depending on the size of the eye, species, and specific frequency (Fig. 19-7). The result is a juxtaposition of the peak on the general region which shows a sensitivity to microwaves (Fig. 19-8).

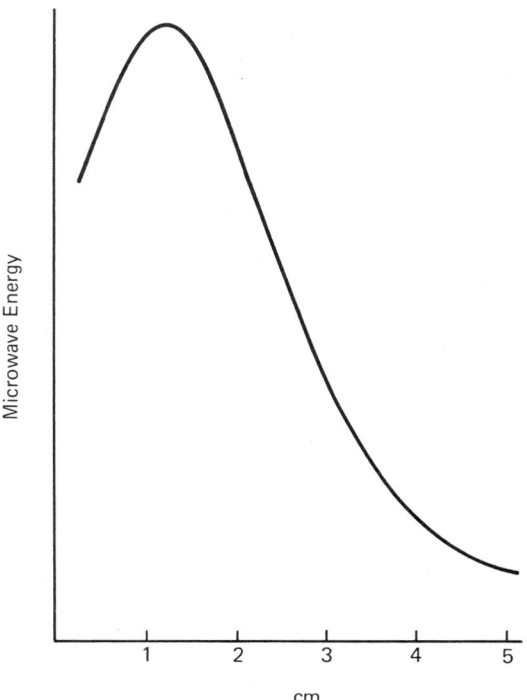

Fig. 19-6. Penetration curve with internal peak.

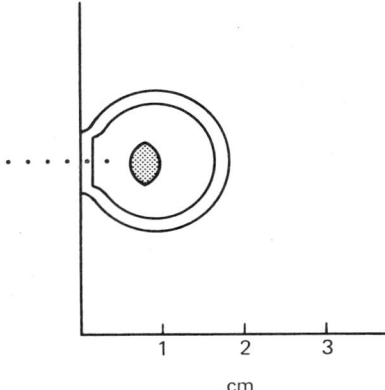

Fig. 19-7. Approximate distance to cataract region.

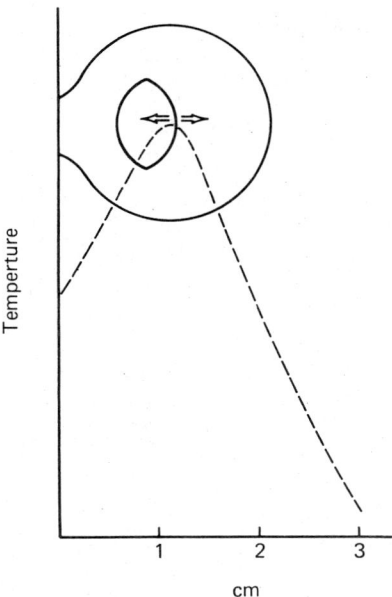

Fig. 19-8. Posterior capsular-subcapsular cortex shown as point of juxtaposition of internal peak and microwave sensitive region of the eye. Maximum is displaced by species, wavelength, and dielectric absorption factors as indicated by arrows, anteriorly or toward the vitreous humor. In the rabbit at 2450 MHz, it is immediately distal to the posterior pole.

In a manner reminiscent of the audio effect, a local deposit of heat at a maximum can induce rapid mechanical processes somewhat like the disintegration of cellular, membranous structures by pressure, and then vacuum forces. Some 10 to 20 dynes/cm^2 are associated with an input of 40 microjoules.

This conclusion is supported by the work of Zaret (1969) who draws the clear distinction between ionizing radiation cataract in the anterior capsular region and the microwave type in the posterior capsular location. However, he assumed a uniform microwave heating in which the iris absorbs the heat in the front of the eye. The contribution of Baillie and his coworkers (1969) has verified the temperature distribution curves obtained in agar gels by Copson (1962) using phantom structures which can be used to relate cataract development to the same juxtaposition of energy peak and sensitive ocular region. Baillie (1969) suggests a 2 to 1 ratio for the energy distributed to the lens substance to that in the vitreous humor substance beyond, and he quotes their specific dielectric constants as

$$\epsilon'_{lens}/\epsilon'_{vitreous\,humor} = 75/35 = 2.14 \qquad (348)$$

The absorption ratio was given as

$$1.9/1 \qquad (349)$$

for lens to vitreous body which supports a selectivity in the microwave energy distribution and a cataractogenesis based on microwave energy density. Computing the energy disposition without these factors tends to displace the maximum distally into the head and to raise the question of possible lethality before cataractogenesis. At 2450 MHz in the rabbit, the power absorption is maximum immediately distal to the posterior pole and falls off quickly in the head.

MIGRATION OF INTERNAL PEAK WITH WAVELENGTH

The difference in cataract location between super, high frequency ionizing radiation and microwave is clear but a question arises about possible relocation of the frequency-dependent peak in various experiments within the microwave region alone. Taflore and Brodwin (1975) computed a hot spot at 40.4°C in the center of a model human eyeball at 100 mW/cm² at 1.5 GHz. Guy et al (1974) put the 915 MHz peak at 1 cm behind the orbit in the brain and the 2450 MHz peak two-thirds of the way to the back of the rabbit eyeball. The model clarifies points of heat conduction and the authors see a resonance shift from the eye "scatterer" at the higher frequency to the whole head at the lower one. Then the peaking at 1 to 3 GHz is at the midpoint of the eyeball. The wavelengths in the medium will be about one tenth of that in space (centimeter waves) based on the dielectric constant of the lens

$$\lambda_m = \lambda_o/\sqrt{\epsilon'} \qquad (350)$$

so that the geometrical considerations will vary greatly in favor of much smaller distances. Similarly, the propagation velocity v_m varies with the conductivity of the medium and this fact changes the wavelength in the medium because

$$f = v_m/\lambda_m \qquad (351)$$

and

$$v = c/\sqrt{\epsilon'} \qquad (352)$$

The higher tissue conductivities abruptly decrease the wavelength in the medium as a consequence. The total electrolytes are known to increase in the lens post irradiation, but there is also hydration so that the effect on the wavelength is mixed.

Then there is considerable interest in the delayed or flywheel effect for "hot" cataractogenesis because the equilibration of the peak could then maintain activation conditions over a broader region of the posterior subcapsular lens even if the peak moved within the eye body. The thermic insult is aggravated by a sustained activation temperature in the poorly thermoregulated volume. It must be recalled that the lens is (on a millimeter scale) far from the nearest vascularization which could, by thermoregulation, reduce the activation period. This contributes to the delayed effect along with the location of the peak under the insulating ocular surface. The surface structures are actually colder because they lose heat to the surroundings so that the peak, if located there, could not have the same effect. The peak during the delay to equilibrium flattens and spreads to affect nearby structures.

A remaining question in the geometrical relationship has to do with comparative morphology. The etiology of cataract is variable with puzzling inconsistencies between individuals and species subject to equivalent exposures. Dr. Piro Kramer *et al* (1978) have helped to explain why cataractogenesis is so variable. Briefly, the answer is that the morphology is also variable. Given a rather uniform EM exposure, the hazard can be much worse for a rabbit or a species with protruding eyes. In the rabbit, the lens uses up almost half the eye volume with its axis from anterior to posterior of about 14 mm. In man and monkeys, on the other hand, the eyeball will be drawn out much like an elongated bullet to about twice this axial length while the lens is thinner. These investigators studied why the monkey was resistant to cataracts when exposed to a resonant slot director backed up with a cavity to model a microwave oven leak. Under these conditions the monkey withstood 500 mW/cm^2 for 60 minutes at 2450 MHz while the rabbit succumbed to cataract after only 180 mW/cm^2 for 140 minutes.

In the monkey, the nasal bridge absorbed considerable energy, protecting the deeper eyes in that the temperature there remained below 42°C. Thus the face was sacrificed to protect the sensory system. The morphology of a resistant individual or species would probably include a prominent brow, deepset eyes, and protruding orbital rims.

POSSIBLE COINCIDENCE OF PEAK AMPLITUDES

Periodicity in the energy distribution curve is also contributed to by reflections from each interface in the tissue pathway, placing peak amplitudes in possible coincidence with the indicated injury. Several lenticular, capsular, and vitreous body interfaces exist which should be considered (Fig. 19-9). It is conceivable that, under some conditions, the reflected peak amplitude from Interface 1 may coincide with the peak in the entering energy distribution (initial peak). When the curve flattens on equilibration, it affects structures which may already be near the trauma temperature. The effect of the rise in ϵ' with wavelength in this region of wavelengths would be to reduce the wavelength in the medium. Then, even if wavelengths in space are larger to begin with, the wavelengths in tissue are less affected than they would be when ϵ' is truly "constant" with changes in wavelength.

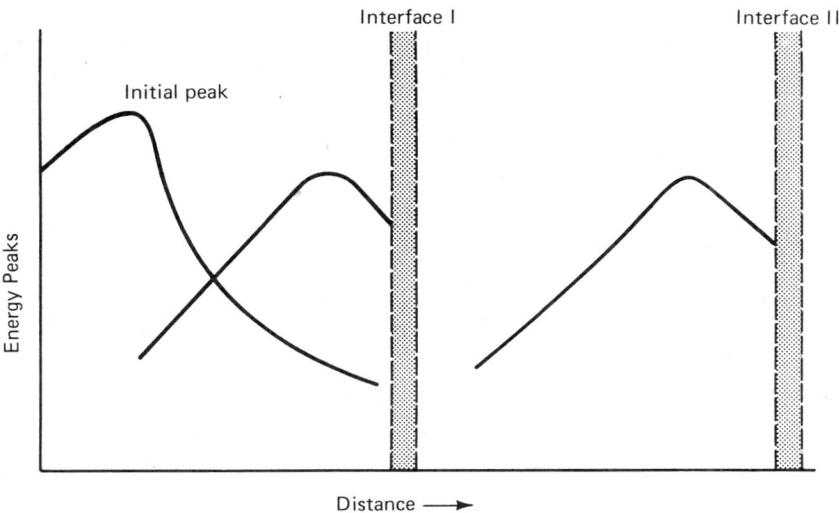

Fig. 19-9. Entrant and reflected energy peaks. Internal multipath radiation may become coincident at sensitive site.

A pulse imposes a higher peak energy density upon this situation which is already trying hard to equilibrate. It therefore aggravates the instabilities that would be created by ordinary continuous wave input. In comparing the pulse process time constant with corresponding continuous wave input, the peak power cannot be attained by the continuous wave power, even working on the average energy flow. Therefore, the two are never equivalent. The stage is then prepared for the internal peak to either cause trauma or to equilibrate toward threshold energy levels for trauma.

THE RESONANCE CONDITION

The pulse pattern is now positioned for possible resonant energy transfer in case the acute injury or energy window is reached. To illustrate this, a short period of pulsing is examined (Fig. 19-10). One

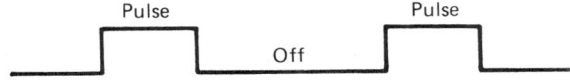

Fig. 19-10. Modulation and peak pulse levels.

pulse may not be cataractogenic, but may produce a yield of defects within a time span that may be regarded as the acute situation. Within this framework, some number of pulses will be sufficient to give the lesion. The cataract is then an accumulation of irreversible defects or defects that are not reparable in the time between pulses.

The analysis is adaptable to one more level of acuteness in the development of the lesion. Should there be a singular moment during the "off" period when the system is particularly sensitive to the radiation input, and if the pulse application is made to coincide with this time, then an acute resonance-related response is obtained. Just as the push applied to a swing is most effective at one particular moment, the resonant structure receives the input like a well-matched junction. The susceptible region is insulted repeatedly at the moment of greatest sensitivity. The internal peak is now well established, and its equilibration broadens the region which it affects. The peak can be utilized to help explain the findings. The sensitivity of the eye may then be explained in terms of spatial (i.e., wavelength) and equilibration constants, pulse time constants, and resonance. Diverse experimental conditions will alter this model. For example, the subject may be in the near or far field of a radiator or located at a microwave focal point which will further delineate the region of maximum effect. Added objects, including dielectric lenses in the near field can induce standing waves between the eye and the object with potentially resonant input between this scatter and the dielectric lens focusing. A dielectric lens acts as a secondary radiator having its own near and far fields. The present model deals with absorbed energy as well as with conditions within the radiating body.

Due to the similarities in structure and the fact that this injury is an endpoint rather than one that depends on the intact organism, the dielectric agar gel phantom or model has a certain validity for simulations in microwave tests. The lenticular substance is also a tight gel, with water held in the network of fibers, and the usual thermoregulation differences are muted in this instance. Further model studies involving the hydration mechanism as an intensifying factor are justified.

TIME CONSTANTS MAY BE UNFAVORABLE FOR REPAIR

Questions which arise in connection with delay periods and exposure levels are: What kind of defects are involved and how does the activation period affect them? In the first place, there is the observation of Carpenter (1968) which showed that a period of seven days between chronic exposures permitted repair. In addition, it is reasonable that the number of defects varies with the exposure. The time periods involved are from one to two weeks after the threshold exposure. When the 280 mW/cm^2 exposure for four minutes is repeated on various days during a two-week period (subthreshold-cumulative) there is a certain repair period between exposures which is protective, not allowing any opacities to develop. This period is about seven days for rabbits. The defects are of two types. There are early, reversible changes such as in lens intumescence, metabolism, and the physical and electrical condition of membranes. The other type of defect in the lens is later and may either persist beyond the repair period or become permanent. An example would be loss of transparency which would be related to the lack of regenerating power in the transparent cells. An early defect may be pushed into the permanent class of defect by repeated insult or by inadequate recovery time between exposures. Some delay is almost the rule in cataractogenesis when opacification and all stages in the process are included.

The latency varies between nonionizing and ionizing radiations. The acute microwave cataract takes one to eight days or an average of 3.5 with threshold exposure. The ionizing radiation type develops in about a month with a dose of a few roentgens in rabbits. This consistent difference suggests a different basis or mechanism for the two classes of cataract. Galactose cataract, like the microwave type, has a rapid onset with opacities developing in two days when the diet is directed toward this metabolic disease. There may be impairments of the lens metabolism, repair malfunctions, or both. It is known, for example, that alloxan-induced diabetes increases the sensitivity of rabbits to cataract. The subthreshold microwave exposures may produce observable lens defects even if no opacity develops. A mature lens may show late type cataract which suggests the chronic weakening of a repair mechanism. The significant common elements would be enzyme activity, phosphatase, ATP concentration, vitamin levels, permeability or membrane integrity, and the presence of important metabolic substances and precursors. The phosphatase activity in the lens, for example, is decreased after microwave exposure (Baillie, 1951 and Kinoshita, 1966). The oxidative metabolism in support of repair is concentrated more in the epithelial layer in the equatorial subcapsular region. This would position the repair processes more conveniently for the ionizing radiation type of lesion and may explain why it takes so much longer to potentiate.

SELF-GENERATING REACTION

The integrity of lens membranes is in doubt after microwave exposure and the stability of the lens protein is a determining factor in recoverability. Hydration, formation of bands and striations, and opacities are all reversible if the stability of lenticular protein is assured. Hydration and ion concentrations have a special significance in the microwave spectrum. Hydration will render the structures more microwave absorptive. Then the equilibrium process would have time constants that parallel dehydration requirements. The time constants for microwave pulsing or relaxation processes will influence the result. Long time periods between exposures would improve the chances for recovery. A super-hydration can favor local dielectric absorption and exposure repetition would provide a precipitating factor at the point of system instability. The time constants for the normalizing versus reexposure processes may then produce a self-generating reaction leading to the lesion.

These cell-level effects are "sub slit lamp" and will not be seen in ordinary examinations. Sectioning and electron microscopy show that, at the cell level, the interdigital process of the membrane is sensitive and it triggers changes by its thickening and increased permeability to vitamin C and glutathione. Another predisposing factor for catastrophe is the location of microvillous processes at the lens suture. There, multimembranous folds occur as the lens fibers abut on a size scale of one micrometer. Then the formation of vacuoles and cavities is an accentuating factor as these form from microwave exposure.

As an indicator, it is found that the vitamin C decrease after microwave exposure apparently is due to the failure of humor pools to replace it. As the lens opacification proceeds, potassium flows out and sodium enters showing a failure of the cation pump. The total electrolytes gain along with the water content. Then, as the cataract develops, more water and still more cations enter. On the repair side, microwaves are seen to inhibit the lens synthesis of DNA and mitosis (Kinoshita, 1966). The DNA synthesis may recover to stimulated levels within 14 days after microwave exposure while the lens epithelium is caused to proliferate.

BIOLOGICAL WARNINGS

After exposure to fields of a few mW/cm^2, visual sensitivity is reduced. Blue color is not as easy to see and white objects are difficult to distinguish. Intraocular pressure varies during the exposure and occupationally exposed persons (after some time has elapsed) may have *eyelid flutter, eye strain, fatigue, lacrimation, outright ocular pain* and *daily episodes of conjunctivitis* (Van Ummersen, 1969). These effects serve as *distinct warnings* that repair mechanisms are under attack and a change in work habits, especially avoiding microwaves for a period of recovery, is clearly indicated. More acutely, there would be lacrimation, constricted pupils, and dilated conjunctival and iris vessels. Before a week there may be turbidity in the anterior chamber and a milky band in the lens posterior cortex.

MICROWAVE OVEN

The question of whether cataracts are due to accumulated injury is not settled, but a multiple exposure will be positive where a single one is negative. At 400 mW/cm^2 for more than 10 exposures, cataracts are produced, but at 400 mW/cm^2, a single exposure can be negative. Then, at 500 mW/cm^2, a single exposure can form an opacity. Assuming that rabbit data can be extrapolated to humans, microwave devices of the type commonly used, the oven for example, are not likely to cause opacities or cataracts. The long term chronic exposure at very low dose may give delayed effects or effects with long latencies. Eye damage may also be related to systemic effects which may or may not precede the cataract. Malfunctioning doors on ovens, of course, are an open invitation to catastrophe.

DIAGNOSTIC INDICATORS

The different latency periods are important diagnostic aids in differentiating microwave and ionizing radiation cataract in view of the fact that days or weeks for the former versus months or years for the latter are required. Caution must be exercised, however, that a comparison is not being made between acute microwave and chronic ionizing radiation cataract. In any case, both begin as vacuole or granular entities in the subcapsular area. The microwave type is in the posterior suture region and although both poles have been implicated in connection with the ionizing type, it is likely that it is anterior. Changes in the whole lens and other structures are more obvious with large exposures and, therefore, more likely for the microwave type which may be observed closer to the exposure event. Vacuoles in the suture area will develop axially with a feathery, striated, or banded appearance, forming a ring opacity. The opacification ring closes on the axis making the cataract. The banded structure of alternately clear and opaque regions in the posterior cortex resembles a ripple effect or polarization-wave phenomenon. It may be associated with alternate concentrations of a metabolite which would normally be distributed uniformly. The bands, double or triple, last about a week or two and are observable with the slit lamp. Their progress in time may be extrapolated to zero to fix the likely exposure in acute cases.

The swollen, hydrated lens fibers are characteristic of both microwave and ionizing cataract, beginning as a tumescent epithelium equatorially. In the nucleus, cell degeneration and fragmentation may occur. The tumescence produces special dielectric absorption which may give microwave selectivity and add to the insult. An early threshold at about one week may parallel the hydration process, that is, added water may increase the absorption of radiation energy.

The action of vitamin C in the lesion-related metabolism is a fascinating matter. It is sensitive to microwave heat and oxidation, and it may disappear on standing for three hours in preparations. It may be involved in the banding and it drops selectively in the lens but not in the aqueous humor, suggesting a transport deficiency. It has been shown by Merola and Kinoshita (1961) and others that the lens ascorbic acid decreases after microwave exposure by about 23% preceding the opacity and prior to the drop in glutathione. As it decreases, the lens epithelium becomes cloudy, especially near the equator. In the ionizing radiation lesion, glutathione is the first to drop. Lens ascorbic acid occurs in microgram amounts in each gram of tissue and it may be protective against this injury. Its fall is delayed along with that of glutathione and thus it is a sequel to this and other cataractogenic injuries, but it precedes opacification. That hypothermia fully protects an animal suggests that there should be a real thermic threshold which can be assumed in safety models. In addition to acute, high power situations, dangerous repetitive exposures to low power would need evaluation and an indicator or accumulator-like loss of ascorbic acid may be both useful and easily measured in animal tests.

Lasers, still very useful (with caution) in repair of detached retinas, are used to evaluate opacities induced by EMR. Ultrashort, picosecond pulses are used in a thickness-measuring configuration for ocular layers and to locate microscopic sites of scattering opacities. Then the size of the scattering particles is susceptible to measurement by the backscattered light. Holography gives 3-dimensional recordings with large depth of field of the microscopic, internal structure of eye parts with a resolution of about six micrometers or less on cataracted eyes. This however is the use of a radiation to measure a radiation effect and is anything but a routine diagnostic aid. As the porcupine said when asked how he manages in close company, "It is a matter of caution, sir, considerable caution."

In microwave cataract, an accidental acute exposure of a few watts per cm^2 may develop the lesion quickly to cataract stage in two months. Late cataract is found connected with both kinds of radiation. For microwaves, it may come from multiple, small exposures over several years and so be more common in veteran workers. Their earliest complaints may be subjective and later some losses in vision may occur. With ionizing radiations, cataract may show the long latency period and be associated with other late effects such as tumor.

The etiology of microwave cataract may involve unilateral exposure in radar technicians as opposed to bilateral cases in other instances. The technicians have considered it safe to boresight transmission lines during assembly and reassembly and usually do it with the same eye. Lacrimation and burning eyes are common complaints with slit lamp observation. Then roughening and thickening of the posterior capsular surface occurs. A frank injury logic would involve a few mW to tens of mw/cm^2 as sufficient to induce the disease. Acute EM injury would not be unlike exposure to flame but with much less pain, even though the eye is pain sensitive. If a person held tools and presented himself as a dipole antenna, the ocular region might be most sensitive even though he felt a minimum of heat sensation. There would be sensation of burning in eyelids, the lens should show swelling at least, and the other irritations mentioned. These technicians, seagoing pilots who are constantly riding radar beams in landing patterns, and who also live near the concentrated EMR, and microwave oven repairmen should be concerned about safety since they probably represent worst cases.

PROBLEMS

1. How can coincident peaks affect cataractogenesis?
2. What determines the wavelength in lens tissue and the wave velocity in the tissue?

3. How can a resonant situation be visualized in tissue damage to a sensitive region?
4. Compare the morphology and chronology of cataract due to ionizing versus nonionizing radiation.
5. What exposure produces microwave opacities or cataract?
6. What is the flywheel effect?
7. Why would 300 nm uV require more exposure than 335 nm for cataracts?
8. What precaution is indicated in cancer termotherapy?

BIBLIOGRAPHY

Appleton, B. and G.C. McCrossen, "Microwave Lens Effects in Humans," *Arch. Ophthalmol.*, Vol. 88, 259-262, 1972.
Appleton, B. *Results of Clinical Surveys for Microwave Ocular Effects*, DHEW Pub. No. FDA 73-8031 (BRH/DBE 73-3), 1973.
Appleton, B. "Investigation of Single-Exposure Microwave Ocular Effects at 3000 MHz.," *Annals of the New York Academy of Sciences*, Vol. 247, 125-131, 1975.
Aslan, E. "Broad-Band Isotropic Electromagnetic Radiation Monitor," *IEEE Trans. Instrument. Measure*, IM-21, 4, 421-424, 1972.
Baillie, H., "Thermal and Nonthermal Cataractogenesis of Microwaves," *Symp. on Biological Effects and Health Implications of Microwave Radiation*, S.F. Cleary, Ed., BRH/DBE 70-2, U.S. Dept. HEW, Sept. 17-19, 60, 1969.
Baillie, H., G. Heaton, and D.K. Pal, "The Dissipation of Microwaves as Heat in the Eye," *Symp. on Biological Effects and Health Implications of Microwave Radiation*, S.F. Cleary, Ed., BRH/DBE, 70-2, U.S. Dept. HEW, Sept. 17-19, 1969.
Bachem, A., "Cataracts," *Am. J. Ophthal.*, Vol. 41, 969, 1956.
Bellows, J.G., *Cataract and Anomalies of the Lens* Mosby, St. Louis, Missouri, 1944.
Bonney, C. H., D, M. Hunter, G. E. Conley, and K. A. Hardy, "Heavy-Ion-Induced Cataractogenesis," *Aviation, Space and Environmental Medicine*, August 1977.
Carpenter, R.L. and C.A. Van Ummersen, "The Action of Microwave Radiation on the Eye," *J. Microwave Power*, Vol. 3, 1, 3-19, 1968.
Carpenter, R.L., G.J. Hagan, and E.S. Ferri, "Use of a Dielectric Lens for Experimental Microwave Irradiation of the Eye," *Annals of the New York Academy of Sciences*, Vol. 247, 142-154, 1975.
Cleary, S., "Microwave Cataractogenesis," *Proceedings of the IEEE*, 68, 1 Jan., 1980.
Cleary, S.F., B.S. Pasternack, and G.W. Beebe, "Cataract Incidence in Radar Workers," *Arch. Environ. Health*, Vol. 11, 179-182, 1965.
Cleary, S.F., and B.S. Pasternack, "Lenticular Changes in Microwave Workers. A Statistical Study," *Arch. Environ. Health*, Vol. 12, 23-29, 1966.
Copson, D.A., *Report, First Seminar on Microwaves*, Raytheon Company, Waltham, Mass., Oct. 7-8, 57-60, 1954.
Copson, D.A., *Microwave Heating*, Avi Publ. Co. Westport, Conn., Chapters 21-22, 1975.
Daily, L., *et al*, *Am. J. Opthalmol.*, Vol. 34, 1301, 1951.
Deficis, J.C., *et al*, "Effect of Electromagnetic Energy on the Formation of Triglycerides Frequency Influence," *J. Microwave Power*, Vol. 11, 2, 1976.
Deficis, A. *et al*, "Variation of Serum Triglycerides Rate Under the Action of Electromagnetic Waves," *J. Microwave Power*, Vol. 11, 2, 1976.
Duke-Elder, W.W., "The Pathological Action of Light Upon the Eye," *Lancet*, I, 1137-1140, 1188-1191, 1250-1254, 1926.
Guy, A. W., J.C. Lin, P. O. Kramer, A. F. Emergy, "Effect of 2450 MHz Radiation on the Rabbit Eye," *IEEE Trans. MTT 23-492*, June 1975.

Hirsch, F.G. and J.T. Parker, "Bilateral Lenticular Opacities Occurring in a Technician Operating a Microwave Generator," *A.M.A. Arch. Ind. Hygiene Occupat. Med.*, Vol. 6, 512-517, 1952.

Hogan, M., J. Alvarado, and J. Weddell, *Histology of the Human Eye*, 638-677, W.B. Saunders Co., Philadelphia, Pa., 1971.

Kinoshita, J.M., L.O. Merola, and I. Dikmak, *Opthalmol.*, Vol. 20, 91, 1966.

Kramar, P., et al, "Theoretical and Experimental Studies of Microwave Induced Cataracts in Rabbits," *Proc. 1973 IEEE-GMTT Int. Microwave Symp.*, 265-267, 1973.

Kramar, P.O., "The Ocular Effects of Microwaves on Hypothermic Rabbits: A Study of Microwave Cataractogenic Mechansims," *Annals of the New York Academy of Sciences*, Vol. 247, 155-165, 1975.

Kramar, P., et al, "Acute Microwave Irradiation and Cataract Formation in Rabbits and Monkeys," *J. of Microwave Power*, Vol. 11, 2, 1976.

Kuck, J.F.R., Jr., *Biochemistry of the Eye*, C.N. Graymore, Ed., Academic Press, New York, 1970.

Kurz, G.H. and R.B. Einaugler, "Cataract Secondary to Microwave Radiation," *Amer. J. Opthalmol.*, Vol. 66, 886, 887, 1968.

Marha, K. "Maximum Admissible Values of HF and UHF Electromagnetic Radiation at Work Places in Czechoslovakia," *Symp. on Biological Effects and Health Implications of Microwave Radiation*, S.F. Cleary, Ed., BRH/DBE 70-2, U.S. Dept. HEW 189, Sept. 17-19, 1969.

McAfee, R.D., L.L. Cazenavette, and M.G. Holland, "Screening for Cataracts Among Veteran Radar Repair and Maintenance Technicians," *Annals of the New York Academy of Sciences*, Vol. 247, 135-141, 1975.

Merola, L.O. and J.H. Kinoshita, "Changes in the Ascorbic Acid Content in Lenses of Rabbit Eyes Exposed to Microwave Radiation," in *Biological Effects of Microwave Radiation*, M.F. Peyton, Ed., Plenum Press, New York, 1961.

Michaelson, S.M., "The Cutaneous Perception of Microwaves," *J. Microwave Power*, Vol. 7, 2, 67-73, 1972.

Milroy, W.C. and S.M. Michaelson, "Microwave Cataractogenesis: A Critical Review of the Literature," *Aerospace Med.*, Vol. 43, 67-75, 1972.

Pitts, D.G., A.P. Cullen, and P.D. Hacker, "Ocular Effects of UV Radiation from 295 to 365 nm, *Invest. Ophth.*, Vis. Sci. 16 (10):932 (1977).

Ready, J.F., *Effects of High Power Laser Radiation*, Academic Press, New York, 1971.

Taflore, A., and M.E. Brodwin, "Computation of EM Fields and Induced Temperatures Within a Model of the Microwave-Irradiated Human Eye," *IEEE, MTT*, 23, 11 Nov. 1975.

Van Ummersen, C.A. and F.C. Cogan, "Effects of Microwave Radiation Lens Epithelial Cells," *Symp. on Biological Effects and Health Implications of Microwave Radiation*, S.F. Cleary, Ed., BRH/DBE 70-2, U.S. Dept. HEW, Sept17-19, 122, 1969.

Weiter, J.J., et al "Ascorbic Acid Changes in Cultured Rabbit Lenses After Microwave Irradiation," *Annals of the New York Academy of Sciences*, Vol. 247, 175-181, 1975.

Williams, D.B., *Inst. of Radio Eng. Trans. Med. Elec.*, p. 14, 1976.

Williams, R.J., "Ultrastructural Changes in the Rabbit Lens Induced by Microwave Radiation," *Annals of the New York Academy of Sciences*, Vol. 247, 166-174, 1975.

Zaret, M., S. Cleary, B. Pasternack and M. Eisenbud, *Occurrence of Lenticular Imperfections in the Eyes of Microwave Workers and Their Association with Environmental Factors*, RADC-TN-61-226, New York University Press, New York, 1961.

Zaret, M. and M. Eisenbud, "Preliminary Results of Studies of the Lenticular Effects of Microwaves Among Exposed Personnel," *Proc. Fourth Ann. Tri-Service Conf. Biol. Effects Microwave Radiation*, 293-308, 1961.

Zaret, M., *Ophthalmic Hazards of Microwave and Laser Environments*, 39th Annual Sci. Meeting Aerospace Med. Ass., 1968.

Zaret, M., *Ophthalmic Hazards of Microwave and Laser Environments*, 40th Annual Sci. Meeting Aerospace Med. Ass., 1969.

Zaret, M., "Cataracts and Avionic Radiations," *British J. of Opthal.*, Vol. 161, June 1977.

20

Quantitative Information

> "For cells, as for societies,
> Time's arrow points ultimately
> To the dust of entropy"
>
> James G. Miller

Systems *theory* is a valuable thought-mode rather than an operating construct that operates in measureable units, while units of information guide and change systems, being vital for their operation. In the two, informational biology ranks as an operator and actively links the primitive information molecule and cell with every evolved cell today and in the future. The tasks of systems theory are analysis, classification, amplification and extrapolation with the multiuse of data having common features. Living *systems* are something else for they operate on and organize material using coded information to do so. Thus informational biology offers a systematic sensitivity gained in these studies which teach the significance of understanding the codes, channels, storage and retrieval of bioinformation. It is necessary to identify the systems that accomplish these things, try to explain the mechanisms, and attempt to count the required units.

The selection among the states or conformations that bodies, systems, objects and events are free to assume, according to a mission, is the most essential decision made in systems. Interactions proceed or are inhibited by the control exerted over interactants so selected. These may be integral and intrasystem when the systems are sufficient unto themselves or intersystematic when depending on satellite, parasitic systems. These decisions involve $\log_2 \Omega$ bits among Ω different states. Among the more important bodies are molecules and the systems are the living ones with their nonliving constructions as satellites, all connected by channel elements. IBEM is a construct introducing a special intersystem framework that explains and enumerates these neglected degrees of freedom in bioorganization. How large Ω can be is related to uncertainty. A large number of possible scenarios makes for much uncertainty as to the outcome, but if the uncertainty goes to zero, the missing information also goes to zero.

IBEM is built around that central concept of a process, involving a system. When extended to multisystems in which informational biology shows the essential communication, the informational intersystem is seen to be related by a channel with characteristic elements.

1. coding
2. flow rates
3. capacity
4. functions
5. conflicts
6. biological hierarchy
7. entropy conditions

The flow of information is inseparable from the parallel energy that is its carrier in the channel. Measured in relation to mass, important constraints on the velocity and the flow will exist unless the mass is reduced to some limit such as that of an electrical charge proceeding at light speed. Molecular codes, coupled into effector loops with catalysis produce competitive rates as in pheromones. EMR vastly improve transmission but originate and terminate in molecular systems as in sunlight activating chlorophyl in photosynthesis. A process signifies a new selection among degrees of freedom for the body or state in a system, behaving according to space or time constants, as in the juxtaposition of peaks for cataract in Chapter 19.

The biological process is emphasized in texts such as *Biological Principles and Processes*, Claude A. Villee and Vincent G. Dethier, 1976, and even more fundamental is the processing of information which describes the pathways of general biology in texts such as *The Study of Life, An Introduction to Biology*, G. H. Orians, 2nd Edition, 1973. These, and this text, followed by *Living Systems*, J. G. Miller, 1977 provide a multidisciplinary exposition.

Miller sees a flow of matter-energy as completing that of information in living systems. While this permits his elaborate scheme of systematic amplification to many levels, actually energy can be seen as arising from operations in matter or EMR and running in parallel with information. In these texts coupling is a valuable concept. It completes the interaction of loops, using vital feedback for control. Then coupling and uncoupling agents point to the controllers in information management in the systems with channels and nets (within their limits) for coordination. At all levels there is transduction, interpretation, discrimination, threshold behavior, phases with meaningful lags and leads, space and time constants, frequency-bandwidth and concepts of conflict, noise, and antinoise. Cell systems are specialized for information and the reader will recall examples of cellular sensory transducers, information storage, coding, channels, decision makers, and associator functions.

Complex living systems need special communcation and control which come from metabolic and regulatory EM information transmitted in tissues from metabolic generators. Examples of communicating codes are:
a) Molecular (Chapter 3).
b) EM pulses for lower and motor-cortex operations (Chapter 15).
c) Symbolic information for higher, temporal, frontal and parietal signals giving awareness and modulations (Chapter 2).
d) Codes for forcing functions, as in the Cole-Cole arc where dielectric constant and loss signify a frequency-coded response (Chapter 11).
e) Codes for information couples, as in the Schwan curve (Chapter 12), which codes EM activations for log ϵ/ϵ_0 effects as a function of the log of frequency.
f) Crisis codes for intuitive survival behavior preempting channels for emergency signals to support great effort of special significance. Higher organisms devote more information processing to this than do lower ones. The ACTH net plays a significant role in the messages.

The independence that photosynthesis brings to organisms contributes to the negentropy of higher organisms. The EMR that are responsible also use sensory processing of EM information which weaves the pattern of associations and memory. These connections sustain and partly liberate the organism from its dependence on germline DNA information. Heredity dominates but experience extends the scope. With less learning and memory capacity, the operations of primitive organisms are closer to patterns made by the original nucleic acids and these systems characteristically produce new generations in minutes and hours. Supporting activity, informational and metabolic is intense; life is in the fast lane reproductively, and organized time and spacewise into meaningful kaleidoscopes of structure which show constancy while changing. Higher organisms exercise choice and even opt to neglect this basic operation, putting other requirements of the culture into the fast lane instead of reproduction.

A channel as a branch or path over which signals may travel, requires activation from input forcing functions and radio and TV provide many examples, usually designed to avoid interference by management of frequency and modulation. Most communication will be multichannel. In biology we see the essential nerve networks, molecular, synthetic, metabolic, catabolic, anabolic, and amphibolic channels, bony conduction, skin channels, and those of membranes, blood, lymph, and ducts. The living channels can interface with artificial media, wires, wireless, and bounding transducers in visual, audio and other channels. They may be in a hierarchy from which significance emerges just as a pattern of spots on TV is successively a box, a pirate chest, one from the pirate Morgan, and finally is seen to be filled with his gold.

The background is a supportive metabolism, normally an excess of energy, which may be enzyme activated, radiant or other forms, all governed as interacting components by information flowing to prescribe recognition on meetings, conformation and cooperation. As an analogy O_2 is also in excess for breathing. Specificity, time and space processing and cycling schedules and regulation are features and the information requires the channel for communication. The channel is not always a nonmaterial one as seen in TV and radio, but is always energized for the transmission as a kind of motivation.

$$c = w \log_2(1 + P/N) \text{ bits/sec.}$$

Shannon's is the limit for a channel and is the bandwidth times the log of 1 plus the ratio of the power of the signal to the white noise present. As c decreases, the efficiency of the information process falls (as it does in aging) until it fails completely in "channel death" (Chapter 4). Normally one expects to find quantal fitting for a channel energy such as in neural pulses, volleys of neurotransmitters and the resulting end plate potentials. Problems in channels bring up the question of electromagnetic radiation intersystem communication which varies over distances that are successfully extrapolated from interatomic to interplanetary. The operations between sources and sinks at velocity approaching c are nearly instantaneous and information interactions at these velocities amount to almost instant sharing or coupling of near and far systems.

There is a relation between perception of microphonics as in Chapter 17 and what is called subliminal perception because both show discrimination at a low level of awareness. The microwave input is above the threshold of the input transducer of the ear but the operations affected are mediated by lower levels in the brain with no effect on the higher ones. The identification of positive channels in this process will be most significant.

Channels have an efficiency factor based on learning, and complex cells like phagocytes learn also to recognize and manage their attack. Channels have a limit for transmitting codes without error subject to constraints afforded by the noise present. Many techniques are available for overcoming the noise barrier including redundant transmission as shown in Chapter 3. In one case, if some messages carry the same information and there are three of them, selection of two that agree can filter out errors. A Pavlovian relation is exposed in information terms when a novel input I is repeatedly associated with a familiar output A, that produces a specific response. That certain input I will become capable of giving the same output as A. A very large multichannel began with the first informational molecule, for example RNA, and continues through every reproduction (reproduction is the transfer of information). Then function and information became coexistent and inseparable.

The energy for a channel varies somewhat independently of the information carried. More energy does not necessarily mean more information. There is, for example, the case studied in Chapter 18 where more energy acts like noise obliterating the information until a saturation of energy produces only the thermal effect. Thus the lumping of EM effects in a power density unit measurement such as watts/cm^2 shows the futility of studying direct microwave interactions in biochannels by the power flux.

The management of information input is constrained by channel capacity, and as shown in Chapter 1, we are subject constantly to *information overload* by media, education and work and the permissible input, without tensing systems inordinately, is surprisingly low. High bioinformation velocity is cellular and molecular with some cells enjoying orders of magnitude more information channel capacity than higher echelons like organs. The number 7 (intensities) is significant as representing the categories in information input or number of intensities that a human distinguishes. For example, a touch cell transmits a maximum of 2.5 bits per second for each information input which is the amount for separating seven or eight categories in the information input. High memory efficiency is associated with higher echelons rather than molecules and cells *per se* and these structures do not guarantee permanency, only entropic decay. Beginnings are made by analysis of nonrandom neural nets, e.g. by finding probabilities in the net operation with counts of contacts among components and reciprocity of contact.

The decay to posentropy occurs in memory attrition. In channel transmissions, lower entropy in the system means a temporary state in which the transmission or storage is possible and the less the entropy, the greater the information. This is how entropy comes to be associated with the amount of information.

As energy flows in a channel, bandwidth gives the amount of multifrequency sensitivity in an addressee. This determines the number of symbols which the channel can transmit and so there is a coordination. Likewise there is a selectivity and fine tuning because certain frequencies are particularly receptive. Thus, access of a receptor to the information may be easy at a resonant frequency and denied at others. For molecular recognition processes the information role is apparent and resonance is a molecular recognition process. Then, since the energy relations in EMR are quantized, one expects to find a quantal fit associated with resonant processes which may be both quantized and resonant or the two phenomena may be identical where resonance gives the quantal fit. Responding systems may thus be selective by virtue of an energy fit at relaxation frequencies as shown in Chapter 11, and a forcing function can be frequency specific, directly, or indirectly, from some other frequency function. Where this is true, the energy is clearly seen as directed by its information content.

An EM wave transmitting through a responding bulk system, which is receptive by virtue of its molecular species, will often resemble a brush that waves through and polarizes or otherwise energizes molecular addressees. With molecular systems responding or relaxing as a dipole ensemble to the sweeping EMR, lag and lead in phase relations communicate further with the system. There may be dissipation and the EM energy is transferred according to the relaxation frequency or appropriate band. Alternatively, the energy is stored in lesser excitation in a form of compliance for retrieval. The energy follows the informing route to its sink and is general and nonspecific, directed by its information which stimulates an addressee system. The latter is specifically responsive rather than a nondefined heat sink. It was shown in Chapter 18 that information is not proportional to energy but instead becomes a thermal noise or similar indication of an excess in the region of absorption near saturation. Such modulations as pulse pattern are undefined in basic energy units and the system response is proportional to the information content of peaks and modulation patterns which have their own quantization as to probabilities and uncertainties. The breaking of codes as was accomplished with the universal genetic one and the probing of intersystem channels from source to sink are objectives in informational biology.

The negentropy concept for information ($H=-S$) is in the sense of improbability (of specified, ordered states). While certainty, predictability, and organization will be the negative of S giving a definition of ordinary as well as of probabilistic information, $H = -S = \Sigma\ P_i$ (log P_i). This groups information with the fundamental states given in equations of state with P, V, T as well, as energy which is ultimately convertible to mass in EMR. Thus the parallel energy and the entropic explanations in many ways keep the information bits together with their energetics.

In life, survival involves a complex struggle (against posentropy in which living systems obtain EM energy as in photosynthesis, which depends on the informing and energizing action of specific light. The photosynthetic organism, while blessed with its own energy sink, survives only for a limited time in darkness. The interaction of molecules is according to the significance of the enzymatic and ambiental situation which to us involved meaning. Information permeates this matrix according to the many functions that reduce uncertainty among the degrees of freedom in expected interactions. There are, then, radiant informing processes, recognition, specificity, coded guidance, regulation and structure with meaningful information built in via bonds and higher molecular organization all measureable in required or stored bits, probably in concentrations on the order of megabits per gram. Then the process such as an input to the skin channel is measureable in bits per second, as is a consequent output from the cortex so that about 2.6 bits/signal are sensed in a cat somatic system (Eccles, 1966). This is an information capacity and depends on the steps in the process. The channel transmits in a somatic, afferent neuron according to the channel rate where a pulse is a bit and about 500 bits per second will flow depending strongly upon the code (Mountcastle et al, 1957). The pulses are observable (in and out) for obtaining the count.

The pulse-bit relation suggests a special minimum stimulus, since less than a whole pulse may go unrecorded. Then neural pulses will be quantized for neurotransmitter volleys. Input adjustments will reflect threshold levels, interact with capacity in the channel and provide discrimination and selectivity. Upward changes in capacity signify a learning process improvement and the input will be subject to impedance balancing, acting as an intelligent gate when mismatched, or easing transfer when matched.

Molecular code constraints will give limits on bits/second/gram producing, as a result, a pressure in evolution for structures capable of ingenious coding techniques, like DNA and the double helix. The EM channel is released from this constraint, but the information organized in molecular compounds takes about 10^{-12} ergs/bit, (Valentinuzzi and Valentinuzzi, 1962) or within an order of magnitude the same energy is needed for one bit in communication channels, when the opposition is white noise at body temperature.

Molecular coding is discussed in Chapters 3 and 4, with its conformation and organization elements. EM transmission velocities are approached by nerves which show great velocities (Chapter 11), and by transmissions in fast sign language. Coded transmissions may also be passed over large distances to specific receptors as in pheromones. In analog codes, some aspect of the signal varies as a function of a certain variable it represents, while in digital codes, the variable represented is broken into discrete units to be counted by some numerical scheme and the signals flow reversibly into numbers. In temporal coding by electric fish, *Brienomyrus brachyisteus triphasic*, the millisecond duration (time domain cues), of a 2-spike electrical discharge is the message signature that identifies the female to the male along with frequency, power and phase in the waveform (Hopkins and Bass, 1981).

The degrees of freedom, m, resemble the playing squares in checkers. To identify a particular status, or square, requires six questions if only binary questions are asked which always split the remaining possibilities in half (see Chapters 1 and 2). The information is given by H, and $m = 2^H$ for unbiased degrees of freedom, and H is $\log_2 m$. The probability is the reciprocal of the number of all alternative states and vice versa. Then H is $\log_2$ of the reciprocal of the probability ($H = \log_2 1/p$). Taking a certain alternative h_i,

$$h_i = \log_2 1/p_i$$

Taking bias into account which favors one state, say weighting it at 90%, p is no longer 0.5 for equal alternatives or a 50-50 chance. The 90% bias case gives,

$$h = \log_2 1/0.9 = \log 1.11 = 0.15 \text{ bits}$$

The other case gives,

$$h = \log_2 1/0.1 = \log 10 = 3.32 \text{ bits}$$

because it is less to be expected and, when it occurs, it represents more information.

The bias will cycle over and over with the information coming out of the probabilities times information. The average here is

$$H = (0.90 \times 0.15) + (0.1 \times 3.32) = 0.47 \text{ bits}$$

Thus, less is discovered on the average, from successive events, when the bias operates than when it does not. One can now have any number of alternatives and probabilities with their p and h values so that a more general average

and
$$H = p_1 h_1 + p_2 h_2 + \ldots p_i h_i \ldots p_m h_m$$
$$H = \Sigma\, p_i h_i$$
$$h = \log 1/p$$
$$H = \Sigma\, p_i \log p_i$$

Then, since $\log 1/x = -\log x$
$$H = -\Sigma\, p_i \log p_i$$

as in Chapter 1.

The information content of a macromolecule, containing x molecular components like amino acids or nucleotides, will depend on the degrees of freedom or alternative selections among the units possible. Suppose there are y alternatives, then z is the power to which 2 must be raised to equal y, or $y = 2^z$ and represents the bits per component. Then the macromolecule will contain zx bits (total components × bits per component) of information. The cell will have the sum of bits for its molecular ensemble and must also have a rate, appropriate for each stage of growth, at which it stores information. Information content is very large and the rate numbers are like those of fast enzyme kinetics. Separately, the cell will depend on its information capacity to identify interactants for control and recognition. Capacity and content produce a degree of organization in information units which is a meaningful taxonomic aid like GC%, and various kinds of typing. Hormonal systems and small signaling units like calcium and adenylates find coded information in reactive sites or in conformation capacity. They are memoryless as compared with nucleic acid, recognition and communication nets. With memories, these can often "learn" by experience with their receptors and generators. One must visualize information channels operating on a rigid schedule, space and timewise, electrochemically in the nerve nets and usually molecularly in hormonal and recognition systems according to the appropriate state in cell physiology.

PROBLEMS

1. Explain how systems theory and information biology are divergent concepts in biophysics.
2. Consider a protein molecule containing 10,000 amino acids of 16 types arranged in a specific order. How many bits could represent the information content?
3. Define reproduction in terms of information.

4. A nucleic acid has a code of four nucleotides and has 5,000 subunits. What is the information content?
5. What is information overload?

BIBLIOGRAPHY

Eccles, J.C., Ed. *Brain and Conscious Experience*, Springer-Verlag, New York, 108, 1966.
Eigen, Manfred, W. Gardiner, P. Schuster, and R. Winkler-Oswatitsch, "The Origin of Genetic Information," *Scientific American*, Vol. 244, 4, p. 88-118, April 1981.
Environmental Protection Agency (United States), Eds. J.B. Kinn and E. Postow, *Index of Publications on Biological Effects of EMR (0-100 GHz)*, Health Effects Research Lab., Res. Triangle Park, NC, 27711, EPA 600/9-81-011, March 1981.
Hopkins C.D., and A.H. Bass, "Temporal Coding of Species Recognition Signals in an Electric Fish," *Science*, Vol. 212, 3 April, 1981.
Miller, James G., *Living Systems*, McGraw-Hill, New York, 1978.
Mountcastle, V.B., P.W. Davier, A.L. Berman, "Response Properties of Neurons of Cats' Somatic Sensory Cortex to Peripheral Stimuli," *J. Neurophysiology*, Vol. 20, 397, 1957.
Orians, Gordon H., *The Study of Life: An Introduction to Biology*, Allyn and Bacon, Boston, 2nd. Ed., 1973.
Valentinuzzi, M. And Valentinuzzi, M.E., "Information Content of Chemical Structures and Some Possible Biological Applications," *Bulletin Math. Biophysics*, 24, 11-27, 1962.
Villee, Claude A. and Vincent G. Dethier, *Biological Principles and Processes*, Saunders, Philadelphia, PA 1976.

Epilogue

It has taken more than a few words to "miniform" this subject down to a ponderable size. Probably only a dedicated specialist would ever attempt to administer this antidote against overspecialization. The antidote is of course, information, with its pathways in biology. Then the new conceptualizations, intersystems and their channels, allow normally scattered systems to be formally associated via the radiations. Perhaps the devotion to these themes is self-evident, but what may not be evident is the effort expended in integrating the widely flung morsels on information scattered through the biological, chemical, physical, engineering, and psychological literature with the writer's own conclusions into a working theory.

Many of the accounts of other work were infuriatingly terse even when the work was highly controversial. It seems as if many want to announce some of their results in a few words that only obliquely describe the real discoveries like Alexander Graham Bell's summoning Watson by phone to see the circuit in the first telephone communication.

Professor Webb was following an alluring trail set up by his observations in 1968 that specific, seemingly related frequencies give either stimulation or inhibition in exposed metabolic systems. The trail turned out to be tortuous with new hypotheses needed at almost every turn. The work independently supported our basic theme and was recognized here. It is also being duplicated and extended in small "pseudopods" elsewhere. On the face of it, it would be extremely fortuitous if Webb picked cellular time schedules that were fitted dimensionally to a frequency series in space and time. Thus much more is needed for full acceptance, yet it will remain true that he was the first to observe these space and phase connected stimulating and inhibiting events.

These studies exposed a real hunger for data on locally generated electromagnetic signals from excited and relaxing molecules at all levels of tissue organization. This is needed to fill in gaps in information biology, the velocity of signals, for example, in signal generation in myelinated versus nonmyelinated neurons, in recognition and attack processes of phagocytosis, photo-oxidation, triplet, singlet and ground states of metabolic atoms and the transport of materials, energy, and information.

The language that is used to describe the knowledge between disciplinary camps is best described as difficult because the writer has to satisfy both sides. This is no more evident than in the meetings and symposia in applied biophysics where biologists jealously guard their priorities and terminologies against equally sanctimonious physicists and engineers. The writer, having lived with both, felt able to travel this two-lane highway at least gingerly.

To the best of our knowledge this work records, for the first time in one place, the intensity of the effort of the Executive Office of the President of the United States in its program for control of electromagnetic pollution of the environment and assessment of biological hazards of nonionizing radiation. The bibliography edited by Kinn and Postow (1981) is a measure of this effort and an excellent addendum to Informational Bioelectromagnetics as it lists 3627 reports on health effects of microwaves. Whether every effort is precisely modeled or not is less significant than the outpouring of biophysical science and the totally new experimental concepts that arise from the intensity of this effort. It is hoped that the reader has been helped in seeing these events in their proper places within the framework of the theme of the book which happens to provide a very appropriate background. This gives a new face to biophysics which has many already, since it clearly shows where physiology contributes and biophysics goes on as in the sections on sensory, feedback, radiation and information intersystems.

These large-scale efforts in science inevitably leave whirlpools of innovation in their wake and this one is no exception. Sophisticated analysis using probing radiations to interrogate metabolic systems concerning their normal/abnormal status is, in a sense, an extension of spectroscopy as suggested by the work. When vitamin C is studied in lens opacities where it decreases after exposure as a function of modulation, it is fairly significant in terms of potential informational analysis. An encyclopedic kind of synthesis is needed to bring out the spinoffs and to add to this lore of basic biology which are ultimate objectives of this book.

The dependence of microwave interaction on tissue water is a potential basis for pulmonary diagnosis and monitoring because the tissues that become edematous show increased reflection and decreased transmission as a function of frequency. Alternatively, samples taken from living tissue may be analyzed. Respiratory movement is followed, without contact, by observing the scattering by the active lung of incident microwaves. These interactions can be projected to make a recorder-enhanced thermograph. One record is a transmission picture with contrasts where the nonionizing radiation is substituted for X-rays. Presently, at least the heart and some lung differences can be seen.

Epilogue

The audio signal with microwaves is a demonstrated event and it could test the function of the part of the ear under examination, such as the vestibule or cochlea, that is being stimulated by it. With Dr. James C. Lin's *Microwave Auditory Effects and Applications* which appeared recently and the definitive work of Dr. Robert M. Lebovitz and others in this field, it is certain that a background exists for exploring the uses in medicine and informational acoustics.

On the side of a safe and better environment in which to live, the reader will have gained a great respect for the bioeffects of EM radiation. He will question whether any ionizing radiations are perfectly safe or if there exists any "threshold" for permissible exposures or absorbed energy. Some combinations he will see as downright lethal—especially any internal radiation of long life. Yet as radiations become more common and pervasive, the process of weighing risks versus benefits becomes routine and the idea of putting down one's roots somewhere may well be subject to reanalysis. The area chosen may mean a constant exposure. It would be better to plan to move once in a while to spread the probabilities, especially for urban dwellers. Airports—especially in the line of the landing pattern, are clearly unhealthy. High rise buildings that put people at the transmitter and receiver levels may give them much more than pacemaker or electronic interference with problems such as skipping an occasional heart beat. Getting "away" on the top of a mountain may mean some dangerous EM neighbors because these are ideal transmitter sites.

The Age of Electronics has transformed us into eager receivers for radio, TV, radar, and other signals. As the transmissions grow in intensity and concentration, we continue at the receiving end with our receivers now doubling as crude field monitoring and interference-indicating devices. Thus the areas free of detectable electromagnetic radiations are those beyond the fringe reception areas, at least for those frequencies. Persons who live in shielded structures and who must use external antennas to receive the signals, are demonstrating the rise of futuristic architectural design for habitable spaces in areas of intense electromagnetic radiations. Most EMR are nonpenetrating, below X-ray wavelengths, for a thickness as little as that of aluminum foil so that even metallic wall paper could be protective.

Working personnel in EM-type activities should favor rotations that spread the risk. Medical surveillance for career people would include tests for changing metabolism, vitality, blood component patterns, eye, fertility, behavior and exposure data. Liver tests are valuable, including any changes in symmetry and size. Many of the hazards converge in special situations like those of pilots, especially helicopter pilots, submarine tower or "sail" watch personnel near the lower mounted radars, and those close to concentrated electronics, or riding radar and radio beams. High voltage transmission line workers have an obvious exposure problem. It is good advice to treat EM sources with the respect given guns which are held with the business end high and never accidentally trained on anyone, excluding for the moment the mischievous or intentional "applications." They are a particular threat because electronic advances and radiation weaponry can be used effectively on either side of law and order.

The 10 mW/cm^2 output limit in occupational or inhabited areas can now be applied with greater ingenuity. Silverman (1980) concluded that there have been enough accidental exposure levels estimated at over 100 mW/cm^2 to indicate that some shipboard personnel are exposed to much more than 10 mW/cm^2 on certain ships at certain times. The gunfire control technicians, and aircraft electronics technicians seem especially in the line of fire, rather then electronics technicians. For most of the Navy, a limit of 10 mW/cm^2, and probably much less, may be acceptable with computer frequency and pattern control.

Dangerous areas need charting with isoexposure lines to indicate degrees of danger. Trainable antennas that exceed the limits should have fail-safe cam cut-outs and access should be limited. Areas labeled zero, low, or high risk are clear, but sometimes a specific hazard needs to be mentioned, such as pacemakers, explosives, or fuels, the last two giving special dangers from accidental electronic detonation or ignition. A special hazard should be recognized with large peak E fields and with certain pulse widths, pulse repetition rates, and pulses of extremely fast rise times. Sometimes these can be avoided completely and replaced by other signals.

Frequency-specific physical therapy can be used with much more expertness than today's devices. Diagnosis by communication with organs and systems of interest, in accord with our basic theme suggests opportunities for much more capable ECG, EEG, electromyography, magnetocularography, magnetoencephalography, and electroretinography. Better prosthetic devices and pacemakers are clear possibilities. There is also the question of EM compatibility on an international scale. Can one country violate the air space of another with waves on an international basis? The answer is probably yes, because nation states appear to be *tolerating* aerial reconnaissance or surveillance of their lands and seas in an ever more precise manner. This could lead to trouble as their well-kept secrets evaporate into common knowledge. We already see electromagnetic countermeasures, constant, secret tampering with the EM environment on a sometimes massive scale. It is obviously necessary to understand this menace lest it lead to international misunderstanding and tension.

To use an extreme case, heart patients could be grouped in a high-rise building and many may be using pacemakers. With the situation, any fields over 10 V/m represent life and death with pacemakers of current design. The 10 mW/cm^2 cw power density represents an E field of about 200 V/m. With pulsed EMR, an average pulse power with a duty factor of 0.001 or less might give a peak E field of over 6 kV/m. Thus the group of heart patients, or perhaps any concentration of the elderly, is susceptible. Of course, the usual effect of the field is interference with a beat, and not with the beating due to the small intersection offered to the beams. The danger is greatest when the PRR > 10 Hz and the level > 10 mW/cm^2. Pacemaker wearers represent a model group showing communication interference with a bioelectric function and the cardiac rhythm is certainly a perfect example. Thus one only needs to define a susceptible, sensitive, section of a population, name the communication interaction and nature of the interference, the maximum permissible level, the effect of pulse, peak and rise time, and have the susceptibles congregated, for a potentially serious situation to exist.

Cardiovascular disease and myocardial infarction may be potentiating elements in electromagnetic sickness. M. Zaret (1977) looked for the cause of the maximum reported rate of heart attacks in the world which exists in the Larelia-Kuopio in Finland. He felt that electromagnetic radiations were implicated as environmental factors, but the area was rural and dense electromagnetic radiations are usually found in *urban* centers. He did find that the area was uncomfortably close to the U.S.S.R. border northwest of Leningrad and the closer the area studied, the more the effect. Either the Finns were in a surveillance zone for Soviet electronics or they were promoting arteriosclerosis by diet and smoking. This is just the type of sensitive group under discussion and the suggestion is that a precipitation of disease can occur. The affected group was widening toward children from the forty-year-olds originally under study when attention to corrective diet and less smoking decreased the male rate for heart attacks by 40%. When superpowers glare at each other through high powered radars, it might be wise for peaceful citizens to live where it is not quite so hot.

Like waves that mutually extinguish each other as do equals and opposites, saturation with EMR obliterates much of the bioinformation content. Thus the flow of information is sensitive to its carrier energy level. When a kilowatt of microwave energy transforms a little scramble of egg and milk into a delicious omelet, the quanta are unconcerned ultimately with the spectroscopy of rotational energies of constituent molecules. These are weak interactions and forces as compared with the dehydrations, denaturations, and changes of state that occur in the milk and egg. But it is only a matter of scale and the weaker interactions can be allowed to surface in more sensitive applications in diagnosis, therapy, and communication, where the noise may be discriminated against.

More dramatically perhaps, a massive irradiation of a vital organ with neutrons and gammas will doom someone from saturation effects in hours, and death from the gastrointestinal syndrome that might have come in a week or two, or hematopoietic death that ensues in three or four weeks would never be less important to the victim.

There is a tendency to deny the subtle effects in favor of the more visible saturation ones. One instance is the battle of words that still rages around the question of the athermic effects of microwaves. By applying masking energy levels, those who subscribe to this error can conclude that there is no evidence of athermic effects. Would they like to have us believe that with the microwaves emitted by ordinary muscle activity, we are simply cooking each other to rare or well-done? Certainly with the neutron bomb, a warrior *could* be cremated, since thermal energy is the ultimate fate of all the radiations and this result would mask all the other subtleties including his death at lower doses. Preferably we might see EMR in a gentle but enormously effective role, guiding metabolic events away from disease and contrary statistics toward health and longevity.

Medicine and science find the radionuclides to be more and more indispensable. These create a new kind of demand, that for good radiobiological housekeeping. This is certainly coming along, but it requires experience that comes only with time and the tuition for this teaching will be expensive. On the good side is the element of strangeness which cautions some to treat radioactive applications with respect and the fact that as more and more professionals are occupationally exposed, they tend to accept less and less exposure. The military requirements seem to be working against this trend. When the Carter administration acted to deter development of the neutron bomb, there was an outcry from Germany and from the United States Secretary of Defense which is significant, because it shows a growth in acceptance of nuclear weapons which will endanger our getting-acquainted time. If the trend continues, there may well be an order to use the weapons within this century. Yet the neutron bomb is more advanced than dirty H and A bombs of the primary type and suggests that we are moving slightly away from the planet-self-distruct situation. While consummate skill exists at military, academic, and industrial levels, the truth is that we have only begun to learn to live with radioactivity. The thorium-impregnated mantle is a case in point. How many unknown sources are around that can distribute body burdens of radionuclides in the manner of hurricane lamps which are found in a majority of recreational homes and used during blackouts for residences and stores? These can use thorium mantles and emit significant radioactive particles which can be absorbed in breathing.

The zero threshold for radiation hazards comes out as a philosophy of *doubling dose* for the normal level of diseases of the hematopoietic system, GI tract, seminiferous epithelium, and epithelium, will be the panic level. This may well bring the end of life as we know it. EM saturation creeping over a megalopolis, ionizing radiation levels on an upturn, and the spectra of mischievous, or warlike use of these agents bring us to the EM frontier.

This cruise through the informational electromagnetics of biology has been an appraisal of the nature and urgency of matters at this EM frontier. Microwaves are clearly a matter of life and death, usually promoting one and sometimes the other, but their very uniqueness in acceptable risk. While it says that any ionizing radiation whatsoever carries the potential for biological harm, nobody is paying any attention except investigators doing research in this area. It is an academic question because it is already impossible to escape (if it were ever

Epilogue

possible) to a zero radiation level. Instead, the upward exposure trend is everywhere evident. As a future threshold for planetary disaster, the the spectrum as commensurate in wavelength with the dimensions of higher organisms gives a commanding position. At any time they can possess a high informational content making them capable of presently crude, but promising for the future, communication with biosystems.

In its role as a political ploy in the Moscow incident, the coverup has the appearance of a clumsy response. Those who made the decision to classify the matter as secret must be exonerated from charges of mischief since many aspects of microwaves and electromagnetic radiations are incompletely understood.

Preliminary answers are affirmative to such questions as:

1. Whether some groups such as children and pregnant women might have special susceptibilities and should avoid more than 0.1 mW/cm^2 (with a safety factor of 10, this equals 10 μW/cm^2).
2. Whether there are:
    a. Molecular level effects
    b. Microscopic level effects
    c. Cumulative effects
    d. Central nervous system and behavioral effects
    e. Genetic effects
    f. Latent, delayed effects

These estimates are, in some cases, educated guesses based on data withheld for security reasons, incomplete, difficult primate experiments, extrapolations that are tenuous, and public health surveys from the Soviet Union which are sketchy or not universally accepted. However, since about 1975, when U.S. scientists began to look at these effects without bias, the tendency is toward more support and better explained experimental results. For example, the work of Dr. Piro *et al* (1978) on cataracts explains how morphology and anatomy explain previously strange inconsistencies in cataractogenesis.

Also, opposite trends are confusing when incidence rises abruptly with age and people exposed in the military, conversely, are young. New criteria are needed for eye tests which can clarify such matters as predisposition, exposure pattern, and the damage criteria. Thus, electron microscopy can detect precataract membrane damage in lens fibers and ultrastructure defects that are not revealed in slit-lamp biomicroscope examination.

An interesting, but probably remote, possibility at this stage would be the occurrence of an all-woman planet due to the bias of the sex ratio as an EMR effect. The females have two X chromosomes while males have an X and a Y chromosome. The XY male produces two kinds of sex cells, one with the X (gynosperm) and the other with the Y or maleness chromosome (androsperm). This is a common basis for the determination of sex which is thus under genetic control of the male. The mechanism is chromosomal and determination is at fertilization. The final product, however, awaits the period of differentiation and all these processes are sensitive to modifications and aberrations. The Y chromosome is vital to sex but relatively poor in information otherwise. In some organisms there is a mechanism for equalizing the number of males and females with feedback to give a controlled system which is called haplodiploidy and is seen in insects such as wasps. If the males arise via parthenogenesis from unfertilized haploid eggs, then the female arises after fertilization and is diploid. When the males decline in number, the females tend to go unfertilized, so that more males are then produced. When the situation is corrected, the excess of males insures fertilization with a resulting surge in females.

The human system is called heterogametic and the probabilities favor a near equalization of males and females. In the author's case, both the maternal and paternal sides have shown a remarkably even distribution, but for the period associated with microwaves, which is about 30 years, his current distribution is weighted 4 to 1 in favor of girls with an overall bias of 6 to 1. Nothing would have been made of this if it hadn't been for the statement in the Russian literature that maleness may be a casualty of EMR and the lighthearted stories about uninterrupted female births in radar plants during the wars. One has to assume that this is no hoax and, ideologies notwithstanding, the Russians are certainly able to tell a boy from a girl. There is no way of knowing whether the exposure needs to be at the time of conception or can be an accumulated predisposition but the latter is more obvious unless Russian electronic plants are "very different." Marha and his group state that morphological changes manifest themselves in an increase in the number of females born and they believe this to be a distinct effect.

In animals, sex ratios may be modified by genetic selection, by frequency of male ejaculations, and possibly by seasonal factors. There is a possibility that morphological differences occur between the human gametes. The androsperm may have a lower specific gravity, larger tail, a smaller, rounder head, greater speed and may be more sensitive to alterations in the medium. The specific gravity difference seems consistent enough to allow gravity separations to be made in rabbit and bull spermatozoa. In human beings, the possibility of differences has been made the basis for predetermining the sex of the child, using a woman's most fertile time when the environment for sperm delivery is less acid to permit androsperm to group in a wave ahead of the gynosperm. Before the egg is ready, the environment is more acid, the waiting period is, according to this explanation, too lengthy for androsperm and females are more likely to be born. This is widely known but incompletely instrumented and therefore is not a trustworthy method for influencing the sex ratio.

The possibility of EM interference with the normal process would be greatest if exposure of the testes resulted in a selective action on the androsperm in spermatogenesis. Many experiments have shown that sperm are sensitive to EMR and probably even to the 1°C rise in temperature in the testes that occurs when special, tight clothing holds the scrotum against the lower abdomen. EMR heating could add to this clothing effect so that less spermatogenesis occurs. However, the missing experimental work as opposed to survey work is a demonstration that the Y chromosome undergoes selective mutation or some form of genetic bias.

Some etiology and pathology can be explained and curves suggested with the keys of informational bioelectromagnetics. To illustrate, take motion sickness, an example that will most benefit sailors, and ask if it might be called an informational "dis-ease." Here, interdisciplinary paths are seen to converge in a meaningful way, because mere mention of this illness calls up a psychological uneasiness in many. Actually, the words act as a cue for 90% of the population and one person on the way to the rail of a heaving boat, or already there, can cause others to join the march or to have to fight down the urge.

Motion sickness is not related to the agitation of abdominal organs nor to the choice of foods on a cruise. It comes about through mismanagement of information, and understanding can lead to control. Experienced sailors and cruising people *adapt* to the motion, which shows that the sensory input *is* capable of being processed without ill effect. For some anatomical reason, the vomiting reflex issues from close proximity to the brain region for processing body attitude information, which suggests a transfer of signals. Normally, the retching signal comes from gastrointestinal disturbances or suggestion. Getting accustomed to boat motion may take a day or two and it is a real adaptation of muscles and the nervous system to the shifting platform offered to the senses at sea. Like shifting to a gyrostabilizer mode, adaptation has the effect of filtering out information that would confuse, such as vision informing one way and the inner ear another. Probably everyone has experienced the sensation of being parked next to another car which starts to back out and of feeling compelled to put a foot on the brake. Out of the corner of an eye, the other car and its field of associated objects is actually *seen* as moving together, but the sensory information *process* says that *you* are moving forward, so you try to stop. Turning and concentrating intently on the other car and its background will dispel the "illusion" of your own car motion, and knowing this could help you to feel more at ease on a boat. The secret lies in managing the confusing, corner-of-your-eye information input.

Sensory input on the body's attitude comes both from vision and the vestibular, inner ear sensors which have a gyroscopic sort of reference, so that the ear informs on the absolute relation of the bodies, the ship, the sailor, the earth, and gravity. When a landlubber is inserted into the wave motion, the information input is somewhat alien to his sensory correlation networks. It is not the motion of the body itself, but the noise and antinoise that determine effects. Duplication of the "dis-ease" in a motion picture theater when the viewer is stationary proves this. Normally, the ear sensors simply confirm the eye-coded information on the scene. Feedback and experience combine to help manage the input and correlation centers process the input in an orderly manner. Experience especially prepares the organism to expect the sensations and all goes well until a strange set of motion information with contradictions is entered, one with no learned correlation process.

At sea, in a gentle roll, the input from side vision, "gyroscopic" ear information, and the straight-on scene contain opposing elements of information with an effect like noise, so that smooth processing requires antinoise capability or biological squelch. In heavy weather, the confusing input is aggravated by more strange bodily attitudes. As the body learns to cope, the new correlations are burned in and are hard to lose even on return to land which seems to roll a bit.

Control is suggested by a sailor with both feet planted firmly on the main deck, outside, with muscle data flowing in as he braces knowingly against each upcoming motion, his eyes and ears getting corroboration over a wide frame of reference, and his brain in a good status for correlation. At the other extreme might be a person being introduced to cruising and glued to the TV below in an aftercabin with the boat pitching on a downwind sail. He would soon head topsides or find a somewhat steady berth amidships probably spread out, looking up, and taking Dramamine if he has guidance, or at the rail if not.

This remedy or its equivalent will function as a biological squelch, reducing the information input and its confusing elements, not unlike the filtering effect of learning by experience. Non-informational explanations of motion sickness are at a disadvantage in pinpointing the cause, suggesting controls, and indicating the role of the several truly helpful remedies.

There is an *information basis* for civilization—the Neolithic period started it with community—implying new levels of communication. Instant replays of Neolithic culture are to be found in aborigine living in Australia, and in some of the groups in New Guinea and Mindanao. Attention to mechanisms and explanations, as opposed to the acceptance of events at face value, seems to accompany step changes in the information basis and at some point formality in communication required language. This was the most significant step change in history and, as a matter of fact, its very origin. It formed a stable link by which a little Sumerian child in 1850 B.C. could talk to President Reagan in 1981 and give him the "base 60" system for seconds, minutes, and degrees. Writing, as always, represented a burst point in information content when communication modes were inadequate for the emotional, business, and physiological record.

Today's globe is *also* informational, showing centers of civilized living, and in these a greater coupling of information is apparent. Like all roads leading to Rome, communication is center-seeking and operating information flows toward the brain. Then, as the tempo of stimulation and response increases, the information loops become more tightly coupled. Culture, then, is evidence of refinement in conditions, leading to growth and progress with manifestations in thought, manners, and taste at the behavioral level. The operations and behavior of people reflect the closeness of coupling in loops which go all the way down to their smallest living subsystem in which systems dedicated to information easily outnumber those devoted to other processing. Then the informing and coupling processer upgrade the organism's survivability to postpone victory for the ashes of entropy.

In describing brain activity, the word operations has little of the limitation implied by behaviorism. It melds well with the explanation of brain activity as a renewing image from multichannel information processing, but the rivers of information that feed the image travel back for about 3×10^9 years, flowing through communicating, molecular codes, programs of many brains, and manifestations in verbal and written records.

The reader has seen how the forcing and informing biofunctions have the effect of coupling the output of three epochs that are upon us—the Age of Nuclear Energy, the Age of Information and the Age of Electronics. This places the subject in a progression of epochs that began with the Stone Age and traveled through the Bronze Age and the Iron Age. Of the three concerning us now, Information is far more fundamental than Electronics, and tightly coupled to Nucleonics.

In the pressure toward improvement of species, it will be the capacity to manage bioinformation that will be the driving force for future civilization. Hints of this came with hybridization and continued more dramatically with recombinant genetics and the synthesis of nucleic acids. The concept of sensory information broadens the definition, but it is still measurable, as quantized visual information illustrates, and as the information content of visual data proves even further. The eye is a biophysical mechanism for coding such information. Less rigorous but not less meaningful would be the evaluation of sensory information, which can range between limits imposed by the capacity to process it. In semantics, the meaning may need constraints so that information may be more specifically examined in operating, living systems. Then the content, coding, and flow in certain channels are important. In the larger sense, information is a resource with a technical aspect and this is the one we see doubling in size every ten years. Other resources are exhaustible, some are renewable, but information has a breeding feature and it is self-amplifying. In use, it interacts with all other resources through their mutual management. The information theme brings us via coding and channels toward an understanding of specificity while energy is less directed, energizing the information channel as intersystems are coupled for desired responses.

Quite suddenly in the last half-decade, an information revolution has taken place in the management of this resource which is more highly charged with disguised meaning than the coming of nuclear power. When it became possible to synthesize workable genes, the implication was to transfer management from a subordinate species to man. Now that genetic sequences can be selected and spliced into receptive nuclear protein, this management becomes functional. Even the incentive has been provided by the Diamond decision in 1980 (Diamond v. Chakrabarty 79-136 U.S. Supreme Court) which blessed forcible entry into the genetic data bank or gene pool by ruling that novel life forms are patentable. Now, instead of new mechanical models, some corporations will unveil new life forms and be able to control their profit potential under the patent umbrella. Information biology is then the foundation for new profit centers for large and small corporations. Political interest runs a parallel course because of the sociopolitico aspect of genetic programming. The new access to living data banks is no less than control over the quality of life, insofar as it can be *seen* to be based on the planet's gene pool. Manipulation within the time frame of laboratory experiments gives biology a new step change that challenges its central tenets. New legal norms must be framed to protect sacred areas or "endangered genetic sequences" threatened by political and corporate influence. This will need information management from the academic world to assure the benefits of this new era while minimizing its pains.

PERSPECTIVES

Briefings on a project like the Pandora one are to aid in decisions and minimize confusion in the minds of those whose intentions are good, and to maximize it in those whose intentions are not good or to redirect them toward acceptable behavior. Thus people may be guided by good or by evil or may be operating *in* or *against* the national interest, whose manifestations can never be constant but must vary with time in history. In the long term, the probability is good that adequate briefing will prevent bad decisions at least when the operator has good intentions. There might then be economy in the national effort since it can be directed particularly at the deviations from good behavior.

The roots of confusion are in the complexities of the bioinforming and forcing functions at molecular and cellular levels. The misunderstandings arise when so few persons are qualified to probe these levels and interpret their observations. At a minimum the analysis requires a broadband specialist who is also a molecular biologist, a specialist type who will remain in short supply indefinitely.

If confusion about electromagnetic radiations is broadcast, it has a bad effect on legitimate applications. When authorities decided to stand like Custer in favor of thermic effects as an explanation for bioresponses, the circumstances did not warrant this heroism since there were no hostilities. Then the concept of "no athermic effects" filtered into circulation in applied science, constraining medical, industrial, and scientific efforts to involve these effects since they did not exist "by edict." After 1970, many were still using the assumption that there were no athermic effects, a ground rule that couldn't be taken seriously. Those in other countries who sought guidance from the superpowers were somewhat better informed due to the fact that they were aware of the opposite approach in Czechoslovakia and Russia.

The reassurance given by the U.S. that 10 mW/cm^2 represented a safe level may have led to global-wide diathermy at excessive levels and may have given us one of our largest studies (Dael 1973) showing the "harmless" nature of such therapy. In these studies, microwave heating was analgesic in obstetric pathology to relieve pain in uterine contrac-

Epilogue 695

tions. Until 1976, 2,000 babies were born healthy with no "untoward" acute or chronic effects from exposure shortly before delivery. Follow-up of the cases reported would settle the question of how harmless this therapy actually was, whether central nervous system function was threatened, and whether more adroit application is required.

As a result of the East-West differences, the U.S. has probably lost points in the cold war. Working on the low-level side, the U.S.S.R. was able to build expertise denied in past times to the U.S. so that, when the latter was able to examine athermic effects, it had to contend with its own barriers. One lesson is that diathermy must be monitored so that others will not use its therapeutic effect to justify overexposure in other applications of EMR. Suppression of bioeffect information at policy levels makes working-level personnel and their supervisors careless or carefree in enforcing housekeeping in terms of EMR, since downgrading hazards to protect free use is the policy. Instead, dangerous devices should be employed so as to endanger only the enemy.

The microwave exposures are, even in the chronic mode, loaded with an informing capability for susceptible subsystems, leaving appropriate responses where the time or space constants dictate an interaction. In comparison, thermal effects should be rather obvious, becoming acute, at higher exposure levels which mask by saturation other communication.

The natural place for top electromagnetic environmental decisions to be made is "where the buck stops being passed"—at the President's desk. Below that level there are conflicting interests, and it is not fair to ask lesser men to make the decisions. A major conflict is between device users and the interference, in the broadest sense, which is caused by the device. The Armed Services unwillingly bear the brunt of the decision. They may be expected to sponsor studies and to develop the most knowledgeable modes of safe operation for devices. Their studies and conclusions must be biased in favor of their missions. Without higher guidance they, in effect, must develop the national posture which has evolved into a working process. A device that is considered vital *to the national interest* is a go-ahead matter, yet to the credit of the Services, the concept of weighing the impact on the environment has developed. Effectively, at the Secretary of Defense level, interference with the mission is barely tolerable since the military responds to orders. This concept is then modifiable by the urgency of the situation. Not tolerable would be interference for example, by any nonmilitary or nonsupporting civilian emissions of electromagnetic radiations in time of attack, and presumably the warring parties would be expected to preempt planetary channels, or try to, as necessary.

The author believes that, prior unfortunate incidents aside, the U.S.A. will easily achieve a balance between legitimate concern for outright damage with associated legal liability or workman's compensation liability and the guarantee that citizens can pursue careers without disagreeable interference from an electromagnetic environment. The quality of life in the U.S.A. will not be downgraded by the malevolent stares of insolent antennas. Where would the President of the U.S. go *today* if a few 60 Hz incandescent lamps were suspicious when the White House was first electrified? Today, White House duties place the President and his staff at a huge EMR focal point which moves whenever they do. Another necessary perspective is to realize that shedding light on the work with 60 Hz radiation powered a glow which spread over the whole earth and freed man from the gas and oil lamps era. If these results left some bioeffect, they were too important to be constrained by biological events and the country decided in favor of progress. In 1940 radar meant survival and therefore, first things first, in the historical scale. Real wisdom is shown by those who see *when* to switch emphasis. They are now seeing low level exposures as unacceptable because this would assume an electromagnetically contaminated environment with continual pressure to regularly raise the exposure permitted as more and more electromagnetic devices become economically significant. Treating the electromagnetic contamination as such, forces a better engineering of the devices and more considerate application of them.

It is obvious that one real key is in the biophysics of microwaves because it has nothing less than a quality factor of life in its hands. Those officials on the griddle who must condone the extension in time of hazardous conditions for radar operators and technicians or microwave emitters depend on this discipline for answers.

The effect of misjudgment is globally costly and it looks as if those who transfixed the U.S. with the 10 mW/cm^2 level caused the nation to become locked into this level when billions of dollars came to be involved in any change of heart. It cost us something too when this human thermoregulating level biased scientists toward denial of athermic effects as an easy "out."

Some companies in weapons sales are also heavily invested in microwave generator sales. This suggests the possibility of a heavy bias and attendant danger to ordinary citizens from multiple exposures unless the firms are able to show equivalent responsibility for biological effects of their devices, beyond the propaganda level. Industrial producers of electromagnetic radiation often find that efficiency and safety go hand-in-hand in the engineering.

As the electromagnetic devices become more advanced in terms of the information they carry or probe, they approach what we have described as direct interaction modes. For example, the AWACS rotating, phased array antenna systems allow very valuable probing. These arrays acquire information, such as offensive vehicle threats or disposition of forces in large operating theaters. They are airborne and global in operation and in ownership but future appraisals will show these probing and processing microwaves to have been crudely employed at the bioinformational and systems level. Even now it is no longer a matter of frequency allocation alone but one of allowable pulse patterns, peak power emission, and interacting systems.

Crowded antennas on shipboard give warships the look of seagoing TV-radio stations, with the crew being pushed farther and farther out from the EMR until it looks like a question of manning the lifeboats or going back to PT boats, where almost all the radiation is outboard.

One MPL in the U.S. is already down to 1 mW/cm^2 for new microwave ovens and edging toward the U.S.S.R. level. Their 1 μW/cm^2 level is no doubt headed upward so that, in the future, a consensus level may be at a midlevel or range centered on 0.1 mW/cm^2 as a maximum flux permitted in inhabited areas. Controlled access areas would then be allowed an MPL of 1 mW/cm^2. Other sensitive situations demanding special protective consideration would be certain pilots who are literally "on a beam," submarine sail watch personnel, research, medical, radiation physicists, and certain electronic technicians.

The situation in regard to judgment at the top level on electromagnetic radiation hazards became both prominent and sensitive around 1970. Then it became a matter of research against the clock for, if the government failed to answer the environmental and injury questions, then almost everyone's ills would have been traced to microwaves or electromagnetic radiations and governments would have been affected beginning in 1972 by issues of sexual libido, health, and work performance of the exposed population.

The cold war has undoubtedly required an aggressive approach to weapon system developments and the experience in Viet Nam is very convincing. The enemy is real and future generations will thank U.S. military leaders who managed to develop systems in the face of determined opposition. However, there is a question of judgment as to the urgency at a given time and as to the rate of weapon system development which should, of course, be commensurate with the threat. Thus there is no need for endangering pilots, engineers and technicians and letting the EM hazard gradually widen its envelope to involve the general population at any time that is not consistent with the urgency and the national security. Those who have pushed too hard, in effect, forced the development of a novel new field of study when biological questions get involved in legal issues. This field, *forensic biophysics*, is emerging now as families, guards, safety officers in the Services, and electronics engineers allege fatal cancers, leukemias, bladder tumors, and coronary attacks as well as lesser injuries in the form of cataracts down to mild disabilities, reduced life span, and foggy vision in airline people and air traffic controllers. Instead of in the laboratory, under calm and unhurried methodology, the issues will be decided in the courts and workmen's compensation hearings.

These decisions will have consequences far beyond the first forensic findings, which will certainly be stormy, as the contending parties deal with matters of etiology or cause and effect, delayed development of injury with mitigating, intervening circumstances, employment of persons already at risk, or negligence in allowing access to areas, warnings, signs in many languages, clear explanation of existing dangers, provision for remote testing of powerful electromagnetic devices, and last but not least, the eternal argument of thermic versus athermic effects.

As this book goes to press in 1981, the Yannon case in New York City provides the first official instance of a lethal chronic exposure to microwaves. Samuel Yannon had working for the New York Telephone Co., tuning the Empire State building's TV microwave antenna since 1954. He began wasting away rapidly, forgetting, losing weight and growing clumsy, so that people, according to his wife, thought him to be her father. Weakened and disabled after 16 years he was obliged to retire, dying four years later of pneumonia, in a Manhattan hospital. By this time he was described as a withered, 60 pound remnant of a human.

The workman's compensation court ruled in favor of the connection between his death and his work. No other chronic and finally lethal exposure is recorded, but an acute exposure in California in front of an aircraft radar is described by the physician, John T. McLaughlin in *California Medicine*, May 1957 under the title "Death from Microwave Radiation," (Radar) who reported on the operation and tests on the victim, a Hughes Aircraft Co. radar repairman. The award to Samuel Yannon's widow, which is probably several million dollars, will assure that RCA company, makers of the antenna, will be more than ever interested in bioelectromagnetic compatibility.

Such interest for a starter will suggest remote servicing for powerful installations and strict ground rules for microwave devices to avoid the threat of costly verdicts in what are essentially engineering malpractice suits.

The world is trying to absorb the enormous load of technology and the injection of phased arrays of radar, neutron bomb technology, and the intensity of modern communication and transportation is simply overpowering in the time frame being applied.

Evolution will henceforth involve a new element of pressure, an electromagnetic environment, because the levels in cities, airports, and many occupational spaces have become biologically significant.

It will be interesting to follow the Pandora story beyond January 12, 1970 when it went underground. The latest known events are those selected by Brodeur including a medical mission to Moscow when Russian counterparts presented Dr. Thomas F. Stossel with a book on hematology as a gesture, followed chronologically by the embassy fire in August 1977. The book was an example of something one scientist would give another scientist who asked questions whose answers should generally be known by those well-instructed in the field. Dr. Stossel, now assistant Secretary of State, must be counted a brave man. Mark Garrison who ran the desk during the Pandora episode is now on the frontier in Moscow as U.S. Ambassador.

BIBLIOGRAPHY

Marha, K., J. Musil, and H. Tuha, *Electromagnetic Fields and the Life Environment*, San Francisco Press, San Francisco, Calif., 1971.
Siverman, C., "Epidemiology Studies of Microwave Effects," *IEEE 68*, Jan. 1920.

Appendix A Abbreviations

alternating current	ac
adenosine diphosphate	ADP
adenosine monophosphate	AMP
adenosine triphosphate	ATP
amplitude modulation	AM
curie	Ci
direct current	dc
deoxyribonucleic acid	DNA
disintegrations per minute	dpm
electrocardiograph	EKG, ECG
electroencephalograph	EEG
electromagnetic	EM
electromagnetic compatibility	EMC
electromagnetic radiation	EMR
electron volt	eV
frequency modulation	FM
gauss	G
gram	g
high frequency	HF
hour	h
infrared	IR
kiloelectronvolt	keV
low frequency	LF
maximum permissible exposure	MPE
maximum permissible level	MPL
megahertz, megacycles per second	MHz, Mc/s

megawatt	MW
messenger RNA	mRNA
meter	m
microgram	μg
microwatt	μW
milligram	mg
megaelectronvolt	MeV
milliliter	ml
millivolt	mV
milliwatt	mW
minute (time)	min
nonionizing radiation	NIR
phase modulation	PM
picofarad	pF
picosecond	psec
power factor	PF
pulse-amplitude modulation	PAM
pulse-code modulation, pulse-count modulation	PCM
pulse-duration modulation	PDM
pulse-position modulation	PPM
pulse-repetition frequency	PRF
pulse-repetition rate	PRR
pulse-time modulation	PTM
pulse-width modulation	PWM
radiation dose, absorbed, 100 ergs/g	rad
radio frequency	RF
relative biological effectiveness	RBE
ribonucleic acid	RNA
roentgen	R
roentgen equivalent man	rem
roentgen equivalent physical	rep
second	sec
signal-to-noise ratio	SNR or S/N
standing wave ratio	SWR
superhigh frequency	SHF
television	TV
tesla	T
transverse electric	TE
transverse electromagnetic	TEM
transverse magnetic	TM

Abbreviations

ultrahigh frequency	UHF
ultraviolet	UV
very-high frequency	VHF
very-low frequency	VLF
voltage standing-wave ratio	VSWR
watt	W
watthour	Wh

Appendix B Useful Factors

$e = 2.7183$

$\ln 10 = 2.3026$

$\pi = 3.1416$

$c = 3 \times 10^{10}$ cm/sec

$h = 6.6249 \times 10^{-27}$ erg sec (Planck)

$k = 1.3805 \times 10^{-16}$ erg/deg (Boltzman) $= 1.3804 \times 10^{-23}$ J/deg

$N = 6.0228 \times 10^{23}$ (Avogadro) molecules/g mole

$e^- = 4.8028 \times 10^{-10}$ esu (electronic charge) $p = 4.8028 \times 10^{-10}$ esu (proton charge)

$m_{e^-} = 9.1082 \times 10^{-28}$ g (electronic mass-rest)

$m_p = 1.6724 \times 10^{-24}$ g (proton mass-rest)

$P = kfE^2 \epsilon'' / \epsilon_o$ (power from EM wave-watts/cm^3)

$E =$ field strength rms volts/cm

$\epsilon'' = \epsilon' \tan \epsilon =$ loss factor

$D = 3\lambda_o / [8.686 \, \pi (\epsilon'/\epsilon_o)^{1/2} \tan \delta]$ (half-power depth in cm)

$\lambda_o =$ free space wavelength, cm

ϵ'/ϵ_o = dielectric constant

wave energy, 10 cm microwave = 1.2×10^{-5} eV

eV = 1.6020×10^{-12} erg

4.187 J/kg/deg = 1 cal/g/deg

4.187 watts = 1 cal/sec

4.187 J = 1 cal

ϵ_o = 8.854×10^{-12} farad/m (free space)

μ_o = $4\pi \times 10^{-7}$ h/m (free space)

$Z = (\mu_o/\epsilon_o)^{1/2}$ = 377 ohms (free space)

Joule = 10^7 ergs

^{12}C atomic mass = 12 atomic mass units = 12 amu
$\qquad\qquad\qquad$ = $12 \times 1.661 \times 10^{-24}$ g

1 horsepower = 746 watts

1 atomic mass unit (amu) = 3.56×10^{-11} cal

1 calorie = 2.62×10^{13} MeV

1 Coulomb = 6.28×10^{18} electrons

1 curie (Ci) = 3.7×10^{10} dps = 2.22×10^{12} dpm

1 rad = 100 ergs/g

1 rep = 93 ergs/g (water or tissue)

1 kilogram (kg) = 2.205 pounds

1 pound = 453.59 g

1 cm = 0.3937 inches

1 angstrom unit (Å) = 10^{-8} cm

erg = 10^{-7} J

Useful Factors

watt = Joule/sec

calorie = 4.187 J = 2.93 × 10^{-4} kWh

kWh = 3.6 × 10^6 J = 3.6 × 10^{13} ergs

kW = 6.24 × 10^{21} eV/sec

BTU = 1055 J

Appendix C Glossary

absorption coefficient Function that characterizes a beam of radiation as it passes through matter. The rate of change in intensity according to the ability to absorb electromagnetic waves.
absorption spectrum Electromagnetic radiation absorbed at a specific wavelength.
acetylcholine (ACh) Derivative of choline which is important because it acts as a chemical transmitter of nerve impulses in the autonomic system. It has been identified in brain tissue. The enzyme cholinesterase hydrolyzes acetylcholine into choline and acetic acid. Necessary in body to prevent acetylcholine poisoning. It functions as the transmitting agent in synaptic conduction.
action potential A series of changes in potential that occurs across the membrane of a neuron following stimulation, produced by ion flux when membrane permeability is changed upon stimulation.
action spectra A particular wavelength for a specialized, selective response.
active transport Transfer of substance into or out of a cell, across the cell membrane against a concentration gradient by a process which requires expenditure of energy.
adaptation Fitness of organism for its environment. Process by which it becomes fit. Characteristic which enables organism to survive in its environment.
adaptive radiation The evolution from a single ancestral species of a variety of species which occupy different habitats.
adenine A purine which is a component of nucleic acids and nucleotides important in energy transfer.
adenosine triphosphate (ATP) An organic compound containing adenine, ribose and three phosphate groups. Present in living cells, and providing energy for a wide variety of life functions.
aerobic Growing or metabolizing in presence of molecular oxygen. Conditions requiring free oxygen.

afferent neuron A neuron that conducts signals toward the central nervous system.

after potential Small potential changes in latter stages of the action potential on a neuron.

agglutination Collecting into clumps of cells or particles formerly distributed in a fluid.

ALCOR Advanced Research Projects Agency, Lincoln Laboratory of Massachusetts Institute of Technology, Coherent Observable Radar for watching reentry vehicles.

allele One of an array of alternative forms of a gene occupying the same locus in homologous chromosomes.

allosteric An enzyme with two distinct receptor sites on the same protein molecule.

allosteric interactions Control of cellular metabolism by conformational changes in functional linkage. Enzymes with separate activating or control sites which interact with or influence each other; producing feedback through the action of the products on the first or allosteric enzyme.

alopecia Loss of hair, for example after exposure to ionizing or possibly nonionizing radiations.

alpha helix A spring-like configuration of protein molecules occurring particularly in globular proteins. Secondary structure of a protein, spiral structure of an amino-acid chain.

alpha particle A helium nucleus emitted spontaneously from natural and manmade radioactive elements, consisting of two protons and two neutrons.

alpha wave Characteristic pattern in a normal electroencephalogram with a frequency near 10 Hz.

alternating current (ac) An electrical current that is continually varying in value and reversing its direction of flow at regular intervals.

alveolus Small saclike dilation.

amino acid Structural unit of proteins. Each amino acid is characterized by the presence of an amino acid group ($-NH_2$) and a carboxyl group ($-COOH$).

amplitude The maximum displacement of an oscillation, vibration or wave from its average value.

anabolism Chemical reaction. Simpler substances are combined to form complex substances, the result is storage of energy and production of new cellular materials and growth.

anaerobic Not requiring the presence of oxygen or growing only in the absence of oxygen.

analog Similar in structure or function but different in elements, conformable, operating with numbers to represent voltages, etc.

anaphase Third stage in mitosis during which chromosomes move to opposite poles.

angstrom unit (Å) Unit of length equal to 10^{-8} cm and used in describing atomic and molecular dimensions.

anterior Head or front end. In erect animals, ventral (stomach) side of body.

antibody Specific protein produced by an organism in response to the introduction of an antigen. Antibodies often tend to combine with the antigens and neutralize their harmful effects.

anticodon Sequence of three nucleotides in transfer RNA that, in the process of protein synthesis, binds to a specific codon in messenger RNA by complementary base pairing. Sometimes opocodon.

antigen Foreign substance, usually protein or protein-polysaccharide, that usually induces antibody formation in a living organism.

antimetabolite Compound that competitively inhibits a specific enzyme or reaction in metabolism because of structural resemblance to natural metabolite.

antinode Any point, line or surface in a stationary-wave system which has maximum amplitude at all times. Usually specified since there can be antinodes of voltage, current, pressure, etc.

antitoxin An antibody which is produced in response to the presence of a toxin and which is capable of neutralizing a toxin.

apoenzyme Protein portion of a conjugated enzyme.

arteriole A minute arterial branch, one proximal to a capillary.

artery A blood vessel that passes away from the heart and transports blood to various parts of the body. Has thick, elastic walls.

atom Smallest part of an element that is capable of undergoing a chemical reaction and that is chemically indestructible and invisible. A structural unit of matter which remains unchanged in chemical reactions except for the loss or gain of electrons.

atomic mass number Total number of protons and neutrons in the nucleus of an atom.

atomic orbital Orbital of electron about the nucleus of an atom.

atrium Chamber giving entrance to another structure or organ. Chamber of the heart receiving blood from veins and pumping it into a ventricle.

autonomic nervous system Portion of the peripheral nervous system which supplies nerves to structures not under voluntary control.

autonomous Existing and functioning independently. A tumor cell that is free of host control or plasmid that replicates independently of the chromosome.

autoradiograph A technique for studying the location of radioactive isotopes in macromolecules and in larger structures. Material to be studied is labeled with a radioactive isotope and is placed in the dark in contact with photographic emulsion. Latent image produced by the radioactivity is subsequently analyzed.

autosome Any chromosome which is not involved in determining sex.

autotrophy Being self-nourishing, manufacturing organic nutrients from inorganic materials.

axon Long extended part of neuron nerve cell that conducts pulses away from the cell body of the neuron.

axoplasm Cytoplasmic fluid of an axon.

bacillus A bacterium with a cylindrical shape. Bacilli represent one of the three major forms of eubacteria, e.g. *Lactobacillus*.

background Radioactivity from other than the source measured or other noise affecting sensory processing.

bacteriophage A virus that infects and kills bacteria.

bacterium A minute, unicellular procaryotic organism characterized by the absence of a formed nucleus.

base Compound that releases hydroxyl ions (OH−) when dissolved in water.

base change Effect on DNA caused by ionization of a base such that the base forms normally forbidden pairs during DNA replication. Results in heritable changes in genetic code.

base deletion Genetic code error caused by complete dissociation of a base from the DNA molecule.

base pair A pair of nitrogenous bases, one purine and one pyrimidine, that join the two parallel strands of the DNA molecule.

basilar membrane Membrane in the cochlea where the organ of Corti rests. This membrane separates cochlear duct from the scala tympani.

benign Being of a mild character and not malignant.

binary Made up of two elements or of two radicals acting as elements.

biologic clocks Activities of plants and animals are adapted to regularly recurring changes in external physical conditions.

biological endpoint A biological response or bioeffect that is definitely associated with an EMR such as a change in the number of lymphocytes. A stimulation in the division of lymphocytes.

bioluminescence Light produced by a chemical process occurring within the living organism. Emission of light by a living organism such as the firefly, certain fungi and many marine forms.

biopole An organism functioning as a dipole in an applied field.

biospectroscopy The spectroscopy of living material.

biosphere Entire zone of land, air, and water at the surface of the earth which is occupied by living things.

bioxidation Electrons removed from an atom or molecule are transferred through electron transmitter systems.

bivalent The structure formed by the association of the members of a pair of homologous chromosomes during meiosis. Having a valence of two.

Boltzmann distribution Mathematical representation which relates the equilibrium spatial distribution of particles to the energy differences of particles at different positions and to the temperature of particles.

bound water Water molecules fixed to a macromolecule.

boutons The presynaptic nerve endings filled with vesicles of transmitter substance.

brachycardia Abnormal, relatively slow heart action whether physiological or pathological.

Bragg's Law Equation for the condition whereby a system of parallel atomic layers in a crystal will reflect a beam of X-rays with maximum intensity.

bremsstrahlung Radiation from decelerating X-rays.

Brownian movement Continuous irregular motion of small particles suspended in a liquid or gaseous medium; in liquid the motion results from the bumping of particles by water molecules.

buffers Substance or mixture of substances which, in a solution, maintains constant hydrogen ion concentration despite addition of comparatively large amounts of acids and bases.

calorie The amount of heat required at a pressure of one atmosphere to raise the temperature of one gram of water one degree centigrade.

calorimetric, calorimetry Measurement of a quantity of energy, by changing it to a thermal form.

capacitance Stored electrical energy from a voltage and ac current at some frequency, in farads. One farad passes one ampere during charging at one volt per second. Depends on the size of the capacitor plates, their separation, and the interplate dielectric.

capillaries Microscopic, thin-walled vessels located in tissues, through whose walls substances pass to the tissue fluid. Interposed between arteries and veins.

capsid Protein shell enclosing nucleic acid in most viruses.

carcinogenic A substance or agent producing or inciting cancer.

carbohydrate Compound of carbon, hydrogen and oxygen, the latter two in the same proportion as in water. As cane sugar, starch and cellulose which are formed in green plants.

catabolism Degrading metabolism involving the release of energy and resulting in the breakdown of complex substances within the organism into simpler compounds.

catalyst Substance which brings about a change in the speed of reaction. Influencing a chemical action without being used up as a result of the reaction. Enzymes.

cataract A clouding of the lens of the eye arising from opacity.

catecholamines Amines containing catechol, e.g. adrenaline, noradrenaline, and dopamine; neurotransmitters.

cation Positively charged ion in a solution; that portion of a compound which, when dissolved (usually in water), tends to flow toward the cathode under the influence of a direct electric current.

central dogma Thesis that RNA is transcripted from DNA only and that protein is translated only from RNA; that the transfer of the genetic information only goes in the direction from DNA to RNA to protein.

central nervous system Brain, brain stem, and the spinal cord.

centriole A small body near the nucleus of a cell which, in the process of cell division, forms the center of the aster rays. Small darkstaining organelle lying near the nucleus in the cytoplasm of animal cells and forming the spindle during mitosis and meiosis.

centromere Region of chromosome which controls its movement during cell division. A specialized region of the chromosome at which the chromatids are joined.

cerebellum A large dorsally projecting part of the brain especially concerned with the coordination of muscles and the maintenance of bodily equilibrium, situated anterior to and above the medulla which it partly overlaps and formed in man of two lateral lobes and a median lobe.

cerebrum Main portion of the vertebrate brain, occupying the upper part of the cranium; united by the corpus callosum; forms the largest part of the central nervous system in man.

chemoreceptor A nervous receptor adapted for excitation by chemical substances. A group of atoms in a cell which fixes chemicals.

chemotherapy Treatment of a disease by means of chemotherapeutic agents.

chlorophyll The pigments which give plants the green color and which are of great importance in changing radiant energy to chemical energy in the process of photosynthesis.

chloroplast A plastid that contains chlorophyll and is the site of photosynthesis and starch formation.

chromatin Filamentous form of DNA which, in cell division, forms the substance of chromosomes. Readily stainable protein of acid character found in nucleus of cells.

chromosomal aberration Abnormal arrangement of normal chromosome complement caused by chromosomal breakage and reunion.

chromosomal break Break in chromosome that serves as a basis for losses of information or repositioning of the physical structure of the chromosomes.

chromosomal exchange Reciprocal translocation of nearly intact chromosome arms. Two breaks, one in each chromosome, are needed for this type of exchange.

chromosome arms Parts of the chromosomes extending from the centromere to the free end.

chromosomes Thread-like structures which carry genetic information and are found in the nuclei of cells. One of several bodies composed of protein and nucleic acid that are found in cell nuclei.

chronon Element controlling periodic phenomena, such as synthesis or division.

cilia Minute hairlike processes often forming part of a fringe, that is capable of a lashing movement. Thread-like appendages by which certain microorganisms propel themselves. Cytoplasmic projections on the free surface of cells.

cingulate cortex The middle part of the cortex between the two hemispheres. Concerned with emotional problems and sometimes with motor and autonomic disturbances.

circadian rhythms Repeats sequences of events which occur at 24 hour intervals.

cis configuration Both mutant alleles of two linked genes or heteroalleles of a cistron are on one homolog and both wild-type alleles are on the other. Cis means having certain atoms or groups on the same side of the molecule.

cistron A section of DNA molecule that specifies the amino acid sequence for a particular polypeptide chain. The genetic unit of biochemical function.

clone Many cells descended from a single ancestral cell by asexual reproduction.

cocci A spheroidal bacterial cell, such as a macrococcus, micrococcus, streptococcus, or gonococcus. Usually less than 1 μm in diameter.

cochlea Cavity of the inner ear, snail shaped. Contains receptors of hearing.

codon Group of three successive nucleotides on a molecule of mRNA which serves as a code for placement of specific tRNA molecules and amino acids into a proper sequence during protein synthesis.

coenzyme Thermostable nonprotein compound that forms the active portion of an enzyme system after recombination with an apoenzyme.

cognitive Information obtained from interaction between incoming radiation and objects.

colinearity Concept that the order of amino acids along a polypeptide chain corresponds to the order of the codons in the cistron specifying that polypeptide.

collagen An albuminoid which is the main supportive protein of connective tissue. Forms gelatin on boiling with water.

collisions Encounters between atoms, molecules, ions, electrons, and so on, that may result in an exchange of energy and change in condition.

colloid A state of matter of single large molecules or aggregations of smaller molecules. Two-phase system in which particles of one phase, ranging in size from 1 to 100 mm, are dispersed in a second phase; a gelatinous material secreted by cuboidal epithelial cells. Colloids have little or no tendency to dialyze and small or no freezing point depression.

commensalism The living together of two organisms, one of which is benefited and the other is neither benefited nor injured.

commensurate interactions A case where a favorable dimensional relationship exists between wavelength and absorbing system.

conditioned reflex A paradigm using an association by pairing of some neutral stimulus such as a sound, with the unconditional stimulus of an inborn reflex, e.g. salivation to food.

cone Photoreceptive cell in the retina of the eye which is particularly sensitive to bright light and functions in color vision.

conformational change Change in a unit such as a protein subunit which will involve cooperative action or many amino acids; when one subunit switches its conformation, it carries the other when the two are well-linked; like sharp responses and switching.

continuous spectrum A spectrum with continuous variation of wavelength from one end to the other, e.g. EMS, X-rays, spectrum from a lamp.

cooperativity Ways in which macromolecular subunits or biosystems link up functionally for purposes of changing to alternative, stable molecular states. Noncooperativity is when binding of one subunit does not affect another subunit for the ligand.

corticosteroid Steroid hormones produced by the cortex of the adrenal gland.

covalent bond A nonionic chemical bond formed by electrons shared between the atoms in a molecule.

crossing over Reciprocal exchange of corresponding segments between two nonsister chromatids of homologous chromosomes.

cybernetics Control systems involving governors, feedback, loops, the processing of information for stabilization or realization of some goal.

cyclic AMP A messenger from hormones to cells called cyclic because phosphate group forms rings with the carbon atoms to which it is attached. Regulatory agent coordinating diverse cell groups. Formed from ATP by adenyl cyclase and hydrolyzed to $5'$-AMP by phosphodiesterase; as a messenger, informs aggregation of free amoeba to form a mature fruiting body (*Dictyostelium discoideum*).

cytochromes Any of several intracellular hemoprotein respiratory pigments that are enzymes, functioning as transporters of electrons to molecular oxygen by undergoing alternate oxidation and reduction. Important in cellular metabolism.

cytokinesis Mitotic division of the cytoplasm portion of the cell.

dead time Length of time following an impulse during which the instrument remains insensitive, unable to record another event. Resolution time which leads to coincidence loss.

decibel The unit of loudness of sound expressing the relationship between two sound levels which are in terms of the logarithm of the ratio of the levels. $10 \log (P_1/P_2)$ for any energy = dB.

degeneracy Existence of two or more synonym codons for a given amino acid. Degeneracy is *complete* or *partial* depending on whether all or some of the possible codons code for amino acids; it is regular if it follows certain rules as distinct from being entirely random.

deletion Loss of genetic information from the chromosome.

demodulate Recovery of electrical signal that represents the information from the carrier wave.

dendrite A short, and usually branched process of a nerve cell which conducts impulses to the cell body.

deuterium Isotope of hydrogen having one proton and one neutron in the nucleus.

diathermy The generation of heat in the body due to absorption in the tissues of high frequency energy passed through them.

dielectric loss Electromagnetic energy dissipation in dielectrics.

diploidy Having the basic chromosome number doubled. The presence of two homologous sets of chromosomes within the nucleus.

discriminator Electronic device that is capable of accepting or rejecting a pulse according to pulse height or voltage.

dispersion Incorporation of the particles of one substance throughout the body of another. A colloid solution.

distal Being farthest from a particular location or point of attachment.

disulfide bond Covalent bond formed between two sulfur atoms. Formed mostly in peptides and proteins between two sulfhydryl groups of two cysteine residues.

DNA (Deoxyribonucleic acid) Localized in cell nuclei. Present in chromosomes. The chemical that contains genetic information, specifically, directions for the synthesis of large and complex protein molecules. Is capable of self-replication. Vehicle for transfer of genetic information from parent to offspring.

doppler effect The apparent change in the measured frequency of a wave pattern caused by movement of the receiving device toward or away from the wave source.

dose Amount of exposure of a biological system to ionizing radiation as in radiotherapy.

dosimetry Measurement of doses.

ectoderm Outermost of the three primitive germ layers of the embryo. From it come the skin and the nervous system.

effective half-life The half-life of a radioisotope in a biological system as a result of the combined effects of the biological half-life and the radiological half-life.

effector Termination of an efferent nerve within a muscle or gland, structures of body by which an organism acts. A metabolite that, when bound to an allosteric enzyme, alters the catalytic activity of the enzyme.

efferent neuron Neuron that conducts signals away from the central nervous system.

elastic collision Collision between particles in which there is no change in the total kinetic energy of the particles.

electroencephalogram (EEG) The tracing of brain waves made by an electroencephalograph.

electrolyte A substance which will conduct an electric current, either in molten state or in solution. Substance which dissolves into electrolytically charged ions when dissolved in water or other polar solvent.

electromagnetic spectrum (EMS) The EM spectrum is an array of forcing functions which could be enlarged at any moment according to the engineering capabilities in designing generators for the frequencies. The spectrum collects all the radiations that have the burden of carrying, containing, or retrieving specific information and the energies that are associated as a function of frequency. Both may be designed to reach transducer molecules or organs playing the role of selective receptors, or to function in the lesser capacity as noise.

electromagnetic wave The passage of energy in mutually perpendicular, transverse (across), electric and magnetic fields, each perpendicular to the direction of propagation. It may be represented physically as a sine wave in which the periodicity is shown by the trigonometric sine function, or by amplitudes plotted, in a clockwise manner, against angles or radians.

electron Elementary particle of nature having a charge of one and a mass of 9.1×10^{-28} grams. The term electron usually implies a negative charge.

electron capture Type of radioactive transformation in which an orbital electron collides with the nucleus and also unites with the nucleus to form a new nuclide having the same mass number but an atomic number diminished by one.

electron transfer system System of enzymes such as the cytochrome system within the mitochondria, which transfer electrons from foodstuff molecules to oxygen.

electron volt The amount of energy gained by an electron in passing from one point to another point that is one volt higher in potential.

Glossary

electrophoresis Migration of suspended solid, liquid or gaseous particles to one of the two electrodes (under the influence of an external emf or as the result of an applied potential difference).

element Type of matter, natural or manmade, which has some number of protons in the atomic nucleus and the same number of electrons circling in orbit. Substance that cannot be decomposed by ordinary chemical methods into simpler substances.

emission The process of ejecting electrons from the surface of a material under the influence of heat, radiation or other causes. Electromagnetic waves radiated by an antenna, a celestial body, tissue quantum source, or machine generator.

emulsion Colloidal mixture of two immiscible fluids, one dispersed in the other in the form of fine droplets.

endergonic The reaction characterized by the absorption of energy. Requires energy to occur.

endocrine Internally secreting. Organ whose function is to secrete into the blood or lymph a substance that has a specific effect on another organ or system.

endoderm Innermost of three primitive germ layers of the embryo. From it come the epithelium of the pharynx, respiratory tract, digestive tract, bladder and urethra.

endoplasmic reticulum Unit membrane network within the cell, linking nucleus, cytoplasm, and cell membrane.

endoskeleton Bony, cartilaginous supporting structures within the body. Supporting framework from within.

energetics The science that deals primarily with energy and its transformations. The total energy transformations of a system.

engram Sensory information processing leaves a residual memory, probably of biochemical changes in RNA. The RNA concentration follows the maturation process, peaking at about 40 and falling off after 60.

ensemble A group of related elements which, when taken together, produce a single effect.

entropy Entropy is approximately the measure of ordering in the state of the ensemble. An ensemble of particles may have a great amount of ordering, or negative entropy. If expelled into a larger volume, the ensemble may become more disordered (positive entropy). A measure of energy not available for mechanical work. According to the equation $H = - \Sigma\ p_i \log p_i$, entropy is related to the information content.

enzyme Any of numerous complex proteins that are produced by living cells and that catalyze specific biochemical reactions at body temperatures. Virtually all chemical reactions in living cells are under enzymatic control.

epidermis Outermost layer of cells of an organism. Outer epithelial layer of the external integument of the animal body that is derived from the embryonic epiblast.

epistemology Nature, grounds and limits of knowledge.

epithelium Layer of tissue that covers the internal and external surface of the body, also the lining of vessels and small cavities. Consists of cells that are joined by small amount of cementing substances.

equilibrium A state in which the forces acting on a system are balanced. A state of adjustment between opposing or divergent influences or elements.

ergodic A process having sequences representative of the whole, being statistically probable that any state will recur.

erythrocytes Red blood corpuscles. Circular, biconcave, disc-shaped, blood cells containing no nucleus. Erythrocytes contain the oxygen-carrying pigment, hemoglobin.

esu Electrostatic unit.

etiology Knowledge dealing with causes; all of the causes of a disease or abnormal condition.

eukaryotic Being applied to organisms that have the nuclei surrounded by membranes, Golgi apparatus, and the mitochondria.

exencephaly Hernia of the brain.

exergonic A biochemical reaction characterized by the liberation of energy.

farad The basic unit of capacitance. A capacitor has a capacitance of one farad when a potential difference of one volt will charge it with one coulomb of electricity.

feedback control The system in which the accumulation of the product of a reaction leads to the decrease in its rate of production, or when deficiency of the product leads to the increase in its rate of production.

ferromagnetic Of or relating to a substance with an abnormally high magnetic permeability, a definite saturation point, and appreciable residual magnetism and hysteresis.

field Area of observation, a realm of knowledge. A space in which a phenomena is to be noted; electric or magnetic field which may be applied to a biosystem as a forcing function.

fission A splitting or breaking up into parts. Asexual reproduction.

fluorescence Instantaneous reemission of light of a greater wavelength than that of the originally absorbed light. Visible radiation emitted by some atoms after the electrons of those atoms have been excited by ultraviolet light.

flux A generalized flow, whether of heat, electricity, fluid or chemical constituents. The number of particles or photons passing through a unit of area per unit of time.

flywheel effect An inertial tendency to continue an effect after the cessation of the motivating influence.

focal point Point along the optical axis of a lens from which rays incident on the lens parallel to the axis diverge, or toward which they converge.

food chain Arrangement of the organisms of an ecological community according to order of predation in which each uses the next, usually lower, member as a food source.

forcing function Energy applied by electromagnetic radiation in the action spectrum of a biological system, informing it to reorder its status. Polarizing influence on an organism that alters its physical, biological, chemical, or psychological state.

forensic biophysics Application of biophysical facts to questions of law.

fovea A small pit, fossa. Indentation in retina of the eye on axis of eyeball, rich in cone cells.

free energy The portion of released energy that is available to do work as a reaction proceeds toward equilibrium.

frequency Number of complete cycles per unit of time for a periodic quantity such as alternating current, sound waves, or vibrating objects. The repetition rate of a periodic motion, measured in Hertz (Hz) or cycles per second for EMR.

frequency response Rating indicating the range over which a circuit or system handles all frequencies uniformly. A measure of capability of an instrument to respond fast enough to changes in the parameter being measured so that it faithfully reproduces the variations in that parameter, in the sense of the process time constant. Sensitivity of a system to a certain frequency.

fusion Joining of light nuclei to form a heavier nucleus. Particle velocities corresponding to millions of degrees are required for this process which is used in the thermonuclear bomb.

galvanometer Instrument for measuring or indicating a small electric current or some function of the current by means of mechanical motion resulting from electromagnetic forces set up by the current.

gamete A germ cell, sperm in the male and egg in the female. Specialized reproductive cells formed by the parental organisms during sexual reproduction.

gamma ray Electromagnetic radiation having its origin in an atomic nucleus. Of shorter wavelength than X-rays and of greater penetrating power; emitted during the decomposition of a radioactive substance. Gamma rays are not deflected by a magnetic field. Used in treatment of cancer.

ganglion A mass of grey nervous tissue which serves as a center of nervous influence. Composed primarily of the cell bodies of neurons.

gauss The cgs electromagnetic unit of flux density. An emf of 1 abvolt will be induced in a conductor 1 cm long moving perpendicular to a flux of 1 gauss at a velocity of 1 cm/sec.

gaussian distribution Bell-shaped distribution of probabilities.

GCA Ground controlled approach radar.

Geiger counter An ionization chamber operating in the Geiger region. Used for detecting and counting ionized particles in the air.

gene Hereditary factor. A self-perpetuating particle by which hereditary characteristics are transmitted. A segment of DNA responsible, in most instances, for the formation of a particular protein or amino acid chain. Elementary unit of the germ plasm regarded as part of the chromosome. Biologic unit of genetic information.

generator Source of an electromagnetic radiation.

genetic equilibrium Situation in which the distribution of alleles in a population is constant in successive generations, unless altered.

genetics The science of heredity.

genome Term used in reference to all of the genes carried by a single gamete, that is, by a single representative of each of all of the chromosome pairs. A complete set of hereditary factors in the haploid assortment of chromosomes, or bacterial DNA.

genotype The total genetic constitution of an organism. Fundamental hereditary constitution.

globulin One class of proteins in blood plasma, some of which function as antibodies. Native protein, insoluble in water and soluble in dilute salt solutions and alkalis.

glycogen Animal starch used as an energy storage substance in higher animals.

glycolysis The digestion of sugar, with metabolic conversion into simpler compounds. Breaking down of the sugar in the body tissues.

glycosidic bond A covalent bond that links sugars together.

gram mole Mass of a substance, in grams, numerically equal to the molecular weight.

grana Small disc-like bodies within chloroplasts that contain alternate layers of chlorophyll, protein and lipid. The functional units of photosynthesis.

half-life The time in which the radioactivity usually associated with a particular radioactive isotope is reduced by one-half through radioactive decay.

haploid Having only one of each of the pair of chromosomes characteristic of the somatic cells of a species.

Hardy-Weinberg Law The relative frequencies of the members of a pair of allelic genes in a population are described by the expansion of the binomial equation $a^2 + 2ab + b^2$.

helix Geometric figure that looks like a spring, being formed by the rotation of a line around, and at a constant distance from, a longitudinal axis.

helix-coil-transition A conformation shift from irregular helix to a regular coil induced by electric field as in poly-γ-benzyl L-glutamate (PBLG) when the field is parallel to the helix axis.

hematopoietic system The blood-making system, including the bone marrow, the spleen, and lymph nodes.

hemoglobin The red, iron-containing protein pigment contained in red blood cells whose function is the transportation of oxygen from the lungs to the various tissues of the body.

heterozygote An organism or cell containing two contrasting alleles of a single gene.

HIPAR High power acquisition radars used in relation to anti-aircraft missiles.

homologous chromosome The two similar chromosomes present in a diploid cell, one being derived from each parent and identical genes or their alleles being located at corresponding positions.

homozygote An organism or cell containing two identical alleles of the gene.

hormones Signaling substances that are produced in cells in one part of the body which diffuse or are transported by the bloodstream to cells in other parts of the body where they regulate and coordinate their activities.

host The organism upon or inside which another organism, being a parasite, perhaps lives and obtains nourishment.

hydrated electron (e^-_{aq}) A reducing agent and free radical, and the most important photochemical product. Polarization product from orientation of water dipoles.

hydrogen bond A relatively weak chemical bond, between two molecules formed when a hydrogen atom is shared between two atoms, usually oxygen atoms. Primary in the structure of nucleic acids and proteins.

hydrolysis Decomposition due to absorption of water. A chemical reaction in which water is one of the reactants. Decomposition of organic compounds by interaction with water, either in the cold, on heating alone, or in the presence of acids or alkalis.

hypertonic Possessing greater osmotic pressure than a standard having a higher than isotonic concentration. Having a greater concentration of solute molecules and a lower concentration of solvent, such as water molecules, and an osmotic pressure greater than that of the solution to which it is compared.

hypothalamus The portion of the diencephalon forming the floor and part of the lateral wall of the third ventricle. Region of forebrain, contains various centers controlling visceral activities, water balance, temperature, sleep, etc.

hypothesis A supposition assumed as a basis of reasoning. Can be further tested by controlled experiments.

hypotonic Pertaining to solutions in which the concentration of solute molecules is lower and a concentration of solvent, such as water, is higher, and thus having an osmotic pressure lower than that of the solution with which it is compared.

immune reaction Production of antibodies in response to the antigens.

immunologic tolerance Ability of an organism to accept cells transplanted from a genetically distinct organism, the results of exposure of the organism to an antigen before it has developed capacity to react, so development of capacity to react might be delayed or postponed indefinitely.

impedance match The condition in which the impedance of a connected system is equal to the internal impedance of the source, thereby giving maximum transfer of energy from source to load.

index of refraction Ratio of speed of light in a vacuum to the speed of light in a particular medium. (n)

induction Process in which a specific structure develops within an organism as a result of chemicals transmitted from another part of the organism. The act by which an object is electrified, magnetized, or given an induced voltage by exposure to a field.

induction coil A device for changing direct current into high-voltage alternating current.

inelastic collision A collision between particles or photons in which there is a change in the total kinetic energy of the particles.

information Ordering influence giving life and continuity, measurable in rigidly prescribed models as bits. This word comes from binary digit and one unit specifies a selection between two probabilities such as on-off, go, no-go, heads-tails.

information/energy sink Place where they disappear or are transformed.

informing function A particular frequency which has a forcing effect on a receptive system's functions.

insulator A device having high electric resistance, used for supporting or separating conductors so as to prevent undesired flow of current from the conductors to other objects.

interferon The protein formed during interaction of animal cells with viruses, that is capable of conferring on fresh animal cells of the same species, resistance to infection with a wide range of viruses.

internode An interannular segment of a nerve fiber. Section of a stem between two nodes.

interphase In the "resting" or interphase nucleus, the chromosomes usually cannot be distinguished individually although chromatin material can be identified.

intersymbol influence Where one symbol has an effect on another in coding; for example in English where q is followed by u or th is followed by e. In biological code, the symbols would be represented by amino acids and bases.

intersystems EM radiations connect systems that are close by or at great distances, and have bioinforming functions.

inversion Segment of a chromosome that has had its gene order reversed relative to the remainder of the chromosome as a result of two chromosomal breaks and rotation through 180°. The reversal of a segment of a chromosome within the chromosome as a whole. The process of producing inverted or scrambled speech. In antenna and propagation theory, the condition when the temperature of the air at high altitudes is higher than that of the surface air or that at some lower altitude.

in vitro In a laboratory, outside the normal environment of a living system.

in vivo Being in a living system.

ionic bonding The last unfilled atomic shell may have only one or two electrons which are easily lost to vacancies in a second atom. The first will be positive since it lost electrons and the gain by the second will cause it to be negative. Oppositely charged atoms such as these can then form ionically bonded compounds.

ionizing Where the energy of an electromagnetic wave is capable of separating electrons from the molecule.

isomer Compounds or substances having molecules with the same chemical formula as one another but different arrangement in structural formulas.

isometric Of equal dimensions. Condition of muscle contraction in which no change in the length of the muscle occurs.

isotopes Atoms of a chemical element having the same atomic number but different atomic masses. One of a group of nuclides of the same element having the same number of protons in the nucleus but differing in the number of neutrons.

isozymes Different molecular forms of proteins with the same enzymatic activity.

joule effect The rate at which heat is produced by the flow of current through a resistance.

karyotype The shapes of individual chromosomes of a set of chromosomes within the nucleus of a cell. The chromosome set characteristics of a given species, 46 for humans.

kinesis Activity of an organism in response to a stimulus, the direction of the response is not controlled by the direction of the stimulus.

kinesthesis The sense which gives us our awareness of the position and movement of the various parts of the body.

kinetics The motion of masses or mass particles in relation to the forces acting on them. That part of mechanics which deals with the effects of forces in changing the motions of bodies.

klystron A vacuum tube for converting direct-current energy by alternately slowing down and speeding up an electron beam, utilizing the transit time between two points to produce a velocity-modulated electron stream that delivers radiofrequency power to a cavity resonator.

knowledge Summary of existing global situation presented as an output from programs to acquire learning.

lac operon Genetic control section. Inhibited by protein, lac processor, so RNA polymerase cannot transcribe. Cyclic AMP activates many operons for transcriptions, while repressors are specific for a gene.

laser Light amplificaton by the stimulated emission of radiation.

LET Linear Energy Transfer.

linkage The inheritance of separate genetic characters in a nonrandom manner. Caused by presence of separate genes on the same chromosome.

load An absorbing substance presented to an applied electromagnetic field.

longitudinal wave Wave whose vibrations are in the direction of the propagation of the wave.

loss Attenuation of electromagnetic energy in an absorbing medium.

luciferin Oxidizable organic substance necessary for the bioluminescence of many organisms. Substrate present in certain organisms capable of bioluminescence, producing light, when acted upon by the enzyme luciferase.

luminescence Emission of radiation, mostly visible light, as a result of absorption of radiant or corpuscular energy or by other causes, such as bioluminescence or chemiluminescence.

lymph The fluid taken up and discharged by the lymphatics. Colorless fluid which is derived from blood plasma and resembles it closely in composition, contains white cells, some of which enter the lymph capillaries from the tissue fluid, others of which are manufactured in the lymph nodes.

lymphatic system Network of vessels that conducts lymph from the body to the bloodstream.

lymphocytes A variety of leukocytes, arising in lymph nodes, with a single nucleus and nongranular protoplasm.

lysis The gradual abatement of a disease. The process of destruction of a cell and liquidation of its contents, for example, by the enzyme lysozyme.

lysosome The intracellular organelles that are present in many animal cells; contain a variety of hydrolytic enzymes, which are released when the lysosome ruptures.

magnetron Cross-field device for the generation of microwaves.

maser Microwave amplification by the stimulated emission of radiation.

mass number Number indicated by the symbol A which represents the total number of nucleons in a nucleus.

mean free path The average distance traveled by a particle between collisions.

medulla Marrow, the deep, inner substance of an organ or part. The most posterior part of the brain, lying next to the spinal cord.

meiosis Process in the maturation of germ cells by which the chromosome number is reduced from diploid to haploid. Kind of nuclear division, usually two successive cell divisions which result in daughter cells with the haploid number of chromosomes of the original cell.

melanin Protein of reddish-brown to black color in hair and epithelial cells. From dark tissues or melanotic tumors.

mesoderm An intermediate layer of cells developing between the ectoderm and endoderm, responsible for bone, muscle, connective tissue, and inner layer of the skin.

messenger RNA (mRNA) Linear form of RNA used to transmit information from nuclear DNA to site of protein synthesis in cytoplasm. A particular kind of ribonucleic acid which is synthesized in the nucleus and passes to the ribosomes in the cytoplasm, combines with RNA in the ribosomes and provides a template for the synthesis of an enzyme or some other specific protein.

metabolism The sum of all the physical and chemical processes by which living organized substance is produced and maintained and by which energy is made available for the use of the organism.

metamorphosis Abrupt transition from one developmental stage to another. Change of structure or shape.

metaphase Phase of cell division in which the chromosome pairs are arranged with their centromers along the equatorial plane.

microcurie One millionth of a curie. Symbol μCi.

micron (μ) One millionth of a meter or one-thousandth of a millimeter.

microtubules Tiny tubes of tubulin protein 250 Å in diameter in eucaryotes universally, in procaryotic spirochetes and gliding bacteria; for flagellar movement, regeneration, transport, mitotic spindle; cold sensitive.

microwaves Electromagnetic radiation in the centimeter and millimeter wave region.

milliroentgen One thousandth of a roentgen. Symbol mr.

mismatch The condition in which the impedance of a source does not match or equal the impedance of the connected load.

mitochondria Small granules or rods in the cytoplasm of cells. Spherical or elongated intracellular organelles which contain the electron transmitter system and certain other enzymes, site of oxidative phosphorylation. Source of cell power.

mitosis Form of cell or nuclear division by means of which each of two daughter nuclei receive the same complement of chromosomes as parent nucleus had.

mode Microwave term specifying the type of oscillation occurring in a line, waveguide, cavity or tube. A complex rhythmic scheme in a system or a basic form or pattern in a system.

modulation The process in which the amplitude, frequency, or phase of a carrier wave is varied with time in accordance with an informing signal.

mole Amount of a chemical compound mass which is equivalent to its molecular weight. The sum of atomic weights of its constituent atoms.

molecule A small mass of matter. An aggregation of atoms composing the smallest unit of a compound possessing its characteristic properties. Smallest particle of a covalently bonded element or compound having the composition and properties of a larger part of the substance.

monochromatic Radiation of only one wavelength.

morphogenesis The differentiating of the developing body into its various organs and parts.

motor unit The skeletal muscle fibers that are stimulated by a single neuron.

MTR Missile Tracking Radar.

multiple alleles Three or more alternate conditions of a single locus which produce different phenotypes.

mutation Abrupt change in the inheritability characteristics of an organism caused by a change in the genetic material of the organism.

myelin The soft material surrounding the axon of a medullated nerve fiber. The fatty material which forms a sheath around the axons of nerve cells in the central nervous system and in certain peripheral nerves.

myofibrils Smallest element of muscle structure capable of contraction, composed of proteins, myosin and actin.

NAD The abbreviation of nicotinamide adenine dinucleotide, a coenzyme that functions as a hydrogen acceptor in biological oxidations.

NADP The abbreviation of nicotinamide adenine dinucleotide phosphate, a coenzyme that functions as a hydrogen acceptor in biological syntheses.

neuroglia The supporting structure of the brain and spinal cord. Connecting and supporting nonnervous cells located in the central nervous sytem surrounding the neurons.

neuron A nerve cell with all its processes, collaterals and terminations. The structural unit of the nervous system.

neurotransmitter Substance secreted by the tip of a neuron which is able to activate a neighboring neuron or muscle.

neutron Electrically uncharged neutral particle of matter along with the protons. Has a mass number of one approximately equal to that of a proton. Existing in the atomic nucleus of all elements except the mass 1 isotope of hydrogen. In the free state (outside the nucleus) it is unstable, having a half-life of about 12 minutes. It decays by the process n = proton + negatron(e^-) + neutrino.

node Where the wave shape crosses the X-axis. Any point, line or surface in a stationary-wave system at which the amplitude of the wave-shaped variable is zero.

noise Approximately defined as any extensive group of frequencies whose causality is not understood and whose occurrence is not well-predicted by the observer. Fluctuation in voltage or other measured information parameter in a system. Signal to noise ratio, shot noise, thermal noise, white noise, noise spectrum.

nuclear isomers Isotopes of elements in different quantum or energy states. Isomers of an isotope will exhibit different properties such as mode of decay.

nucleolus Spherical, stainable body found inside the cell nucleus. Composed primarily of ribonucleic acid and believed to be the site of synthesis of ribosomes. Visible during interphase.

nucleoside A purine or pyrimidine base that is attached to a ribose or a deoxyribose sugar.

nucleotide A nucleoside-phosphoric acid complex composed of a purine or pyrimidine base, a 5-carbon sugar (ribose or deoxyribose) and a phosphate group (PO_4). A basic unit from which DNA and RNA are made up.

nuclide Any one of the more than one thousand species of atoms characterized by the number of protons and neutrons in the nucleus.

operant conditioning Preparation of an animal by training it to respond to a cue for some operation. Responses such as touching a switch by a rat to obtain a pleasurable sensation that operates on the animal to produce reinforcement.

operator site An entity postulated to account for the control of operon. Adjacent to structural genes in operon and is believed to be the site on DNA to which repressor molecules are bound, inhibiting the synthesis of mRNA by the genes in adjacent operon.

operon A unit consisting of a number of structural genes, or cistrons, that control a series of biochemical reactions in a biosynthetic pathway, an operator gene controlling the activity of the structural genes, and a promoter region lying between the operator and the first structural gene.

orbit The closed path or track of a body in space repeated periodically.

organ of Corti The organ that changes sound signals into nerve impulses; located on the basilar membrane inside the cochlea.

organelle One of the specialized structures within a cell, mitochondria, chloroplasts.

orthogenesis Evolution progressing in a given direction.

osmosis The passage of fluid through a membrane resulting from differences in concentration of solutions on either side of the membrane.

osmotic pressure The pressure needed to prevent water from passing into a solution from which it is separated by a membrane permeable only to water.

ossicle One of the small bones in the middle ear, the malleus, the incus, or the stapes.

oval window Membrane between the middle ear and inner ear against which the stapes transmits pressure signals which are carried into the cochlea through a fluid.

pacemaker An electrical device worn by those who may suffer from skipped heartbeats. A center for program control if a program for aging is assumed. The part whose rate of reaction sets the pace for a series of interrelated reactions.

parameter An arbitrary constant whose value characterizes a member of a system (as a function of curves); any of a set of physical properties whose values determine the characteristics or behavior of something.

parenchyma The essential or functional elements of an organ. Plant cells that are relatively unspecialized, being thin-walled, containing chlorophyll and are typically loosely packed, function in photosynthesis and in storage of nutrients.

permeant Ionic or other species that pass through a membrane such as sodium or potassium.

phagocyte A cell specialized for eliminating damaged material, toxins, and harmful bacteria.

phase meter An instrument for measuring the difference in phase between two alternating quantities of the same frequency.

phase relation The tendency for time factors in a forcing function to be synchronous or asynchronous with temporal events in the informed system.

phase transitions Change in state of molecule, helix-coil changes, hysteresis, one-to-many input-output cooperative shifts.

phenotype The sum total of an organism's observable or measurable characteristics produced by its genotype's interaction with the environment.

pheromone Receptor, such as olfactory bombykol molecule, pheromone receptor involves transduction for odor sensation with membrane conductance and depolarization; receptor cell senses one molecule of attractant.

phosphorescence The decay of a molecule in the triplet state to the ground state. Emission of light without heat.

photo cell A semiconductor device in which charged carriers are released to flow as a current when activated by incident light.

photolysis Decomposition by light. The splitting of a molecule under the action of light.

photon A quantum of electromagnetic radiation.

photoperiodism Physiologic response of animals and plants to variations of light and darkness.

photosynthesis The process by which plants synthesize carbohydrates from carbon dioxide and water, using the radiant energy from the sun captured by the chlorophyll tissue of green plant cells.

phototropism The response in growth of an organism to light.

phylogeny The evolutionary history of a group of organisms. Inversions and translocations can be used to trace phylogeny.

pi electrons The mobile electrons that are located in a system of conjugated single and double bonds which are not associated with a single atom or bond but with the conjugated system as a whole.

piezoelectric Having the ability to generate a voltage when mechanical force is applied, and the reverse.

pituitary The small gland that lies below the hypothalamus of the brain to which it is attached by a narrow stalk. The anterior lobe forms in the embryo as an outgrowth of the roof of the mouth. The posterior lobe grows down from the floor of the brain.

plane wave A wave whose wavefront is a plane surface.

ploidy The number of sets of chromosomes in a cell.

point mutations A change in the genetic code by addition or removal of a nucleotide, transposition of the code or substitution of nucleotides.

polar body A small cell that consists of practically nothing but a nucleus, formed during oogenesis, maturing of the egg, and appears as a speck at the animal pole of the egg.

polarization In optics the act or process of making light or other radiation vibrate in a definite form so that the paths of the vibrations in a plane perpendicular to the ray are straight lines, circles or ellipses giving plane polarization, circular polarization, or elliptical polarization. Orientation of a system due to a forcing function in biology, psychology, or social sciences.

polarizer A Nicol prism or other device for polarizing light. Specifically, the first of two prisms used for the purpose, and which receives the light as it enters.

polymer Large molecule formed when several monomer units are joined chemically.

polyploids The organisms that have more than two haploid sets of chromosomes; for example triploid, tetraploid and so on.

polysome A linear array of a group of ribosomes held together by a molecule of messenger RNA. The active site of cellular protein synthesis.

porphyrin A complex organic compound containing four pyrrole nuclei linked through methylene carbons; a component of hemoglobin and chlorophyll.

pH (potential of Hydrogen) Negative logarithm of the hydrogen ion concentration by which the degree of acidity or alkalinity of a fluid may be expressed.

power density An intensity of electromagnetic radiation or flux, usually in watts per square centimeter.

precursor A substance that precedes another substance in a metabolic pathway. A substance from which another is synthesized.

propagation The travel of electromagnetic or sound waves through a medium. Propagation does not refer to the flow of current in the ordinary sense.

prophase The initial stage of mitosis. Characterized by the formation of chromosomes from the chromatin in the nucleus of a cell.

protein A nitrogen-containing compound found in animal and vegetable tissues. Macromolecules containing carbon, hydrogen, oxygen, nitrogen and usually sulfur and phosphorus, composed of chains of amino acids bound in peptide bonds. One of the principal compounds found in all cells.

proton A particle present in the nuclei of all atoms having a positive charge, equal to that of an electron and a mass number of 1.

proximal Nearest or nearer to the body or to the point of attachment.

quantum Finite unit or bundle of radiant energy emitted when an electron moves to the next inner orbit or energy level of an atom. Quantum jump. A discrete packet of electromagnetic radiation. Quantum of charge (e^-), quantum number.

quantum mechanics Physical theory dealing with atomic structure in terms of quantities that can actually be measured. It embraces matrix mechanics, the transformation theory of Jordan and Dirac, and the wave mechanics of Shrödinger. The motion of atomic particles emerge as quantized, even though the deterministic position of the particles is replaced by probabilistic distribution.

quantum number One of a set of integral or half-integral numbers, one for each degree of freedom which determine the state of an atomic system in terms of constants of nature.

quantum theory The concept that energy is radiated in discrete units of energy called quanta.

quenching Process of limiting the discharge of an ionization detector, either externally by momentary reduction in applied potential to the tube through suitable electronic circuitry, or internally by introduction of a quenching agent like butane or chlorine. Quickly reducing thermal energy level.

radiation absorbed dose The basic unit of absorbed dose of ionizing radiation. One rad is equal to the absorption of 100 ergs of energy per gram of matter.

radiation syndrome A set of biological responses associated with a certain kind and intensity of radiation.

radioactive decay When the activity of radioactive material is decreased by time.

radiolysis The breakdown of a chemical such as water by radiation.

radiosensitive Biological sensitivity of material to one or more of the radiations.

radiotherapy The use of radiation in curing.

receptor Atoms of a molecule that, when acted upon by atoms of another molecule, produce the chemical transformations in cell protoplasm. A sensory nerve ending that responds to a given type of stimulus. A compound present in the target cell of a hormone that specifically takes up and binds that hormone.

recessive genes The genes which do not express their phenotype unless carried by both members of a set of homologous chromosomes. Genes which produce their effect only when homozygous.

recombinant Exhibiting genetic recombination by transformation, transduction, or conjugation of bacteria.

redox potential (oxidation-reduction potential) Potential difference set up at an inert electrode immersed in a reversible oxidation-reduction system; the measure of the state of oxidation of the system, volts on scale of electronegativity.

redundance The quality of a message that is longer than its minimum effective length.

reflex A reflected action especially such as movement occurring in response to specific stimulation. An inborn, automatic but involuntary response to a given stimulus which is determined by the anatomic relations of the involved neurons, the functional units of the nervous system.

refraction The bending of a ray of light, heat, sound, or a radio wave or other radiant energy passing obliquely from one medium to another in which the velocity of propagation is different.

refractory period The time after the stimulation of a biosystem during which it is no longer able to respond.

relative biological effectiveness (RBE) =

$$\frac{\text{Dose in rads to produce effect with therapy X-rays}}{\text{Dose in rads to produce effect with radiation under investigation}}$$

relaxation time The time interval between the time of maximum tension of a contracted muscle and the time at which the resting state is restored. Time constant. Behavior with time on withdrawal of EM field or forcing function, like a broadened resonance.

resolution May be temporal or spatial. Ability to separate or distinguish the identity of two objects or time signals that almost coincide with the use of an instrument.

resonant Conditions established by a biosystem and a particular frequency which results in the easing in of energy. Also circuits may be resonant with special current or voltage effects. Resonances may be mechanical, chemical, acoustic as well as combined.

reticulum The network of fibrils or filaments within cells or in the intracellular matrix.

rhizome The subterranean root stem of a plant.

rhodopsin The light-sensitive, purple-red pigment of the outer segment of retinal rods, which undergoes a chemical reaction triggered by light which stimulates the receptor cell to send an impulse to the brain, resulting in the sensation of light.

ribonuclease An enzyme capable of destroying ribonucleic acid.

ribonucleic acid (RNA) A polymer of nucleotides connected via a phosphate-ribose backbone, involved in the synthesis of proteins. Present in both nucleus and cytoplasm.

ribosomes Small bodies, usually attached to the endoplasmic reticulum within a cell, rich in RNA and are the site of protein synthesis.

roentgen The quantity of X- or gamma radiation, with associated corpuscular emission per 0.001293 grams of air (i.e. 1 ml at 0°C and 760 mm) producing, in air, ions carrying 1 electrostatic unit of quantity of electricity of either sign.

REM (roentgen equivalent man) A quantity of radiation absorbed by man producing an effect equivalent to the absorption of 1 roentgen of X- or gamma radiation.

REP (roentgen equivalent physical) The amount of ionizing radiation capable of producing 1.615×10^{12} ion pairs per gram of tissue or the amount absorbed by tissue up to 93 ergs per gram. This unit is used in particular to measure beta radiation.

rotation absorption Electromagnetic energy (usually microwave) absorbed by a molecular system causing an excitation in the rotary mode for some element of a system. Quantum rotation means rotational energy states at only fixed levels with relaxation giving emission of a quantum.

saccharide One of a series of carbohydrates including the sugars.

scintillator A counter employing a phosphor, photomultiplier tube and associated circuits for the detection of radiation.

self-absorption Effect of finite thickness of a radioactive material in which the material itself absorbs part of the emitted radiation.

self-replication A direction by one molecule for the construction of a second molecule exactly like the first. DNA is a self-replicating molecule; a double structure, each half of which can direct the formation of its complement.

semiconductor One of three classes of materials, the other two being conductors and insulators. Closely related to dielectrics.

senescence The aging of a biological system in the sense of a program but without necessarily involving a life-spanning, for example menapause for the reproductive system.

sequence A set of ordered elements as in playing cards of the same suit by rank.

sertoli cells Testicular cells which, with the system of Leydig and spermatozoon cells, offer a logic for development of entire testicular structure.

sialic acid N-acylneuraminic acid or any other derivatives of its alcoholic hydroxyl groups.

sodium pump The mechanism that restores the ion concentration characteristics of the resting state to neurons after a depolarization pulse has passed along the neuron.

spallation A nuclear reaction in which light particles are ejected as the result of bombardment (as by high-energy protons).

specific ionization The number of ion pairs produced per unit of distance along the track of an ionizing particle.

spectrin Human erythrocyte protein on inner cell membrane surface constitutes 20% by weight of the total membrane protein.

spectrometer An instrument that disperses radiation and measures the deviations or wavelengths corresponding to the spectral lines; usually used in analysis.

spectroscope Any of several instruments used for spreading individual wavelengths in a beam of radiation so that they may be observed with the eye as the resulting spectrum.

spindle A fibrous structure that is visible within the cell during division, controlling chromosome movement.

squelch A filtration process for noise in signal managment often by narrowband or exclusion methods. Antinoise measures in a biosystem.

standing wave A wave reflected at both ends along its direction of propagation characterized by points along the wave (nodes) where the minimum disturbance of the medium occurs and by points (antinodes) midway between the nodes where the maximum disturbance occurs.

stationary Statistical situations where the probability distributions and the moments associated with them are the same and without effects from the time of measurement.

stochastic A process which involves a random variable, chance, and probability. Like tossing coins.

strain A population of genetically identical cells; a clone.

structural gene Section of a DNA molecule that codes for a particular enzyme.

substrate Substance on which an enzyme acts.

superoxide dismutase The enzyme for the detoxification of a superoxide in biosystems.

synapsis The exact pairing of homologous chromosomes during early stages of meiosis.

tachycardia Speeding of heartbeat.

tesla current A high-frequency current produced by a voltage that is fairly high but is intermediate between that for Oudin and that for D'Arsonval currents. Used in electrobiology. Nicoli Tesla also developed ac current and experimented with biosphere-modifying currents and charges.

tetraploid Individual or cell that has four sets of chromosomes.

thermal neutrons Neutrons that have energies corresponding to room temperature, approximately 0.025 eV which is the kinetic energy of a molecule at about 300°K.

thermic effect An effect which is primarily related to the higher temperature experienced by a system. Opposed to athermic effects in which the temperature rise is not considered adequate to explain the response. Some responses are a combination as when a heat pulse in a cool environment distracts an animal from a task or causes him to lose a cue.

thermonuclear reaction A nuclear reaction. The activation energy for the reaction is provided by the thermal agitation of the reacting nuclei.

time constants The measure of the time required for a capacitor to charge or discharge in a circuit. It is numerically equal in seconds to the product of the resistance in megohms and capacity in microfarads. Also active processes, relaxations, responses and reactions.

tracer An isotopic tracer is an isotope that is used to tag or follow a chemical reaction or process such that its location and concentration can later be determined, for example, radioactive tracer ^{32}P or deuterium with its greater mass.

transducer A device by means of which energy may flow from one or more transmission systems to one or more other transmission systems, by means of a change in its form, for example, electromechanical transducer.

transfer RNA (tRNA) A form of RNA which serves as an adaptor molecule in the synthesis of proteins by delivering amino acids to the peptide.

transformation The transfer of genetic information in the form of DNA, between related bacteria, for example, the transformable pneumococcus can become virulent by this process.

transition element Any of various metallic elements (e.g. chromium, iron, or nickel) that have valence electrons in two shells instead of in only one—also called transition metal; being transitional between the more highly electropositive and the less highly electropositive elements.

translocation The change in position of a chromosome segment to another part of the same chromosome or to a different chromosome. There is reciprocal translocation where there is the exchange of segments between two chromosomes.

transmitter substances Chemical mediator for conducting nerve impulse across the synaptic gap between axon terminals and the postsynaptic membrane. Norepinephrine, for affective behavior, and about six others give function linkage in psychopharmacology studies of molecular neurobiology.

transmodal transduction Chemical energy or energy-rich molecules as ATP or cyclic AMP matched to biosystems for action; muscle contractibility, sensory information processing via the five senses, their receptors and the action potentials that result; pheromone attractant in olfaction, quantum basis for vision.

triplet code The sequence of three nucleotides which comprise the codons, units of genetic information in DNA that specify the order of amino acids in a peptide chain.

triplet state The state that results when an electron is activated by absorbing a photon, moves to an outer orbit of higher energy and pairs with an electron of like spin.

tritium A hydrogen isotope of mass three. Its nucleus contains one proton and two neutrons.

TTR Target tracking radar.

uniport Membrane transport system which transports molecule without necessarily having to balance the action with reverse transport.

unit membrane Membrane that is incorporated in the structure of many cell organelles that consist of double-layered phospholipid molecules.

vacuo Vacuum, the absence of matter.

valence Of orbital electrons which participate in chemical reactions through their loss, gain or sharing.

velocity A vector quantity denoting the speed and specific direction of a linear motion.

velocity modulation Modulation of other system by controlling the velocity. Modification of the velocity of the electrons in a beam. Variations of electron velocity are utilized to convert direct-current beam energy into radio-frequency energy.

viable Capable of living, growing, developing or functioning.

viable cell count The count of living organisms only.

virus A minute infectious agent having a nucleic acid core and a protein shell.

vitreous Being glasslike or hyaline, designates the vitreous body of the eyes that contains clear transparent jelly called vitreous humor which fills the posterior part of the eyeball.

wave A propagated disturbance, usually periodic. A cyclic disturbance that propagates through the medium without moving the particles, of the medium, along with it; chemical wave, change along a membrane or fiber, self-propagating, not with change in membrane potential, communicating as in phototaxis where the intersystem information leaves off and the message moves to locomotive system in bacterium; spreading phenomena with a concentration wave.

wavelength The distance between two successive peaks of the same polarity in a wave.

wave media Elastic media such as water and air will propagate waves but EM waves provide their own medium—their electric and magnetic fields, so that they propagate well in a vacuum. The energy passing through the wave train need not disturb or translocate the medium substance as when a boat pitches up and down as water waves pass underneath.

work function The energy required to free an electron. Energy in excess of the work function w imparts a velocity to the electron where $e^-_{energy} = \frac{1}{2}mv^2$. The photoelectric effect of Einstein is then described by $\frac{1}{2}mv^2 = hf - w$.

xerophthalmia A type of blindness that is characterized by an abnormally dry, lusterless and horny layer of epithelium over the cornea. Due to deficiency in vitamin A.

X-rays Electromagnetic radiation in the region below 100 angstroms. Similar to light but having much shorter wavelengths. Usually generated by accelerating electrons to high velocity and suddenly stopping them by collision with a metal target.

X-ray wavelength Determined by the magnitude of the nuclear change. The greater the atomic number of an element, the less the wavelength of the X-rays characteristic of it.

Zeeman effect The splitting of a spectral line by a strong magnetic field.

zygote The diploid cell produced at fertilization by the union of the haploid gametes.

Index

α-ketobutyrate, 152
ACTH, 174
ATP, 249
absorption coefficients, 220
absorption spectroscopy, 255
accessary pigments, 309
Acetabularia, 195
acetyl coenzyme-A, 146, 155
acetylcholinesterase, 634
acoustics, 456, 497
action spectra, 550
active transport, 182
acute microwave exposure, 507
adaptations, 310
adenylate cyclase, 634
adenylates, 155
Adey's trigger conditions, 557
adjuvents, 158
aerobiosis, 405
aging, 166, 210
    cataract, 656
    deficiency of L-dopamine, 173
    diabetes, 170
    eye lens crystalline fibers, 170
    fungi, 167
    pituitary hormone, 174
    programmed, 176
    spanning, 167
    vitamin C, 170
    yeast, 167
Alber, Werner, 421
allostery, 148
allowable current gradient, 593
alphanumeric logic, 118
amelanotic melanoma, 250
amino acids, 121
    most probable conformation, 156
amplification, 5, 6, 79, 118, 145, 153, 157
anaerobiosis, 410
anathonemus petersii, 423
anemia, 341
animal behavior, 511-512
anomalous dispersion, 469
ant orientation, 314
antibodies
    amplification, 157
    cooperative effort, 157
    phase, 157
antigens, 158
    adjuvents, 158
    determinants, 158
    haptens, 158
antinoise, 176, 208
Apis mellifers, 314
Arbacia punctulata, 418
Ashby, W.R., 3
aspartate transcarbamulase, 152
athermic effects of microwaves, 687
athermic exposure, 511
atmospheric scatter, 313
atomic and thermonuclear
    explosions, 378
atomic structures
    information dense regions, 134
audio effect, 622
auditory sensitivity, 639
automata, 17
autoradiographs, 290, 388

B cells, 171
B system lymphocytes, 161
backscatter, 312
bacteria, 405, 546, 548
bacterial photosynthesis, 250
ball lightning, 125
bands, 251
   spectral, 252
bee orientation, 314
Beecher, William, 505
binary system, 74
biochemical feedback, 38
biodielectrics, 475
bioelectric gradient, 192
biohazards, 277-278
bioinformation management, 17
   systems, 12
biological dosimeter, 349
biological feedback and
   information, 4
biological identification by
   transferrin, 271
biological material, sources for EMR, 296
biological squelch, 207
biological warnings, 668
biomolecular receptivity, 210
   specific frequency, 210
biomolecular time processes, 270
bionics, 79-80
biopole, 314, 644
biosolid state, 236
biospectroscopy, 236
biosquelch process, 176
biosystems, reaction to strong
   magnetic fields, 249
biradical, 247
bird behavior, 508
bird migration, 314
bits per immune response, 165
blood changes, 349
blood conductivity, 457
blood groups, 159
blood pattern, 578
blood pressure, 647
blue light, 314
   scatter, 314
   sea color, 314
   sky, 314
Bohr magneton, 264
bone fracture, healing, 192, 564
bony structures as protection, 643
Boolean algebra, 65
bound water, 270, 411-412, 489
Brachydamia rerio, 565

brain
   as a regulator, 201
   calcium efflux, 555
   enzyme inactivation, 633
brainwaves, 20, 23, 245
broad-band absorptions, 422
Brodeur, P., 505

calcium ion, 142
   switch, 144
cancer, 279
capacitor model, 431
carcinogens, 249
cardiovascular symptoms, 506
Carnot cycle, 200
carrier wave, 554
catalysis, 152
cataract
   aging, 656
   microwave, 656
cataractogenesis curve, 657
cell division, 511
cells
   comparing normal and tumor, 552
   contact inhibition, 193
   reaction to strong magnetic fields, 249
cellular information, 13
channel, 674
channel death, 168
Chart of the Nuclides, 368-376
chemical bonds in DNA, 96
chemiluminescence, 399
chemoreceptors, 156
chemotactic morphogenesis, 190
Chiroptera, 178
chromatophore, 552
chromosome aberrations, 565
chronic exposure, 507
chronon theory, 195
circadian cycles, 428
circuit parameters, 598
clone, 173
   antibody cell line, 173
cloning of cultured cells, 192
closed and open loops, 5, 17
code degeneracy, 108-109
codons, symbolic for species, 97
coenzymes, 146
cognitive energy, 231
coherence, 312, 480
coherent waves
   interference, 286
coincidence of peak amplitudes, 664
Cole, Kenneth, 417

Index

Cole-Cole arc, 471
colinearity, 110
colloidal suspension, 460
commensurate interaction, 302
communicating codes, 675
communication link, 458
communications theory and coding, 63
compartments, 412
competitive analogs, 154
Compton scattering, 302
Computer and the Brain, The, 3
computer functions, organismic, 41
computers, 77
concentration gradient, 186
condensed mixtures, 474
conformers, 156
conforming bounds, 142
conforming protein, 114
conjunction, AND operation, 69
contact inhibition, 194
continuous spectrum, 250
continuum of biological effects, 538
control systems, 200
cooperative effort, 136
Cotton-Mouton effect, 273
coupled enzyme sequences, 162
coupling, 545
   biological coupled EM wave modulation, 511
   in bacterial cell development, 186
crisis management, 150
critical mass, 11
crossed field effect, 513
crossing over in chromosomes, 98-99
crosslinking, 170
culture injury, 194
cybernetic analysis, 7, 17
cybernetics, 4, 12, 13, 27, 239
   cellular information, 13
cyclic holograms, 619
cytoplasmic inheritance, 98

DNA, 93
   amount, 550
   and neural nets, 86
   chemical bonds, 96
   codons, 97
   evolutionary advantage over RNA, 129
   guanine-cytosine percentage, 97
   information, 549
   polymerases, 171
   synthesis in bacteria, 125
   X-ray crystallography, 101
Debye, P., 218

decay scheme, 376
decimal prefixes, 291
deformation, 148
deformational change, 139
degeneracy, 176
dehydrogenases, 140
del Castillo, Jose, 419
densely ionizing radiation, 406
depolarization of populations, 246
deprogramming, 181
   influence on pacemakers, 181
dermal shell, microwave significance, 630
Design for a Brain, 3
determinants, 158
diabetes, 170
diagnostic indicators, 669
diamagnetism, 263
diathermy, 695
Dictyostelium discordium, 190
dielectric constant, 238, 246, 247, 255
   of water, 134
dielectric mobility, 637
dielectromagnetics, 237
differentiation, cellular, 189
diffraction, 250, 313
   grating, 250, 313
dipoles, 442
   ensemble, 431
   orientation, 219
direct and indirect action, 408
direct information transfer, 523
direct mode, 546
discontinuous spectrum, 250
displacement, 212, 246
dissociation energies, 292
dopamine, 173
   as pacemaker, 174
dosimeters, 386
Drosophila, 129, 418

E. coli, see *Escherichia coli*
EEGons, 560
EM field signature, 562
EM signature, 554
EMR, penetration into leaf, 287
ESR, 249, 250
   biohazards, 277
   see also electron spin resonance
ear anatomy, 614
earth's electrical field, 539
Edman degradator, 113
eigenvalue, 258
Einstein, Albert, 258
elastic photon-molecule collisions, 259

elastic wave quanta, 307
elasticity, 174
electric eel, 200
electric fish, 679
electroencephalography, 142
electromagnetic energy, microwave, 262
electromagnetic environment, 307
electromagnetic pollution, 684
electromagnetic radiation
    equivalents, 302
electromagnetic spectrum, 4
electromechanical transducer, 307
electron paramagnetic
    resonance, (EPR), 247
electron-positron pair production, 302
electron spin resonance, 247, 260
    polarization, 261
    transferrin, 268
    see also ESR
electron transport system, 249
embryonic development, 347
energy, 18
    and wavelength, 303
    cognitive, 231
    internal peaks, 602
    levels, 292
    radio frequency — microwave, 265
    saturation levels, 307
energy of parallel and antiparallel
    subgroups, 261
ensemble, 82
entropy, 5, 47
environment radioactivity, 332
environmental fields, 511
enzymatic regulation, 4
enzymes, 186
    coupled sequences, 162
    functional metallic centers, 249
    reverse transcriptase, 188
equivalent circuit, 424
    parameters, 598
equivalent quanta, 229
ergodic processes, 83
Erlich ascites tumor cells, 167
error catastrophe, 171
erythroblastosis, 159
erythrocytes, 152
Escherichia coli, 411, 418, 548
    T_4 phage, 124
etiolated plants, 310
evoked biopotentials, 144
evolution, 162, 238
excitable membranes, 418

exencephalia, 347
exponential growth, 119
eye lens crystalline fibers, 170
eye-pupil reflex, 34, 40
eye structures
    human, 652
    rabbit, 652

Faraday effect, 272
Fatt, Paul, 419
feedback, 31, 143, 149, 152
    biochemical, 38
    machines, 244
feedback information and
    communication, 17
fields, 595
    near and far, 595
filled wavefuide, 607
final-product-sensitive-type enzyme, 147
flag pathways, 5
fluorescent wavelengths, 254
flywheel effect, 628, 659
forcing functions, 213, 246, 294
free radicals, 175, 247, 249, 393, 402
    "heavy", 250
    in biology, 248
    organic, 248
frequency, geometrical fit, 626
frequency-dependent impedance, 420
frequency-dependent response, 128
frequency effects, 401
frequency shift, 258
Fricke, Hugo, 417
fringing field, 606
fructose, 149
functional linkages, 404
fungi, 167

gas phase reactions, 232
gel formation, 170
gene mutation, 119
general systems analysis, 17
genetic information, membrane
    integrity of, 169
geometrical cross section, 561
Gibbs free energy, photosynthesis, 51
global symptoms, 506
glucocorticoids, 577
gluconeogenesis, 149
Goffman, W. 22
Gordon, S.V., 506
Great Ideas in Information Theory,
    Language and Cybernetics, 3

Index 741

growth modification, 563
guanine-cytosine percentage, 97
Gymnarchus niloticus, 423

habitable space, in areas of intense electromagnetic radiations, 685
hair cell displacements, 617
hairpin loop, 170
haptens, 158
hazardous conditions of radiation, 605
"heavy" free radical, 250
HeLa cells, 167
helical director, 422
helices, 175
helix-forming information, 165
hemoglobin, 246
Herpes virus, 566
Hertzian waves, 293
hierarchies of systems, 509
histological changes, 346
hologram, 286, 288
holographic microscope, 287
human blood groups, 159
  erythroblastosis, 159
  Rh compatability, 159
human body, as a microwave source, 297
human reasoning or operating link, 2
hydrated electron, 394
hydroelectric transduction, 613
hydrosols, scatter, 314
hyperthermia, 569
hysteresis, 144

identification, 5
immune response, bits per, 165
immunoglobulin, 157
  "tuning fork" configuration, 158
immunology, methods in, 160
immunoresponse, 157, 576
  B system lymphocytes, 161
  macrophage, 161
  T system lymphocytes, 161
  transplant rejection, 159
immunosystem, information status, 339
impedance, 5, 417, 422
inadvertent radiators, 594
inelastic collisions, 259
information, 17, 18, 33
  and antigen activity, 338
  cellular, 13
  counterflowing, 164
  features in organisms, 84
  helix-forming, 165
  level of radiation received, 128
  polarization, 313, 315
  status of the immunosystem, 339
  wave mechanisms, 19
  also see type of information
information content, 47, 680
  entropy, 47
  thermobiodynamics, 47
information couple, 308
information density, 207
information loops, 33
information management, 22
  in aging, 166
information overload, 677
information-related reactions, 172
information retrieval, 290
information theory, 164
  in hearing, 85
information transmission, modulation, and noise, 6
informational biomolecules, 271
informational direction, 145
informational molecules, 146, 163
  alternate pathways, 146
  coenzymes, 146
  key intermediates, 146
  mass action, 146
  noise, 146
informational probing, 282
infrared radiation of human skin, 295
infrared sensing, 45
insect pupae, 506
insect spiracles, 232
insects, 566
intelligible probe, 261
interaction of polarities, 314
interface
  man and machine, 1
  primary, 211
interference, 313
interference pattern, 310
interferometer, 312
intersystem coupling with chlorophyll, 308
intersystems, 424
  interactions, 282
  vision, 44
  vitamin A, 44
intersystems and nervous system connections, 42
inverse square law effects, 330
ionization potentials, 292
ionization tracks, 399

ionizing versus nonionizing radiation, 8
isosteric control, 152
isotope dilution, 389

jaundice, diagnosis of, 249
Justesen, D.R., 648

Katz, Bernard, 419
Keilin, David, 253
Kerr effect, 273
knowledge and data, 20
knowledge and data bases, 17
Kolmogoroff, A.N., 67

L-isoleucine synthesis and feedback, 147
lactate dehydrogenase, 201, 279
Larmor precession frequency, 264
lasers, 11, 288, 289, 478
   and acoustics, 288
   microprobes, 260
   sources, 294
Law of LeChatelier, 46
leghorn hen, 643
lenticular structures, 653
life expectancy, 168, 340
   ling span, 168
   zero error syntehsis, 169
lifespan, 170
ligands, 145
Lin, James C., 685
linear growth, 119
lines, spectral, 252
Lipmann, F., 421
living systems, 673
logical degeneracy, 109
long span, 168
longevity, 177
lung-trachea, 459
lymphocytes, 576
lysozyme molecule, 138

macrophages, 157
magnetic components, 538
magnetic fields, 244, 592
   strong, 249
magnetic moment, 247
magnetoculogram, 244
malignancy, 194
marine animals, 314
Markoff process, 83
maser, 11, 289, 479
   sources, 294
mass spectroscopy, 253
maximum exposure to radiation, 515

maximum safe levels of
   electromagnetic energy, 8
Maxwell, J.C., 301, 303
mechanical control systems, 31
"medium", 290
   propagation, 290
membrane, 182, 245
membrane dynamics, 142
membrane integrity, 169
   enzyme release, 170
   of genetic information, 169
membrane models, 422
memory, 143, 173, 189
memory cells, 171
   T and B cells, 171
menopause, 174
metabolic bridges, 553
metabolic effect, 557
metalloenzymes, 404
microbeam, 288
microorganisms, 549
microwave biology, 590
microwave cataract, 656
mcriowave cold war, 524
microwave diathermy, 507
microwave frequency for reinforcement
   of electron split, 264
microwave hyperthermia, 288
microwave information, 78
   amplitude and code, 79
microwave oven, 668
microwave quanta, 593
microwave sickness, 506
microwave spectral applications, 296
microwave thermal stress, 632
microwave thermography, 288
microwave ultrasound, 307
microwaves
   geometrical relationships, 305
   interference information on
      physiological events, 312
   plasma, 402
   significance of the dermal shell, 630
migration of internal peak with
   wavelength, 663
Miledi, Ricardo, 419
mind versus brain, 62
Mini-15, 113
minimum conformational
   free energy, 156
mitosis, 211
mitotic delay, 511
mobility effects, 638
mobility in differentiation, 191

mode, 303
modeling, 199
  natural, 199
modeling and programs, 17
models, 25
modulation information, 554
molecular bioengineering, 98
molecular biology, 272
  central dogma of, 163
molecular polarization, 273
molecular recognition, 427
molecules
  excited singlet state, 254
  excited triplet state, 254
  singlet ground state, 254
  vibrational energy state, 254
Moore, J.W., 419
morphogenesis, 183
mosaics, 347
motion sickness, 691
multi-path knowledge, 5
multiple shift key, 511
multiradiation effects, 212
multiwavelength signals, 283
mutagenic mutations, 164
mutations, 239
myoglobin, 266

$NADH_2$, 249
NMR, biohazards, 277
NAND operation, 71
natural protection, 642
natural selection, 18, 123
nearest neighbor analysis, 111
nervous system, 346
neuro-cybernetic operations, 30
neurotransmitter enzymes, 143
neurotransmitters, 171, 634
neutrino, 549
Nils Bohr model of the atom, 257
Nitella, 418
nitrosamines, 279
noise, 146, 169, 175, 197, 532
  from radiomimetic chemicals, 169
  shot, 127
  thermal, 127
  white, 127
noise saturation, 578
nonregenerating cells, 170
NOR operation, 71
nuclear magnetic relaxation, 270
  and biomolecular time processes, 270
  timed, 270
nuclear magnetic resonance
  spectroscopy, 268
nuclear reactions, 365
nucleotide triplets, 108
nucleotides, 95
nuclide therapy, 336
numeric and alphanumeric logic, 65
nutrition, 180

octopus, 200
1,6-diphosphatase, 149
operations research, 17, 39, 40
operator, 154
operon, 154, 183
optical density, 222
optical dielectric constant, 430
OR operation, class sum, 69
order of gene mutation, 550
organic free radicals, 248
organic radicals, 249
organism
  digital/analog computing, 17
organizational hierarchy, 228
Origin of the Species, 18
ostrich, 200
overlap regions, 296
overlapping code, 108
overloading, 125
oxygen effects on radiation, 340, 407
oxygen sensitivity, 130

p-n junction, 297
paramagnetic susceptibility, 247
paramecia, 197
Parsegian, V.L., 30
passive transport, 182
pathogen, 157
  probability of success, 157
pearl chain effect, 639
penetration, 548
penetration curve with internal peak, 661
penetration of EMR into leaf, 287
peptide bonds, scission of, 128
periodic patterns, 187
periodic system, 237
periodicity, 30
phagocytes, 157, 405
phase relations, 189, 197, 429
phenylketonuria, 279
phosphodiesterase, 634
phosphorescent emission bands, 254
photoelectrons, 302
photons at 60 Hz, 306
photoperiodism, 195
photopreprotection, 289

photoreactivation, 100, 289
photosynthesis, 19
phototaxis, 19, 310
phylogeny, 642
Physarum polycephalum, 567
plane polarizers, 312
plant and animal uptake, 337
ploidy, 209
polarization, 6, 165, 187, 206, 211, 216, 237, 244, 246, 261, 312, 393, 403, 404, 423, 596
   and displacement, 212
   information, 313, 315
   model, 238
   molecular, 273
   plane polarizers, 312
polarization force, 622
polarize responses, 5
polarized growth, 188
polarized infrared, 318
polarized ultraviolet, 318
polarizer, discriminator for information, 317
Polaroid lens, 312
polymers, 476
population-dependent cell functions, 193
power density
   corrective measures, 540
   decibels, 540
power density unit, 515
Poynting's vector, 516
primary interface, 211
principal wave, 303
   higher mode, 303
   TEM, 303
probing structures, hazards, 271
profile of the energy source, 8
programmed aging, 176
Programs of the Brain, 27
prolactin hormone, 200
propagation, 5, 6
properdin, 338
proportional control, 31
protein, sensitivity of, 212
protection, 601
   and reflection, 601
protective compounds, 410
protective functions of skin, 631
push and pull cooperation, 155
pyrimidine base synthesis, 152
pyruvate, 145

quantal fit, 677
quantizing, 187
quantum, 258
quantum fitting, 230
quantum mechanics relations at atomic level, 304
quarter-wave transformer, 455

RNA
   base-paired loops, 112
   recognition, 114, 116
   sequential transcription, 196
   special diversity, 112
   viral takeover, 114
radial growth, 187
radiation, 5, 18
   cataractogenesis, 654
   densely ionizing, 406
   doubling dose, 341
   dual nature, 300
   energy and wavelength, 306
   in prehistoric times, 127
   in role of noise, 125
   intensity, 258
   interaction, primary and secondary, 290
   ionizing, 296
   lower dose rates, 341
   nonionizing, 296
   oxygen effects, 340
   pulse, 430
   sensitivity to, 329
   sparsely ionizing, 406
   syndrome, 340
radioimmunoreactions, 335
radiolysis, 247, 396
radiomimetic chemicals, 169
Raman, C.V., 257
Raman effect, 272
Raman information shift, 257
Raman laser, infrared, 260
Raman shift, 259
Raman spectral lines, 259
   horizontal and vertical components, 259
   polarized light, 259
rapid rise in temperature, 620
rate acceleration, 271
Rayleigh scatter, 272
reaching reflex, 32
recognition, 338
redox systems, electron transfer, 248
redundancy, 5, 80, 109, 129, 169, 176
reflected and plane wave preparations, 590
reflection, 251, 601

Index

and protection, 601
reflectometer, 312
reflex
   eye-pupil, 34
   reaching, 32
refraction, 251
refractive index, 430
refractometer, 313
regulation, 5
rejection mechanism, 172
relaxation, 238, 423
relaxation and reactions, fast, constant, or slow, 250
relaxation frequencies, 677
repair time constants, 659
replicon, 196
reproduction, 675
resolution, 251, 287
resonance, 5, 215, 216, 238, 263, 264, 427, 549
resonant effects, 638
reverse transcriptase, 188
Rh compatability, 159
Rhodospirillum rubrum, 187
ribosomal access time, 116
rod and cone transducers, 302
rotational absorption, biohazards, 278
rotational energy, 635

Sadcikova, M., 506
salamander, water drive, 200
sand dollar, 418
saturation, 201
saturation energy levels, 307
scaling, from small animals to humans, 506
scatter, 231, 250, 258, 272, 313
   atmospheric, 313
   plane polarized, 313
   Rayleigh, 272
   X-ray, 103
Schwartz, Mischa, 6
scission of peptide bonds, 128
self-folding helix, 156
self-generating reaction, 667
self-organization, 5, 6
   enzyme, 140
semiconductor, 400
senescence, 174
   and aging, 174
sensitivity of protein, 212
sensitization, 648
sequence analysis, 113
   Edman degradator, 113

Mini-15, 113
sequential RNA transcription, 196
serine, used in cloning, 192
servomechanism, 40
   and self-regulation, 17
   eye-pupil reflex, 40
shark, 200
shifted frequency, gain or loss in energy, 301
shock wave mechanism, 620
shock waves, 508
sickle cell anemia, 119
Simon, Peter, 421
simple harmonic motion, 30, 309
simultaneous cell growth and function, 184
Singh, J., 3
skin, protective functions, 631
sleep and awareness, 62
Smith, Hamilton O., 421
social movements, 246
sorting out, 189
space frequency, 292
   angular, 292
spanning, 167
sparsely ionizing radiation, 406
spatial thermic conditions, 559
specific absorption rates, 517
spectral bands, 252
spectral information analysis, 282
spectral lines, 252
spectral reconnaissance, 288
spectral signatures, 282
spectrophotometry, 255
spectroscopy, 236
   absorption, 255
spectrum
   continuous, 250
   discontinuous, 250
Spirillum volutans, 405
splitting effect, at precession frequency, 265
splitting factor, 261
squid axon, 199, 418
squid field monitor, 244
standing waves, 310
Staphylococcus aureus, 572
Stark effect, 274
stationary processes, 83
stem cells, 167
stereochemical behavior, 133
stimulation versus inhibition, 566
stochastic process, 83
Streptomyces griseus, 129

Streufert, S., 13
sulfhydryl, as protection, 249
superconducting magnets, 264
swallowing reflex, 647
symbolic information, intersymbol, 80
symbolic logic of dendrites, 84
synchronous population, 573
systems, 24
systems stabilization, 36, 37, 38
systems theory, 5, 673
Szent-Gyorgyi, A., 13

T cells, 171
T system lymphocytes, 161
TEM, 303
target theory, 408
Terrapene carolina, 178
Testudo sumeiri, 178
thermal pollution, 283
thermobiodynamics, 47
This Cybernetic World, 30
3D forms, 142
thymus gland, 171
time constant, 224
time domain reflectometry, 519
time sequence sensing, 289
tissue
    ensemble, 294
    renewing, 177
tissue culture, 200
tracer studies, 235
transducers, 29
transduction, 5, 198
transformation, 198
transition metal, 249
transplant rejection, 159
transport system
    active, 182
    passive, 182
triplet state, 262
tumor cells, effect of oxygen, 570
tumor-sepcific antibodies, 573
tumor virus, 194

U.S.S.R. permissible exposure, 515
unit membrane, 418
unpaired electrons, 260

vacuum ultraviolet, 252
vasodilation, 643
velocity, 290
vibrating charge carriers, 126
viruses, 119
vitamin E, 179
von Neumann, John, 3

water binding, 270
    proton relaxation, 270
wave
    dimensions, 303
    mechanisms, 19
    notation, 187
    number, 293
        linked to the energy of the wave, 293
    particle nature, 231
"white source" 257
whole body, energy transfer to man, 513
Wiener, Norbert, 67
work function, 302

X-ray crystallography of DNA, 101
X-ray diffraction, 289
X-Y sex determination, 178
Xeroderma pigmentosum, 176

Y chromosome, 159
yeast, 167
Yockey, H.P., 13
Young, J.Z., 26, 419

Zeeman effect, 274-275
zeitgebers, 197

DATE DUE			